Human BIOLOGY

Human BIOLOGY

fourth edition

Sylvia S. Mader

WCB Wm. C. Brown Publishers

Dubuque, IA Bogota Boston Buenos Aires Caracas Chicago
Guilford, CT London Madrid Mexico City Sydney Toronto

Book Team

Editor *Colin Wheatley*
Developmental Editor *Kristine Noel*
Production Editor *Sue Dillon*
Designer *Christopher E. Reese*
Art Editor *Kathleen Huinker Timp*
Photo Editor *Lori Hancock*
Permissions Coordinator *Vicki Krug*
Visuals/Design Developmental Coordinator *Donna Slade*

Wm. C. Brown Publishers
A Division of Wm. C. Brown Communications, Inc.

Vice President and General Manager *Beverly Kolz*
Vice President, Publisher *Kevin Kane*
Vice President, Director of Sales and Marketing *Virginia S. Moffat*
Vice President, Director of Production *Colleen A. Yonda*
National Sales Manager *Douglas J. DiNardo*
Marketing Manager *Craig Johnson*
Advertising Manager *Janelle Keeffer*
Production Editorial Manager *Renée Menne*
Publishing Services Manager *Karen J. Slaght*
Permisions/Records Manager *Connie Allendorf*

Wm. C. Brown Communications, Inc.

President and Chief Executive Officer *G. Franklin Lewis*
Corporate Senior Vice President, President of WCB Manufacturing *Roger Meyer*
Corporate Senior Vice President and Chief Financial Officer *Robert Chesterman*

Cover Credit © KOTOH/ZEFA/H. Armstrong Roberts, Inc.

Copyedited by Laura Beaudoin

A Times Mirror Company

Library of Congress Catalog Card Number: 93–073907

ISBN 0–697–15956–6 (Paper) 0–697–15956–4 (Case)

Printed in the United States of America by Wm. C. Brown Communications, Inc., 2460 Kerper Boulevard, Dubuque, IA 52001

10 9 8 7 6 5 4 3 2

for my family

Brief Contents

Contents

Introduction

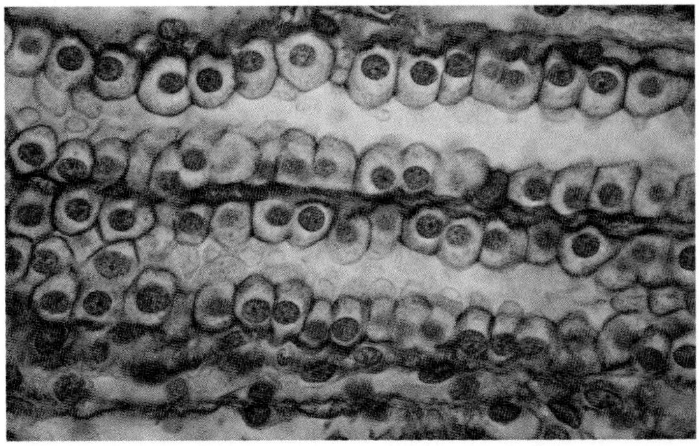

Part ONE

Chapter 1

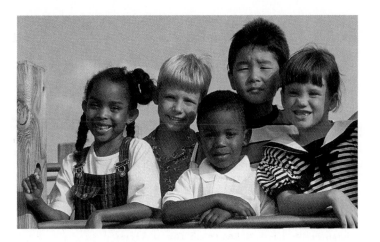

Preface

Human Biology is suitable for use in one-semester biology courses that emphasize human physiology and the relationship of humans to other living things. All students should leave college with a firm grasp of how their bodies normally function and how the human population can become more fully integrated into the biosphere. This knowledge can be applied daily and helps ensure our continued survival as individuals and as a species.

In keeping with its 2 major themes, the new edition of *Human Biology* has 2 types of readings; most chapters contain both a *Health Focus* and an *Ecology Focus*. The Health Focus readings are designed to help students cope with some particular health problem, such as how to design a healthy diet, how to protect vision and hearing, and how to do a shower test for breast cancer or testicular cancer. The Ecology Focus readings draw attention to some particular environmental problem, such as the relationship between ozone holes and skin cancer, the danger of carbon monoxide poisoning, and the possible ecological effects of biotechnology. The *Instructor's Manual* offers suggestions as to how the readings can be used in the classroom to stimulate discussion, think critically, or improve students' communication skills. Each reading is considered separately, and a variety of activities are offered to meet the current needs of instructors.

The fourth edition of *Human Biology* has the same style and organization as the previous edition. Each chapter presents the topic clearly, simply, and distinctly so that students will feel capable of achieving an adult level of understanding. Detailed, high-level scientific data and terminology are not included because I believe that true knowledge consists of working concepts rather than technical facility.

Many chapters have a new title, and the heads within each chapter have been rewritten to spark student interest. Each chapter has been carefully revised, sometimes completely rewritten, to improve student understanding. The following chapters in particular are significantly changed.

Introduction,"A Human Perspective," was rewritten to feature an ecological orientation and to improve the portion devoted to the scientific method.

Chapter 6, "Composition and Function of Blood," is a new chapter. All coverage of immunity now appears in the next chapter on immunity.

Chapter 9, "Urinary System and Excretion," has been reorganized and simplified in keeping with adopters' suggestions.

Chapter 16, "Development and Aging," now has an even greater human emphasis.

The "AIDS Supplement" was updated completely to take into account the very latest information concerning this topic.

Part Five, "Human Genetics": All chapters (17, "Chromosomal Inheritance"; 18, "Genes and Medical Genetics"; and 19, "DNA and Biotechnology") have been simplified and rewritten to strengthen their human approach. My intent was to encourage instructors to include these modern and relevant topics in their curriculums. Chapter 20, "Cancer," received special attention. It now stresses general concepts and reviews all modern ideas about the characteristics, causes, treatment, and prevention of cancer. A new appendix gives specific information about the most frequent types of cancer in humans.

As before, *Human Biology* has a beautiful illustration program with many new or improved 4-color illustrations. Color screens have sometimes been added to make illustrations visually more appealing and interesting. In addition, many figures in the text have been revised to make drawings of structures more clear or to increase readability of labels.

New to this edition are multimedia figures, which are also illustrated in computer animations that accompany the text. These are cellular or organismal processes that are best understood by viewing a dynamic simulation.

The chapter summaries have been lengthened and improved by the addition of text art. The pedagogy, which remains strong, is reviewed in the portion of the preface addressed "To the Student."

Organization of the Text

Introduction: A Human Perspective

The Introduction lays a foundation and provides a rationale for the text as a whole. It discusses the biological characteristics of humans and includes an Ecology Focus reading on the preservation of the tropical rain forests.

This chapter also defines and explains the scientific method in more depth than formerly.

Part One: Human Organization

Part One presents principles needed for the parts that follow. It includes chemistry, cell structure and function, and body organization. Homeostasis, which is introduced in this part, is a recurring theme in the chapters that follow.

Part Two: Maintenance of the Body

Part Two covers those systems of the human body that could be described as vegetative, and it explains how they contribute to homeostasis.

Chapter 6, "Composition and Function of Blood," was rewritten and all topics pertinent to immunity now appear in the following chapter. The chapter on the urinary system and excretion was simplified to increase student understanding. A new illustration consolidates nephron anatomy into just one illustration.

Part Three: Integration and Coordination in Humans

The control of homeostasis by the nervous and hormonal systems is found in Part Three. It also includes the musculoskeletal system and sense organs. By adopter request, coverage of the brain has been increased and each part is considered in more depth.

Part Four: Human Reproduction

Part Four deals with human reproduction and development. The new chapter, entitled "Development and Aging," replaces the old development chapter. The new chapter is better organized and simpler. It has more of a human emphasis in that only human development is now considered.

AIDS Supplement

The "AIDS Supplement" gives statistics about this pandemic condition and reviews information about an HIV infection—its stages, treatment, and prevention. The supplement has been updated in keeping with the most current information available.

Part Five: Human Genetics

The transference of traits from one generation to the next and associated topics are considered in Part Five.

Every chapter in the part has been simplified and given more of a human orientation. This up-to-date and modern topic should be a part of every curriculum, and it is my hope that instructors will want to make room for these chapters in their syllabi.

The cancer chapter was rewritten to emphasize general principles and the causes, treatment, and prevention of cancer. Every effort was made to present this high-interest topic at a level appropriate for students using the text. The Health Focus for the chapter explains a shower test for breast cancer and for testicular cancer.

Part Six: Human Evolution and Ecology

The chapters in this part cover the principles of evolution and ecology. The section of chapter 21, "Evolution," that pertains to human evolution now considers primates only in general terms and stresses hominid evolution.

Chapter 22, "Ecosystems," explains the principles of chemical cycling and energy flow through an ecosystem. Chapter 23, "Population Concerns," considers human population growth and modern ecological concerns such as global warming, tropical rain forest destruction, and the growing ozone depletion.

Readings

Two types of readings are included in the text. The Health Focus readings give practical information concerning some particular topic of interest, such as how to do a breast exam for cancer, when not to give blood, and proper nutrition. The Ecology Focus readings draw attention to some particular environmental problem, such as the need to preserve tropical rain forests, the relationship between ozone holes and skin cancer, and the possible ecological effects of biotechnology. The *Instructor's Manual* explains how to use the readings in the classroom.

Appendices and Glossary

The appendices contain optional information. Appendix A is an expanded table of chemical elements; appendix B is a new presentation of the metric system, which contains practice exercises; appendix C reviews most of the drugs of abuse; and appendix D gives pertinent information about the most frequent types of cancer. Appendix E gives the answers to the objective questions found at the end of each chapter.

The glossary defines all the boldface terms in the text. These terms are the ones most necessary for making the study of biology successful. Terms that are difficult to pronounce have a phonetic breakdown.

Additional Aids

Transparencies

A set of 200 transparency acetates accompanies the text. Most feature key illustrations and are in full color.

Student Study Art Notebook

This notebook contains a copy of all the illustrations available as transparencies with the text. These are the very same text illustrations that are used for overhead projection in the lecture hall. When the instructor shows a transparency, students can put notes in the space provided.

Bio Sci II Videodisc

This critically acclaimed laser disk, produced by Videodiscovery, features more than 12,000 still and moving images. The disk can be used for lecture support, individual student study, or student group activity. Bio Sci II Videodisc is accompanied by Bio Sci II Stacks Software. This software, using Macintosh HyperCard or IBM Linkway, puts the entire textual database for Bio Sci II at the instructor's or student's fingertips while simultaneously controlling the videodisc. The *Instructor's Manual* that accompanies *Human Biology* contains a bar code directory that correlates specific frames on the laser disk to illustrations in the text.

Instructor's Manual/ Test Item File

The *Instructor's Manual/Test Item File,* revised by the author, is designed to assist instructors as they plan and prepare for classes using *Human Biology.* The first part of the *Instructor's Manual* pertains to the text chapters and the second part is the *Test Item File.*

The *Instructor's Manual* has a completely new format to better serve the needs of today. Each chapter in the manual begins with a list of behavioral objectives, and the questions in the *Test Item File* are sequenced according to these behavioral objectives. Instructors can decide which behavioral objectives are best suited to their particular course and then easily locate the appropriate test questions.

New to this edition, the *Instructor's Manual* gives suggestions for student activities, and these suggestions are centered around the Heath Focus and Ecology Focus readings. These activities are designed to increase the communication skills of students. Particularly useful to previous adopters, the *Instructor's Manual* reviews changes in organization and content made in the fourth edition. It also gives suggested answers to critical thinking questions and writing across the curriculum questions found at the end of each text chapter.

The *Test Item File* for each chapter contains approximately 60 objective test questions and several essay questions. Some questions require more thought than others, and the questions are coded as to level of difficulty. These same questions are found in the computerized version of the test item file.

Visuals Testbank

This testbank contains black and white versions of the same 200 illustrations available as transparencies. The labels are deleted and copies can be run off for student quizzing or practice.

Study Guide

To ensure close coordination with the text, the author has written the *Study Guide* that accompanies the text. Each text chapter has a corresponding *Study Guide* chapter that includes a listing of behavioral objectives, study questions, and a chapter test. Answers to the study questions and the chapter tests are provided to give students immediate feedback.

The behavioral objectives in the *Study Guide* are the same as those in the *Instructor's Manual,* and the practice test questions in the *Study Guide* are sequenced and keyed to these objectives. Instructors who make their choice of behavioral objectives known to the students can thereby direct student learning in an efficient manner. Instructors and students who make use of the *Study Guide* should find that student performance increases dramatically.

Laboratory Manual

The author has also written the *Laboratory Manual to accompany Human Biology.* With few exceptions, each chapter in the text has an accompanying laboratory exercise in the manual (some chapters have more than one accompanying exercise). In this way, instructors are better able to emphasize particular portions of the curriculum if they wish. The 20 laboratory sessions in the manual are designed to further help students appreciate the scientific method and to learn the fundamental concepts of biology and the specific content of each chapter. All exercises have been tested for student interest, preparation time, and feasibility.

Laboratory Resource Guide

More extensive information regarding preparation is found in the *Laboratory Resource Guide.* The guide includes suggested sources for materials and supplies, directions for making up solutions and otherwise setting up the laboratory, expected results for the exercises, and suggested answers to all questions in the laboratory manual. It is available for free to all adopters of the laboratory manual.

Videotapes and CD-ROM

New to this edition are videotapes and a CD-ROM disk, which depict certain biological processes in dynamic animations. The multimedia figures are identified with a videotape or CD-ROM icon. The multimedia figures are also listed on the inside front cover of this book.

Micrograph Slides

This ancillary provides 35mm slides of many photomicrographs and all electron micrographs in the text.

Computer Software

Lecture Software

This software contains all the major heads of the text. It provides a framework for lecture notes because they can be entered into the outline wherever desired. The lecture software is available to adopters upon request.

WCB Microtest

Printing an exam yourself requires access to a personal computer—IBM (DOS or Windows) that uses 3.5" or 5.25" diskettes, or a Macintosh. Diskettes are available through your local Wm. C. Brown sales representative or by phoning Educational Services at 1–800–228–0459. You will receive complete instructions for making up an exam.

Wm. C. Brown Publishers also provides support services via mail (Judi David, Wm. C. Brown Publishers, 2460 Kerper Blvd., Dubuque, IA 52001), or by phone (1–800–258–2385) to assist in the use of the text generator software, as well as the creation and printing of tests if you do not have access to a suitable computer.

The Gundy-Weber Knowledge Map of the Human Body

This 13-disk, Mac-Hypercard program, by G. Craig Gundy, Weber State University, is for use by instructors and students alike. It features computer graphics, animations, labeling exercises, self-tests, and practice questions to help students examine the systems of the human body. Contact your local Wm. C. Brown sales representative or call 1–800–351–7671.

The Knowledge Map Diagrams

1 Introduction, Tissues, Integument System (0-697-13255-2)

2 Viruses, Bacteria, Eukaryotic Cells (0-697-13257-9)

3 Skeletal System (0-697-13258-7)

4 Muscle System (0-697-13259-5)

5 Nervous System (0-697-13260-9)

6 Special Senses (0-697-13261-7)

7 Endocrine System (0-697-13262-5)

8 Blood and the Lymphatic System (0-697-13263-3)

9 Cardiovascular System (0-697-13264-1)

10 Respiratory System (0-697-13265-X)

11 Digestive System (0-697-13266-8)

12 Urinary System (0-697-13267-6)

13 Reproductive System (0-697-13268-4)

Demo (0-697-13256-0)

Complete Package (0-697-13269-2)

ACKNOWLEDGEMENTS

William J. Brett
Indiana State University

Lisa Danko
Mercyhurst College

Martin E. Hahn
William Paterson College

Patricia Matthews
Grand Valley State University

Patricia P. Rosen
Moorhead State University

To the Student

*H*uman Biology includes a number of aids that will help you study biology successfully and enjoyably. These features are explained here by referring to appropriate text pages.

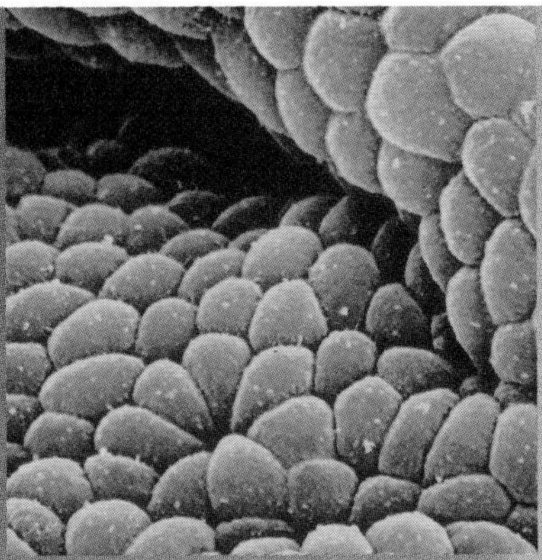

Chapter 4

Digestive System and Nutrition

The discovery of digestive enzymes involved the use of a human as the experimental material. William Beaumont was an American doctor who had a French-Canadian patient, Alexis St. Martin. St. Martin had been shot in the stomach, and when the wound healed, he was left with a fistula, or opening, that allowed Beaumont to collect gastric (stomach) juices and to look inside the stomach to see what was going on there. Beaumont was able to determine that the muscular walls of the stomach contract vigorously and mix food with juices that are secreted whenever food enters the stomach. He found that gastric juice contains hydrochloric acid (HCl) and a substance (enzyme) active in digestion. He concluded that gastric juices are produced separately from the protective mucous secretions of the stomach. Beaumont's work, which was very carefully and painstakingly done, pioneered the study of the physiology of digestion.

Chapter Concepts

1. The human digestive system is an extended tube with specialized parts between 2 openings, the mouth and the anus. 68
2. The products of digestion are small molecules, such as amino acids and glucose, that can cross cell membranes. 68
3. The liver and the pancreas are the accessory organs of digestion because their secretions assist the digestive process. 75
4. The digestive enzymes are specific and have an optimum temperature and pH. 77
5. Proper nutrition supplies the body with energy and nutrients, including the essential amino acids and fatty acids, and all vitamins and minerals. 79–86

Chapter Outline

Chapter Openers

Each chapter begins with an introductory page that discusses a topic of interest and provides a listing of chapter contents and a chapter outline. The concepts provide a framework for the content of the chapter and the outline includes the first and second level heads of the chapter.

Boldfaced Terms

Terms that are pertinent to the topic being discussed appear in boldfaced print and are defined in context. They also appear in the glossary where hard-to-pronounce terms are accompanied by a phonetic breakdown.

Text Line Art

Graphic diagrams placed within textual passages help clarify difficult concepts and enhance learning.

Respiratory Tract

During **inspiration** (breathing in) and **expiration** (breathing out), air is conducted toward or away from the lungs by a series of cavities, tubes, and openings, listed in order in table 8.1 and illustrated in figure 8.2.

As air moves in along the air passages, it is filtered, warmed, and moistened. Filtering is accomplished by coarse hairs and cilia in the region of the nostrils and by cilia alone in the rest of the nose and the trachea. In the nose, the hairs and the cilia act as a screening device. In the trachea, cilia beat upward, carrying mucus, dust, and occasional bits of food that "went down the wrong way" into the pharynx, where the accumulation can be swallowed or expectorated. The air is warmed by heat given off by the blood vessels lying close to the surface of the lining of the air passages, and it is moistened by the wet surface of these passages.

Conversely, as air moves out during expiration, it cools and loses its moisture. As the air cools, it deposits its moisture on the lining of the windpipe and the nose, and the nose may even drip as a result of this condensation. The air still retains so much moisture, however, that upon expiration on a cold day, it condenses and forms a small cloud.

Air is warmed, filtered, and moistened as it moves from the nose toward the lungs.

Each portion of the respiratory tract has its own structure and function, as described in the sections that follow.

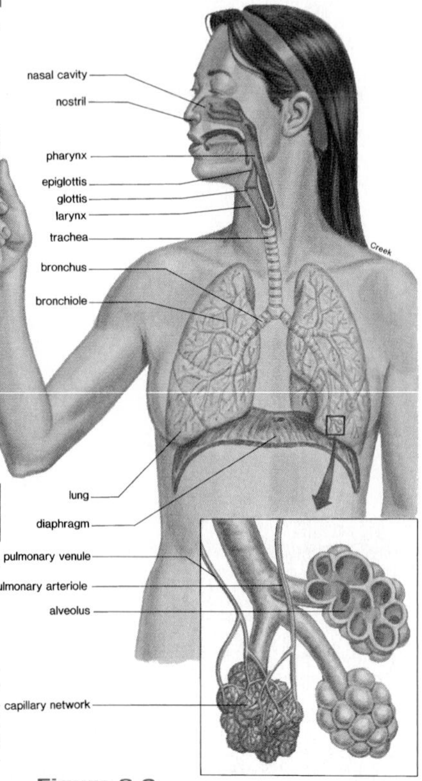

Figure 8.2

The human respiratory tract, with an enlargement showing the internal structure of a lung portion. Gas exchange occurs between the alveoli and blood within the surrounding capillary network. Notice that the pulmonary arteriole carries deoxygenated blood (colored blue) and the pulmonary venule carries oxygenated blood (colored red) (see p. 100).

The Nose: Two Cavities

The nose contains 2 *nasal cavities*, which are narrow canals with convoluted lateral walls separated from one another by a wall composed of bone and cartilage. Special ciliated cells in the narrow upper recesses of the nasal cavities (see fig. 12.3) act as odor receptors. Nerves lead from these cells to the brain, where the impulses generated by the odor receptors are interpreted as smell.

The tear (lacrimal) glands drain into the nasal cavities by way of tear ducts. For this reason, crying produces a runny nose. The nasal cavities also communicate with the cranial sinuses, air-filled, mucous membrane-lined spaces in the skull. If these membranes are inflamed due to a cold or an allergic reaction, mucus can accumulate in the sinuses, causing a sinus headache.

The nasal cavities empty into the nasopharynx, the upper portion of the pharynx. The *eustachian tubes* lead from the nasopharynx to the middle ears (see fig. 12.11).

The nasal cavities, which receive air, open into the nasopharynx.

CHAPTER EIGHT *Respiratory System* 147

Breathing is more continuously necessary than eating. While it is possible to stop eating altogether for several days, it is not possible to remain alive for longer than several minutes without breathing. The normal breathing rate is about 14 to 20 times per minute. The more energy expended, the greater the breathing rate. The average young adult male utilizes about 250 ml of oxygen per minute in a basal, or restful, state. Exercise and the digestion of food raise the need for oxygen. The average amount of oxygen needed with mild exercise is 500 ml of oxygen per minute.

Breathing supplies the body with the oxygen (O_2) needed for aerobic cellular respiration, as indicated in the following equation:

38 ADP + 36 Ⓟ → 36 ATP

$$C_6H_{12}O_6 + 6\,O_2 \longrightarrow 6\,H_2O + 6\,CO_2$$

As glucose is broken down to water and carbon dioxide, ATP molecules are formed.

Breathing not only draws oxygen into the body, it also pushes carbon dioxide out. It is only the first step of respiration, which can be said to include the following steps (fig. 8.1):

1. **Breathing:** entrance or exit of air into and out of the lungs;
2. **External and internal respiration:** External respiration is the movement of the gases oxygen (O_2) and carbon dioxide (CO_2) between air and blood; internal respiration is the movement of these gases between blood and tissue fluid.
3. **Aerobic cellular respiration:** production of ATP in cells.

Figure 8.1

Respiration is divided into 4 components: breathing brings oxygen into the lungs; external respiration is the exchange of gases between lungs and blood; internal respiration is the exchange of gases between blood and tissues; and aerobic cellular respiration is the production of ATP in cells—an oxygen-requiring process.

Table 8.1
Path of Air

STRUCTURE	DESCRIPTION	FUNCTION
Nasal cavities	Hollow spaces in nose	Filter, warm, and moisten air
Pharynx	Chamber behind oral cavity and between nasal cavity and larynx	Connection to surrounding regions
Glottis	Opening into larynx	Passage of air into larynx
Larynx	Cartilaginous organ that contains vocal cords (voice box)	Sound production
Trachea	Flexible tube that connects larynx with bronchi (windpipe)	Passage of air to bronchi
Bronchi	Major divisions of trachea that enter lungs	Passage of air to each lung
Bronchioles	Branched tubes that lead from the bronchi to the alveoli	Passage of air to each alveolus
Lungs	Soft, cone-shaped organs that occupy a large portion of the thoracic cavity	Gas exchange

146 *Maintenance of the Body* PART TWO

In-Chapter Summaries

Summary statements are placed strategically within the body of each chapter. These give periodic reinforcement of the information presented. Such statements are highlighted for easy identification.

Tables

Numerous, strategically placed tables list and summarize important information, making it readily accessible for efficient study.

Ecology Focus

Our Crops Deserve Better

When plants photosynthesize, they convert carbon dioxide from the air and water from the soil into a carbohydrate, namely glucose. Carbohydrates are organic food for plants and all other living things. Animals, including humans, either eat plants directly or eat animals that have eaten plants. Plants have the metabolic capability to modify and change carbohydrates into all the other types of organic molecules they require. This makes it possible for vegetarians to not only acquire carbohydrates but also amino acids from plants. The usual slimness of vegetarians testifies that most plants are not an abundant source of fat, and this fact can only make the vegetarian diet even more appealing.

Plant cell metabolism utilizes most of the same vitamins and minerals that humans require. Tables 4.6 and 4.7 show that plants are an important source of vitamins and minerals as well as organic food. Minerals are found in the soil in low concentrations, but plants are able to take them up and concentrate them. As the root system of a plant grows, it branches and branches again so that the roots are exposed to a tremendous amount of soil. It has been estimated that a rye plant has roots totaling about 900 kilometers in length. Water and minerals enter the roots by diffusion, but eventually active transport is used to concentrate minerals within the organs of a plant. A plant uses a great deal of ATP for active transport.

We are lucky that plants can concentrate minerals, for animals, like ourselves, often are dependent on them for supplying such elements as potassium for cardiac contraction, phosphorus for strong bones, and manganese for metabolic reactions. Once plants have taken up minerals, they are often incorporated into other molecules, including amino acids, phospholipids, and nucleotides.

Figure 4A

Proper management allows
many years. Mismanagemer
conditions and the loss of la
photograph, crops are being
sand, and therefore, land pro

ability of plants to carry on photosynthesis (p. 471). Ground-level ozone in smog destroys leaves and roots. Acid rain (see Ecology Focus, chapter 1, p. 17), which causes minerals to be leached from the soil, can even cause the death of plants. Global warming may cause the extinction of many plant species. Global warming is due to the buildup of carbon dioxide in the atmosphere from the combustion of gasoline, coal, and oil. Carbon dioxide acts like the glass of a greenhouse—it allows the sun's rays to pass through but traps the heat and doesn't allow it to escape. An increase in the average annual temperature is predicted.

The best climate for growing crops may shift to more rocky northern terrains with a resultant loss in crop yield. These problems don't begin to compare to the reduction in crop yield due to

Unfortunately, t
is affected by the ve
that affect the health
already mentioned th
crease in UV radiatio
tion of ozone will a

Ecology Focus Readings

These readings draw attention to a significant environmental problem. The example on the left examines the impact of environmental quality on crop production. Other readings discuss such topics as the need to preserve tropical rain forests, the relationship between ozone holes and skin cancer, and the possible ecological effects of biotechnology.

Health Focus

Weight Loss the Healthy Way

For a woman 19 to 22 years, 5 feet 4 inches tall, who exercises lightly, 2,100 Kcal* per day are normally recommended. For a man the same age, 5 feet 10 inches tall, who exercises lightly, the recommendation is 2,900 Kcal. Those who wish to lose weight need to reduce their caloric intake and/or increase their level of exercise. Exercising is a good idea, because to maintain good nutrition the intake of calories per day should probably not go below 1,200 Kcal a day. Also, for the reasons discussed in this chapter, carbohydrates should still make up at least 50% of these calories; proteins should be no more than 25% and the rest can be fats. A deficit of 500 Kcal a day (through intake reduction or increased exercise) is sufficient to lose a pound of body fat in a week. Once you realize that a diet needs to be judged according to the principles of adequacy of nutrients; balance in regard to carbohydrates, proteins, and fats; moderation in number of calories; and variety of food sources, it is easy to see that many of the diets and gimmicks people use to lose weight are bad for their health. Unhealthy approaches include the following:

Pills

The most familiar pills, and the only ones approved by the FDA, are those that claim to suppress the appetite. They may actually work at first, but the appetite soon returns to normal and weight lost is regained. Then the user has the problem of trying to get off the drug without gaining more weight. Other types of pills are under investigation and sometimes can be obtained illegally. But, as yet, there is no known drug that is both safe and effective for weight loss.

Low-Carbohydrate Diets

The dramatic weight loss that occurs with a low-carbohydrate diet is not due to a loss of fat; it is due to a loss of muscle mass and water. Glycogen and important minerals are also lost. When a normal diet is resumed, so is the normal weight.

Liquid Diets

Despite the fact that liquid diets provide proteins and vitamins, the number of Kcal is so restricted that the body cannot burn fat quickly enough to compensate and muscle is still broken down to provide energy. A few people on this regime have died, probably because even the heart muscle was not spared by the body.

Single-Category Diets

These diets rely on the intake of only one kind of food, either a fruit or vegetable or rice alone. However, no single type of food provides the balance of nutrients needed to maintain health. Some dieters on strange diets suffer the consequences—in one instance an individual lost hair and fingernails.

*A *calorie* is a standard way to measure energy; it is the amount of heat necessary to raise the temperature of one gram of water one degree Centigrade. Food energy is measured in *kilocalories* (thousands of calories), abbreviated Kcal.

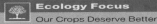

Questions to Ask About a Weight-Loss Diet

1. Does the diet have a reasonable number of Kcal?
   10 Kcal per pound of current weight is suggested. In any case, no fewer than 1,000–1,200 Kcal for a normal-sized person.
2. Does the diet provide enough protein?
   For a woman 120 lb, 44 g protein each day is recommended. For a man 154 lb, 56 g is recommended. More than twice this amount is too much.
   For reference, 1 c milk and 1 oz meat each has 8 g protein.
3. Does the diet provide too much fat?
   No more than 20%–30% of total Kcal is recommended.
   For reference, a pat of butter has 45 Kcal. 1 g fat = 9 Kcal.
4. Does the diet provide enough carbohydrates?
   100 g = 400 Kcal is the very least recommended per day; 50% of total Kcal should be carbohydrates.
   For reference, a slice of bread contains 14 g of carbohydrates.
5. Does the diet provide a balanced assortment of foods?
   The diet should include breads, cereals, legumes; vegetables (especially dark-green and yellow ones); low-fat milk products; and meats or a meat substitute.
6. Does the diet make use of ordinary foods that are available locally?
   Diets should not require the purchase of unusual or expensive foods.

From W. N. Whitney, et. al., Understanding Nutrition, 5th ed. Copyright © 1990 West Publishing Company, St. Paul, MN.

Health Focus Readings

These readings give practical information about a particular health concern, such as a healthy approach to weight loss, shown here. Others discuss how to do a breast exam for cancer, when not to give blood, and proper nutrition.

Multimedia Figures

Certain illustrations, identified by a videotape or CD-ROM icon, can be studied further by viewing an animated videotape or CD-ROM. Ask your instructor about availability.

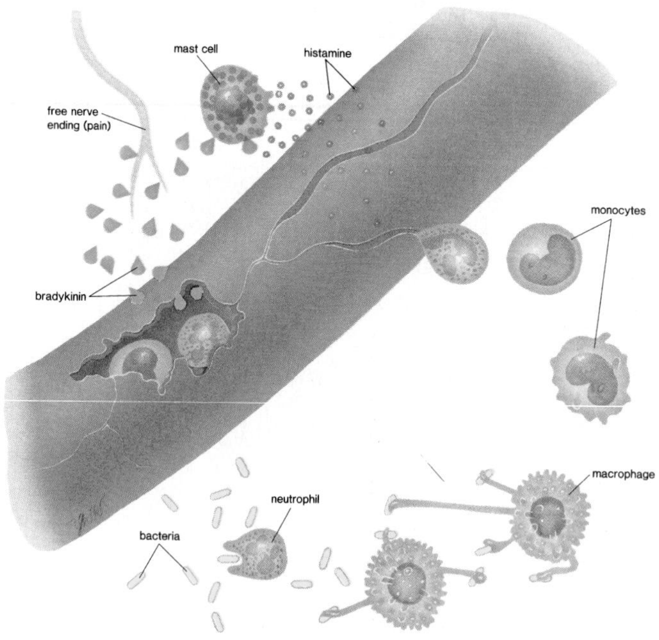

Figure 7.3

Inflammatory reaction. When a blood vessel is injured, mast cells release histamine, which dilates blood vessels, and bradykinin, which stimulates the pain nerve endings. Neutrophils and monocytes congregate at the injured site and squeeze through the capillary wall. The neutrophils begin to phagocytize bacteria. The monocytes become macrophages, large cells that are especially good at phagocytosis and also stimulate other white blood cells to action.

passes by way of blood to the red bone marrow, where it stimulates the production and the release of white blood cells, usually neutrophils.

As the infection is being overcome, some neutrophils die. These, along with dead tissue, cells, bacteria, and living white blood cells, form **pus**, a thick, yellowish fluid. Pus indicates that the body is trying to overcome the infection.

> The inflammatory reaction is a "call to arms"—it marshals phagocytic white blood cells to the site of bacterial invasion.

> Characteristics of B Cells:
> - Antibody-mediated immunity
> - Produced and mature in bone marrow
> - Direct recognition of antigen
> - Clonal expansion produces antibody-secreting plasma cells as well as memory cells

Antibodies and Antigens Often Form Complexes

The most common type of antibody (IgG) is a Y-shaped protein molecule with 2 arms. Each arm has a "heavy" (long) chain and a "light" (short) chain of amino acids. These chains have *constant regions,* where the sequence of amino acids is set, and *variable regions,* where the sequence of amino acids varies (fig. 7.6). The constant regions are not identical among all the antibodies. Instead, they are the same within different classes of antibodies (table 7.1). The variable regions form an antigen-binding site, and their shape is specific to a particular antigen. The antigen combines with the antibody at the antigen-binding site in a lock-and-key manner.

The antigen-antibody reaction can take several forms, but quite often the antigen-antibody reaction produces complexes of antigens combined with antibodies. Such an antigen-antibody complex, sometimes called the immune complex, marks the antigen for destruction by other forces. For example, the complex may be engulfed by neutrophils or macrophages or it may activate complement. Complement makes microbes more susceptible to phagocytosis, as discussed.

Classes of Antibodies: Improving the Odds

There are 5 different classes of circulating antibodies (table 7.1). IgG antibodies are the major type in blood, and lesser amounts are also found in lymph and tissue fluid. IgG antibodies attack microbes and their toxins. A toxin is a specific chemical produced, for example by bacteria, that is poisonous to other living things. IgM antibodies are a cluster of 5 of the Y-shaped structures shown in figure 7.6*a.* These antibodies appear soon after an infection begins and disappear before it is over. IgA antibodies contain 2 Y-shaped structures and are the main type of antibody found in bodily secretions. They attack microbes and their toxins before they reach the bloodstream. IgD antibodies are identical to the membrane-bound antibodies on B cells. IgE antibodies are responsible for allergic reactions and are discussed later (p. 140).

Figure 7.6

Antigen-antibody reaction. **a.** An IgG antibody contains 2 heavy (long) amino acid chains and 2 light (short) amino acid chains arranged to give 2 variable regions, where a particular antigen is capable of binding with the antibody. **b.** Quite often the antigen-antibody reaction produces complexes of antigens combined with antibodies.

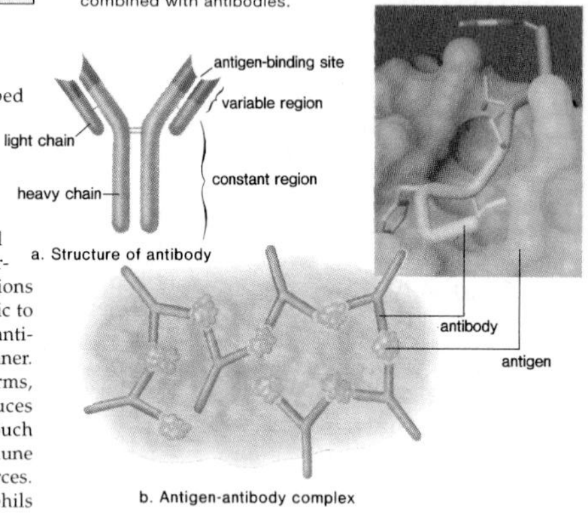

> An antigen combines with an antibody at the antigen-binding site in a lock-and-key manner. The reaction can produce antigen-antibody complexes, which contain several molecules of antibody and antigen.

Table 7.1

Antibodies

Classes	Presence	Function
IgG	Main antibody type in circulation	Attacks microbes* and bacterial toxins; enhances phagocytosis
IgM	Antibody type found in circulation, largest antibody	Activates complement; clumps cells
IgA	Main antibody type in secretions such as saliva and milk	Attacks microbes and bacterial toxins
IgD	Antibody type found as a membrane-bound receptor	Functions unknown
IgE	Antibody type found as membrane-bound receptor on basophils in blood and on mast cells in tissues	Responsible for allergic reactions

*Viruses and bacteria

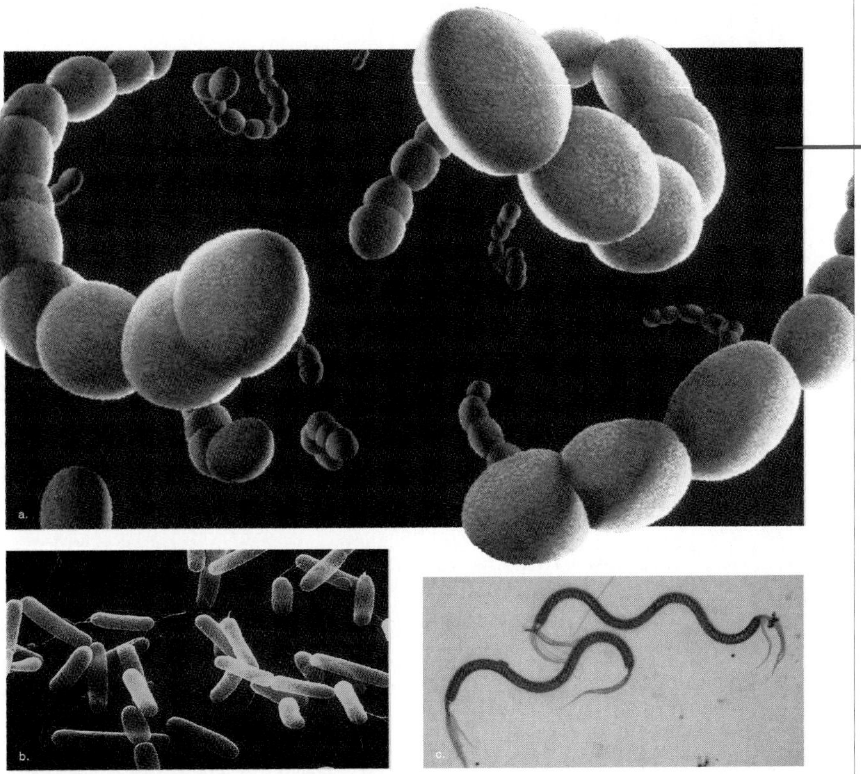

Figure 15.5

Scanning electron micrographs of bacteria. **a.** Spherical-shaped bacteria. **b.** Rod-shaped bacteria. **c.** Spiral-shaped bacteria that use flagella for locomotion. See figure 15.7 for a generalized drawing of a bacterium.

Bacterial in Origin

Although **bacteria** are generally larger than viruses, they are still microscopic. As a result, it is not always obvious that they are abundant in the air, water, and soil and on most objects. It has even been suggested that the combined weight of all bacteria would exceed that of any other type of organism on earth. Bacteria occur in 3 basic shapes (fig. 15.5): round, or spherical (coccus); rod (bacillus); and spiral (a curved shape called a spirillum). Some bacteria can locomote by means of flagella.

Dramatic Visual Program

Colorful, informative photographs and illustrations support the learning program by sparking interest and providing insight into the topics being discussed.

Chapter Summaries

A summary reviews the material of the chapter and may be read before or after you have studied the chapter. The addition of illustrations reinforces the important concepts of the chapter.

Deafness: Two Major Types

There are 2 major types of deafness: *conduction deafness* and *nerve deafness*. Conduction deafness can be due to a congenital defect, such as that which occurs when a pregnant woman contracts German measles during the first trimester of pregnancy. (For this reason every female should be immunized against rubella before the childbearing years.) Conduction deafness can also be due to infections that have caused the ossicles to fuse, restricting their ability to magnify sound waves. Because respiratory infections can spread to the ear by way of the eustachian tubes, every cold and ear infection should be taken seriously.

Nerve deafness most often occurs when cilia on the sense receptors within the cochlea have worn away. Since this can happen with normal aging, old people are more likely than younger persons to have trouble hearing. However, as discussed in the Health Focus in this chapter, nerve deafness also occurs when people listen to loud music amplified to or above 130 decibels. Because the usual types of hearing aids are not helpful for nerve deafness, it is wise to avoid subjecting the ears to any type of continuous loud noise. Costly cochlear implants, which stimulate the auditory nerve directly, are available, but those who have these electronic devices report that the speech they hear is like that of a robot.

SUMMARY

All receptors are the first part of a reflex arc. Each type is sensitive to a particular kind of stimulus in the external or internal environment. When stimulation occurs, receptors initiate nerve impulses that are transmitted to the spinal cord and/or brain. Only when nerve impulses reach the cerebrum are we conscious of sensation. The cerebrum can be mapped according to the type of sensation felt and the localization of sensation.

The skin contains receptors for touch, pressure, pain, and temperature (hot and cold). Taste and smell are due to chemoreceptors that are stimulated by chemicals in the environment. The taste buds contain cells in contact with nerve fibers, while olfactory cells are specialized nerve fibers. The former end in microvilli and the latter end in cilia.

Vision is dependent on the eye, the optic nerve, and the visual cortex of the cerebrum. The eye has 3 layers. The outer layer, the sclera, can be seen as the white of the eye; it also becomes the transparent bulge in the front of the eye called the cornea. The rods, receptors for vision in dim light, and the cones, receptors that depend on bright light and provide color and detailed vision, are located in the retina, the inner layer of the eyeball. The cornea, the humors, and especially the lens bring the light rays to focus on the retina. To see a close object, accommodation occurs as the lens rounds up. Due to the optic chiasma, both sides of the brain must function together to give three-dimensional vision.

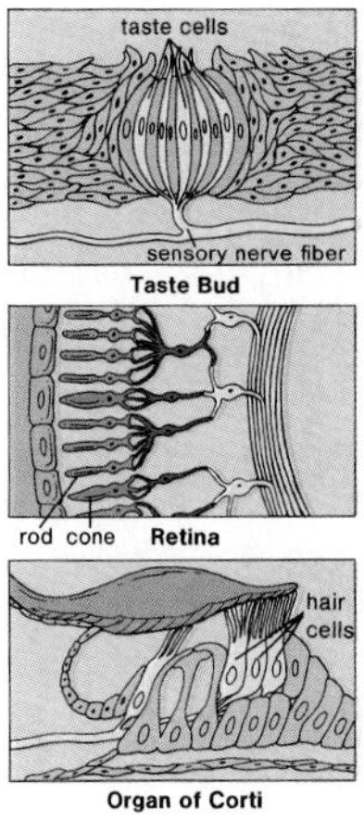

The ear also contains receptors for our sense of balance. Dynamic equilibrium is dependent on the stimulation of hair cells within the ampullae of the semicircular canals. Static equilibrium relies on the stimulation of hair cells by otoliths within the utricle and the saccule.

Hearing is a specialized sense dependent on the ear, the auditory nerve, and the auditory cortex of the cerebrum. The ear is divided into 3 parts: outer, middle, and inner. The outer ear consists of the pinna and the auditory canal, which direct sound waves to the middle ear. The middle ear begins with the tympanic membrane and contains the ossicles (hammer, anvil, and stirrup). The hammer is attached to the tympanic membrane, and the stirrup is attached to the oval window, which is covered by membrane. The inner ear contains the cochlea, the semicircular canals, plus the utricle and saccule. The outer and middle portions of the ear simply convey and magnify the sound waves that strike the oval window. Its vibrations set up pressure waves within the cochlea, which contains the organ of Corti, consisting of hair cells with the tectorial membrane above. When the cilia of the hair cells strike this membranae, nerve impulses are initiated that finally result in hearing.

Study and Thought Questions

The study questions are page-referenced and allow you to test your understanding of the chapter's concepts. The critical thinking questions challenge you to go one step further and use the information in a new and different manner. The ethical issue questions ask you to consider the social implications of knowledge or biological advances.

Writing Across the Curriculum

It is generally recognized that students need an opportunity to practice writing in all of their courses. These questions are thought questions that can best be answered by writing a short paragraph.

Objective Questions

Fill-in-the blank questions allow you to quiz yourself prior to taking an examination. This section also often asks you to label an illustration from the chapter. Answers to the objective questions are found in appendix E.

Study and Thought Questions

1. List and discuss the receptors of the skin. (p. 228)
2. Discuss the chemoreceptors. (pp. 229–230)
3. Describe the anatomy of the eye (p. 230), and explain focusing and accommodation. (pp. 232–233)

   Critical Thinking Devise a categorization for the parts of the eye listed in table 12.2. Justify your system.
4. Describe sight in dim light. What chemical reaction is responsible for vision in dim light? (p. 234) Discuss color vision. (p. 235)
5. Relate the need for corrective lenses to 3 possible eye shapes. Discuss bifocals. (p. 236)
6. Describe the anatomy of the ear (pp. 238–239) and how we hear. (pp. 242–243)
7. Describe the role of the utricle, the saccule, and the semicircular canals in balance. (p. 242)

   Critical Thinking Both balance and hearing utilize mechanical stimulation and hair cells. Why does one result in the sense of balance and the other in the sense of hearing?
8. Discuss the 2 causes of deafness, including why younger people frequently suffer a hearing loss. (p. 244)

   Ethical Issue What should be done, if anything, to protect teenagers who play music so loudly that their hearing is endangered? Explain your answer.

Writing Across the Curriculum

1. In general, list the steps, from sense organ to brain, by which sensation occurs. Discuss the weaknesses and strengths of depending on such a method for knowledge of the outside world.
2. Question 1 assumes that the brain is responsible for sensation. Based on figure 12.5, why might there be some integration of data in the eye itself? Does this fact lend strength or weakness to the way by which we see?
3. What do the retina and organ of Corti have in common?

Objective Questions

1. Vision, hearing, taste, and smell do not occur unless nerve impulses reach the proper portion of the ______.
2. Taste cells and olfactory cells are ______ because they are sensitive to chemicals in the air and in food.
3. The receptors for vision, the ______ and the ______, are located in the ______, the inner layer of the eye.
4. The cones give us ______ vision and work best in ______ light.
5. Vision in dim light is dependent on the presence of ______ in the rods.
6. The lens ______ for viewing close objects.
7. People who are nearsighted cannot see objects that are ______. A ______ lens restores this ability.
8. The ossicles are the ______, the ______, and the ______.
9. The semicircular canals are involved in our sense of ______.
10. The organ of Corti is located in the ______ canal of the ______.
11. Label this diagram of the eye.

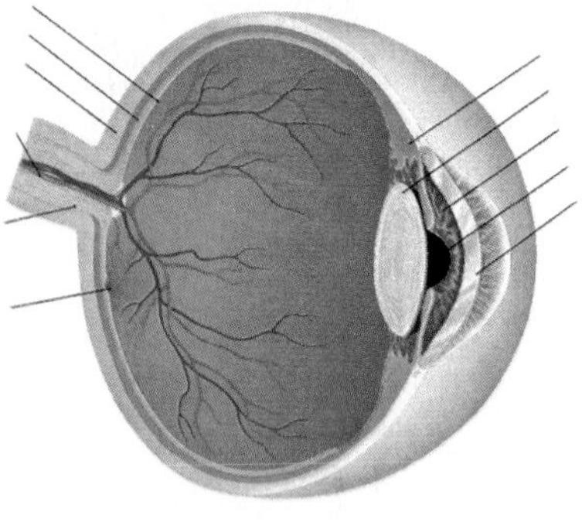

Selected Key Terms

allantois (ah-lan-to-is) An extraembryonic membrane that serves as a source of blood vessels for the umbilical cord. 318

amniocentesis The removal of a small amount of amniotic fluid to examine the chromosomes and the enzymatic potential of fetal cells. 324

amnion (am´ne-on) One of the extraembryonic membranes; a fluid-filled sac around the embryo. 318

blastocyst An early stage of embryonic development that consists of a hollow ball of cells. 322

chorion (ko´re-on) An extraembryonic membrane that forms an outer covering around the embryo and contributes to the formation of the placenta. 318

conception Fertilization and implantation of an embryo resulting in pregnancy. 318

embryo The organism in its early stages of development; first week to 2 months. 322

embryonic development The period of development from the first week to 2 months, during which time all major organs form. 321

episiotomy (ĕ-piz´e-c procedure perfo childbirth in wh the vagina is en tearing. 329

extraembryonic mem that are not a pa but are necessar existence and he embryo. 318

fertilization The union of a sperm nucleus and an egg nucleus, which creates the zygote with the diploid number of chromosomes. 317

fetal development The period of development from the ninth week through birth. 321

gestation period (jes-ta´shun) The period of development measured from the start of the last menstrual cycle until birth; in humans, typically 280 days. 318

implantation The attachment and penetration of the embryo to the lining of the uterus (endometrium). 318

lanugo (lah-nu´go) Short, fine hair that is present during the later portion of fetal development. 327

morula (mor´u-lah) An early stage in development in which the embryo consists of a mass of cells, often spherical. 322

prolactin (PRL) A hormone secreted by the anterior pituitary that stimulates the production of milk from the mammary glands. 330

umbilical cord Cord through which blood vessels pass, connecting the fetus to the placenta. 319

umbilicus The scar left after the umbilical cord stump shrivels and falls off. 330

vernix caseosa (ver´niks ka˝se-o´-sah) Cheeselike substance covering the skin of the fetus. 327

yolk sac An extraembryonic membrane that serves as the first site for blood cell formation. 318

zygote (zi´gōt) Diploid cell formed by the union of 2 gametes; the product of fertilization. 318

Selected Key Terms

A list of boldfaced terms from the chapter appears in the chapter summary section. Each term is accompanied by its definition and, for hard-to-pronounce terms, a phonetic breakdown.

Further Readings for Part Four

Anderson, R. M., and May, R. M. May 1992. Understanding the AIDS pandemic. *Scientific American.*

Aral, S. L., and Holmes, K. K. February 1991. Sexually transmitted diseases in the AIDS era. *Scientific American.*

Eigen, M. July 1993. Viral quasispecies. *Scientific American.*

Fischetti, V. A. February 1992. Streptococcal M protein. *Scientific American.*

Gilbert, S. 1991. *Developmental biology.* 3d ed. Sunderland, Mass.: Sinauer Associates.

Hampton, J. K., Jr. 1991. *Biology of human aging.* Dubuque, Iowa: Wm. C. Brown Publishers.

Kalil, R. E. December 1989. Synapse formation in the developing brain. *Scientific American.*

Kart, C. S., et al. 1992. *Human aging and chronic disease.* Boston: Jones and Bartlett.

Kimura, D. September 1992. Sex differences in the brain. *Scientific American.*

Mills, J., and Masur, H. August 1990. AIDS-related infections. *Scientific American.*

Moore, K.L. 1988. *Essentials of human* embryology. Toronto: B.C. Decker, or St. Louis: Mosby Year Book.

Murray, A. W., and Kirschner, M. W. March 1991. What controls the life cycle. *Scientific American.*

Nathans, J. February 1989. The genes for color vision. *Scientific American.*

National Research Council, Institute of Medicine. 1990. *Developing new contraceptives: Obstacles and Opportunities.* Washington, D.C.: National Academy Press.

Nilsson, L. 1977. *A child is born.* Rev. ed. New York: Delacorte Press.

Rietschel, E. T., and Brade, H. August 1992. Bacterial endotoxins. *Scientific American.*

Further Readings

A list of readings suggests references that can be consulted for further study of topics covered. The sources listed were chosen for readability and accessibility.

Introduction

A Human Perspective

This book has 2 primary functions. The first is to explore human anatomy and physiology so you will know how the body functions. The second is to take a look at human evolution and ecology so you will understand the place of humans in nature. Both the human body and the environment are self-regulating systems that can be thrown out of kilter by misuse and mismanagement. An appreciation of the delicate balance present in both systems provides the perspective from which future decisions can be made. It is hoped that adequate information will better enable you to keep your body and the environment healthy.

Introductory Concepts

1 Human beings have characteristics in common with all living things. 3

2 The scientific method is the process by which scientists gather information about the material world. 7

3 All persons have to be prepared to use scientific information to make value judgments. 9

Introductory Outline

Figure I.1

Classification and evolution of humans. **a.** Living things are classified into 5 major kingdoms and humans are in the animal kingdom. **b.** Human beings are most closely related to the other vertebrates shown. The evolutionary tree of life has many branches; the vertebrate line of descent is just one of many.

Kingdom	Representative Organisms
animals (Animalia)	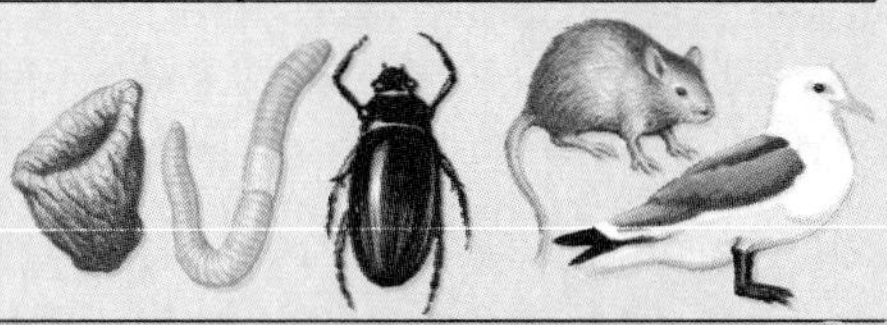 Invertebrates (sponges, worms, insects) Vertebrates (fishes, amphibia, birds, mammals)
plants (Plantae)	 Mosses, ferns, trees, flowering plants
fungi (Fungi)	 Molds, mushrooms
protists (Protista)	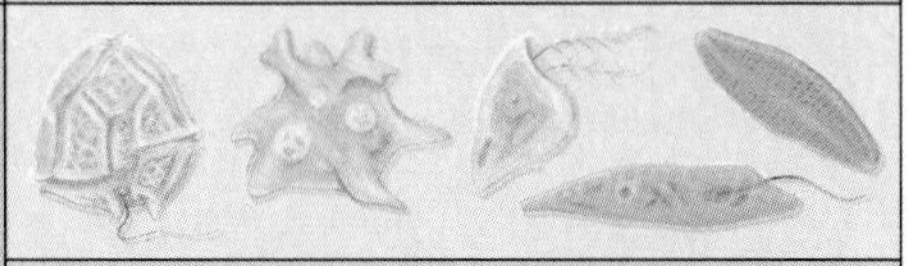 Protozoa, algae, slime molds
monerans (Monera)	 Bacteria

a. Survey of living things

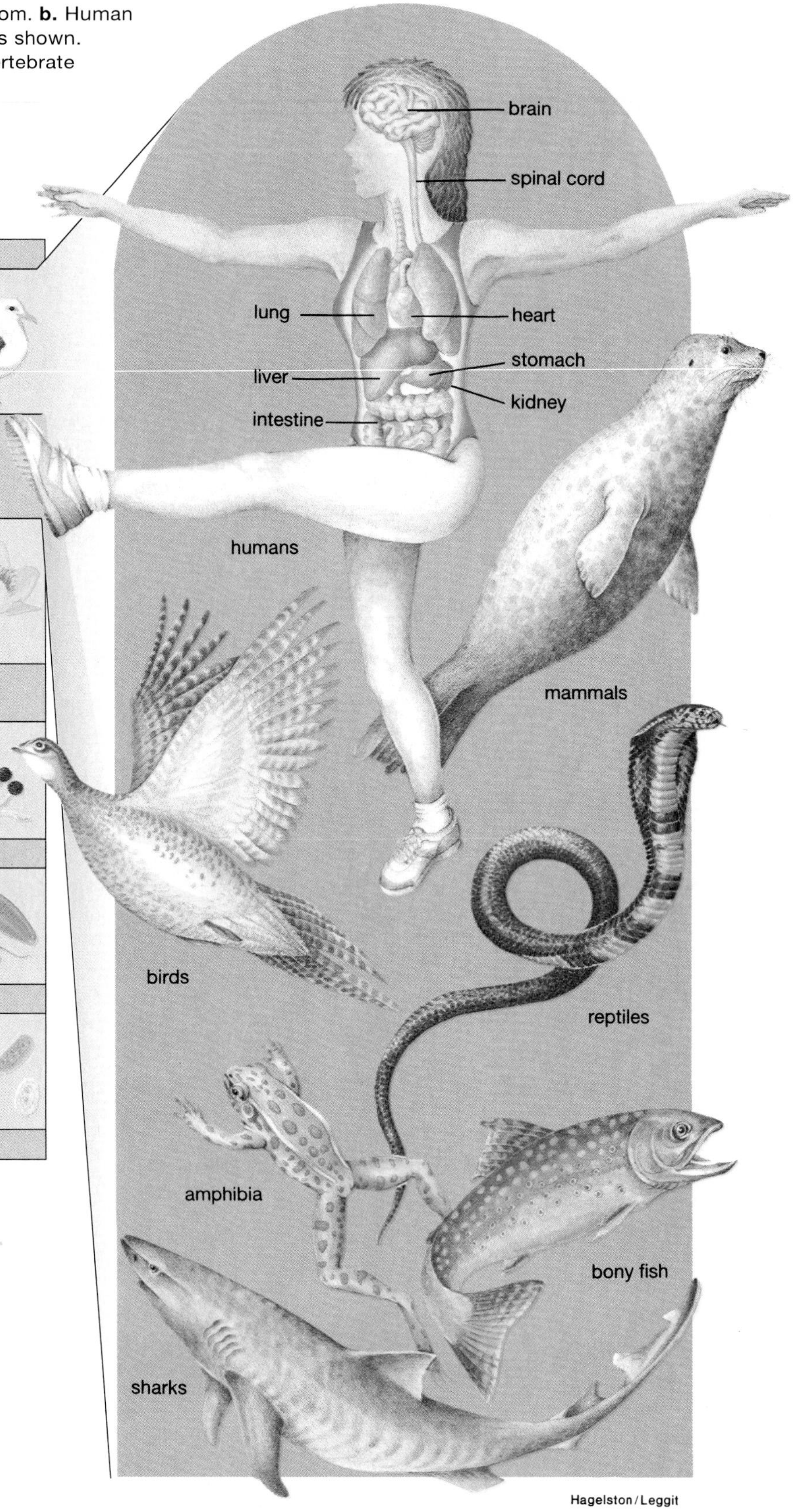

b. Survey of vertebrates

Biologically Speaking

You are about to launch on a study of human biology. Before you begin, it is appropriate to define humans by looking at their biological characteristics. Biologically speaking:

Human beings are a product of an evolutionary process. Life has a history that began with the evolution of the first cell(s) about 3.5 billion years ago. Thereafter, living things became increasingly complex—humans are among the most complex animals to have evolved. Criteria such as degree of complexity, motility, life cycle, and mode of nutrition are used by biologists to classify living things into 5 major groups called **kingdoms** (fig. I.1). *Humans are vertebrates in the animal kingdom.* **Vertebrates** have a nerve cord that is protected by a vertebral column whose repeating units (the vertebrae) indicate that we and other vertebrates are segmented animals. Segmentation leads to the specialized structures we observe in humans. Among the vertebrates, we are most closely related to the apes, from whom we are distinguished by our highly developed brains, completely upright stance, and the power of creative language. It is possible to trace human ancestry from the first cell through a series of prehistoric ancestors until the evolution of modern-day humans.

The presence of the same types of chemicals tells us that *human beings are related to all other living things.* DNA is the genetic material, and ATP is the energy currency in the cells of all living things, for example. It is even possible to do research with bacteria (kingdom Monera) and have the results apply to humans. For example, it is common practice today to test food additives first by applying the chemicals to bacterial culture plates. If the chemical causes mutations in bacteria, it is considered unsafe for human consumption.

Human beings are a part of the biosphere. All living things are a part of the **biosphere,** a network of life that spans the surface of the earth. In any portion of the biosphere, such as a particular forest or pond, the various populations interact with one another and with the physical environment. They form an **ecosystem** in which chemicals cycle and energy flows. As figure I.2 shows, all living things are dependent upon solar energy and upon plants, which use this energy to convert chemicals into a form that is usable by all living things, including humans.

Human beings are a product of a cultural heritage. We are born without knowledge of civilized ways of behavior, and we gradually acquire these by adult instruction and imitation of role models. Unfortunately, it is our cultural inheritance that makes us think we are separate from nature and makes it difficult for us to see that actually we are part of the biosphere (fig. I.3).

Humans have modified the biosphere greatly. Wherever humans settle, they first clear the area to grow crops; then they build houses on what was once farmland; and finally, they convert small towns into cities. With each step, fewer and fewer of the original organisms remain, until at last there are primarily humans and their domesticated plants and animals.

The Ecology Focus boxes found throughout the text emphasize how important it is that we do all we can now to preserve the biosphere, because only then can we be assured that we will continue to exist. Presently, there is great concern about preserving the **biodiversity** (the wide range of living things) in the world's tropical rain forests, as discussed in the introductory Ecology Focus. The preservation of biodiversity is extremely important, but we should also be aware that tropical rain forests perform services for us. For example, they act like a giant sponge and absorb carbon dioxide, a pollutant that pours into the atmosphere from the burning of fossil fuels such as oil and coal. An increased amount of carbon dioxide in the atmosphere is expected to cause an increase in the average daily temperature. Carbon dioxide also combines with water and oxygen to form carbonic acid, which contributes to acid rain.

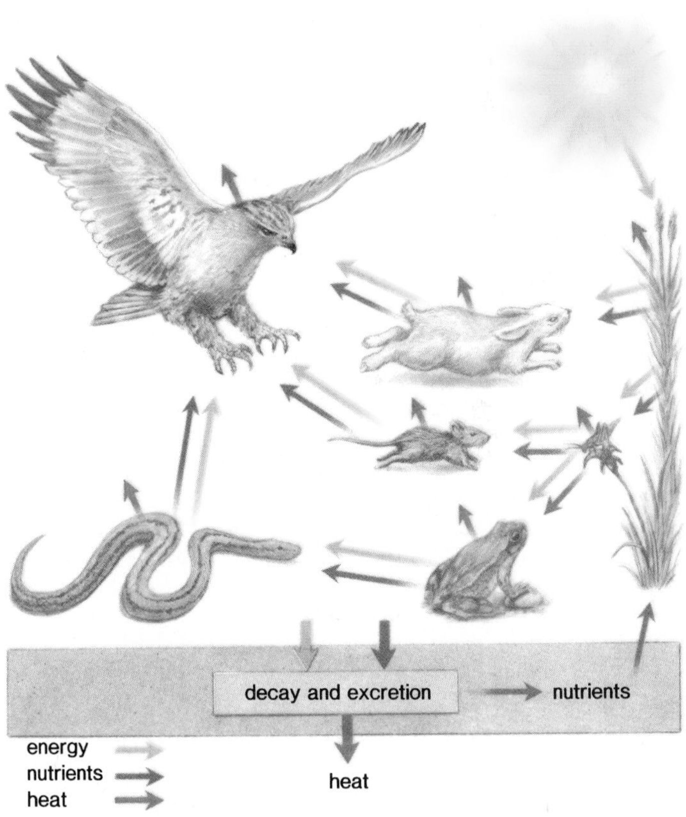

Figure I.2

Within an ecosystem, chemicals cycle (*see grey arrows*); plants make and use their own organic food, and this becomes food for several levels of animal consumers, including humans. When these organisms die, their inorganic remains are used by plants as they produce organic food. Energy flows (*see yellow arrows*); solar energy used by plants to produce organic food is eventually converted to heat by all members of an ecosystem (therefore a constant supply of solar energy is required for life to exist).

Ecology Focus

Tropical Rain Forests: Can We Live Without Them?

So far, only about 1.7 million species of organisms have been discovered and named. Two-thirds of the plant species, 90% of the nonhuman primates, 40% of birds of prey, and 90% of the insects live in the tropics. Many more species of organisms are estimated to exist but have not yet been discovered (perhaps as many as 30 million), and these are believed to live in the tropical rain forests that occur in a green belt spanning the planet on both sides of the equator.

Tropical forests cover 6%–7% of the total land surface of the earth—an area roughly equivalent to our contiguous 48 states. Every year humans destroy an area of forest equivalent to the size of Oklahoma. At this rate, these forests and the species they contain will disappear completely in just a few more decades. Even if only the forest areas now legally protected survive, 56%–72% of all tropical forest species would still be lost.

The loss of tropical rain forests results from an interplay of social, economic, and political pressures. Many people already live in the forest, and as their numbers increase, more of the land is cleared for farming. People move to the forests because internationally financed projects build roads and open the forests up for exploitation. Small-scale farming accounts for about 60% of tropical deforestation, and this is followed by commercial logging, cattle ranching, and mining (fig. IA). International demand for timber promotes destructive logging of rain forests in Southeast Asia and South America. The market for low-grade beef encourages their conversion to pastures for cattle. The lure of gold draws miners to rain forests in Costa Rica and Brazil.

The destruction of tropical rain forests gives only short-term benefits but is expected to cause long-term problems. The forests act like a giant sponge, soaking up rainfall during the wet season and releasing it during the dry season. Without them, a regional yearly regime of flooding followed by drought is expected to destroy property and reduce agricultural harvests. Worldwide, there could be changes in climate that would affect the entire human race. On the other hand, the preservation of tropical rain forests will result in benefits. For example, the rich diversity of plants and animals would continue to exist for scientific and pharmacological study. One-fourth of the medicines we currently use come from tropical rain forests. For example, the rosy periwinkle from Madagascar has produced 2 potent drugs for use against Hodgkin disease, leukemia, and other blood cancers. It is hoped that many of the still-unknown plants will provide medicines for other human ills.

Studies show that if the forests were used as a sustainable source of nonwood products, such as nuts, fruits, and latex rubber, they would generate as much or more revenue while continuing to perform their various ecological functions. And biodiversity could still be preserved. Brazil is exploring the concept of "extractive reserves," in which plant and animal products are harvested, but the forest itself is not cleared. Ecologists have also proposed "forest farming" systems, which mimic the natural forest as much as possible while providing abundant yields. But for such plans to work maximally, the human population size and the resource consumption per person must be stabilized.

Preserving tropical rain forests is a wise investment. Such action promotes the survival of most of the world's species—indeed, the human species, too.

Human beings reproduce and grow. Reproduction and growth are fundamental characteristics of all living things. When living things **reproduce,** they create a copy of themselves and assure the continuance of the species. (A species is a type of living thing.) Human reproduction requires that sperm contributed by the father fertilize an egg contributed by the mother. Growth occurs as the resulting cell develops into the newborn. Development includes all the changes that occur from the fertilized egg to death and, therefore, all the changes that occur during childhood, adolescence, and adulthood.

Human beings are highly organized. A **cell** is the basic unit of life, and human beings are multicellular since they are composed of many types of cells. Like cells form tissues, and tissues make up organs. Each type of organ is a part of an organ system (see fig. I.1). The different systems perform the specific functions listed in table I.1. Together, the organ systems maintain **homeostasis,** an internal environment for cells that only varies within certain limits. For example, cells require a constant supply of nutrients and they give off waste products. The digestive system takes in nutrients, and the circulatory system distributes

Figure IA

Species in jeopardy. **a.** A rare mountain gorilla overlooks its Rwanda rain forest habitat. **b.** *(below, right)* Deforestation by small-scale farming, logging, ranching, and mining lays waste to large portions of many of the world's rain forests, endangering species known and unknown and causing climate changes that could affect the entire human population.

Figure I.3

Biosphere versus cultural inheritance. Human beings are a part of the biosphere. Their cultural inheritance, represented here by a wedding, sometimes interferes with this recognition and prevents them from acting in ways that protect and preserve the biosphere, represented here by a forest.

these to the cells. The waste products are excreted by the excretory system. The work of the nervous and endocrine systems is critical because they coordinate the functions of the other systems. The concept of homeostasis is discussed further in chapter 3, and how the systems maintain homeostasis is discussed in chapters 4–15.

Homeostasis is maintained when we are healthy; illness indicates that the body is not in balance. Throughout the text Health Focus boxes include information about what we can do to help the organ systems maintain homeostasis.

> Like all living things, humans are a product of the evolutionary process and also have a cultural heritage. Like all living things, we produce and grow and are highly organized. Our organization maintains homeostasis and therefore the health of the body.

Table I.1

Human Organ Systems

System	Function
Digestive	Converts food particles to nutrient molecules
Circulatory	Transports nutrients to and wastes from cells
Immune	Defends against disease
Respiratory	Exchanges gases with the environment
Excretory	Eliminates metabolic wastes
Nervous	Regulates systems and internal environment
Musculoskeletal	Supports and moves organism
Endocrine	Regulates systems and internal environment
Reproductive	Produces offspring

How Biologists Work

Biology is a science, and as such it is a process resulting in a body of knowledge that allows us to understand and manipulate the material world. Scientists, including biologists, often employ the methodological approach described in figure I.4 to investigate a problem. For example, suppose biologists wish to determine if a new artificial sweetener (termed sweetener S) is a safe food additive. First, they would study any previous **data,** or objective information, concerning sweeteners and on the basis of this information, they might formulate a hypothesis that sweetener S is a safe food additive. Next, the biologists might design the experiment described in figure I.5 to test the hypothesis.

Experimental group: 50% of diet is sweetener S
Control group: diet contains no sweetener S

A **control group** goes through all the steps of the experiment except the one being tested.

No doubt the biologists would keep the 2 groups of mice in separate cages in the laboratory. A laboratory setting is preferred because it is here that all aspects of an experiment such as this can be controlled. Both groups of mice would be kept under the same experimental conditions and fed the same diet except for the presence or absence of sweetener S. At the end of the experiment, both groups of mice are to be examined for bladder cancer. Let's suppose that 50% of the mice in the experimental group

Figure I.4

The scientific method often includes these steps.

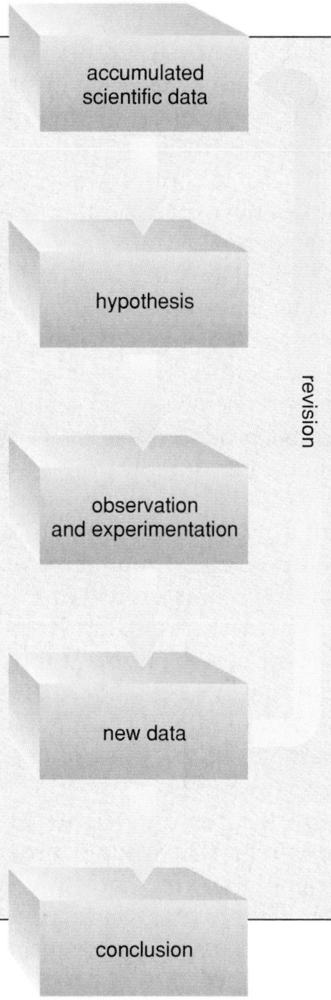

Figure I.5

Design of a controlled experiment. From *left* to *right*: genetically identical mice are divided randomly into the control group and the experimental group. All groups are exposed to the same environmental conditions, such as housing, temperature, and water supply. Only the experimental group is subjected to the test. At the end of the experiment, all mice are examined for bladder cancer.

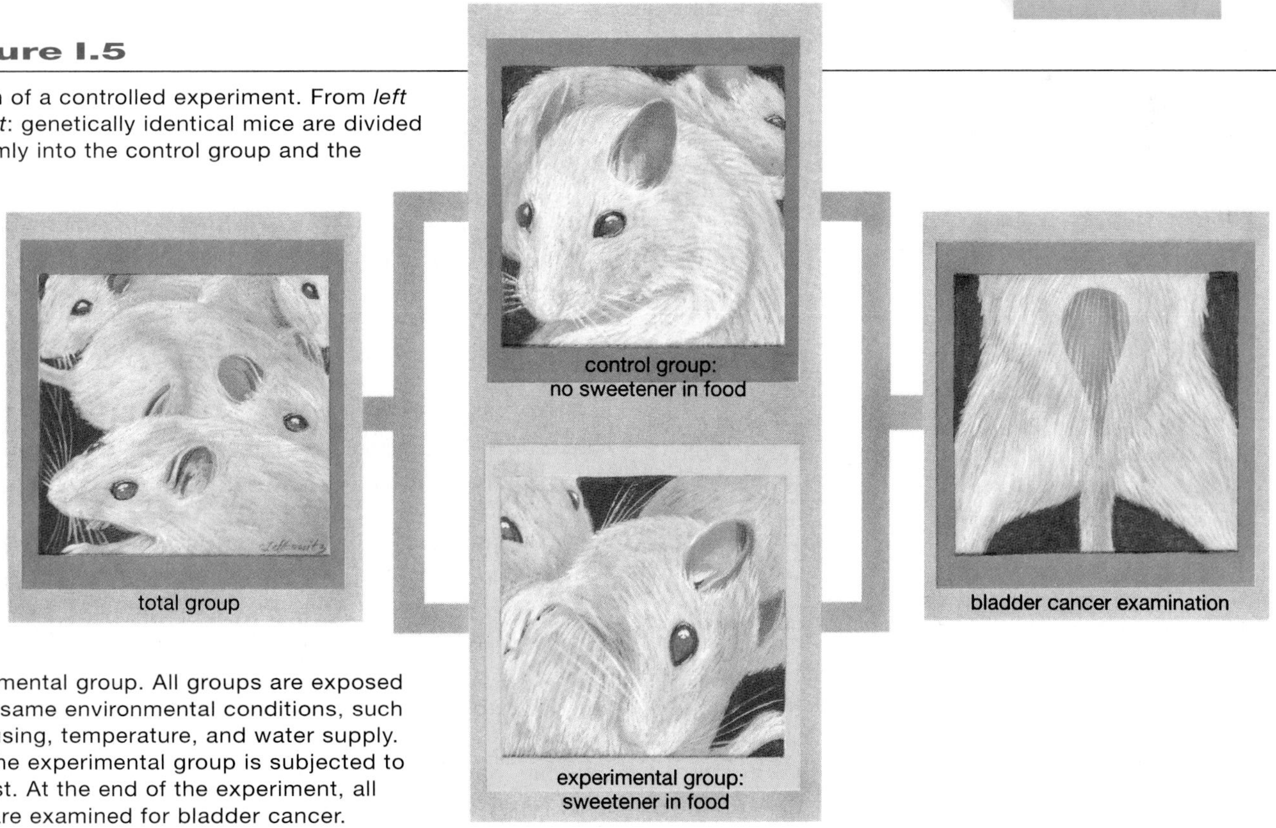

are found to have bladder cancer while none in the control group have bladder cancer. The results of this experiment do not support the hypothesis that sweetener S is a safe food additive.

Use of a control group gives greater validity to the results of the experiment. In the described experiment, the control group did not show cancer, but the experimental group did. Now, we can conclude that sweetener S in the diet must have brought about the cancer because this was the only difference between the 2 groups. Suppose, however, that both the experimental and the control group showed bladder cancer. Now, we cannot conclude that sweetener S in the diet produced the effect—that is, bladder cancer. There must be some other factor(s) causing the cancer.

Notice that while it is possible for an experiment to prove a hypothesis false, an experiment cannot prove a hypothesis true. If neither group had showed cancer we could only conclude that on the basis of this particular experiment, it seems that sweetener S is a safe food additive. Also, any experiment must be repeatable. Other scientists using the same design and carrying out the experiment under the same conditions are expected to get the same results. If they do not, the original experiment is called into question.

At this point the biologists might decide to do more experiments. They might now wish to hypothesize that sweetener S is safe if the diet contains less than 50% sweetener S. They might decide to feed sweetener S to groups of mice at ever-greater concentrations:

Group 1: diet contains no sweetener S (control group)

Group 2: 5% of diet is sweetener S

Group 3: 10% of diet is sweetener S

↓

Group 11: 50% of diet is sweetener S

The hypothetical results of this experiment are shown in figure I.6. Scientists prefer mathematical data because such data are highly objective and not subject to individual interpretation. When mathematical data are presented in the form of a table or a graph, the data are more clearly understood by most. This time, the biologists might conclude that sweetener S is safe within certain limits, that is, from 0%–10% of diet can be sweetener S. As the intake of sweetener S increases beyond this amount, there is an ever-greater chance of bladder cancer.

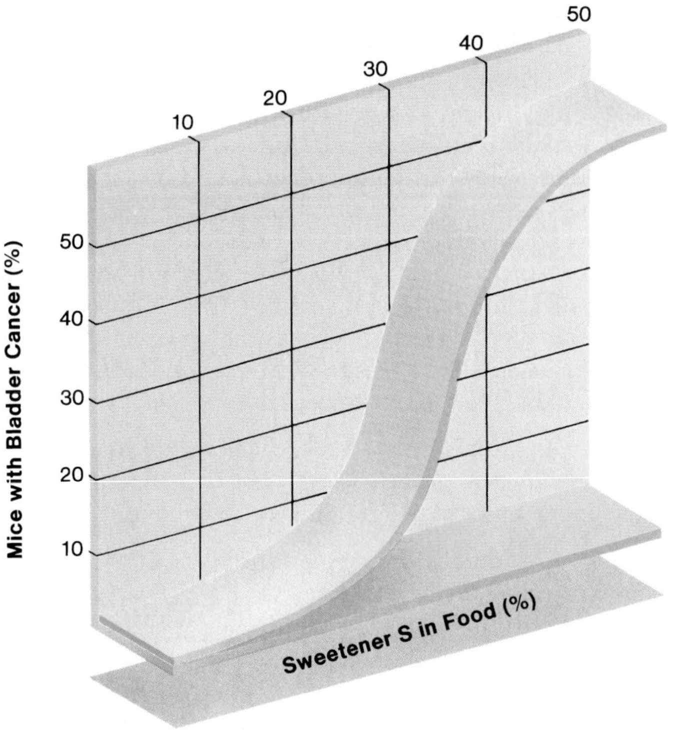

Figure I.6

Presenting the data. Scientists often acquire mathematical data that they report in the form of tables or graphs. Mathematical data are more decisive and objective than subjective observations. The data in this instance suggest that there is an ever-greater chance of bladder cancer if the food is more than 10% sweetener S.

Biological Concepts and Theories

The ultimate goal of science is to understand the natural world in terms of concepts, which are interpretations that take into account the results of many experiments and observations. These concepts are stated as theories. In a movie, a detective might claim to have a theory about a crime, or you might say that you have a theory about the win-loss record of your favorite baseball team. But in science, the word **theory** is applied to those hypotheses that are supported by many observations and are considered valid by an overwhelming majority of scientists. For this reason, biologists refer to the theory of evolution; it enables scientists to understand the history of life, the variety of living things, and the anatomy, the physiology, and the development of organisms—and even their behavior.

Our Social Responsibility

Science has improved our lives in many ways. The most obvious examples are in the field of medicine. The discovery of antibiotics such as penicillin and the vaccines for polio, measles, and mumps has increased the average life span by decades. Cell biology research is helping us understand the mechanisms that cause cancer. Genetic research has produced new strains of agricultural plants that ease the burden of feeding our burgeoning world population.

Science has also produced disturbing conclusions and has fostered technologies such as offshore oil rigs that have proven to be ecologically disastrous if not controlled properly. Too often we blame science and think that scientists are duty bound to pursue only those avenues of research that will not conflict with our system of values and/or result in environmental degradation, particularly when applied in an irresponsible manner. Yet science, by its very nature, is impartial and simply attempts to study natural phenomena. Science does not make ethical or moral decisions.

To take a biological example, consider that biologists now understand how female hormones function in the body and have developed RU-486, a medication that brings about an abortion. They alone are not obligated to decide how society should make use of this medication. This is the task of all members of society. In this and all other instances, all men and women have the responsibility to decide how best to use scientific knowledge so that it benefits the human species and all living things. Therefore, this text frequently emphasizes the application of scientific, especially biological, information to everyday human concerns.

> Scientists make use of the scientific method to discover information about the material world. It is the task of all persons to use this information as they make value judgments about their own lives and about the environment.

Summary

Human beings, just like other organisms, are a product of the evolutionary process. They are members of the animal kingdom, and are the vertebrates most closely related to the apes. Like other living things, human beings reproduce, are highly organized, and maintain a fairly constant internal environment. They are members of the biosphere but have a cultural heritage that sometimes hinders the realization of their place in nature.

When studying the world of living things, biologists and other scientists use the scientific method, which consists of the following steps:

> data, hypothesis, observation and →
> experimentation, new data, conclusions

It is the responsibility of all to make ethical and moral decisions about how best to make use of the results of scientific investigations.

Study and Thought Questions

1. Name 5 characteristics of human beings and discuss each one. (p. 3)
2. Name the 5 kingdoms of classification, and name types of organisms in each kingdom. (p. 2)

   Critical Thinking Do the characteristics you listed in question 1 apply to other living things?
3. Give evidence that human beings are related to all other living things. (p. 3)
4. Human beings are dependent upon what services performed by plants? (p. 3)

   Critical Thinking When plants make their own food they take up carbon dioxide and give off oxygen. Considering this, what other service is performed by plants?
5. What is homeostasis and how is it maintained? Choose one organ system and tell how it helps maintain homeostasis. (pp. 4, 6)
6. Name the steps of the scientific method and discuss each one. (p. 7)

   Ethical Issue We are constantly bombarded by information from all sides. Is there anything special or unique about scientific information based on scientific data compared to conclusions that are not based on such data?
7. How do you recognize a control group, and what is its purpose in an experiment? (pp. 7, 8)
8. What is our social responsibility in regard to scientific findings? (p. 8)

Writing Across the Curriculum

1. In what way are human beings a part of the biosphere despite having a cultural heritage?
2. Write a letter to a friend explaining why he or she should be interested in preserving the tropical rain forests.

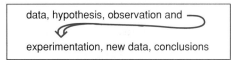

Objective Questions

1. Human beings are termed __________ because they have a backbone.
2. Human beings are dependent upon plants because they use __________ energy to make food.
3. To reproduce is to make a __________ of one's self.
4. Aside from our biological heritage, we also receive a __________ heritage in large part from our parents but also from society as a whole.
5. Human beings are made up of cells; therefore they are said to be ___________.
6. The __________ group in an experiment does not undergo the treatment being tested.
7. After considering previous findings, a scientist formulates a __________, which will be tested by observations and experimentation.
8. __________ has a responsibility to decide how scientific knowledge should be used.

Selected Key Terms

biodiversity The wide range of living things. 3

biosphere That part of the earth's surface and atmosphere where living organisms exist. 3

cell The structural and functional unit of an organism; the smallest structure capable of performing all the functions necessary for life. 4

control group In experimentation, a sample group that undergoes all the steps in the experiment except the one being tested. 7

data Experimentally derived facts. 7

ecosystem A setting in which populations interact with each other and with the physical environment. 3

homeostasis The maintenance of the internal environment, such as temperature, blood pressure, and other body conditions, within narrow limits. 4

kingdom A classification category into which organisms are placed: Monerans, Protists, Fungi, Plants, and Animals. 3

reproduce To make a copy similar to oneself; for example, bacteria dividing to produce more bacteria, or egg and sperm joining to produce offspring in more advanced organisms. 4

theory A concept supported by a large number of conclusions drawn by using the scientific method. 8

vertebrate Animal possessing a backbone composed of vertebrae. 3

Part ONE

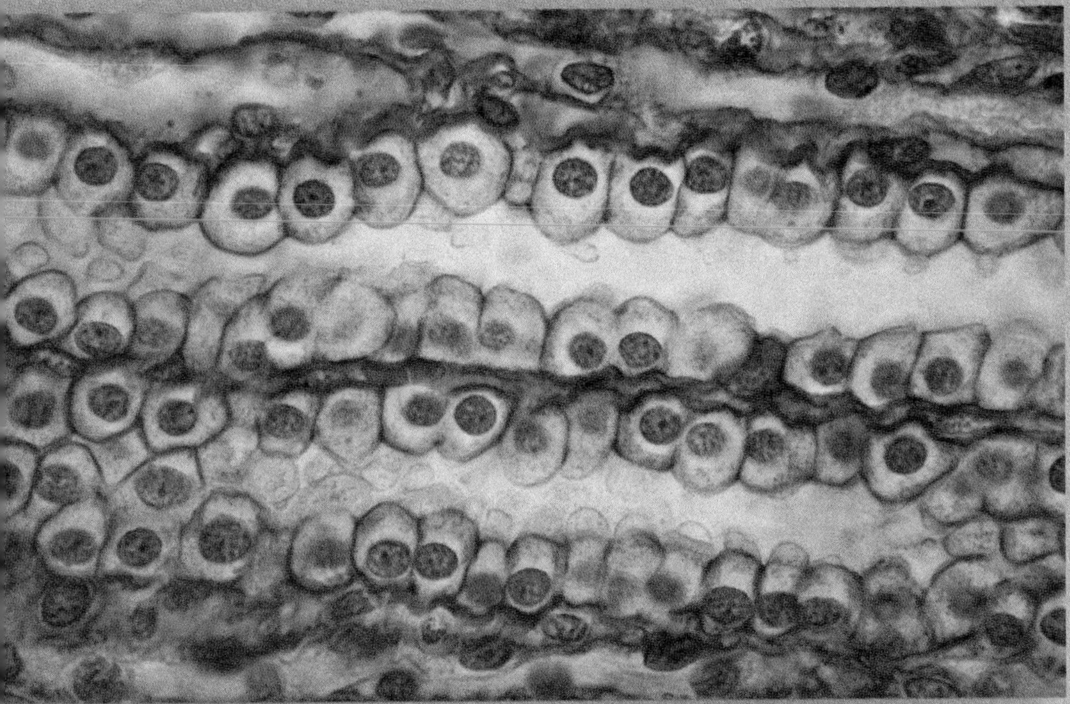

Human Organization

The human body is composed of cells, the smallest units of life. An understanding of cell structure, physiology, and biochemistry serves as a foundation for understanding how the human body functions.

Principles of inorganic and organic chemistry are discussed before a study of human cell structure is undertaken. The human cell is bounded by a membrane and contains organelles, many of which are also membranous. Membranes regulate entrance and exit of molecules and help cellular organelles carry out their functions.

The many cells of the body are specialized into tissues that are found within the organs of the various systems of the body. All body systems help maintain a dynamic constancy of the internal environment so that proper physical conditions exist for each cell.

Amino acid crystal

Chapter 1

Chemistry of Life

It is not always easy to understand that human beings, and indeed all living and nonliving things, are composed of chemicals. After all, it is not possible to see the chemicals that make up the body. However, a few minutes' reflection regarding the dietary needs of the body usually convinces us that we are indeed made of chemicals. For example, calcium is needed to maintain the bones, iron is necessary to prevent anemia, and adequate amino acid intake is required to build muscles. Chemicals are specifically arranged to form the constituents of the cells making up the body.

Chapter Concepts

Chapter Outline

Elements and Atoms

All living and nonliving things are composed of **elements,** which are basic substances that cannot be broken down further into simpler substances. Considering the variety of living and nonliving things in the world, it's quite remarkable that there are only 90 naturally occurring elements. It is even more surprising that over 90% of the human body is composed of just 3 elements: carbon, oxygen, and hydrogen. Every element has a name and a symbol; for example, the symbol for calcium is Ca (fig. 1.1). Although calcium is not one of the most abundant atoms in cells, it has many important functions. It is necessary for strong bones and plays a role in nervous conduction and muscular contraction.

Figure 1.1

Elements and atoms. **a.** The atomic symbol, atomic number, and atomic weight are given for the common elements in living things. **b.** The symbol for carbon is C and the atomic number is 6. The atomic number tells the number of protons, which is equal to the number of electrons in an electrically neutral atom. Protons and neutrons are located in the nucleus of an atom. They account for the weight of an atom: the usual weight of carbon is 12. Electrons are located in the shells. The first shell of any atom can contain up to 2 electrons; thereafter, each shell of the atoms noted in the table can contain up to 8 electrons. Notice that carbon has 4 electrons in the outermost shell.

Common Elements in Living Things

Element	Atomic Symbol	Atomic Number	Atomic Weight	Comment
hydrogen	H	1	1	These elements make up most biological molecules.
carbon	C	6	12	
nitrogen	N	7	14	
oxygen	O	8	16	
phosphorus	P	15	31	
sulfur	S	16	32	
sodium	Na	11	23	These elements occur mainly as dissolved salts.
magnesium	Mg	12	24	
chlorine	Cl	17	35	
potassium	K	19	39	
calcium	Ca	20	40	

a.

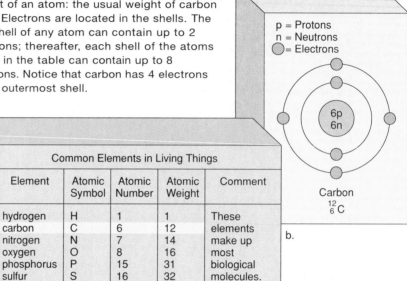

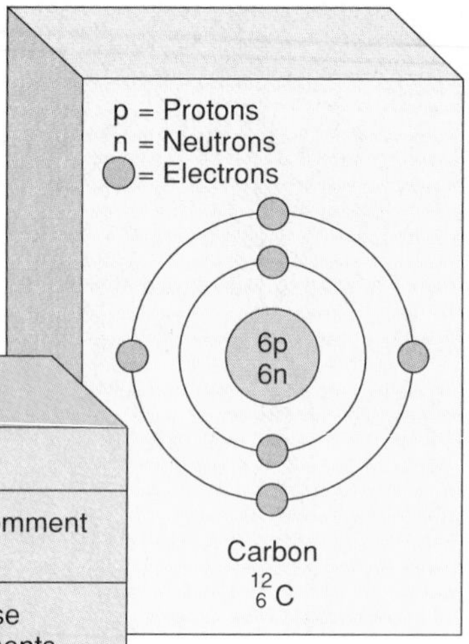

b.

Atoms Have Structure

An **atom** is the smallest unit of an element that still retains the chemical and physical properties of an element. While it is possible to split an atom by physical means, an atom is the smallest unit to enter into chemical reactions. For our purposes, it is satisfactory to think of each atom as having a central nucleus, where subatomic particles called **protons** and **neutrons** are located, and shells, where **electrons** orbit about the nucleus. Two important features of protons, neutrons, and electrons are their weight and charge.

Name	Charge	Weight
Electron	One negative unit	Almost no weight
Proton	One positive unit	One atomic unit
Neutron	No charge	One atomic unit

The structure of the carbon atom is given in figure 1.1*b*. The further the shell from the nucleus, the higher the energy level and the faster the electrons move. The chemical properties of an atom depend largely on the number of electrons in the outer shell.

Isotopes are atoms of a particular element that differ only by the number of neutrons they contain. For example, all carbon atoms have 6 protons and 6 electrons; most carbon atoms have 6 neutrons but others have more or less than that number. Certain isotopes called *radioactive isotopes* undergo disintegration or decay. As they decay, they emit radiation, which may be detected using a special counter or scanner. Radioactive isotopes are widely used in biological and medical research; for example, because the thyroid gland uses iodine (I), it is possible to administer a dose of radioactive iodine and then observe later that the thyroid has taken it up (fig. 1.2).

Molecules and Compounds

Atoms often bond with each other to form a chemical unit called a **molecule.** Some molecules contain more than one kind of atom, and in those instances, a molecule is a part of a compound. A **compound** has many copies of the same type of molecule. The formula for a molecule tells the proportion of the different atoms in the molecule.

$CaCl_2$

In the compound calcium chloride, each molecule has 1 calcium atom and 2 chlorine atoms.

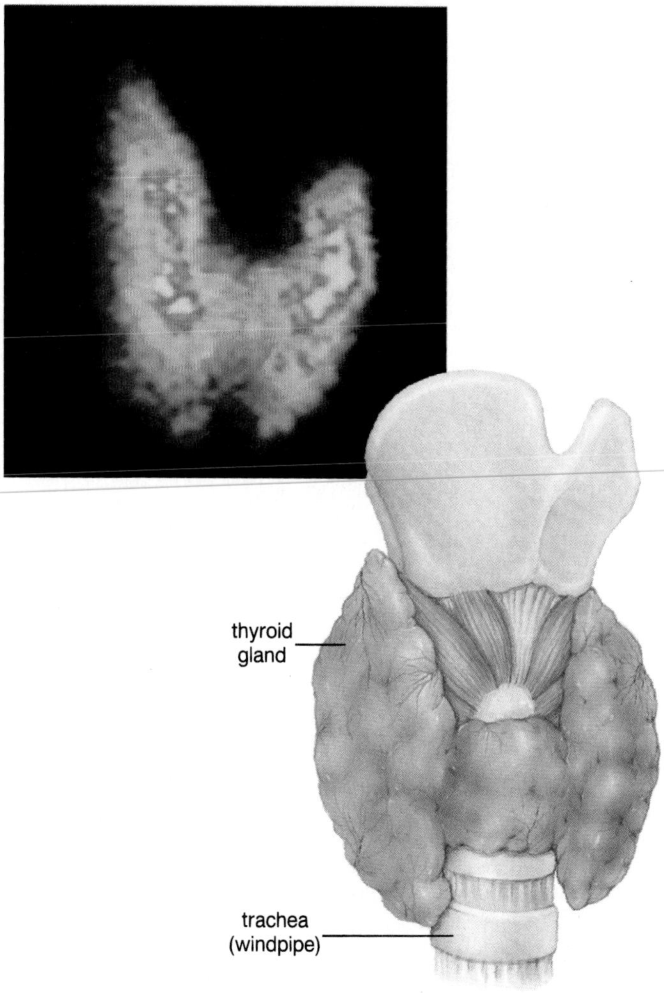

Figure 1.2

Use of radioactive iodine. A scan of the thyroid gland 24 hours after the patient was administered radioactive iodine. The thyroid gland is located at the base of the neck.

Atom-Atom Interactions

Compounds and/or molecules result when atoms react with one another. Usually, reactions between atoms involve the electrons in their outer shells. The *octet rule,* based on chemical findings, states that atoms react with one another to achieve 8 electrons in their outer shells. Hydrogen (H) is one exception to this rule. In hydrogen, the first shell, which is complete with 2 electrons only, is the outer shell.

In one type of reaction, there is a transfer of an electron(s) from one atom to another to form a molecule. Such atoms are thereafter called **ions,** and the reaction is called *an ionic reaction.* For example, figure 1.3 depicts a reaction between sodium (Na) and chlorine (Cl) in which chlorine takes an electron from sodium. Now the sodium ion (Na^+) carries a positive charge, and the chloride ion (Cl^-) carries a negative charge. Notice that a negative charge indicates that the ion has more electrons (–) than protons (+), and a positive charge indicates that the ion has more protons (+) than electrons (–). Oppositely charged ions are attracted to one another, and this attraction is called an **ionic bond.**

In another type of reaction, atoms form a molecule by sharing electrons. The bond that forms between these atoms is called a **covalent bond.** For example, when oxygen reacts with 2 hydrogen atoms, water (H_2O) is formed (fig. 1.4).

Sometimes the atoms in a covalently bonded molecule share electrons evenly, but in water the electrons spend more time circling the larger oxygen than the smaller hydrogens. Therefore, there is a slight positive charge on the hydrogen atoms and a slight negative charge on the oxygen atom. For this reason, water is called a *polar molecule,* and hydrogen bonding occurs between water molecules (fig. 1.5). A **hydrogen bond** occurs whenever a

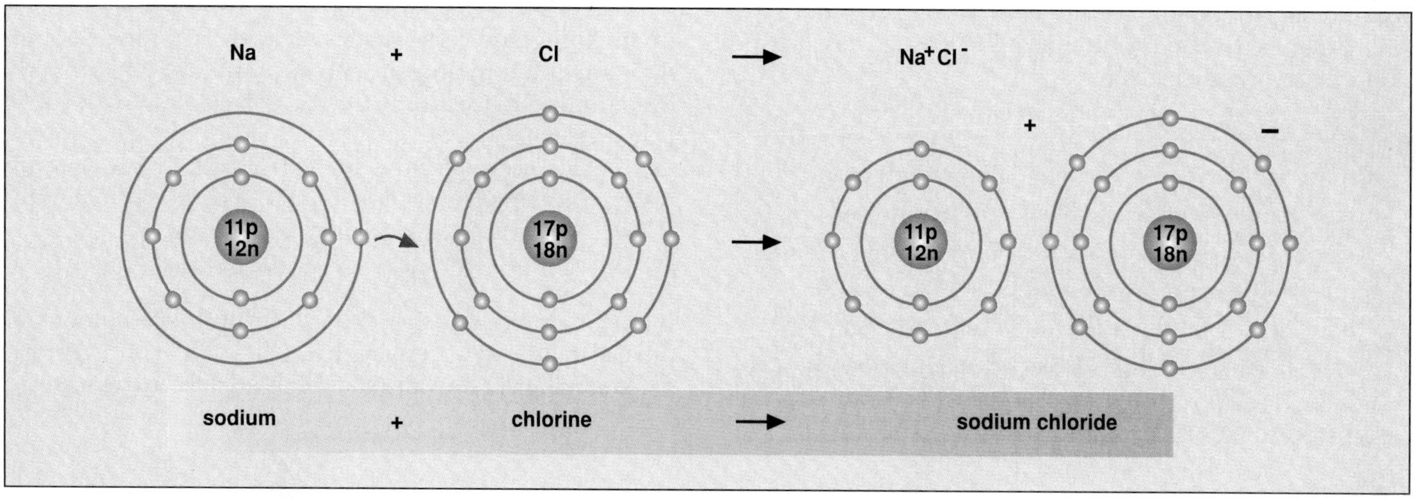

Figure 1.3

Formation of the salt sodium chloride. During this ionic reaction, an electron is transferred from the sodium atom to the chlorine atom. Each resulting ion carries a charge as shown. Most people use the term *salt* to refer only to sodium chloride, but chemists use the term to refer to similar combinations of positive and negative ions.

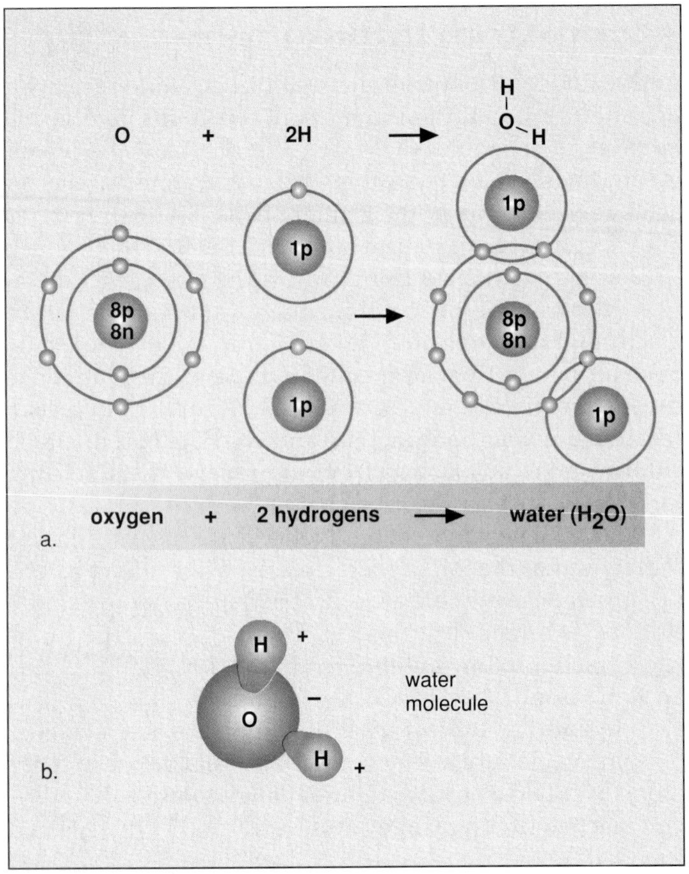

Figure 1.4

Water molecule. **a.** Following a covalent reaction, oxygen is sharing electrons with 2 hydrogen atoms. **b.** Three-dimensional model of water molecule showing that water molecules are polar; each hydrogen (H) carries a partial positive charge, and each oxygen (O) carries a partial negative charge.

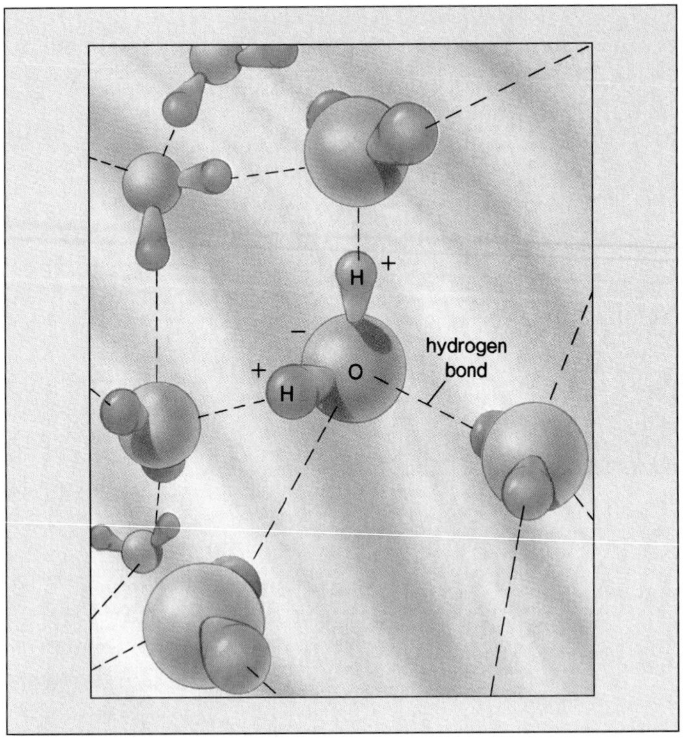

Figure 1.5

Hydrogen bonding between water molecules. The polarity of the water molecules brings about hydrogen bonding between the molecules in the manner shown. The dashed lines represent hydrogen bonds. Polarity causes water to be an excellent solvent; it dissolves various chemical substances, particularly other polar molecules. Hydrogen bonding allows water to absorb a great deal of heat and still remain liquid; therefore, water is liquid at body temperature. Eventually, however, water vaporizes; when humans sweat, body heat is used to vaporize water.

partially positive hydrogen is attracted to a partially negative atom in another molecule. The hydrogen bond is represented by a dashed line in figure 1.5 because it is a weak bond that is easily broken.

Atoms react with one another to form molecules. In one type of reaction, positively and negatively charged ions are formed when an electron(s) is (are) transferred from one atom to another. In another type of reaction, the atoms share electrons in a molecule where the atoms are covalently bonded to one another.

Acids and Bases

Living things are very sensitive to acids and bases because they affect the hydrogen ion concentration $[H^+]$ in body fluids and cells. Cells only function properly when the $[H^+]$ is near normal.

Acids are molecules that dissociate or break up in water, releasing hydrogen ions (H^+). For example, hydrochloric acid (HCl) dissociates in this manner:

$$HCl \rightarrow H^+ + Cl^-$$

Bases are molecules that either take up hydrogen ions or dissociate in water, releasing hydroxide ions (OH^-). For example, sodium hydroxide (NaOH) dissociates in this manner:

$$NaOH \rightarrow Na^+ + OH^-$$

Ecology Focus

The Harm Done by Acid Rain

Normally, rainwater has a pH of about 5.6 because the carbon dioxide in the air combines with water to give a weak solution of carbonic acid. Rain falling in the northeastern United States and southeastern Canada now has a pH between 5.0 and 4.0. Remember that a pH of 4 is 10 times more acidic than a pH of 5 to appreciate the increase in acidity this represents.

There is very strong evidence that this observed increase in rainwater acidity is a result of the burning of fossil fuels such as coal and oil, as well as gasoline derived from oil. When fossil fuels are burned, sulfur dioxide (SO_2) and nitrogen oxides (NO_x) are produced, and they combine with water vapor in the atmosphere to form acids. These acids return to earth as rain or snow, a process properly called wet deposition but more often called acid rain. Dry particles of sulfate salts (e.g., $MgSO_4$) and nitrate salts (e.g., Na_2NO_3) descend from the atmosphere during dry deposition.

Unfortunately, regulations that require the use of tall smokestacks on burners to reduce local air pollution only cause pollutants to be carried far from their place of origin. Acid deposition in southeastern Canada is due to the burning of fossil fuels in factories and power plants located in the Midwest. Tensions between Canada and the United States have been eased by the 1990 Clean Air Act, which called for a 10-million-ton reduction in sulfur dioxide emissions and a 2-million-ton reduction in nitrogen oxide emissions in the United States by the year 2000. Thereafter, sulfur dioxide emissions will be capped and no increase in total emission will be allowed. Canada agreed to make comparable reductions in its emissions of air pollutants, some of which find their way to the northeastern United States. In 1991, the 2 countries signed a formal air pollution agreement.

Acid rain adversely affects lakes, particularly in areas where the soil is thin and lacks limestone (calcium carbonate, $CaCo_3$), a buffer to acid rain. Acid rain leaches aluminum (Al) from the soil, carries it into the lakes, and converts mercury (Hg) deposits in lake bottom sediments to soluble methyl mercury, a toxin. Lakes not only become more acidic, they also show accumulation of toxic substances. In Norway and Sweden, at least 16,000 lakes contain no fish and an additional 52,000 lakes are threatened by the change in pH levels. In Canada, some 14,000 lakes are almost fishless and an additional 150,000 are in peril because of excess acidity. In the United States, about 9,000 lakes (mostly in the Northeast and Upper Midwest) are threatened, one-third of them seriously.

In forests, acid rain weakens trees because it leaches away nutrients and releases aluminum. By 1988, most spruce, fir, and other conifers atop North Carolina's Mount Mitchell were dead from being bathed in ozone and acid fog for years. The soil was so acidic that new seedlings could not survive. Nineteen countries in

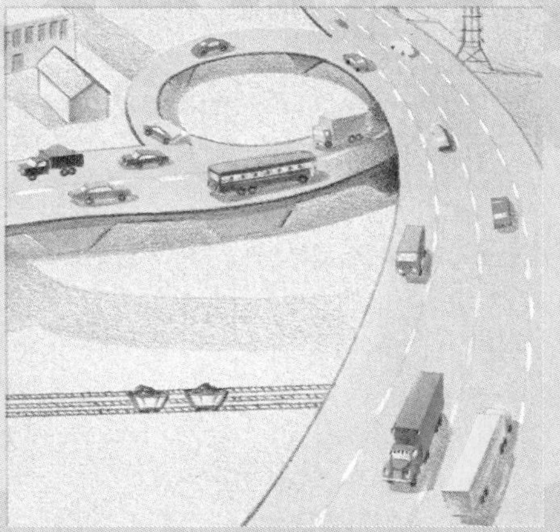

Europe have reported woodland damage ranging from 5%–15% of the forested area in Yugoslavia and Sweden to 50% or more in the Netherlands, Switzerland, and the former West Germany. More than one-fifth of Europe's forests are now damaged.

These aren't the only effects of acid rain. Other effects include reduction of agricultural yields, damage to marble and limestone monuments and buildings, and even illnesses in humans. Acid rain has been implicated in the increased incidence of lung cancer and possibly colon cancer in residents of the East Coast. Tom McMillan, Canadian Minister of the Environment, says that acid rain is "destroying our lakes, killing our fish, undermining our tourism, retarding our forests, harming our agriculture, devastating our heritage and threatening our health."

There are, of course, things that can be done.

1. Whenever possible, use alternative energy sources, such as solar, wind, hydropower, and geothermal energy.
2. Use low-sulfur coal or remove the sulfur impurities from coal before it is burned.
3. Require people to use mass transit rather than drive their own automobiles.
4. Reduce our energy needs through other means of energy conservation.

These measures and possibly others could be taken immediately. It is only necessary for us to determine that they are worthwhile.

Sources: G. Tyler Miller, Living in the Environment, *1993, Wadsworth Publishing Company, Belmont, CA, 1992; and Lester R. Brown, et al.,* State of the World, *W. W. Norton and Company, Inc., New York, NY.*

pH Scale: [H+] Can Be Measured

The **pH scale** is used to indicate the acidity and basicity (alkalinity) of a solution.[1] When a water molecule dissociates it releases an equal number of hydrogen ions (H^+) and hydroxide ions (OH^-).

$$H—O—H \rightarrow H^+ + OH^-$$

The portion of water molecules that dissociates is 10^{-7}(or 0.0000001), which is the source of the pH value for neutral solutions. Neutral solutions have a pH of 7. In other words, the pH scale was devised to simplify discussion of the hydrogen ion concentration [H^+].

The pH scale ranges from 0 to 14 (fig. 1.6). Any number below pH 7 is an acid pH, and any number above pH 7 is a basic pH. As we move down the pH scale, each unit has 10 times the number of H^+ as the previous unit, and as we move up the scale each unit has 10 times the number of OH^- as the previous unit.

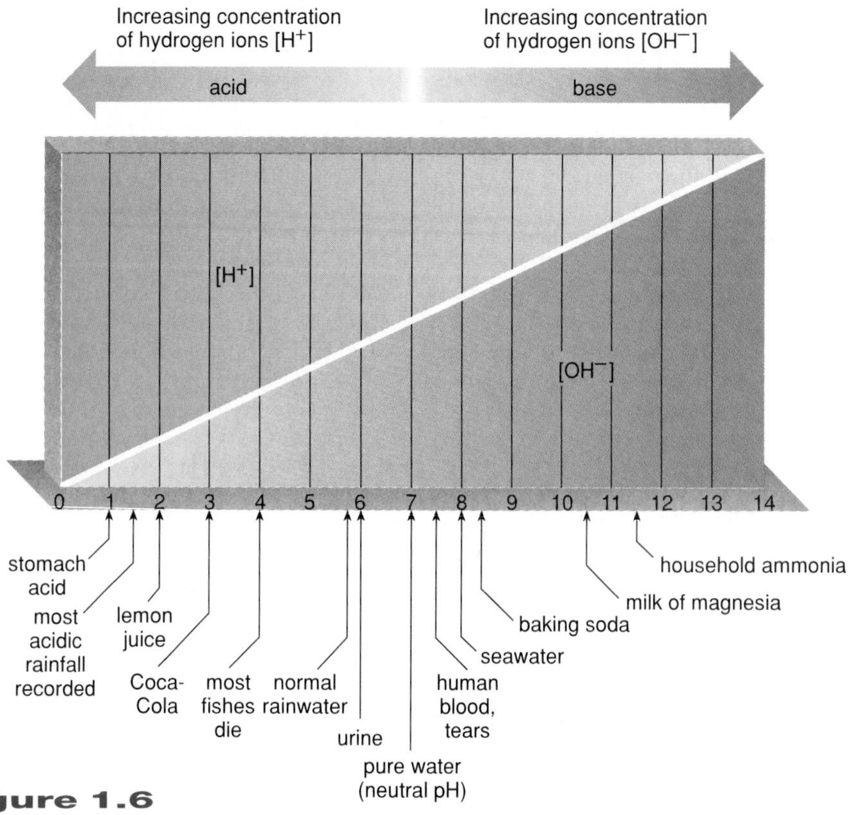

Figure 1.6

The pH scale. Any pH above 7 is basic, while any pH below 7 is acidic. The proportionate amount of hydrogen ions (H^+) to hydroxide ions (OH^-) is indicated by the diagonal line. The pH of some common substances is given.

In living things, pH needs to be maintained within a narrow range or there are health consequences. The Ecology Focus for this chapter discusses the environmental consequences of rain and snow becoming more acidic.

Buffers: [H+] Can Be Controlled

The pH of our blood when we are healthy is always about 7.4. **Buffers,** chemicals or combinations of chemicals that take up excess hydrogen ions (H^+) or hydroxide ions (OH^-), help to keep the pH constant.

When an acid is added to a buffered solution, a buffer takes up excess hydrogen ions, and when a base is added to a buffered solution, a buffer takes up excess hydroxide ions. Therefore, the pH changes minimally whenever a solution is buffered.

> Acids have a pH that is less than 7, and bases have a pH that is greater than 7. The presence of buffers helps keep the pH of body fluids constant at about neutral, or pH 7, because a buffer can absorb both hydrogen and hydroxide ions.

1. pH is defined as the negative logarithm of the hydrogen ion concentration [H^+].

Organic Molecules

Table 1.1 contrasts inorganic molecules with organic molecules. Both types of molecules are necessary to the proper functioning of the human body. The chemistry of carbon accounts for the formation of the very large number of organic compounds we associate with living organisms. Carbon has 4 electrons in its outer shell, and this allows it to bond with as many as 4 other atoms. Carbon can also share 2 pairs of electrons with another atom, giving a double covalent bond. Usually carbon bonds to hydrogen, oxygen, nitrogen, or another carbon atom. The ability of carbon to bond to itself makes carbon chains of

Table 1.1

Inorganic Versus Organic Molecules

Inorganic Molecules	Organic Molecules
Usually contain positive and negative ions	Always contain carbon and hydrogen
Usually ionic bonding	Always covalent bonding
Always contain a small number of atoms	May be quite large with many atoms
Often associated with nonliving elements	Associated with living organisms

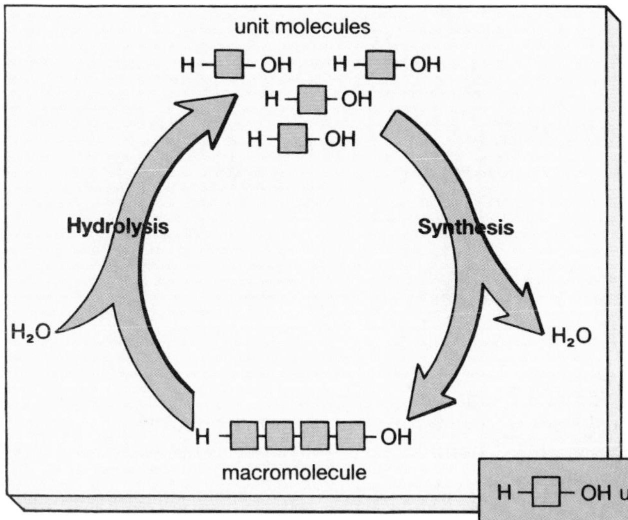

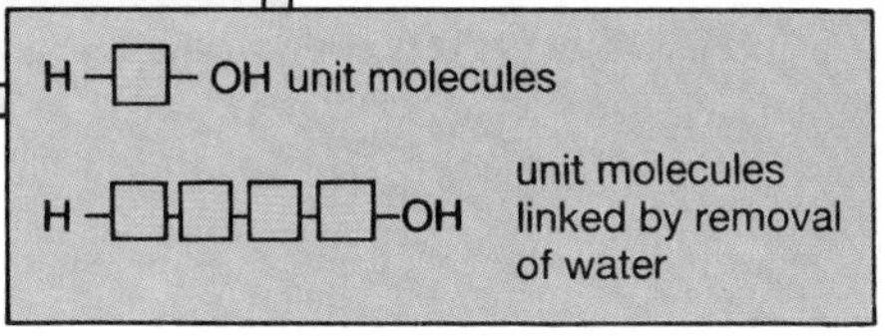

Figure 1.7

Synthesis and hydrolysis. When synthesis of a macromolecule (polymer) occurs, unit molecules join as water is released. When hydrolysis of a macromolecule occurs, water is added as unit molecules are released.

various lengths and shapes possible. For example, carbon-to-carbon bonding sometimes results in ring compounds of biological significance.

Carbon chains make up the skeleton, or backbone, of organic molecules. Organic molecules all have a carbon backbone, but in addition they may have different functional groups. *Functional groups* are clusters of atoms with a certain pattern that always behave in a certain way. For example, the functional group $-\mathrm{C}\begin{smallmatrix}=\mathrm{O}\\ -\mathrm{OH}\end{smallmatrix}$ is an acidic group because it releases hydrogen ions (H^+), causing a decrease in pH. Functional groups determine the particular characteristics of organic molecules because they all have a carbon backbone.

Small organic molecules are often unit molecules of **macromolecules** such as the biomolecules—proteins, polysaccharides, lipids, and nucleic acids—found in cells. These are the macromolecules and unit molecules we will study:

Macromolecules	Unit Molecule(s)
Protein	Amino acid
Polysaccharide	Monosaccharide
Lipid	Glycerol and fatty acid (for fat)
Nucleic acid	Nucleotide

A covalent bond that joins 2 unit molecules together within a macromolecule is created after the removal of hydrogen (—H) from one molecule and a hydroxyl group (—OH) from the next molecule. As water forms, dehydration **synthesis** occurs; dehydration because water is removed and synthesis because a bond is made. Macromolecules, called *polymers* when they are chains of unit molecules, can be broken down in a manner opposite to synthesis: the addition of water leads to the disruption of the bonds linking the unit molecules together. During this process, called **hydrolysis,** one molecule takes on hydrogen and the next takes on a hydroxyl group (fig. 1.7). This is a hydrolysis reaction because water (hydro) is used to break (lyse) a bond.

Dehydration synthesis (removal of water) joins monomers to form macromolecules, and hydrolysis (the addition of water) breaks macromolecules apart into unit molecules.

Proteins

Proteins are large polymers whose main function can be structural. For example, in humans, keratin is a protein that makes up hair and nails, and collagen is a protein found in all types of connective tissue, including ligaments, cartilage, bone, and tendons. The muscles contain proteins (actin and myosin), which account for their ability to contract (fig. 1.8).

Some proteins function as **enzymes,** which are necessary contributors to the chemical workings of the cell and therefore of the body. Enzymes are organic catalysts that speed up chemical reactions. They work so quickly that a reaction that could take several hours or days without an enzyme takes only a fraction of a second when the enzyme is present.

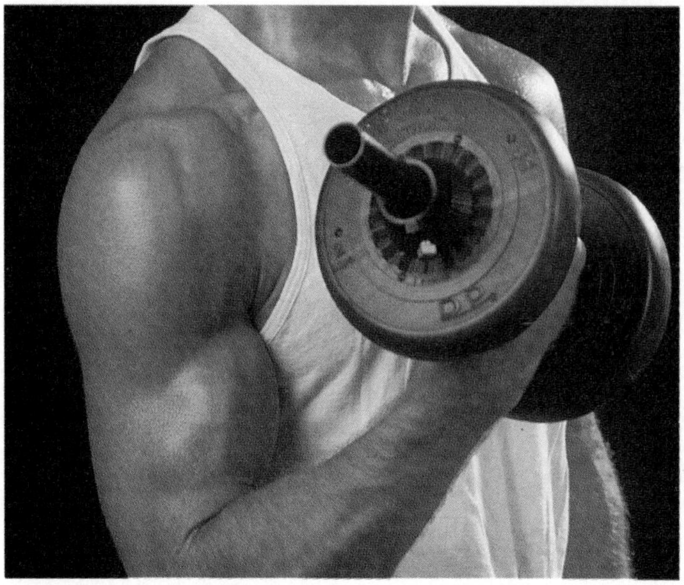

Figure 1.8

The well-developed muscles of an athlete. Muscle cells contain many protein molecules. A major concern today is steroid use to promote the buildup of muscles. This practice can be detrimental to health (p. 262).

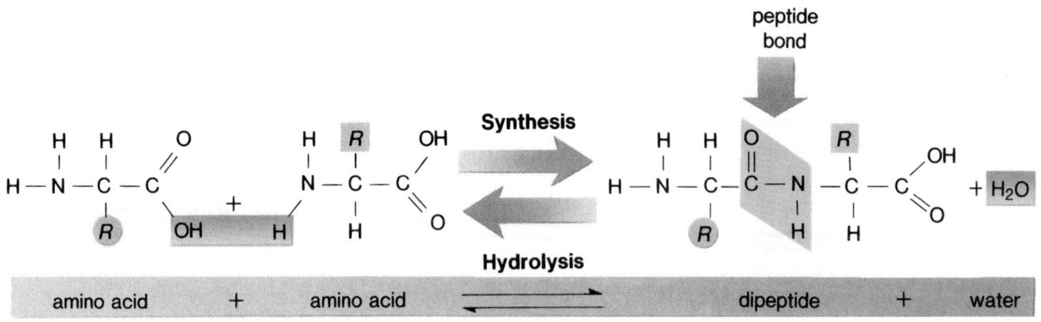

Figure 1.9

Formation of a dipeptide. On the left-hand side of the equation, there are 2 different amino acids, as signified by the difference in the *R*-group notations. As the peptide bond forms, water is given off—the water molecule on the right-hand side of the equation is derived from components removed from the amino acids on the left-hand side. During hydrolysis, water is added to the dipeptide and the peptide bond is broken, releasing the amino acids.

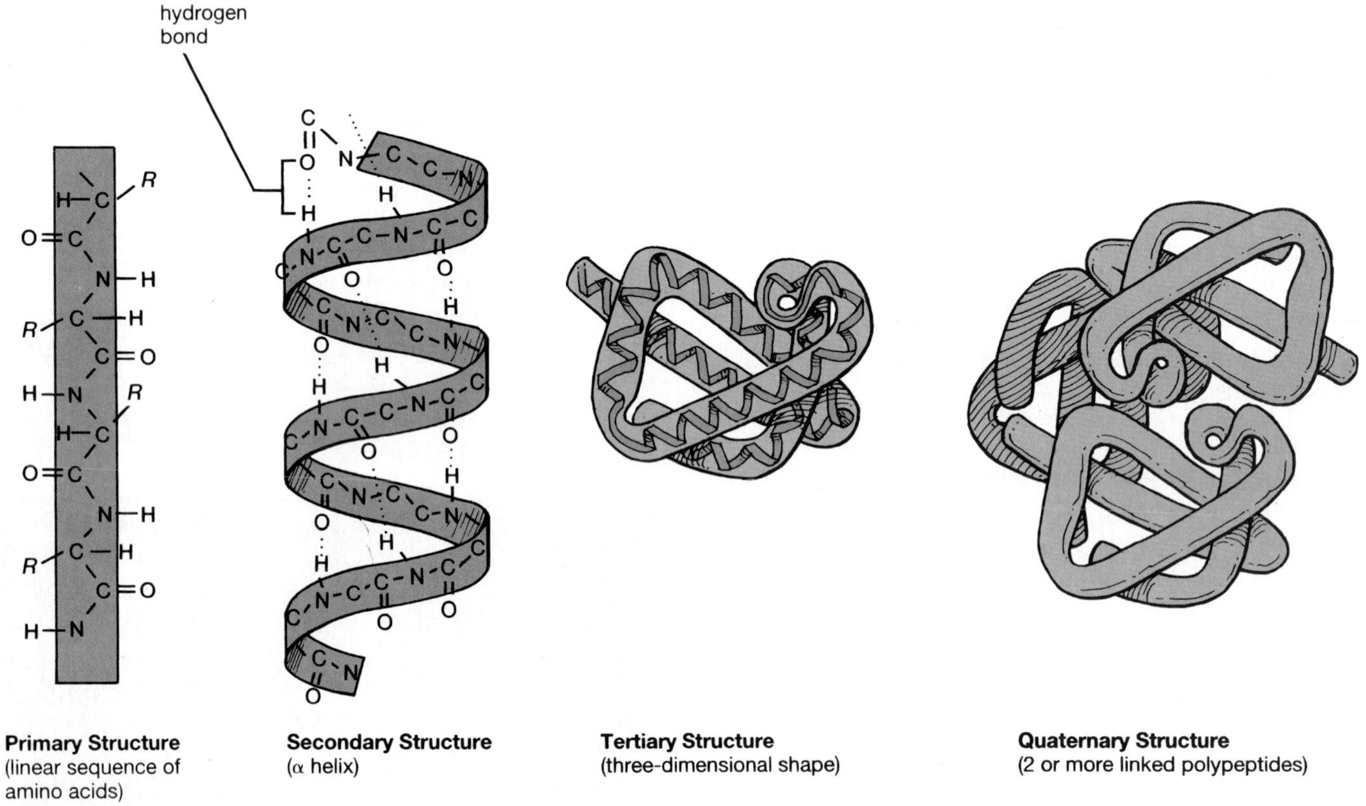

Figure 1.10

Levels of protein structure. Primary structure of a protein is the order of the amino acids; secondary structure is often an alpha (α) helix, in which hydrogen bonding occurs along the length of a polypeptide, as indicated by the dotted lines; in globular proteins, the tertiary structure is the twisting and the turning of the helix, which takes place because of bonding between the *R* groups; and the quaternary structure occurs when a protein contains 2 or more linked polypeptides. Enzymes are globular proteins, but some proteins, like muscle proteins, are fibrous.

Amino Acids Form Polypeptides

The unit molecules found in proteins are called **amino acids.** The name *amino acid* refers to the fact that the molecule has 2 functional groups, an *amino acid* and an *acid group.*

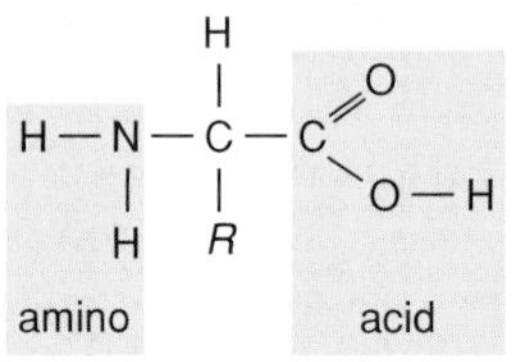

R = remainder of molecule

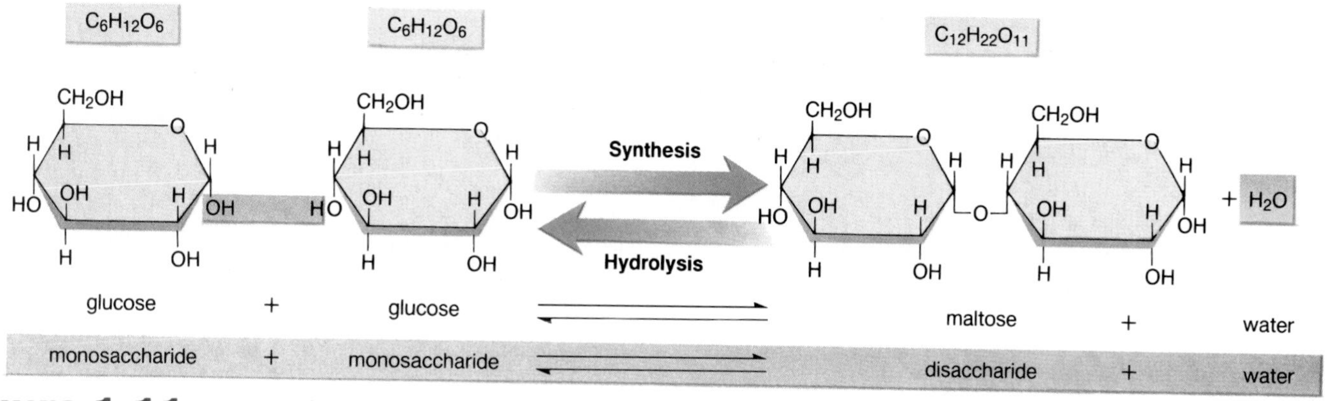

Figure 1.11

Synthesis and hydrolysis of maltose, a disaccharide containing 2 glucose units. During synthesis, a bond forms between the 2 glucose molecules as the components of water are removed. During hydrolysis, the components of water are added as the bond is broken.

Amino acids differ from one another by their *R groups*, the portion of the molecule that varies from one amino acid to the other. The *R* group can be a single hydrogen atom or a hydrocarbon chain that is straight, branched, or even forming a ring. Some *R* groups contain a functional group such as a hydroxyl group (—OH).

The bond that joins 2 amino acids together is called a **peptide bond.** As you can see in figure 1.9, when synthesis occurs, the acid group of one amino acid reacts with the amino group of another amino acid and water is given off. A dipeptide results when 2 amino acids join; a **polypeptide** is a string of amino acids joined by peptide bonds. All polypeptides have a backbone consisting of N—C—C—N—C—C—N; they differ by the sequence of the attached *R* groups. A polypeptide can contain hundreds, even thousands, of amino acids, and a protein consists of one or more polypeptides (fig. 1.10).

Proteins Have Levels of Structure

Proteins are said to have 4 levels of structure: primary, secondary, tertiary, and quaternary. The *primary structure* is simply the sequence, or order, of the different amino acids. Any number of the 20 different amino acids may be joined in various sequences, and each type of protein has its own particular sequence. The resulting chain is like a necklace comprised of up to 20 different types of beads that reoccur and are linked in a set way.

The *secondary structure* is the usual orientation of the amino acid chain. One common arrangement of the chain is the alpha (α) helix, a right-handed spiral held in place by hydrogen bonding between members of the various peptide bonds. The *tertiary structure* of a protein refers to its final three-dimensional shape. In a structural protein like collagen, the helical chains lie parallel to one another, but in enzymes the helix bends and twists in different ways. The final shape of a protein is maintained by various types of bonding between the *R* groups. Covalent, ionic, and hydrogen bonding are all seen.

If a protein has more than one polypeptide chain, each chain has its own primary, secondary, and tertiary structures. Within the protein, these separate chains are arranged to give a fourth level of structure termed the *quaternary structure.* Figure 1.10 illustrates some of the main features of protein chemistry.

> An amino acid is the unit molecule for peptides and proteins. Proteins have both structural and metabolic[2] functions in the human body.

Carbohydrates

Carbohydrates are characterized by the presence of H—C—OH groupings in which the ratio of hydrogen atoms to oxygen atoms is approximately 2:1. Since this ratio is the same as the ratio in water, the compound's name—hydrates of carbon—is very appropriate. If the number of carbon atoms in the compound is low (from about 3 to 7), the carbohydrate is a monosaccharide. Larger carbohydrates are created by joining together monosaccharides in the manner described in figures 1.7 and 1.11.

Sugars Are Many

As their name implies, **monosaccharides** are sugars having only one unit. These compounds are often designated by the number of carbons they contain; for example, pentose sugars have 5 carbons, and hexose sugars have 6 carbons. **Glucose** is a 6-carbon sugar, with the structural formula shown in figure 1.11. Although there are other monosaccharides with the molecular formula $C_6H_{12}O_6$, in this text we will use the molecular formula $C_6H_{12}O_6$ to mean glucose, since glucose is the most common 6-carbon monosaccharide found in cells. Cells use glucose as a direct energy source.

2. Metabolism is all the chemical reactions that occur in a cell.

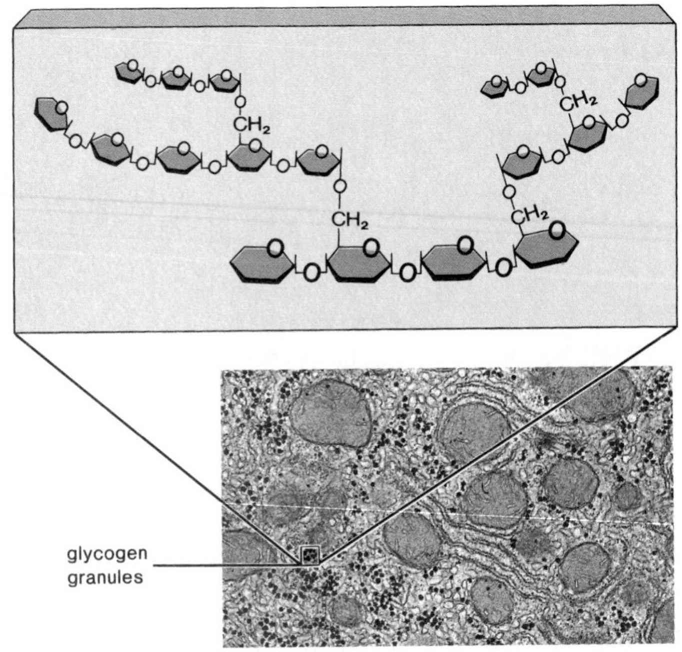

Figure 1.12

Glycogen structure and function. Glycogen is a highly branched polymer of glucose molecules. The branching allows breakdown to proceed at several points simultaneously. The electron micrograph shows glycogen granules in liver cells. Glycogen is the storage form of glucose in animals.

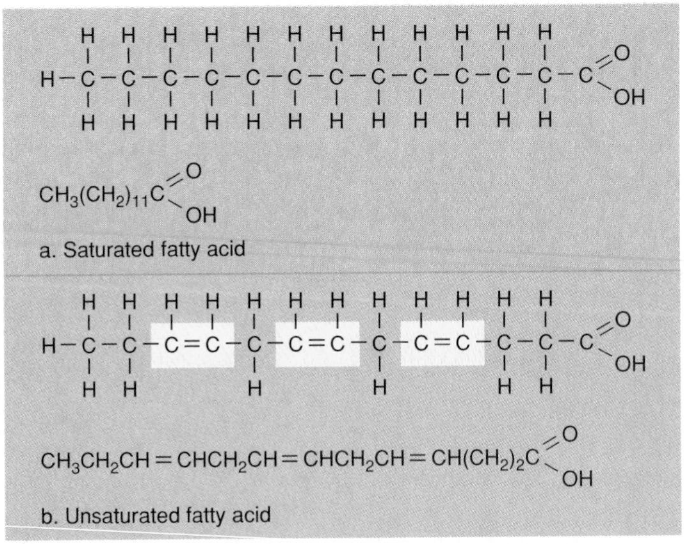

Figure 1.13

Saturated fatty acid versus unsaturated fatty acid. Fatty acids are either saturated (have no double bonds between carbon atoms) or unsaturated (have double bonds between carbon atoms). **a.** In a saturated fatty acid, the carbon atoms (C) carry all the hydrogen atoms (H) possible. **b.** In this unsaturated fatty acid, there are 3 double bonds.

The term **disaccharide** tells us that there are 2 monosaccharide units joined in the compound. When 2 glucose molecules join, maltose is formed (fig. 1.11). You may also be interested to know that when glucose and another monosaccharide, fructose, are joined, the disaccharide called *sucrose* is formed. Sucrose is derived from plants and is commonly used at the table to sweeten foods. Another sugar you may have heard of is lactose, or milk sugar, which is composed of glucose and another monosaccharide, galactose.

Polysaccharides: Animal and Plant

A **polysaccharide** is a carbohydrate that contains a large number of monosaccharide molecules. Three polysaccharides are common in animals and plants: glycogen, starch, and cellulose. All of these are polymers of glucose. Even though all 3 chains contain only glucose, they are distinguishable from one another.

Glycogen, the storage form of glucose in humans, is a molecule with many side branches (fig. 1.12). After eating, the liver stores glucose as glycogen; in between eating, the liver releases glucose so that the blood concentration of glucose is always near 0.1% (100 mg/100 ml).

Starch and *cellulose* are found in plants. Plants store glucose as starch, a polymer similar in structure to glycogen except that it has few side branches. Starch is an important source of glucose energy in our diet because it can be hydrolyzed to glucose by digestive enzymes. Athletes sometimes favor a meal rich in pasta for this reason. In cellulose, a common structural compound in plants, the glucose units are joined by a slightly different type of linkage compared to that of glycogen and starch. For this reason we are unable to digest cellulose (often called *fiber*), and it passes through our digestive tract. As discussed in the Health Focus for this chapter, fiber in the diet promotes good health for any number of reasons.

> In this text, $C_6H_{12}O_6$ stands for glucose, which is converted to glycogen and serves as an important stored energy source for humans.

Lipids

Many **lipids** are nonpolar and therefore are insoluble in water. This is true of fats, the most familiar lipids, such as lard, butter, and oil, which are used in cooking or at the table. In the body, fats serve as long-term energy sources and as insulation against the cold. Adipose (fat) tissue is composed of cells that contain many molecules of neutral fat.

Fats and Oils Are Energy Rich

A neutral (nonpolar) fat contains 2 types of unit molecules: **glycerol** and **fatty acids.** Each fatty acid is composed of a long chain of carbon atoms with hydrogens attached and ends in an acid group (fig. 1.13). The length of the hydrocarbon chain can vary, and a fatty acid can be either *saturated*

Health Focus

Health Effects of Fiber

Cereals, vegetables, and fruits all contain fiber, substances that cannot be broken down by human digestive enzymes. Bran is the chief fiber constituent of grains such as oat or wheat (fig. 1A). Oat bran is a gummy soluble fiber, while wheat bran is a rigid insoluble fiber. These 2 types of fiber function differently in the digestive tract.

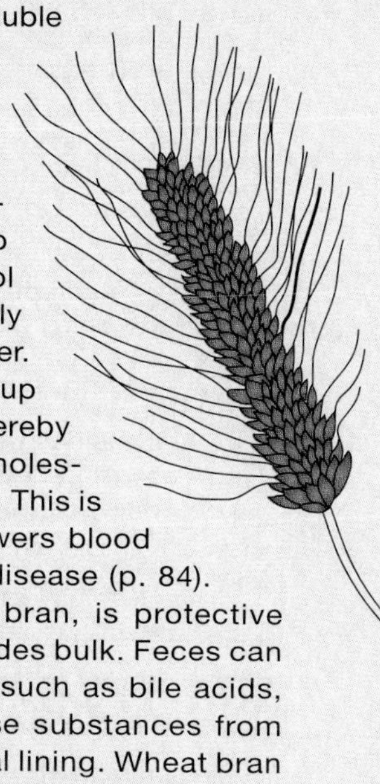

Bile produced in the liver is delivered to the small intestine, where it functions in the digestive process. It so happens that the liver uses cholesterol to make bile acids, which are normally absorbed for recycling back to the liver. Soluble fiber, such as oat bran, soaks up bile acids, preventing recycling and thereby causing the liver to break down more cholesterol in order to make more bile acids. This is the manner in which soluble fiber lowers blood cholesterol and helps prevent heart disease (p. 84).

Insoluble fiber, such as wheat bran, is protective against colon cancer because it provides bulk. Feces can contain cancer-causing substances, such as bile acids, but regular elimination prevents these substances from remaining in contact with the intestinal lining. Wheat bran also reduces the absorption of dietary calcium, and this too may be protective against colon cancer. European studies show that people who eat calcium-rich dairy products have lower rates of colon cancer. Altogether, foods that contain fiber have been found to have the following benefits.*

- *Weight control.* A diet high in fibrous foods can promote weight loss, if those foods displace concentrated fats and sweets. This is possible because fibrous foods offer less energy per bite than concentrated fats and sweets, thus providing fullness before too much energy is taken in.
- *Constipation and diarrhea relief.* Some fibers attract water into the digestive tract, thus softening the stools. Others help to solidify watery stools. By the one mechanism, they help relieve constipation, and by the other, they help relieve diarrhea.
- *Hemorrhoid prevention.* Softer and larger stools ease elimination for the rectal muscles and reduce the pressure in the lower bowel, creating less likelihood that rectal veins will swell.
- *Appendicitis prevention.* Fiber helps prevent compaction of the intestinal contents, which could obstruct the appendix and permit bacteria to invade and infect it.
- *Diverticulosis prevention.* Fiber exercises the muscles of the digestive tract so that they retain their health and tone and resist bulging out into the pouches characteristic of diverticulosis.

Figure 1A

Foods on the right are rich in soluble fiber, which lowers blood cholesterol. Foods on the left are rich in insoluble fiber, which maintains regularity. Foods in the middle contain both types of fiber.

wheat bran	navy beans	oat bran
wheat products	green beans	carrots
brown rice	kidney beans	barley
cooked lentils	green beans	apples

◄——————————————►

Rich in insoluble fiber — Rich in soluble fiber

- *Colon cancer prevention.* Some fibers speed up the passage of food materials through the digestive tract, thus shortening the "transit time" and helping to prevent exposure of the tissue to cancer-causing agents in food. Some fibers bind bile and carry it out of the body; this is also thought to reduce cancer risk.
- *Blood lipid and cardiovascular disease control.* Some fibers bind lipids such as bile and cholesterol and carry them out of the body with the feces so that the blood lipid concentrations are lowered, and possibly the risk of heart and artery disease as a consequence.
- *Blood glucose and insulin modulation.* Monosaccharides absorbed from some complex carbohydrates, in the presence of fiber, produce a moderate insulin response and an even rise in blood glucose concentrations. Insulin levels are high in obesity, cardiovascular disease, and diabetes (type II), so this effect of fiber may be beneficial in all 3 diseases.
- *Diabetes control.* Thanks to its effects on blood glucose concentrations, high-fiber foods help to manage diabetes. Persons with mild cases of diabetes, given high-fiber diets, have been able to reduce their insulin doses.

There has been a great deal of emphasis placed on using bran to supply fiber needs, but actually this may not be the wisest course of action. There are a variety of foods that contain fiber while providing other nutrients as well (fig. 1A). To ensure that you get the full range of fiber benefits listed, it is best to include different types of whole grains, vegetables, and fruits in your diet each day and not rely on any particular purified form of fiber.

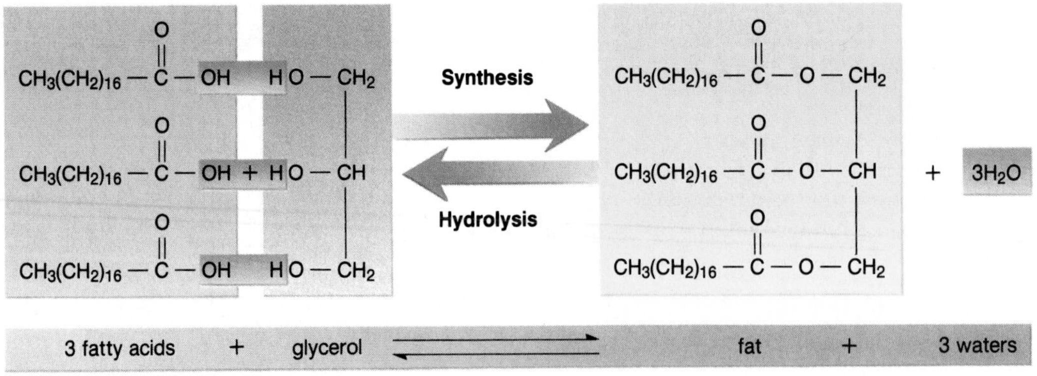

Figure 1.14

Synthesis and hydrolysis of a neutral (nonpolar) fat. Three fatty acids plus glycerol react to produce a fat molecule and 3 water molecules. A fat molecule plus 3 water molecules react to produce 3 fatty acids and glycerol.

or *unsaturated.* Saturated fatty acids have no double bonds between the carbon atoms. The carbon chain is saturated, so to speak, with all the hydrogens that can be held. Unsaturated fatty acids have double bonds in the carbon chain wherever the number of hydrogens is less than 2 per carbon atom. Unsaturated fatty acids are most often found in vegetable oils and account for the liquid nature of these oils. Vegetable oils are hydrogenated to make margarine. Polyunsaturated margarine still contains a large number of unsaturated, or double, bonds.

Glycerol is a compound with 3 H—C—OH groups attached by way of the carbon atoms. When fat is formed, by dehydration synthesis, the —OH groups react with the acid portions of 3 fatty acids so that 3 molecules of water are formed. The reverse of this reaction represents hydrolysis of the fat molecule into its separate components (fig. 1.14).

Other Lipids Are Important, Too

Other lipids include the phospholipids and steroids. Phospholipids, as their name implies, contain a phosphate group:

```
      O
      ||
 ⁻O—P—O⁻
      |
     ⁻O
```

Essentially, phospholipids are constructed like neutral fats except one of the hydrocarbon chains contains a phosphate group or a grouping that contains both phosphate and nitrogen. This portion of the molecule becomes a polar "head," and the other 2 hydrocarbon chains become the nonpolar "tails." The cell membrane contains 2 layers of phospholipid molecules; the polar heads are directed outward and the tails are directed inward (fig. 2.2).

Steroids have an entirely different structure from neutral fats. Like cholesterol, a molecule you may have heard about because it is implicated in cardiovascular disease, other steroids have 4 fused carbon rings. The sex hormones are steroids, and male and female sex hormones have the same basic structure but slightly different attached groups.

> Lipids include nonpolar fats, long-term energy storage molecules that form from glycerol and 3 fatty acids, and the related phospholipids, which have a charged group. The steroids have an entirely different structure, similar to that of cholesterol.

Nucleic Acids

Nucleic acids are huge, macromolecular compounds with very specific functions in cells; for example, the genes are composed of a nucleic acid called **DNA (deoxyribonucleic acid).** DNA has the ability to replicate, that is, to make a copy of itself. It also controls protein synthesis. Another important nucleic acid, **RNA (ribonucleic acid),** works in conjunction with DNA to bring about protein synthesis.

Both DNA and RNA are polymers of nucleotides and therefore are chains of nucleotides joined together. Just like the other synthetic reactions we have studied in this section, when these units are joined together to form nucleic acids, water molecules are removed.

> Both DNA and RNA are composed of nucleotides. DNA makes up the genes and along with RNA controls protein synthesis.

Nucleotides Have Three Parts

Every **nucleotide** is a molecular complex of 3 types of unit molecules: phosphate, a 5-carbon sugar, and a nitrogen-containing base. In DNA the sugar is deoxyribose and in RNA the sugar is ribose, and this difference accounts for their respective names. There are 4 different types of bases in DNA and RNA: adenine (A) and guanine (G) have a double ring; thymine (T) and cytosine (C) have a single ring (fig. 1.15). In RNA, the base uracil (U) is used in place of the base thymine. (These structures are called bases because they have basic characteristics that raise the pH of a solution.)

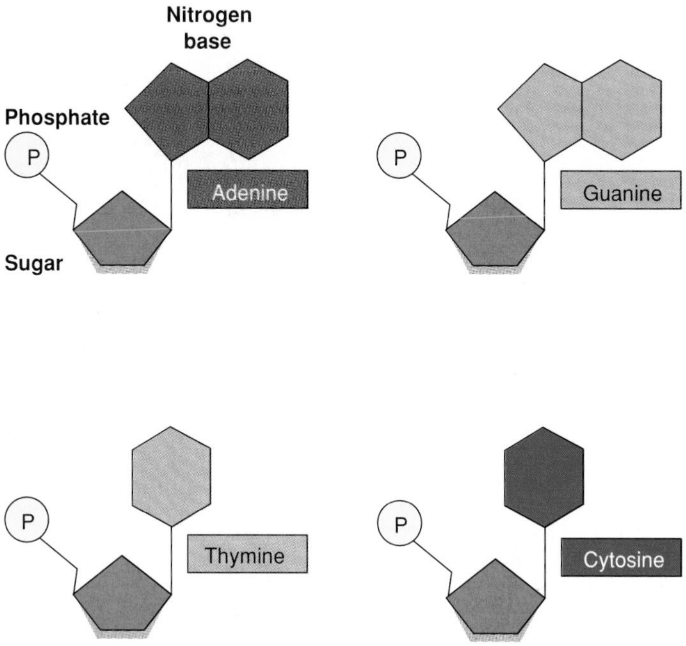

Figure 1.15

Nucleotides in DNA. Each nucleotide contains a phosphate, a sugar, and a nitrogen base. There are 4 bases: adenine (A), guanine (G), thymine (T), and cytosine (C), which differ by the number of rings and/or the atoms attached to the rings.

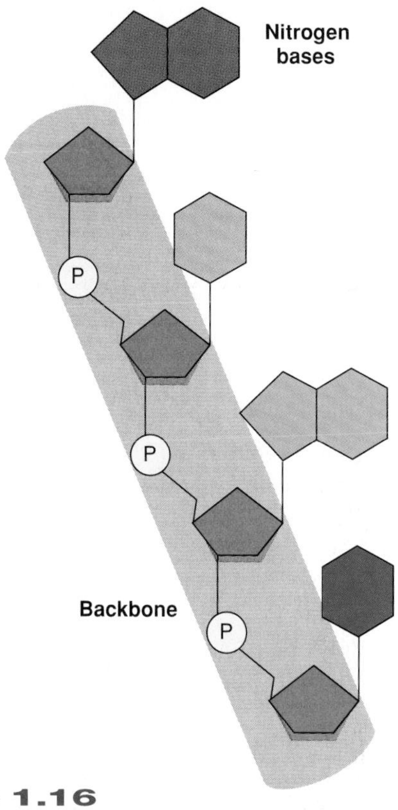

Figure 1.16

Generalized nucleic acid strand. Nucleic acid polymers contain a chain of nucleotides. Each strand has a backbone made of sugar and phosphate molecules. The bases project to the side.

Nucleic Acid Strands Have Backbones

When nucleotides join, they form a linear polymer, called a strand, in which the so-called backbone is made up of phosphate-sugar-phosphate-sugar. The bases project to one side of the backbone. RNA is single-stranded (fig. 1.16), but DNA is double-stranded. The 2 strands of DNA twist about one another in the form of a **double helix** (fig. 1.17*a* and *b*). The 2 strands are held together by hydrogen bonds, which are represented by dashed lines between the bases. The base T is always paired with the base A and the base G is always paired with the base C because their chemical structures are complementary. Therefore, this is called *complementary base pairing.*

If we unwind the DNA helix, it resembles a ladder in which the sugar phosphate backbone makes up the sides of the ladder and the hydrogen-bonded bases make up the rungs, or steps, of the ladder (fig. 1.17*c*). Notice that the 2 strands run in opposite directions. The 5′ end of one strand is opposite the 3′ end of the other strand. The designation 5′ refers to the fifth carbon of ribose and the 3′ end refers to the third carbon of ribose.

Table 1.2

DNA Structure Compared to RNA Structure

	DNA	RNA
Sugar	Deoxyribose	Ribose
Bases	Adenine, guanine, thymine, cytosine	Adenine, guanine, uracil, cytosine
Strands	Double-stranded with base pairing	Single-stranded
Helix	Yes	No

DNA has a structure like a twisted ladder: sugar-phosphate backbones make up the sides of the ladder, and hydrogen-bonded bases make up the rungs of the ladder. The base A is always paired with the base T, and the base C is always paired with the base G. RNA differs from DNA in several respects (table 1.2).

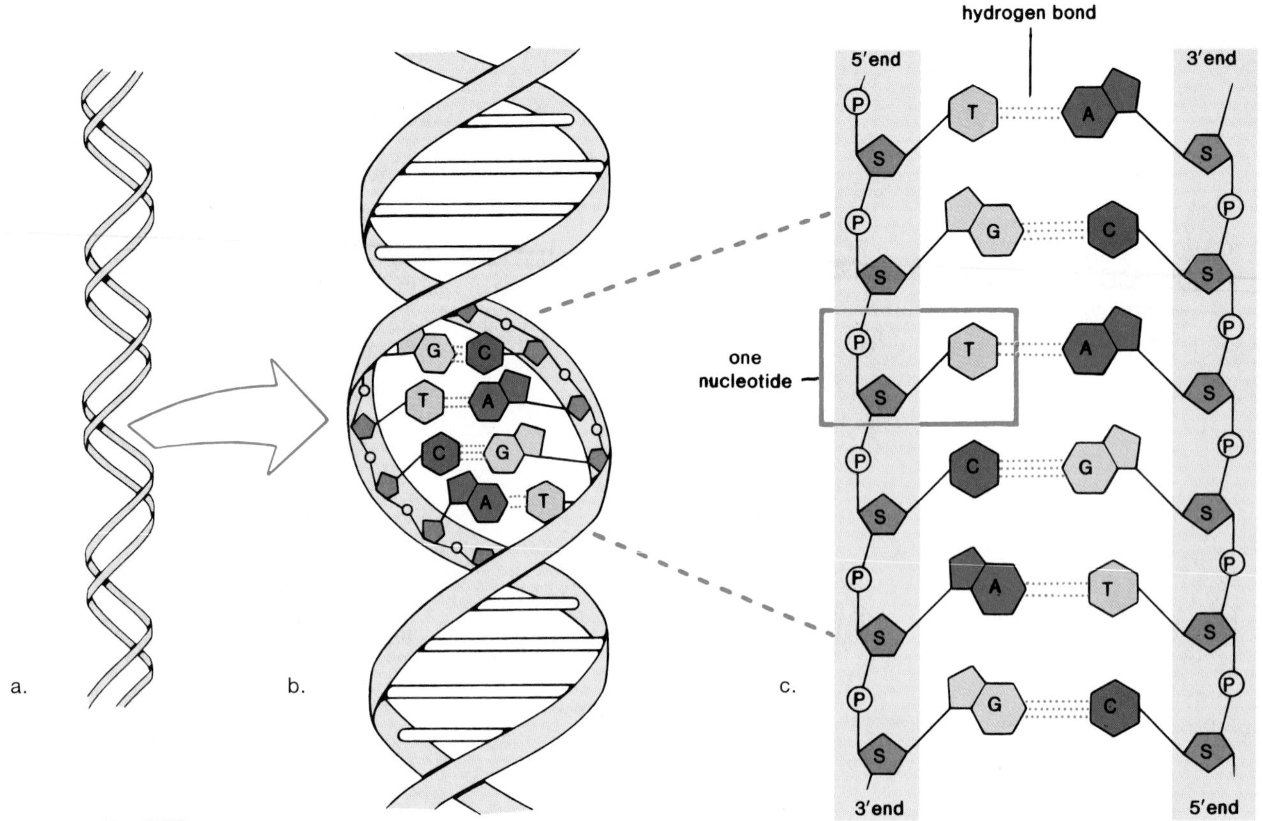

Figure 1.17

Overview of DNA structure. **a.** Double helix. **b.** Complementary base pairing. **c.** Ladder configuration. Notice that the uprights are composed of sugar and phosphate molecules and the rungs are complementary paired bases.

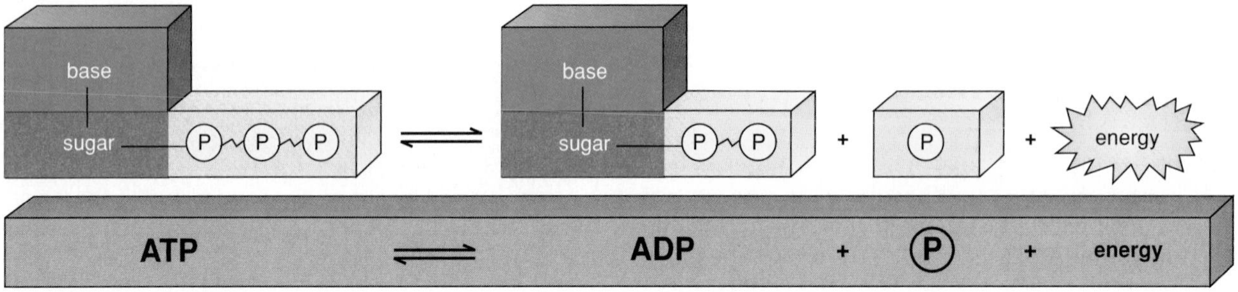

Figure 1.18

ATP reaction. ATP is an energy carrier in cells because it has 2 high-energy phosphate bonds (indicated in the figure by wavy lines). When cells require energy, the last phosphate bond is broken, and energy is made available for cellular work.

ATP for Energy

ATP (adenosine triphosphate) is a very special type of nucleotide (fig. 1.18). It is composed of the base adenine and the sugar ribose (together called adenosine) and 3 phosphate groups. The wavy lines in the formula for ATP indicate it is a high-energy molecule. When the phosphate bonds are broken, a significant amount of energy is released. Because of this property, ATP is the energy currency of cells; when cells "need" something, they "spend" ATP.

ATP is used in body cells for synthetic reactions, active transport, nervous conduction, and muscle contraction. When energy is required for these processes, the end phosphate group is removed from ATP, breaking down the ATP molecule to **ADP (adenosine diphosphate)** and Ⓟ (phosphate) (fig. 1.18).

ATP is the energy molecule of cells because it contains high-energy phosphate bonds that release energy when they are broken.

Summary

All matter is made up of atoms, each having a weight that is dependent on the number of protons and neutrons in the nucleus and chemical properties that are dependent on the number of electrons in the outermost shell. Atoms react with one another to form molecules. In ionic reactions, one atom gives electrons to another and in covalent reactions, atoms share electrons.

Table 1.3

Biomolecules

Macromolecules	Unit Molecule	Usual Atoms
Protein	Amino acid	C, H, O, N, S
Carbohydrate, e.g., glycogen	Glucose	C, H, O
Lipid	Glycerol and fatty acids	C, H, O
Nucleic acid	Nucleotide	C, H, O, N, P

Water, acids, and bases are important inorganic compounds. Water has a neutral pH; acids decrease and bases increase the pH of water. The organic molecules of interest are proteins, carbohydrates, lipids, and nucleic acids, each of which has a particular unit molecule(s) (table 1.3). Dehydration synthesis joins unit molecules and hydrolysis releases them. Some proteins are enzymes; carbohydrates serve as immediate energy sources; and fats are a long-term energy source for the individual. Nucleic acids are of 2 types, DNA and RNA. DNA is the genetic material, and both of these have functions related to protein synthesis, which will be discussed in chapter 19.

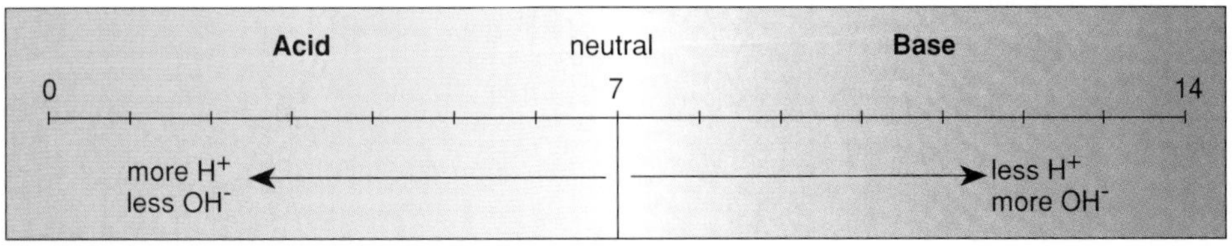

Study and Thought Questions

1. Describe the composition of an atom, and give the weight and charge of an atom's components. (p. 14)
2. Give an example of an ionic reaction, and define the term *ion*. Define the term *ionic bond*. (p. 15)

   Ethical Issue Should humans be described as chemical and physical machines? For some, this description increases the possibility of curing human ills, but for others, it detracts from the beauty and wonder of life.
3. Give an example of a covalent reaction, and define the term *covalent bond*. (p. 15)
4. On the pH scale, which numbers indicate an acidic solution? a basic solution? Why? (p. 18)
5. What are buffers, and why are they important to life? (p. 18)

   Critical Thinking Carbonic acid (H_2CO_3) is a buffer that releases H^+. Will carbonic acid release H^+ when the blood turns acidic or when the blood turns basic?
6. Name 4 general differences between inorganic and organic molecules. (p. 18)
7. Explain synthesis by dehydration and breakdown by hydrolysis of organic compounds. (p. 19)
8. Describe the primary, secondary, and tertiary structures of proteins. What functions do proteins serve in the body? (p. 21)

   Critical Thinking Actin and myosin are the proteins found in muscles. How do you predict they differ structurally?
9. Name a monosaccharide, a disaccharide, and a polysaccharide, and state appropriate functions. What is the most common unit molecule for these? (p. 21)
10. What type of molecules react to form a neutral fat? Explain the difference between a saturated and an unsaturated fatty acid. (p. 22)

    Critical Thinking Oleic and linoleic acid are both unsaturated fatty acids. How might they differ structurally?
11. Name several types of lipids, and state their functions. (p. 24)
12. What are the 2 types of nucleic acids in cells, and what are their functions? What is the unit molecule of a nucleic acid? (p. 24)
13. Name 4 differences between DNA and RNA. (p. 26)

Writing Across the Curriculum

1. The human body is composed of organic molecules but also contains inorganic molecules. Argue for the importance of inorganic molecules in the body.
2. Compare the structure of a glycogen molecule to a protein molecule. Relate the structure of these molecules to their function.
3. How is water created and used in the body's chemical reactions?

Objective Questions

1. ________________ are the smallest units into which matter can be chemically broken.
2. Isotopes differ by the number of ______________ in the nucleus.
3. The 2 primary types of reactions and bonds are ________________ and ________________ .
4. A type of weak bond, called ____________ bonding, exists between water molecules.
5. Acidic solutions contain more ______________ ions than basic solutions, and they have a __________ pH.
6. ________________ speed up chemical reactions.
7. The primary structure of a protein is the sequence of ______________; the secondary structure is very often a ______________; the tertiary structure is the final______________ of the protein.
8. Glycogen is a polymer of _____________, molecules that serve to give the body immediate __________.
9. A neutral fat hydrolyzes to give one_____________ molecule and 3_______________ molecules.
10. The genes are composed of_____________, a nucleic acid made up of_____________ joined together.
11. Label this diagram of synthesis and hydrolysis:

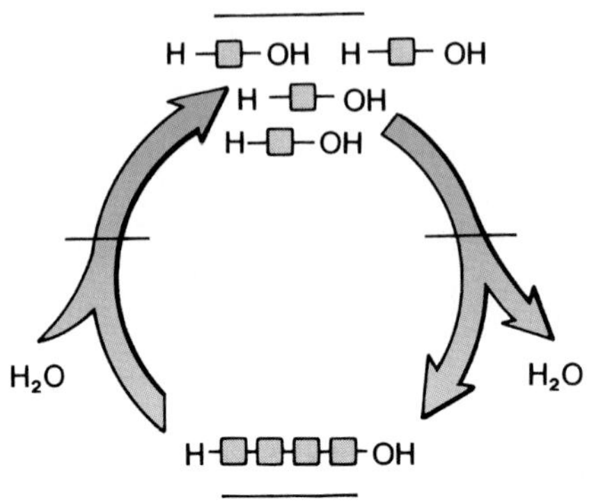

Selected Key Terms

acid A solution in which pH is less than 7; a substance that contributes or liberates hydrogen ions (protons) in a solution. 16

amino acid Unit molecule of a protein that takes its name from the fact that it contains an amino group (—NH_2) and an acid group (—COOH). 20

atom Smallest unit of matter that cannot be divided by chemical means. 14

ATP (adenosine triphosphate) A compound containing 3 phosphate groups, 2 of which are high-energy phosphates. It is the "common currency" of energy for most cellular processes. 26

base A solution in which pH is greater than 7; a substance that contributes or liberates hydroxide ions (OH^-) in a solution; alkaline; opposite of acidic. Also, a term commonly applied to one of the components of a nucleotide. 16

buffer A substance or compound that prevents large changes in the pH of a solution. 18

DNA (deoxyribonucleic acid) A nucleic acid found in the cells; the genetic material that directs protein synthesis in cells. 24

electron A subatomic particle that has almost no weight and carries a negative charge; orbits in a shell about the nucleus of an atom. 14

enzyme An organic catalyst that speeds up a specific reaction or a specific type of reaction in cells. 19

hydrogen bond A weak attraction between a hydrogen atom carrying a partial positive charge and an atom of another molecule carrying a partial negative charge. 15

ion An atom or group of atoms carrying a positive or negative charge. 15

isotope One of 2 or more atoms with the same atomic number that differs in the number of neutrons and therefore in weight. 14

lipid One of a class of organic molecules that are insoluble in water; notably fats, oils, and steroids. 22

neutron A subatomic particle that has a weight of one atomic mass unit, carries no charge, and is found in the nucleus of an atom. 14

nucleic acid A large organic molecule made up of nucleotides joined together; for example, DNA and RNA. 24

peptide bond The covalent bond that joins 2 amino acids. 21

pH scale A measure of the hydrogen ion concentration; any pH below 7 is acid and any pH above 7 is basic. 18

polypeptide A macromolecule composed of many amino acids linked by peptide bonds. 21

polysaccharide A macromolecule composed of many units of sugar. 22

protein One of a class of organic compounds that are composed of either one or several long polypeptides. 19

proton A subatomic particle found in the nucleus of an atom that has a weight of one atomic mass unit and carries a positive charge; a hydrogen ion. 14

RNA (ribonucleic acid) A nucleic acid found in cells that assists DNA in controlling protein synthesis. 24

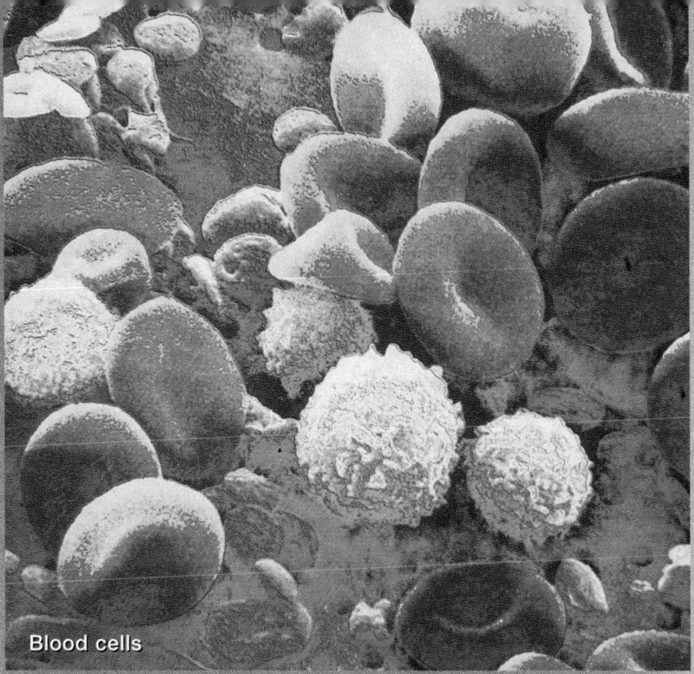

Chapter 2

Cell Structure and Function

We are multicellular animals. The cell is the fundamental unit of our bodies, and it is at the cellular level that we must understand health and disease. Because cells are microscopic, it is sometimes hard to imagine that it is not the intestine or heart that is causing the difficulty—it is the cells that make up the intestine or heart. Nowhere is this more evident than when we study the cause of cancer of the uterus, the lungs, the colon, and so forth. In all cases, cancer is characterized by uncontrolled growth of cells due to irregularities of cell structure and function.

Chapter Concepts

1 The amount of detail microscopes allow us to see varies from one instrument to another. 30

2 The cell membrane regulates the entrance and exit of molecules into and out of the cell. 32

3 A cell is highly organized and contains organelles that carry out specific functions. 33

4 The nucleus, a centrally located organelle, controls the metabolic functioning and structural characteristics of the cell. 34

5 A system of membranous canals and vacuoles works to produce, store, modify, transport, and digest macromolecules. 36

6 Mitochondria are organelles concerned with the conversion of glucose energy into ATP molecules. 37

7 The cell has a cytoskeleton composed of microtubules and filaments; the cytoskeleton gives the cell a shape and allows it and its organelles to move. 37

8 The cytoplasm contains metabolic pathways, each a series of reactions controlled by enzymes. 39

Chapter Outline

Microscopy: Reveals Cell Structure

Your body and all living things are made up of cells. Human cells come in many different shapes and size, but no matter what the shape or the size, each one has the same basic organization, which can be revealed by microscopy.

Two types of microscopes are commonly used to look at cells. *Light microscopes* utilize light to view the object, and electron microscopes utilize electrons. A *transmission electron microscope* gives us an image of the interior of the object, while a *scanning electron microscope* provides a three-dimensional view of the surface of an object (fig. 2.1).

The useful magnification of a transmission electron microscope (100,000×) is much greater than the useful magnification of a light microscope (1,000×) because the resolving power is much greater. Resolving power is the capacity to distinguish between 2 points. If 2 points are seen as separate, then the image appears more detailed than if the 2 points are seen as one point. At the very best, a light microscope can distinguish 2 points separated by 200 nm (nanometer = 1×10^{-6} mm), but the transmission electron microscope can distinguish 2 points separated by about 0.2 nm.

In the case of scanning electron microscopy, a narrow beam of electrons is scanned over the surface of the specimen, which has been coated with a thin metal layer.

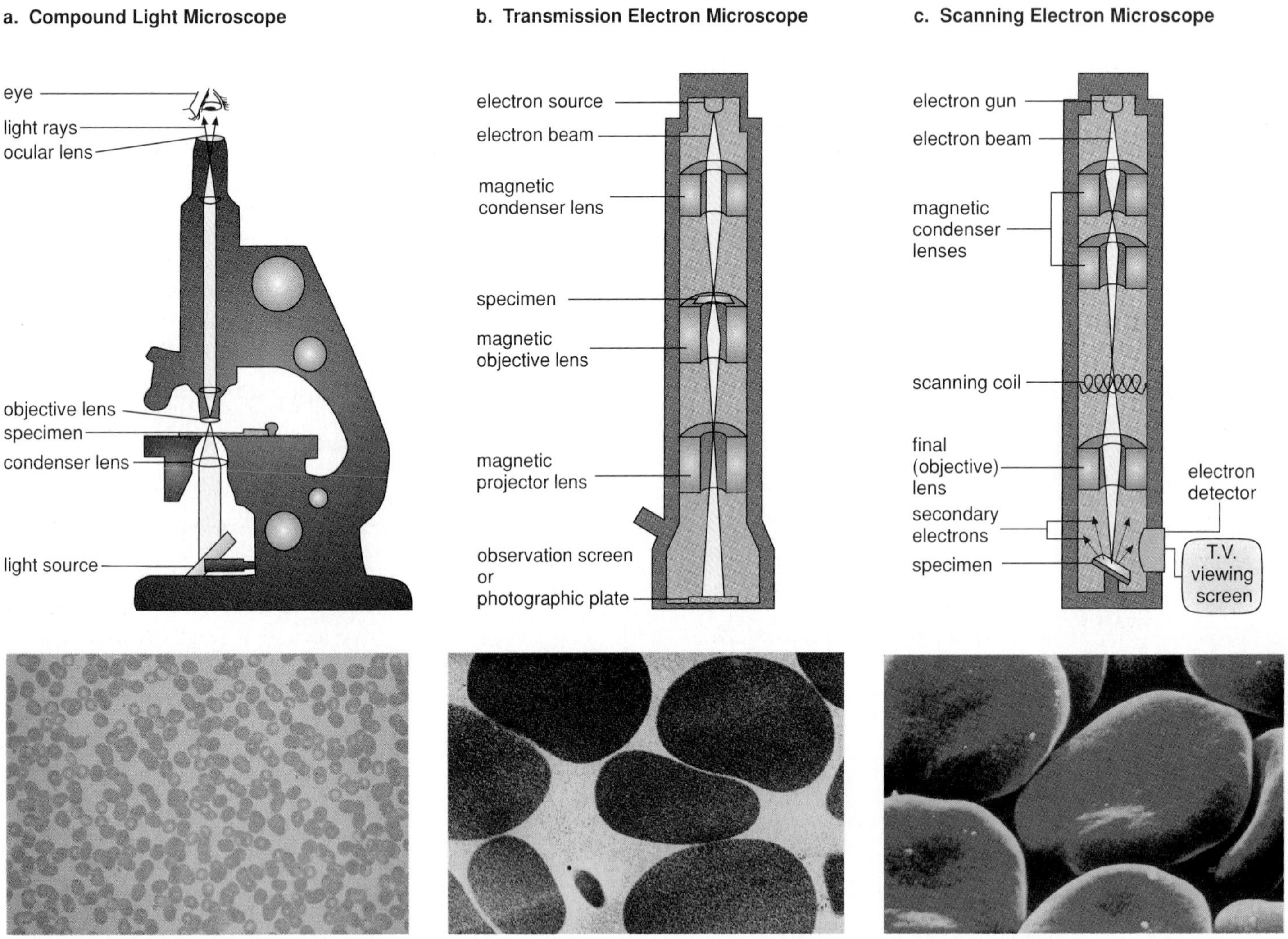

Figure 2.1

a. Microscopes. Photomicrograph of red blood cells. A light microscope uses light to view a specimen. The specimen can be living if it is thin enough to allow light to pass through. A light microscope does not magnify or distinguish as much detail as the electron microscope. **b.** Transmission electron micrograph (TEM) of red blood cells. The transmission electron microscope uses electrons to "view" the specimen. The specimen is nonliving and must be thin enough to allow electrons to pass through. The magnification and amount of detail seen is far greater than with the light microscope. **c.** Scanning electron micrograph (SEM) of red blood cells. The scanning microscope scans the surface of the specimen with an electron beam. The secondary electrons given off are collected and the result is a three-dimensional image of the specimen.

The metal gives off secondary electrons, which are collected and produce a television-type picture of the specimen's surface on a screen.

There are new types of scanning microscopes called scanning-probe microscopes. In one type, laser light is focused on a tiny probe that is pressed against an organic polymer. Reflected light tells the position of the probe as it moves up and down the polymer, and these data allow a computer to give a three-dimensional picture of a single biological molecule. The probe moves up and down in response to electron movements and atomic forces that exist between the probe and the material.

A picture obtained by using the light microscope sometimes is called a photomicrograph, and a picture resulting from the use of electron microscopes is called a transmission electron micrograph (TEM) or a scanning electron micrograph (SEM), depending on the type of microscope used.

> Electron micrographs have helped biologists develop an understanding of cell structure.

The Cell Membrane

An animal cell is surrounded by an outer **cell membrane.** The cell membrane marks the boundary between the outside of the cell and the inside of the cell, termed the **cytoplasm.** Cell membrane integrity and how it functions are necessary to the life of the cell.

The cell membrane is a phospholipid bilayer with attached or embedded proteins. The structure of a phospholipid is such that the molecule has a polar head and nonpolar tails (fig. 2.2*a*). The polar heads being charged are hydrophilic (water loving) and face outward, where they are likely to encounter a watery environment. The nonpolar tails are hydrophobic (water hating) and face inward, where there is no water. When phospholipids are placed in water, they naturally form a circular bilayer because of the chemical properties of the heads and the tails. At body temperature, the phospholipid bilayer is a liquid; it has the consistency of olive oil, and the proteins are able to change their position by moving laterally. The fluid-mosaic model, a working description of membrane structure, suggests that the protein molecules have a changing pattern (form a mosaic) within the fluid phospholipid bilayer (fig. 2.2*b*).

Short chains of sugars are attached to the outer surface of some protein and lipid molecules (called glycoproteins and glycolipids, respectively). It is believed that these carbohydrate chains specific to each cell mark it as belonging to a particular individual and account for such characteristics as blood type or why a patient's system sometimes rejects an organ transplant. Other glycoproteins have a special configuration that allows them to act as a receptor for a chemical messenger like a hormone. Some cell membrane proteins form channels through which certain substances can enter cells or are **carriers** involved in the passage of molecules through the membrane.

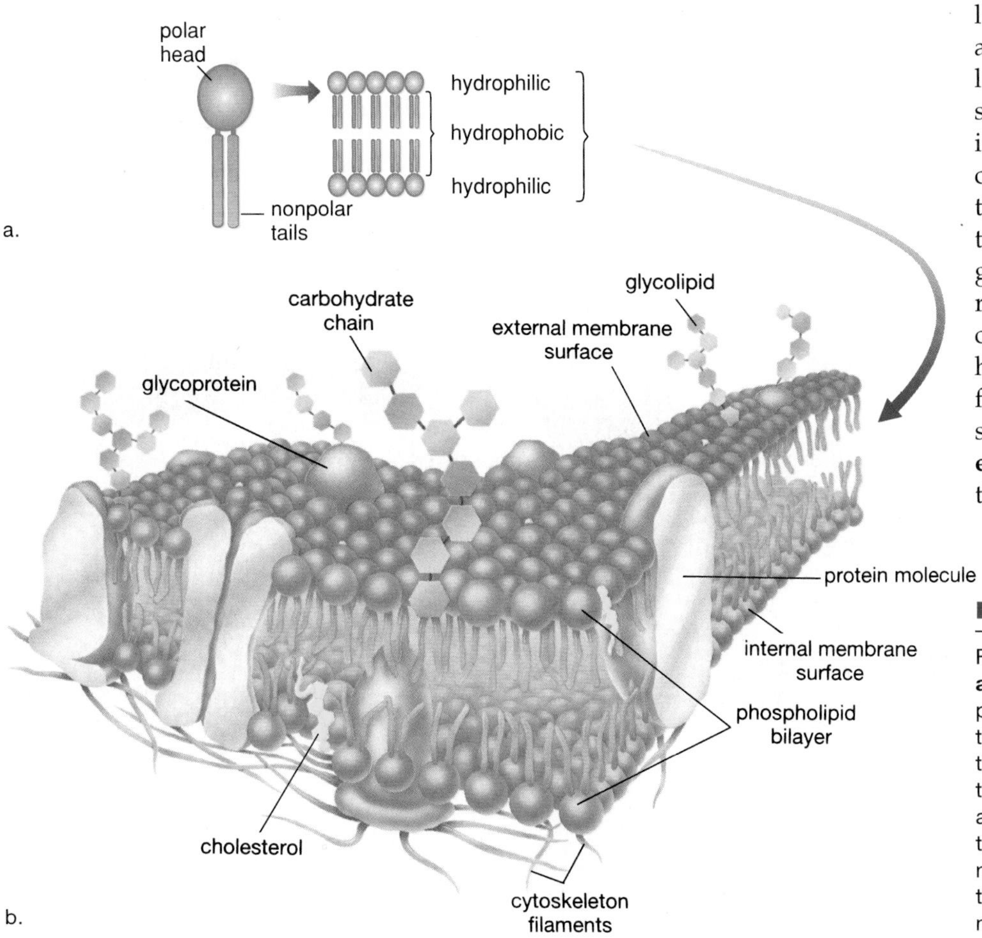

Figure 2.2

Fluid-mosaic model of the cell membrane. **a.** The membrane is composed of a phospholipid bilayer. The polar heads of the phospholipids are at the surfaces of the membrane; the nonpolar tails make up the interior of the membrane. **b.** Proteins are embedded in the membrane. Some of these function as receptors for chemical messengers, as conductors of molecules through the membrane, and as enzymes in metabolic reactions. Carbohydrate chains of glycolipids and glycoproteins are involved in cell-to-cell recognition.

Table 2.1

Passage of Molecules into and out of Cells

Name	Direction (by concentration)	Requirements	Examples
Diffusion	Higher → lower	Concentration gradient	Lipid-soluble molecules, water, and gases
Transport			
Facilitated	Higher → lower	Carrier plus concentration	Sugars and amino acids
Active	Lower → higher	Carrier plus ATP	Sugars, amino acids, and ions
Exocytosis	Lower → higher	Vesicle release	Secretion
Endocytosis	Lower → higher	Vesicle formation	Phagocytosis

How the Cell Membrane Functions

The cell membrane allows only certain molecules to enter and exit the cytoplasm freely; therefore, the cell membrane is said to be **selectively permeable.** Small molecules that are lipid soluble, such as oxygen and carbon dioxide, can pass through the membrane easily. Certain other small molecules, like water, are not lipid soluble but still cross the membrane passively by moving through a protein channel. Still other molecules and ions require the use of a carrier to enter a cell.

> The cell membrane, composed of phospholipid and protein molecules, is selectively permeable and regulates the entrance and exit of molecules and ions from the cell.

Lipid-soluble molecules, including gases and water, pass through the membrane by **diffusion** (table 2.1), the movement of molecules from the area of greater concentration to the area of lesser concentration until they are equally distributed. To illustrate diffusion, imagine opening a perfume bottle in the corner of a room. The smell of the perfume soon permeates the room because the molecules that make up the perfume move to all parts of the room. Another example is putting a tablet of dye into water. The water eventually takes on the color of the dye as the tablet dissolves.

Osmosis is the diffusion of water across a cell membrane. It occurs whenever there is an unequal concentration of water on either side of a selectively permeable membrane. Normally, body fluids are **isotonic** to cells (fig. 2.3)—there is an equal concentration of substances (solutes) and water (solvent) on both sides of the cell membrane and cells maintain their usual size and shape.

If red blood cells are placed in a **hypotonic** solution, which has a higher concentration of water (lower concentration of solute) than do the cells, water enters the cells and they swell to bursting. On the other hand, if red blood cells are placed in a **hypertonic** solution, which has a lower concentration of water (higher concentration of solute) than do the cells, water leaves the cells and they shrink.

These changes have occurred due to osmotic pressure. *Osmotic pressure* is the force exerted on a selectively permeable membrane because water has moved from the area of higher to lower concentration of water (higher concentration of solute).

Most solutes do not simply diffuse across a cell membrane; rather, they are transported by means of protein carriers within the membrane. During **facilitated transport,** a molecule (e.g., an amino acid or glucose) is transported across the cell membrane from the side of higher concentration to the side of lower concentration. The cell does not need to expend energy for this type of transport because the molecule is moving in the normal direction.

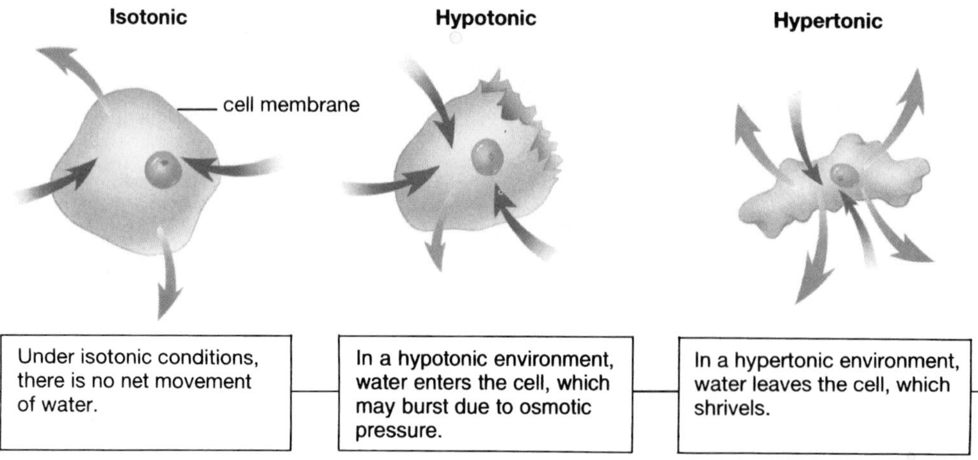

Isotonic
Hypotonic
Hypertonic
cell membrane
Under isotonic conditions, there is no net movement of water.
In a hypotonic environment, water enters the cell, which may burst due to osmotic pressure.
In a hypertonic environment, water leaves the cell, which shrivels.

Figure 2.3

Osmosis. The arrows indicate the movement of water.

During **active transport** a molecule is moving contrary to the normal direction; that is, from lower to higher concentration. For example, iodine collects in the cells of the thyroid gland; sugar is completely absorbed from the gut by cells that line the digestive tract; and sodium (Na^+) is sometimes almost completely withdrawn from urine by cells lining kidney tubules. Active transport requires a protein carrier and the use of cellular energy obtained from the breakdown of ATP (p. 26). When ATP is broken down, energy is released and in this case the energy is used by a carrier to carry out active transport.

Cellular Organelles

The cell contains a number of **organelles,** small bodies with specific structures and functions that are located in the cytoplasm (fig. 2.4 and table 2.2). These help the cell carry out its many activities.

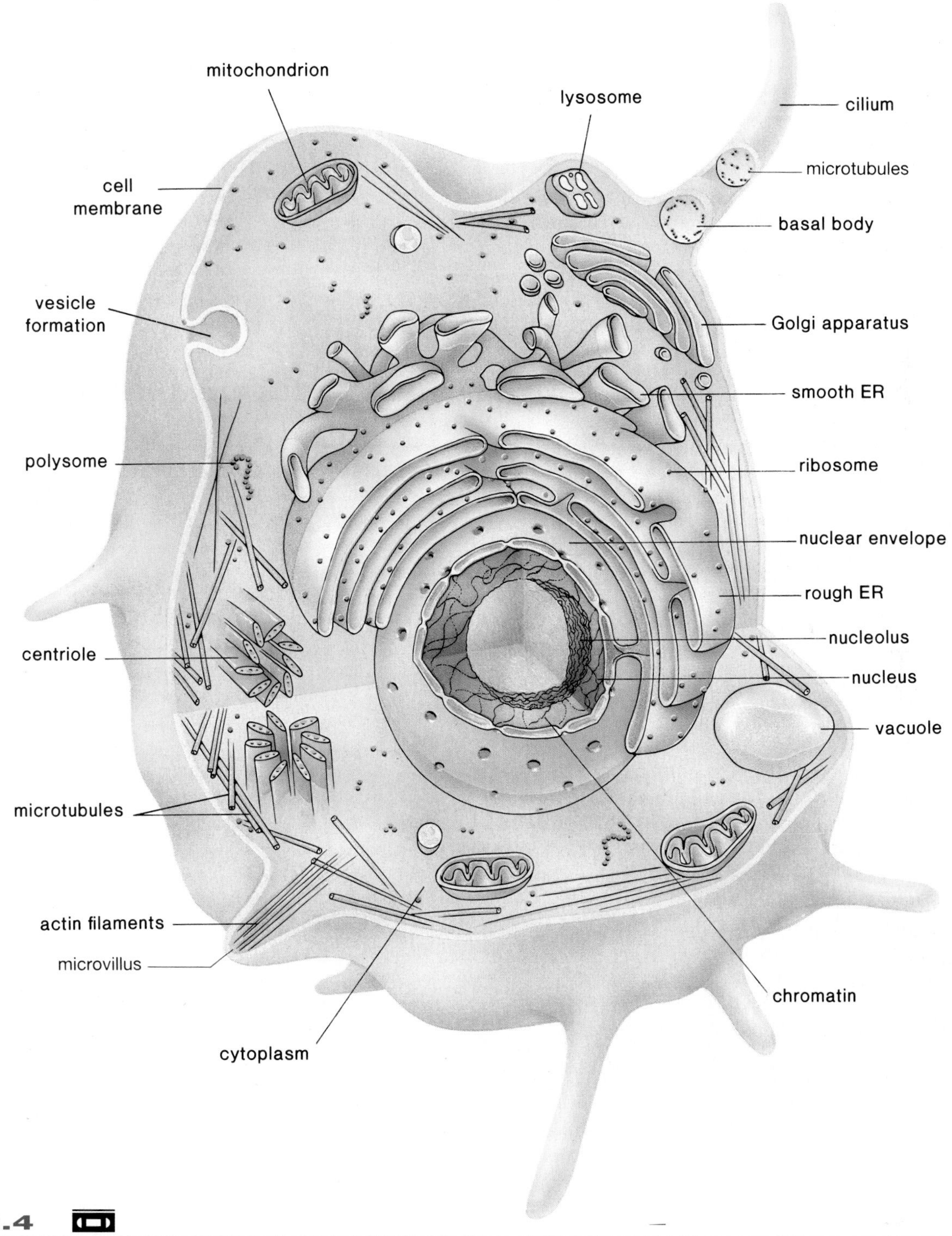

Figure 2.4

Animal cell. This generalized representation is based on electron micrographs.

Table 2.2

Cellular Organelles (Simplified)

Name	Structure	Function
Cell membrane	Bilayer of phospholipid and globular proteins	Passage of molecules into and out of cell
Nucleus	Nuclear envelope surrounds chromatin, nucleolus, and nucleoplasm	Control of cell
Nucleolus	Concentrated area of RNA in the nucleus	Ribosome formation
Chromatin (chromosomes)	Composed of DNA and protein	Directs protein synthesis
Endoplasmic reticulum	Folds of membrane forming flattened channels and tubular canals	Transport by means of vesicles
Rough	Studded with ribosomes	Protein synthesis
Smooth	Having no ribosomes	Lipid and carbohydrate synthesis
Ribosome	RNA and protein in 2 subunits	Protein synthesis
Golgi apparatus	Stack of membranous saccules	Processing, packaging, and secretion
Vacuole and vesicle	Membranous sacs	Containers of material
Lysosome	Membranous container of hydrolytic enzymes	Intracellular digestion
Mitochondrion	Inner membrane (cristae) within outer membrane	Aerobic cellular respiration and ATP synthesis
Cytoskeleton	—	Cell shape and subcellular movement
Actin filament	Actin protein	Member of cytoskeleton
Microtubule	Tubulin protein	Member of cytoskeleton
Centriole	9 + 0 pattern of microtubules	Organization of microtubules; associated with cell division
Cilium and flagellum	9 + 2 pattern of microtubules	Movement of cell

Nucleus: Controller of the Cell

The **nucleus,** the largest organelle found within the cell, is enclosed by a double-layered **nuclear envelope** that is actually continuous with the endoplasmic reticulum, discussed in the following paragraphs. As illustrated in figure 2.5, there are pores, or openings, in the envelope through which large molecules pass from the *nucleoplasm,* the fluid portion of the nucleus, to the cytoplasm or vice versa.

The nucleus is of primary importance in the cell because it is the control center that oversees the metabolic functioning of the cell and ultimately determines the cell's characteristics. Within the nucleus there are masses of threads called **chromatin,** so called because they take up stains and become colored. Chromatin is indistinct in the nondividing cell, but it condenses to rodlike structures called **chromosomes** at

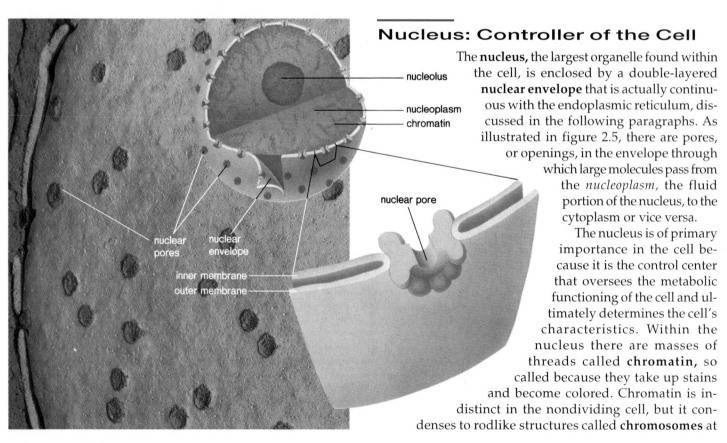

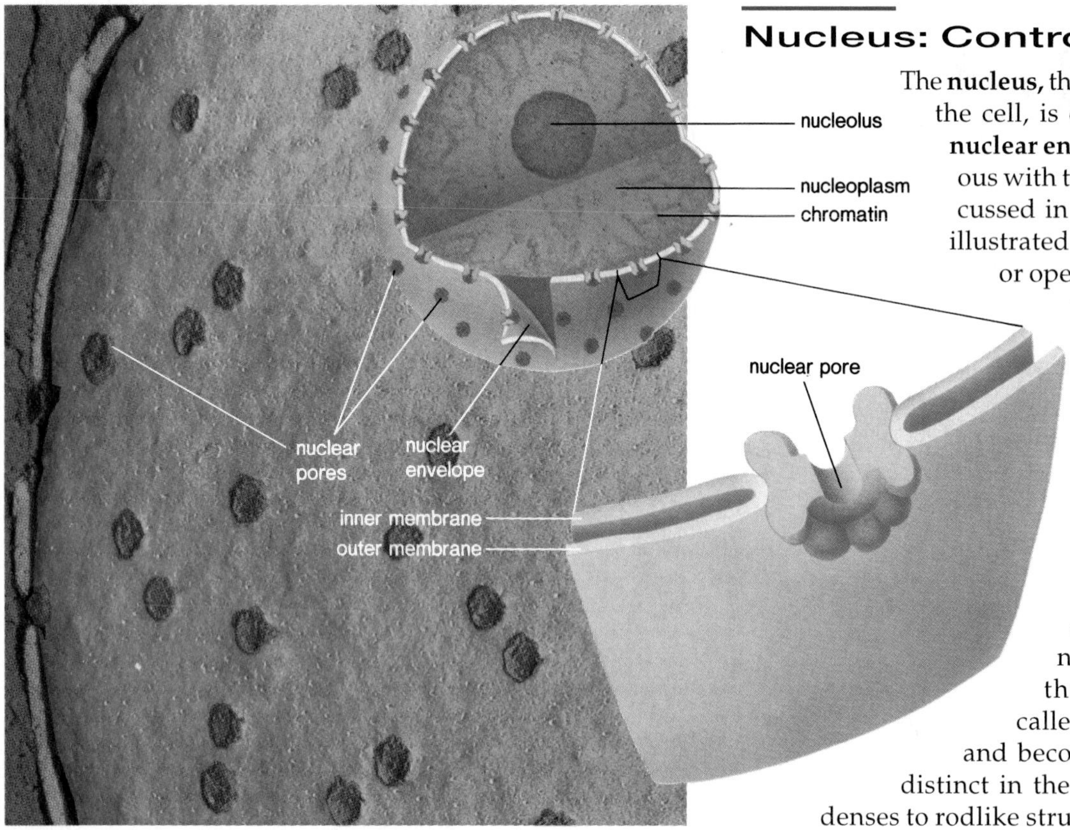

Figure 2.5

Anatomy of the nucleus. The nucleoplasm contains chromatin. Chromatin has a special region called the nucleolus, which is where rRNA is produced. The nuclear envelope contains pores, as shown in this micrograph of a freeze-fractured nuclear envelope. Each pore is lined by a complex of 8 protein granules.

the time of cell division. Chemical analysis shows that chromatin, and therefore chromosomes, contain DNA along with certain proteins and some RNA. This is not surprising because we already know that chromosomes contain the genes and that the genes are composed of DNA, the hereditary material. DNA, with the help of RNA, directs protein synthesis within the cytoplasm, and it is this function that allows DNA to control the cell. (DNA and RNA are discussed in chapter 1.)

> The nucleus contains chromatin, which condenses into chromosomes just prior to cell division. Chromosomes contain DNA that, with the help of RNA, directs protein synthesis in the cytoplasm. Another type of RNA, rRNA, is made within the nucleolus before migrating to the cytoplasm, where it is incorporated into ribosomes.

One or more **nucleoli** are present in the nucleus. These dark-staining bodies are actually specialized parts of chromatin in which a type of RNA called ribosomal RNA (rRNA) is produced. Ribosomal RNA joins with proteins to form subunits that become part of the ribosomes.

Ribosomes and Endoplasmic Reticulum: Protein Makers and Transporters

Ribosomes look like small, dense granules in low-power electron micrographs (fig. 2.6*a*), but higher resolution micrographs show that each contains 2 subunits (fig. 2.6*c*). The subunits, each of which has a particular mix of rRNA and proteins, are produced in the nucleus, but they are not assembled into one ribosome until they reach the cytoplasm.

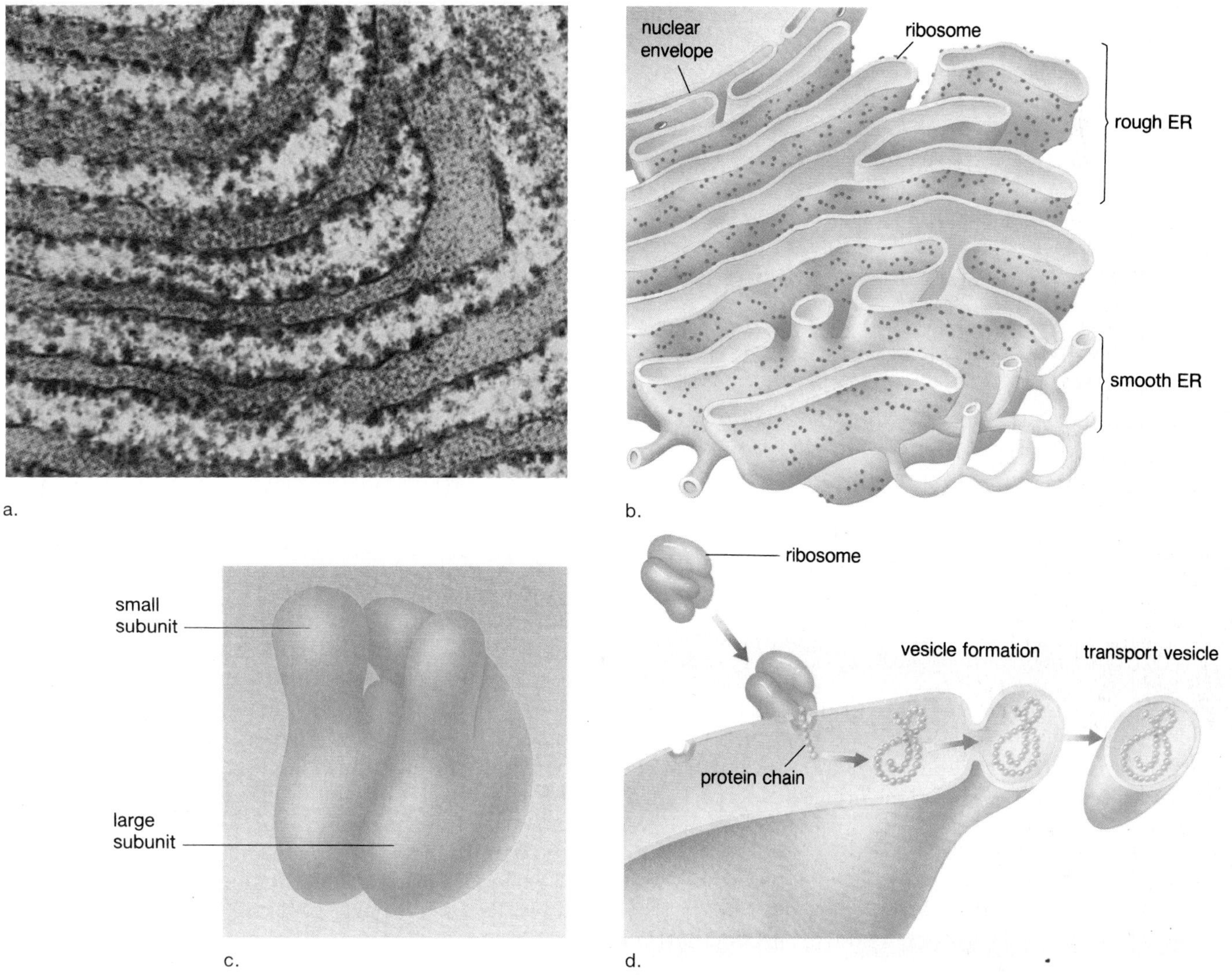

Figure 2.6

Endoplasmic reticulum. **a.** Electron micrograph of rough ER shows a cross section of many flattened vesicles with ribosomes attached to the sides that abut the cytoplasm. **b.** Drawing that shows the 3 dimensions of the ER. **c.** Model of a single ribosome illustrates that each is actually composed of 2 subunits. **d.** Method by which the ER acts as a transport system.

Protein synthesis occurs at the ribosomes, which can lie free within the cytoplasm. In the cytoplasm several ribosomes, each producing the same type protein, are arranged in a functional group called a *polysome*. Most likely, these proteins are for use inside the cell.

Ribosomes can be attached to the **endoplasmic reticulum (ER),** a membranous system of tubular canals that is continuous with the nuclear envelope and branches throughout the cytoplasm. If ribosomes are attached to the ER, it is called **rough ER.** Rough ER makes proteins for export from the cell. The proteins enter the lumen (interior space) of the rough ER (fig. 2.6*d*) and proceed to the lumen of the **smooth ER,** so called because no ribosomes are attached to it. A small vesicle then pinches off from the smooth ER. Most vesicles formed in this way move through the cytoplasm to the Golgi apparatus, where the proteins are received and are processed further. This is how the ER serves as a transport system.

Smooth ER also has specific functions in different cells. Sometimes it specializes in the production of lipids. For example, smooth ER is abundant in cells of the testes and the adrenal cortex, both of which produce steroid hormones. In muscle cells, smooth ER acts as a storage area for calcium ions that are released when contraction occurs. In the liver, it is involved in the detoxification of drugs, including alcohol. It is quite possible that drugs are detoxified within structures called peroxisomes, membrane-bound vesicles often attached to smooth ER that contain enzymes capable of breaking down of various substances.

Golgi Apparatus: Processing, Packaging, and Secretion

The **Golgi apparatus** is named for the person who first discovered its presence in cells (fig. 2.7). It is composed of a stack of about a half-dozen or more saccules that look like hollow pancakes. In human cells, one side of the stack (the inner face) is directed toward the nucleus, and the other side of the stack (the outer face) is directed toward the cell membrane. Vesicles, which are tiny membranous sacs, occur at the edges of the saccules.

The Golgi apparatus contains enzymes that modify proteins, particularly by changing the chain of sugars that makes them glycoproteins. The inner face of the Golgi apparatus receives protein-filled vesicles that bud from the ER. Some biologists believe that these fuse to form a saccule that remains as a part of the Golgi until the proteins are repackaged in new vesicles. Others believe that vesicles from the ER proceed directly to the outer face of the Golgi where protein modification is carried out in its saccules. The vesicles that leave the Golgi move to different locations in the cell (fig. 2.7). Some of the vesicles proceed to the cell membrane, where they discharge their protein contents in a process called exocytosis (see table 2.1). Because this is *secretion,* it often is said that the Golgi apparatus is involved in processing, packaging, and secretion. Some of the other vesicles that leave the Golgi apparatus contain proteins for cell membrane inclusion.

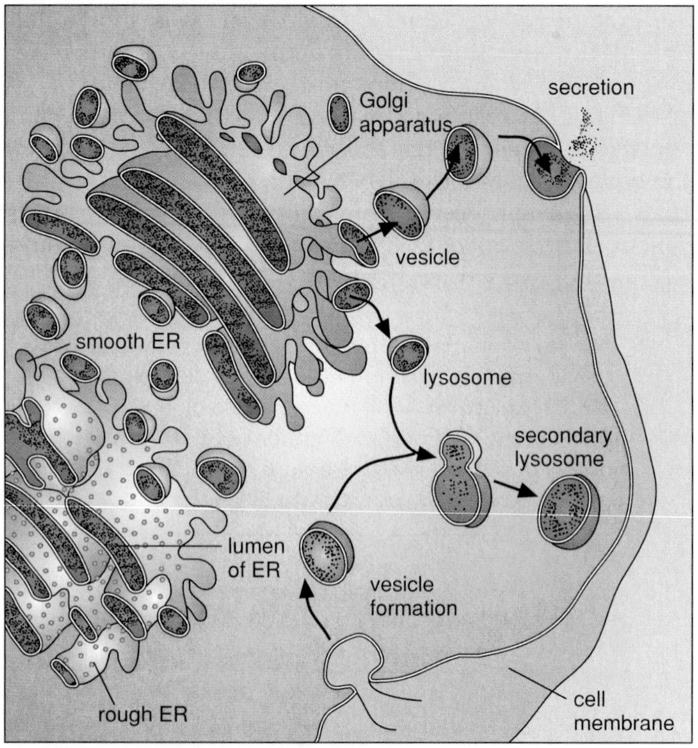

Figure 2.7

Golgi apparatus function. The Golgi apparatus receives vesicles from the smooth ER and thereafter forms at least 2 types of vesicles, lysosomes and secretory vesicles. Lysosomes contain hydrolytic enzymes that can break down macromolecules. Vesicles bringing macromolecules into a cell sometimes join with lysosomes, forming structures called secondary lysosomes. Thereafter, the molecules are digested. The secretory vesicles formed at the Golgi apparatus discharge their contents at the cell membrane.

Lysosomes: Cellular Digestion

Lysosomes (fig. 2.7) are vesicles formed by the Golgi apparatus that contain *hydrolytic enzymes,* which digest macromolecules. Macromolecules are sometimes brought into a cell by vesicle formation at the cell membrane, a process called endocytosis (see table 2.1). A lysosome can fuse with an endocytic vesicle and can digest its contents into simpler molecules that then enter the cytoplasm. Some white blood cells defend the body by engulfing bacteria, a process that involves vesicle formation at the cell membrane. When lysosomes fuse with these vesicles, the bacteria are digested. It should come as no surprise, then, that even parts of a cell are digested by its own lysosomes (called autodigestion because *auto = self*). Normal cell rejuvenation most likely takes place in this way, but autodigestion is also important during development. For example, the fingers of a human embryo are at first webbed, but they are freed from one another following lysosomal action.

Occasionally, a child is born with a metabolic disorder involving a missing or inactive lysosomal enzyme. In these cases, the lysosomes fill to capacity with macromolecules that cannot be broken down. The best known of these lysosomal storage disorders is Tay-Sachs disease, discussed on page 366.

> The ER is a membranous system of tubular canals that can be smooth or rough. Proteins synthesized at the rough ER are processed and are packaged in vesicles by the Golgi apparatus. Some vesicles discharge their contents at the cell membrane, and some are lysosomes that digest any material enclosed therein.

Mitochondria: Energy Converters

A **mitochondrion** is bounded by a double membrane (fig. 2.8). The inner membrane is folded to form little shelves, called *cristae,* which project into the *matrix,* an inner space filled with a gel-like fluid.

Mitochondria produce ATP molecules. Every cell uses a certain amount of ATP energy to synthesize molecules, but many cells use ATP to carry out their specialized function. For example, muscle cells use ATP for muscle contraction that produces movement, and nerve cells use it for the conduction of nerve impulses so we are aware of our environment.

Mitochondria often are called the powerhouses of the cell: just as a powerhouse burns fuel to produce electricity, the mitochondria oxidize glucose products to produce ATP molecules. In the process, mitochondria use up oxygen and give off carbon dioxide and water. The oxygen you breathe in enters cells and then the mitochondria; the carbon dioxide you breathe out is released by the mitochondria. Because gas exchange is involved, it is said that mitochondria carry on **aerobic cellular respiration.** A shorthand way to indicate the chemical transformation associated with cellular respiration is:

> ATP
>
> carbohydrate + oxygen ⟶ carbon dioxide + water
>
> Read as follows: As carbohydrate is broken down to carbon dioxide and water, ATP molecules are built up.

> Mitochondria are the sites of aerobic cellular respiration, a process that provides ATP energy molecules to the cell.

a.

b.

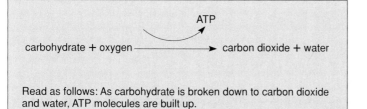

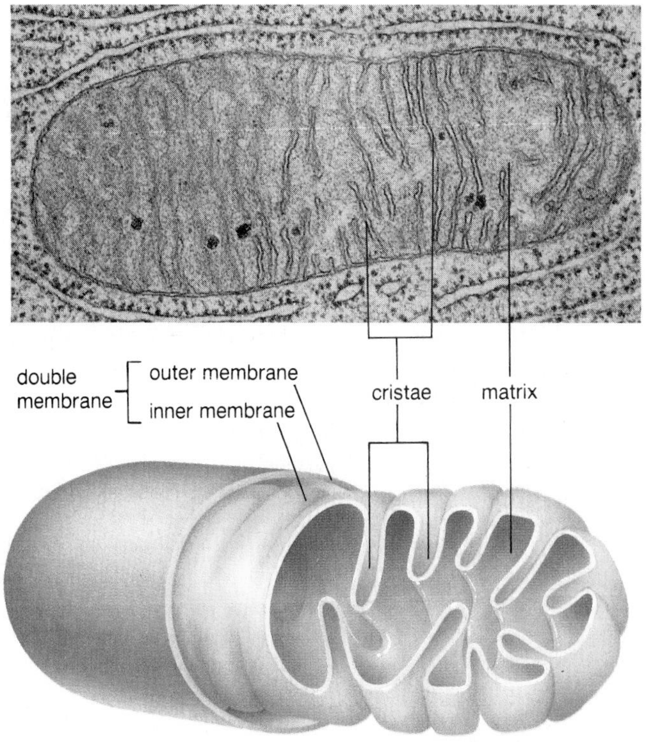

Figure 2.8

Mitochondrion structure. **a.** Electron micrograph. **b.** Generalized drawing in which the outer membrane and a portion of the inner membrane have been cut away to reveal the cristae.

Cytoskeleton: Cell Shape and Movement

Several types of filamentous protein structures form a **cytoskeleton** that helps maintain the cell's shape and either anchors the organelles or assists their movement as appropriate (fig. 2.9). The cytoskeleton includes microtubules and actin filaments.

Microtubules are shaped like thin cylinders and are several times larger than actin filaments. Each cylinder contains 13 rows of tubulin, a globular protein, arranged in a helical fashion. Remarkably, microtubules can assemble and disassemble. In many cells, the regulation of microtubule assembly is under the control of a microtubule organizing center (MTOC), which lies near the nucleus. Microtubules radiate from the MTOC, helping to maintain the shape of the cell and acting as tracts along which organelles move. It is well known that during cell division, microtubules form spindle fibers, which assist the movement of chromosomes (p. 352).

Actin filaments are long, extremely thin fibers that usually occur in bundles or other groupings. Actin filaments have been isolated from various types of cells, especially those in which movement occurs. Microvilli, which project from certain cells and can shorten and extend, contain actin filaments. Actin filaments, like microtubules, can assemble and disassemble.

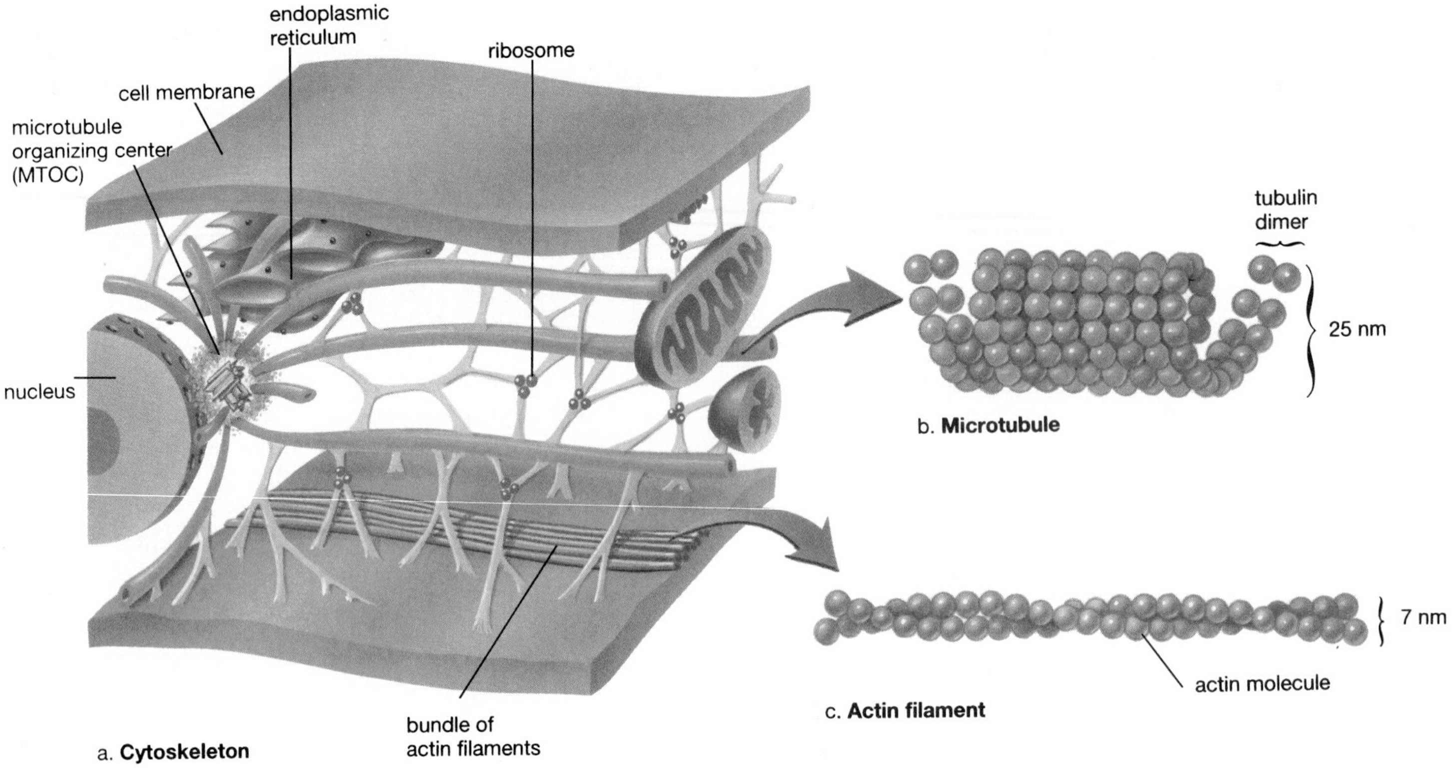

Figure 2.9

The cytoskeleton. **a.** The cytoskeleton both anchors the organelles and allows them to move. **b.** A microtubule is a cylinder composed of tubulin protein molecules. Microtubules radiate from the microtubule organizing center (MTOC) and assist intracellular movement of organelles. **c.** An actin filament contains molecules of the protein actin. Bundles of actin filaments lie close to the cell membrane. Actin filaments also form networks within the cytoplasm.

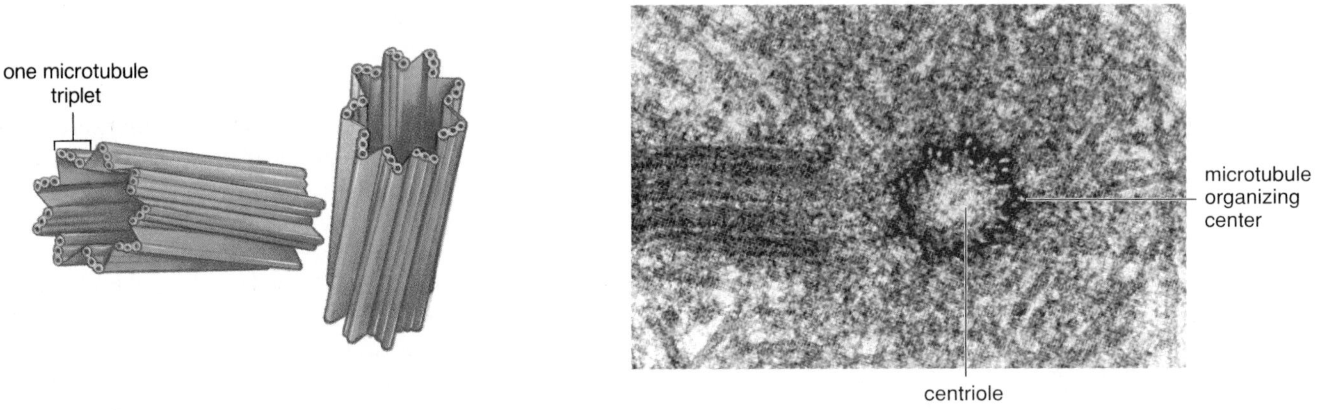

Figure 2.10

Centrioles. Centrioles are composed of 9 microtubule triplets. They lie at right angles to one another within the microtubule organizing center, which is believed to assemble microtubules at the time of cell division.

> The cytoskeleton contains actin filaments and microtubules. Actin filaments, thin actin strands, and microtubules, 13 rows of tubulin protein molecules arranged to form a hollow cylinder, maintain the shape of the cell and also direct the movement of cell parts.

Centrioles: Microtubules Also

Centrioles are short cylinders with a 9 + 0 pattern of microtubules. There are 9 outer microtubule triplets and no center microtubules (fig. 2.10). There is always one pair of centrioles lying at right angles to one another near the nucleus (see fig. 2.4). Before a cell divides, the centrioles duplicate, and the members of the new pair are also at right angles to one another.

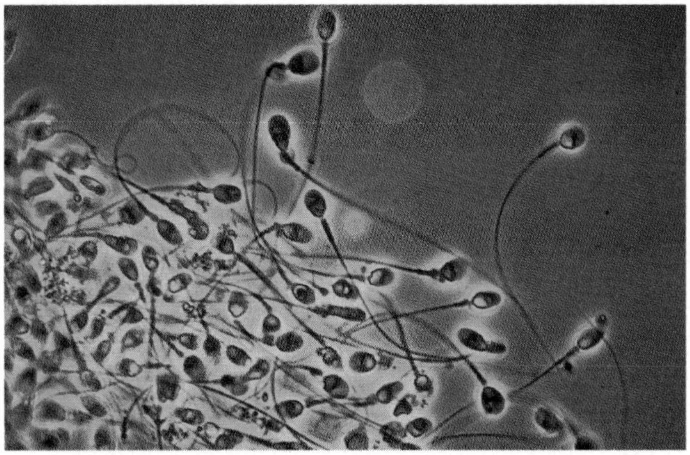

Figure 2.11

Sperm cells use long, whiplike flagella to move about.

Centrioles are part of a microtubule organizing center that also includes other proteins and substances. Microtubules begin to assemble in the center, and then they grow outward, extending through the entire cytoplasm. Centrioles also give rise to basal bodies that direct the formation of cilia and flagella. In addition, centrioles may be involved in other cellular processes that use microtubules, such as movement of material throughout the cell or the formation of the spindle (p. 352). Their exact role in these processes is uncertain, however.

Cilia and Flagella: Making Cells Move

Cilia and **flagella** are hairlike projections of cells that can move either in an undulating fashion, like a whip, or stiffly, like an oar. Cells that have these organelles are capable of self-movement or moving material on the surface of the cell. For example, sperm cells, carrying genetic material to the egg, move by means of flagella (fig. 2.11). The cells that line our respiratory tract are ciliated (see fig. 3.2). These cilia sweep debris trapped within mucus back up the throat, and this action helps keep the lungs clean.

Each cilium and flagellum has a basal body lying in the cytoplasm at its base. **Basal bodies,** like centrioles, have a 9 + 0 pattern of microtubule triplets. They are believed to organize the structure of cilia and flagella even though cilia and flagella have a 9 + 2 pattern of microtubules. In cilia and flagella, there are 9 microtubule doublets surrounding 2 central microtubules. This arrangement is believed to be necessary to their ability to move.

> Centrioles have a 9 + 0 pattern of microtubules and give rise to basal bodies that organize the 9 + 2 pattern of microtubules in cilia and flagella. Centrioles may be connected in some way to the origination of microtubules and to the spindle fibers that are seen during cell division.

Cellular Metabolism

Cellular **metabolism** includes all the chemical reactions that occur in a cell. Quite often these reactions are organized into metabolic pathways.

$$A \xrightarrow{1} B \xrightarrow{2} C \xrightarrow{3} D \xrightarrow{4} E \xrightarrow{5} F \xrightarrow{6} G$$

The letters, except *A* and *G*, are *products* of the previous reaction and the *reactants* for the next reaction. *A* represents the beginning reactant(s), and *G* represents the end product(s). The numbers in the pathway refer to different enzymes. *Every reaction in a cell requires a specific enzyme.* In effect, no reaction occurs in a cell unless its enzyme is present. For example, if enzyme number 2 in the diagram is missing, the pathway cannot function; it will stop at *B*. Since enzymes are so necessary in cells, their mechanism of action has been studied extensively.

> Metabolic pathways contain many enzymes that perform their reactions in a sequential order.

Enzymes and Coenzymes

When an enzyme speeds up a reaction, the reactant(s) that participates in the reaction is called the enzyme's *substrate(s).* Enzymes are often named for their substrate(s) (table 2.3). Enzymes have a specific region, called an **active site,** where the substrates are brought together so that they can react. An enzyme's specificity is caused by the shape of the active site, where the enzyme and its substrate(s) fit together in a specific way, much as the pieces of a jigsaw puzzle fit together (fig. 2.12). After one reaction is complete, the product or products are released, and the enzyme is ready to catalyze another reaction. This can be summarized in the following manner:

$$E + S \rightarrow ES \rightarrow E + P$$

(where E = enzyme, S = substrate, ES = enzyme-substrate complex, and P = product).

Environmental conditions such as an incorrect pH or high temperature can cause an enzyme to become denatured. A denatured enzyme no longer has its usual shape and is therefore unable to speed up its reaction.

Table 2.3

Enzymes Named for Their Substrates

Substrate	Enzyme
Lipid	Lipase
Urea	Urease
Maltose	Maltase
Ribonucleic acid	Ribonuclease
Lactose	Lactase

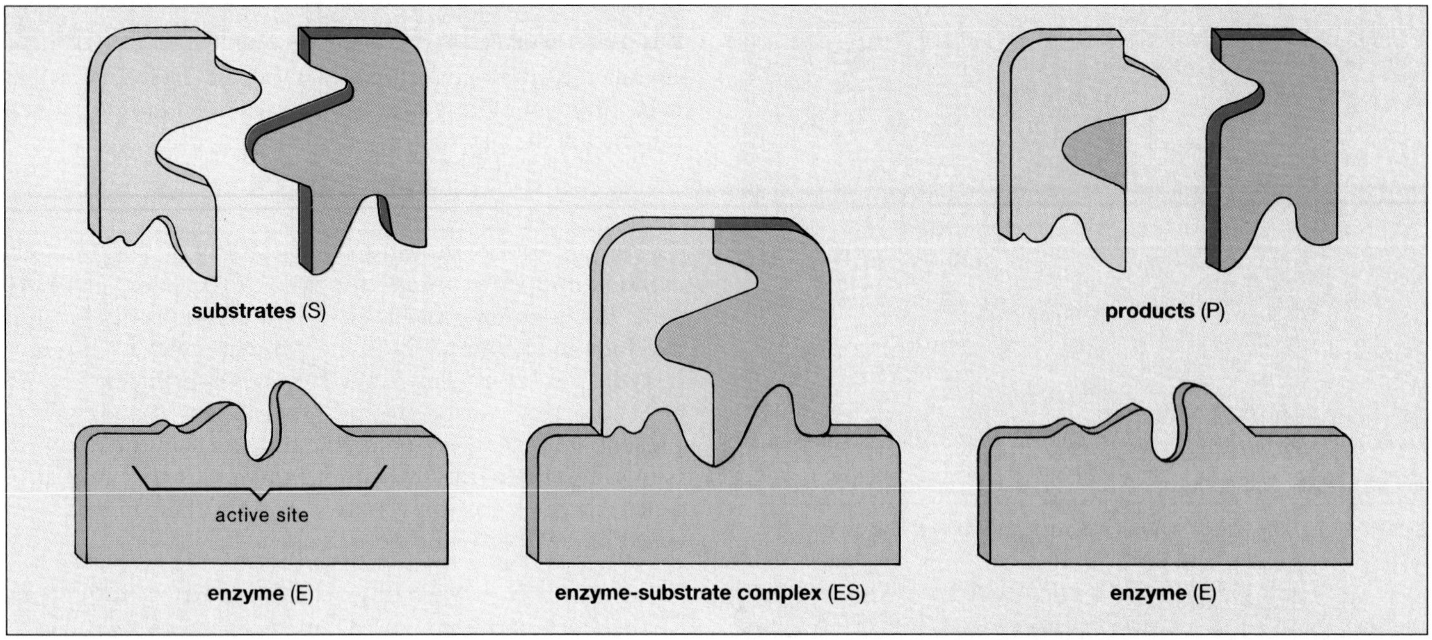

Figure 2.12

Enzymatic action. An enzyme has an active site where the substrates and enzyme fit together in such a way that the substrates are oriented to react. Following the reaction, the products are released.

Many enzymes have **coenzymes,** nonprotein portions that assist the enzyme and may even accept or contribute atoms to the reaction. It is of interest that vitamins and metals such as copper and zinc are often components of coenzymes. The vitamin niacin is a part of the coenzyme NAD, which removes hydrogen (H) atoms from substrates and therefore is called a *dehydrogenase.* NAD that is carrying hydrogen atoms is written as $NADH_2$ because NAD removes 2 hydrogen atoms at a time.

Hydrogen atoms are sometimes removed by NAD when molecules are broken down. As we shall see, the removal of hydrogen atoms releases energy that can be used for ATP buildup.

> Enzymes are specific because they have an active site that accommodates their substrates. Enzymes often have organic, nonprotein helpers called coenzymes. NAD is a dehydrogenase, a coenzyme that removes hydrogen from substrates.

Cellular Respiration: ATP Buildup

Cellular respiration is an important part of cellular metabolism because it accounts for ATP buildup in cells. Cellular respiration includes *aerobic* (requires oxygen) cellular respiration and fermentation, an *anaerobic* (does not require oxygen) process.

Aerobic Cellular Respiration: Oxygen Needed

During *aerobic cellular respiration,* glucose ($C_6H_{12}O_6$) is broken down to carbon dioxide (CO_2) and water (H_2O). Even though it is possible to write an overall equation for the process, it does not occur in one step. Glucose breakdown requires 3 subpathways: *glycolysis,* the *Krebs cycle,* and the *electron transport system* (fig. 2.13). The location of these subpathways is as follows:

Glycolysis—occurs in the cytoplasm, outside a mitochondrion

Krebs cycle—occurs in the matrix of a mitochondrion

Electron transport system—occurs on the cristae of mitochondrion

Each arrow you see in glycolysis and the Krebs cycle (fig. 2.13) represents a different enzyme, and the letters represent the product of the previous reaction and the substrate for the next reaction. Notice how each pathway resembles a conveyor belt in which a beginning substrate continuously enters at the start and, after a series of reactions, end products leave at the termination of the belt. It is important to realize, too, that all 3 pathways occur at the same time. They can be compared to the inner workings of a watch, in which all parts are synchronized.

It is possible to relate the reactants and products of the overall equation for aerobic cellular respiration to the subpathways:

1. Glucose, a C_6 molecule, is to be associated with **glycolysis,** the breakdown of glucose to 2 molecules of pyruvate (pyruvic acid), a C_3 molecule. During

WANT TO DO WELL ON TESTS AND EXAMS?

Wm. C. Brown Publishers now has developed a **TAKE-HOME TUTOR** that will help you achieve your goals in the classroom and prepare for careers in the nursing and allied health sciences. The ***BIOLOGY LEARNING SUPPORT PACKAGE*** is a complete set of additional study aids designed to facilitate learning and achievement. The package includes:

Student Study Guide
ISBN 0-697-16236-2
For each text chapter there is a corresponding study guide chapter—including *suggestions for review, objective questions,* and *essay questions*—that reviews the major chapter concepts.

Is Your Math Ready for Biology?
ISBN 0-697-22677-8
This unique booklet provides a *diagnostic test* that measures your math ability. Part II of the booklet provides *helpful hints* on the necessary math skills needed to successfully complete a biological science course.

How to Study Science
ISBN 0-697-14474-7
This excellent workbook offers you helpful suggestions for meeting the considerable challenges of a college science course. It offers tips on *how to take notes, how to get the most out of laboratories,* as well as *how to overcome "science anxiety."* The book's unique design helps you develop critical- thinking skills, while facilitating careful note taking.

A Life Science Lexicon
ISBN 0-697-12133-X
This portable, inexpensive reference will help you quickly master the vocabulary of the life sciences. Not a dictionary, it carefully explains the rules of *word construction and derivation,* and gives *complete definitions* of all important terms.

Animated Illustrations of Physiological Processes Videotape
ISBN 0-697-21512-0
This videotape *animates 13 illustrations of complex physiological processes.* These colorful, moving visuals make challenging concepts easier to understand.

Explorations in Human Biology CD-ROM
ISBN 0-697-22959-9
This interactive software, for use with Macintosh and IBM Windows, consists of 16 different animated modules that cover *key topics* discussed in a human biology course.

BE YOUR OWN TUTOR!

To get your own **TAKE-HOME TUTOR**, just ask your bookstore manager. If it's unavailable through your bookstore, call toll-free 800-338-5578 (in the United States), 800-268-4178 (in Canada).

WCB **Wm. C. Brown Publishers**
A Division of Wm. C. Brown Communications, Inc.

Mader tht 15956

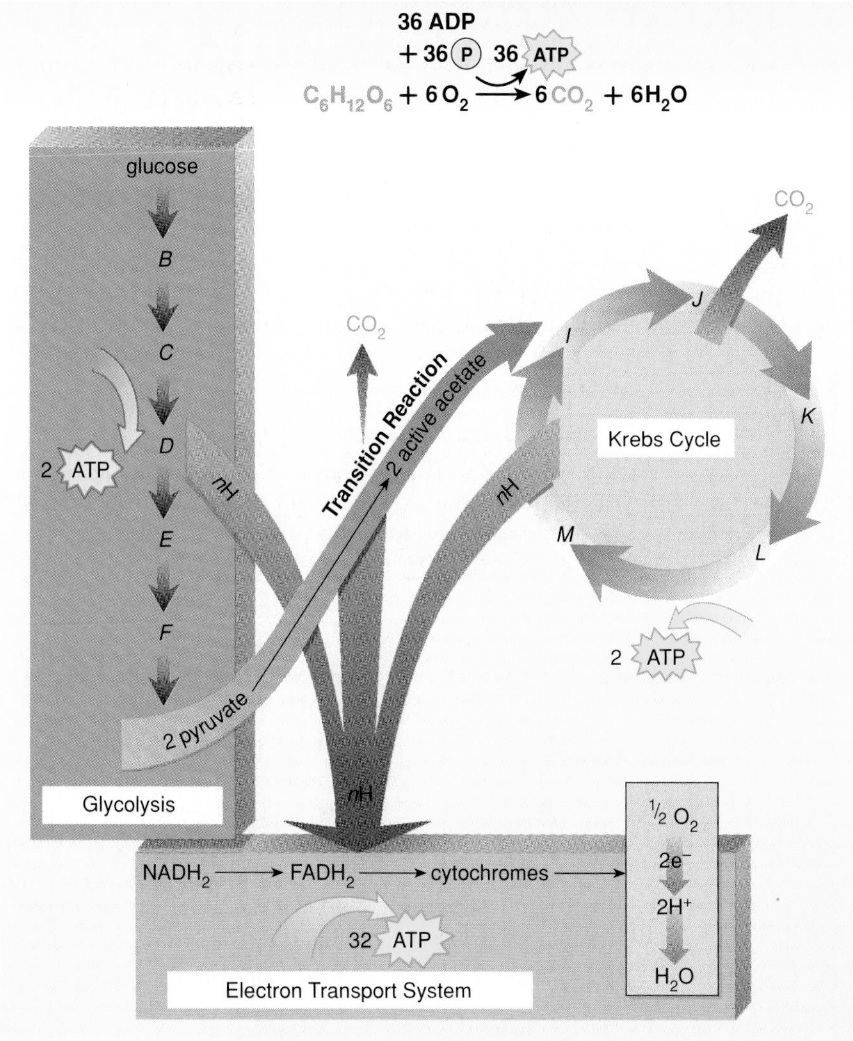

Figure 2.13

Aerobic cellular respiration. The overall reaction shown at the top actually requires 3 subpathways: glycolysis, the Krebs cycle, and the electron transport system. As the reactions occur, a number of hydrogen (nH) atoms and carbon dioxide (CO_2) molecules are removed from the various substrates. Oxygen (O_2) acts as the final acceptor for hydrogen atoms ($2e^- + 2H^+$) and becomes water (H_2O).

glycolysis, energy is released as hydrogen (H) atoms are removed. This energy is used to form 2 ATP molecules as described in figure 1.18.

2. Carbon dioxide, CO_2, is to be associated with the transition reaction and the Krebs cycle, both of which occur in mitochondria. During the transition reaction, pyruvate is converted to active acetate (AA), a C_2 molecule, after CO_2 comes off. Because the transition reaction occurs twice per glucose molecule, 2 molecules of CO_2 are released. Hydrogen (H) atoms are also removed at this time.

AA enters the **Krebs cycle,** a cyclical series of reactions that give off CO_2 molecules and produce one ATP molecule. Since the Krebs cycle occurs twice per glucose molecule, altogether 4 CO_2 and 2 ATP are produced per glucose molecule. Hydrogen (H) atoms are also removed as the Krebs cycle occurs.

3. Oxygen, O_2, and water, H_2O, are to be associated with the **electron transport system.** The electron transport system begins with $NADH_2$, the coenzyme that carries most of the hydrogen (H) atoms to it, but after that it consists of molecules that carry electrons. Electrons are removed from the hydrogen atoms, leaving behind hydrogen ions (H^+), and then the electrons are passed from one molecule to another until the electrons are received by oxygen. At this point, $2H^+$ combine with ionic oxygen to give water. As the electrons are passed down the system, energy is released to allow the buildup of ATP.

4. ATP is to be associated with glycolysis, the Krebs cycle, and the electron transport system. Usually 32 ATP are produced altogether by the electron transport system.

Table 2.4 summarizes the discussion of aerobic cellular respiration.

Table 2.4

Overview of Aerobic Cellular Respiration

Name of Pathway	Result
Glycolysis	Removal of H from substrates produces 2 ATP
Transition reaction	Removal of H from substrates releases 2 CO_2
Krebs cycle	Removal of H from substrates releases 4 CO_2 Produces 2 ATP after 2 turns
Electron transport system	Accepts H from other pathways and passes electrons on to O_2, producing H_2O Produces 32 ATP

Aerobic cellular respiration requires glycolysis, which takes place in the cytoplasm; the Krebs cycle, which is located in the matrix of the mitochondria; and the electron transport system, which is located on the cristae of the mitochondria.

Fermentation: Oxygen Not Used

Fermentation is an anaerobic process. When oxygen is not available to cells, the electron transport system soon becomes inoperative because oxygen is not present to accept electrons. In this case, most cells have a safety valve so that some ATP can still be produced. Glycolysis operates as long as it is supplied with "free" NAD; that is, NAD that can pick up hydrogen atoms. Normally, $NADH_2$ takes hydrogens to the electron transport system and thereby becomes "free" of hydrogen atoms. However, if the system is not working due to lack of oxygen, $NADH_2$ passes its hydrogen atoms to pyruvate as shown in the following reaction:

$$\begin{array}{ccc} NADH_2 & \curvearrowright & NAD \\ \text{pyruvate} & \longrightarrow & \text{lactate} \end{array}$$

The Krebs cycle and electron transport system do not function as part of fermentation. When oxygen is available again, lactate (lactic acid) can be converted back to pyruvate and metabolism can proceed as usual.

Fermentation takes less time than aerobic cellular respiration, but since glycolysis alone is occurring, it produces only 2 ATP per glucose molecule. Also, fermentation results in the buildup of lactate. Lactate is toxic to cells and causes muscles to cramp and fatigue. If fermentation continues for any length of time, death follows.

It is of interest to know that fermentation takes its name from yeast fermentation. Yeast fermentation produces alcohol and carbon dioxide (instead of lactate). When yeast is used to leaven bread, it is the carbon dioxide that produces the desired effect. When yeast is used to produce alcoholic beverages, it is the alcohol that humans make use of.

> Fermentation is an anaerobic process, a process that does not require oxygen but produces very little ATP per glucose molecule and results in lactate or alcohol and carbon dioxide buildup.

Summary

A cell is surrounded by a cell membrane, which regulates the entrance and exit of molecules. Some molecules, such as water and gases, diffuse through the membrane. The direction in which water diffuses is dependent on its concentration within the cell compared to outside the cell.

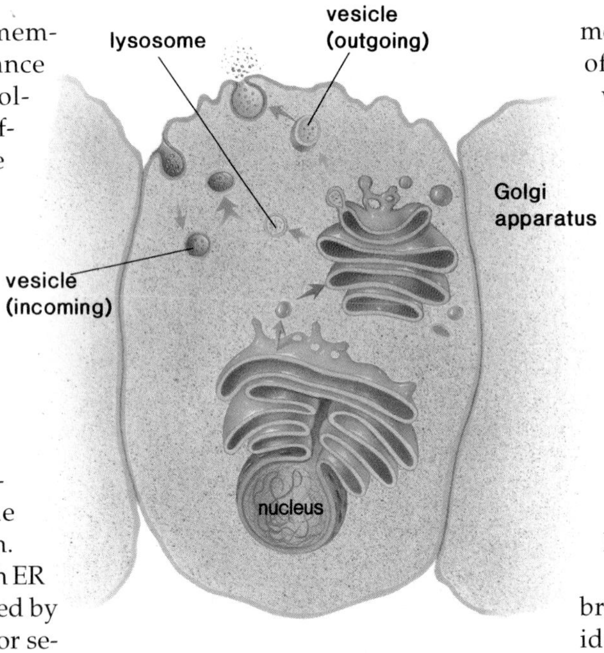

Table 2.2 lists the cell organelles we have studied in the chapter. The nucleus is a large organelle of primary importance because it controls the rest of the cell. Within the nucleus lies the chromatin, which condenses to become chromosomes during cell division.

Proteins are made at the rough ER before being modified and packaged by the Golgi apparatus into vesicles for secretion. During secretion, a vesicle discharges its contents at the cell membrane. Golgi-derived lysosomes fuse with incoming vesicles to digest any material enclosed within, and lysosomes also carry out autodigestion of old parts of cells.

Mitochondria are the powerhouses of the cell. During the process of aerobic cellular respiration, mitochondria convert carbohydrate energy to ATP energy.

Microtubules and actin filaments make up the cytoskeleton, which maintains the cell's shape and permits movement of cell parts. Centrioles are a part of the microtubule organizing center, which is associated with the formation of microtubules in general and the spindle that appears during cell division. Centrioles also produce basal bodies that give rise to cilia and flagella.

Cellular metabolism uses pathways in which a series of reactions proceed in an orderly, step-by-step manner. Each of these reactions requires a specific enzyme. Sometimes enzymes require coenzymes, nonprotein portions that participate in the reaction. NAD is a coenzyme.

Aerobic cellular respiration (the breakdown of glucose to carbon dioxide and water) includes 3 pathways: glycolysis, the Krebs cycle, and the electron transport system. If oxygen is not available in cells, the electron transport system is inoperative and fermentation (an anaerobic process) occurs. Fermentation makes use of glycolysis only, plus one more reaction in which pyruvate is reduced to lactate.

Study and Thought Questions

1. Describe the structure and biochemical makeup of a membrane. (p. 31)
2. What are the 3 mechanisms by which substances enter and exit cells? Define isotonic, hypertonic, and hypotonic solutions. (p. 32)
3. Describe the nucleus and its contents, including the terms *DNA* and *RNA* in your description. (p. 34)

   Critical Thinking Devise a controlled experiment to show that the nucleus is necessary to the continued life of a cell.
4. Describe the structure and function of endoplasmic reticulum. Include the terms *rough* and *smooth ER* and *ribosomes* in your description. (p. 35)

   Ethical Issue The consumption of alcohol increases the amount of smooth ER in liver cells to the point that the cells do not function properly. Should alcoholics and others who abuse their bodies be barred from any future universal health care programs?
5. Describe the structure and function of the Golgi apparatus. Mention vesicles and lysosomes in your description. (p. 36)
6. Describe the structure of mitochondria, and relate this structure to the pathways of aerobic cellular respiration (p. 37)
7. Describe the composition of the cytoskeleton. (p. 37)
8. Describe the structure and function of centrioles, cilia, and flagella. (pp. 38, 39)

   Critical Thinking Movement requires energy. The microtubules in cilia interact to produce movement. What molecule do you predict is involved in this interaction?
9. Discuss and draw a diagram for a metabolic pathway. Discuss and give a reaction to describe the specificity theory of enzymatic action. Define coenzyme. (p. 39)
10. Name and describe the events within the 3 subpathways that make up aerobic cellular respiration. Why is fermentation wasteful and potentially harmful to the human body? (p. 40)

    Critical Thinking In cells, fatty acids are broken down directly to active acetate molecules, which enter the Krebs cycle. Considering that a glucose molecule results in 2 active acetates and a fatty acid of 18 carbons results in 9 active acetates, explain why fats are long-term storage forms of energy in cells.

Writing Across the Curriculum

1. What is the advantage of organelles in complex cells?
2. Present evidence that the cell is dynamic rather than static as it appears to be in drawings and micrographs.
3. Show that human beings have a cellular basis by describing the function of the oxygen we breathe in and the origination of the carbon dioxide we breathe out.

Objective Questions

For questions 1–5, match the organelles in the key to their functions.

Key: **a.** mitochondria
b. nucleus
c. Golgi apparatus
d. rough ER
e. centrioles

1. packaging and secretion __________
2. cell division __________
3. powerhouses of the cell __________
4. protein synthesis __________
5. control center for cell __________
6. Microtubules and actin filaments are a part of the __________, the framework of the cell that provides its shape and regulates movement of organelles.
7. Water enters a cell when it is placed in a __________ solution.
8. Substrates react at the __________, located on the surface of their enzyme.
9. During aerobic cellular respiration, most of the ATP molecules are produced at the __________, a series of carriers located on the __________ of mitochondria.
10. Fermentation of a glucose molecule produces only __________ ATP compared to the __________ ATP produced by aerobic cellular respiration.

11. Label only the parts of the cell that are involved in protein synthesis and modification. Explain your choices.

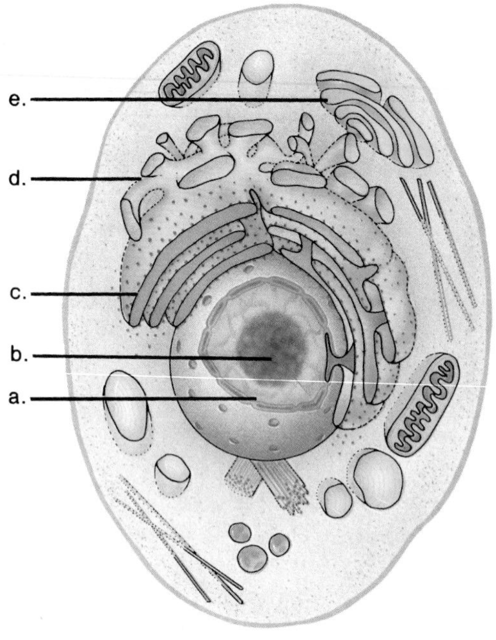

12. Complete the following diagram by labeling the pathways and by adding ATP, CO_2, O_2, and H_2O where needed.

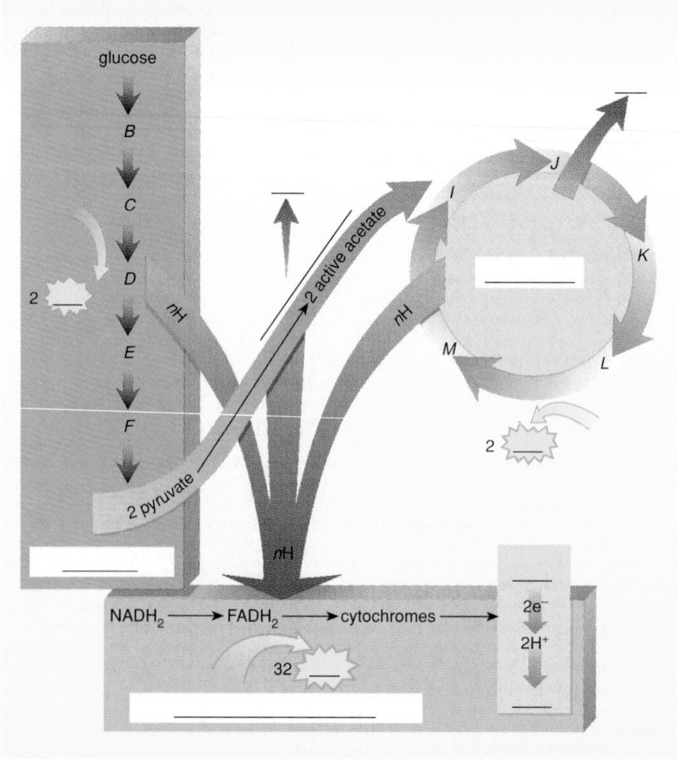

SELECTED KEY TERMS

actin filament An extremely thin fiber found within the cytoplasm that is composed of the protein actin and is involved in the maintenance of cell shape and the movement of cell contents. 37

active site The region on the surface of an enzyme where the substrate binds and where the reaction occurs. 39

centriole (sen´tre-ōl) A short, cylindrical organelle in animal cells that contains microtubules in a 9 + 0 pattern and that is associated with the formation of basal bodies and the formation of the spindle during cell division. 38

chromosome Rod-shaped body in the nucleus seen during cell division; contains the hereditary units or genes. 34

coenzyme A nonprotein molecule that aids the action of an enzyme. 40

cytoplasm Cellular region inside cell membrane but excluding the nucleus. 31

cytoskeleton Filamentous protein structures found throughout the cytoplasm that help maintain the shape of the cell, anchor the organelles, and allow the cell and its organelles to move. 37

electron transport system A series of molecules within the inner mitochondrial membrane that pass electrons one to the other from a higher energy level to a lower energy level; the energy released is used to build ATP. 41

endoplasmic reticulum (ER) (en-do-plaz´-mik rĕ-tik´u-lum) A complex system of tubules, vesicles, and sacs in cells, sometimes having attached ribosomes. Rough ER has ribosomes; smooth ER does not. 36

fermentation Anaerobic breakdown of carbohydrates that results in organic end products such as alcohol and lactic acid. 42

glycolysis (gli-kol´i-sis) A metabolic pathway found in the cytoplasm that participates in aerobic cellular respiration and fermentation. It converts glucose to 2 molecules of pyruvate. 40

Golgi apparatus (gol´je) An organelle that consists of concentrically folded membranes and functions in the processing, packaging, and secretion of cellular products. 36

Krebs cycle A cyclical metabolic pathway found in the matrix of mitochondria that participates in aerobic cellular respiration; breaks down C_2 (2-carbon) groups to carbon dioxide and hydrogen atoms. 41

lysosome (li´so-sōm) An organelle in which digestion takes place due to the action of powerful hydrolytic enzymes. 36

microtubule A organelle composed of 13 rows of globular proteins; found in multiple units in several other organelles, such as the centriole, cilia, and flagella, as well as spindle fibers. 37

mitochondrion (mi´to-kon´dre-on) An organelle in which cellular respiration produces the energy molecule, ATP. 37

nucleolus (nu-kle´-o-lus) (pl., nucleoli) An organelle found inside the nucleus; a special region of chromatin containing DNA that produces RNA for ribosome formation. 35

nucleus A large organelle containing the chromosomes and acting as a control center for the cell; center of an atom. 34

organelle Specialized structures within cells (e.g., nucleus, mitochondria, and endoplasmic reticulum). 33

ribosome (ri´bo-sōm) Minute particle, found attached to the endoplasmic reticulum or loose in the cytoplasm, that is the site of protein synthesis. 35

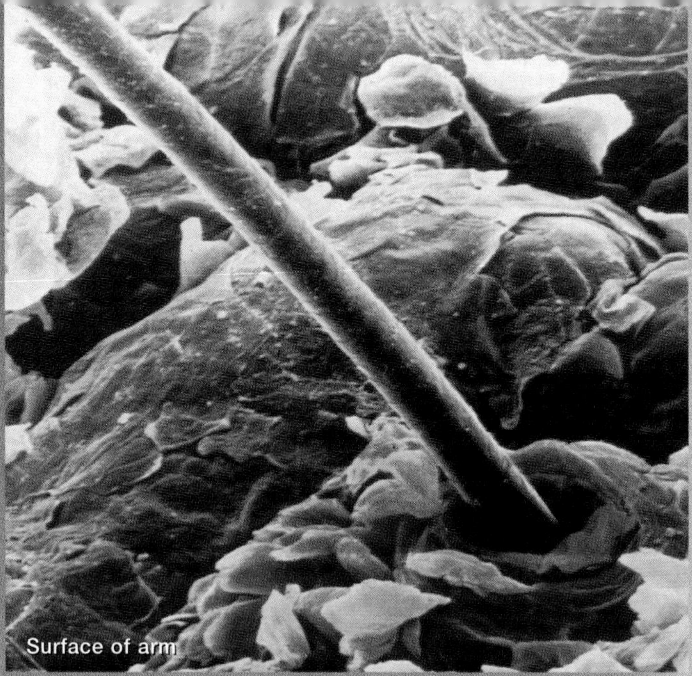

Chapter 3

Human Organization

It's all very well to think of humans as being multicellular animals, but actually the cells are differentiated and have specific structures and functions. Like cells form tissues that are grouped into muscular tissue, nervous tissue, connective tissue, and epithelial tissue. We all know that muscular tissue contracts, nervous tissue conducts, and connective tissue binds, but what about epithelial tissue? Epithelial tissue forms a protective covering on the exterior of the body and all its various orifices. Because of epithelial tissue, the body is like a citadel that forbids the entrance of microorganisms.

Chapter Concepts

1. Human tissues can be categorized into 4 major types: epithelial, connective, muscular, and nervous tissues. 46
2. Organs usually contain several types of tissues. For example, although skin is composed primarily of epithelial tissue and connective tissue, it also contains muscle and nerve fibers. 52
3. Organs are grouped into organ systems, each of which has specialized functions. 56
4. Humans exhibit a marked ability to maintain a relatively constant internal environment. 57
5. All organ systems contribute to homeostasis. 57

Chapter Outline

The human body has levels of organization (fig. 3.1). Cells of the same type are joined to form a tissue. Different tissues are found in an organ, and various types of organs are arranged into an organ system. Finally, the organ systems make up the organism.

Epithelial Tissue

Epithelial tissue, also called epithelium, consists of tightly packed cells that form a continuous sheet over the entire body surface and most of the body's inner cavities. On the external surface, it forms a covering that protects the body from injury and drying out. On internal surfaces, epithelial tissue may be specialized for other functions in addition to protection; for example, it secretes mucus along the digestive tract; it sweeps up impurities from the lungs by means of hairlike extensions called cilia; and it efficiently absorbs molecules from kidney tubules because of fine cellular extensions called microvilli.

There are 3 types of epithelial tissue (fig. 3.2): **squamous epithelium,** which *is composed of flattened cells;* **cuboidal epithelium,** which contains *cube-shaped cells;* and **columnar epithelium,** in which *the cells resemble pillars or columns.* An epithelium can be simple or *stratified.* Simple means the tissue has a single layer of cells, and **stratified** means the tissue has layers piled one on top of the other. The outer layer of skin is stratified squamous epithelium. One type of columnar epithelium is *pseudostratified*—it appears to be layered, but actually true layers do not exist because each cell touches a basement membrane. The basement membrane joins an epithelium to underlying connective tissue. Three types of junctions join the tightly packed epithelial cells, as described in figure 3.3.

An epithelium sometimes secretes a product, in which case it is known as glandular. A **gland** can be a single epithelial cell, as in the case of the mucus-secreting goblet cells, found within the columnar epithelium lining the digestive tract, or a gland can contain many cells. Glands that secrete their product into ducts are called **exocrine glands,** and those that secrete their product directly into the bloodstream are called **endocrine glands.**

Epithelial tissue forms a protective covering on the exterior of the body and lines all the cavities or lumens within the body.

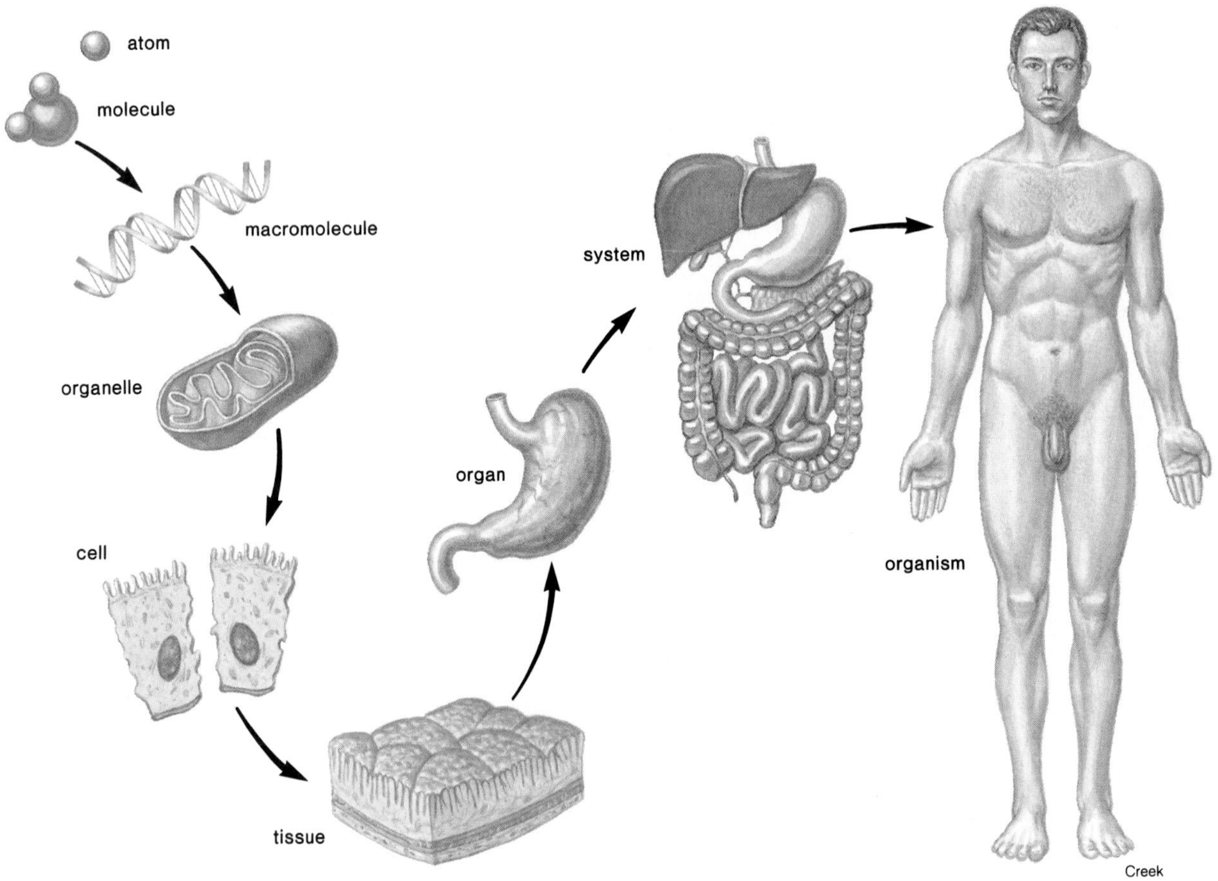

Figure 3.1

Levels of organization in the human body. Cells are composed of molecules, tissues are made up of cells, organs are composed of tissues, and the organism contains organ systems.

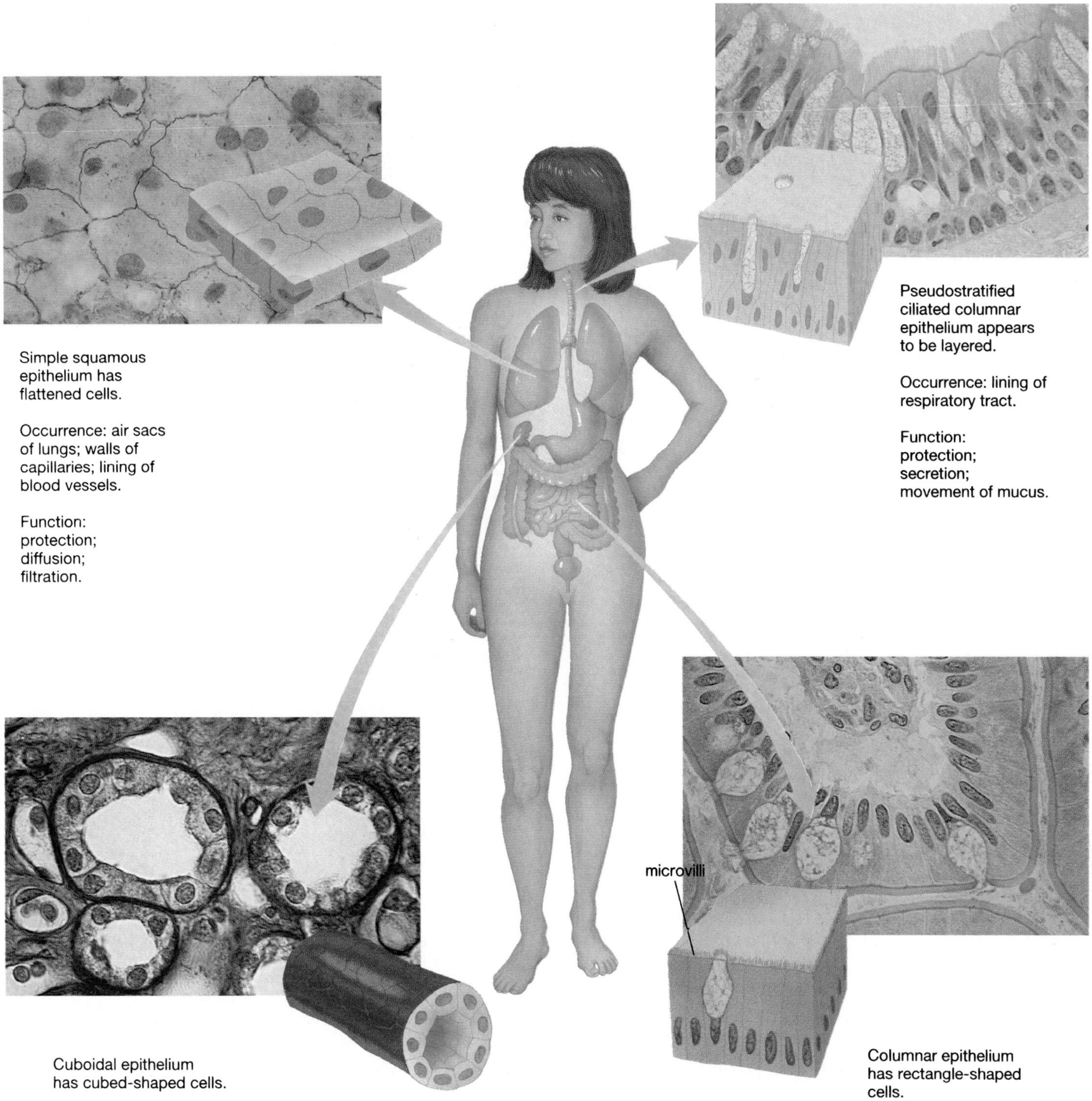
Simple squamous epithelium has flattened cells.
Occurrence: air sacs of lungs; walls of capillaries; lining of blood vessels.
Function: protection; diffusion; filtration.
Pseudostratified ciliated columnar epithelium appears to be layered.
Occurrence: lining of respiratory tract.
Function: protection; secretion; movement of mucus.
microvilli
Cuboidal epithelium has cubed-shaped cells.
Occurrence: surface of ovaries, lining of kidney tubules.
Function: protection; secretion; absorption.
Columnar epithelium has rectangle-shaped cells.
Occurrence: lining of uterus and intestine.
Function: protection; secretion; absorption.

Figure 3.2

Epithelial tissues. The 3 types of epithelial tissues—squamous, cuboidal, columnar—are named for the shape of their cells. They all have a protective function and other functions as noted.

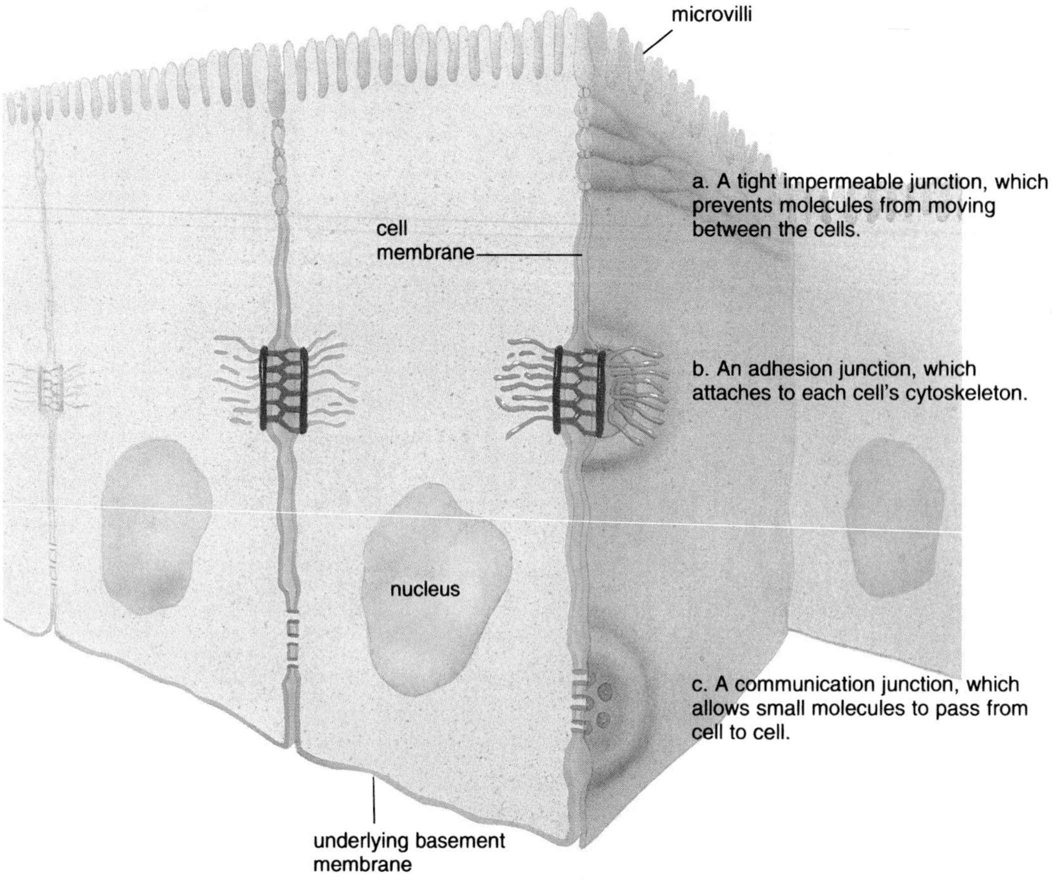
microvilli
a. A tight impermeable junction, which prevents molecules from moving between the cells.
cell membrane
b. An adhesion junction, which attaches to each cell's cytoskeleton.
nucleus
c. A communication junction, which allows small molecules to pass from cell to cell.
underlying basement membrane

Figure 3.3

Junctions between epithelial and other types of cells.
a. Tight, impermeable junctions are useful, for example, between intestinal cells so that liquid from the intestine cannot pass between the cells and must pass into the cells.
b. Adhesion junctions are useful, for example, between the squamous epithelial cells of the skin, which must withstand considerable stretching and mechanical stress.
c. Communication junctions are useful, for example, when cells must work in conjunction. They are found between muscle cells, which must contract in unison.

Connective Tissue

Connective tissue binds structures together, provides support and protection, fills spaces, stores fat, and forms blood cells (table 3.1). As a rule, connective tissue cells are separated widely by a **matrix,** in this instance a noncellular material found between cells. The matrix of connective tissue may have fibers of 2 types. White fibers contain collagen, a substance that gives them flexibility and strength. Yellow fibers contain elastin, a substance that is not as strong as collagen but is more elastic.

Table 3.1

Connective Tissue

Type	Function	Location
Loose connective tissue	Binds organs	Beneath the skin; beneath most epithelial layers
Adipose tissue	Insulates; stores fat	Beneath the skin; around the kidneys
Fibrous connective tissue	Binds organs	Tendons; ligaments
Cartilage		
Hyaline cartilage	Supports; protects	Ends of bones; nose; rings in walls of respiratory passages
Elastic cartilage	Supports; protects	External ear; part of the larynx
Fibrocartilage	Supports; protects	Between bony parts of backbone and knee
Bone	Supports; protects	Bones of skeleton
Blood	Transports gases, nutrients, and wastes about body; infection fighting; blood clotting	Blood vessels

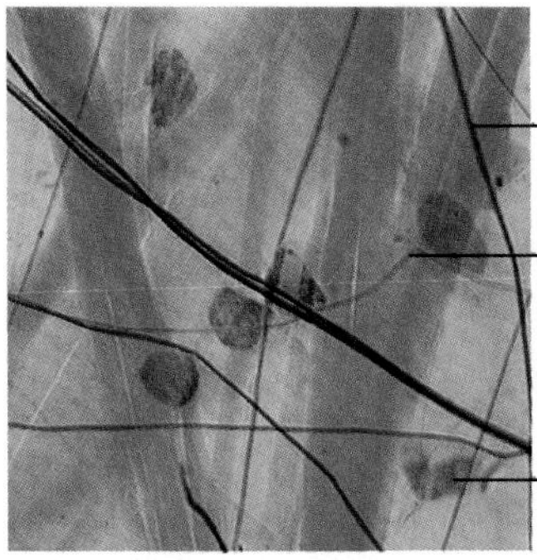
elastic fiber
collagen fiber
fibroblast
Loose connective tissue
has space between components.
Occurrence:
under skin and most epithelial layers
Function:
supports and binds organs

a.

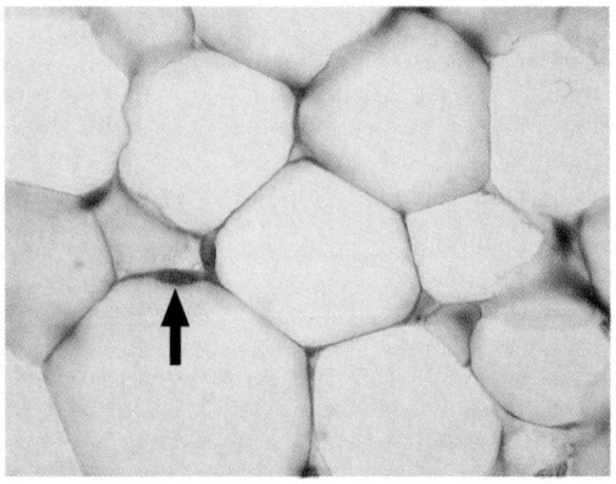
Adipose tissue
cells are filled with fat.
Occurrence:
under skin; around organs
Function:
insulates; stores fat

b.

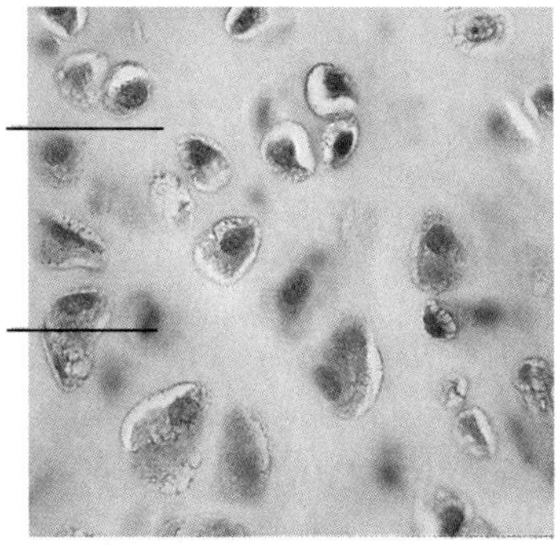
Hyaline cartilage
has cells in a lacuna.
Occurrence:
ends of bones, nose, walls of respiratory passages
Function:
supports, protects
matrix
cells within a lacuna

c.

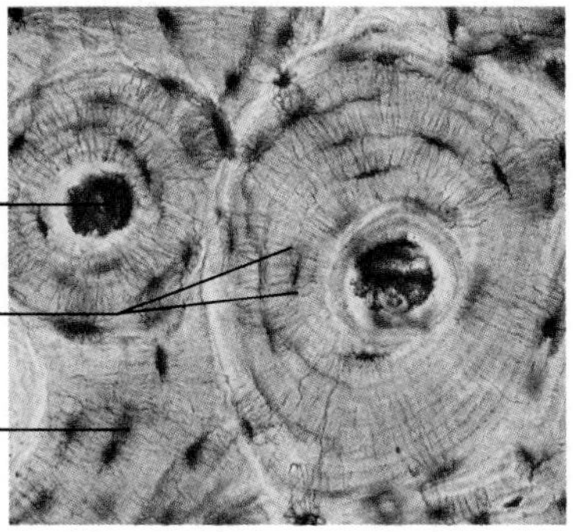
Compact bone
has cells in concentric rings.
Occurrence:
bones of skeleton
Function:
supports, protects
Haversian canal
canaliculi
osteocyte within a lacuna

d.

Figure 3.4

Connective tissue examples. **a.** In loose connective tissue, cells called fibroblasts are separated by a jellylike matrix, which contains both collagen and elastic fibers. **b.** Adipose tissue cells have nuclei (*arrow*) pushed to one side because the cells are filled with fat. **c.** In hyaline cartilage, the flexible matrix is glassy in appearance. **d.** In compact bone, the hard matrix contains calcium salts. The concentric rings of cells in lacunae are part of a Haversian system, an elongated cylinder with central canals that contains blood vessels and nerve fibers.

Loose Connective Tissue: Two Types of Fibers

Loose connective tissue binds structures (fig. 3.4*a*). The cells of this tissue, which are mainly *fibroblasts,* are located some distance from one another and are separated by a jellylike matrix that contains many white collagen fibers and yellow elastic fibers. The collagen fibers occur in bundles and are strong and flexible. The elastic fibers form networks that when stretched return to their original length. Loose connective tissue commonly lies beneath epithelial layers. In certain instances, epithelium and its underlying connective tissue form body membranes (p. 57). In addition, **adipose tissue** (fig. 3.4*b*) is a type of loose connective tissue in which the fibroblasts enlarge and store fat and in which the intercellular matrix is reduced.

Fibrous Connective Tissue: Tendons and Ligaments

Fibrous connective tissue contains many collagenous fibers that are packed closely together. This type of tissue has more specific functions than loose connective tissue. For example, **tendons** are straps of fibrous connective tissue which connect muscles to bones, and **ligaments** connect bones to other bones at joints. Tendons and ligaments take a long time to heal following an injury because their blood supply is relatively poor.

> Loose connective tissue and fibrous connective tissue, which bind body parts, differ according to the type and the abundance of fibers in the matrix.

Cartilage: Less Rigid Matrix

In **cartilage,** the cells lie in small chambers called **lacunae** (singular, lacuna), separated by a matrix that is solid yet flexible. Unfortunately, because this tissue lacks a direct blood supply, it heals very slowly. There are 3 types of cartilage, distinguished by the type of fiber in the matrix.

Hyaline cartilage (fig. 3.4*c*), the most common type of cartilage, contains only very fine collagenous fibers. The matrix has a milk glass appearance. Hyaline cartilage is found in the nose, at the ends of the long bones and the ribs, and in the supporting rings of the windpipe. The fetal skeleton also is made of this type of cartilage. Later, the cartilaginous fetal skeleton is replaced by bone.

Elastic cartilage has more elastic fibers than hyaline cartilage. For this reason, it is more flexible and is found, for example, in the framework of the outer ear.

Fibrocartilage has a matrix containing strong collagenous fibers. Fibrocartilage is found in structures that withstand tension and pressure, such as the pads between the vertebrae in the backbone and the wedges found in the knee joint. This tissue tends to degrade as a person ages.

Bone: Rigid Matrix

Bone is the most rigid connective tissue. It consists of an extremely hard matrix of calcium salts deposited around protein fibers. The minerals give bone rigidity, and the protein fibers provide elasticity and strength, much as steel rods do in reinforced concrete.

The shaft of a long bone is compact bone (fig. 3.4*d*). In **compact bone,** bone cells (osteocytes) are located in lacunae that are arranged in concentric circles around tiny tubes called Haversian canals. Nerve fibers and blood vessels are in these canals. The latter bring the nutrients that allow bone to renew itself. The nutrients can reach all of the cells because there are minute canals (canaliculi) containing thin processes of the osteocytes that connect the cells with one another and with the Haversian canals.

The ends of a long bone contain **spongy bone** (see fig. 11.1), which has an entirely different structure from compact bone. Spongy bone contains numerous bony bars and plates separated by irregular spaces. Although lighter than compact bone, spongy bone still is designed for strength. Just as braces are used for support in buildings, the solid portions of spongy bone follow lines of stress.

> Cartilage and bone are support tissues. Cartilage is more flexible than bone because the matrix is rich in protein and not calcium salts like that of bone.

Blood: Liquid Matrix

Blood (fig. 3.5) is a connective tissue in which the cells are separated by a liquid called plasma, the contents of which are listed in table 3.2. Blood cells are of 2 types: red blood cells **(erythrocytes),** which carry oxygen, and white blood cells **(leukocytes),** which aid in fighting infection. Also present in plasma are *platelets,* which are important to the initiation of blood clotting. Platelets are not complete cells; rather, they are fragments of giant cells originating in the bone marrow.

Blood is unlike other types of connective tissue in that the intercellular matrix (i.e., plasma) is not made by the cells. Plasma is a mixture of different types of molecules and ions that enter the blood at various locations (table 3.2). Some people do not classify blood as connective tissue; instead, they suggest a separate tissue category for blood called vascular tissue.

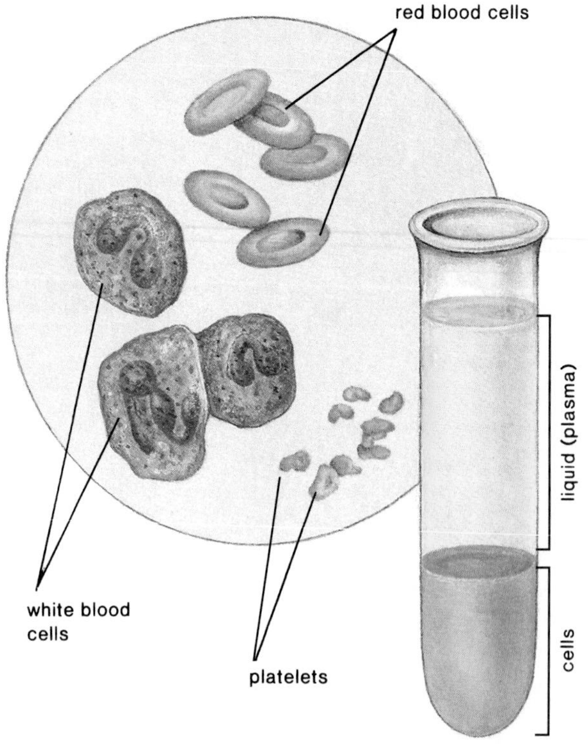

Figure 3.5

Blood, a liquid tissue. Blood is classified as connective tissue because the cells are separated by a matrix—plasma. Plasma, the liquid portion of blood, contains several types of cells (red blood cells and white blood cells) and cell fragments (platelets).

Table 3.2
Blood Plasma

WATER (92% OF TOTAL)	
SOLUTES (8% OF TOTAL)	
Inorganic ions (salts)	Na^+, Ca^{++}, K^+, Mg^{++}; Cl^-, HCO_3^-, $HPO_4^=$, $SO_4^=$
Gases	O_2, CO_2
Plasma proteins	Albumin, globulins, fibrinogen
Organic nutrients	Glucose, fats, phospholipids, amino acids, etc.
Nitrogenous waste products	Urea, ammonia, uric acid
Regulatory substances	Hormones, enzymes

> Blood is a connective tissue in which the matrix is plasma.

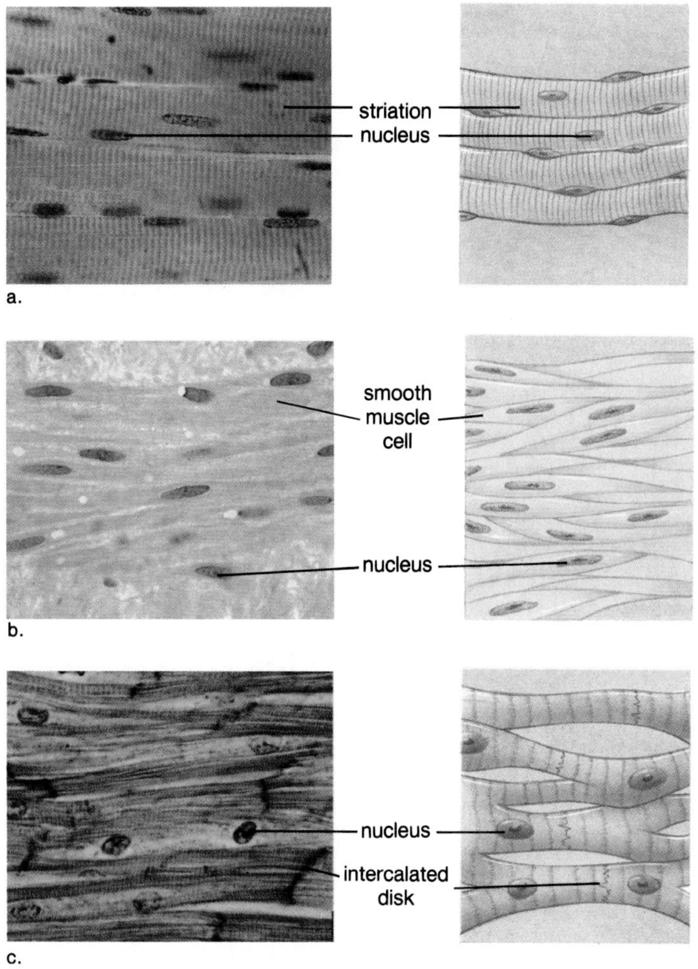

Figure 3.6

Muscular tissue. **a.** Skeletal muscle is voluntary and striated. **b.** Smooth muscle is involuntary and nonstriated. **c.** Cardiac muscle is involuntary and striated. The cell membranes of the branching cells fit together at intercalated disks.

Muscular Tissue

Muscular (contractile) tissue is composed of cells that are called *muscle fibers.* Muscle fibers contain actin filaments and myosin filaments, whose interaction accounts for the movements we associate with animals. There are 3 types of vertebrate muscles: *skeletal, smooth,* and *cardiac.*

Skeletal muscle is attached via tendons to the bones of the skeleton (fig. 3.6*a*); it moves body parts. It is under voluntary control and contracts faster than all the other muscle types. Skeletal muscle fibers are cylindrical and quite long—sometimes they run the length of the muscle. They arise during development when several cells fuse, giving one muscle fiber with multiple nuclei. The nuclei are located at the periphery of the fiber, just inside the cell membrane. Skeletal muscle fibers are **striated.** Light and dark bands run perpendicular to the length of the fiber. These bands are due to the placement of actin filaments and myosin filaments in the cell.

Smooth (visceral) muscle is so named because its fibers lack striations. The spindle-shaped fibers form layers in which the thick middle portion of one fiber is opposite the thin ends of adjacent fibers. Consequently, the nuclei form an irregular pattern in the tissue (fig. 3.6*b*). Smooth muscle is not under voluntary control and therefore is said to be involuntary. Smooth muscle, found in walls of viscera (intestine, stomach, and other internal organs) and blood vessels, contracts more slowly than skeletal muscle but can remain contracted for a longer time. When the smooth muscle of the intestine contracts, it moves the food along, and when the smooth muscle of the blood vessels contracts, it constricts the blood vessels, helping to raise the blood pressure.

Cardiac muscle, which is found only in the wall of the heart, is responsible for the heartbeat, which pumps blood. Cardiac muscle seems to combine features of both smooth muscle and skeletal muscle (fig. 3.6*c*). Its fibers have striations like skeletal muscle, but the contraction of the heart is involuntary for the most part. Cardiac muscle fibers also differ from skeletal muscle fibers in that they have a single, centrally placed nucleus. These fibers are branched and seemingly fused one with the other, and the heart appears to be composed of one large interconnecting mass of muscle fibers. Actually, cardiac muscle fibers are separate and individual, but they are bound end to end at intercalated disks, areas where folded cell membranes between 2 fibers contain desmosomes and gap junctions.

All muscular tissue contains actin filaments and myosin filaments; these form a striated pattern in skeletal and cardiac muscle but not in smooth muscle.

Nervous Tissue

The brain and the spinal cord contain conducting cells termed neurons. A **neuron** is a specialized cell that has 3 parts: (1) a *dendrite,* which conducts impulses (sends a message) to the cell body; (2) the *cell body,* where the nucleus is located; and (3) an *axon,* which conducts impulses away from the cell body (fig. 3.7).

When axons and dendrites are long, they are called *nerve fibers.* Outside the brain and the spinal cord, nerve fibers are bound by connective tissue to form **nerves.** Nerves conduct impulses from sense organs to the spinal cord and the brain, where the phenomenon called sensation occurs. They also conduct nerve impulses away from the spinal cord and the brain to the muscles, causing them to contract, and to the glands, causing them to secrete.

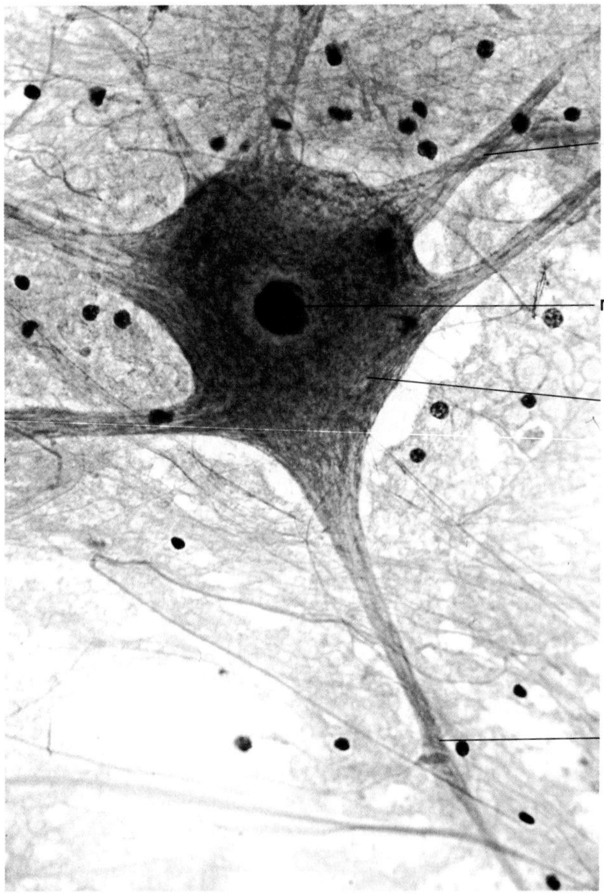

Figure 3.7

A neuron. Conduction of the nerve impulse is dependent on neurons, each of which has the 3 parts indicated. A dendrite takes nerve impulses to the cell body, and an axon takes them away from the cell body. The nucleus is in the cell body.

In addition to neurons, nervous tissue contains *neuroglial cells.* These cells maintain the tissue by supporting and protecting neurons. They also provide nutrients to neurons and help to keep the tissue free of debris.

Skin as an Organ

We will consider the skin as an example of an organ, a structure that is composed of 2 or more tissues (fig. 3.8). Some people like to call the skin the *integumentary system,* especially since it cannot be placed in one of the other organ systems; however, the skin does not really have distinct organs. The skin covers the body, protecting underlying parts from physical trauma, microbial invasion, and water loss. Skin also helps to regulate body temperature (p. 60), and because it contains receptors, the skin helps us to be aware of our surroundings and to communicate with others by touch.

Skin Layers: Structure and Function

The **epidermis** is the outer, thinner layer of the skin. It is made up of stratified squamous epithelium, which is continually produced by a bottom layer of cells termed basal cells. Newly formed cells push to the surface and then gradually flatten and harden. Eventually, the cells die and are sloughed off. Hardening is caused by cellular production of a waterproof protein called *keratin.* When you have *dandruff,* the rate of keratinization is 2 or 3 times the normal rate in certain areas of the scalp. Over much of the body, keratinization is minimal, but the palm of the hand and the sole of the foot have a particularly thick outer layer of dead keratinized cells arranged in spiral and concentric patterns. These patterns form fingerprints and footprints.

Specialized cells in the epidermis called *melanocytes* produce melanin, the pigment responsible for skin color. When you sunbathe, the melanocytes become more active, producing melanin in an attempt to protect the skin from the damaging effects of the ultraviolet (UV) radiation in sunlight.

> The epidermis, the outer layer of skin, is made up of stratified squamous epithelium. New cells, continually produced in the innermost layer of the epidermis, push outward, become keratinized, die, and are sloughed off.

The **dermis,** a layer of fibrous connective tissue, is thicker than the epidermis. It contains elastic fibers and collagen fibers. The collagen fibers form bundles, which interlace with each other and run, for the most part, parallel to the skin surface. As a person ages and is exposed to the sun, the number of these fibers decreases, and those remaining have characteristics that make the skin less supple and prone to wrinkling.

There are several types of structures in the dermis. A hair shaft, except for the root, contains dead and hardened epidermal cells; the root is alive and resides at the base of a *hair follicle.* Each follicle has one or more *oil* (sebaceous) *glands,* which secrete sebum, an oily substance that lubricates the hair and the skin. Particularly on the nose and the cheeks, the sebaceous glands may fail to discharge, and the secretions collect and form "whiteheads" or "blackheads." The color of blackheads is due to oxidized sebum. If pus-inducing bacteria also are present, a boil or a pimple may result.

The *arrector pili muscle,* a smooth muscle, is attached to the hair follicle in such a way that when contracted, the muscle causes the hair to "stand on end." When you are frightened or are cold, goose bumps develop due to the contraction of these muscles.

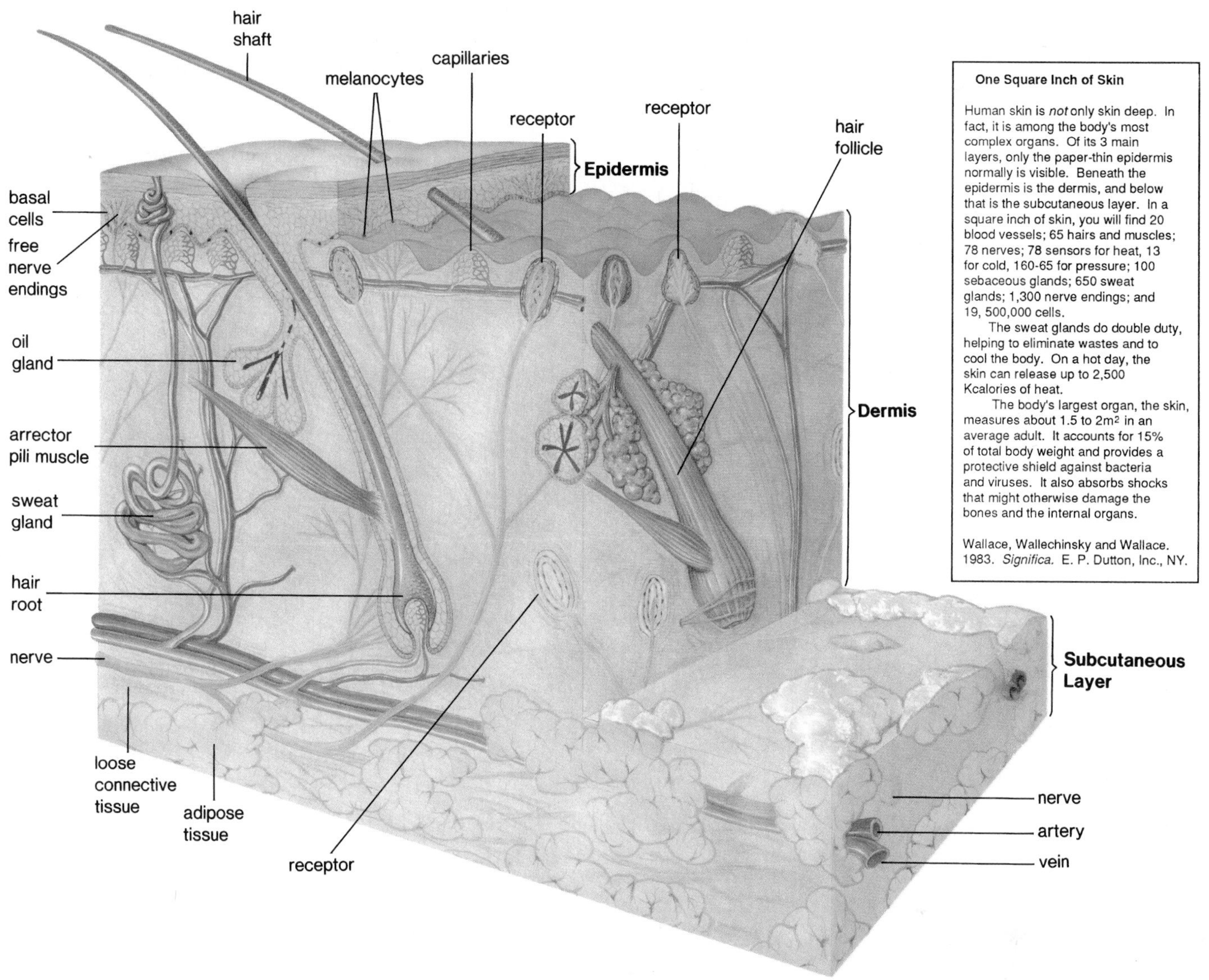

One Square Inch of Skin

Human skin is *not* only skin deep. In fact, it is among the body's most complex organs. Of its 3 main layers, only the paper-thin epidermis normally is visible. Beneath the epidermis is the dermis, and below that is the subcutaneous layer. In a square inch of skin, you will find 20 blood vessels; 65 hairs and muscles; 78 nerves; 78 sensors for heat, 13 for cold, 160-65 for pressure; 100 sebaceous glands; 650 sweat glands; 1,300 nerve endings; and 19, 500,000 cells.

The sweat glands do double duty, helping to eliminate wastes and to cool the body. On a hot day, the skin can release up to 2,500 Kcalories of heat.

The body's largest organ, the skin, measures about 1.5 to 2m² in an average adult. It accounts for 15% of total body weight and provides a protective shield against bacteria and viruses. It also absorbs shocks that might otherwise damage the bones and the internal organs.

Wallace, Wallechinsky and Wallace. 1983. *Significa.* E. P. Dutton, Inc., NY.

Figure 3.8

Human skin anatomy. Skin contains these layers: epidermis and dermis, plus a subcutaneous layer, which lies beneath the dermis.
Source: Wallace, Wallechinsky, and Wallace, 1983, Significa. *E. P. Dutton, Inc., NY.*

Sweat (sudoriferous) *glands* are quite numerous and are present in all regions of skin. A sweat gland begins as a coiled tubule within the dermis, but then it straightens out near its opening. Some sweat glands open into hair follicles, and others open onto the surface of the skin.

Small receptors are present in the dermis. *Receptors* are specialized nerve endings, present in all organs and parts of the body, that respond to either external or internal stimuli as appropriate. In the skin there are different receptors for touch, pressure, pain, and temperature. The fingertips contain the most touch receptors, and these add to our ability to use our fingers for delicate tasks. The dermis also contains nerve fibers and blood vessels. When blood rushes into these vessels, a person blushes, and when blood is reduced in them, a person turns "blue."

The dermis, composed of fibrous connective tissue, lies beneath the epidermis. It contains hair follicles, sebaceous glands, and sweat glands. It also contains receptors, blood vessels, and nerve fibers.

The **subcutaneous layer,** which lies below the dermis, is composed of loose connective tissue, including adipose tissue. Adipose tissue helps to insulate the body from either gaining heat from the outside or losing heat from the inside. A well-developed subcutaneous layer gives a rounded appearance to the body. Excessive development of this layer accompanies obesity. The subcutaneous layer also attaches the skin to the underlying muscles.

Ecology Focus

Ozone Depletion Threatens the Biosphere

The earth's atmosphere is divided into layers. The troposphere envelopes us as we go about our day-to-day lives. When ozone (O_3) is present in the troposphere (called ground-level ozone), it is considered a pollutant because it adversely affects a plant's ability to grow and our ability to breathe oxygen (O_2). In the stratosphere, some 50 kilometers above the earth, ozone forms a shield that absorbs the ultraviolet (UV) rays of the sun so that they do not strike the earth.

UV radiation causes mutations that can lead to skin cancer and can make the lens of the eyes develop cataracts. It also is believed to adversely affect the immune system and our ability to resist infectious diseases. Crop and tree growth is impaired, and UV radiation also kills off small plants (phytoplankton) and tiny shrimp-like animals (krill) that sustain oceanic life. Without an adequate ozone shield, our health and food sources are threatened.

Depletion of the ozone layer within the stratosphere in recent years is, therefore, of serious concern. It became apparent in the 1980s that some worldwide depletion of ozone had occurred and that there was a severe depletion of some 40%–50% above the Antarctic every spring. Severe depletions of the ozone layer are commonly called "ozone holes." Detection devices now tell us that there is an ozone hole above the Arctic as well, and ozone holes could also occur within northern and southern latitudes, where many people live. Whether or not these holes develop depends on prevailing winds, weather conditions, and type of particles in the atmosphere. A United Nations Environment Program report predicts a 26% rise in cataracts and nonmelanoma skin cancers for every 10% drop in the ozone level. A 26% increase translates into 1.75 million additional cases of cataracts and 300,000 more nonmelanoma skin cancers every year, worldwide.

The cause of ozone depletion can be traced to the release of chlorine atoms (Cl) into the stratosphere (fig. 3A). Chlorine atoms combine with ozone and strip away the oxygen atoms, one by one. One atom of chlorine can destroy up to 100,000 molecules of ozone before settling to the earth's surface as chloride years later. These chlorine atoms come from the breakdown of chlorofluorocarbons (CFCs), chemicals much in use by humans. The best known CFC is Freon, a heat transfer agent found in refrigerators and air conditioners. CFCs are also used as cleaning agents and foaming agents during the production of styrofoam found in coffee cups, egg cartons, insulation, and paddings. Formerly, CFCs were used as propellants in spray cans, but this application is now banned in the United States and several European countries.

Most countries of the world have agreed to stop using CFCs by the year 2000, but the United States is halting production by 1995. Scientists are now searching for CFC substitutes that will not release chlorine atoms (nor bromine atoms) to harm the ozone shield. In the meantime, there are certain steps we can take.

- Make sure your auto repair shop recycles the coolant—rather than venting it to the air—whenever you have your air conditioner worked on.
- If you buy a new refrigerator, have the coolant removed from your old refrigerator by a professional before discarding it.
- Do not use foam insulation or foam-padded mattresses and furniture.
- Do not use halon fire extinguishers or spot removers that contain 1,1,1-trichlorethane, because this compound also destroys ozone.

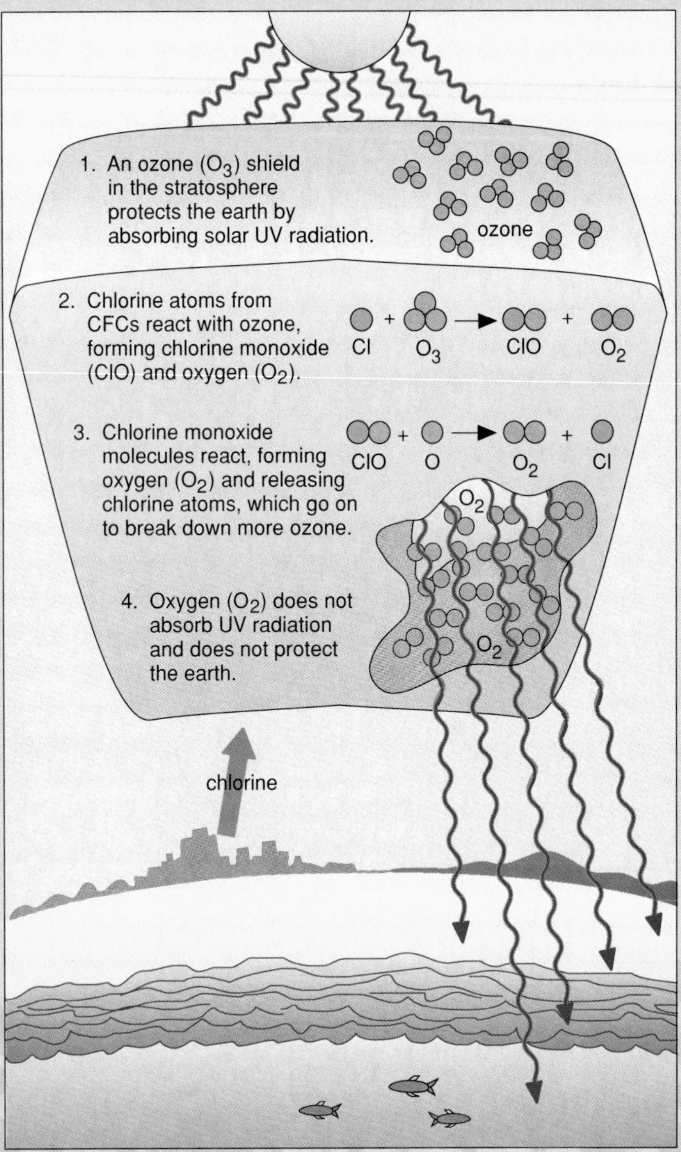

1. An ozone (O_3) shield in the stratosphere protects the earth by absorbing solar UV radiation.
ozone
2. Chlorine atoms from CFCs react with ozone, forming chlorine monoxide (ClO) and oxygen (O_2).
Cl + O_3 → ClO + O_2
3. Chlorine monoxide molecules react, forming oxygen (O_2) and releasing chlorine atoms, which go on to break down more ozone.
ClO + O → O_2 + Cl
4. Oxygen (O_2) does not absorb UV radiation and does not protect the earth.
O_2
O_2
chlorine

Figure 3A

The development of an ozone hole due to the release of chlorine atoms from CFCs.

Health Focus

Start Saving Your Skin

a.

City	UV Index
Anchorage	100
Seattle	477
Minneapolis	570
Boston	591
Chicago	637
New York City	639
Philadelphia	656
Washington, D.C.	683
Columbus	698
St. Louis	714
San Francisco	715
Boise	715
Los Angeles	824
Dallas	871
Atlanta	875
Las Vegas	876
Phoenix	889
Denver	951
Houston	999
Miami	1028
Honolulu	1147

b.

Figure 3B

Sunbathing is even more dangerous today because of ozone depletion. **a.** Sunbathing on the beach. **b.** The UV index measures the relative amount of solar UV radiation for cities. An ozone hole will increase each city's index by about 20%; as the index rises, so does the incidence of skin cancer.

In Victorian days, Caucasian women carried parasols to keep their skin fair. But early in this century, some fair-skinned people began to prefer the golden-brown look, and they took up sunbathing as a way to achieve a tan. A few hours after exposure to the sun, pain and redness due to dilation of blood vessels occurs. Tanning occurs when melanin granules increase and melanocytes rise to the surface of the skin as a way to prevent any further damage by ultraviolet (UV) rays. The sun gives off 2 types of UV rays: UV-A rays and UV-B rays. UV-A rays penetrate the skin deeply, affect connective tissue, and cause the skin to sag and wrinkle. UV-A rays are also believed to potentiate the effect of the UV-B rays, which are the cancer-causing rays. UV-B rays are more prevalent at midday.

Skin cancer is categorized as either nonmelanoma or melanoma. Nonmelanoma skin cancers are of 2 types. Basal cell carcinoma, the most common type, begins when UV radiation causes epidermal basal cells to form a tumor, while at the same time suppressing the immune system's ability to detect the tumor. The signs of a tumor are varied. They include an open sore that will not heal, a recurring reddish patch, a smooth, circular growth with a raised edge, a shiny bump, or a pale mark. About 95% of patients are easily cured, but reoccurrence is common.

Squamous cell carcinoma begins in the epidermis proper. Squamous cell carcinoma is 5 times less common than basal cell carcinoma, but it is more likely to spread to nearby organs and the death rate is about 1% of cases. The signs of squamous cell carcinoma are the same as that for basal cell carcinoma, except that it may also show itself as a wart that bleeds and scabs.

Melanoma that starts in the melanocytes has the appearance of an unusual mole. Unlike a dark mole that is circular and confined, melanoma moles look like spilled ink spots. A variety of shades occur in the same mole, and they can itch, hurt, or feel numb. The skin around the mole turns grey, white, or red. Melanoma is most apt to appear in persons who have fair skin, particularly if they have suffered occasional severe burns as children. The chance of melanoma increases with the number of moles a person has. Most moles occur before the age of 14 and their appearance is linked to sun exposure. Melanoma rates have risen since the turn of the century, but the incidence has doubled in the last decade. Now about 32,000 cases of melanoma are diagnosed each year, and one in 5 persons diagnosed dies within 5 years.

Since the incidence of skin cancer is related to UV exposure, scientists have developed a UV index to determine how powerful the sun is at different cities in the United States (fig. 3B). The index assigns a baseline level of 100 to Anchorage, Alaska. In general, the more southern the city, the higher the UV index, and the greater the risk of skin cancer.

Looking ahead, we should note that for every 10% decrease in the ozone layer, the UV index per city rises by 13%–20%, and the chance of skin cancer also then rises. Even if you live in Anchorage, you should take the following steps to protect yourself from the sun.

To prevent the possible occurrence of skin cancer, do the following:

- Use a broad-spectrum sunscreen, meaning that it protects you from both UV-A and UV-B radiation with an SPF (sun protection factor) of at least 15. (This means, for example, that if you usually burn after a 20-minute exposure, it will take 15 times that duration before you will burn.)
- Wear protective clothing. Choose fabrics with a tight weave and wear a wide-brimmed hat. A baseball cap does not protect the rims of the ears.
- Stay out of the sun particularly between the hours of 10 A.M. and 3 P.M. Avoiding the sun in the midday hours cuts your annual exposure by as much as 60%.
- Wear sunglasses that have been treated to absorb both UV-A and UV-B radiation. Otherwise, sunglasses can expose your eyes to more damage than usual because pupils dilate in the shade.
- Avoid tanning machines. Although most tanning devices use high levels of only UV-A, the deep layers of the skin become more vulnerable to UV-B radiation when you are later exposed to the sun.

Skin Damage

Serious skin damage occurs when the skin is burned. A burn can affect one or all of the layers of the skin. A first-degree burn, which affects only the epidermal layer, is characterized by redness, pain, and swelling. As with sunburn, the skin usually peels in a few days. A second-degree burn, which affects both the epidermis and dermis, usually causes pain and blistering. A third-degree burn is not painful because nerve endings have been destroyed. It is most serious because it leaves underlying body parts with no protection and subject to infection. When a person is burned over a large portion of the body, it is sometimes difficult to find enough skin to make autografting (a graft from skin remaining) possible. Under these circumstances, physicians can now make use of artificial skin, consisting of 2 layers. The inner layer is a lattice made from shark cartilage and collagen fibers from cowhide. The outer layer is a rubberlike silicone plastic. After the artificial skin is sewn in place, the lattice is slowly digested away and replaced by the patient's own cells. At that time, the silicone layer can be safely removed.

Sun exposure damages the skin in ways we have already mentioned. The sun's UV rays cause sunburn, wrinkling of the skin, and skin cancer. This is of particular concern now because the *ozone shield* is being depleted for reasons discussed in the Ecology Focus for this chapter. The ozone shield, a 32–48-kilometer layer of ozone in the atmosphere, absorbs the sun's UV rays and prevents them from reaching the earth. Depletion of the ozone shield will make sun exposure even more dangerous, and the incidence of skin cancer is expected to increase. The Health Focus for this chapter tells you how to protect your skin from the sun's damaging rays.

Studying Organ Systems

Each human system has a specific location within the body. The central nervous system is located in a *dorsal* (towards the back) cavity (fig. 3.9). The brain is protected by the skull, and the spinal cord, which gives off spinal nerves, is protected by the vertebrae. Other internal organs are found within a *ventral* (front) body cavity. This cavity is divided by a muscular diaphragm that assists breathing. The heart, a pump

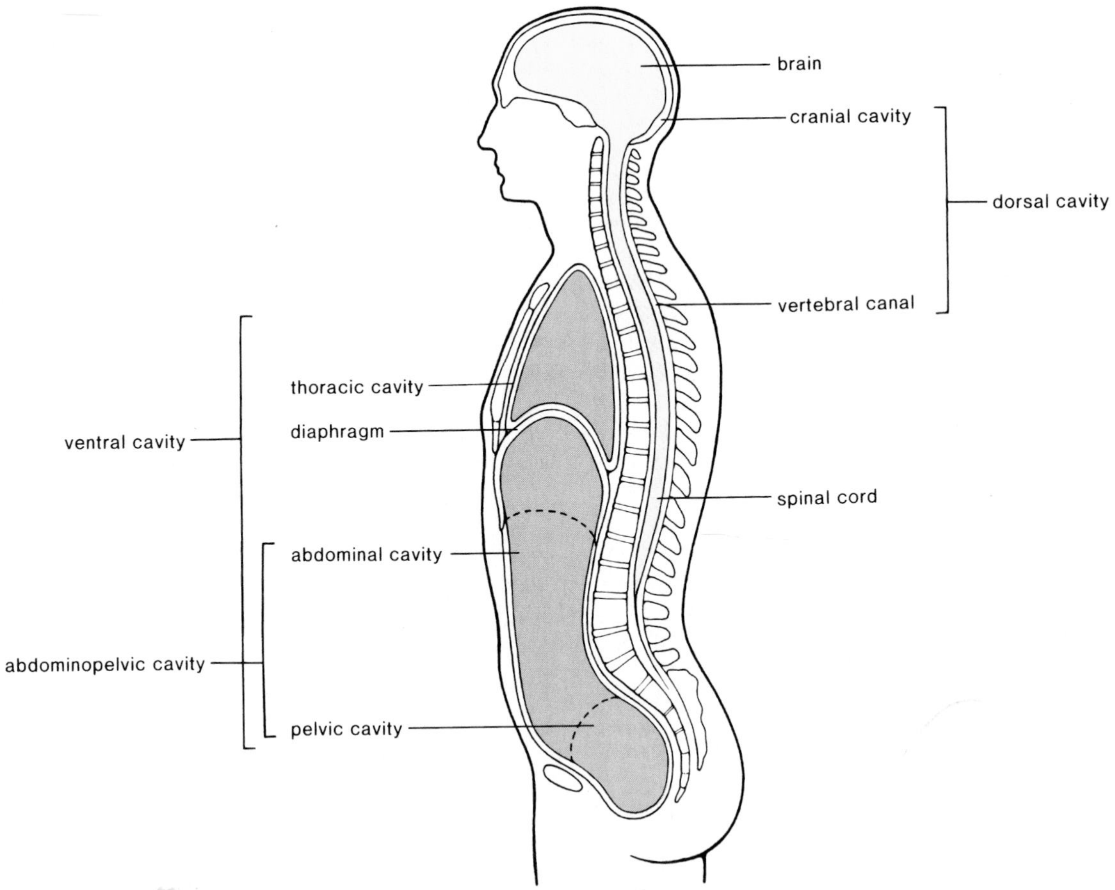

Figure 3.9

Organs are located in cavities of the human body. Notice that we are obviously vertebrates since we have a backbone composed of vertebrae.

for the closed circulatory system, and the lungs are located in the upper *thoracic* (chest) cavity. The major portion of the digestive system and the kidneys is located in the *abdominal* cavity; much of the reproductive system, the urinary bladder, and the terminal portion of the digestive system are located in the *pelvic* cavity.

The musculoskeletal system is also an internal organ system. The skeleton provides the surface area for attachment of striated muscles, which are well developed and powerful. The musculoskeletal system makes up most of the body weight and is specialized for locomotion.

Body Membranes: Two-Part Lining

The term *membrane* on the organ level generally refers to a thin lining or covering that is composed of a layer of epithelial tissue overlying a layer of loose connective tissue. Mucous membrane lines the organs of the respiratory and digestive systems. This type of membrane, as its name implies, secretes mucus. Serous membrane lines enclosed cavities and covers the organs that lie within these cavities, such as the heart, lungs, and kidneys. This type of membrane secretes a watery lubricating fluid.

> There are numerous cavities within the body. These cavities are usually lined with membrane made of a layer of epithelial tissue and a deeper layer of loose connective tissue.

Homeostasis

Homeostasis means that the internal environment remains relatively constant regardless of the conditions in the external environment. For example:

1. Blood glucose concentration remains at about 0.1%.
2. The pH of the blood is always near 7.4.
3. Blood pressure in the brachial artery remains near 120/80.
4. Body temperature remains around 37° C.

The ability of the body to keep the internal environment within a certain range allows humans to live in a variety of habitats, such as the arctic regions, deserts, or the tropics.

The internal environment of the body includes a tissue fluid that bathes all of the tissues. The composition of tissue fluid must remain constant if cells are to remain alive and healthy. Tissue fluid is created when water (H_2O), oxygen (O_2), and nutrient molecules leave a capillary (the smallest of the blood vessels). Tissue fluid is cleansed when water, carbon dioxide (CO_2), and other waste molecules enter a capillary from the fluid (fig. 3.10). Tissue fluid remains constant only as long as blood composition remains constant. Although we are accustomed to using the word *environment* to mean the external environment of the body, it is important to realize that it is the internal environment of tissues that is ultimately responsible for our health and well-being.

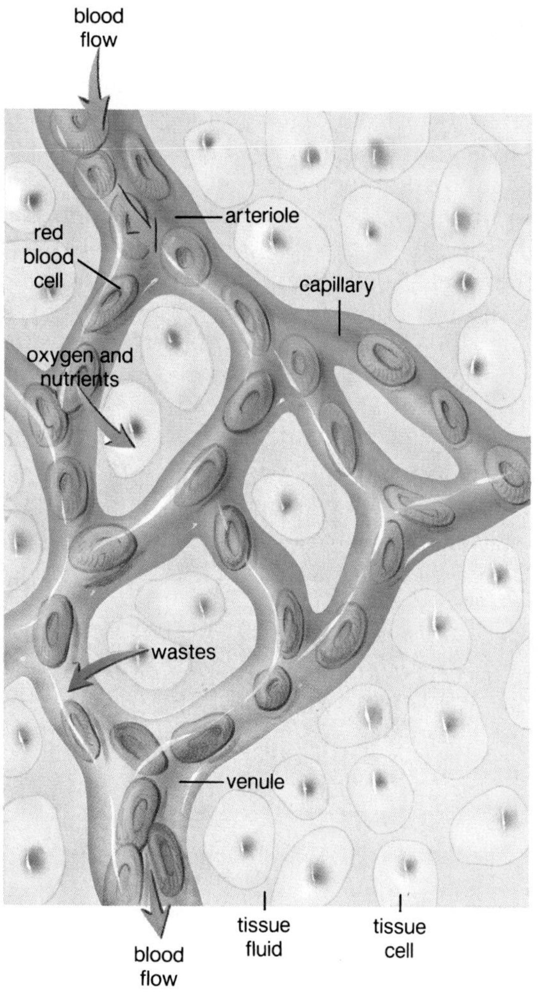

Figure 3.10

Formation of tissue fluid from blood. The internal environment of the body is the blood and the tissue fluid. Tissue cells are surrounded by tissue fluid, which is continually refreshed because nutrient molecules constantly exit the bloodstream and waste molecules continually enter the bloodstream.

> The internal environment of the body consists of blood and tissue fluid, which bathes the cells.

Most systems of the body contribute to maintenance of homeostasis. The digestive system takes in and digests food, providing nutrient molecules that enter the blood and replace the nutrients that are constantly being used up by the body cells. The respiratory system adds oxygen to the blood and removes carbon dioxide. The amount of oxygen taken in and carbon dioxide given off can be increased to meet bodily needs. The chief regulators of blood plasma composition, however, are the liver and the kidneys. They monitor and then alter the chemical composition of plasma as required. Immediately after glucose enters the blood from digested food, it can be removed by the liver for storage as glycogen. Later, glycogen can be broken down to replace the glucose used by the body cells; in this way, the glucose composition of the blood remains constant in between eating. The liver also removes toxic chemicals, such as ingested alcohol and drugs, and produces urea, a nitrogenous waste excreted by the kidneys. The kidneys also excrete water and salts, substances that can affect the pH of the plasma.

> All of the systems of the body (fig. 3.11) contribute to homeostasis; that is, maintenance of the relative constancy of the internal environment.

The endocrine system, consisting of glands that secrete hormones, and the nervous system are involved in the regulation of the other systems of the body. Therefore, these systems are ultimately in control of homeostasis. The endocrine system is slower acting than the nervous system, which rapidly brings about a particular response.

Previously, we mentioned that the liver is involved in homeostasis because it stores glucose as glycogen. But actually there is a hormone produced by an endocrine gland that regulates storage of glucose by the liver. When the glucose content of the blood rises after eating, the pancreas secretes insulin, a hormone that causes the liver to store glucose as glycogen. Now the glucose level falls and the pancreas no longer secretes insulin. This is called control by **negative feedback** because the response (low blood glucose) negates the original stimulus (high blood pressure). In some instances, an endocrine gland is sensitive to the blood level of a hormone whose concentration it regulates. For example, the pituitary gland produces a hormone that stimulates the thyroid gland to secrete its hormone. When the blood level of this hormone rises to a certain level, the pituitary gland no longer stimulates the thyroid gland.

A negative feedback system can regulate itself because it has a sensing device, which is a mechanism by which the system detects a particular condition. For example, consider the feedback mechanism that functions to maintain the room temperature of a house. In this feedback

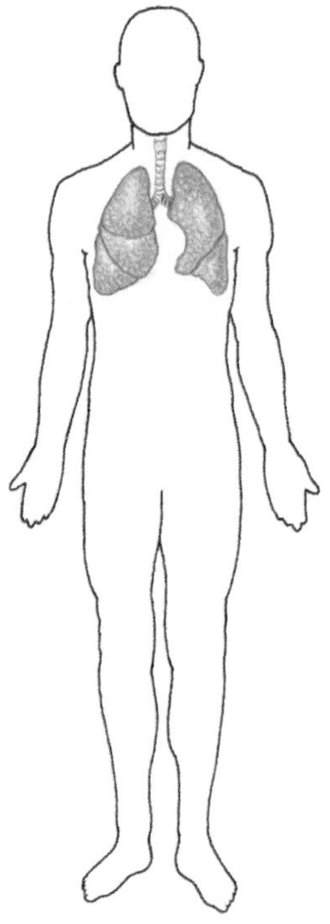

Respiratory system function: Gaseous exchange between external environment and blood

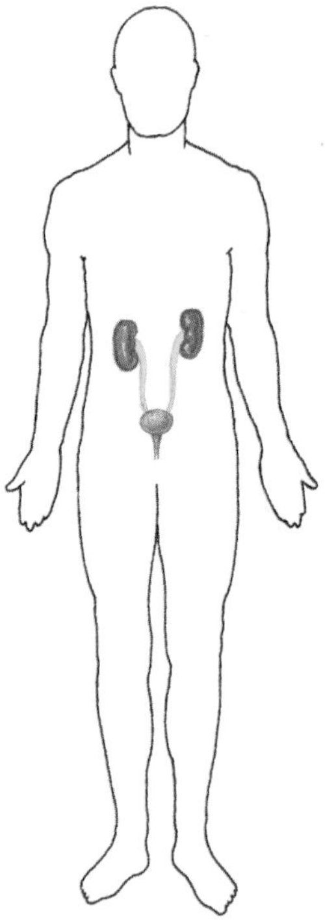

Urinary system function: Filtration of blood; maintenance of volume and chemical composition of the blood

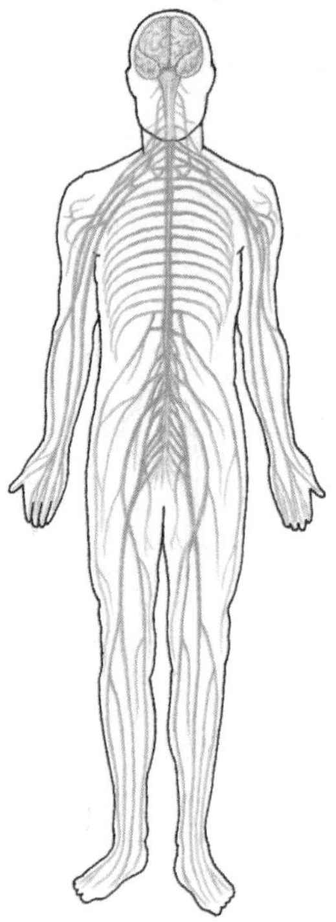

Nervous system function: Regulation of all body activities; learning and memory

Figure 3.11

Organ systems of the human body.

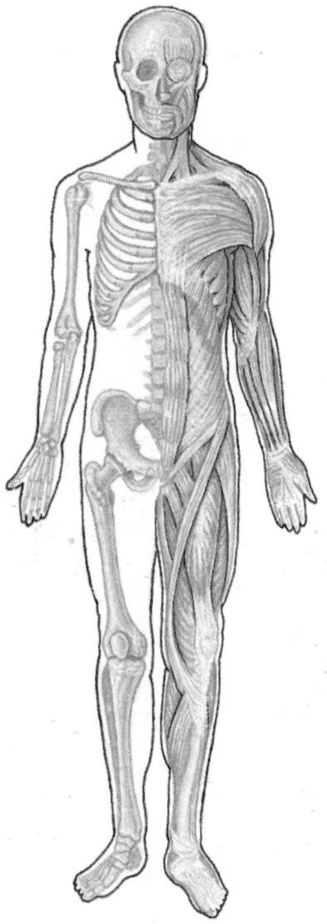

Musculoskeletal system function: Body form and movement; skeleton produces blood cells and muscles produce body heat

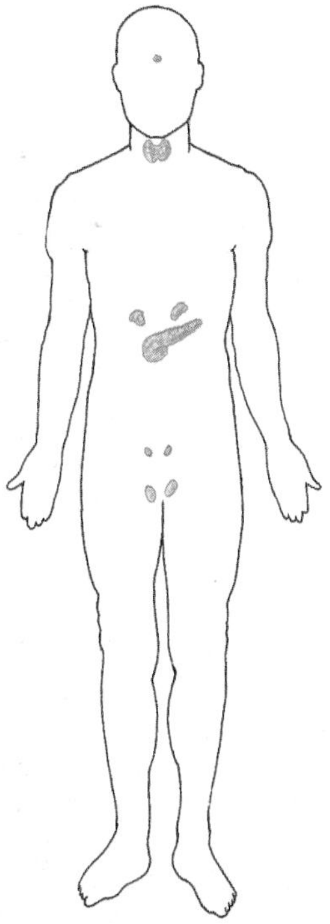

Endocrine system function: Secretion of hormones for regulation of systems and internal environment

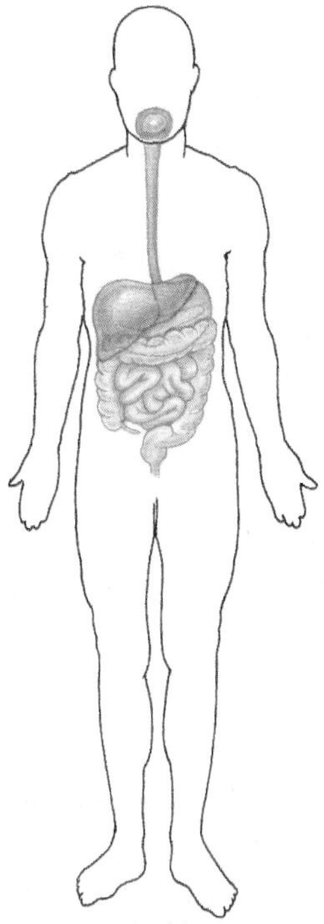

Digestive system function: Breakdown and absorption of food materials; elimination of wastes

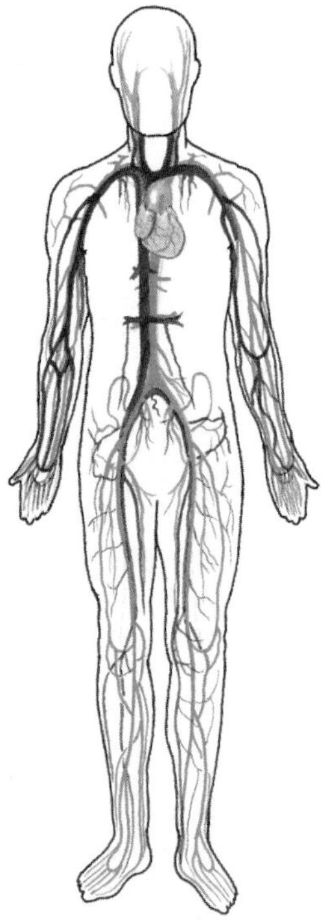

Circulatory system function: Transport of life-sustaining materials to body cells; removal of metabolic wastes from cells

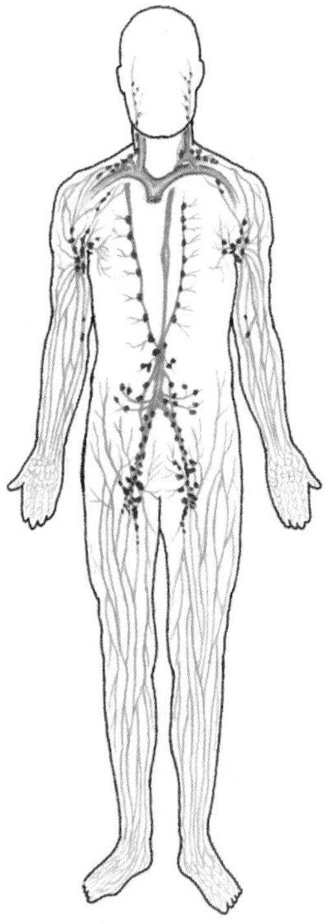

Lymphatic system function: Body immunity; absorption of fats; drainage of tissue fluid

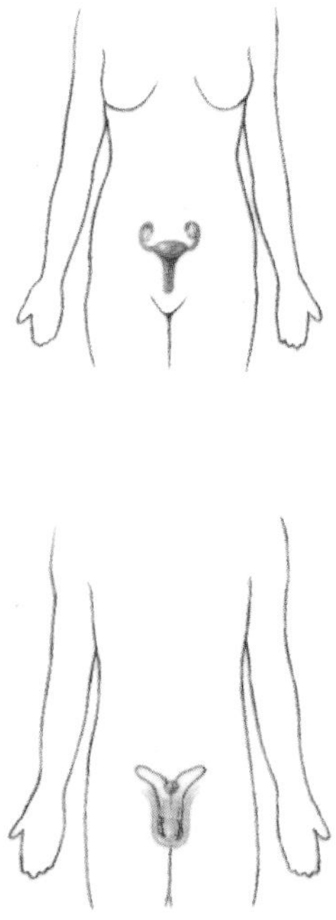

Reproductive system function: Production of gametes (sperm and egg); male transfers sperm to female where fertilization, implantation, and development of embryo and fetus take place.

system, the thermostat is a device that is sensitive to room temperature. The furnace produces heat, and when the temperature of a room reaches a certain point, the thermostat signals a switching device that turns the furnace off. On the other hand, when the temperature falls below that indicated on the thermostat, it signals the switching device, which turns the furnace on again.

Figure 3.12*a* shows that in the body there are receptors that fulfill the role of sensing devices. When a receptor is stimulated, it signals a regulator center that then turns on an effector. The effector brings about a response that negates the original conditions that stimulated the receptor. In the absence of suitable stimulation, the receptor no longer signals the regulator center.

Figure 3.12*b* gives an actual example involving the nervous system. When blood pressure rises, receptors signal a regulator center, which then sends out nerve impulses to the arterial walls, causing them to relax, and the blood pressure now falls. Therefore, the receptors are no longer stimulated and the system shuts down. Notice that negative feedback control results in a fluctuation above and below a mean. We will now explore this concept in regard to control of body temperature in the next section.

Controlling Body Temperature

Control by negative feedback results in fluctuation between 2 levels, as illustrated by maintenance of body temperature (fig. 3.13).

The receptor and the regulator center for body temperature are located in the hypothalamus. The receptor is sensitive to the temperature of the blood, and when the temperature falls below normal, the regulator center directs (via nerve impulses) the blood vessels of the skin to constrict. This conserves heat. Also, the arrector pili muscles pull hairs erect, and a layer of insulating air is trapped next to the skin. If body temperature falls even lower, the regulator center sends nerve impulses to the skeletal muscles, and shivering occurs. Shivering generates heat, and gradually body temperature rises to 37° C and perhaps higher.

During the period of time that the body temperature is normal, the receptor and the regulator center are not active, but once body temperature is higher than normal, they are reactivated. When this happens, the regulator center directs the blood vessels of the skin to dilate. This allows more blood to flow near the surface of the body, where heat can be lost to the environment. The regulator center activates the sweat glands, and the evaporation of sweat helps to lower body temperature. Gradually, body temperature decreases to 37° C and perhaps lower. Once body temperature is below normal, the cycle begins again.

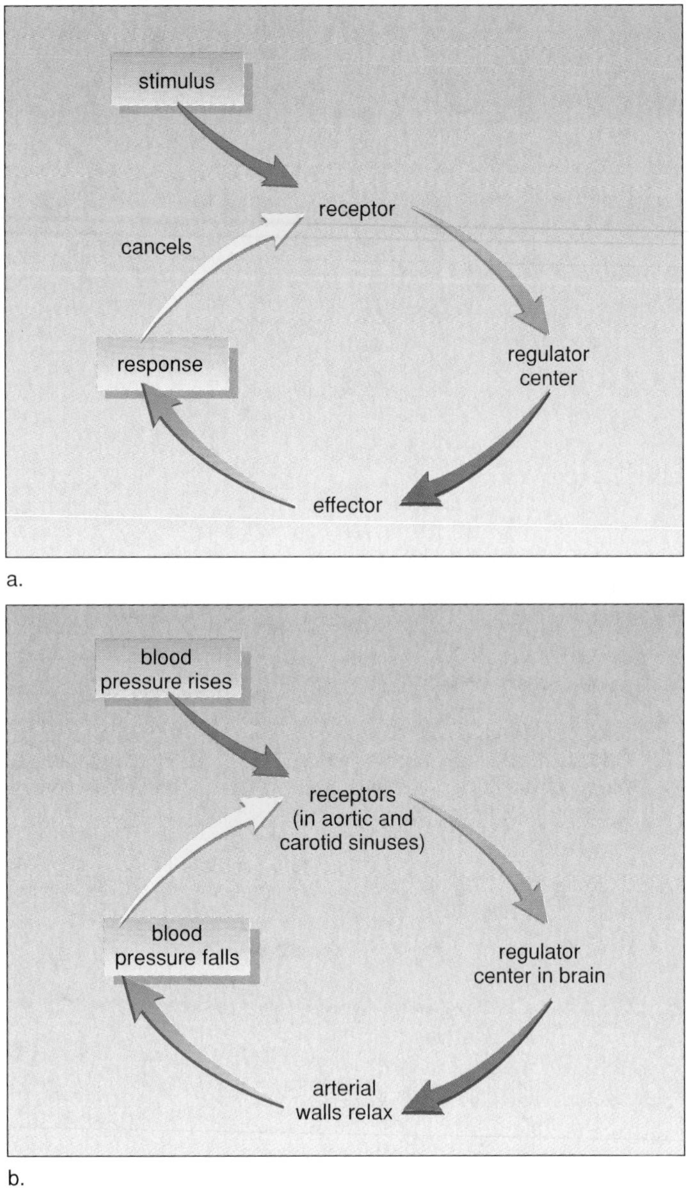

Figure 3.12

Negative feedback control. **a.** A stimulus causes a receptor to signal a regulator center in the brain. The regulator center signals effectors to respond, and the response cancels the stimulus. **b.** For example, when blood pressure rises, special receptors in blood vessels signal a particular center in the brain. The brain signals the arteries to relax, and blood pressure falls.

> Homeostasis of internal conditions is a self-regulatory mechanism that results in slight fluctuations above and below a mean. For example, body temperature rises above and drops below a normal temperature of 37° C.

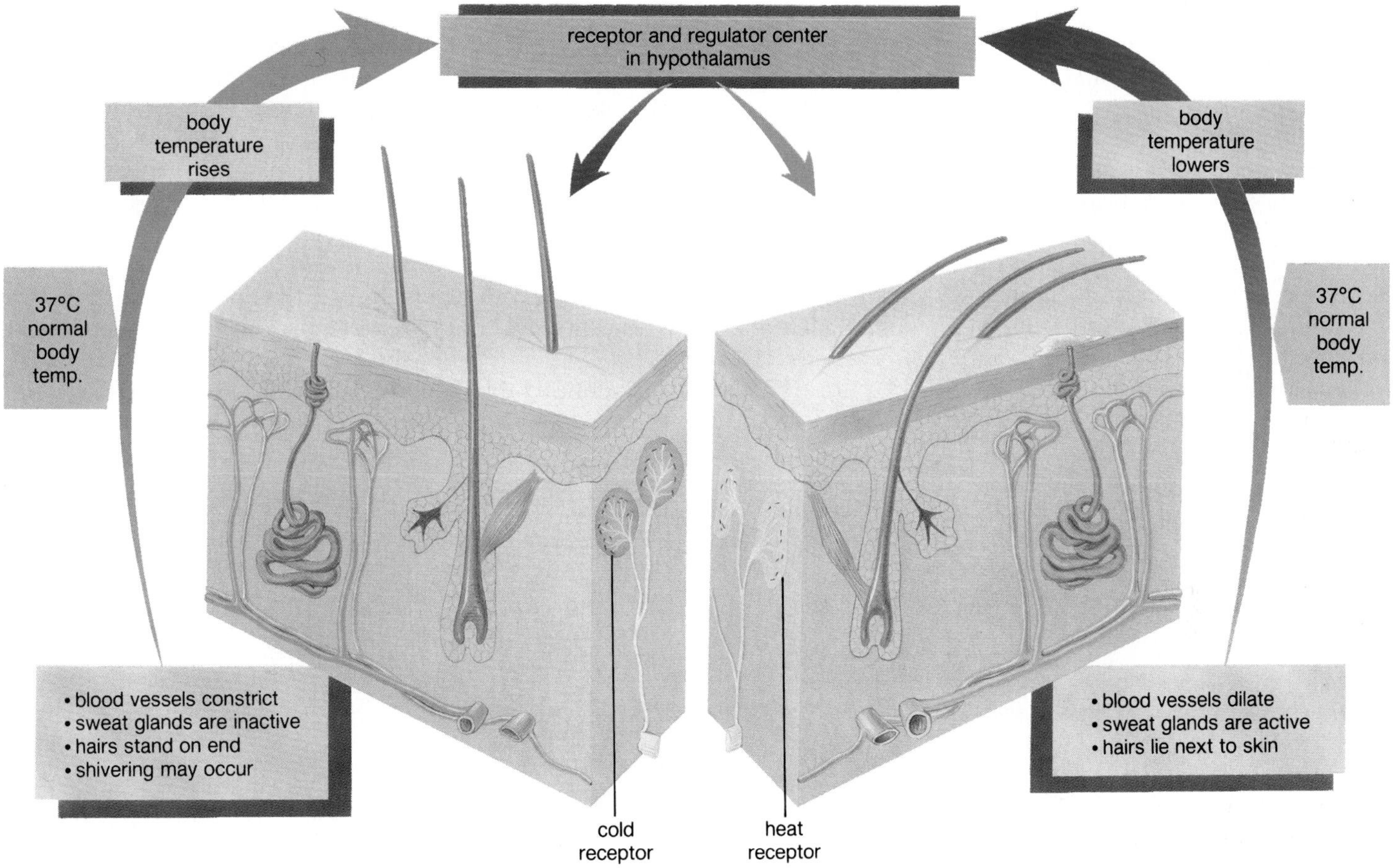

Figure 3.13

Temperature control. When the body temperature rises, the regulator center directs the blood vessels to dilate and the sweat glands to be active. The body temperature lowers. Then, the regulator center directs the blood vessels to constrict, hairs to stand on end, and even shivering to occur if needed. The body temperature rises again. Because the regulator center is activated only by extremes, the body temperature fluctuates above and below normal.

SUMMARY

Human tissues are categorized into 4 major types. Epithelial tissue covers the body and lines its cavities. Connective tissue often binds body parts. Contraction of muscular tissue permits movement of the body and its parts. Nerve impulses conducted by neurons within nervous tissue help to bring about coordination of body parts.

Tissues join to form organs, each with a specific function. Skin is a 2-layered organ that waterproofs and protects the body. The epidermis contains a germinal layer that produces new epithelial cells that become keratinized as they move toward the surface. The dermis, a largely fibrous connective tissue, contains epidermally derived glands and hair follicles, nerve endings, and blood vessels. Encapsulated nerve endings form receptors for touch, pressure, and temperature; free nerve endings register pain. Sweat glands and blood vessels help control body temperature. A subcutaneous layer, which is made up of loose connective tissue containing adipose cells, lies beneath the skin and fastens it to muscle.

Organs are located in cavities of the body. In humans, the brain and spinal cord are located in a dorsal cavity and most other internal organs are located in a ventral cavity that contains the thoracic, abdominal, and pelvic cavities.

Organs are grouped into organ systems. All organ systems contribute to the constancy of tissue fluid and blood. Special contributions are made by the liver, which keep blood glucose constant, and the kidneys, which regulate the pH. The hormonal and nervous systems regulate the other systems. Both of these are controlled by a feedback mechanism, which results in fluctuation above and below the desired level.

Epithelial Tissue
(e.g., cuboidal epithelium)

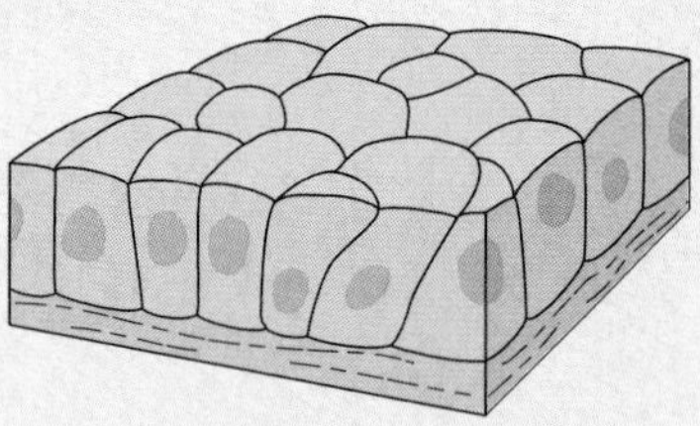

- named for shape of cell
- located on body surface or lining a cavity
- various functions, including protection

Connective Tissue
(e.g., loose connective tissue)

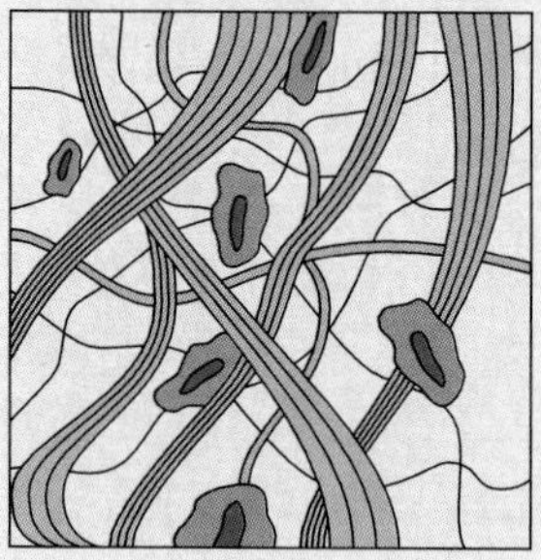

- cells separated by a matrix that contains fibers
- located between and around body parts
- functions to support, protect, and hold things together

Muscular Tissue
(e.g., cardiac muscle)

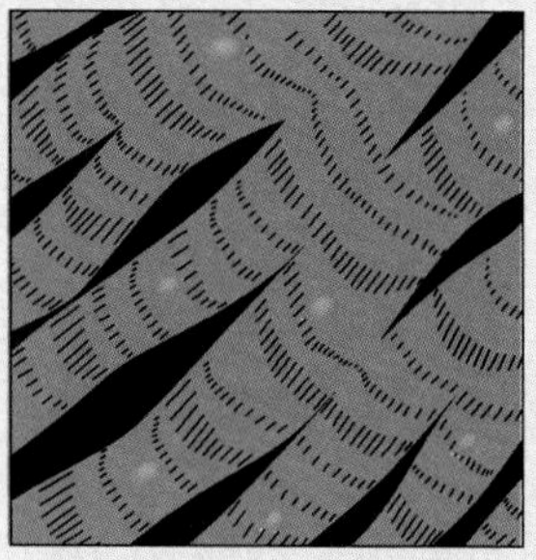

- cells contain contractile filaments
- located in skeletal muscles and organ walls
- functions to move the body and body parts

Nervous Tissue
(e.g., neuron)

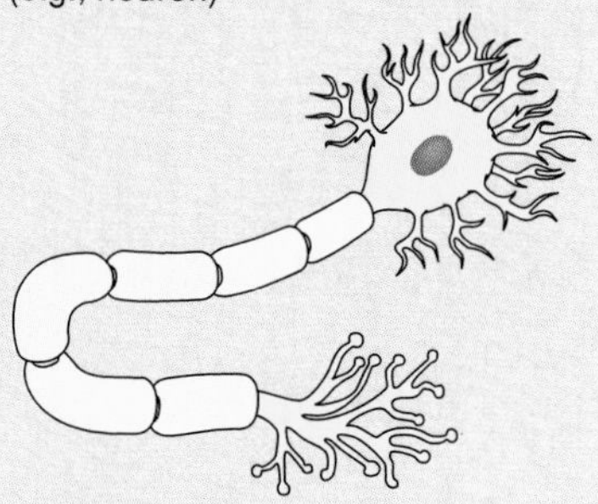

- cells have nerve fibers
- located in brain, spinal cord, and ganglia
- functions to transmit nerve impulses

Study and Thought Questions

1. State in order the levels of organization of the human body. (p. 46)
2. Name the 4 major types of tissues. (pp. 46–52)
3. What are the functions of epithelial tissue? Name the different kinds, and give a location for each. (p. 46)

   Critical Thinking Since structure suits function, why is it appropriate for the epithelium of the windpipe to have cilia and the epithelium of the small intestine to have microvilli?
4. What are the functions of connective tissue? Name the different kinds, and give a location for each. (p. 48)
5. What are the functions of muscular tissue? Name the different kinds, and give a location for each. (p. 51)
6. Nervous tissue contains what type of cell? Which organs in the body are made up of nervous tissue? (p. 51)

   Critical Thinking Contrast the shape of a muscle fiber with a neuron to show that structure suits function.
7. Describe the structure of skin, and state at least 2 functions of this organ. (p. 52)
8. In general terms, describe the location of the human organ systems. (p. 56)
9. Distinguish between cell membrane and body membrane. (p. 57)
10. What is homeostasis, and how is it achieved in the human body? (p. 57)

    Critical Thinking Often homeostasis results in fluctuation above and below a mean. Explain why this is the case for body temperature.

Writing Across the Curriculum

1. Sometimes, the skin is said to be the "integumentary system." Which structures should be included in the integumentary system aside from the skin?
2. Classification schemes are arbitrary in nature. Make up a new classification system for the tissues described in this chapter.
3. Explain why it is beneficial for humans to have homeostatic control systems.

Objective Questions

1. The tissue covering all body surfaces and lining cavities is called ______________.
2. The epithelium that lines most digestive system organs is composed of ______________ tissue.
3. The noncellular material that separates connective tissue cells is called a ______________.
4. Adipose tissue is a type of ______________ tissue where ______________ is stored in fibroblasts.
5. ______________ connect muscles to bones.
6. Blood is a ______________ tissue in which the matrix is ______________.
7. Cardiac muscle is ______________ but involuntary.
8. Nervous tissue is composed of what types of cells? ______________
9. Skin has these layers: epidermis and ______________, plus a subcutaneous layer.
10. Outer skin cells are filled with ______________, a waterproof protein.
11. Homeostasis is the maintenance of the relative ______________ of the internal environment, that is, the blood and ______________ fluid.
12. Give the name, the location, and the function for each of these tissues.
    - a. Type of epithelial tissue
    - b. Type of muscular tissue
    - c. Type of connective tissue

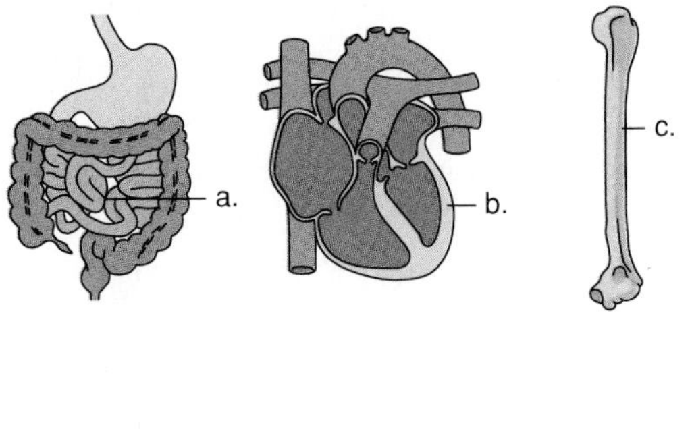

Selected Key Terms

bone Connective tissue in which the cells lie within lacunae embedded in a hard matrix of calcium salts deposited around protein fibers. 50

cartilage A connective tissue in which the cells lie within lacunae embedded in a flexible matrix. 49

compact bone Bone that contains Haversian systems cemented together. 50

connective tissue A type of tissue characterized by cells separated by a matrix that often contains fibers. 48

dermis (der´mis) The thick skin layer that lies beneath the epidermis. 52

epidermis (ep˝ ĭ-der´mis) The outer skin layer composed of stratified squamous epithelium. 52

epithelial tissue (ep˝ ĭ-the´le-al tish´u) A type of tissue that lines cavities and covers the external surface of the body. 46

hyaline cartilage (hi´ah-lĭn kar´tĭ-lij) Cartilage composed of very fine collagenous fibers and a matrix of a milk glass appearance. 49

lacuna (lah-ku´nah) A small pit or hollow cavity, as in bone or cartilage, where a cell or cells are located. 49

ligament Fibrous connective tissue that joins bone to bone. 49

muscular (contractile) tissue A type of tissue that contains cells capable of contracting; skeletal muscles are attached to the skeleton, smooth muscle is found within walls of internal organs, and cardiac muscle comprises the heart. 51

negative feedback A self-regulatory mechanism that is activated by an imbalance and results in a fluctuation about a mean. 58

neuron (nu´ron) Nerve cell that characteristically has 3 parts: dendrites, cell body, and axon. 51

spongy bone Porous bone found at the ends of long bones. 50

stratified Layered; stratified epithelium contains several layers of cells. 46

striated Having bands; cardiac and skeletal muscle are striated with bands of light and dark. 51

subcutaneous layer (sub˝ ku-ta´ne-us la´er) A tissue layer found in vertebrate skin that lies just beneath the dermis and tends to contain adipose tissue. 53

tendon Strap of fibrous connective tissue that joins muscle to bone. 49

Further Readings for Part One

Allen, R. D. February 1987. The microtubule as an intracellular engine. *Scientific American.*

Atkins, P. W. 1987. Molecules. New York: *Scientific American.*

Becker, W. M., and Deamer, D. W. 1991. *The world of the cell.* 2d ed. Redwood City, Calif.: Benjamin/Cummings Publishing.

Bretscher, M. S. October 1985. The molecules of the membrane. *Scientific American.*

Devlin, T. 1992. *Textbook of biochemistry: With clinical correlations.* 3d ed. New York: John Wiley & Sons, Inc.

Doolittle, R. F., and Bork, P. October 1993. Evolutionary mobile molecules in proteins. *Scientific American.*

Glover, D. M., et al. June 1993. The centrosome. *Scientific American.*

Goulding, M. March 1993. Flooded forests of the Amazon. *Scientific American.*

Hoffman, R. February 1993. How should chemists think? *Scientific American.*

Hole, J. W. 1990. *Human Anatomy and Physiology.* 5th ed. Dubuque, Iowa: Wm. C. Brown Publishers.

Holloway, M. July 1993. Sustaining the Amazon. *Scientific American.*

Lienhard, G. E., et al. January 1992. How cells absorb glucose. *Scientific American.*

Linder, M. E., and Gilman, A. G. July 1992. G Proteins. *Scientific American.*

McIntosh, J., and McDonald, L. October 1989. The mitotic spindle. *Scientific American.*

May, R. M. October 1992. How many species inhabit the earth? *Scientific American.*

Murray, A. W., and Kirschner, M. W. March 1991. What controls the life cycle. *Scientific American.*

Neher, E. and Sakmann, B. March 1992. The patch clamp technique. *Scientific American.*

Olson, A. J., and Goodsell, D. S. November 1992. Visualizing biological molecules. *Scientific American.*

Pennisi, E. February 27, 1993. Cancer linked to aging DNA repair ability. *Science News.*

Sadava, D. 1993. *Cell biology: Organelle structure and function.* Boston: Jones and Bartlett Publishers.

Sharon, N., and Lis, H. January 1993. Carbohydrates in cell recognition. *Scientific American.*

Todorov, I. N. December 1990. How cells maintain stability. *Scientific American.*

Welch, W. J. May 1993. How cells respond to stress. *Scientific American.*

Wickramasinghe, H. K. October 1989. Scanned-probe microscopes. *Scientific American.*

Part TWO

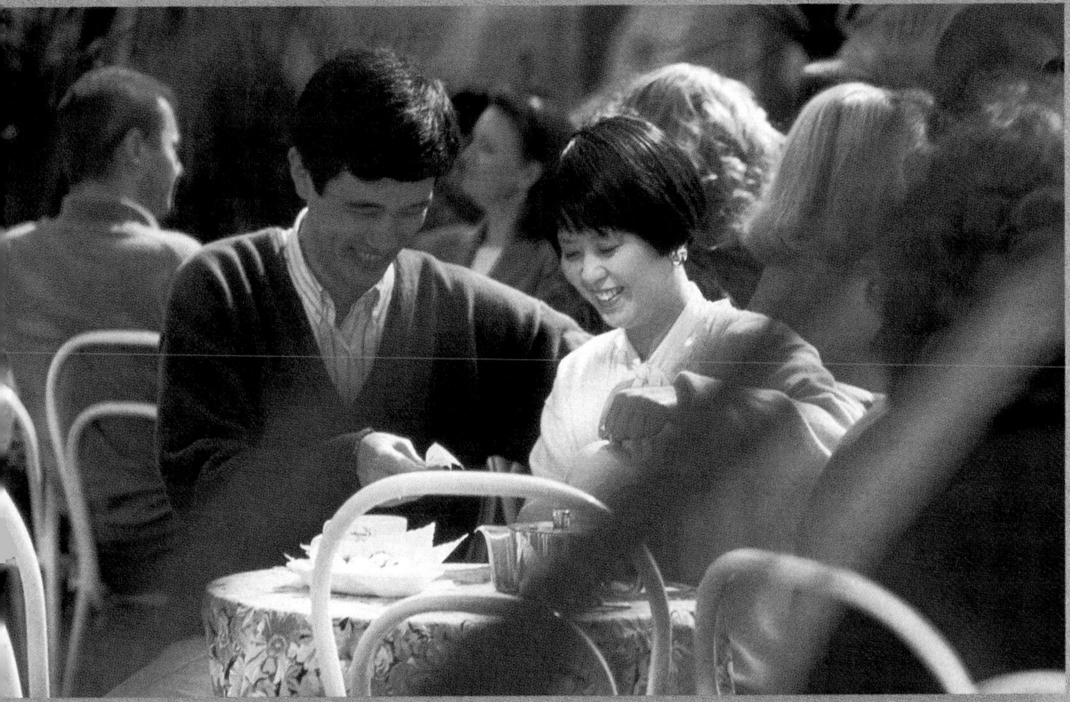

Maintenance of the Body

All of the systems of the body help maintain homeostasis, the dynamic equilibrium of the internal environment. Our internal environment is the blood within blood vessels and the fluid that surrounds the cells of the tissues. The heart pumps the blood and sends it in vessels to the tissues, where exchange of materials occurs with tissue fluid. The composition of blood tends to remain relatively constant as a result of the actions of the digestive, respiratory, and excretory systems. Nutrients enter the blood at the small intestine, external gas exchange occurs in the lungs, and waste products are excreted at the kidneys. The immune system prevents microorganisms from taking over the body and interfering with its proper functioning.

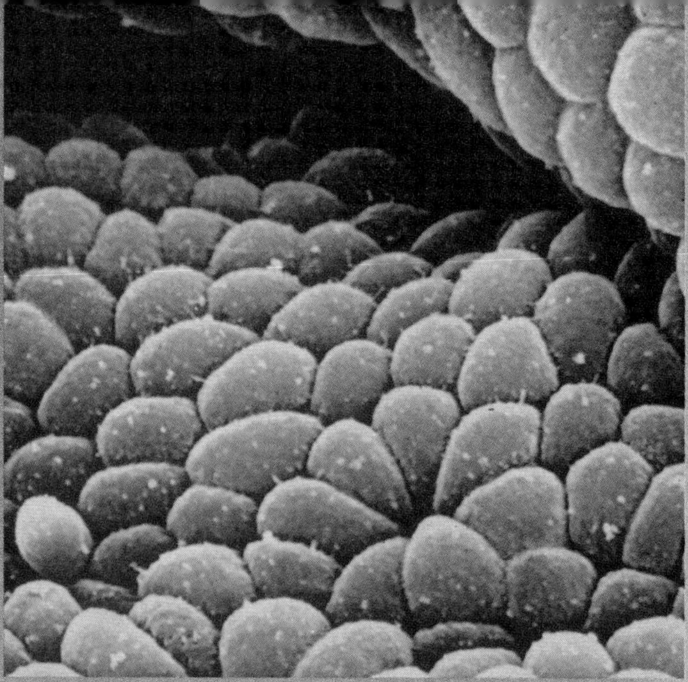

Chapter 4

Digestive System and Nutrition

The discovery of digestive enzymes involved the use of a human as the experimental material. William Beaumont was an American doctor who had a French-Canadian patient, Alexis St. Martin. St. Martin had been shot in the stomach, and when the wound healed, he was left with a fistula, or opening, that allowed Beaumont to collect gastric (stomach) juices and to look inside the stomach to see what was going on there. Beaumont was able to determine that the muscular walls of the stomach contract vigorously and mix food with juices that are secreted whenever food enters the stomach. He found that gastric juice contains hydrochloric acid (HCl) and a substance (enzyme) active in digestion. He concluded that gastric juices are produced separately from the protective mucous secretions of the stomach. Beaumont's work, which was very carefully and painstakingly done, pioneered the study of the physiology of digestion.

Chapter Concepts

1. The human digestive system is an extended tube with specialized parts between 2 openings, the mouth and the anus. 68
2. The products of digestion are small molecules, such as amino acids and glucose, that can cross cell membranes. 68
3. The liver and the pancreas are the accessory organs of digestion because their secretions assist the digestive process. 75
4. The digestive enzymes are specific and have an optimum temperature and pH. 77
5. Proper nutrition supplies the body with energy and nutrients, including the essential amino acids and fatty acids, and all vitamins and minerals. 79–86

Chapter Outline

Digestion takes place within a tube, called the digestive tract, which begins with the mouth and ends with the anus (table 4.1 and fig. 4.1). Digestion of food in humans is an extracellular (outside the cell) process. Digestive enzymes are secreted by the tract or by glands that lie nearby. Food is never found within these *accessory glands*, only within the tract itself.

While strictly speaking, the term *digestion* means the breakdown of food by enzymatic action, in this text the term is expanded to include both physical and chemical processes that reduce food to small absorbable molecules. Only small molecules can cross cell membranes and be absorbed by the intestine lining. Many of us think that since we eat meat (protein), potatoes (carbohydrate), and butter (fat), these are the substances that nourish our bodies. Instead, it is, for example, the amino acids from the protein and the sugars from the carbohydrate that are absorbed and are transported throughout the body to nourish the cells. Any component of food, such as cellulose, a plant material that cannot be digested to small molecules, leaves the intestine as waste.

Table 4.1

Path of Food

Organ	Function
Mouth	Reception and chewing of food; digestion of starch
Esophagus	Passageway
Stomach	Storage and mechanical breakdown of food; acidity kills bacteria; digestion of protein
Small intestine	Digestion of all foods; absorption of nutrients
Large intestine	Absorption of water (and some vitamins); storage of nondigestible remains
Anus	Defecation

The Digestive System

The functions of the digestive system are to ingest the food, to digest it to small molecules that can cross cell membranes, to absorb these nutrient molecules, and to eliminate nondigestible wastes.

Mouth: Food Receiver

The oral cavity of the mouth receives food when humans eat. Most people enjoy eating largely because they like the taste of food. Receptors called taste buds are found primarily on the tongue. These are activated by the presence of food in the mouth. Taste buds initiate nerve impulses, which travel by way of cranial nerves to the brain. Still, what we call taste is largely due to stimulation of olfactory (smell) receptors in the nose. If you have a cold and your nose is blocked, food has little taste. The sense of taste is discussed further in chapter 12.

With our teeth we chew food into pieces convenient for swallowing. During the first 2 years of life, the 20 deciduous, or baby, teeth appear. These are eventually replaced by 32 adult

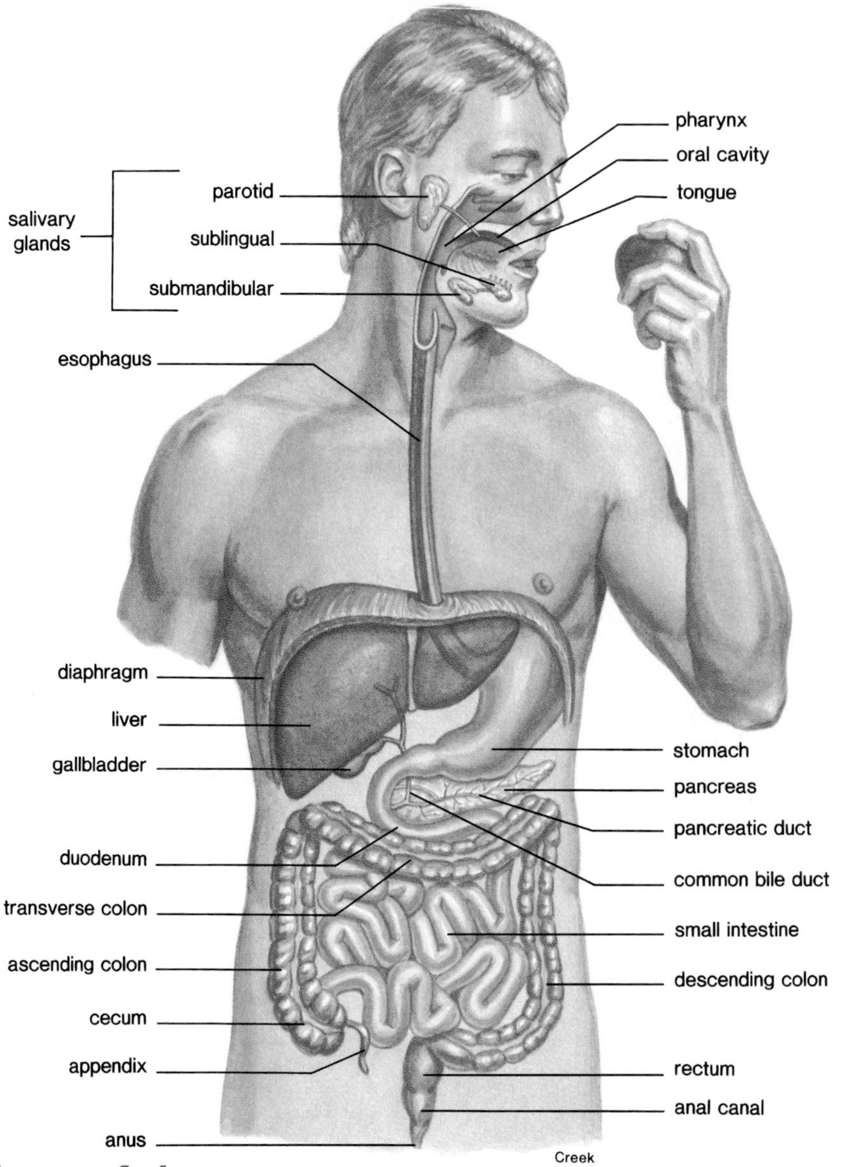

Figure 4.1

Trace the path of food from the mouth to the anus. The large intestine consists of the transverse, ascending, and descending colons, plus the rectum and anal canal. Note the placement of the accessory organs of digestion, the liver and the pancreas.

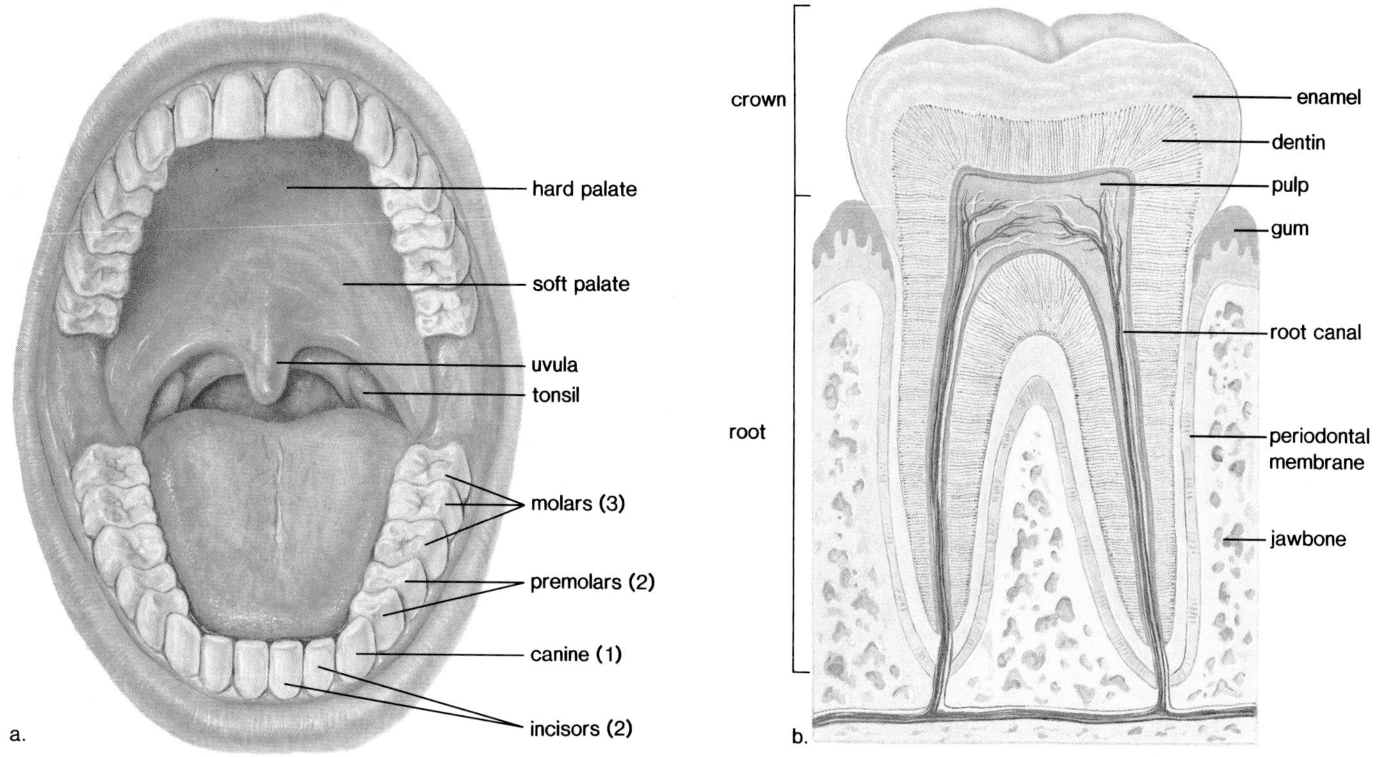

Figure 4.2

Mouth and teeth. **a.** The chisel-shaped incisors bite; the pointed canines tear; the fairly flat premolars grind; and the flattened molars crush food. The last molar, called a wisdom tooth, may fail to erupt, or if it does, it is sometimes crooked and useless. Often dentists recommend the extraction of the wisdom teeth. **b.** Longitudinal section of a tooth. The crown is the portion that is sometimes replaced by a dentist. When a root canal is done, the nerves are removed. When the periodontal membrane is inflamed, the teeth can loosen.

teeth (fig. 4.2). Each tooth has 2 main divisions, a crown and a root. The crown has a layer of enamel, an extremely hard outer covering of calcium compounds; dentin, a thick layer of bonelike material; and inner pulp, which contains the nerves and the blood vessels. Dentin and pulp are also found in the root.

Tooth decay, or *caries,* commonly called a cavity, occurs when the bacteria within the mouth metabolize sugar and give off acids, which erode the tooth. Two measures can prevent tooth decay: eating a limited amount of sweets and daily brushing and flossing of teeth. It also has been found that fluoride treatments, particularly in children, can make the enamel stronger and more resistant to decay. Gum disease is more apt to occur with aging. Inflammation of the gums (gingivitis) can spread to the periodontal membrane (fig. 4.2), which lines the tooth socket. A person then has **periodontitis,** characterized by a loss of bone and loosening of the teeth so that extensive dental work may be required. Stimulation of the gums in a manner advised by your dentist is helpful in controlling this condition.

The roof of the mouth separates the nasal cavities from the oral cavity. The roof has 2 parts: an anterior **hard palate** and a posterior **soft palate** (fig. 4.2*a*). The hard palate contains several bones, but the soft palate is composed largely of muscle. The soft palate ends in the *uvula,* a suspended process often mistaken by the layperson for the tonsils. In fact, the tonsils are at the sides of the oral cavity, at the base of the tongue, and in the nose (called adenoids). The tonsils play a minor role in protecting the body from disease-causing organisms.

The 3 pairs of **salivary glands** are exocrine glands that send their juices (saliva) by way of ducts to the mouth (fig. 4.1). The *parotid glands* lie at the sides of the face immediately below and in front of the ears. These glands swell when a person has the mumps, a viral infection most often seen in children. Each parotid gland has a duct, which opens on the inner surface of the cheek at the location of the second upper molar. The *sublingual glands* lie beneath the tongue, and the *submandibular glands* lie beneath the posterior floor of the oral cavity. The ducts from these glands open into the mouth under the tongue. You can locate all these openings if you use your tongue to feel for small flaps on the inside of your cheek and under your tongue. An enzyme within saliva begins the process of digesting starch.

The tongue, which is composed of striated muscle with an outer layer of mucous membrane, mixes the chewed food with saliva. It then forms this mixture into a mass called a *bolus* in preparation for swallowing.

> The salivary glands send saliva into the mouth, where the teeth chew the food and the tongue forms it into a bolus for swallowing.

Pharynx: A Crossroad

Swallowing occurs in the **pharynx,** a region between the mouth and the esophagus, a long muscular tube (fig. 4.3). Swallowing is a *reflex action,* that is, it is usually performed automatically, without conscious thought. During swallowing, food normally enters the esophagus because the air passages are blocked. Unfortunately, we have all had the unpleasant experience of having food "go the wrong way." The wrong way may be either into the nasal cavities or into the trachea (windpipe). If it is the latter, coughing will most likely force the food up out of the trachea and into the pharynx again. Usually during swallowing, the soft palate moves back to cover the opening to the *nasopharynx,* which leads to the nasal cavities. The trachea moves up under the **epiglottis,** which then covers the **glottis,** the opening to the larynx (voice box). The up and down movement of the *Adam's apple,* a part of the larynx, is easy to observe when a person eats.

The air passage and the food passage cross in the pharynx. When you swallow, the air passage usually is blocked off, and food must enter the esophagus.

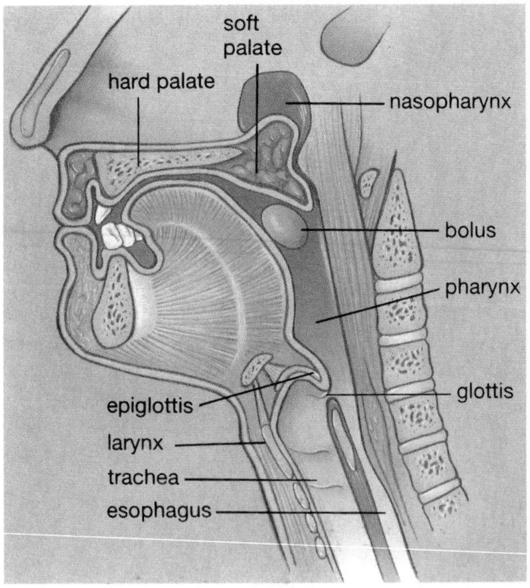

Figure 4.3

Swallowing. When food is swallowed, the soft palate covers the nasopharyngeal openings and the epiglottis covers the glottis forcing the bolus to pass down the esophagus. Therefore, you do not breathe when swallowing.

Esophagus: Food Conductor

The **esophagus,** which passes through the thoracic cavity and into the abdominal cavity, conducts the bolus from the pharynx to the stomach. The wall of the esophagus in the abdominal cavity is representative of the digestive tract in general (fig. 4.4). A *mucosa* lines the **lumen** (space within the tube); this is followed by a *submucosa* of loose connective tissue, which contains nerve and blood vessels; a muscularis, a smooth muscle layer with both circular and longitudinal muscles; and finally, a *serosa* of connective tissue.

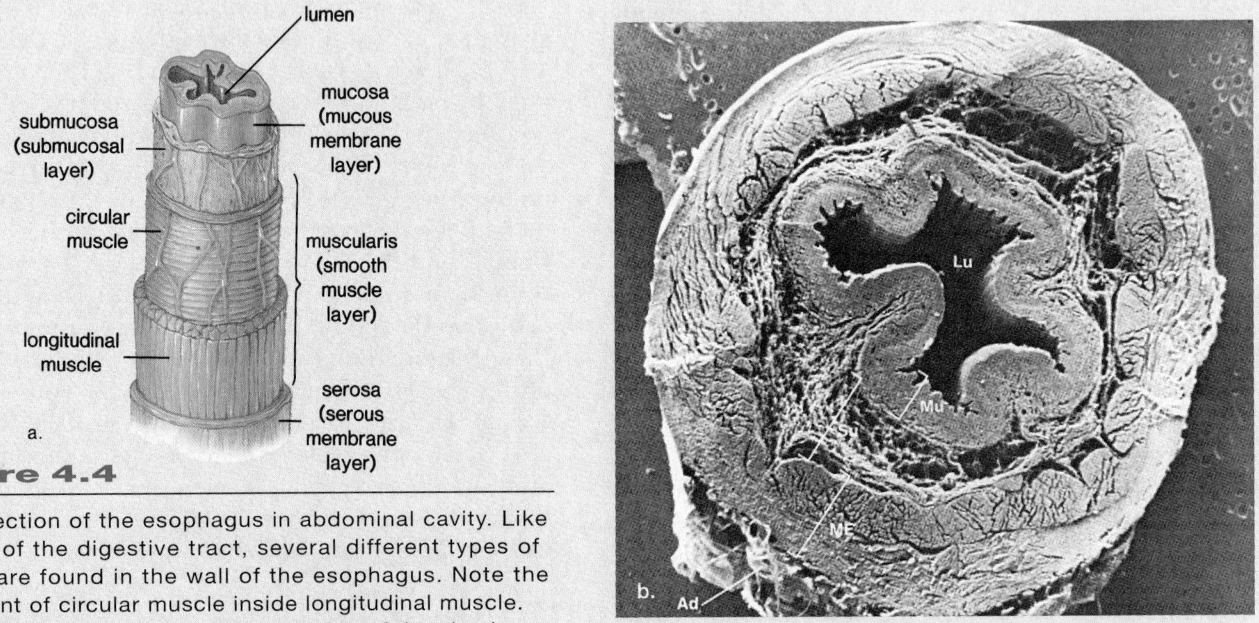

Figure 4.4

Cross section of the esophagus in abdominal cavity. Like the rest of the digestive tract, several different types of tissues are found in the wall of the esophagus. Note the placement of circular muscle inside longitudinal muscle. This arrangement ensures that the action of the circular muscle does not interfere with that of the longitudinal muscle. **a.** Diagrammatic drawing. **b.** Scanning electron micrograph (Lu = central lumen; Mu = mucosa; Su = submucosa; ME = muscularis externa; and Ad = adventitia, or serosa).

Micrograph from R. G. Kessel and R. H. Kardon, Tissues and Organs: A Text Atlas of Scanning Electron Microscopy, *1979 W. H. Freeman and Company.*

A rhythmic contraction of the digestive tract, called **peristalsis** (see fig. 4.8), pushes the food along. Occasionally, peristalsis begins even though there is no food in the esophagus. This produces the sensation of a lump in the throat.

The entrance of the esophagus to the stomach is marked by a constrictor called the lower esophageal sphincter, although the muscle in this sphincter is not as developed as in a true sphincter. **Sphincters** are muscles that encircle tubes and act as valves; tubes close when sphincters contract, and they open when sphincters relax. When food is swallowed, the sphincter relaxes, allowing the bolus to pass into the stomach. Normally, the lower esophageal sphincter prevents the acidic contents of the stomach from entering the esophagus. Heartburn, which feels like a burning pain rising up into the throat, occurs when some of the stomach contents escape into the esophagus. When vomiting occurs, a reverse peristaltic wave causes the sphincter to relax, and the contents of the stomach are propelled upward through the esophagus.

Stomach: Food Storer and Grinder

The **stomach** (fig. 4.5) is a thick-walled, J-shaped organ that lies on the left side of the body beneath the diaphragm. The stomach is continuous with the esophagus above and the duodenum of the small intestine below. The stomach stores food. The wall of the stomach has deep folds, which disappear as the stomach fills to an approximate capacity of 1 liter. Its muscular wall churns, mixing the food with gastric juice. The term *gastric* always refers to the stomach.

The columnar epithelial lining of the stomach has millions of gastric pits, which lead into **gastric glands**. The gastric glands produce gastric juice. Gastric juice contains an enzyme called pepsin, which digests protein, hydrochloric acid (HCl), and mucus. The gastric glands also produce *gastrin*, a hormone. HCl causes the stomach to have a high acidity of about pH 2, and this is beneficial because it kills most bacteria present in food. Although HCl does not digest food, it does break down the connective tissue of meat and activates pepsin.

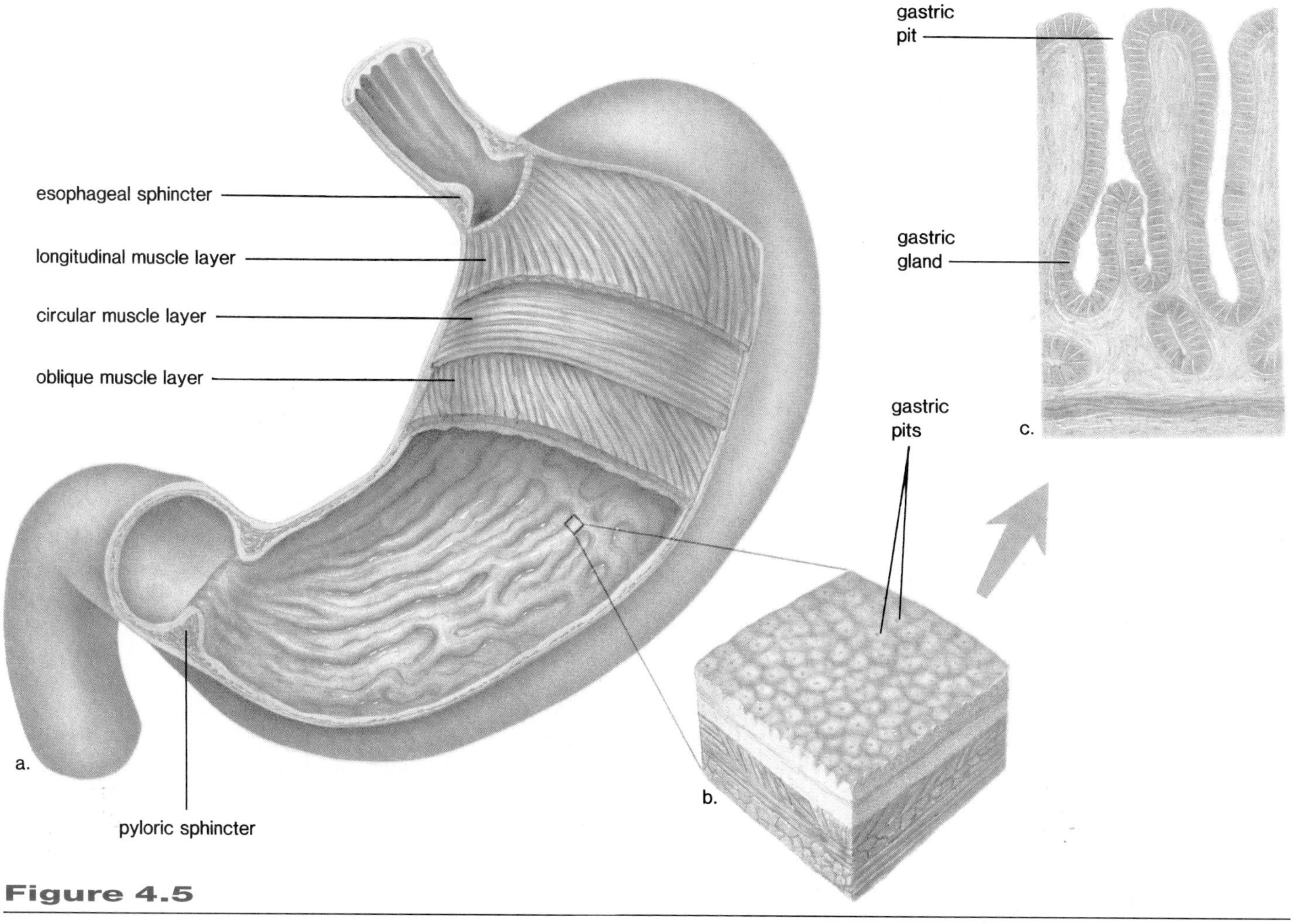

Figure 4.5

Stomach anatomy. **a.** The stomach has 3 muscle layers, whose contraction helps mix food, and the interior has deep folds that can expand to allow more food to be stored. **b.** The mucosa of the stomach contains gastric glands in which individual cells secrete either mucus or HCl plus the digestive enzyme, pepsin.

Table 4.2

Hormones of the Digestive Tract

Hormone	Source and Secretion	Function
Gastrin	Stomach wall, in response to the presence of food rich in protein	Stimulates gastric glands to increase their secretory activity
Secretin	Intestinal wall, in response to acid chyme entering the small intestine	Stimulates pancreas to secrete sodium bicarbonate
Cholecystokinin (CCK)	Intestinal wall, in response to food rich in fat	Stimulates pancreas to secrete digestive enzymes; stimulates gallbladder to contract and release bile

Gastrin is released from gastric glands when there is protein in the stomach. Gastrin goes into the bloodstream, and when it circulates back to the stomach, the gastric glands continue to secrete (table 4.2).

Normally, the wall of the stomach is protected by a thick layer of mucus secreted by goblet cells in its lining. If by chance HCl penetrates this mucus, autodigestion of the wall can begin, and an ulcer results. An ulcer is an open sore in the wall caused by the gradual disintegration of tissue. It is believed that the most frequent cause of an ulcer is oversecretion of gastric juice due to too much nervous stimulation; persons under stress tend to have a greater incidence of ulcers. However, there is now evidence that a bacterial (*Helicobacter pylori*) infection may impair the ability of epithelial cells to produce the protective mucus.

Alcohol is absorbed in the stomach, but there is no absorption of food substances. Normally, the stomach empties in about 2–6 hours. When food leaves the stomach, it is a pasty material called *chyme.* Chyme leaves the stomach and enters the small intestine by way of the *pyloric sphincter.* The pyloric sphincter repeatedly opens and closes, allowing chyme to enter the small intestine in small squirts only.

> The stomach can expand to accommodate large amounts of food. When food is present, the stomach churns, mixing food with acidic gastric juice.

Small Intestine: Food Processor and Absorber

The **small intestine** gets its name from its small diameter (2.5 cm,[1] compared to that of the large intestine, which is 6.5 cm in diameter), but perhaps it should be called the long intestine because it averages about 6.0 m in length compared to the large intestine, which is about 1.5 m long. The small intestine receives bile from the gallbladder and secretions from the pancreas, chemically and mechanically breaks down chyme, absorbs nutrient molecules, and transports undigested material to the large intestine.

The first 25 cm of the small intestine is called the **duodenum.** Duodenal ulcers sometimes occur because the gastric juice within chyme digests the intestinal wall in this region. Ducts from the gallbladder and the pancreas join to form a common duct, which enters into the duodenum (see fig. 4.1). The juices of the small intestine are normally basic because pancreatic juice, which enters the duodenum, contains sodium bicarbonate ($NaHCO_3$). This normally neutralizes the acidity of chyme from the stomach. The wall of the small intestine contains fingerlike projections called **villi** (fig. 4.6). Because they are so numerous, the villi give the intestinal wall a soft, velvety appearance. Each villus has an outer layer of columnar epithelium and contains blood vessels and a small lymphatic vessel called a **lacteal.** The lymphatic system is an adjunct to the circulatory system—its vessels carry a fluid called lymph to the circulatory veins.

The cell membranes of epithelial cells making up the villi contain microvilli, fingerlike projections. In electron micrographs, the microvilli give the cells a fuzzy border, collectively called a brush border. The microvilli bear the intestinal digestive enzymes, which are therefore referred to as brush-border enzymes. These enzymes finish the digestion of chyme to small molecules that can be absorbed. The microvilli greatly increase the surface area of the small intestine for absorption of nutrients.

The intestinal wall also secretes the hormones secretin and cholecystokinin (CCK) (see table 4.2). These hormones travel in the bloodstream and stimulate the pancreas to secrete its juices; CCK also stimulates the gallbladder to contract and release bile.

Absorption of nutrient molecules occurs across the wall of each villus and continues until all small molecules have been absorbed. Therefore, absorption is an active process involving active transport of molecules across cell membranes and requiring an expenditure of cellular energy. Sugars and amino acids cross the columnar epithelial cells to enter the blood. The components of fats rejoin in epithelial cells and are packaged as lipoprotein droplets, which enter the lacteals.

> The small intestine is specialized to absorb the products of digestion. It is quite long (6.0 m) and has fingerlike projections called villi, where nutrient molecules are absorbed into the circulatory (glucose and amino acids) and lymphatic systems (fats).

1. See Appendix B, Metric System, p. 478.

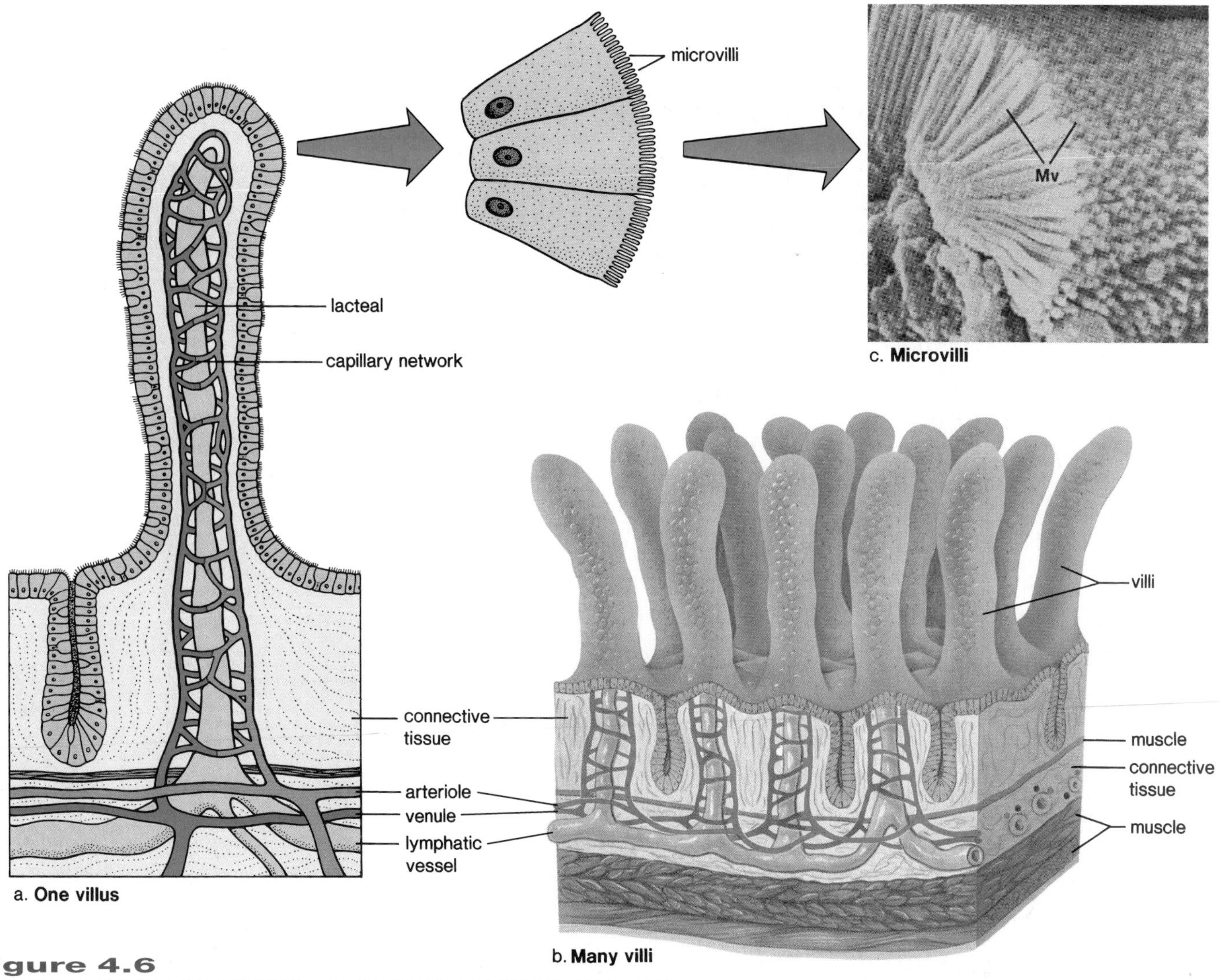

Figure 4.6

Anatomy of intestinal lining, which absorbs the products of digestion. **a.** A villus contains blood vessels and a lacteal of the lymphatic system. **b.** The many villi that project from the intestinal wall increase its surface area. **c.** A villus is itself covered by microvilli (Mv), as shown in this transmission electron micrograph.

Micrograph from R. G. Kessel and R. H. Kardon, Tissues and Organs: A Text Atlas of Scanning Electron Microscopy, *1979 W. H. Freeman and Company.*

Large Intestine: Water and Salt Absorber

The **large intestine,** which includes the cecum, the colon, the rectum, and the anal canal, is larger in diameter than the small intestine (6.5 cm compared to 2.5 cm). The large intestine absorbs water, salts, and some vitamins. It also stores nondigestible material until it is defecated (expelled) at the anus.

The *cecum,* which lies below the entrance of the small intestine, is the blind end of the large intestine. The cecum has a small projection called the vermiform **appendix** (*vermiform* means wormlike) (fig. 4.7). In humans, the appendix, like the tonsils, may play a role in immunity. This organ is subject to inflammation, a condition called

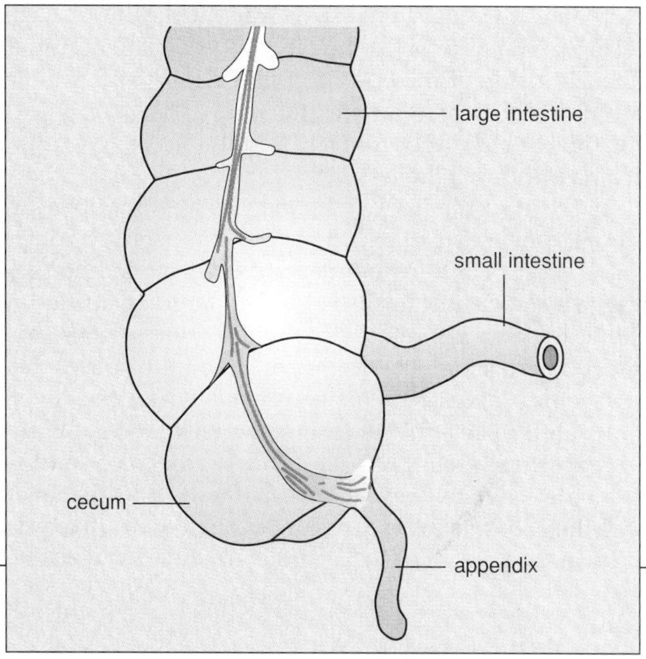

Figure 4.7

The anatomical relationship between the small intestine and the ascending colon. The cecum is the blind end of the ascending colon. The appendix is attached to the cecum.

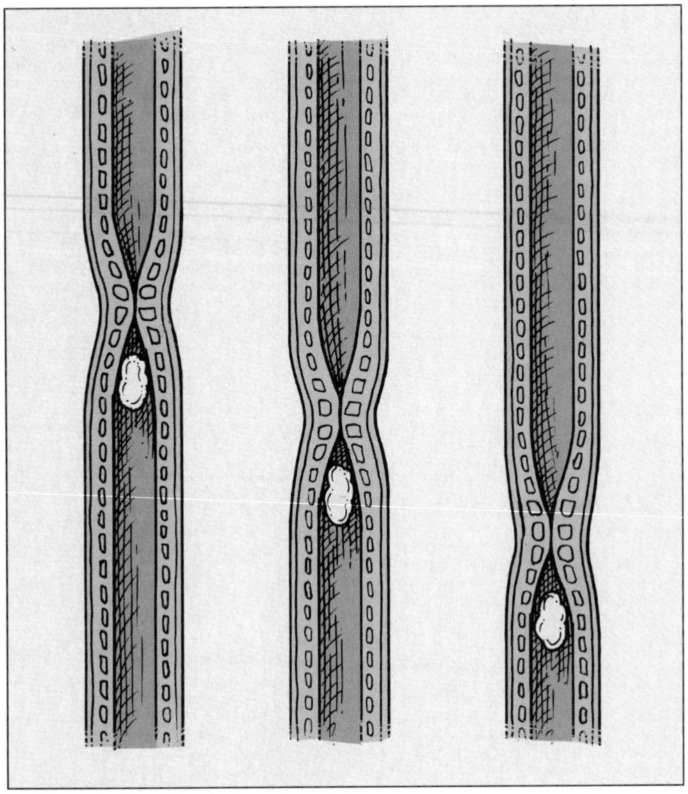

Figure 4.8

Peristalsis in the digestive tract. Rhythmic waves of muscle contraction move material along the digestive tract. The 3 drawings show how a peristaltic wave moves through a single section of intestine over time.

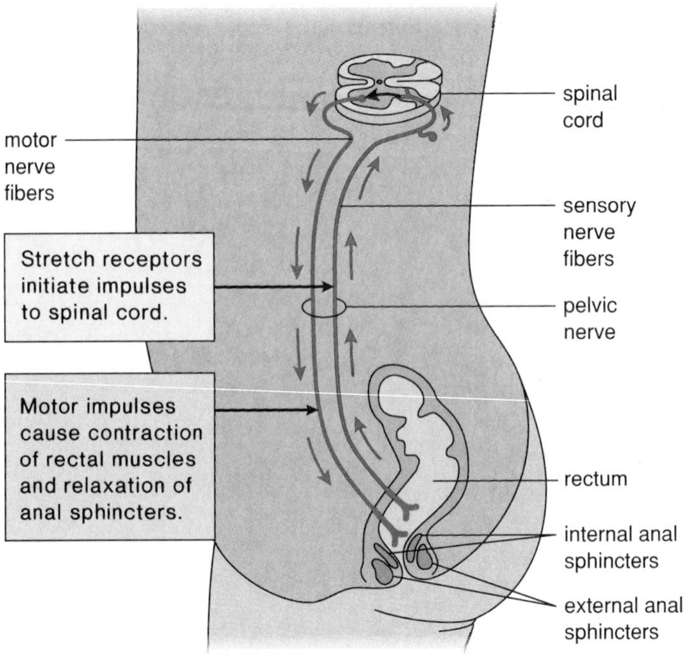

Figure 4.9

Defecation reflex. The accumulation of feces in the rectum causes it to stretch, which initiates a reflex action resulting in rectal contraction and expulsion of the fecal material.

appendicitis. If inflamed, it is wise to remove the appendix before the fluid content rises to the point that the appendix bursts, a situation that can lead to peritonitis, a generalized infection of the serosa of the abdominal cavity. Peritonitis can lead to death.

Peristalsis (fig. 4.8), which begins in the esophagus, occurs along the entire digestive tract, including the colon. The **colon** has 3 parts: the *ascending colon,* which goes up the right side of the body to the level of the liver; the *transverse colon,* which crosses the abdominal cavity just below the liver and the stomach; and the *descending colon,* which passes down the left side of the body to the *rectum,* the last 20 cm of the large intestine. The rectum opens at the **anus,** where *defecation,* the expulsion of *feces,* occurs (fig. 4.9). Feces contains nondigestible remains, bile pigments, which account for its color, and large quantities of bacteria, which account for its smell. Normally, these are noninfectious bacteria that live off any food substances that have not been digested. For many years, it was believed that facultative bacteria (bacteria that can live with or without oxygen), such as *Escherichia coli,* were the major inhabitants of the colon, but new culture methods show that over 99% of the colon bacteria are obligate anaerobes (bacteria that die in the presence of oxygen). Not only do the bacteria break down nondigestible material, they also produce some vitamins and other molecules that can be absorbed and used by us. In this way, they perform a service for us.

Water is considered unsafe for swimming when the coliform bacterial count reaches a certain level. A high count is an indication that a significant amount of feces has entered the water. The more feces present, the greater the possibility that infectious bacteria are also present.

The colon is subject to the development of polyps, small growths arising from the epithelial lining. Polyps, whether benign or cancerous, can be removed individually. If colon cancer is detected while still confined to a polyp, the expected outcome is a complete cure. Some investigators believe that dietary fat increases the likelihood of colon cancer because dietary fat causes an increase in bile secretion. It could be that intestinal bacteria convert bile salts into substances that promote the development of cancer. On the other hand, fiber in the diet seems to inhibit the development of colon cancer. Dietary fiber absorbs water and adds bulk, thereby diluting the concentration of bile salts and facilitating the movement of substances through the intestine. Regular elimination (defecation) reduces the time that the colon wall is exposed to any cancer-promoting agents in feces.

Diarrhea and Constipation

Two common everyday complaints associated with the large intestine are *diarrhea* and *constipation.* The major causes of diarrhea are infection of the lower tract and nervous stimulation. In the case of infection, such as food poisoning caused by eating contaminated food, the intestinal wall becomes irritated and peristalsis increases (see fig. 4.8). As a protective measure, water is not absorbed, and the diarrhea that results rids the body of the infectious organisms. In nervous diarrhea, the nervous system stimulates the intestinal wall and diarrhea results. Prolonged diarrhea can lead to dehydration because of water loss and to disturbances in the heart's contraction due to an imbalance of salts in the blood.

When a person is constipated, the feces are dry and hard. One reason for this condition is that socialized persons have learned to inhibit defecation to the point that the desire to defecate is ignored. Two components of the diet can help to prevent constipation: water and fiber (roughage). Water intake prevents drying out of the feces, and fiber provides the bulk needed for elimination. The frequent use of laxatives is discouraged. If, however, it is necessary to take a laxative, a bulk laxative is the most natural because, like fiber, it produces a soft mass of cellulose in the colon. Lubricants, like mineral oil, make the colon slippery, and saline laxatives, like milk of magnesia, act osmotically—they prevent water from being absorbed and may even cause water to enter the colon, depending on the dosage. Some laxatives are irritants; they increase peristalsis to the degree that the contents of the colon are expelled.

Chronic constipation is associated with the development of hemorrhoids, a condition that is discussed on page 105.

> The large intestine does not produce digestive enzymes; it does absorb water and salts. In diarrhea, too little water has been absorbed by the large intestine; in constipation, too much water has been absorbed.

Two Accessory Organs

The pancreas and the liver are accessory organs of digestion. Figure 4.1 shows how ducts conduct pancreatic juice from the pancreas and bile from the liver to the duodenum.

The Pancreas: Secretes Digestive Enzymes

The **pancreas** lies deep in the abdominal cavity, resting on the posterior abdominal wall. It is an elongated and somewhat flattened organ that has both an endocrine and an exocrine function. We now are interested in its exocrine function—most of its cells produce pancreatic juice, which contains enzymes for digestion of carbohydrate, protein, and fat. In other words, the pancreas secretes enzymes for the digestion of all types of food. The enzymes travel by way of ducts to the duodenum of the small intestine (see fig. 4.1). Pancreatic secretion is controlled by the hormones secretin and CCK (see table 4.2).

The Liver: Makes Bile

The **liver,** which is the largest gland in the body, lies mainly in the upper right section of the abdominal cavity, under the diaphragm.

The liver produces up to 1,500 ml of bile each day. This bile is sent by way of bile ducts to the gallbladder, where it is stored. The **gallbladder** is a pear-shaped, muscular sac attached to the undersurface of the liver. Here, water is absorbed, and bile becomes a thick, mucuslike material. Bile leaves the gallbladder, when stimulated to do so by CCK (see table 4.2), and proceeds to the duodenum via ducts (see fig. 4.1).

Bile is a yellowish green fluid because it contains the bile pigments bilirubin and biliverdin, which are derived from the breakdown of hemoglobin, the pigment found in red blood cells. Bile also contains bile salts (derived from cholesterol), which emulsify fat in the duodenum of the small intestine. When fat is emulsified, it breaks up into droplets, which can be acted upon by a digestive enzyme from the pancreas (p. 77).

The liver acts as the gatekeeper to blood. Once nutrient molecules have been absorbed by the small intestine, they enter the **hepatic portal vein,** pass through the blood vessels of the liver, and then enter the hepatic vein (fig. 4.10).

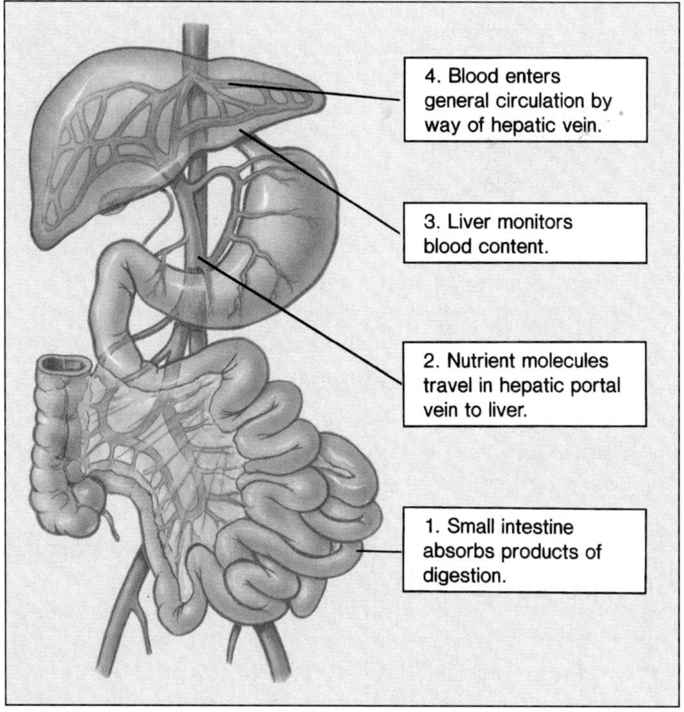
4. Blood enters general circulation by way of hepatic vein.
3. Liver monitors blood content.
2. Nutrient molecules travel in hepatic portal vein to liver.
1. Small intestine absorbs products of digestion.

Figure 4.10

Hepatic portal system. The hepatic portal vein takes the products of digestion from the digestive system to the liver, where they are processed before entering the circulatory system proper.

The liver removes poisonous substances from the blood and works to keep the contents of the blood constant. For example, the liver removes excess glucose present in the hepatic portal vein and stores it as glycogen:

$$\text{glucose} \longrightarrow \text{glycogen} + H_2O$$

Between eating periods, when the glucose level of the blood falls below 0.1%, glycogen is broken down to glucose, which enters the hepatic vein. In this way, the glucose content of the blood remains near 0.1%. It is interesting to note that glycogen sometimes is called animal starch because both starch and glycogen are made up of glucose molecules and are forms of stored energy (p. 22).

If, by chance, the supply of glycogen or glucose runs short, the liver converts amino acids to glucose molecules:

$$\text{amino acids} \longrightarrow \text{glucose} + \text{amino groups}$$

Recall that amino acids contain nitrogen in the form of amino groups, whereas glucose contains only carbon, oxygen, and hydrogen. Therefore, before amino acids can be converted to glucose molecules, **deamination,** the removal of amino groups from the amino acids, must take place. By an involved metabolic pathway, the liver converts these amino groups to urea:

$$H_2N - \overset{\overset{\displaystyle O}{\|}}{C} - NH_2$$

Urea is the usual nitrogenous waste product of humans; after its formation in the liver, it is excreted by the kidneys.

The liver also makes plasma proteins from amino acids. These proteins are not used as food for cells; rather, they serve important functions within the blood itself.

Altogether, we have mentioned the following functions in the liver:

1. Converts hemoglobin from red blood cells to breakdown products (bilirubin and biliverdin) excreted along with bile salts in bile.
2. Produces bile, which is stored in the gallbladder before entering the small intestine, where bile salts emulsify fats.
3. Detoxifies the blood by removing and metabolizing poisonous substances.
4. Stores glucose as glycogen after eating and breaks down glycogen to glucose to maintain the glucose concentration of blood between eating periods.
5. Produces urea from the breakdown of amino acids.
6. Makes plasma proteins from amino acids.

> Two accessory organs of digestion, the pancreas and the liver, send secretions to the duodenum via ducts. The pancreas produces pancreatic juice, which contains enzymes for the digestion of carbohydrate, protein, and fat. The liver produces bile, which is stored in the gallbladder and is used to emulsify fats.

Figure 4.11

Gallstones. After removal, this gallbladder was cut open to show its contents—numerous gallstones. The dime was added later to indicate the size of the stones.

Jaundice, hepatitis, and cirrhosis are 3 serious diseases that affect the entire liver and hinder its ability to repair itself. Therefore, they are life-threatening diseases. When a person becomes jaundiced, there is a yellowish tint to the skin and whites of the eyes. Bilirubin has been deposited in the skin due to an abnormally large amount in the blood. In *hemolytic jaundice,* red blood cells are broken down in abnormally large amounts; in *obstructive jaundice,* the bile duct is blocked or the liver cells are damaged. Obstructive jaundice often occurs when crystals of cholesterol come out of solution and form gallstones. The stones may be so numerous that passage of bile along a bile duct is blocked, and the gallbladder must be removed (fig. 4.11).

Jaundice can also result from *viral hepatitis,* a collective term that includes several types of hepatitis. Hepatitis A is most often caused by eating shellfish from polluted waters. Other types are commonly spread by blood transfusions, kidney dialysis, or injection with inadequately sterilized needles. Hepatitis can also be acquired by sexual contact. There is a vaccine for hepatitis B, and this should reduce the incidence of this disease.

Cirrhosis is a chronic disease of the liver in which the organ first becomes fatty. Liver tissue is then replaced by inactive fibrous scar tissue. In alcoholics, who often develop cirrhosis of the liver, the condition most likely is caused by the excessive amounts of alcohol (a toxin) the liver is forced to break down (p. 200).

> The liver is a very critical organ. Any malfunction is a matter of considerable concern.

Digestive Enzymes

The digestive enzymes are **hydrolytic enzymes,** which catalyze breakdown by the introduction of water at specific bonds (see fig. 1.7). Digestive enzymes are like other enzymes we have studied (p. 39). They are proteins with a particular shape that fits their substrate. They also have an optimum pH, which maintains their shape, thereby enabling them to speed up their specific reaction.

The various digestive enzymes are present in the digestive juices mentioned previously. We now consider how carbohydrates, proteins, and fats, the major components of food, are digested.

Starch digestion begins in the mouth. Saliva from the salivary glands has a neutral pH and contains **salivary amylase,** the first enzyme to act on starch:

$$\text{starch} + H_2O \xrightarrow{\text{salivary amylase}} \text{maltose}$$

In this equation, salivary amylase is written above the arrow to indicate that it is neither a reactant nor a product in the reaction. It merely speeds the reaction in which its substrate, starch, is digested to many molecules of maltose. Although maltose molecules cannot be absorbed by the intestine, additional digestive action in the small intestine converts maltose to glucose.

Protein digestion begins in the stomach. Gastric juice secreted by gastric glands has a very low pH—about 2—because it contains hydrochloric acid (HCl). Pepsinogen, a precursor that is converted to the enzyme **pepsin** when exposed to HCl, is also present in gastric juice. Pepsin acts on protein to produce peptides:

$$\text{protein} + H_2O \xrightarrow{\text{pepsin}} \text{peptides}$$

Peptides vary in length, but they always consist of a number of linked amino acids. Peptides are usually too large to be absorbed by the intestine lining, but later they are broken down to amino acids in the small intestine.

Starch, proteins, and fats are all digested in the small intestine. Pancreatic juice, which enters the duodenum, is basic because it contains sodium bicarbonate ($NaHCO_3$). It also contains enzymes for the digestion of all types of food. One pancreatic enzyme, **pancreatic amylase,** digests starch:

$$\text{starch} + H_2O \xrightarrow{\text{pancreatic amylase}} \text{maltose}$$

Another pancreatic enzyme, **trypsin,** digests protein:

$$\text{protein} + H_2O \xrightarrow{\text{trypsin}} \text{peptides}$$

Trypsin is secreted as trypsinogen, which is converted to trypsin in the duodenum.

Lipase, a third pancreatic enzyme, digests fat droplets after they have been emulsified by bile salts:

$$\text{fat} \xrightarrow{\text{bile salts}} \text{fat droplets}$$

$$\text{fat droplets} + H_2O \xrightarrow{\text{lipase}} \text{glycerol} + \text{fatty acids}$$

The end products of lipase digestion, glycerol and fatty acid molecules, are small enough to cross the cells of the intestinal villi, where absorption takes place. As mentioned previously, glycerol and fatty acids enter the cells of the villi, and within these cells, they are rejoined and packaged as lipoprotein droplets, which enter the lacteals (see fig. 4.6).

Peptidases and **maltase,** 2 enzymes present in the mucosa of the intestinal villi (see fig. 4.6), complete the digestion of protein and starch to small molecules that cross into the cells of the villi. Peptides, which result from the first step in protein digestion, are digested to amino acids by peptidases:

$$\text{peptides} + H_2O \xrightarrow{\text{peptidases}} \text{amino acids}$$

Maltose, which results from the first step in starch digestion, is digested to glucose by maltase:

$$\text{maltose} + H_2O \xrightarrow{\text{maltase}} \text{glucose}$$

Other disaccharides, each of which has its own enzyme, are digested in the small intestine. The absence of any one of these enzymes can cause illness. For example, many people, including as many as 75% of African Americans, cannot digest lactose, the sugar found in milk, because they do not produce lactase, the enzyme that converts lactose to its components, glucose and galactose. Drinking untreated milk often gives these individuals the symptoms of *lactose intolerance* (diarrhea, gas, cramps), caused by a large quantity of undigested lactose in the intestine. In most areas, it is possible to purchase milk made lactose-free by the addition of synthetic lactase.

Table 4.3 lists some of the major digestive enzymes produced by the digestive tract, salivary glands, or the pancreas. Each type of food is broken down by specific enzymes.

> Digestive enzymes present in digestive juices break down food to the nutrient molecules: glucose, amino acids, fatty acids, and glycerol. The first 2 are absorbed into the blood capillaries of the villi and the last 2 re-form within epithelial cells and enter the lacteals as lipoprotein droplets.

Table 4.3

Major Digestive Enzymes

Food	Digestion	Enzyme	Optimum pH	Produced By	Site of Action
Starch	Starch + H_2O → maltose	Salivary amylase	Neutral	Salivary glands	Mouth
		Pancreatic amylase	Basic	Pancreas	Small intestine
	Maltose + H_2O → glucose*	Maltase	Basic	Intestine	Small intestine
Protein	Protein + H_2O → peptides	Pepsin	Acidic	Gastric glands	Stomach
		Trypsin	Basic	Pancreas	Small intestine
	Peptides + H_2O → amino acids*	Peptidases	Basic	Intestine	Small intestine
Fat	Fat droplets + H_2O → glycerol + fatty acids*	Lipase	Basic	Pancreas	Small intestine

Absorbed by villi.

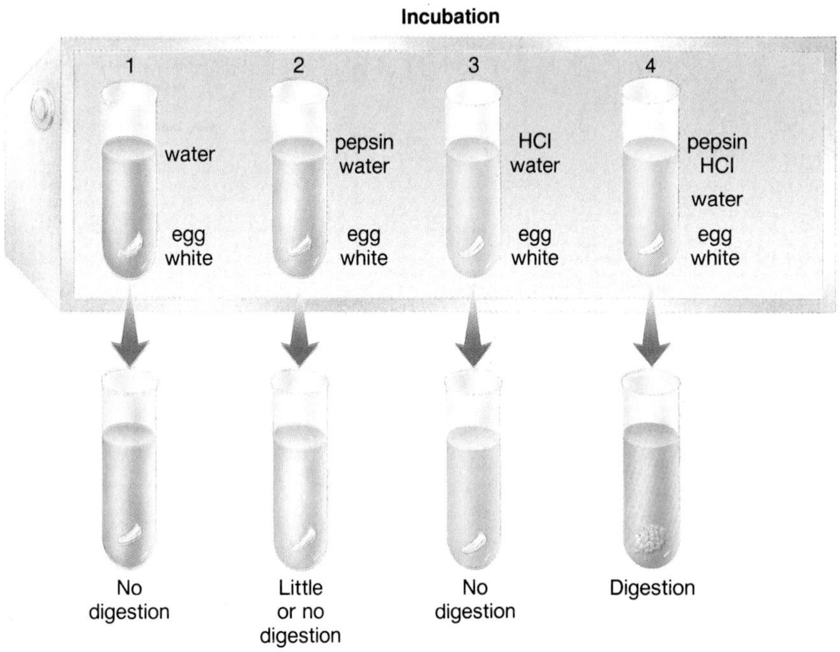

Figure 4.12

An experiment to demonstrate that enzymes digest food when the environmental conditions are correct. Tube 1 lacks the enzyme pepsin, and no digestion occurs; tube 2 has too high a pH, and little or no digestion occurs; tube 3 has the proper pH because of the presence of hydrochloric acid, but still no digestion occurs because the enzyme is missing; tube 4 contains the enzyme, and the environmental conditions are correct for digestion.

Best Conditions for Digestion

Laboratory experiments can define the necessary conditions for digestion (fig. 4.12). For example, the following 4 test tubes can be prepared and observed for the digestion of egg white, or the protein albumin.

1. H_2O + a small sliver of egg white (protein)
2. Pepsin + H_2O + a small sliver of egg white
3. HCl + H_2O + a small sliver of egg white
4. Pepsin + HCl + H_2O + a small sliver of egg white

All tubes now are placed in an incubator at body temperature for at least one hour. At the end of this time, we can predict that tube 4 will show the best digestive action because the environmental conditions are appropriate. Tube 3 does not contain the enzyme (pepsin) and tube 2 has too high a pH (HCl is lacking), so these 2 tubes are expected to show little or no digestion. Tube 1 is a control tube, and no digestion is expected to occur in this tube. This experiment shows that for digestion to occur, the enzyme, as well as the substrate and HCl, must be present.

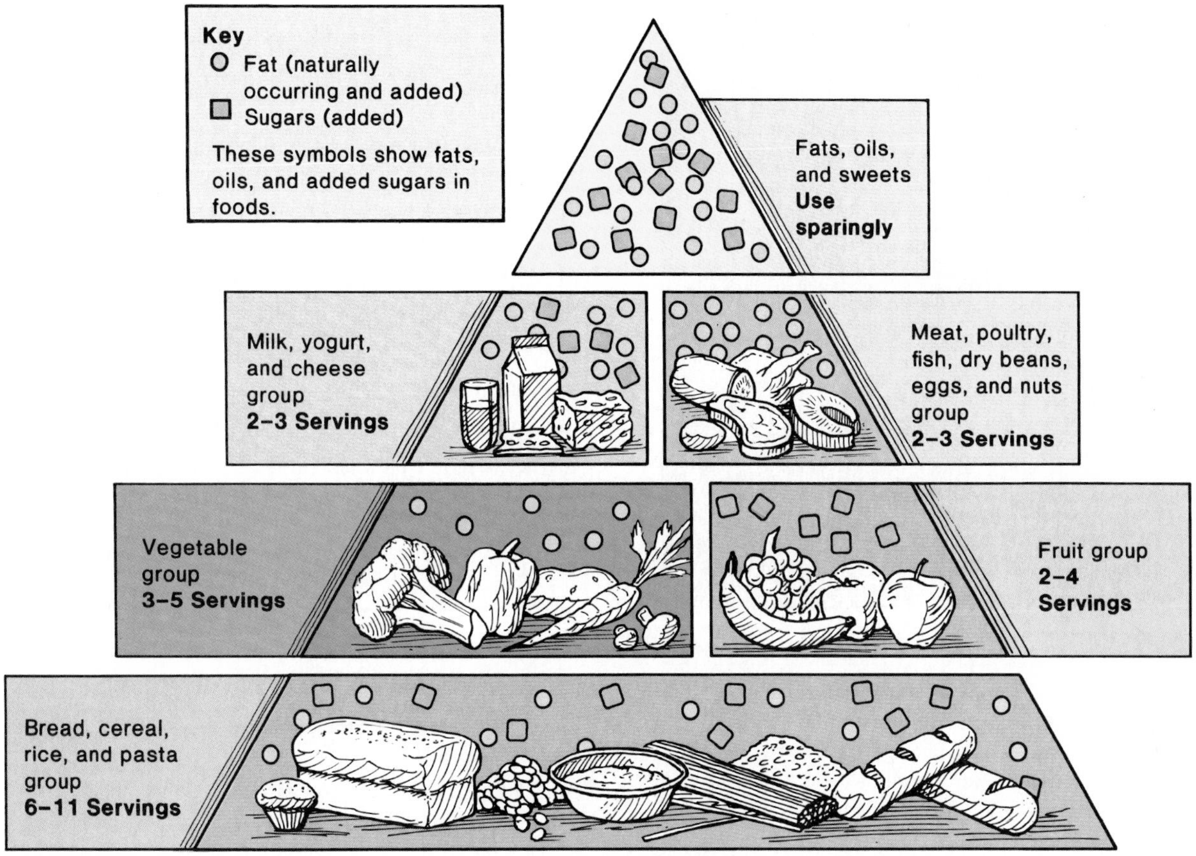

Figure 4.13 **Food Guide Pyramid:** A Guide to Daily Food Choices

Ideal American diet. The U.S. Department of Agriculture uses a pyramid to show the ideal American diet because it emphasizes the importance of including grains in the diet and the undesirability of eating fats, oils, and sweets.

Source: U.S. Department of Agriculture.

Nutrition

The body requires many different types of organic molecules and a smaller number of various types of inorganic ions and compounds from the diet each day. Nutrition involves an interaction between food and the living organism, and a **nutrient** is a substance that the body uses for the maintenance of health. For the diet to contain all the essential nutrients, it must be balanced. A *balanced diet* includes a variety of foods proportioned as shown in figure 4.13.

Carbohydrates: Direct Energy Source

The quickest, most readily available source of energy for the body is carbohydrates, which can be complex (i.e., polysaccharides), as in breads and cereals, or simple, as in candy, ice cream, and soft drinks. As mentioned previously, starches are digested to glucose, which is stored by the liver in the form of glycogen. Between eating periods, the blood glucose level is maintained at about 0.1% by the breakdown of glycogen or by the conversion of amino acids to glucose. While body cells can utilize fat as an energy source, brain cells require glucose. If necessary, amino acids are taken from the muscles—even from heart muscle. To avoid this situation, it is suggested that the daily diet contain at least 100 grams of carbohydrate, even when an individual is dieting. The Health Focus for this chapter concerns dieting.

> Carbohydrates are needed in the diet to maintain the blood glucose level.

Actually, the dietary guidelines produced jointly by the U.S. Department of Agriculture and the Department of Health and Human Services recommend that we increase the proportion of carbohydrates per total energy content of the diet:

	Typical Diet (%)	Recommended Diet (%)
Proteins	12	12
Carbohydrates	46	58
Fats	42	30

Figure 4.14

Complex carbohydrates. To meet our energy needs, dieticians recommend complex carbohydrates like those shown here rather than simple carbohydrates like candy and ice cream. The latter are more likely to cause weight gain than the recommended complex ones displayed.

Table 4.4

Reducing Dietary Sugar

To reduce dietary sugar, the following suggestions are recommended.

1. Eat fewer sweets, such as candy, soft drinks, ice cream, and pastry.
2. Eat fresh fruits or fruits canned without heavy syrup.
3. Use less sugar—white, brown, or raw—and less honey and syrups.
4. Avoid sweetened breakfast cereals.
5. Eat less jelly.
6. Drink pure fruit juices, not imitations.
7. When cooking, use spices like cinnamon instead of sugar to flavor foods.
8. Do not put sugar in tea or coffee.

Further, it is assumed that these carbohydrates are complex and not simple (fig. 4.14). Simple carbohydrates (e.g., sugars) are labeled "empty calories" by some dieticians because they contribute to energy needs and weight gain and are not part of foods that supply other nutritional requirements. Table 4.4 gives suggestions on how to cut down the consumption of dietary sugar (simple carbohydrates).

In contrast to simple sugars, complex carbohydrates are likely to be accompanied by a wide range of other nutrients and by fiber, which is nondigestible plant material. The Health Focus in chapter 1 (p. 23) discussed the benefits of fiber. Insoluble fiber, such as that found in wheat bran, has a laxative effect and therefore may reduce the risk of colon cancer. Soluble fiber, such as that found in oat bran, may possibly reduce cholesterol in blood because it combines with bile acids and cholesterol in the intestine and prevents them from being absorbed. While the diet should have an adequate amount of fiber, a high-fiber diet can be detrimental. Some evidence suggests that the absorption of iron, zinc, and calcium is impaired by a diet too high in fiber.

> Complex carbohydrates, along with fiber, are considered beneficial to health.

Proteins: Supply Building Blocks

Foods rich in protein include red meat, fish, poultry, dairy products, legumes, nuts, and cereals. Following digestion of protein, amino acids enter the bloodstream and are transported to the tissues. Ordinarily, amino acids are not used as an energy source. Most are incorporated into structural proteins found in muscles, skin, hair, and nails. Others are used to synthesize such proteins as hemoglobin, plasma proteins, enzymes, and hormones.

Adequate protein formation requires 20 different types of amino acids. Of these, 9 are required from the diet because the body is unable to produce them. These are termed the **essential amino acids.** The body produces the other 11 amino acids by simply transforming one type into another type. Some protein sources, such as meat, are *complete;* they provide all 20 types of amino acids. Vegetables and grains supply us with amino acids, but each vegetable or grain alone is an *incomplete* protein source because at least one of the essential amino acids is absent. Absence of one essential amino acid prevents utilization of the other 19 amino acids. However, it is possible to combine foods to acquire all the essential amino acids. For example, the combinations of cereal with milk or beans with rice provide all the essential amino acids.

> A complete source of protein is absolutely necessary to ensure a sufficient supply of the essential amino acids.

Health Focus

Weight Loss the Healthy Way

For a woman 19 to 22 years, 5 feet 4 inches tall, who exercises lightly, 2,100 Kcal* per day are normally recommended. For a man the same age, 5 feet 10 inches tall, who exercises lightly, the recommendation is 2,900 Kcal. Those who wish to lose weight need to reduce their caloric intake and/or increase their level of exercise. Exercising is a good idea, because to maintain good nutrition the intake of calories per day should probably not go below 1,200 Kcal a day. Also, for the reasons discussed in this chapter, carbohydrates should still make up at least 50% of these calories; proteins should be no more than 25% and the rest can be fats. A deficit of 500 Kcal a day (through intake reduction or increased exercise) is sufficient to lose a pound of body fat in a week. Once you realize that a diet needs to be judged according to the principles of adequacy of nutrients; balance in regard to carbohydrates, proteins, and fats; moderation in number of calories; and variety of food sources, it is easy to see that many of the diets and gimmicks people use to lose weight are bad for their health. Unhealthy approaches include the following:

Pills

The most familiar pills, and the only ones approved by the FDA, are those that claim to suppress the appetite. They may actually work at first, but the appetite soon returns to normal and weight lost is regained. Then the user has the problem of trying to get off the drug without gaining more weight. Other types of pills are under investigation and sometimes can be obtained illegally. But, as yet, there is no known drug that is both safe and effective for weight loss.

Low-Carbohydrate Diets

The dramatic weight loss that occurs with a low-carbohydrate diet is not due to a loss of fat; it is due to a loss of muscle mass and water. Glycogen and important minerals are also lost. When a normal diet is resumed, so is the normal weight.

Liquid Diets

Despite the fact that liquid diets provide proteins and vitamins, the number of Kcal is so restricted that the body cannot burn fat quickly enough to compensate and muscle is still broken down to provide energy. A few people on this regime have died, probably because even the heart muscle was not spared by the body.

Single-Category Diets

These diets rely on the intake of only one kind of food, either a fruit or vegetable or rice alone. However, no single type of food provides the balance of nutrients needed to maintain health. Some dieters on strange diets suffer the consequences—in one instance an individual lost hair and fingernails.

*A *calorie* is a standard way to measure energy; it is the amount of heat necessary to raise the temperature of one gram of water one degree Centigrade. Food energy is measured in *kilocalories* (thousands of calories), abbreviated Kcal.

Questions to Ask About a Weight-Loss Diet

1. Does the diet have a reasonable number of Kcal?

   10 Kcal per pound of current weight is suggested. In any case, no fewer than 1,000–1,200 Kcal for a normal-sized person.

2. Does the diet provide enough protein?

   For a woman 120 lb, 44 g protein each day is recommended. For a man 154 lb, 56 g is recommended. More than twice this amount is too much.

   For reference, 1 c milk and 1 oz meat each has 8 g protein.

3. Does the diet provide too much fat?

   No more than 20%–30% of total Kcal is recommended.

   For reference, a pat of butter has 45 Kcal. 1 g fat = 9 Kcal.

4. Does the diet provide enough carbohydrates?

   100 g = 400 Kcal is the very least recommended per day; 50% of total Kcal should be carbohydrates.

   For reference, a slice of bread contains 14 g of carbohydrates.

5. Does the diet provide a balanced assortment of foods?

   The diet should include breads, cereals, legumes; vegetables (especially dark-green and yellow ones); low-fat milk products; and meats or a meat substitute.

6. Does the diet make use of ordinary foods that are available locally?

   Diets should not require the purchase of unusual or expensive foods.

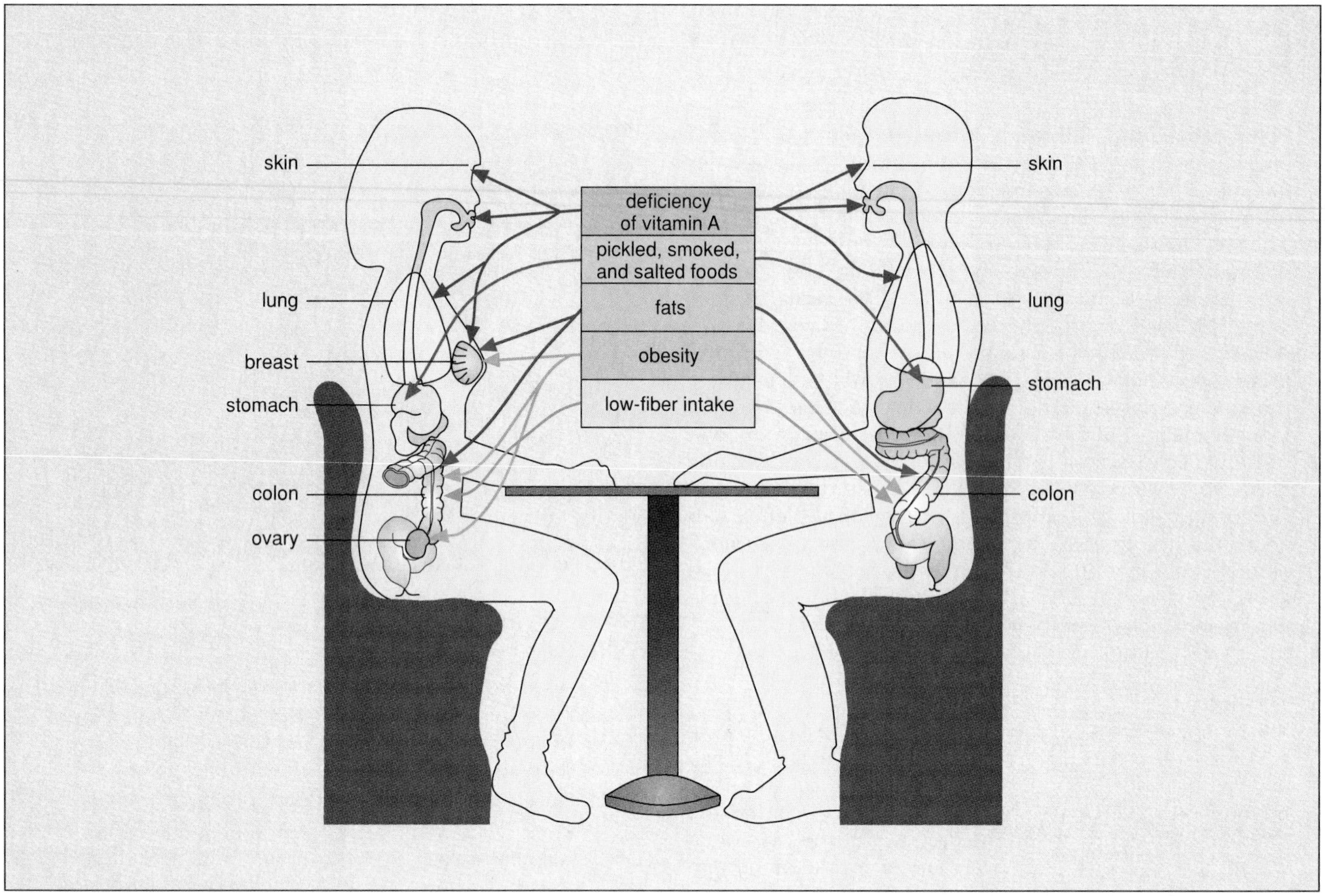

Figure 4.15

Diet and cancer. Evidence is growing to suggest these dietary factors cause the development of cancer in these organs. Among the primary cancer fighters in foods are vitamins A and C and selenium, a trace mineral.

Amino acids are not stored in the body, and a daily supply is needed. However, it does not take very much protein to meet the daily requirement. Two servings of meat a day is usually enough.

Consuming more meat than recommended can actually be detrimental. Calcium loss in the urine has been noted when dietary protein intake is over twice the RDA. Some meats (e.g., hamburger) are high in protein but also fat. Everything considered, it is probably a good idea to depend on protein from plant origins (e.g., whole-grain cereals, dark breads, legumes) to a greater extent than is the custom in this country.

Lipids: High Energy Source

Our discussion of lipids is divided into 2 parts: fats and cholesterol.

Fats: Beware High-Fat Diets

Fats are present not only in butter, margarine, and oils but also in many foods high in protein. After being absorbed, the products of fat digestion are packaged as lipoproteins and enter the lacteals. Later, they are transported by blood to the tissues. The liver can alter ingested fats to suit the body's needs, except it is unable to produce the fatty acid linoleic acid. Since this is required for phospholipid production, linoleic acid is considered an essential fatty acid. Although fats have the highest caloric content, they should not be avoided entirely because they do contain this essential fatty acid.

While we need to be sure to ingest some fat to satisfy our need for linoleic acid, recent dietary guidelines (p. 79) suggest that we should reduce the amount of fat per total energy content of the diet from 40% to 30%. Dietary fat has been implicated in cancer of the colon, pancreas, ovary, prostate, and breast (fig. 4.15).

Table 4.5

Reducing Lipids

To reduce dietary fat, the following suggestions are recommended.

1. Choose lean red meat, poultry, fish, or dry beans and peas as a protein source.
2. Trim fat off meat and remove skin from poultry before cooking.
3. Cook meat or poultry on a rack so that fat drains off.
4. Broil, boil, or bake rather than fry.
5. Limit your intake of butter, cream, hydrogenated oils, shortenings, and coconut and palm oils.*
6. Use herbs and spices to season vegetables instead of butter, margarine, or sauces. Use lemon juice instead of salad dressing.
7. Drink skim milk instead of whole milk, and use skim milk in cooking and baking.
8. Eat nonfat and low-fat foods.

To reduce dietary cholesterol, the following suggestions are recommended.

1. Avoid cheese, egg yolks, and liver. Preferably, eat white fish, poultry, and shellfish.
2. Substitute egg whites for egg yolks in both cooking and eating.
3. Include soluble fiber in the diet. Oat bran, oatmeal, beans, corn, and fruits such as apples, citrus fruits, and cranberries are high in soluble fiber.

*Although coconut and palm oils are from plant sources, they are saturated fats.

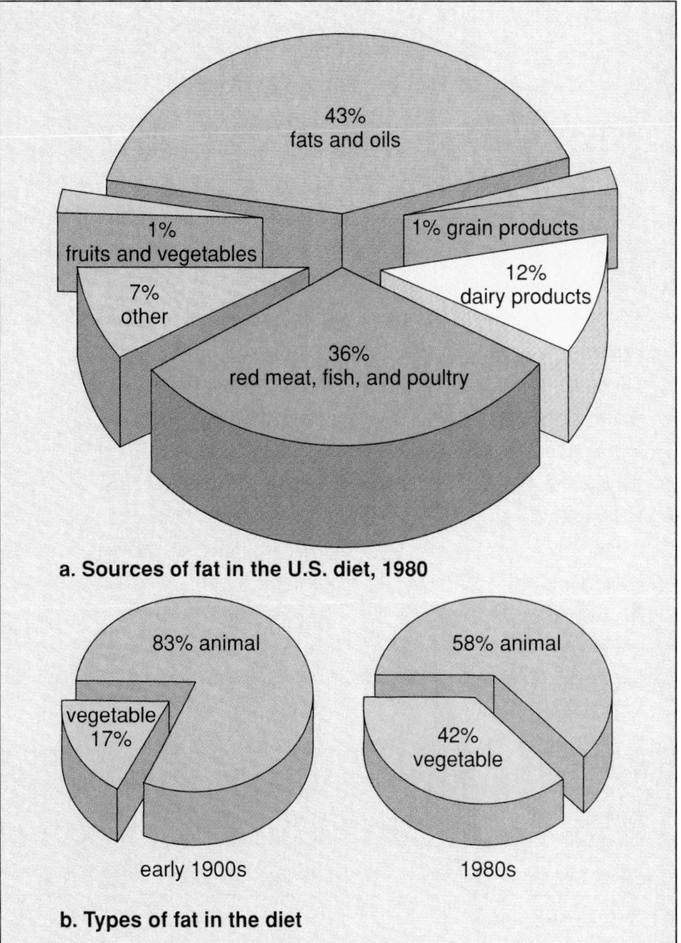

Figure 4.16

Fat content in the diet. **a.** Americans acquire 43% of fat intake from animal fats, such as butter, and from plant oils, such as cooking oils; 36% comes from red meat, poultry, and fish; and lesser amounts come from the sources shown. **b.** The amount of fat acquired from vegetable sources is now larger than it was in the early 1900s.

There is evidence that a high-fat diet increases chances of cancer development.

Fat is the component of food that has the highest energy content (9 Kcal/gram compared to 4 Kcal/gram for carbohydrates). Raw potatoes, which contain roughage, have about 0.9 Kcal per gram, but when they are cooked in fat, the number of Kcal jumps to 6 Kcal per gram. Another problem for those trying to limit their caloric intake is that fat is not always highly visible: butter melts on toast or potatoes. Table 4.5 gives suggestions for cutting down on the amount of fat in the diet.

As a nation, we have increased our consumption of fat from plant sources and have decreased our consumption from animal sources, such as red meat and butter (fig. 4.16). Most likely, this is due to recent publicized studies linking diets high in saturated fats and cholesterol to hypertension and heart attack.

Cholesterol and Heart Disease

According to the National Heart, Lung, and Blood Institute, a blood cholesterol level of 240 mg/100 ml or higher is associated with the development of cardiovascular disease. If the cholesterol level is this high, additional testing can determine how much of each of 2 important subtypes of cholesterol is in blood. Cholesterol is carried from the liver to the cells (including the endothelium of the arteries) by plasma proteins called *low-density lipoproteins* (*LDLs*) and is carried away from the cells to the liver by *high-density lipoproteins (HDLs).* Therefore, LDL is the type of lipoprotein that apparently contributes to formation of plaque (p. 102), which can clog the arteries, while HDL protects against the development of clogged arteries.

Table 4.6

Vitamins: Their Role in the Body and Their Food Sources

Vitamin	Major Role in Body	Good Food Sources
Fat Soluble		
Vitamin A	Vision, health of skin, hair, bones, and sex organs	Deep green or yellow vegetables, dairy products
Vitamin D	Health of bones and teeth	Dairy products, tuna, eggs
Vitamin E	Strengthening of red blood cell membrane	Green leafy vegetables, whole grains
Vitamin K	Clotting of blood, bone metabolism	Green leafy vegetables, cabbage, cauliflower
Water Soluble		
Thiamine (B_1)	Carbohydrate metabolism	Pork, whole grains
Riboflavin (B_2)	Energy metabolism	Whole grains, milk, green vegetables
Niacin (B_3)	Energy metabolism	Organ meats, whole grains
Pyroxidine (B_6)	Amino acid metabolism	Meats, fish, whole grains
Vitamin B_{12}	Red blood cell formation	Meats and dairy foods
Biotin	Carbohydrate metabolism	Eggs, most foods
Folic acid	Formation of red blood cells, DNA and RNA	Green leafy vegetables, nuts, whole grains
Pantothenic acid	Energy metabolism	Most foods
Vitamin C	Collagen formation	Citrus fruits, tomatoes

From David C. Nieman, et al., Nutrition. *Copyright © 1990 Wm. C. Brown Communications, Inc., Dubuque, IA. Reprinted by permission.*

Table 4.7

Minerals: Their Role in the Body and Their Food Sources

Mineral	Major Role in Body	Good Food Sources
Macrominerals		
Calcium (Ca)	Strong bones and teeth, nerve conduction	Dairy products, leafy green vegetables
Phosphorus (P)	Strong bones and teeth	Meat, dairy products, whole grains
Potassium (K)	Nerve conduction, muscle contraction	Many fruits and vegetables
Sodium (Na)	Nerve conduction, pH balance	Table salt
Chloride (Cl)	Water balance	Table salt
Magnesium (Mg)	Protein synthesis	Whole grains, green leafy vegetables
Microminerals		
Zinc (Zn)	Wound healing, tissue growth	Whole grains, legumes, meats
Iron (Fe)	Hemoglobin synthesis	Whole grains, legumes, eggs
Fluorine (F)	Strong bones and teeth	Fluoridated drinking water, tea
Copper (Cu)	Hemoglobin synthesis	Seafood, whole grains, legumes
Iodine (I)	Thyroid hormone synthesis	Iodized table salt, seafood

From David C. Nieman, et al., Nutrition. *Copyright © 1990 Wm. C. Brown Communications, Inc., Dubuque, IA. Reprinted by permission.*

A diet low in saturated fat and cholesterol decreases the blood cholesterol level (LDL level) in some individuals. The suggestions in table 4.5 promote this type of diet.

Vitamins and Minerals: Required in Small Amounts

For good health the diet should contain a variety of vitamins and minerals. The Ecology Focus for this chapter discusses how we are ultimately dependent on plants to meet these needs.

Vitamins are organic compounds (other than carbohydrate, fat, and protein) that the body is unable to produce but uses for metabolic purposes. Many vitamins are portions of coenzymes, or enzyme helpers. For example, niacin is part of the coenzyme NAD (p. 40) and riboflavin is part of another dehydrogenase, FAD. Coenzymes are needed in only small amounts because each can be used over and over again. An exception is vitamin A. Vitamin A is a precursor for "visual purple," a pigment needed in the cells of the eye for night vision.

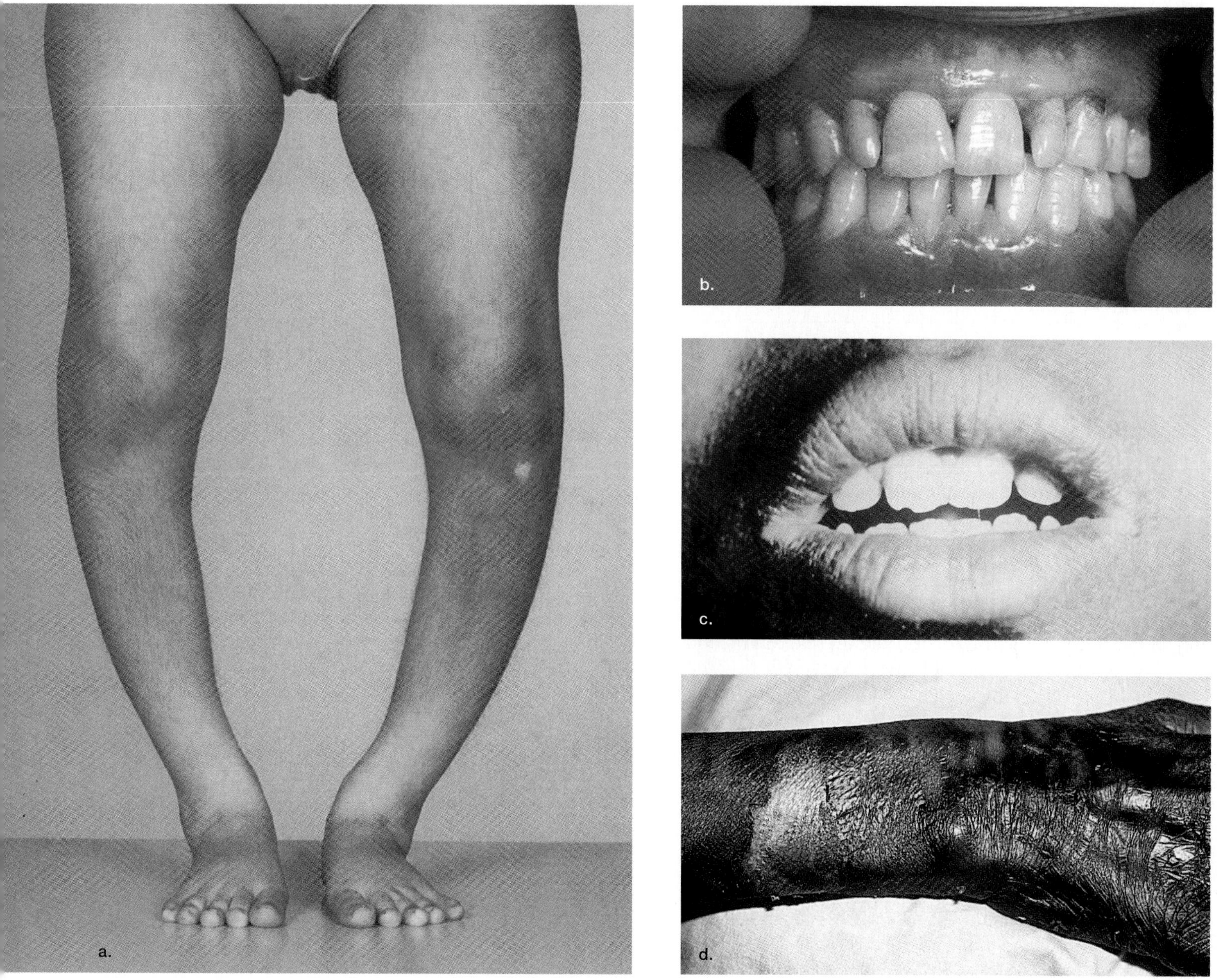

Figure 4.17

Illnesses due to vitamin deficiency. **a.** Bowing of bones (rickets) due to vitamin D deficiency. **b.** Bleeding of gums (scurvy) due to vitamin C deficiency. **c.** Fissures of lips (cheilosis) due to riboflavin (vitamin B_2) deficiency. **d.** Dermatitis (pellagra) of areas exposed to light due to niacin (vitamin B_3) deficiency.

If vitamins are lacking, various symptoms develop (fig. 4.17). Although many substances are advertised as vitamins, in reality there are only 13 vitamins (table 4.6). In general, carrots, squash, turnip greens, and collards are good sources of vitamin A. Citrus fruits and other fresh fruits and vegetables are natural sources of vitamin C. Sunshine and fortified milk are primary sources of vitamin D, and whole grains are good sources of the B vitamins.

The National Academy of Sciences suggests that we eat more fruits and vegetables to acquire a good supply of vitamins C and A because these 2 vitamins may help to prevent cancer. Nevertheless, the intake of excess vitamins by way of pills is discouraged because this practice can possibly lead to illness. For example, excess vitamin C can cause kidney stones, and also excessive vitamin C is converted to oxalic acid, a molecule that is toxic to the body. Vitamin A taken in excess over long periods can cause hair loss, bone and joint pains, and loss of appetite. Excess vitamin D can cause an overload of calcium in the blood, which in children leads to loss of appetite and retarded growth. Megavitamin therapy always should be supervised by a physician.

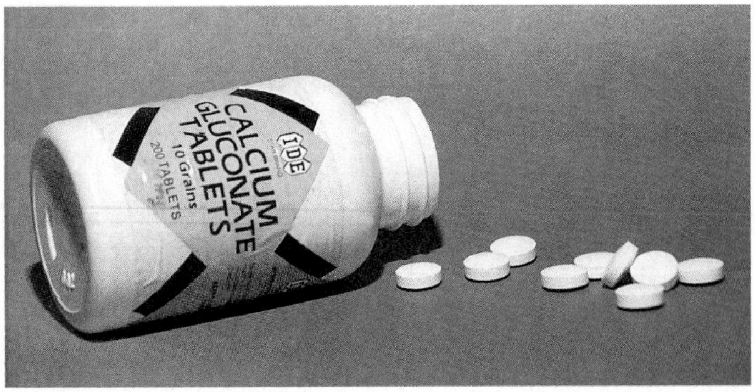

Figure 4.18

Calcium in the diet. Many over-the-counter calcium supplements now are available to boost the amount in the diet.

Table 4.8

Reducing Dietary Sodium

To reduce dietary sodium, the following suggestions are recommended.

1. Use spices instead of salt to flavor foods.
2. Add little or no salt to foods at the table, and add only small amounts of salt when you cook.
3. Eat unsalted crackers, pretzels, potato chips, nuts, and popcorn.
4. Avoid hot dogs, ham, bacon, luncheon meat, smoked salmon, sardines, and anchovies.
5. Avoid processed cheese and canned or dehydrated soups.
6. Avoid prepared catsup and horseradish.
7. Avoid brine-soaked foods, such as pickles and olives.

In addition to vitamins, various **minerals** are required by the body (table 4.7). Minerals are divided into the macrominerals, which are recommended in amounts more than 100 mg per day, and the microminerals (trace elements), which are recommended in amounts less than 20 mg per day. The macrominerals sodium, magnesium, phosphorus, chlorine, potassium, and calcium are constituents of cells and body fluids and are structural components of tissues. For example, calcium is needed for the construction of bones and teeth and for nerve conduction and muscle contraction.

The microminerals seem to have very specific functions. For example, iron is needed for the production of hemoglobin, and iodine is used in the production of thyroxin, a hormone produced by the thyroid gland. As research continues, more and more elements are added to the list of microminerals considered to be essential. During the past 3 decades, for example, very small amounts of molybdenum, selenium, chromium, nickel, vanadium, silicon, and even arsenic have been found to be essential to good health.

Occasionally, individuals do not receive enough iron (especially women), calcium, magnesium, or zinc in their diet. Adult females need more iron in the diet than males (18 mg compared to 10 mg) because they lose hemoglobin each month during menstruation. Stress can bring on a magnesium deficiency, and due to its high-fiber content, a vegetarian diet may make zinc less available to the body. However, a varied and complete diet usually supplies the RDAs for minerals.

Calcium and Bone Disease

There has been much public interest in the dietary addition of calcium supplements (fig. 4.18) to counteract osteoporosis, a degenerative bone disease that afflicts an estimated one-fourth of older men and one-half of older women in the United States. These individuals have porous bones, which break easily because they lack sufficient calcium. Studies have shown, however, that calcium supplements cannot prevent osteoporosis after menopause, even when the dosage is 3,000 mg a day. In postmenopausal women, bone-eating cells called osteoclasts are known to be more active than bone-forming cells called osteoblasts (p. 209). Until now, the most effective defense against osteoporosis in older women has been estrogen replacement and exercise, which encourage the work of osteoblasts. Recently, however, studies have shown that the drug etidronate disodium, which inhibits osteoclast activity, is effective in osteoporotic women when administered at the proper dosage.

Women can guard against osteoporosis when they are older by forming strong, dense bones when they are young. Eighteen-year-old women on the average get only 679 mg of calcium a day, when the dietary requirement is 800 mg (or 1,000 mg according to the National Institutes of Health [NIH]). They should consume more calcium-rich foods, such as milk and dairy products. Taking calcium supplements may not be as effective; a cup of milk supplies 270 mg of calcium, while a 500-mg tablet of calcium carbonate provides only 200 mg. The excess supplemental calcium is not taken up by the body, as it is not in a form that is *bioavailable*. However, an excess of bioavailable calcium can lead to kidney stones.

> Dietary calcium and exercise, plus estrogen therapy if needed, are the best safeguards against osteoporosis.

Too Much Sodium

The recommended amount of sodium intake per day is 500 mg, although the average American takes in 4,000–4,700 mg every day. In recent years, this imbalance has caused concern because high sodium intake has been linked to hypertension (high blood pressure) in some people. About one-third of the sodium we consume occurs naturally in foods; another one-third is added during commercial processing; and we add the last one-third either during home cooking or at the table in the form of table salt.

Clearly, it is possible for us to cut down on the amount of sodium in the diet. Table 4.8 gives recommendations for doing so.

> Excess sodium in the diet can lead to hypertension; therefore, excess sodium intake should be avoided.

Ecology Focus

Our Crops Deserve Better

When plants photosynthesize, they convert carbon dioxide from the air and water from the soil into a carbohydrate, namely glucose. Carbohydrates are organic food for plants and all other living things. Animals, including humans, either eat plants directly or eat animals that have eaten plants. Plants have the metabolic capability to modify and change carbohydrates into all the other types of organic molecules they require. This makes it possible for vegetarians to not only acquire carbohydrates but also amino acids from plants. The usual slimness of vegetarians testifies that most plants are not an abundant source of fat, and this fact can only make the vegetarian diet even more appealing.

Plant cell metabolism utilizes most of the same vitamins and minerals that humans require. Tables 4.6 and 4.7 show that plants are an important source of vitamins and minerals as well as organic food. Minerals are found in the soil in low concentrations, but plants are able to take them up and concentrate them. As the root system of a plant grows, it branches and branches again so that the roots are exposed to a tremendous amount of soil. It has been estimated that a rye plant has roots totaling about 900 kilometers in length. Water and minerals enter the roots by diffusion, but eventually active transport is used to concentrate minerals within the organs of a plant. A plant uses a great deal of ATP for active transport.

We are lucky that plants can concentrate minerals, for animals, like ourselves, often are dependent on them for supplying such elements as potassium for cardiac contraction, phosphorus for strong bones, and manganese for metabolic reactions. Once plants have taken up minerals, they are often incorporated into other molecules, including amino acids, phospholipids, and nucleotides.

Figure 4A

Proper management allows land to be productive for many years. Mismanagement can lead to desert conditions and the loss of land fertility. In this photograph, crops are being covered by windblown sand, and therefore, land productivity will be reduced.

Unfortunately, the health of plants is affected by the very same pollutants that affect the health of humans. We have already mentioned that the expected increase in UV radiation due to the depletion of ozone will adversely affect the ability of plants to carry on photosynthesis (p. 471). Ground-level ozone in smog destroys leaves and roots. Acid rain (see Ecology Focus, chapter 1, p. 17), which causes minerals to be leached from the soil, can even cause the death of plants. Global warming may cause the extinction of many plant species. Global warming is due to the buildup of carbon dioxide in the atmosphere from the combustion of gasoline, coal, and oil. Carbon dioxide acts like the glass of a greenhouse—it allows the sun's rays to pass through but traps the heat and doesn't allow it to escape. An increase in the average annual temperature is predicted.

The best climate for growing crops may shift to more rocky northern terrains with a resultant loss in crop yield. These problems don't begin to compare to the reduction in crop yield due to soil erosion, however. Soil erosion occurs when the wind blows and the rain washes away the topsoil because farmers have plowed the land in straight rows and/or left it without adequate cover. The U.S. Department of Agriculture estimates that erosion is causing a steady drop in the productivity of farmland equivalent to the loss of 1.25 million acres per year. Soil erosion can be halted by the implementation of proven techniques (e.g., contour farming, drip irrigation, no-till farming) that might also halt desertification, the further degradation of the land to desert conditions (fig. 4A).

Considering how dependent we are on plants, we must remember that plants as well as animals, including humans, are affected when the quality of the environment is reduced.

Eating Disorders

Authorities recognize 3 primary eating disorders; obesity, bulimia, and anorexia nervosa. Although they exist in a continuum as far as body weight is concerned, there is much overlap between them.

Obesity is defined as a body weight of more than 20% above the ideal weight. It most likely is caused by a combination of factors, including endocrinal, metabolic, and social factors. The social factors include the eating habits of other family members. Obese individuals need to consult a physician if they want to bring their body weight down to normal and to keep it there permanently.

> Obesity has many complex causes that possibly can be detected by a physician.

Bulimia can coexist with either obesity or anorexia nervosa. People, usually young women, who are afflicted have the habit of eating to excess and then purging themselves by some artificial means, such as vomiting or laxatives. These individuals usually are depressed, but whether the depression causes or is caused by the bulimia cannot be determined. While individual psychological help does not seem to be effective, there is some indication that group therapy helps. The possibility of a hormonal disorder has not been ruled out, however.

Anorexia nervosa is diagnosed when an individual is extremely thin but still claims to "feel fat" and continues to diet. It is possible these individuals have an incorrect body image that makes them think they are fat. It is also possible they have various psychological problems, including a desire to suppress their sexuality. Menstruation ceases in very thin women.

> Both bulimia and anorexia nervosa are serious disorders that require the assistance of competent medical personnel.

Summary

The human digestive tract consists of the mouth, pharynx, esophagus, stomach, small intestine, and large intestine. Only these structures actually contain food, but the salivary glands, liver, and pancreas supply substances that aid in the digestion of food. In the mouth, food is chewed and acted upon by salivary amylase before it is swallowed. Peristaltic action then moves the food along the esophagus to the stomach, a J-shaped muscular organ that churns the food. Here, the gastric glands produce pepsin, an enzyme that breaks down protein, and HCl, an acid.

In contrast, the small intestine has a basic pH because it receives pancreatic juice, which contains sodium bicarbonate. In the small intestine, fat is emulsified by bile salts to fat droplets before being acted upon by pancreatic lipase. Protein is digested by pancreatic trypsin, and starch is digested by pancreatic amylase. Intestinal enzymes present on the surface of the intestinal mucosa finish the digestion of proteins and carbohydrates. The action of digestive enzymes is summarized in table 4.3.

The walls of the small intestine have fingerlike projections called villi, within which are blood capillaries and a lymphatic lacteal. Amino acids and glucose enter the blood; glycerol and fatty acids, which are rejoined and packaged as lipoproteins by epithelial cells, enter the lacteal. The blood from the small intestine moves into the hepatic portal vein, which goes to the liver, an

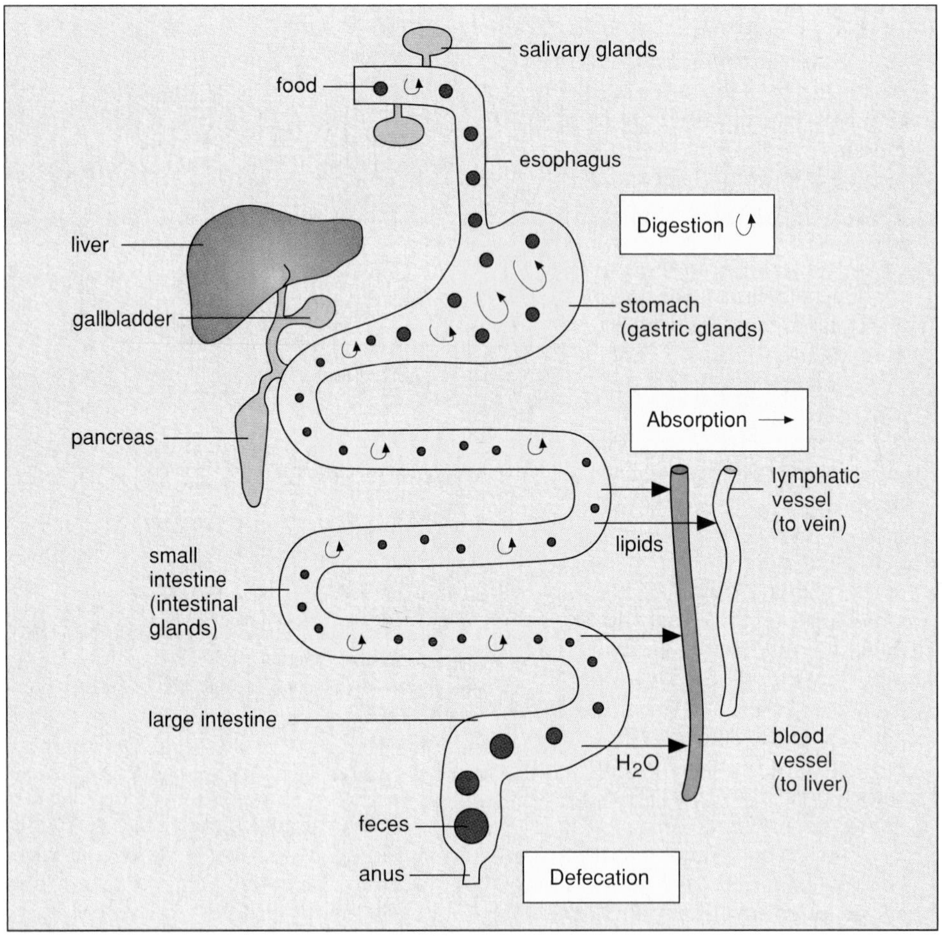

organ that monitors blood composition and, for example, maintains a constant level of glucose in the blood.

The only material that passes from the small intestine to the large intestine is nondigestible material. The large intestine absorbs water from this material and contains a large population of bacteria that can use it as food. In the process, the bacteria produce vitamins that can be absorbed and used by our bodies. They also produce the odor characteristic of feces, which pass out of the body during defecation, a process controlled by reflex action. Feces contain nondigestible substances and bacteria.

Digestive enzymes function according to the lock-and-key theory of enzymatic action and have the usual enzymatic properties. They are specific to their substrate and speed up specific reactions at body temperature and optimum pH. The nutrients released by the digestive process should provide us with an adequate amount of energy, essential amino acids and fatty acids, and all necessary vitamins and minerals. To bring this about, the diet should be balanced.

Study and Thought Questions

1. List the parts of the digestive tract, anatomically describe them, and state the contribution of each to the digestive process. (p. 68)

   Critical Thinking How does the structure of each organ mentioned suit its function?
2. Discuss the absorption of the products of digestion into the lymphatic and circulatory systems. (p. 72)
3. What functions do intestinal bacteria perform? (p. 74)
4. What are gastrin, secretin, and CCK? Where are they produced? What are their functions? (p. 72)
5. List the accessory glands, and describe the part they play in the digestion of food. (p. 75)

   Critical Thinking Hormones help coordinate body activities. Apply this statement to the action of the hormones produced by the digestive tract.
6. List 6 functions of the liver. How does the liver maintain a constant glucose level in blood? (p. 75)
7. What is jaundice? cirrhosis of the liver? (p. 76)
8. Discuss the digestion of starch, protein, and fat, listing all the steps that occur to bring about digestion of each of these. (p. 77)

   Critical Thinking Would pepsin be able to digest starch? Explain in reference to figure 4.13.
9. Give reasons why carbohydrates, proteins, fats, vitamins, and minerals are all necessary for good nutrition. (pp. 79–86)

   Ethical Issue A proper diet can apparently help prevent heart disease and cancer, the leading causes of death in the United States. Should nutrition be a required subject for all students graduating from college? Why or why not?
10. Name and discuss 3 eating disorders. (p. 88)

Writing Across the Curriculum

1. All systems of the body contribute to homeostasis. How does the digestive system contribute?
2. Reexamine figure 4.1 of the digestive tract, and give examples to show that the structure of individual portions along the tract suit their particular functions.
3. Name 2 ways your body would be affected if your pancreas ceased to function.

Objective Questions

1. In the mouth, salivary ______________ digests starch to ______________.
2. When swallowing, the ______________ covers the opening to the larynx.
3. The ______________ takes food to the stomach, where ______________ is primarily digested.
4. The gallbladder stores ______________, a substance that ______________ fat.
5. The pancreas sends digestive juices to the ______________, the first part of the small intestine.
6. Pancreatic juice contains ___________ for digesting protein, ___________ for digesting starch, and ___________ for digesting fat.
7. Whereas pepsin prefers a _________ pH, the enzymes found in pancreatic juice prefer a _________ pH.
8. The products of digestion are absorbed into the cells of the ___________, fingerlike projections of the intestinal wall.
9. After eating, the liver stores glucose as __________.
10. The diet should include a complete protein source, one that includes all the ___________.

11. Label this diagram of the digestive tract, and give a function for each organ.

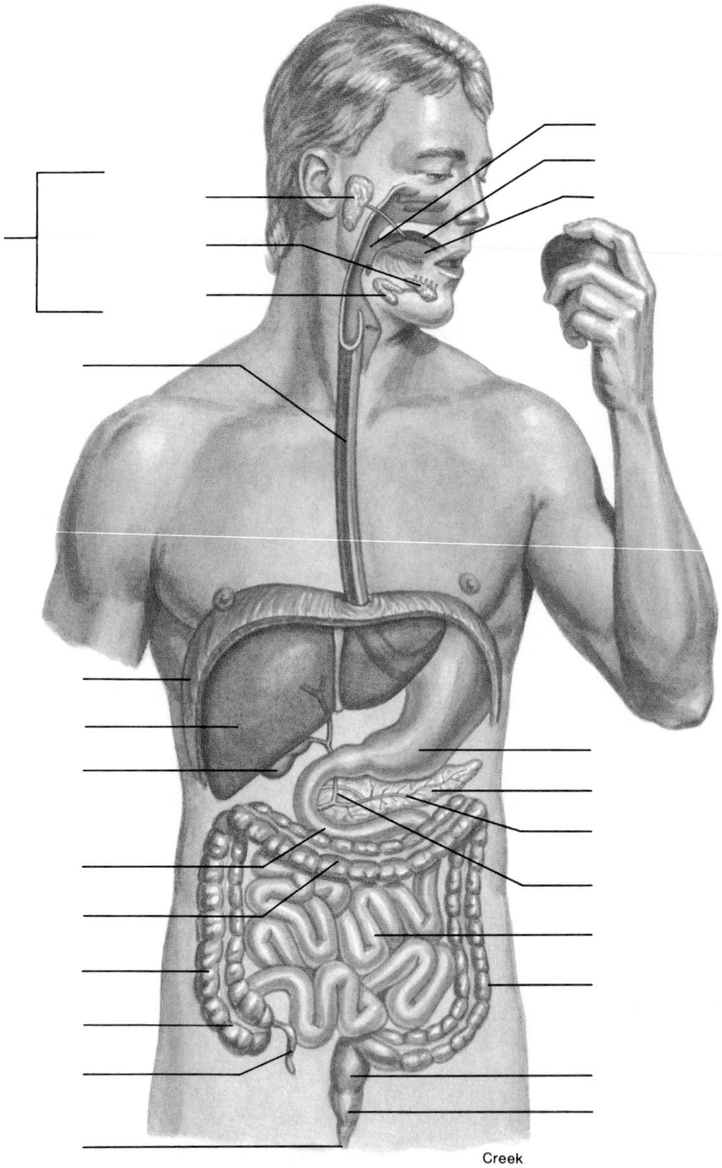

12. Predict and explain the results of this experiment for each test tube.
    a. Tube 1: water, bile salts, oil
    b. Tube 2: water, bile salts, pepsin, oil
    c. Tube 3: water, bile salts, pancreatic lipase, oil

Selected Key Terms

bile A secretion of the liver that is temporarily stored in the gallbladder before being released into the small intestine, where it emulsifies fat. 75

colon The large intestine. 74

epiglottis A structure that covers the glottis during the process of swallowing. 70

gallbladder A saclike organ associated with the liver that stores and concentrates bile. 75

gastric gland Gland within the stomach wall that secretes gastric juice. 71

glottis Slitlike opening to the larynx between the vocal cords. 70

hard palate Bony, anterior portion of the roof of the mouth, which contains several bones. 69

hydrolytic enzyme An enzyme that catalyzes a reaction in which the substrate is broken down with the addition of water. 77

lipase A fat-digesting enzyme secreted by the pancreas. 77

lumen The cavity inside any tubular structure, such as the lumen of the digestive tract. 70

pepsin A protein-digesting enzyme secreted by gastric glands. 77

peristalsis (per″ i-stal′ sis) A rhythmic contraction that serves to move the contents along in tubular organs, such as the digestive tract. 71

pharynx (far′ ingks) A common passageway (throat) for both food intake and air movement. 70

salivary gland A gland associated with the mouth that secretes saliva. 69

soft palate Entirely muscular posterior portion of the roof of the mouth. 69

sphincter A muscle that surrounds a tube and closes or opens the tube by contracting and relaxing. 71

trypsin A protein-digesting enzyme secreted by the pancreas. 77

villus (vil′ us) (pl. villi) Fingerlike projection from the wall of the small intestine that functions in absorption. 72

vitamin Essential requirement in the diet, needed in small amounts. They are often part of coenzymes. 84

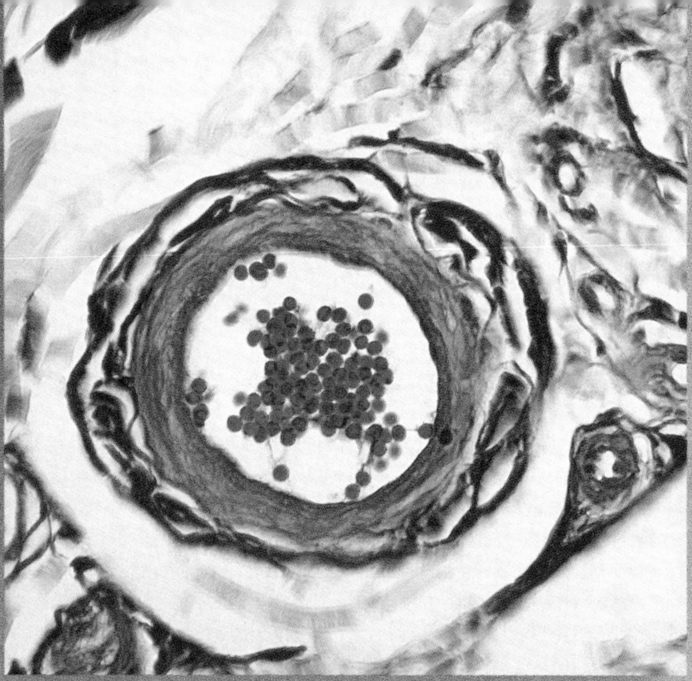

Chapter 5

Circulatory System

Life is dependent on the proper functioning of the coronary blood vessels, the little tubes that serve the needs of cardiac muscle. A common circulatory problem today is blockage of the coronary arteries so that they are unable to function properly. Although coronary artery disease (CAD) develops slowly over the years, a heart attack may come on quite suddenly. Evidence is growing that coronary artery disease may be preventable in part, but there is no quick fix. A lifetime of devotion to these little vessels is required, and good health habits are a necessity. A now-famous study for over 38 years of 6,000 residents in the city of Framingham, Massachusetts, has helped investigators determine that cigarette smoking, elevated blood cholesterol, and the presence of hypertension all predispose an individual to CAD. Other important factors are lack of exercise, obesity, stress, diabetes, and a family history of coronary artery disease.

Chapter Concepts

Chapter Outline

Blood moving through tubular vessels brings our cells their daily supply of nutrients, such as amino acids and glucose, and takes away their wastes, such as carbon dioxide. The heart keeps blood moving along its predetermined circular path. Circulation of blood is so important that if the heart stops beating for only a few minutes, death results.

Blood Vessels

The circulatory system has 3 types of blood vessels: the **arteries** (and arterioles), which carry blood away from the heart; the **capillaries,** which permit exchange of material with the tissues; and the **veins** (and venules), which return blood to the heart (fig. 5.1).

Arteries and Arterioles: Away from the Heart

Arteries have thick walls (fig. 5.1*b*). The walls have an inner membranous layer called endothelium, which contains squamous epithelial cells; a thick middle layer of elastic tissue and smooth muscle; and an outer fibrous connective tissue layer. An artery can expand to accommodate the sudden increase in blood volume after each heartbeat, even though the walls are so thick that they are supplied with blood vessels.

Arterioles are small arteries just visible to the naked eye. The middle layer of arterioles has some elastic tissue but is composed mostly of smooth muscle, the fibers of which encircle the arteriole. If these muscle fibers contract, the lumen of the arteriole gets smaller; if the fibers relax, the lumen of the arteriole enlarges. Whether arterioles are constricted or dilated affects blood pressure. The greater the number of vessels dilated, the lower the blood pressure.

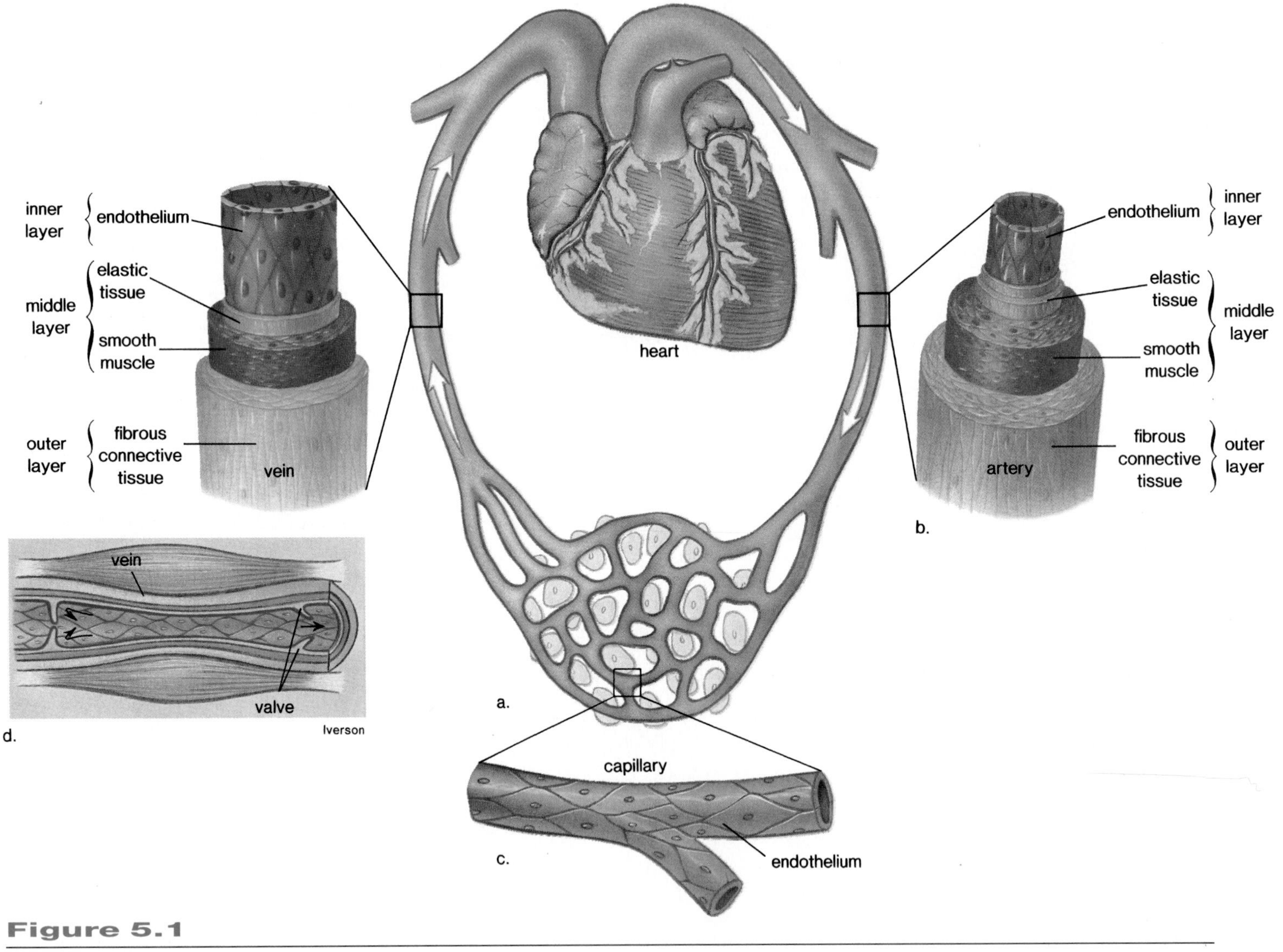

Figure 5.1

Blood vessels. **a.** Blood leaving the heart moves from an artery to arterioles to capillaries to venules and then returns to the heart by way of a vein. **b.** Arteries have well-developed walls with a thick middle layer of elastic tissue and smooth muscle. **c.** Capillary walls are one cell thick. **d.** Veins have flabby walls, particularly because the middle layer is not as thick as in arteries. Veins have valves, which point toward the heart.

Capillaries: Exchange Takes Place

Arterioles branch into capillaries. Each capillary is an extremely narrow, microscopic tube with one-cell-thick walls composed only of endothelium (fig. 5.1c). *Capillary beds* (networks of many capillaries) are present in all regions of the body; consequently, a cut to any body tissue draws blood. The capillaries are a very important part of the human circulatory system because an exchange of nutrient and waste molecules takes place across their thin walls. Oxygen and nutrients diffuse out of a capillary into the tissue fluid that surrounds cells, and carbon dioxide diffuses into the capillary (see fig. 6.7). Some water also leaves a capillary; any excess is picked up by lymphatic vessels, which return it to the blood circulatory system. The lymphatic system is discussed in chapter 7.

Since the capillaries serve the cells, the heart and the other vessels of the circulatory system can be thought of as the means by which blood is conducted to and from the capillaries. Only certain capillaries are open at any given time. Shunting of blood is possible because each capillary bed has a thoroughfare channel that allows blood to go directly from arteriole to venule (fig. 5.2). After eating, blood is shunted through the muscles of the body and diverted to the digestive system. This is why swimming after a heavy meal may cause cramping.

Veins and Venules: To the Heart

Veins and venules take blood from the capillary beds to the heart. First, the **venules** (small veins) drain blood from the capillaries and then join to form a vein. The walls of venules (and veins) have the same 3 layers as arteries, but the middle layer is poorly developed and therefore the walls are thinner (see fig. 5.1d). Veins often have **valves,** which allow blood to flow only toward the heart when open and prevent the backward flow of blood when closed.

At any given time, more than half of the total blood volume is in the veins and the venules. If a loss of blood occurs, for example due to hemorrhaging, nervous stimulation causes the veins to constrict, providing more blood to the rest of the body. In this way, the veins act as a blood reservoir.

Arteries and arterioles carry blood away from the heart; veins and venules carry blood to the heart; and capillaries join arterioles to venules.

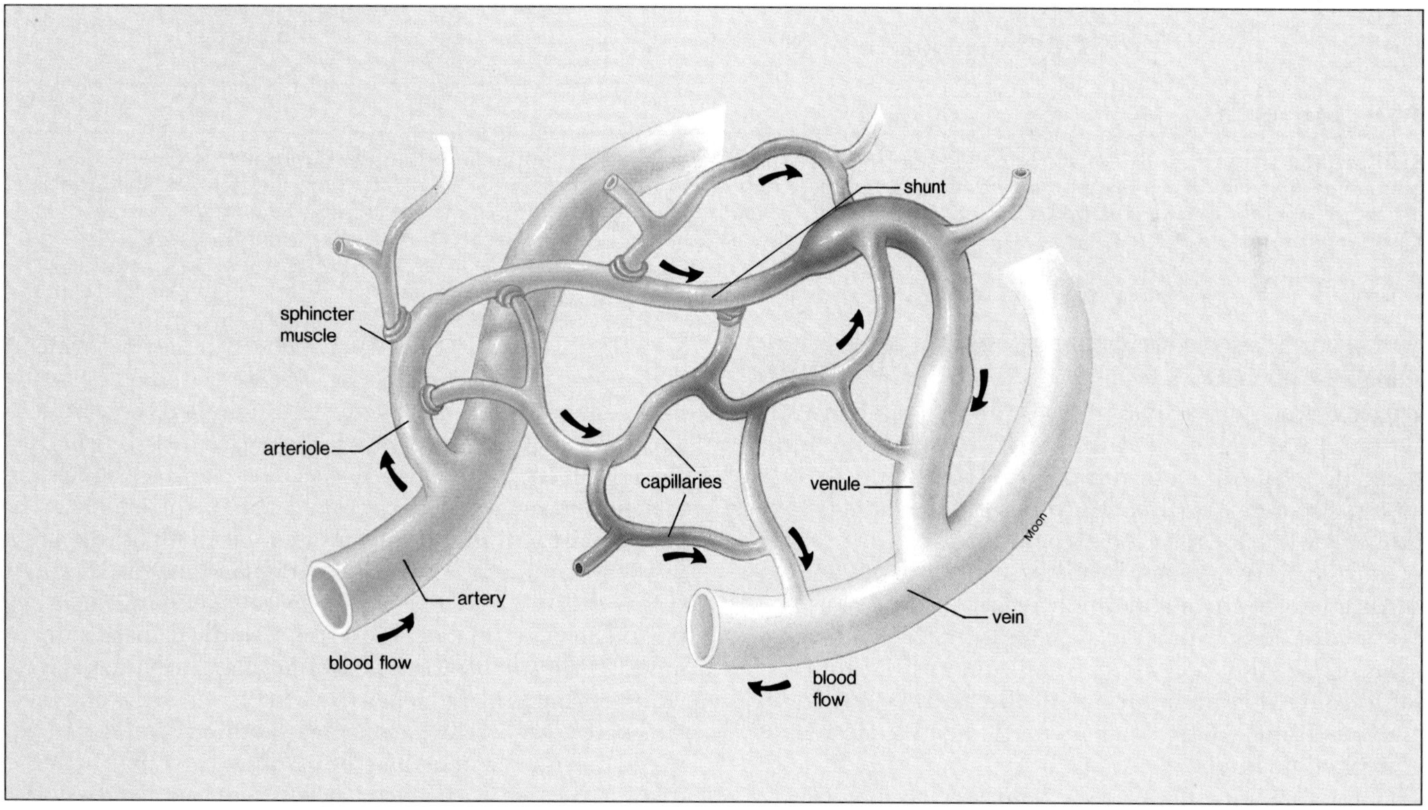

Figure 5.2

Anatomy of a capillary bed. Capillary beds form a maze of vessels that lie between an arteriole and a venule. Blood can move directly between the arteriole and the venule by way of a shunt. When sphincter muscles are closed, blood flows through the shunt. When sphincter muscles are open, the capillary bed is open, and blood flows through the capillaries. Except in the lungs, as blood passes through a capillary, it gives up its oxygen (O_2). Therefore, blood goes from being oxygenated in the arteriole (red color) to being deoxygenated (blue color) in the vein.

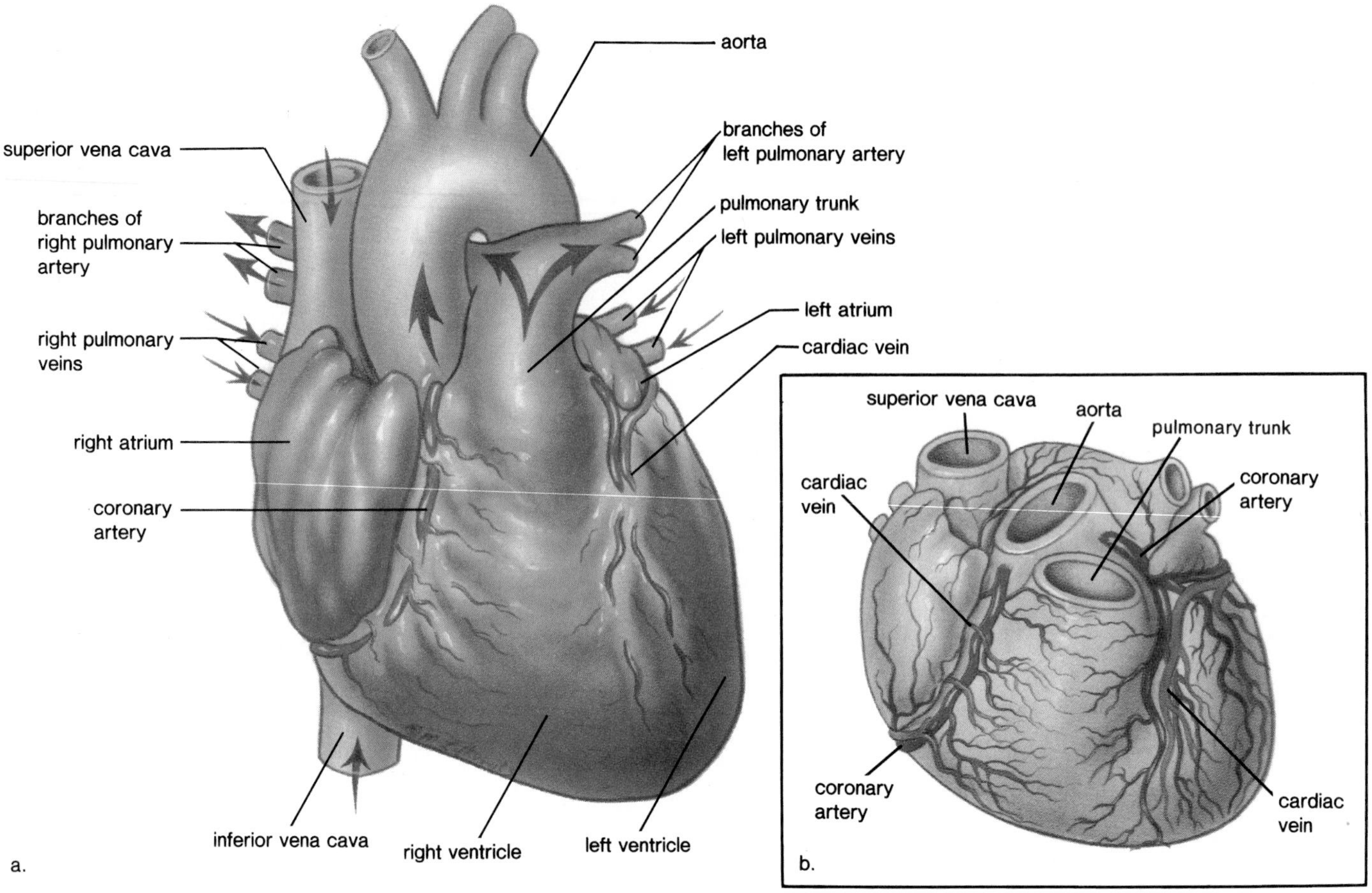

Figure 5.3

External heart anatomy. **a.** The venae cavae bring deoxygenated blood to the right side of the heart from the body, and the pulmonary arteries take this blood to the lungs. The pulmonary veins bring oxygenated blood from the lungs to the left side of the heart, and the aorta takes this blood to the body. **b.** The coronary arteries and cardiac veins pervade cardiac muscle. The coronary arteries are the first blood vessels to branch off the aorta. They bring oxygen and nutrients to cardiac cells.

The Heart: Pumps Blood

The **heart** is a cone-shaped, muscular organ about the size of a fist (fig. 5.3). It is located between the lungs directly behind the sternum (breastbone) and is tilted so that the apex is directed to the left. The major portion of the heart, called the **myocardium,** consists largely of cardiac muscle tissue. The muscle fibers of the myocardium are branched and tightly joined to one another. The heart lies within the pericardium, a thick, membranous sac that contains a small quantity of lubricating liquid. The inner surface of the heart is lined with endocardium, which consists of connective tissue and endothelial tissue.

Internally, a wall called the **septum** separates the heart into a right side and a left side (fig. 5.4). The heart has 4 chambers: 2 upper, thin-walled atria (sing., **atrium**), sometimes called auricles, and 2 lower, thick-walled **ventricles.** The atria are much smaller and weaker than the muscular ventricles, but they hold the same volume of blood.

The heart also has valves, which direct the flow of blood and prevent its backward movement. The valves that lie between the atria and the ventricles are called the **atrioventricular valves.** These valves are supported by strong fibrous strings called chordae tendineae. The chordae, which are attached to muscular projections of the ventricular walls, support the valves and prevent them from inverting when the heart contracts. The atrioventricular valve on the right side is called the tricuspid valve because it has 3 cusps, or flaps. The valve on the left side is called the bicuspid (or the mitral) because it has 2 flaps. There are also **semilunar valves,** which resemble half moons, between the ventricles and their attached vessels. The pulmonary semilunar valve lies between the right ventricle and the pulmonary trunk. The aortic semilunar valve lies between the left ventricle and the aorta.

Humans have a 4-chambered heart (2 atria and 2 ventricles). A septum separates the right side from the left side.

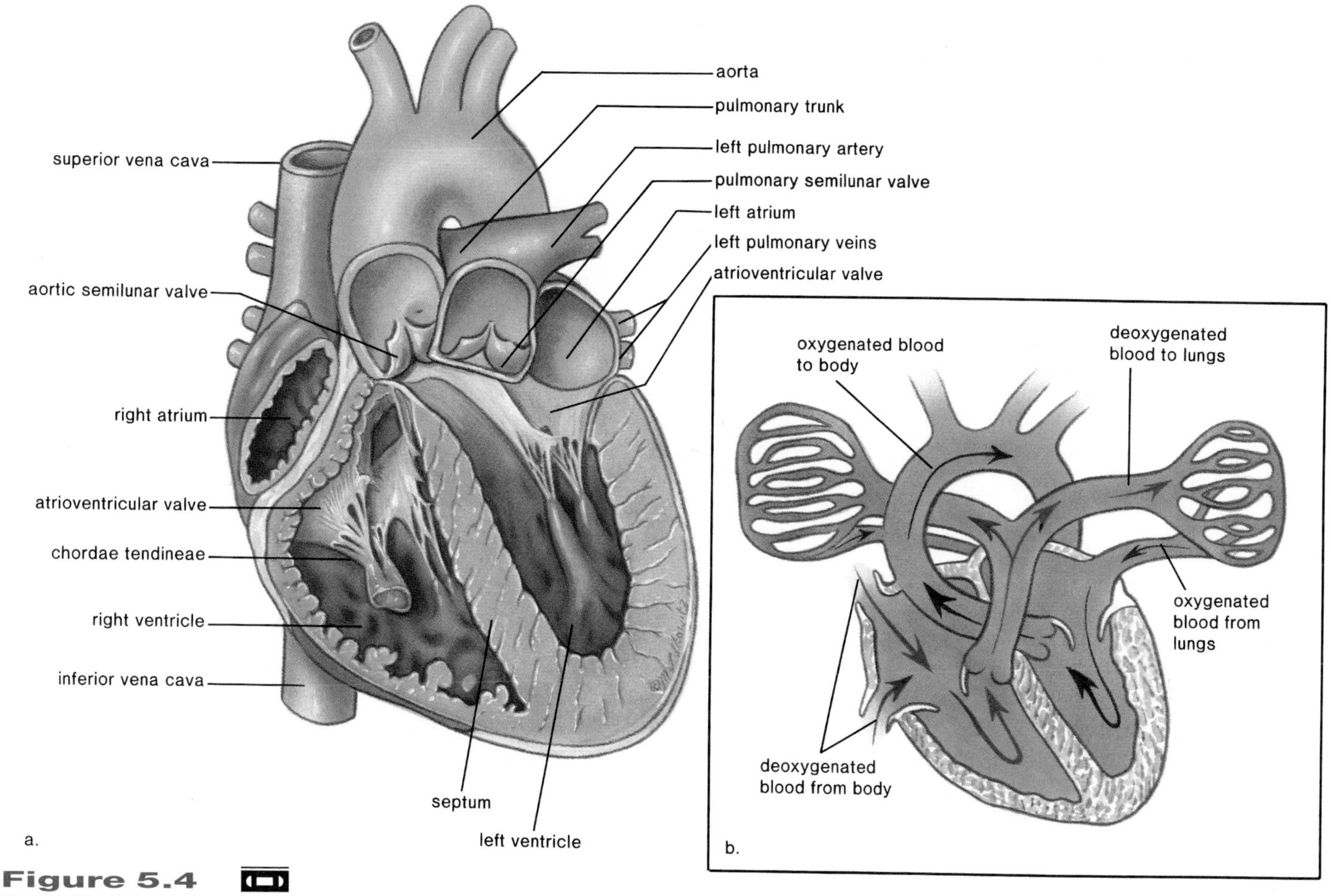

Figure 5.4

Internal view of the heart. **a.** The right side of the heart contains deoxygenated blood. The venae cavae empty into the right atrium, and the pulmonary trunk leaves the right ventricle. The left side of the heart contains oxygenated blood. The pulmonary veins enter the left atrium, and the aorta leaves from the left ventricle. **b.** This diagrammatic representation of the heart allows you to trace the path of the blood. On the right side of the heart: venae cavae, right atrium, right ventricle, pulmonary arteries to lungs. On the left side of the heart: pulmonary veins, left atrium, left ventricle, aorta to body. Restate this and put in the names of the valves where appropriate.

Path of Blood Through the Heart

We can trace the path of blood through the heart (fig. 5.4) in the following manner:

The superior (anterior) **vena cava** and the inferior (posterior) vena cava, both carrying deoxygenated blood (low in oxygen and high in carbon dioxide), enter the right atrium.

The right atrium sends blood through an atrioventricular valve (the tricuspid valve) to the right ventricle.

The right ventricle sends blood through the pulmonary semilunar valve into the pulmonary trunk and the **pulmonary arteries** to the lungs.

The **pulmonary veins**, carrying oxygenated blood (high in oxygen and low in carbon dioxide) from the lungs, enter the left atrium.

The left atrium sends blood through an atrioventricular valve (the bicuspid or mitral valve) to the left ventricle.

The left ventricle sends blood through the aortic semilunar valve into the **aorta** to the body proper.

From this description, you can see that deoxygenated blood never mixes with oxygenated blood and that blood must go through the lungs in order to pass from the right side to the left side of the heart. In fact, the heart is a *double pump* because the right side of the heart sends blood through the lungs, and the left side sends blood throughout the body. Since the left ventricle has the harder job of pumping blood to the entire body, its walls are thicker than those of the right ventricle, which pumps blood to the lungs.

> The right side of the heart pumps blood to the lungs, and the left side of the heart pumps blood throughout the body.

Heartbeat: Every 0.85 Seconds

Each heartbeat is called a cardiac cycle (fig. 5.5). First, the 2 atria contract at the same time; then the 2 ventricles contract at the same time. Then all chambers relax. The word **systole** refers to contraction of heart muscle, and the word **diastole** refers to relaxation of heart muscle. The heart contracts, or beats, about 70 times a minute, and each heartbeat lasts about 0.85 sec. A normal adult rate can vary from 60 to 100 beats per minute.

When the heart beats, the familiar lub-dub sound occurs as the valves of the heart close. The lub is caused by vibrations occurring when the atrioventricular valves close and the ventricles contract. The dub is heard when the semilunar valves close. A heart murmur, or a slight slush sound after the lub, is often due to ineffective valves, which allow blood to pass back into the atria after the atrioventricular valves have closed. Rheumatic fever resulting from a bacterial infection is one cause of a faulty valve, particularly the bicuspid valve. If operative procedures are unable to open and/or restructure the valve, it can be replaced with an artificial valve.

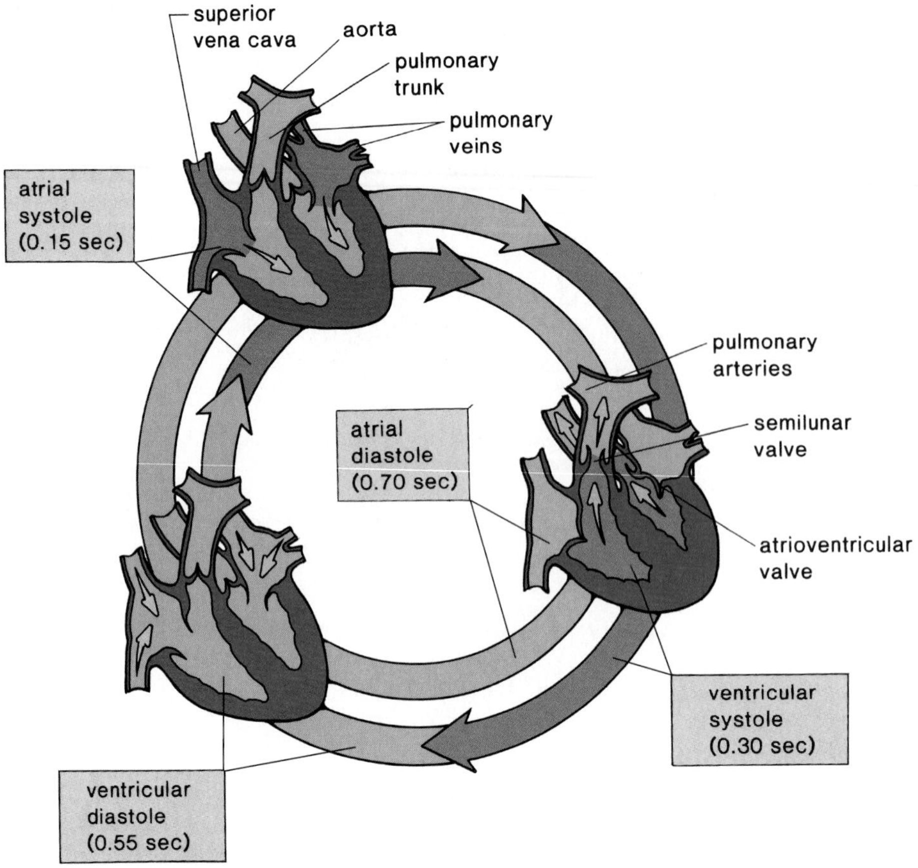

Figure 5.5

Cardiac cycle at 70 beats per minute. During systole the 4 chambers of the heart contract; the atria contract for 0.15 sec and the ventricles contract for 0.30 sec. During diastole the chambers rest. Note when the semilunar and atrioventricular valves are open or closed.

The surge of blood entering the arteries causes their elastic walls to stretch, but then they almost immediately recoil. This alternating expansion and recoil of an arterial wall can be felt as a **pulse** in any artery that runs close to the body's surface. It is customary to feel the pulse by placing several fingers on a radial artery, which lies near the outer border of the palm side of the wrist. A carotid artery, on either side of the trachea in the neck, is another accessible location to feel the pulse. Normally, the pulse rate indicates the rate of the heartbeat because the arterial walls pulse whenever the left ventricle contracts.

Conduction System: Intrinsic Control of Heartbeat

Nodal tissue is a unique type of tissue located in 2 regions of the heart, which has both muscular and nervous characteristics. The **SA** (sinoatrial) **node** is found in the upper dorsal wall of the right atrium; the other, the **AV** (atrioventricular) **node,** is found in the base of the right atrium very near the septum (fig. 5.6*a*). The SA node initiates the heartbeat and automatically sends out an excitation impulse every 0.85 sec; this causes the atria to contract. When the impulse reaches the AV node, the AV node signals the ventricles to contract by way of large fibers terminating in the more numerous and smaller Purkinje fibers. The SA node is called the **pacemaker** because it usually keeps the heartbeat regular. If the SA node fails to work properly, the heart still beats, but slower (40 to 60 beats per minute). To correct this condition, it is possible to implant an artificial pacemaker, which automatically gives an electric shock to the heart every 0.85 sec. The heart then beats regularly again.

With the contraction of any muscle, including the myocardium, ionic changes occur; these can be detected with electrical recording devices. The pattern that results, called an **electrocardiogram** (ECG or EKG) (fig. 5.6*b*), has an atrial phase and a ventricular phase. The first wave in the electrocardiogram, called the *P* wave, represents the excitation and contraction of the atria. The second wave, or the *QRS* wave, occurs just prior to ventricular contraction. The third, or *T,* wave is caused by the recovery of the ventricles. An examination of the electrocardiogram indicates whether the heartbeat has a normal or an irregular pattern.

Ventricular fibrillation caused by uncoordinated contraction of the ventricles is of special interest because it can be caused by an injury or drug overdose. It is the most common cause of sudden cardiac death in a seemingly healthy person. Once the ventricles are fibrillating, they have to be defibrillated by applying a strong electric current for a short period of time. Then the SA node may be able to reestablish a coordinated beat.

> The conduction system of the heart includes the SA node, the AV node, and the Purkinje fibers. With an ECG, it is possible to determine if the conduction system, and therefore the beat of the heart, is regular.

Nervous Control of the Heartbeat: Extrinsic Control

The rate of the heartbeat is also under nervous control. A cardiac center in the medulla oblongata (p. 195) of the brain can alter the beat of the heart by way of the *autonomic system* (p. 193). This system has 2 divisions: the *parasympathetic system,* which promotes those functions we tend to associate with normal activities, and the *sympathetic system,* which brings about those responses we associate with times of stress. For example, the parasympathetic system causes the heartbeat to slow down, and the sympathetic system causes the heartbeat to speed up and produce a stronger beat. Various factors, such as the relative need for oxygen or blood pressure, determine which of these systems is activated.

> The heart rate is regulated largely by the autonomic nervous system.

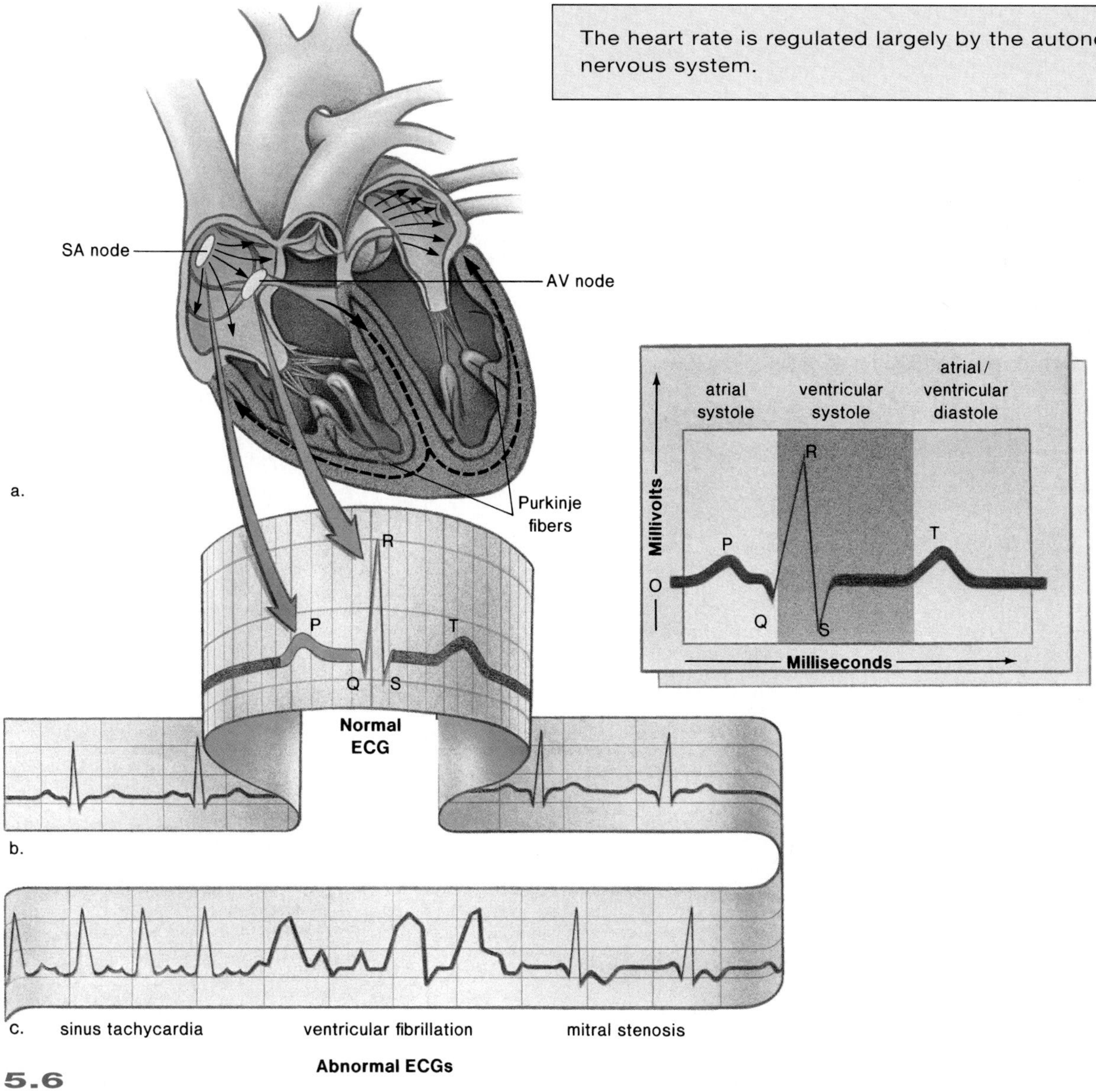

Figure 5.6

Control of the cardiac cycle. **a.** The SA node sends out a stimulus, which causes the atria to contract. When this stimulus reaches the AV node, it signals the ventricles to contract by way of the Purkinje fibers. **b.** A normal ECG indicates that the heart is functioning properly. The *P* wave occurs just prior to atrial contraction; the *QRS* wave occurs just prior to ventricular contraction; and the *T* wave occurs when the ventricles are recovering from contraction. **c.** Abnormal ECGs: sinus tachycardia is an abnormally fast heartbeat due to a fast pacemaker; ventricular fibrillation is an irregular heartbeat due to irregular stimulation of the ventricles; and mitral stenosis occurs because the bicuspid (mitral) valve is obstructed.

Blood Pressure and Swiftness of Flow

Blood pressure is the pressure of blood against the wall of a blood vessel. A sphygmomanometer is used to measure blood pressure, as described in figure 5.7. The highest arterial pressure, called the *systolic pressure,* is reached during ejection of blood from the heart. The lowest arterial pressure is called the *diastolic pressure.* Diastolic pressure occurs while the heart ventricles are relaxing. Normal resting blood pressure for a young adult is said to be 120 mm of mercury (Hg) over 80 mm, or simply 120/80. The higher number is the systolic pressure, and the lower number is the diastolic pressure. Actually, 120/80 is the expected blood pressure in the brachial artery of the arm; blood pressure decreases with distance from the left ventricle (fig. 5.8). Blood pressure is, therefore, higher in the arteries than in the arterioles. Further, there is a sharp drop in blood pressure when the arterioles reach the capillaries. The decrease can be correlated with the increase in the total cross-sectional area of the vessels as blood moves through arteries, arterioles, and then into capillaries. There are more arterioles than arteries and many more capillaries than arterioles.

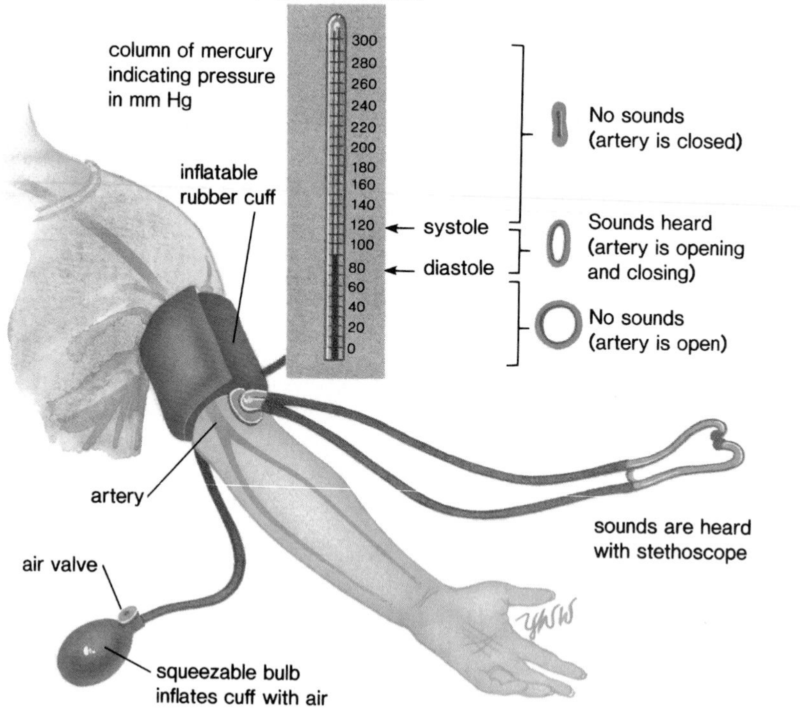

Figure 5.7

Determination of blood pressure using a sphygmomanometer. The technician inflates the cuff with air, gradually reduces the pressure, and listens with a stethoscope for the sounds that indicate blood is moving past the cuff in an artery. This is systolic blood pressure. The pressure in the cuff is further reduced until no sound is heard, indicating that blood is flowing freely through the artery. This is diastolic pressure.

The velocity of blood flow varies in different parts of the circulatory system. Blood pressure accounts for the velocity of the blood flow in the arterial system and therefore, as blood pressure decreases due to the increased cross-sectional area of the arterial system, so does velocity. Blood moves more slowly through the capillaries than it does through the aorta. This is important because the slow progress allows time for the exchange of molecules between blood in the capillaries and the surrounding tissues.

Blood pressure cannot account for the movement of blood through the venules and the veins since they lie on the other side of the capillaries. Instead, movement of blood through the venous system is due to skeletal muscle contraction. When the skeletal muscles contract, they put pressure against the weak walls of the veins. This causes blood

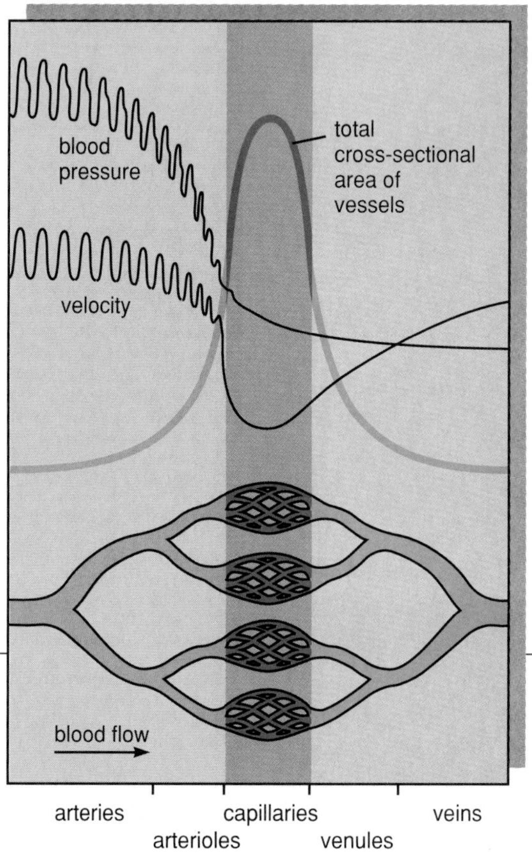

Figure 5.8

Diagram illustrating how velocity and blood pressure are related to the total cross-sectional area of blood vessels. Capillaries have the greatest cross-sectional area and blood is under the least pressure and has the least velocity. Skeletal muscle contraction, not blood pressure, accounts for the velocity of blood in the veins.

to move past a *valve* (fig. 5.9). Once past the valve, blood cannot return. The importance of muscle contraction in moving blood in the venous system can be demonstrated by forcing a person to stand rigidly still for a number of hours. Frequently, fainting occurs because blood collects in the limbs, robbing the brain of oxygen. In this case, fainting is beneficial because the resulting horizontal position aids in getting blood to the head.

Blood flow gradually increases in the venous system (see fig. 5.8) due to a progressive reduction in the cross-sectional area as small venules join to form veins. The 2 venae cavae together have a cross-sectional area only about double that of the aorta. The blood pressure is lowered in the thoracic cavity whenever the chest expands during inspiration. This also aids the flow of venous blood into the thoracic cavity because blood flows in the direction of reduced pressure.

> Blood pressure accounts for the flow of blood in the arteries and the arterioles; skeletal muscle contraction accounts for the flow of blood in the venules and the veins.

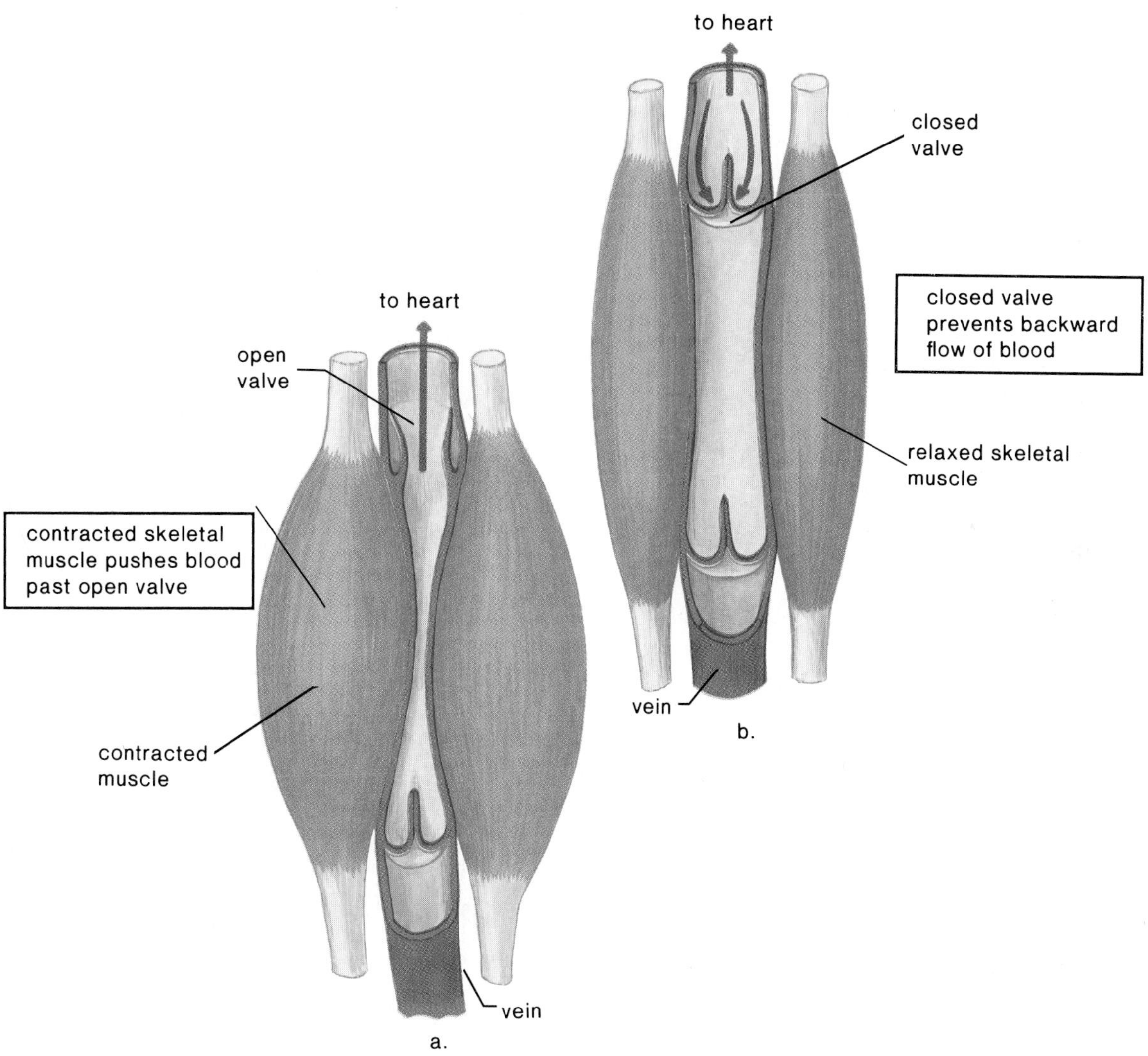

Figure 5.9

Skeletal muscle contraction moves blood in veins. **a.** Muscle contraction exerts pressure against the vein, and blood moves past the valve. **b.** Blood cannot flow back once it has moved past the valve.

Vascular Pathways

The cardiovascular system, which is represented in figure 5.10, includes 2 circuits: the **pulmonary circuit,** which circulates blood through the lungs, and the **systemic circuit,** which serves the needs of body tissues.

Pulmonary Circuit: Through the Lungs

The path of blood through the lungs can be traced as follows. Blood from all regions of the body first collects in the right atrium and then passes into the right ventricle, which pumps it into the pulmonary trunk. The pulmonary trunk divides into the right and left *pulmonary arteries,* which branch as they approach the lungs. The arterioles take blood to the pulmonary capillaries, where carbon dioxide is given off and oxygen is picked up. Blood then enters the pulmonary venules, which lead back through the *pulmonary veins* to the left atrium. Since blood in the pulmonary arteries is deoxygenated but blood in the pulmonary veins is oxygenated, it is not correct to say that all arteries carry oxygenated blood and all veins carry deoxygenated blood. It is just the reverse in the pulmonary circuit.

> The pulmonary arteries take deoxygenated blood to the lungs, and the pulmonary veins return oxygenated blood to the heart.

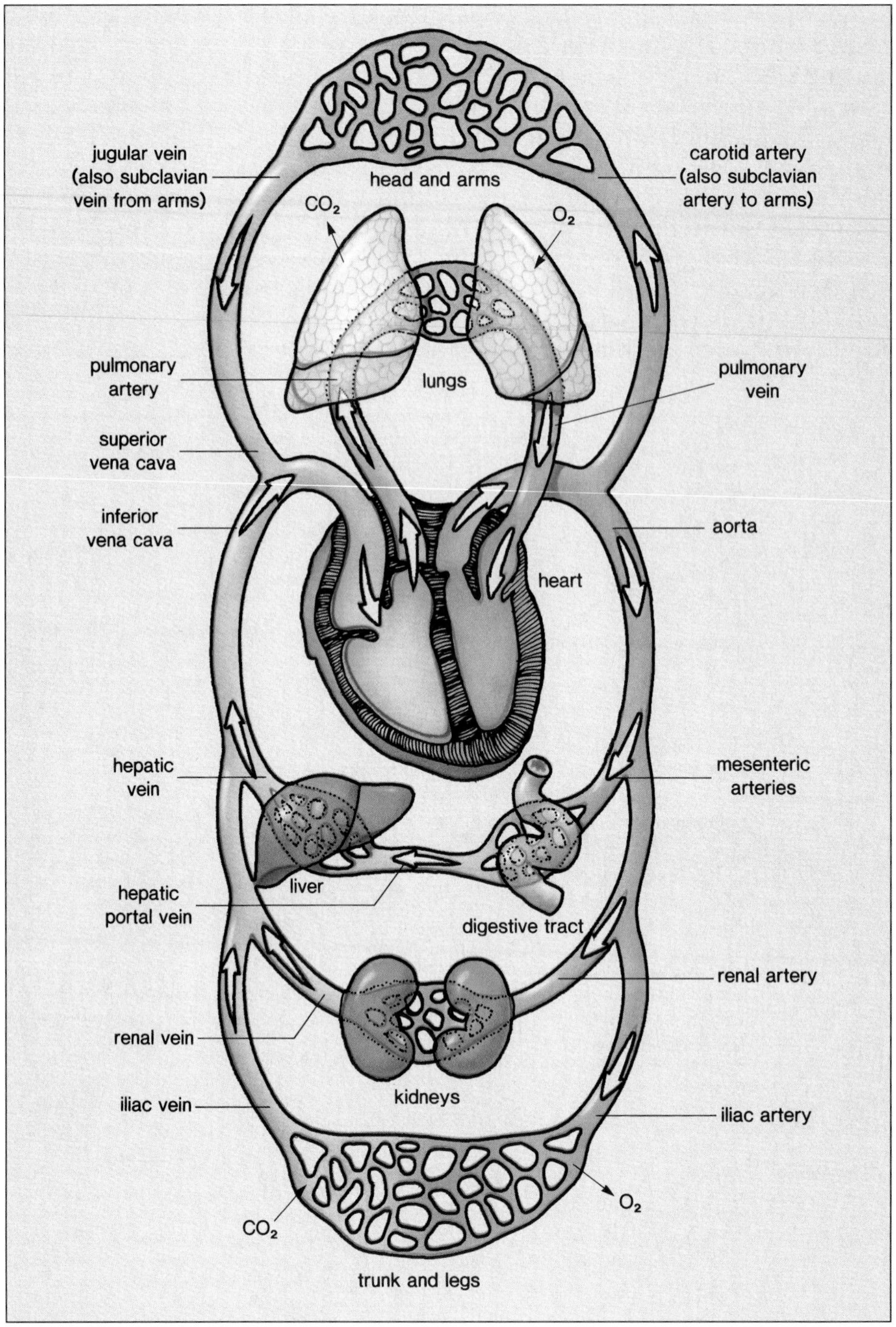

Figure 5.10

Cardiovascular system. The blue-colored vessels carry deoxygenated blood, and the red-colored vessels carry oxygenated blood; the arrows indicate the flow of blood. Compare this diagram, useful for learning to trace the path of blood, to figure 5.11 to realize that both arteries and veins go to all parts of the body. Also, there are capillaries in all parts of the body. No cell is located far from a capillary.

Systemic Circuit: Serving the Body

The systemic circuit includes all of the other arteries and veins shown in figures 5.10 and 5.11. The largest artery in the systemic circuit is the *aorta,* and the largest veins are the *superior* and *inferior venae cavae.* The superior vena cava collects blood from the head, the chest, and the arms, and the inferior vena cava collects blood from the lower body regions. Both enter the right atrium. The aorta and the venae cavae serve as the major pathways for blood in the systemic circuit.

The path of systemic blood to any organ in the body begins in the left ventricle, which pumps blood into the aorta. Branches from the aorta go to the major body regions and

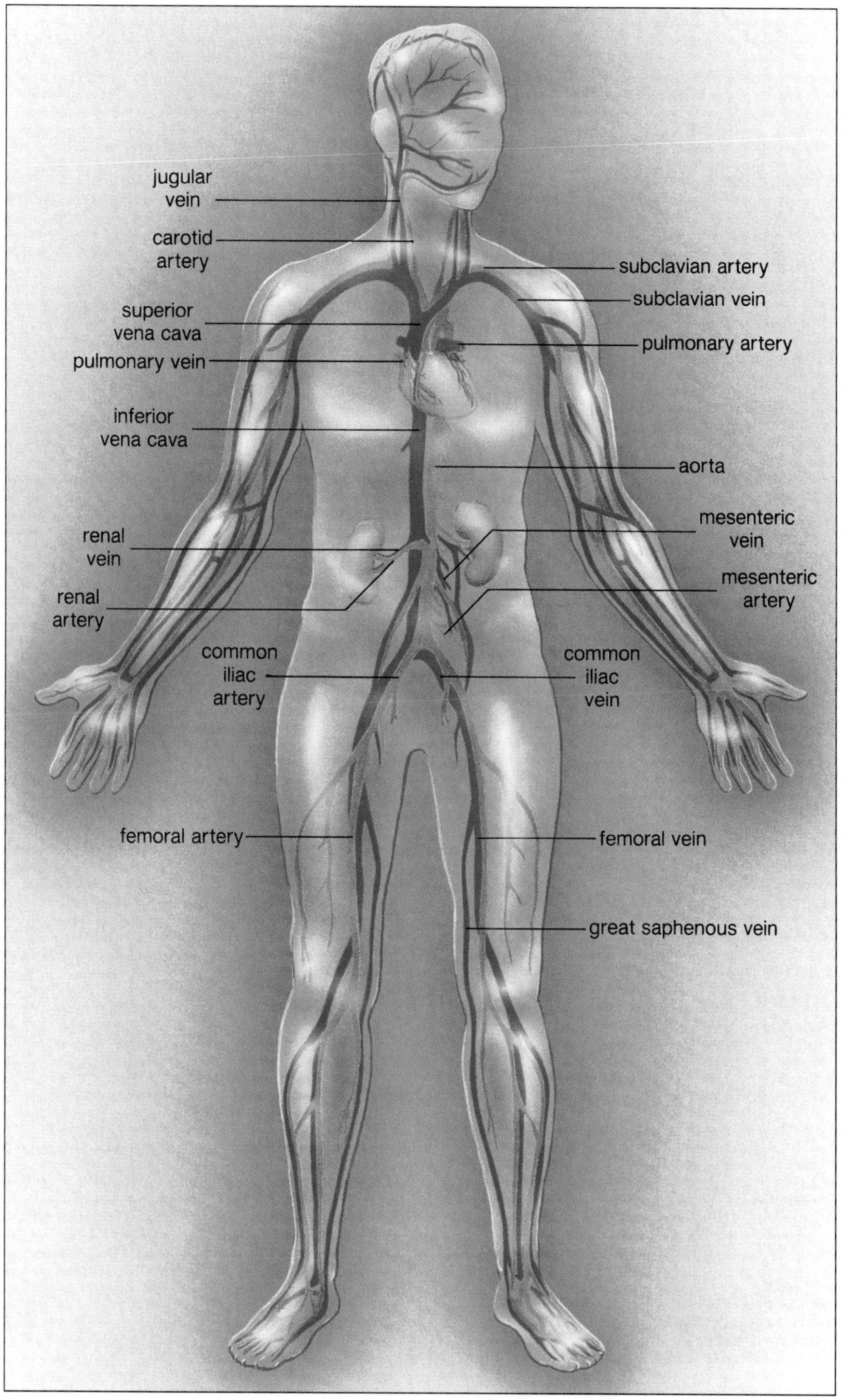

Figure 5.11

Human circulatory system. A more realistic representation of major blood vessels in the body shows that arteries and veins go to all parts of the body. The superior and inferior venae cavae take their names from their relationship to which organ?

organs. For example, the path of blood to the kidneys can be traced as follows:

> left ventricle—aorta—renal artery—renal arterioles, capillaries, venules—renal vein—inferior vena cava—right atrium

To trace the path of blood to any organ in the body, you need only mention the aorta, the proper branch of the aorta, the organ, and the vein returning blood to the vena cava. In most instances, the artery and the vein that serve the same organ are given the same name (fig. 5.11). In the systemic circuit, unlike the pulmonary system, arteries contain oxygenated blood and have a bright red color, but veins contain deoxygenated blood and appear a dark purplish color.

The **coronary arteries** (see fig. 5.3*b*), which are a part of the systemic circuit, are extremely important because they serve the heart muscle itself. (The heart is not nourished by the blood in its chambers.) The coronary arteries are the first branch off the aorta. They originate just above the aortic semilunar valve and they lie on the exterior surface of the heart, where they divide into diverse arterioles. The coronary capillary beds join to form venules. The venules converge to form the cardiac veins, which empty into the right atrium. The coronary arteries have a very small diameter and may become blocked, as discussed on page 103.

The body has a portal system called the *hepatic portal system,* which is associated with the liver. A portal system begins and ends in capillaries; in this instance, the first set of capillaries occurs at the villi of the small intestine and the second occurs in the liver. Blood passes from the capillaries of the intestinal villi into venules that join to form the *hepatic portal vein,* a vessel that connects the villi of the intestine with the liver. The *hepatic vein* leaves the liver and enters the inferior vena cava (see fig. 4.10).

While figure 5.10 is helpful in tracing the path of blood, remember that all parts of the body receive both arteries and veins, as illustrated in figure 5.11.

> The systemic circuit takes blood from the left ventricle of the heart to the right atrium of the heart. It serves the body proper.

Health Focus

Prevention of Cardiovascular Disease

All of us can take steps to prevent the occurrence of cardiovascular disease, the most frequent cause of death in the United States. There are genetic factors that predispose an individual to cardiovascular disease, such as family history of heart attack under age 55, male gender, and ethnicity (African Americans are at greater risk). Those with one or more of these risk factors need not despair, however. It only means that they need to pay particular attention to these guidelines for a heart-healthy life-style.

THE DON'TS

Smoking

Hypertension is well recognized as a major contributor to cardiovascular disease. When a person smokes, the drug nicotine, present in cigarette smoke, enters the bloodstream. Nicotine causes the arterioles to constrict and the blood pressure to rise. Restricted blood flow and cold hands are associated with smoking by most people. More serious is the need for the heart to pump harder to propel the blood through the lungs at a time when the oxygen-carrying capacity of the blood is reduced. (How smoking reduces the oxygen-carrying capacity is explained in the Ecology Focus for chapter 6.)

Obesity

Hypertension also occurs more often in persons who are obese—those who are more than 20% above recommended weight. More tissues require servicing, and the heart sends the extra blood out under greater pressure in those who are overweight. Since it is very difficult for obese individuals to lose weight, it is recommended that weight control be a lifelong endeavor. Even a slight decrease in weight can bring with it a reduction in hypertension. A 4.5-kilogram weight loss doubles the chance that blood pressure can be normalized without drugs.

THE DO'S

Healthy Diet

It was once thought that a low-salt diet was protective against cardiovascular disease, and it still may be in certain persons. Theoretically, hypertension occurs because the more salty the blood the greater the osmotic pressure and the higher the water content. Ways to reduce salt in the diet are given in table 4.8. In recent years, the emphasis has switched to a diet low in saturated fats and cholesterol as protective against cardiovascular disease. Ways to reduce fat and cholesterol in the diet are given in table 4.5. Cholesterol is ferried in the blood by 2 types of plasma proteins called LDL (low-density lipoprotein) and HDL (high-density lipoprotein). LDL (called "bad" lipoprotein) takes cholesterol to the tissues from the liver, and HDL (called "good" lipoprotein) transports cholesterol out of the tissues to the liver. When the LDL level in blood is abnormally high or the HDL level is abnormally low, cholesterol accumulates in the cells. When cholesterol-laden cells line the arteries, plaque develops, which interferes with circulation (fig. 5A).

The National Heart, Lung, and Blood Institute recommends that everyone know his or her blood cholesterol level. Individuals with a high blood cholesterol level (240 mg/100 ml) should be further tested to determine what their LDL blood cholesterol level is. If the LDL blood cholesterol level is 160 mg/100 ml or higher and/or if the total-to-HDL cholesterol ratio is higher than 4.5, the person is considered at risk.

If diet alone does not improve the blood cholesterol levels and the LDL versus HDL levels, then other measures are needed. Aside from drugs, exercise is sometimes effective.

Exercise

Those who exercise are less apt to have cardiovascular disease. One study found that moderately active men who spent an average of 48 minutes a day on a leisure-time activity such as gardening, bowling, or dancing had one-third fewer heart attacks than peers who spent an average of only 16 minutes each day. Exercise helps to keep weight under control, reduces hypertension, and may help minimize stress. One physician recommends that his cardiovascular patients walk for one hour, 3 times a week, and in addition they are to practice meditation and yogalike stretching and breathing exercises to reduce stress.

Circulatory Disorders

Cardiovascular disease (CVD) is the leading cause of untimely death in the Western countries. Modern research efforts have resulted in improved diagnosis, treatment, and prevention. This section discusses the range of advances that have been made in these areas. The Health Focus for this chapter emphasizes how to prevent CVD from developing in the first place.

Cardiovascular disease is the number-one killer in the Western world.

Hypertension: Silent Killer

It is estimated that about 20% of all Americans suffer from *hypertension,* which is high blood pressure indicated by a blood pressure reading. Women of any age are considered to have hypertension if their blood pressure reading is 160/95 or above. For a man under age 45, a reading above 130/90 is hypertensive, and beyond age 45, a reading above 140/95 is considered hypertensive. While both systolic and diastolic pressures are considered important, it is the diastolic pressure that is emphasized when medical treatment is being considered.

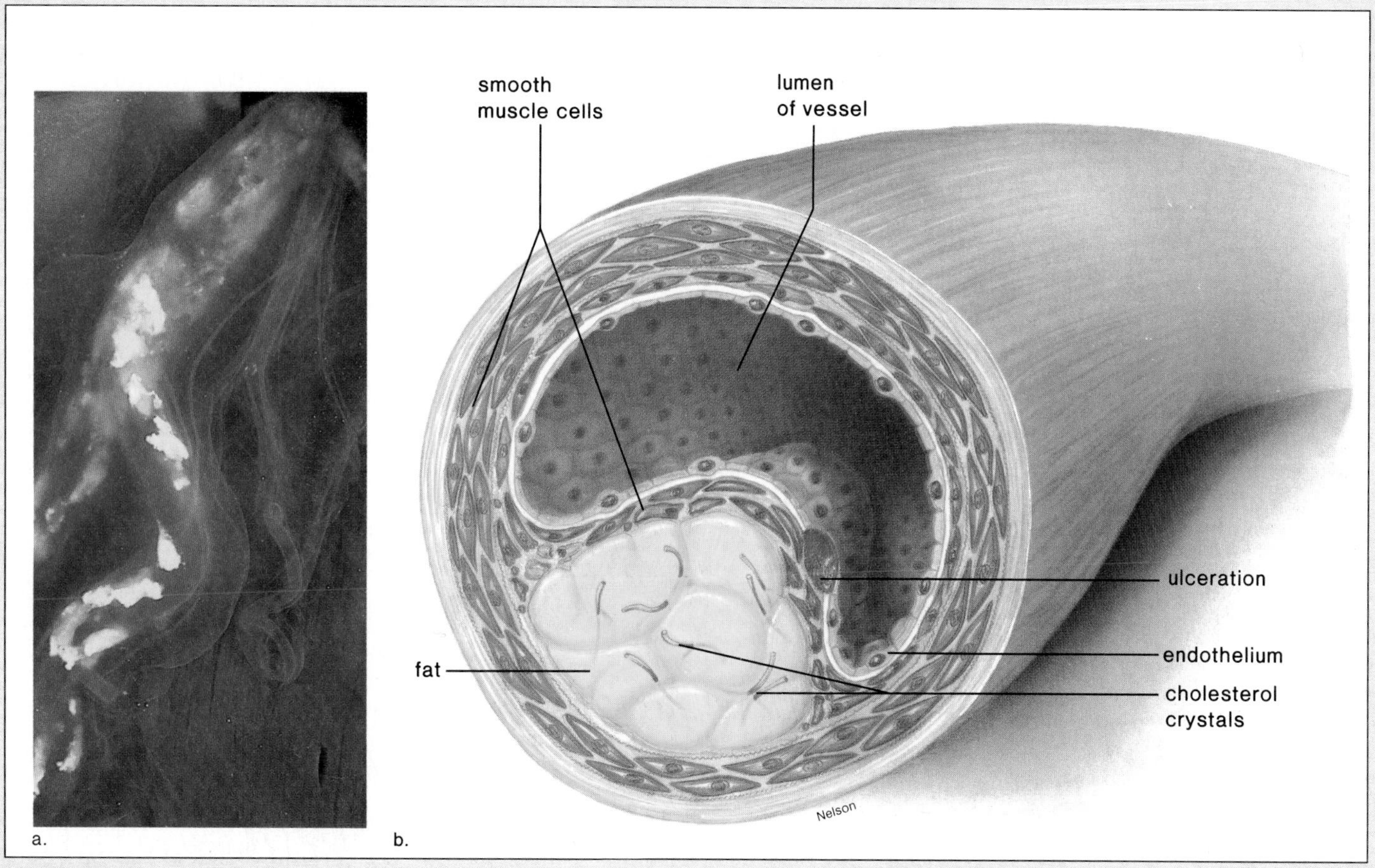

Figure 5A

Plaque. **a.** Plaque (yellow) in the coronary artery of a heart patient. **b.** Cross section of plaque shows its composition and indicates how it bulges out into the lumen of an artery, obstructing blood flow.

Hypertension is sometimes called a silent killer because it may not be detected until a stroke or heart attack occurs. Therefore, it is important to have regular blood pressure checks and to adopt a life-style that protects against the development of hypertension. The Health Focus for this chapter includes the good health habits that lower the risk of cardiovascular disorders.

Atherosclerosis: Fatty Arteries

Hypertension also is seen in individuals who have *atherosclerosis* (formerly called arteriosclerosis), an accumulation of soft masses of fatty materials, particularly cholesterol, beneath the inner linings of arteries. Such deposits are called *plaque,* and as it develops, plaque tends to protrude into the vessel and interfere with the flow of blood. Atherosclerosis begins in early adulthood and develops progressively through middle age, but symptoms may not appear until an individual is 50 or older. To prevent its onset and development, the American Heart Association and other organizations recommend a diet low in saturated fat and cholesterol, as also discussed in the Health Focus for this chapter.

Plaque can cause a clot to form on the irregular arterial wall. As long as the clot remains stationary, it is called a *thrombus*, but when and if it dislodges and moves along with the blood, it is called an *embolus*. If *thromboembolism* is not treated, complications can arise, as mentioned in the following section.

> Development of atherosclerosis, which is associated with a high blood cholesterol level, can lead to thromboembolism.

Stroke and Heart Attack: Lack of Oxygen

Both strokes and heart attacks are associated with hypertension and atherosclerosis. A cardiovascular accident (CVA), also called a *stroke*, occurs when a portion of the brain dies due to a lack of oxygen. A stroke, characterized by paralysis or death, often results when a small arteriole bursts or is blocked by an embolus. A person sometimes is forewarned of a stroke by a feeling of numbness in the hands or the face, difficulty in speaking, or temporary blindness in one eye. A myocardial infarction (MI), also called a *heart attack*, occurs when a portion of the heart muscle dies because of a lack of oxygen. Due to atherosclerosis, the coronary artery may be partially blocked. The individual may then suffer from *angina pectoris*, characterized by a radiating pain in the left arm. Nitroglycerin or related drugs dilate blood vessels and help relieve the pain. When a coronary artery is completely blocked, perhaps because of thromboembolism, a heart attack occurs.

> Stroke and heart attack are associated with both hypertension and atherosclerosis.

Dissolving Blood Clots

Medical treatment for thromboembolism includes 2 drugs that can be given intravenously to dissolve a clot: streptokinase, normally produced by bacteria, and tPA, which is genetically engineered (see p. 391). Both drugs convert plasminogen, a molecule found in blood, into plasmin, an enzyme that dissolves blood clots. In fact, tPA, which stands for *tissue plasminogen activator*, is the body's own way of converting plasminogen to plasmin. Streptokinase and tPA are used particularly when it is known that a clot is present, but they must be used quickly to prevent permanent heart damage.

If a person has symptoms of angina or a thrombolytic stroke, then an anticoagulant drug, such as aspirin, may be given. Aspirin reduces the stickiness of platelets and therefore lowers the probability that a clot will form. There is evidence that aspirin protects against first heart attacks, but there is no clear support for taking aspirin every day to prevent strokes in symptom-free people. Physicians warn that long-term use of aspirin might have harmful effects, including bleeding in the brain.

Clearing Clogged Arteries

Surgical procedures are available to clear clogged arteries. In *angioplasty*, a cardiologist threads a plastic tube into an artery of an arm or a leg and guides it through a major blood vessel toward the heart. When the tube reaches the region of plaque in a coronary artery (see fig. 5.3), a balloon attached to the end of the tube is inflated, forcing the vessel open. The problem with this procedure is the vessel may not remain open, and worse, it may cause clots to form. Various alternatives, including a laser technique, are available.

Each year thousands of persons have *coronary bypass* surgery. During this operation, surgeons take a segment of another blood vessel from the patient's body and stitch one end to the aorta and the other end to a coronary artery past the point of obstruction (fig. 5.12).

Once the heart is exposed, some physicians use lasers to open up clogged coronary vessels. Presently, this technique is used in conjunction with coronary bypass operations, but eventually it may be possible to use lasers independently, without opening the thoracic cavity.

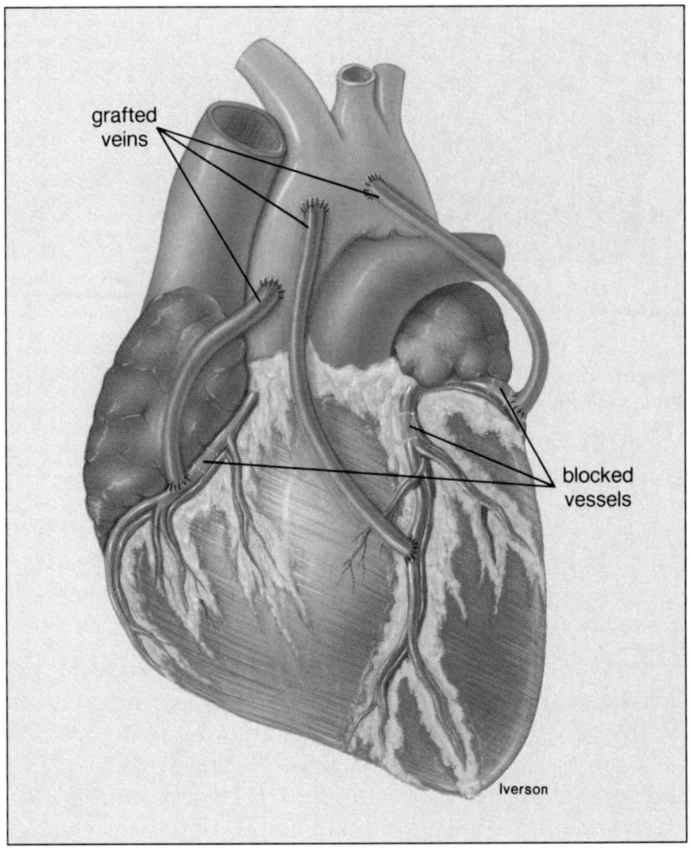

Figure 5.12

Coronary bypass operation. During this operation, the surgeon grafts segments of another vessel, usually a vein, between the aorta and the coronary vessels, bypassing areas of blockage. Patients who are ill enough to require surgery often receive 2 or 3 bypasses in a single operation.

Donated and Artificial Hearts

Persons with weakened hearts eventually may suffer from *congestive heart failure,* meaning the heart no longer is able to pump blood adequately and blood backs up in the heart and lungs. These individuals, depending on their age, are candidates for a donor heart transplant. The difficulties with a donor heart transplant are, first, one of availability and, second, the tendency of the body to reject foreign organs.

Sometimes, it is possible to repair the heart instead of replacing it. For example, a back muscle can be wrapped around a heart too weak to pump adequately. An artificial pacemaker causes the muscle to contract regularly and help pump the blood.

On December 2, 1982, Barney Clark became the first person to receive a Jarvik-7, an artificial heart. The Jarvik-7, no longer approved for experimental use in humans, is driven by bursts of air received from a large external machine. Today, investigators are experimenting with a totally internal electric heart. Radio signals transmit power from a portable battery pack through the skin to an orange-sized mechanical heart consisting of 2 motor-driven plastic sacs.

Veins: Dilated and Inflamed

Varicose veins are abnormal and irregular dilations in superficial (near the surface) veins, particularly those in the lower legs. Varicose veins in the rectum, however, are commonly called piles, or more properly, *hemorrhoids.* Varicose veins develop when the valves of the veins become weak and ineffective due to the backward pressure of blood. The problem can be aggravated when venous blood flow is obstructed by crossing the legs or by sitting in a chair so that its edge presses against the back of the knees.

Phlebitis, or inflammation of a vein, is a more serious condition, particularly when a deep vein is involved. Blood in the inflamed vessel may clot, in which case thromboembolism occurs. An embolus that originates in a systemic vein eventually may come to rest in a pulmonary arteriole, blocking circulation through the lungs. This condition, termed *pulmonary embolism,* can result in death.

Summary

Blood vessels include arteries (and arterioles), which carry blood away from the heart; capillaries, where exchange of molecules with the tissues occurs; and veins (and venules), which return blood to the heart.

The movement of blood in the circulatory system is dependent on the beat of the heart. During the cardiac cycle, the SA node (pacemaker) initiates the beat and causes the atria to contract. The AV node picks up the stimulus and initiates contraction of the ventricles. The heart sounds, lub-dub, are due to the closing of the atrioventricular valves (and ventricular contraction), followed by the closing of the semilunar valves.

Blood pressure accounts for the flow of blood in the arteries, but because blood pressure drops off after the capillaries, it cannot cause blood flow in the veins. Skeletal muscle contraction pushes blood past a venous valve, which then shuts, preventing backward flow. The velocity of blood flow is slowest in the capillaries, where exchange of nutrient and waste molecules takes place.

The circulatory system is divided into the pulmonary circuit and the systemic circuit. In the pulmonary circuit,

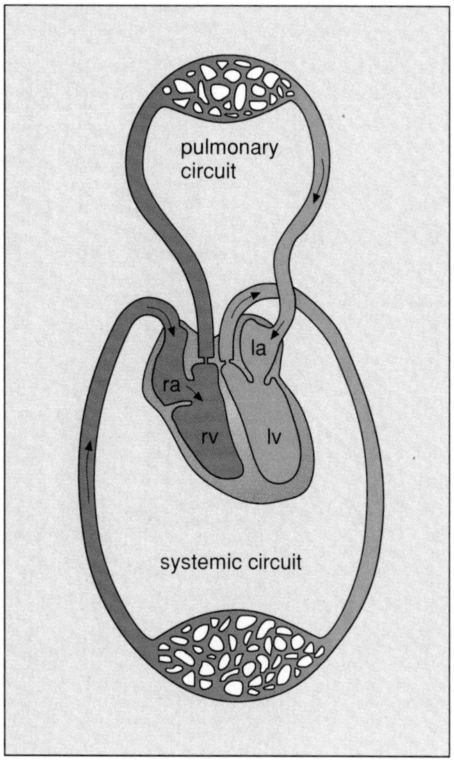

the pulmonary artery takes blood from the right ventricle to the lungs, and the pulmonary veins return it to the left atrium. To trace the path of blood in the systemic circuit, start with the aorta from the left ventricle. Follow its path until it branches to an artery going to a specific organ. It can be assumed that the artery divides into arterioles and capillaries and that the capillaries lead to venules. The vein that takes blood to the vena cava most likely has the same name as the artery. In the adult systemic circuit, but not in the pulmonary circuit, the arteries carry oxygenated blood and the veins carry deoxygenated blood.

Hypertension and atherosclerosis are 2 circulatory disorders that lead to heart attack and to stroke. Medical and surgical procedures are available to control cardiovascular disease, but the best policy is prevention by following a heart-healthy diet, getting regular exercise, maintaining a proper weight, and not smoking cigarettes.

Study and Thought Questions

1. What types of blood vessels are there? Discuss their structure and function. (p. 92)
2. Trace the path of blood in the heart, mentioning the vessels attached to and the valves within the heart. (p. 95)
3. Describe the cardiac cycle (using the terms *systole* and *diastole*), and explain the heart sounds. (p. 96)
4. Describe the cardiac conduction system and an ECG. Tell how an ECG is related to the cardiac cycle. (p. 96)
5. Trace the path of blood in the pulmonary circuit as it travels from and returns to the heart. (p. 100)

   Critical Thinking During fetal development, a blood vessel connects the pulmonary trunk to the aorta. Why is this arrangement appropriate?
6. Trace the path of blood from the mesenteric arteries to the aorta, indicating which of the vessels are in the systemic circuit and which are in the pulmonary circuit. (p. 101)
7. What is blood pressure, and where in the body is the average normal arterial blood pressure about 120/80? (p. 98)
8. In which type of vessel is blood pressure highest? lowest? Velocity is lowest in which type of vessel, and why is it lowest? Why is this beneficial? What factors assist venous return of the blood? (p. 98)
9. What is atherosclerosis? (p. 103) Name 2 illnesses associated with hypertension and thromboembolism. (p. 104)

   Ethical Issue Apparently, aspirin is helpful to individuals who already have heart disease; others should be cautious about taking aspirin regularly. Should the government regulate aspirin advertisers or must consumers beware?
10. Discuss the medical and surgical treatment of cardiovascular disease. (p. 104)

Writing Across the Curriculum

1. During fetal development, a blood vessel connects the pulmonary trunk to the aorta. What effect does this have on circulation to the lung? Under these circumstances, would the aorta carry fully oxygenated blood?
2. Assume that emboli are most apt to lodge in a capillary. If an embolus has formed in the carotid artery, in which organ would you expect it to lodge? Where would you expect it to lodge if it has formed in the hepatic portal vein?
3. Heart disease seems to run in families. One day it may be possible to determine who is genetically predisposed to heart disease. Should such information be a part of the public record or should it be confidential? Discuss the consequences of each position.

Objective Questions

1. Arteries are blood vessels that take blood ________________ from the heart.
2. When the left ventricle contracts, blood enters the ________________.
3. The pulmonary veins carry blood ______________ in oxygen.
4. The right side of the heart pumps blood to the ________________.
5. The __________ node is known as the pacemaker.
6. The arteries that serve the heart are the ________________ arteries.
7. The pressure of blood against the walls of a vessel is termed ________________.
8. Blood moves in the arteries due to __________ and in the veins due to ________________.
9. Varicose veins develop when ___________ become weak and ineffective.
10. Reducing the amount of _______________ and _______________ in the diet reduces the chances of plaque buildup in the arteries.
11. Label this diagram of the heart.

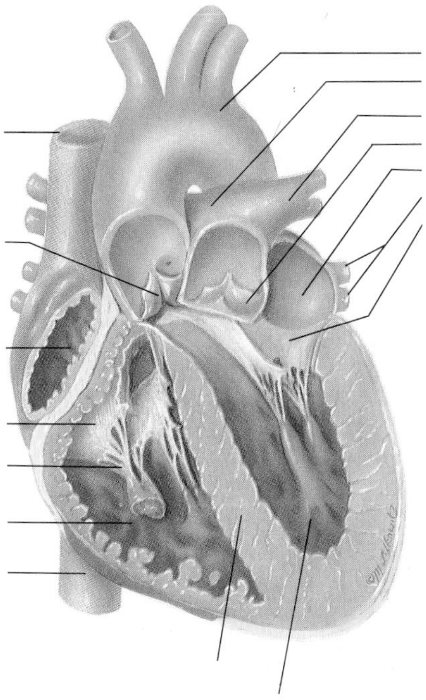

Selected Key Terms

aorta (ā-or´tah) Major systemic artery that receives blood from the left ventricle. 95

arteriole (ar-te´re-ōl) Vessel that takes blood from an artery to capillaries. 92

artery Vessel that takes blood away from the heart to arterioles; characteristically possessing thick elastic and muscular walls. 92

atrium (a´tre-um) Chamber; particularly an upper chamber of the heart lying above the ventricles; either the right atrium or the left atrium. 94

AV node A small region of neuromuscular tissue that transmits impulses received from the SA node to the ventric-ular walls. 96

capillary (kap´ĭ-lar″e) Microscopic vessel connecting arterioles to venules through the thin walls of which molecules either exit or enter the blood. 92

coronary artery Artery that supplies blood to the wall of the heart. 101

diastole (di-as´to-le) Relaxation of a heart chamber. 96

pulmonary circuit That part of the circulatory system that takes deoxygenated blood to and oxygenated blood away from the gas-exchanging surfaces in the lungs. 100

SA node Small region of neuromuscular tissue in the atria that initiates the heartbeat. Also called the pacemaker. 96

systemic circuit That part of the circulatory system that serves body parts other than the gas-exchanging surfaces in the lungs. 100

systole (sis´to-le) Contraction of a heart chamber. 96

valve Membranous extension of a vessel or the heart wall that opens and closes, ensuring one-way flow; common to the systemic veins, the lymphatic veins, and the heart. 93

vein Vessel that takes blood to the heart from venules; characteristically having nonelastic walls. 92

vena cava (ve'nah ka'vah) A large systemic vein that returns blood to the right atrium of the heart; either the superior or inferior vena cava. 95

ventricle Cavity in an organ, such as a lower chamber of the heart; either the right ventricle or the left ventricle; or the ventricles of the brain. 94

venule Vessel that takes blood from capillaries to a vein. 93

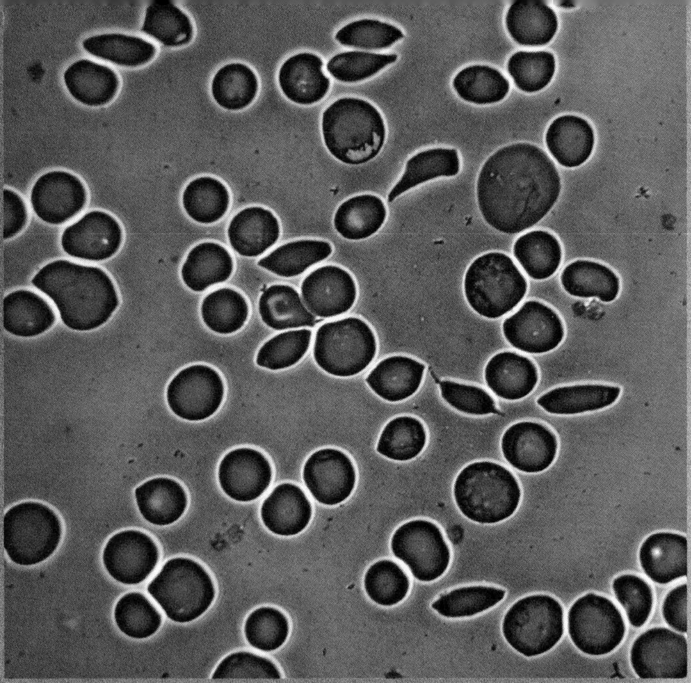

Chapter 6

Composition and Function of Blood

In persons with sickle-cell disease, the red blood cells aren't biconcave disks like normal red blood cells; they are irregular. In fact, many are sickle-shaped. The defect is caused by an abnormal hemoglobin that piles up inside the cells. Because the sickle-shaped cells can't pass along narrow capillary passageways like disk-shaped cells, they clog the vessels and break down. This is why persons with sickle-cell disease suffer from poor circulation, anemia, and poor resistance to infection. Internal hemorrhaging leads to further complications, such as jaundice, episodic pain of the abdomen and joints, and damage to internal organs. The importance of a normal structure and function of blood components is dramatically illustrated by sickle-cell disease.

Chapter Concepts

Chapter Outline

Formed Elements	Function and Description	Source
Red blood cells (erythrocytes)  4 million–6 million per mm³ blood	Transport O_2 and help transport CO_2 7–8 μm in diameter Bright-red to dark-purple bioconcave disks without nuclei	Bone marrow
White blood cells (leukocytes)	Fight infection	Bone marrow
Granular leukocytes		
 Basophil 20–50 per mm³ blood	10–12 μm in diameter Spherical cells with lobed nuclei; large, irregularly shaped, deep-blue granules in cytoplasm	
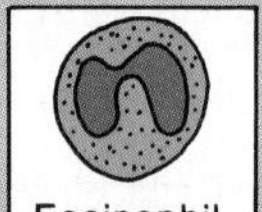 Eosinophil 100–400 per mm³ blood	10–14 μm in diameter Spherical cells with bilobed nuclei; coarse, deep-red, uniformly sized granules in cytoplasm	
 Neutrophil 3,000–7,000 per mm³ blood	10–14 μm in diameter Spherical cells with multilobed nuclei; fine, pink granules in cytoplasm	
Agranular leukocytes		
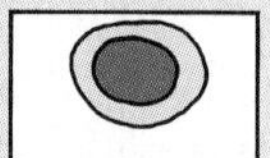 Lymphocyte 1,500–3,000 per mm³ blood	5–17 μm in diameter (average 9–10 μm) Spherical cells with large, round nuclei	
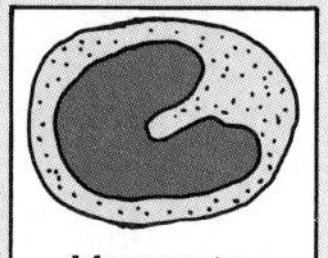 Monocyte 100–700 per mm³ blood	14–24 μm in diameter Large spherical cells with kidney-shaped, round, or lobed nuclei	
Platelets (thrombocytes)  250,000–500,000 per mm³ blood	Initiate clotting 2–4 μm in diameter Disk-shaped cell fragments with no nuclei; purple granules in cytoplasm	Bone marrow

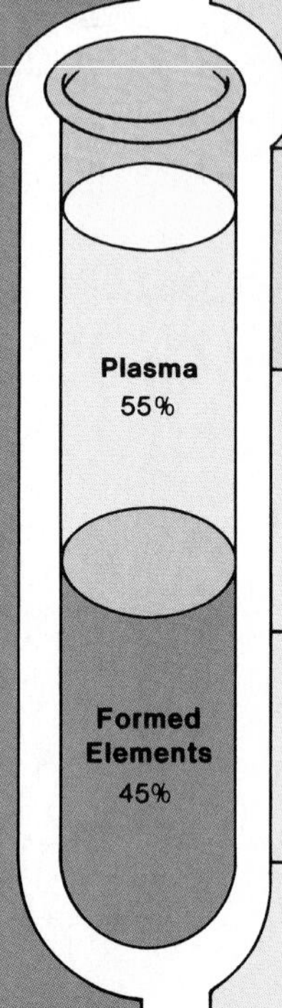

Plasma	Function	Source
Water (90–92% of plasma)	Maintains blood volume; transports molecules	Absorbed from intestine
Plasma proteins (7–8% of plasma)	Maintain blood osmotic pressure and pH	Liver
Albumin	Maintain blood volume and pressure	
Fibrinogen	Clotting	
Globulins	Transport; fight infection	
Salts (less than 1% of plasma)	Maintain blood osmotic pressure and pH; aid metabolism	Absorbed from intestinal villi
Gases		
Oxygen	Cellular respiration	Lungs
Carbon dioxide	End product of metabolism	Tissues
Nutrients Fats Glucose Amino acids	Food for cells	Absorbed from intestinal villi
Urea	Nitrogenous waste	Liver
Hormones, vitamins, etc.	Aid metabolism	Varied

*with Wright's stain

Figure 6.1

Composition of blood. When blood is transferred to a test tube and is prevented from clotting, it forms 2 layers. The transparent yellow top layer is plasma, the liquid portion of blood. The formed elements are in the bottom layer. The tables describe these components in detail.

Blood has the following functions: (1) transport of molecules such as nutrients, hormones, respiratory gases, and wastes about the body; at capillaries, blood supplies nutrients to cells and removes their wastes; (2) regulation of body temperature and pH; (3) clotting to prevent an excessive loss of blood; (4) fighting infections. This function of blood is discussed in the next chapter.

If blood is transferred from a person's vein to a test tube and is prevented from clotting, it separates into 2 layers (fig. 6.1). The lower layer consists of red blood cells (erythrocytes), white blood cells (leukocytes), and blood platelets (thrombocytes). Collectively, these are called the **formed elements.** Formed elements make up about 45% of the total volume of whole blood. The upper layer is plasma, which contains a variety of inorganic and organic molecules dissolved or suspended in water. Plasma accounts for about 55% of the total volume of whole blood.

The functions of blood contribute to homeostasis. Cells are surrounded by tissue fluid whose composition must be kept relatively constant or the cells cease to function in an effective manner. Only if the composition of blood is within the range of normality can tissue fluid also have the correct composition. All of the functions of blood are necessary to keep tissue fluid constant.

Blood functions to maintain homeostasis so that the internal environment of cells (tissue fluid) remains relatively constant.

Red Blood Cells

Red blood cells (erythrocytes) are small biconcave disks that lack a nucleus when mature. They occur in great quantity; there are 4 to 6 million red blood cells per mm^3 of whole blood. Each red blood cell is packed full of the respiratory pigment **hemoglobin;** the absence of a nucleus provides more space for hemoglobin.

Hemoglobin

Hemoglobin is a respiratory pigment that carries oxygen. Hemoglobin combines with oxygen to form oxyhemoglobin. Oxyhemoglobin gives blood its bright red color. When blood has less oxygen, it has a dark bluish color. Each red blood cell contains about 200 million hemoglobin molecules. If this much hemoglobin were suspended within the plasma rather than enclosed within the cells, blood would be so thick the heart would have difficulty pumping it.

Each hemoglobin molecule contains 4 polypeptide chains that make up the protein globin, and each chain is associated with heme, a complex iron-containing group (fig. 6.2). The iron portion of hemoglobin acquires oxygen in the lungs and gives it up in the tissues. Plasma carries only about 0.3 ml of oxygen per 100 ml, but whole blood carries 20 ml of oxygen per 100 ml. This shows that hemoglobin increases the oxygen-carrying capacity of blood more than 60 times. Hemoglobin also assists in the transport of carbon dioxide, which is discussed in chapter 8.

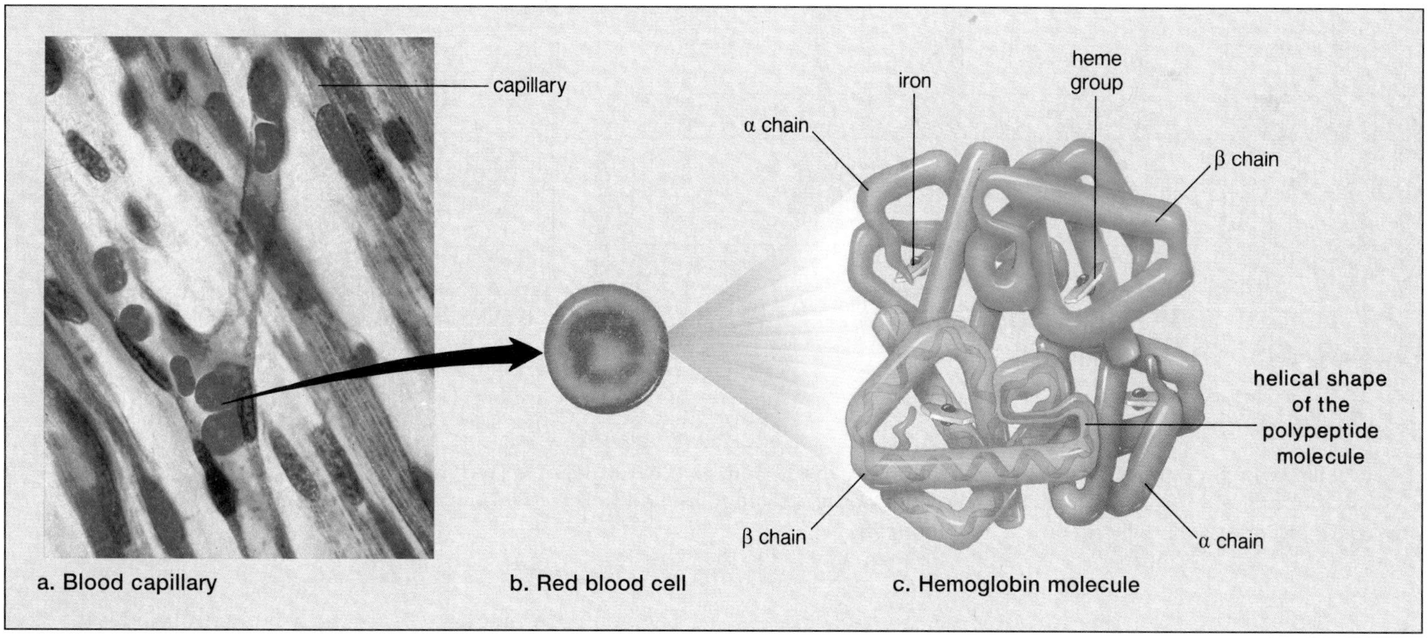

Figure 6.2

Physiology of red blood cells. **a.** Red blood cells move single file through the capillaries. **b.** Each red blood cell is a biconcave disk containing many molecules of hemoglobin, the respiratory pigment. **c.** Hemoglobin contains 4 polypeptide chains, 2 of which are alpha (α) chains and 2 of which are beta (β) chains. The plane in the center of each chain represents an iron-containing heme group. Oxygen combines loosely with iron when hemoglobin is oxygenated.

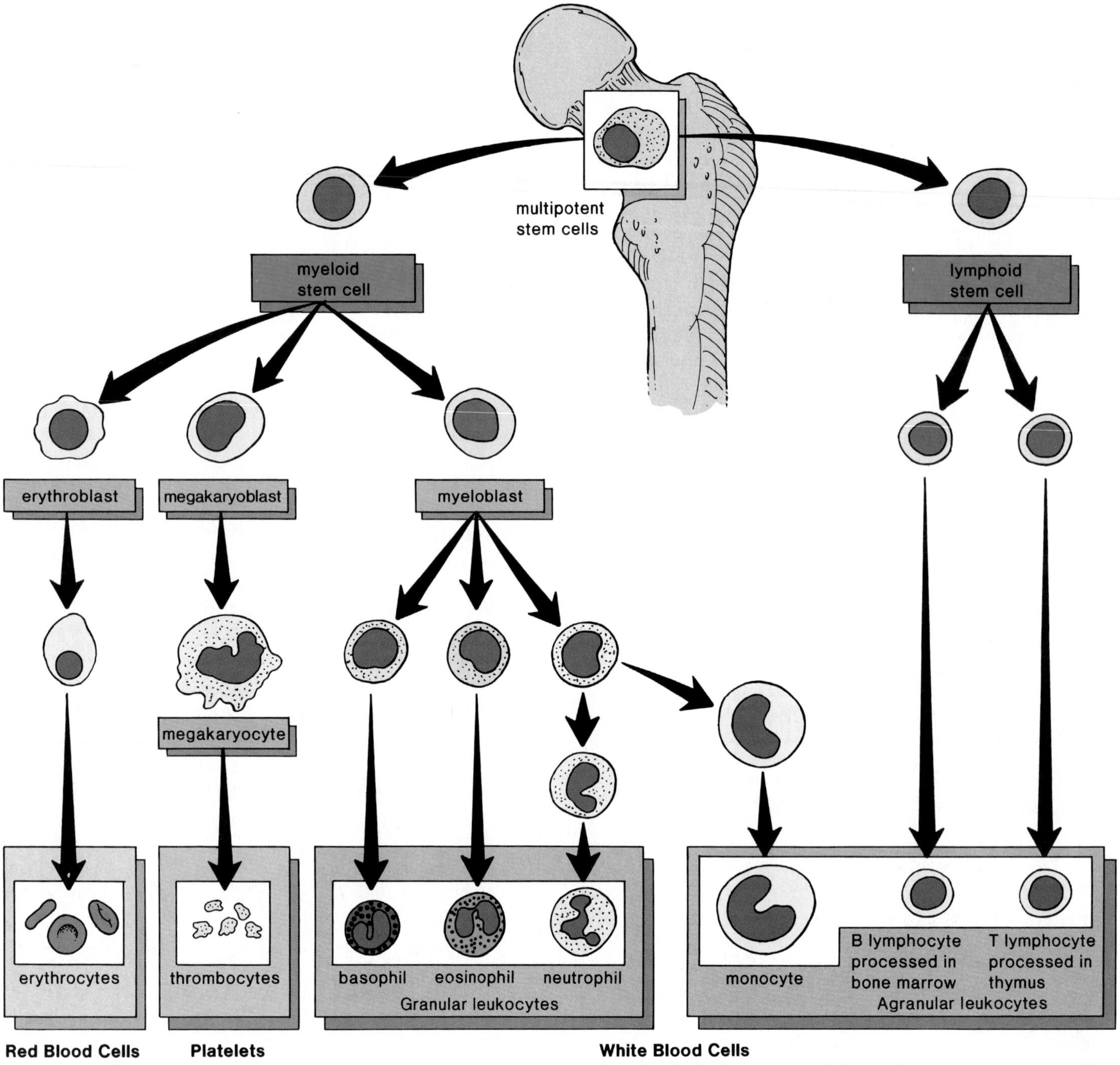

Figure 6.3

Blood cell formation in red bone marrow. Multipotent stem cells give rise to specialized stem cells. The myeloid stem cell gives rise to still other cells, which become red blood cells, platelets, and all the white blood cells except lymphocytes. The lymphoid stem cell gives rise to the lymphocytes.

Although the iron portion of hemoglobin carries oxygen, the equation for oxygenation of hemoglobin is usually written as

$$Hb + O_2 \underset{\text{tissues}}{\overset{\text{lungs}}{\rightleftarrows}} HbO_2$$

The hemoglobin on the right, which is combined with oxygen, is called oxyhemoglobin. *Oxyhemoglobin* has a bright red color. The hemoglobin on the left, which has given up oxygen to tissue fluid, is called deoxyhemoglobin. *Deoxyhemoglobin* is a dark purplish color.

Oxygen is transported from the lungs to the tissues in combination with hemoglobin, a pigment found in red blood cells.

Carbon monoxide, present in combustion gases from automobiles, furnaces, and stoves, combines with hemoglobin more readily than does oxygen, and it stays combined for several hours, making hemoglobin unavailable for oxygen transport. In New York City traffic, the blood concentration of carbon monoxide has been shown to reach 5.8%, a dangerous level when compared to the 1.5% that physicians consider safe. The Ecology Focus for this chapter discusses this component of air pollution.

Life Cycle of Red Blood Cells

In infants, red blood cells are produced in the red bone marrow of all bones, but in adults production primarily occurs in the red bone marrow of the skull bones, ribs, sternum, vertebrae, and pelvic bones.

The number of red blood cells produced increases whenever arterial blood carries a reduced amount of oxygen, as happens when an individual first takes up residence at a high altitude. Under these circumstances, the kidneys (and probably other organs as well) release *erythropoietin,* a molecule that is carried in blood to red bone marrow, where it stimulates further production of red blood cells. Erythropoietin, now mass-produced through biotechnology (p. 390), is sometimes abused by athletes in order to raise their red blood cell counts.

All blood cells, including erythrocytes, are formed from special red bone marrow cells called *stem cells* (fig. 6.3). A stem cell is ever capable of dividing and producing new cells that differentiate. Myeloid stem cells produce erythroblasts, which undergo various stages of maturation to become red blood cells. As they become mature, red blood cells lose their nucleus and acquire hemoglobin (fig. 6.4). Possibly because they lack a nucleus, red blood cells only live about 120 days. They are destroyed in the liver and spleen, where they are engulfed by macrophages, which are large phagocytic cells. It is estimated that about 2 million red blood cells are destroyed per second, and therefore an equal number must be produced to keep the red blood cell count in balance.

When red blood cells are broken down, the hemoglobin is released. The iron is recovered and is returned to the bone marrow for reuse. The heme portion of the molecule undergoes chemical degradation and is excreted by the liver in the bile. These are the bile pigments bilirubin and biliverdin, which contribute to the color of feces. Chemical breakdown of heme is also what causes a bruise of the skin to change color from red/purple to blue to green to yellow.

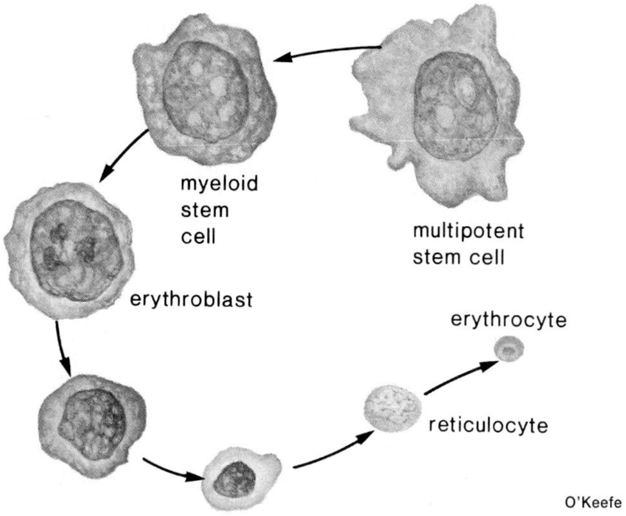

Figure 6.4

Maturation of red blood cells (erythrocytes). Red blood cells are made in the bone marrow, where stem cells continuously divide. During the maturation process, a red blood cell loses its nucleus, gains hemoglobin, and gets much smaller.

Red blood cells are produced in red bone marrow after stem cells divide to form cells that undergo maturation in stages. Red blood cells live only 120 days and are destroyed by phagocytic cells in the liver and spleen.

Anemia

When there is an insufficient number of red blood cells or the cells do not have enough hemoglobin, the individual suffers from **anemia** and has a tired, run-down feeling. In iron-deficiency anemia, the hemoglobin blood level is low. It may be that the diet does not contain enough iron. Certain foods, such as raisins, fortified cereals, and liver, are rich in iron, and the inclusion of these in the diet can help to prevent this type of anemia.

In another type of anemia, called pernicious anemia, the digestive tract is unable to absorb enough vitamin B_{12}, found in dairy products, fish, eggs, and poultry. This vitamin is essential to the proper formation of red blood cells; without it, immature red blood cells tend to accumulate in the bone marrow in large quantities. A special diet and administration of vitamin B_{12} by injection is an effective treatment for pernicious anemia.

The opening paragraph for this chapter described sickle-cell disease, which is accompanied by a type of anemia called hemolytic anemia because the red blood cells have a tendency to rupture. *Hemolysis* is the rupturing of red blood cells. Hemolytic disease of the newborn, which is discussed at the end of this chapter (p. 122), is also a type of hemolytic anemia.

Illness (anemia) results when the blood has too few red blood cells and/or not enough hemoglobin.

Ecology Focus

Carbon Monoxide, A Deadly Poison

Carbon monoxide (CO) is an air pollutant that primarily comes from the incomplete combustion of natural gas and gasoline. Figure 6A shows that transportation contributes most of the carbon monoxide to our cities' air. But power plants, factories, waste incineration, and home heating also contribute to the carbon monoxide level. Cigarette smoke contains carbon monoxide and is delivered directly to the smoker's blood and also to nonsmokers nearby.

Because carbon monoxide is a colorless, odorless gas, people can be unaware that it is affecting their systems. But it binds to iron 200 times more tightly than oxygen. Hemoglobin contains iron and so does cytochrome oxidase, the carrier in the electron transport system that passes electrons on to oxygen. When these molecules bind to carbon monoxide preferentially, they cannot perform their usual functions. The end result is that delivery of oxygen to mitochondria is impaired and so is the functioning of mitochondria.

Flushed red skin, especially on facial cheeks, is a first sign of carbon monoxide poisoning because hemoglobin bound to carbon monoxide is a brighter red than oxygenated hemoglobin. Marked euphoria, then sleepiness, coma, and death follow. Removing a person from the carbon monoxide source is not sufficient treatment because carbon monoxide, unlike oxygen, remains tightly bound to iron for many hours. A transfusion of red blood cells will help to increase the carrying capacity of the blood, and pure oxygen given under pressure will displace some carbon monoxide. Despite good medical care, some people still die each year from CO poisoning.

The level of carbon monoxide in polluted air may not be sufficient to kill people, but it does interfere with the ability of the body to function properly. The elderly and those with cardiovascular disease are especially at risk. One study found that even levels once thought to be safe can cause angina patients to experience chest pains because oxygen delivery to the heart is reduced.

We can all help prevent air pollution and reduce the amount of carbon monoxide in the air by doing the following:

- Don't smoke (especially indoors).
- Walk or bicycle instead of driving.
- Use public transportation instead of driving.
- Heat your home with solar energy—not furnaces.
- Support the development of more efficient automobiles, factories, power plants, and home furnaces.
- Support the development of alternative fuels. When the gas hydrogen is burned, for example, the result is water, not carbon monoxide and carbon dioxide.

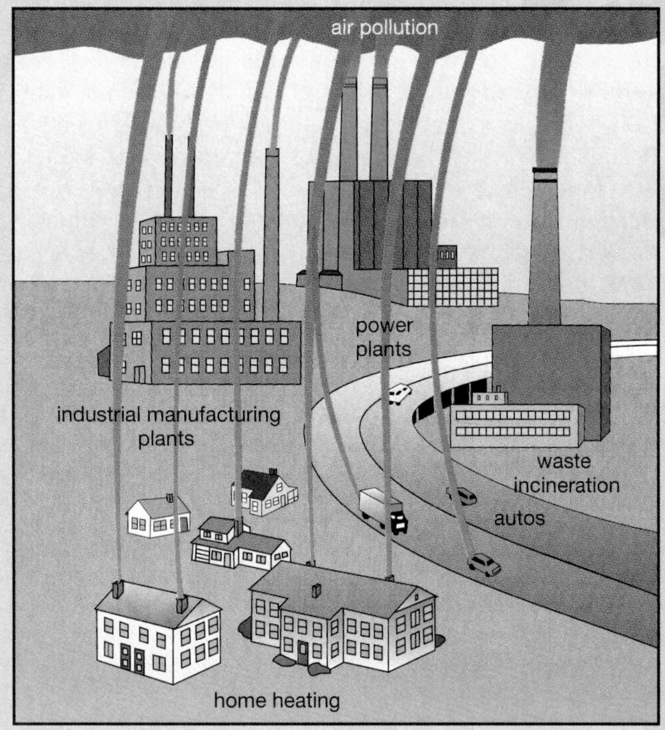

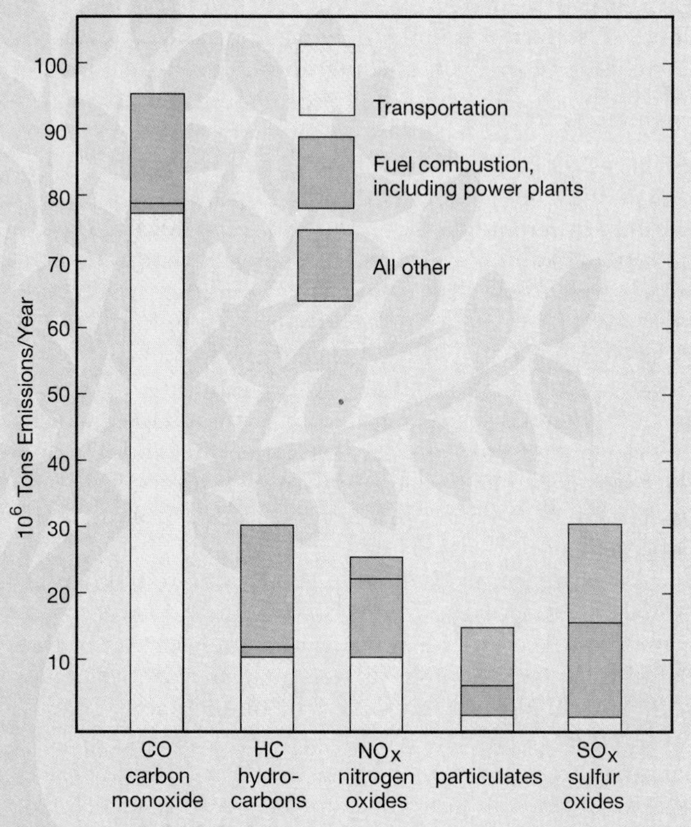

Figure 6A

Air pollution. Air pollutants enter the atmosphere from the sources noted.

White Blood Cells

White blood cells (leukocytes) differ from red blood cells in that they are usually larger, have a nucleus, lack hemoglobin, and without staining, appear white in color. White blood cells are not as numerous as red blood cells. There are only 5,000–11,000 per mm^3 of blood. White blood cells fight infection in ways that are discussed at greater length in the next chapter, which concerns immunity.

White blood cells are derived from stem cells in the red bone marrow and they, too, undergo several maturation stages (see fig. 6.3). There is an increase in the production of white blood cells whenever the body is invaded by **microbes** (infectious agents such as viruses, bacteria, and fungi). Each type of white blood cell seems capable of producing specific growth factors that circulate back to the bone marrow and stimulate their increased production.

Red blood cells are confined to the blood, but white blood cells are also found in lymph and tissue fluid. They are able to squeeze through pores in the capillary wall (fig. 6.5). Some are always found in small numbers in the tissues, but their number greatly increases when there is an infection. Many white blood cells live only a few days—they probably die while engaging microbes. Others live months or even years.

> White blood cells fight infection. They defend us against microbes that have invaded the body.

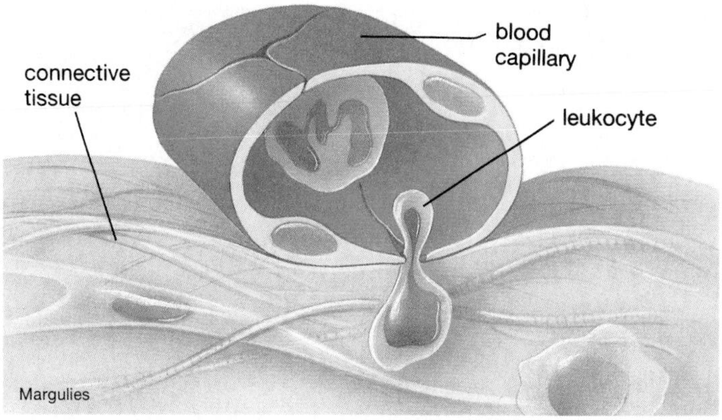

Figure 6.5

White blood cells can squeeze between the pores in a capillary wall and enter the connective tissue outside the bloodstream.

Types of White Blood Cells

It is possible to divide white blood cells into the **granular leukocytes** and the **agranular leukocytes.** Both types of cells have granules in the cytoplasm surrounding the nucleus, but they are more prominent in granular leukocytes. The granules contain various enzymes and antibiotic-like proteins, which help white blood cells defend the body. There are 3 types of granular leukocytes and 2 types of agranular leukocytes. They differ somewhat by the size of the cell and the shape of the nucleus.

Granular Leukocytes

Neutrophils are the most abundant of the white blood cells (see fig. 6.1). They have a multi-lobed nucleus joined by nuclear threads; therefore, they are also called *polymorphonuclear.* They have granules that do not significantly take up the stain eosin, a pink to red stain, or a basic stain that is blue to purple. (This accounts for their name, neutrophil.) Neutrophils are the first type of white blood cells to respond to an infection, and they engulf bacteria and cell debris during **phagocytosis.**

Eosinophils have a bilobed nucleus, and their granules take up eosin and become a red color. (This accounts for their name, eosinophil.) Not much is known specifically about the function of eosinophils, but they are known to increase in number when there is a parasitic worm infection.

Basophils have a U-shaped or lobed nucleus. Their granules take up the basic stain and become a dark blue color. (This accounts for their name, basophil.) Basophils enter the tissues and are believed to become mast cells, which release the histamine associated with allergic reactions. Histamine dilates blood vessels and causes contraction of smooth muscle.

Agranular Leukocytes

The agranular leukocytes (monocytes and lymphocytes) typically have a spherical or kidney-shaped nucleus. **Monocytes** are the largest of the white blood cells, and after taking up residence in the tissues, they differentiate into even larger macrophages (fig. 6.6). Macrophages phagocytize microbes and stimulate other white blood cells to defend the body.

The **lymphocytes** are of 2 types, B lymphocytes and T lymphocytes. B lymphocytes are responsible for antibody-mediated immunity—that is, they produce antibodies, proteins that combine with *antigens,* molecules that mark microbes, and substances for destruction by the immune system. T lymphocytes are responsible for cell-mediated immunity—that is, they directly destroy any cell that bears antigens. B lymphocytes and T lymphocytes are discussed more fully in the next chapter.

> White blood cells are divided into the granular leukocytes and the agranular leukocytes. Each type of white blood cell has a specific role to play in defending the body against disease.

Leukemia

In leukemia, an abnormally large number of immature white blood cells fill the red bone marrow and prevent red blood cell development. Anemia results and the immature white cells offer little protection from disease. The cause of leukemia, a type of cancer, is unknown, but proper chemotherapy has been most successful, particularly in acute childhood leukemia.

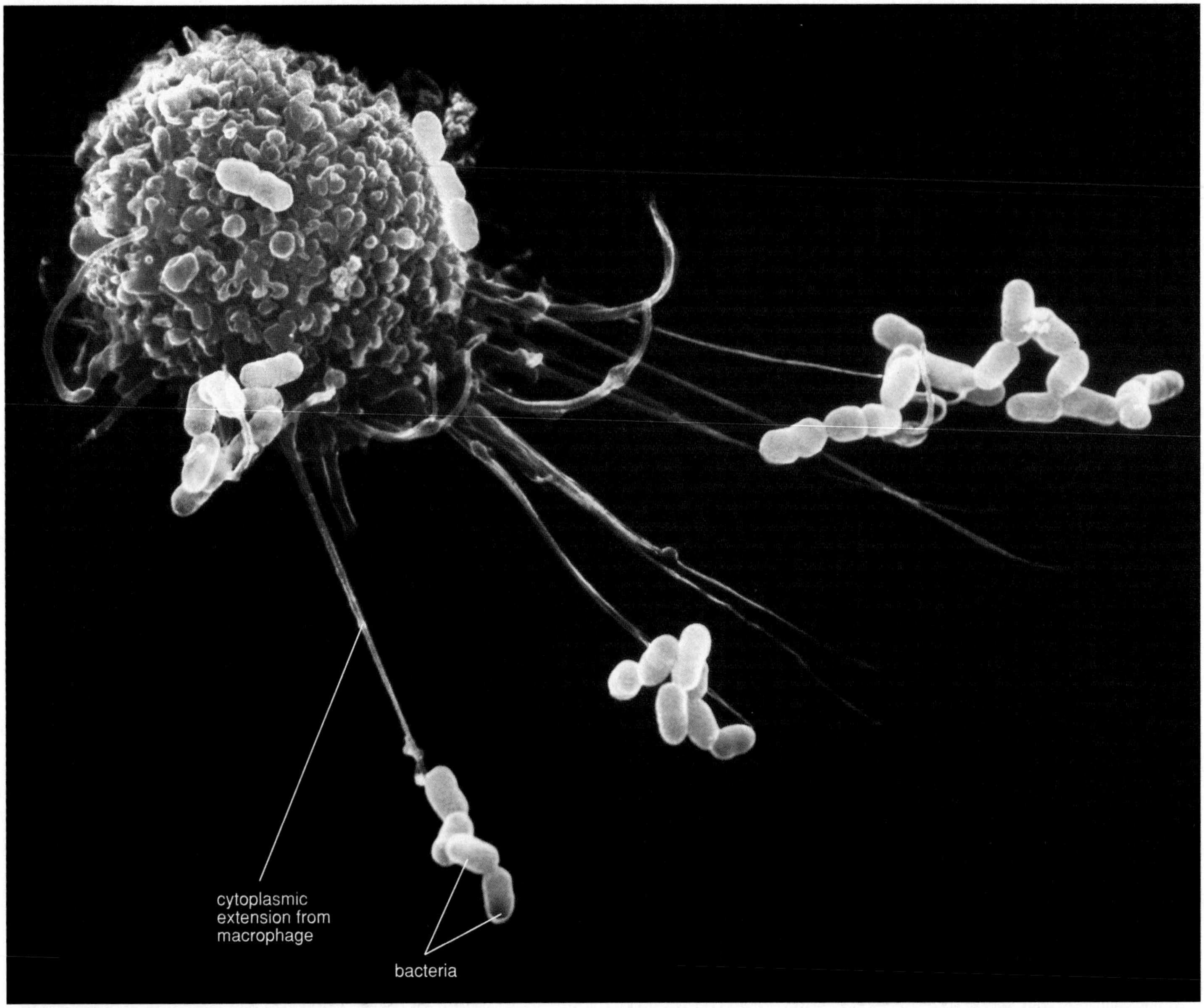

Figure 6.6

Macrophage (red) engulfing bacteria (green). Monocyte-derived macrophages are the body's scavengers. They engulf microbes and debris in the body's fluids and tissues.

Plasma

Plasma is the liquid portion of blood; over 90% of plasma is water. However, there are also various salts (ions) and organic molecules in plasma. The salts, which are simply dissolved in plasma, help maintain the pH and osmotic pressure of the blood. The small organic molecules like glucose, amino acids, and urea can also dissolve in plasma. Glucose and amino acids are food for cells; urea is a waste product on its way to the kidneys for excretion. The large organic molecules in plasma include hormones and the plasma proteins.

Plasma Proteins

Plasma proteins make up 7%–8% of plasma. Most plasma proteins are made by the liver. Exceptions are antibodies and the protein-type hormones. The plasma proteins are able to take up and release hydrogen ions; therefore, they help maintain the pH of blood around neutral pH. They also help maintain the osmotic pressure, which helps keep water in the blood. Certain plasma proteins combine with and transport large organic molecules in blood. For example, albumin transports the molecule bilirubin,

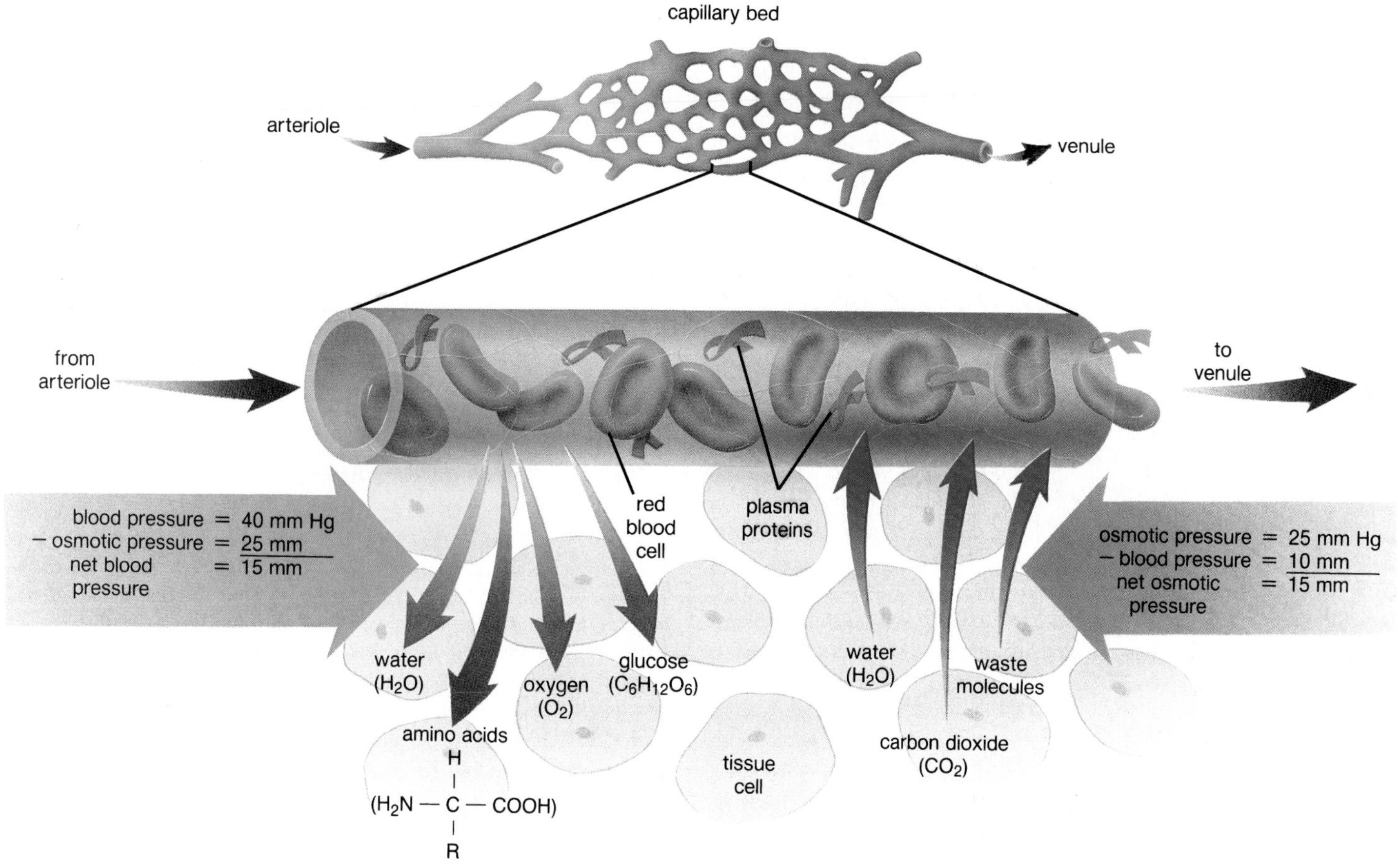

Figure 6.7

A capillary, illustrating the exchanges that take place and the forces that aid the process. At the arterial end of a capillary, the blood pressure is higher than the osmotic pressure, therefore water (H_2O) and nutrients tend to leave the bloodstream. In the midsection, molecules including oxygen (O_2) follow their concentration gradients. At the venous end of a capillary, the osmotic pressure is higher than the blood pressure, therefore water and wastes tend to enter the bloodstream. Notice that the red blood cells and the plasma proteins are too large to exit a capillary.

a breakdown product of hemoglobin. Lipoproteins, whose protein portion is a type of plasma protein called globulin, transport cholesterol. There are 3 types of globulins designated alpha, beta, and gamma globulins. Antibodies are gamma globulins. Other plasma proteins also have specific functions. Fibrinogen is necessary to blood clotting, for example.

Exchange of Materials with Tissue Fluid

The internal environment consists of blood and tissue fluid. The composition of tissue fluid stays relatively constant because of exchanges with blood in the region of capillaries (fig. 6.7). Water makes up a large part of tissue fluid, and any excess is collected by lymphatic capillaries, which are always found near cardiovascular capillaries.

Arterial End of the Capillary

When arterial blood enters the tissue capillaries, it is bright red because red blood cells are carrying oxygen. It is also rich in nutrients, which are dissolved in the plasma. At the arterial end of the capillary, blood pressure (40 mm Hg) is higher than the osmotic pressure of the blood (25 mm Hg). Blood pressure, you recall, is created by the pumping of the heart; the osmotic pressure is caused by the presence of salts and, in particular, by the plasma proteins that are too large to pass through the wall of the capillary. Since the blood pressure is higher than the osmotic pressure, fluid together with nutrients (glucose and amino acids) exit from the capillary. This is a *filtration* process because large substances, such as red blood cells and plasma proteins, remain, but small substances, such as water and nutrient molecules, leave the capillaries. Tissue fluid, created by this process, consists of all the components of plasma except the proteins.

Midsection

Along the length of the capillary, molecules follow their concentration gradient as diffusion occurs. Diffusion, you recall, is the movement of molecules from an area of greater concentration to an area of lesser concentration. The area of greater concentration for nutrients and oxygen is always blood because after these molecules have passed into the tissue fluid, they are taken up and metabolized by the tissue cells. The cells use glucose ($C_6H_{12}O_6$) and oxygen (O_2) in the process of aerobic cellular respiration, and they use amino acids for protein synthesis. Following aerobic cellular respiration, the cells give off carbon dioxide (CO_2) and water (H_2O). Carbon dioxide and waste products of metabolism leave the cell by diffusion. Since tissue fluid is always the area of greater concentration for these waste materials, they diffuse into the capillary.

Oxygen and nutrient molecules (e.g., glucose and amino acids) exit a capillary near the arterial end; waste molecules (e.g., carbon dioxide) enter a capillary near the venous end.

Venous End of the Capillary

At the venous end of the capillary, blood pressure is much reduced (10 mm Hg), as can be verified by reviewing figure 6.7. However, there is no reduction in osmotic pressure (25 mm Hg), which tends to pull fluid into the capillary. As water enters a capillary, it brings with it additional waste molecules. Blood that leaves the capillaries is deep purple in color because red blood cells contain reduced hemoglobin.

Retrieving fluid by means of osmotic pressure is not completely effective. There is always some fluid that is not picked up at the venous end. This excess tissue fluid enters the lymphatic capillaries (fig. 6.8). Lymph is tissue fluid contained within lymphatic vessels. The lymphatic system is a one-way system, and lymph is returned to the systemic venous blood when the major lymphatic vessels enter the subclavian veins (p. 129). The lymphatic system is discussed in greater detail in chapter 7.

> Lymphatic capillaries lie in close proximity to blood capillaries, where they collect excess tissue fluid.

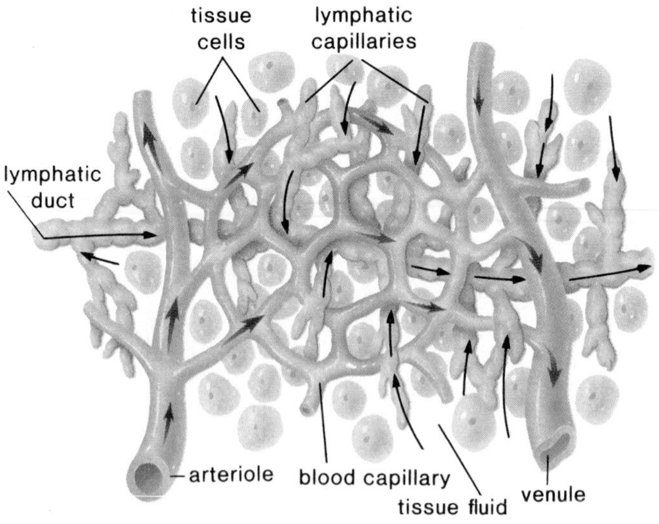

Figure 6.8

Lymphatic capillaries. Arrows indicate that lymph is formed when lymphatic capillaries take up excess tissue fluid. Lymphatic capillaries lie near blood capillaries.

Blood Clotting

When an injury to a blood vessel occurs, **clotting,** or coagulation of blood, takes place. Platelets are necessary to blood clotting.

Platelets

Platelets (thrombocytes) result from fragmentation of certain large cells, called megakaryocytes, in the bone marrow (see fig. 6.3). These cells are produced at a rate of 200 billion a day, and the blood contains 200,000–500,000 per mm^3. Platelets clump together to plug blood vessel breaks when an injury occurs, and they also start the clotting process.

Steps in Blood Clotting

There are at least 12 clotting factors in the blood that participate in the formation of a blood clot. We will discuss the roles played by platelets, prothrombin, and fibrinogen. **Fibrinogen** and **prothrombin** are proteins manufactured and deposited in blood by the liver. Vitamin K, found in green vegetables and formed by intestinal bacteria, is necessary for the production of prothrombin, and if by chance this vitamin is missing from the diet, hemorrhagic disorders develop.

When a blood vessel in the body is damaged, platelets clump at the site of the puncture and partially seal the leak. They and the injured tissues release a clotting factor called prothrombin activator, which converts prothrombin to thrombin. This reaction requires calcium (Ca^{++}) ions. **Thrombin,** in turn, acts as an enzyme that severs 2 short amino acid chains from each fibrinogen molecule. These activated fragments then join end to end, forming long

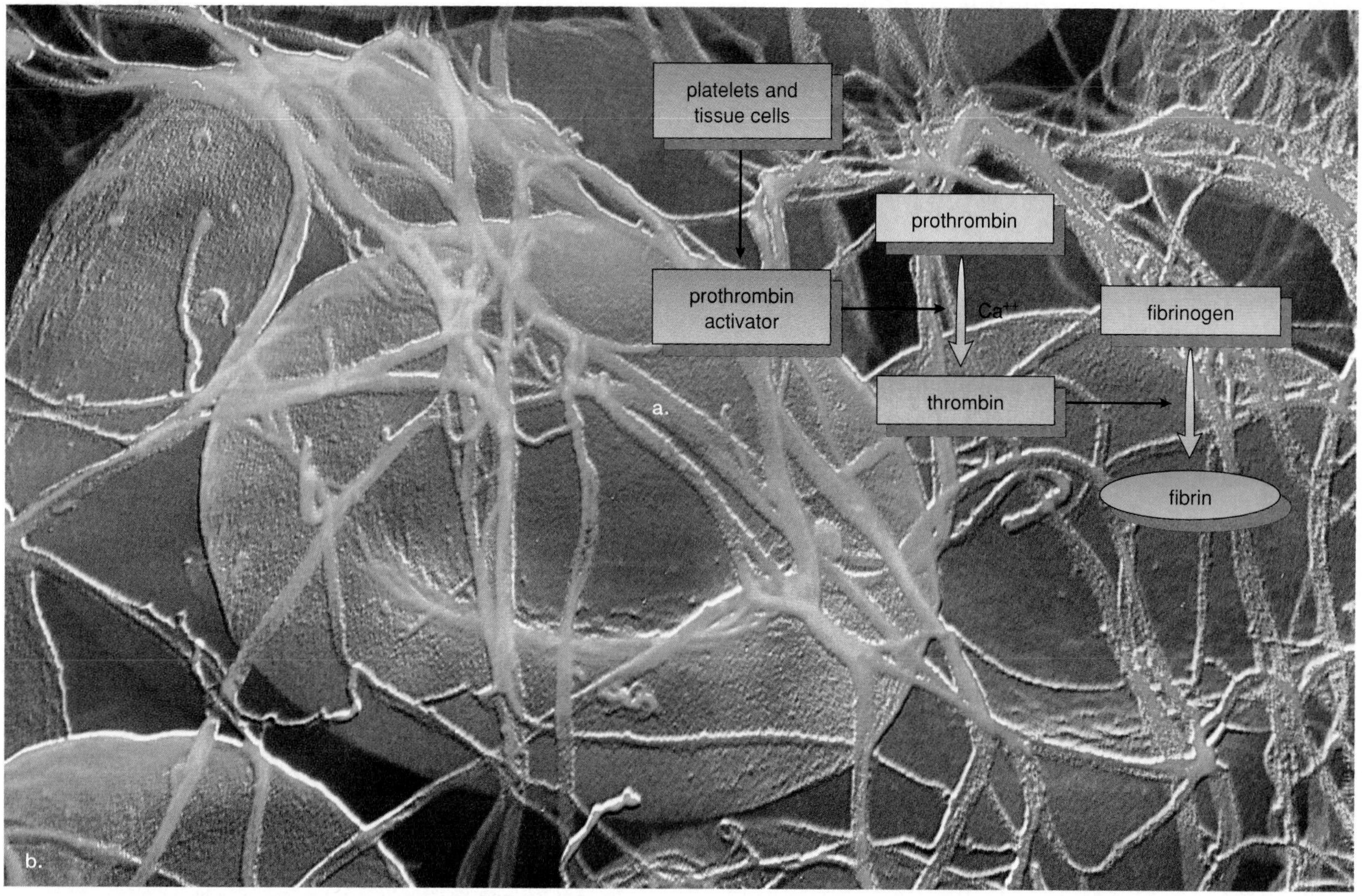

Figure 6.9

Blood clotting. **a.** Prothrombin and fibrinogen are components of blood. Prothrombin activator is an enzyme that speeds up the conversion of prothrombin to thrombin. Thrombin is an enzyme that speeds up the conversion of fibrinogen to fibrin. **b.** Scanning electron micrograph showing a red blood cell (erythrocyte) caught in the fibrin threads of a clot.

threads of **fibrin.** Fibrin threads wind around the platelet plug in the damaged area of the blood vessel and provide the framework for the clot. Red blood cells also are trapped within the fibrin threads; these cells make a clot appear red. The steps necessary for blood clotting upon injury are quite complex and are summarized in figure 6.9.

A fibrin clot is only temporarily present. As soon as blood vessel repair is initiated, an enzyme called plasmin destroys the fibrin network and restores the fluidity of plasma. This is a protective measure because, as discussed in chapter 5, a blood clot can act as a thrombus or an embolus. A blood clot interferes with circulation and can cause the death of tissues in the area.

If blood is allowed to clot in a test tube, a yellowish fluid develops above the clotted material. This fluid is called **serum,** and it contains all the components of plasma except fibrinogen. Table 6.1 reviews the many different terms we have used to refer to various body fluids related to blood.

Table 6.1

Body Fluids

Name	Composition
Blood	Formed elements and plasma
Plasma	Liquid portion of blood
Serum	Plasma minus fibrinogen
Tissue fluid	Plasma minus most proteins
Lymph	Tissue fluid within lymphatic vessels

Health Focus

What To Know When Giving Blood

The Procedure

After you register to give blood, you are asked private and confidential questions about your health history and your life-style, and any questions you may have are answered.

Your temperature, blood pressure, and pulse are checked, and a drop of your blood is tested.

You will have several opportunities prior to giving blood and even afterwards to let Red Cross officials know whether your blood is safe to give to another person.

All of the supplies, including the needle, are sterile and are used only once—for YOU. You *cannot* get infected with HIV (the virus that causes AIDS) or any other disease from donating blood.

When the actual donation is started, you may feel a brief "sting." The procedure takes about 10 minutes, and you will have given about a pint of blood. Your body replaces the liquid part (plasma) in hours and the cells in a few weeks.

After you donate, you are given a card with a number to call if you decide after you leave that your blood may not be safe to give to another person.

An area is provided in which to relax after donating blood. Most people feel fine while they give blood and afterward. A few may have an upset stomach, a faint or dizzy feeling, or a bruise, redness, and pain where the needle was. Very rarely, a person may faint, have muscle spasms, and/or suffer nerve damage.

Your blood is tested for syphilis, AIDS antibodies, hepatitis, and other viruses. You are notified if tests give a positive result. Your blood won't be used if it could make someone ill.

The Cautions

DO NOT GIVE BLOOD

if you have:

- ever had hepatitis;
- had malaria or have taken drugs to prevent malaria in the last 3 years;
- been treated for syphilis or gonorrhea in the last 12 months.

if you have AIDS or one of its symptoms:

- unexplained weight loss (4.5 kilograms or more in less than 2 months);
- night sweats;
- blue or purple spots on or under skin;
- long-lasting white spots or unusual sores in mouth;
- lumps in neck, armpits, or groin for over a month;
- diarrhea lasting over a month;
- persistent cough and shortness of breath;
- fever higher than 37° C for more than 10 days.

DO NOT GIVE BLOOD

if you are at risk for AIDS, that is, if you have:

- taken illegal drugs by needle, even once;
- taken clotting factor concentrates for a bleeding disorder such as hemophilia;
- tested positive for any AIDS virus or antibody;
- been given money or drugs for sex, since 1977;
- had a sexual partner within the last 12 months who did any of the above things;
- (for men): had sex *even once* with another man since 1977; within the last 12 months had sex with a female prostitute;
- (for women): had sex with a male or female prostitute within the last 12 months; *or* had a male sexual partner who had sex with another man *even once* since 1977.

DO NOT GIVE BLOOD to find out whether you test positive for antibodies to the viruses (HIV) that cause AIDS. Although the tests for HIV are very good, they aren't perfect. HIV antibodies may take weeks to develop after infection with the virus. If you were infected recently, you may have a negative test result yet be able to infect someone. **It is for this reason that you must not give blood if you are at risk of getting AIDS or other infectious diseases.**

Courtesy of the American Red Cross.

Blood Typing

Blood typing involves the 2 types of molecules called antigens and antibodies. As mentioned previously, when a foreign substance acting as an *antigen* enters the body an *antibody* reacts with it. The membrane of red blood cells contains possible antigens for a recipient of transfused blood, and the recipient's plasma may contain antibodies that will react with them. Such reactions are life threatening and should be avoided; hence the need to type blood.

ABO System

There are many different systems for typing the blood of humans, but the most common system is the ABO system. In the ABO system, it is determined whether type A or type B antigens are on the red blood cells. For example, if a person has type A blood, the A antigen is on his or her red blood cells. This molecule is not an antigen to this individual, although it can be an antigen to a recipient of his or her blood.

In the simplified ABO system, there are 4 types of blood: A, B, AB, and O (table 6.2). Type O blood has neither the A antigen nor the B antigen on red blood cells; the other types of blood are designated by the antigen(s) present on red blood cells.

Within the plasma, there are antibodies to the antigens that are *not* present on the person's red blood cells. Therefore, for example, type A blood has an antibody called anti-B in the plasma. Type AB blood has neither anti-A nor anti-B antibodies because both antigens are on the red blood cells. This is reasonable because if these antigens were present, **agglutination,** or clumping of red blood cells, would occur. Agglutination of red blood cells can cause blood to stop circulating in small blood vessels, and this leads to organ damage. It also is followed by hemolysis, which brings about the death of the individual. For a recipient to receive blood from a donor, the recipient's plasma must not have an antibody that causes the donor's cells to agglutinate. For this reason, it is important to determine each person's blood type. Figure 6.10 demonstrates a way to use the antibodies derived from plasma to determine the blood type. If clumping occurs after a sample of blood is exposed to a particular antibody, the person has that type of blood.

> In the ABO system, there are 4 types of blood: A, B, AB, and O. Type O blood has neither the A antigen nor the B antigen on red blood cells; the other types of blood are designated by the antigen(s) present on red blood cells.

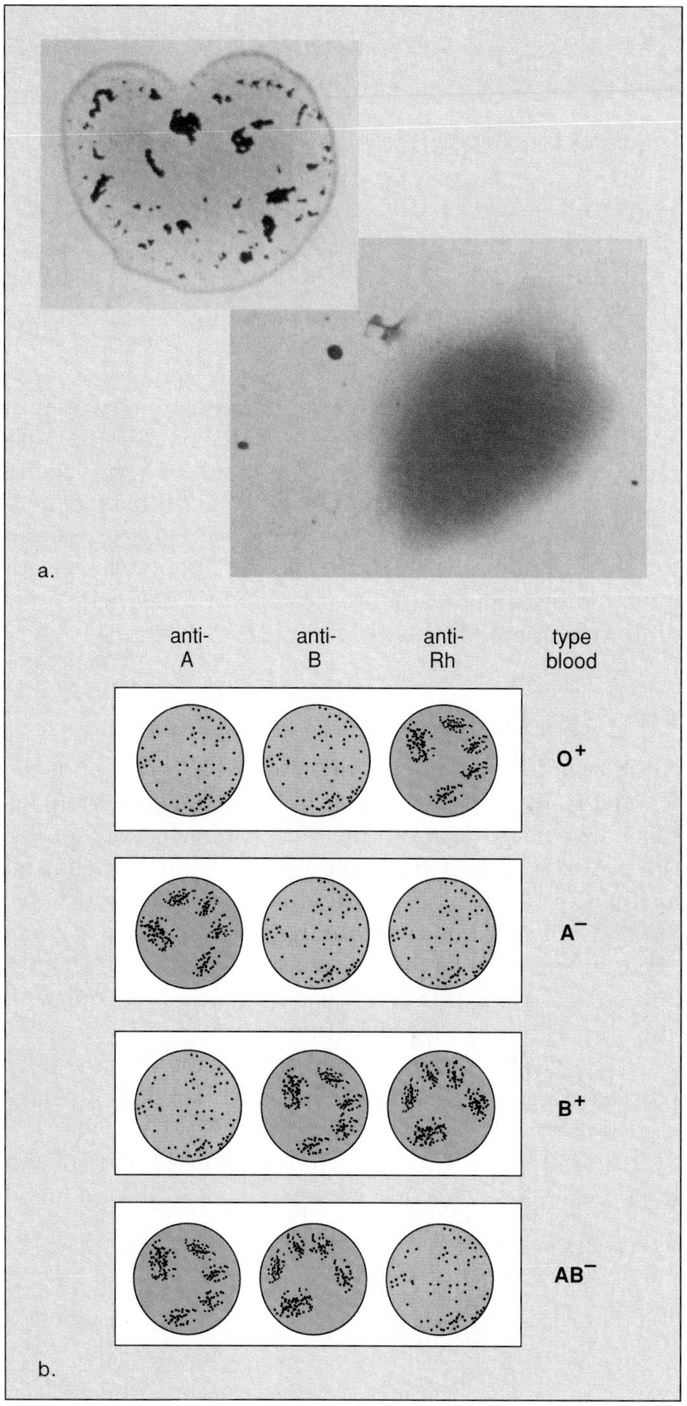

Figure 6.10

Blood typing. The standard test to determine ABO and Rh blood type consists of putting a drop of anti-A antibodies, anti-B antibodies, and anti-Rh antibodies separately on a slide. To each of these 3 antibody solutions, a drop of the person's blood is added. **a.** If agglutination occurs, as seen in the top photo, the person has this antigen on red blood cells. **b.** Several possible results.

Table 6.2

The ABO System

Blood Type	Antigen on Red Blood Cells	Antibody in Plasma	% U.S. African American	% U.S. Caucasian	% U.S. Asian	North American Indian
A	A	Anti-B	27	41	28	8
B	B	Anti-A	20	9	27	1
AB	A, B	None	4	3	5	0
O	None	Anti-A and anti-B	49	47	40	92

Today, blood transfusions are a matter of concern not only because blood types should match, but also because each person wants to receive blood that is of good quality and free of infectious agents. Blood is tested for the more serious agents such as those that cause AIDS, hepatitis, and syphilis. Donors can help protect the nation's blood supply by knowing when not to give blood. This is the topic of this chapter's Health Focus.

Rh System

Another important antigen in matching blood types is the **Rh factor.** Eighty-five percent of the U.S. population has this particular antigen on the red blood cells and are Rh^+ (Rh positive). Fifteen percent do not have this antigen and are Rh^- (Rh negative). Rh^- individuals normally do not have antibodies to the Rh factor, but they may make them when exposed to the Rh factor. It is possible to use anti-Rh antibodies for blood testing. When Rh^+ blood is mixed with anti-Rh antibodies, agglutination occurs.

The designation of blood type usually also includes whether the person has the Rh factor (Rh^+) or does not have the Rh factor (Rh^-) on the red blood cells.

During pregnancy (fig. 6.11), if the mother is Rh^- and the father is Rh^+, the child may be Rh^+. The Rh^+ red blood cells may begin leaking across the placenta into the mother's circulatory system, as placental tissues normally break down before and at birth. The presence of these Rh antigens causes the mother to produce anti-Rh antibodies. In this or a subsequent pregnancy with another Rh^+ baby, anti-Rh antibodies produced by the mother (but usually not anti-A and anti-B antibodies discussed earlier) may cross the placenta and destroy the child's red blood cells. This is called hemolytic disease of the newborn (HDN) because hemolysis continues after the baby is born. Due to red blood cell destruction followed by heme breakdown, bilirubin rises in the blood. Excess bilirubin can lead to brain damage and mental retardation or even death.

The Rh problem has been solved by giving Rh^- women an RH immunoglobulin injection either midway through the first pregnancy or no later than 72 hours after giving birth to any Rh^+ child. This injection contains anti-Rh antibodies that attack any of the baby's red blood cells in the mother's blood before these cells can stimulate her immune system to produce her own antibodies. The injection is not beneficial if the woman has already begun to produce antibodies; therefore, the timing of the injection is most important.

The possibility of hemolytic disease of the newborn exists when the mother is Rh^- and the father is Rh^+.

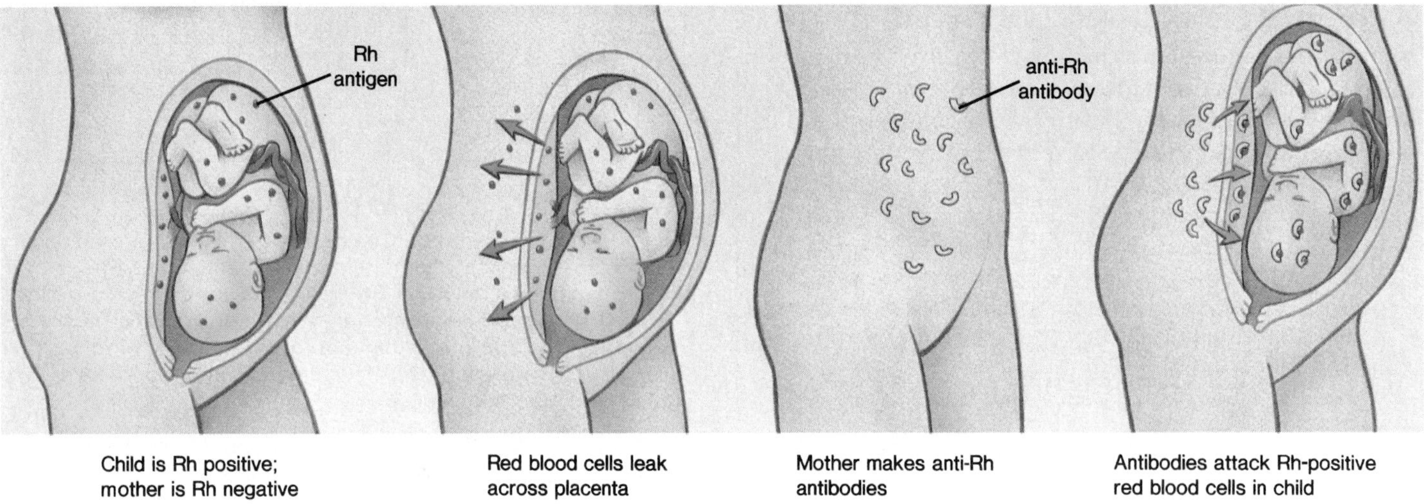
Rh antigen
anti-Rh antibody
Child is Rh positive; mother is Rh negative
Red blood cells leak across placenta
Mother makes anti-Rh antibodies
Antibodies attack Rh-positive red blood cells in child

Figure 6.11

The development of hemolytic disease of the newborn.

SUMMARY

Blood is composed of 2 parts—formed elements and plasma—and has several functions, all of which contribute to homeostasis. Blood transports, regulates body temperature and pH, clots, and fights infections.

Red blood cells are small biconcave disks that lack a nucleus. They are made in the bone marrow and destroyed in the liver and spleen. The production of red blood cells is controlled by oxygen concentration of the blood. When the oxygen concentration decreases, more red blood cells are produced. Red blood cells contain hemoglobin, the respiratory pigment, which combines with oxygen and transports it to the tissues.

White blood cells are larger than red blood cells, they have a nucleus, and they are white unless stained. Like red blood cells, they are produced in the red bone marrow. White blood cells are divided into the granular leukocytes and the agranular leukocytes. The granular leukocytes have conspicuous granules; in eosinophils granules are red when stained with eosin, and in basophils granules are blue when stained with a basic dye. The granules in neutrophils don't take up either dye significantly. Neutrophils are the most plentiful of the white blood cells, and they are able to phagocytize microbes. Many neutrophils die within a few days when they are fighting an infection. The agranulocytes include the lymphocytes and the monocytes. The lymphocytes function in specific immunity and are discussed in the next chapter. The monocytes become large phagocytic cells of great significance. They engulf worn-out red blood cells and microbes at a ferocious rate.

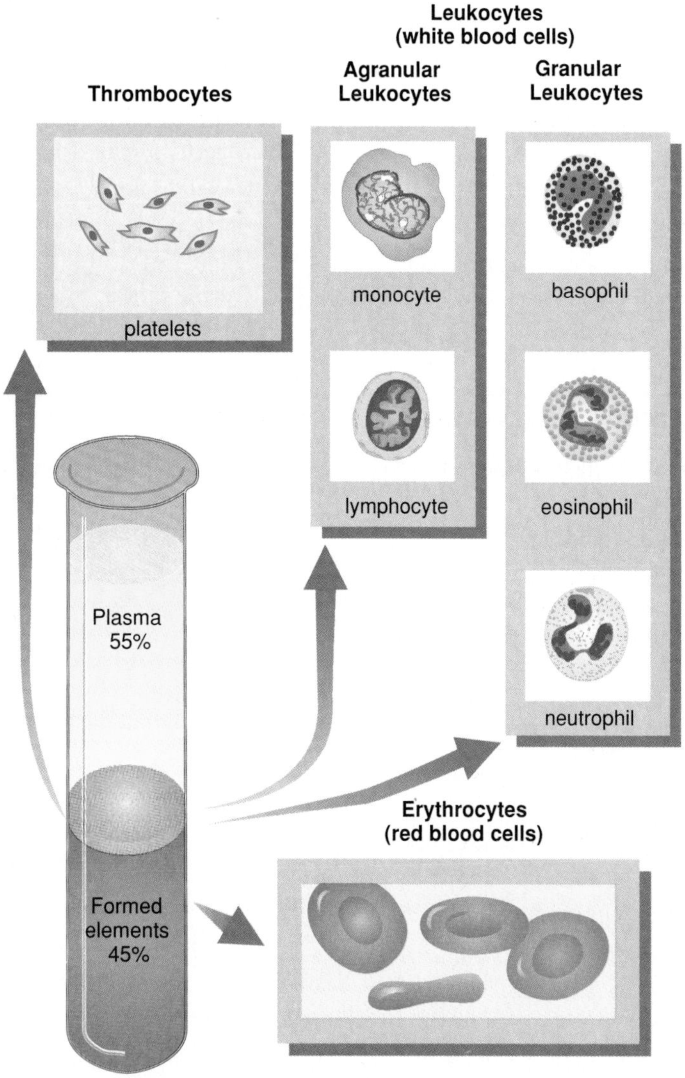

Plasma is mostly water (over 90%) and the plasma proteins (about 10%). Among the plasma proteins, albumin fulfills the general functions of maintaining osmotic pressure, regulating pH, and transporting molecules. Other plasma proteins have specific functions: fibrinogen and prothrombin are necessary to blood clotting and antibodies are gamma globulins.

Small organic molecules like glucose and amino acids are dissolved in plasma and serve as nutrients for cells. They leave the blood at capillaries. At the arterial end of a capillary, blood pressure is greater than osmotic pressure; therefore, water leaves the capillary. In the midsection, oxygen and nutrients diffuse out of the capillary; carbon dioxide and other wastes diffuse into the capillary. At the venous end, osmotic pressure created by the presence of proteins exceeds blood pressure, causing water to enter the capillary.

When there is a break in a blood vessel, the platelets clump to form a plug. Blood clotting itself requires a series of enzymatic reactions involving blood platelets, prothrombin, and fibrinogen. In the final reaction, fibrinogen becomes fibrin threads entrapping cells. The fluid that escapes from a clot is called serum and consists of plasma minus fibrinogen.

Red blood cells of an individual are not necessarily received without difficulty by another individual. For example, the membranes of red blood cells may contain type A, B, AB, or no antigens. In the plasma there are 2 possible antibodies: A or B. If the corresponding antigen and antibody are put together, clumping, or agglutination, occurs; in this way, the blood type of an individual may be determined in the laboratory. After determination of the blood type, it is theoretically possible to decide who can give blood to whom. For this, it is necessary to consider the donor's antigens and the recipient's antibodies.

Another important antigen is the Rh antigen. This particular antigen must be considered in the transfusing of blood; it also is important during pregnancy because an Rh^- mother may form antibodies to the Rh antigen after the birth of the child who is Rh^+. These antibodies can cross the placenta to destroy the red blood cells of any subsequent Rh positive child.

Study and Thought Questions

1. State the 2 main components of blood, and give the functions of blood. (p. 111)
2. What is hemoglobin, and how does it function? (p. 111)

   Critical Thinking The structure of hemoglobin allows it to combine with hydrogen ions (H^+) in the tissues. Why is this appropriate?
3. Describe the life cycle of red blood cells, and tell how the production of red blood cells is regulated. (p. 113)
4. Name the 5 types of white blood cells; describe the structure and give a function for each type. (p. 115)

   Critical Thinking GM-CSF (granulocyte-macrophage colony-stimulating factor) is a hormone that stimulates the production of certain blood cells. Use figure 6.3 to predict which blood cells, and explain your answer.
5. List the major components of plasma and discuss each. Name the primary plasma proteins, and give a function for each. (p. 116)
6. What forces operate to facilitate exchange of molecules across the capillary wall? (pp. 117–118)
7. Name the steps that take place when blood clots. Which substances are present in blood at all times, and which appear during the clotting process? (pp. 118–119)
8. What are the 4 ABO blood types? For each, state the antigen(s) on the red blood cells and the antibody(ies) in the plasma. (p. 121)

   Critical Thinking Type AB blood is sometimes called the universal recipient, meaning that persons with type AB blood can receive blood from any individuals with any other ABO blood type. Explain this terminology.
9. Explain why a person with type O blood cannot receive a transfusion of type A blood. (p. 121)
10. Problems can arise during childbearing if the mother is which Rh type and the father is which Rh type? Explain why this is so. (p.122)
11. Define blood, plasma, tissue fluid, lymph, and serum. (p. 119)

Writing Across the Curriculum

1. Humans have exchange areas with the external environment in order to maintain their internal environment. Refer to figure 5.10, and choose 3 exchange areas and discuss each.
2. Blood coagulation is a series of enzymatic reactions that lead from one to the other. Of what value is such a complicated system?
3. Explain the reason hemolytic disease of the newborn occurs and what medical steps are usually taken to prevent it.

Objective Questions

1. The liquid part of blood is called ___________.
2. Red blood cells carry ___________, and white blood cells ___________.
3. Hemoglobin that is carrying oxygen is called ___________.
4. Human red blood cells lack a ___________ and only live about ___________ days.
5. When a blood clot occurs, fibrinogen has been converted to ___________ threads.
6. The most common granular leukocyte is the ___________, a phagocytic white blood cell.
7. B lymphocytes produce ___________ that react with antigens.
8. At a capillary, ___________, ___________, and ___________ leave the arterial end, and ___________ and ___________ enter the venous end.
9. Type AB blood has the antigens ___________ and ___________ on red blood cells and ___________ antibodies in plasma.
10. Hemolytic disease of the newborn can occur if the mother is ___________ and the father is ___________.
11. Add these labels to this diagram of a capillary: arterial end, plasma proteins, venous end, oxygen, nutrients, carbon dioxide, water.

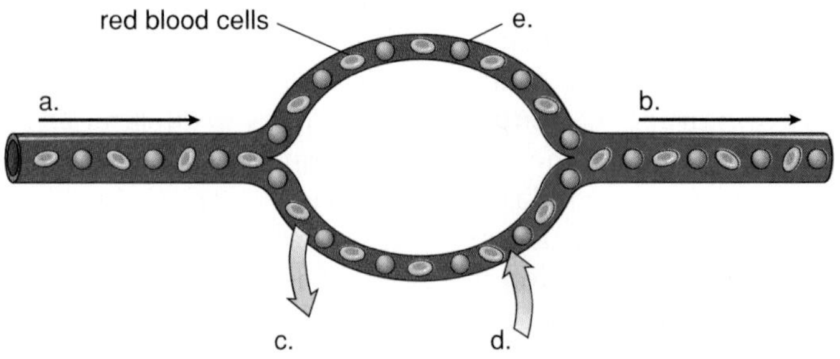

Selected Key Terms

agglutination (ag-gloo″ tĭ-na-shun) Clumping of cells, particularly in reference to red blood cells involved in an antigen-antibody reaction. 121

agranular leukocyte (ah-gran′ u-lar loo′ ko-sīt) White blood cell that does not contain distinctive granules. 115

clotting Process of blood coagulation, usually when injury occurs. 118

fibrinogen Plasma protein that is converted into fibrin threads during blood clotting. 118

formed element A constituent of blood that is either cellular (red blood cells and white blood cells) or at least cellular in origin (platelets). 111

granular leukocyte White blood cell that contains distinctive granules. 115

hemoglobin An iron-containing protein molecule in red blood cells that combines with and transports oxygen. 111

microbe Microscopic infectious agent, such as a fungus, a bacterium, or a virus. 115

neutrophil Granular leukocyte that is the most abundant of the white blood cells; first to respond to infection. 115

phagocytosis (fag″ o-si-to′ sis) The taking in of bacteria and/or debris by engulfing; cell eating. 115

plasma The liquid portion of blood consisting of all components except the formed elements. 116

platelet A cell fragment that is necessary to blood clotting; thrombocyte. 118

prothrombin Plasma protein that is converted to thrombin during the process of blood clotting. 118

serum Light-yellow liquid left after clotting of blood; blood minus formed elements and fibrinogen. 119

thrombin An enzyme that converts fibrinogen to fibrin threads during blood clotting. 118

Chapter 7

Lymphatic System and Immunity

The immune system protects us from disease, but unfortunately it also accounts for the unpleasant symptoms of hay fever, asthma, and other allergic reactions. Almost any substance that is foreign to the body can theoretically provoke an allergic reaction. Dust, milk, molds, eggs, and pollen are common causes, however. After repeated exposure to an allergen, the body begins to produce antibodies of a special type (IgE), which attach to the membrane of the mast cell. The next time the allergen appears, it combines with the IgE antibodies and the mast cells to release histamine and other substances into the body. Now the allergy makes itself known by such symptoms as red eyes, a runny nose, labored breathing, and mood and sleep disturbances. The cure, repeated injections of the allergen into the bloodstream, is often the treatment for an allergy. This doesn't make sense unless you know that this causes the buildup of IgG antibodies. When these antibodies combine with the allergen instead of IgE, the allergic response doesn't occur.

Chapter Concepts

1 The lymphatic vessels form a one-way system, which transports lymph from the tissues and fat from the lacteals to certain cardiovascular veins. 129

2 Immunity consists of nonspecific and specific defenses to protect the body against disease. 130

3 Nonspecific defenses consist of barriers to entry, the inflammatory reaction, and protective proteins. 130

4 Specific defenses require 2 types of lymphocytes, B lymphocytes and T lymphocytes. 133

5 Induced immunity for medical purposes involves the use of vaccines to achieve long-lasting immunity and the use of antibodies to provide temporary immunity. 138

6 While immunity preserves our existence, it also is responsible for certain undesirable effects, such as tissue rejection, allergies, and autoimmune diseases. 142

Chapter Outline

The Lymphatic System

The **lymphatic system** consists of lymphatic vessels and the lymphoid organs (fig. 7.1). This system, which is closely associated with the cardiovascular system, has 3 main functions: (1) lymphatic vessels take up excess tissue fluid and return it to the bloodstream; (2) lymphatic capillaries absorb fats at the intestinal villi and transport them to the bloodstream (p. 72); and (3) the lymphatic system helps to defend the body against disease.

Lymphatic Vessels: One-Way Transport

Lymphatic vessels are quite extensive; every region of the body is richly supplied with lymphatic capillaries. The construction of the larger lymphatic vessels is similar to that of cardiovascular veins, including the presence of valves. Also, the movement of lymph within lymphatic vessels is dependent upon skeletal muscle contraction. When the muscles contract, lymph is squeezed past a valve, which closes and prevents lymph from flowing backwards.

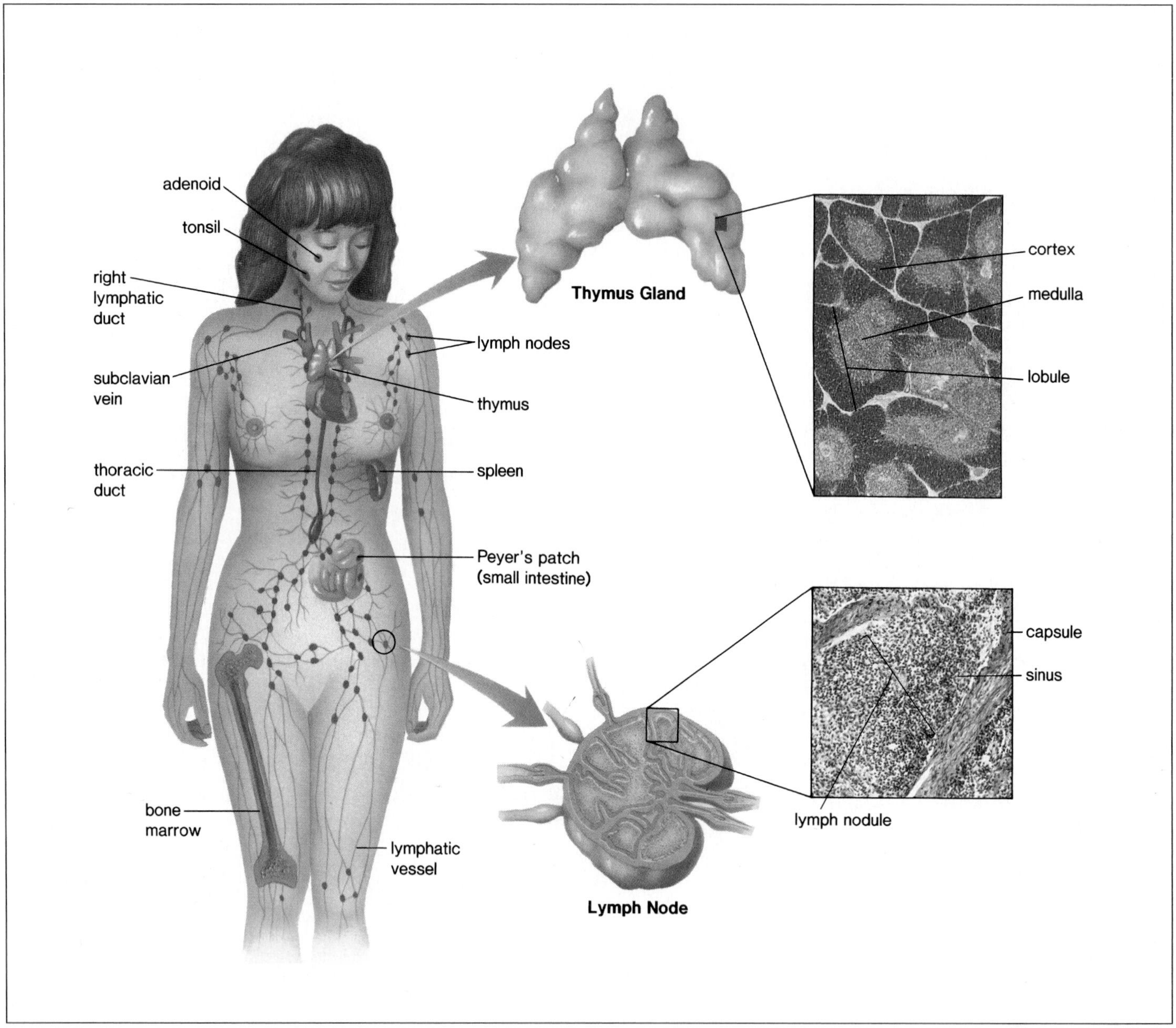

Figure 7.1

Lymphatic system. The lymphatic vessels drain excess fluid from the tissues and return it to the cardiovascular system. The thoracic duct and the right lymphatic duct are the major lymphatic vessels. Adenoids, tonsils, lymph nodes, thymus, spleen, Peyer's patches, lymphatic vessels, and bone marrow are also part of this system.

The lymphatic system is a one-way transport system that begins with lymphatic capillaries. These capillaries take up fluid that has diffused from and has not been reabsorbed by the blood capillaries (see fig. 6.8). Once tissue fluid enters the lymphatic vessels, it is called **lymph.** The lymphatic capillaries join to form lymphatic vessels, which merge before entering one of 2 ducts: the thoracic duct or the right lymphatic duct.

The *thoracic duct* is much larger than the right lymphatic duct. It serves the lower extremities, the abdomen, the left arm, and the left side of the head and the neck. In the thorax, the thoracic duct enters the left subclavian vein. The *right lymphatic duct* serves only the right arm and the right side of the head and the neck. It enters the right subclavian vein. The subclavians are cardiovascular veins in the shoulder regions. They join the superior vena cava, which empties into the heart.

Edema: Too Much Fluid

Edema is localized swelling caused by the accumulation of tissue fluid. Tissue fluid accumulates if too much of it is being made and/or not enough of it is being drained away. Pulmonary edema is a life-threatening condition associated with congestive heart failure. Due to a weak heart, blood backs up in the pulmonary circuit and causes an increase in pulmonary venous blood pressure, which leads to excess tissue fluid. Excess tissue fluid causes poor gas exchange, and the individual may suffocate.

During a breast cancer operation, lymph nodes and lymphatic vessels are sometimes removed to prevent the spread of cancer. Without these lymphatic vessels, tissue fluid collects and edema results. In the tropics, infection of lymphatic vessels by a parasitic worm can result in elephantiasis, a condition in which a limb swells to elephantine proportions.

> The lymphatic system is a one-way transport system. Lymph flows from a capillary to ever-larger lymphatic vessels and finally to a lymphatic duct, which enters a subclavian vein.

Lymphoid Organs: Assist Immunity

The lymphoid organs of special interest are the bone marrow, the lymph nodes, the spleen, and the thymus (see fig. 7.1).

Red bone marrow is the site of origination for all types of blood cells, including the 5 types of white blood cells, which function in immunity (fig. 7.2). The marrow contains stem cells that are ever capable of dividing and producing cells that go on to differentiate into the various types of blood cells. All the white blood cells originate from the same multipotent stem cell but eventually have their own particular stem cell (see fig. 6.3). In a child, most bones have red bone marrow, but in an adult it is present only in the bones of the skull, the sternum (breast bone), the ribs, the clavicle, the pelvic bones, and the vertebral column (p. 211). The red bone marrow consists of a network of connective tissue fibers, called reticular fibers, which are produced by cells called reticular cells. These and the stem cells and their progeny are packed about thin-walled sinuses (open spaces) filled with venous blood. Differentiated blood cells enter the bloodstream at these sinuses.

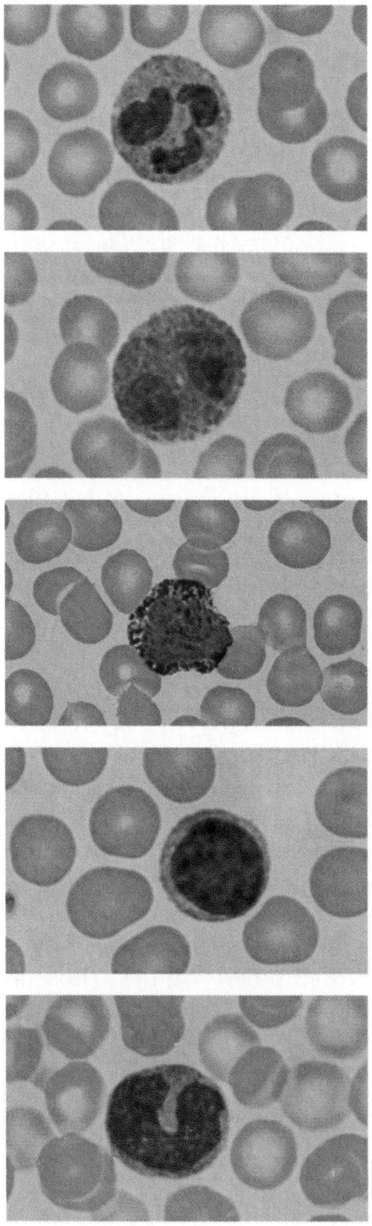

Neutrophil
40 – 70%
Phagocytizes primarily bacteria

Eosinophil
1 – 4%
Phagocytizes and destroys antigen-antibody complexes

Basophil
0 – 1%
Congregates in tissues; releases histamine when stimulated

Lymphocyte
20 – 45%
B type produces antibodies in blood and lymph:
T type kills virus-containing cells

Monocyte
4 – 8%
Becomes macrophage—phagocytizes bacteria and viruses

Figure 7.2

There are 5 types of white cells, which differ according to structure and function. The frequency of each type cell is given as a percentage of the total.

Lymph nodes, which are small (about 1–25 mm), ovoid, or round structures, occur at certain points along lymphatic vessels. A lymph node has a fibrous connective tissue capsule. Connective tissue also divides a node into nodules. Each nodule contains a sinus filled with many lymphocytes and macrophages. As lymph passes through the sinuses, the macrophages purify it of infectious organisms and any other debris.

While nodules usually occur within lymph nodes, they can also occur singly or in groups. The *tonsils,* located in back of the mouth on either side of the tongue, and the adenoids, located on the posterior wall above the border of the soft palate, are composed of partly encapsulated lymph nodules. Also, nodules called *Peyer's patches* are found within the intestinal wall.

The lymph nodes occur in groups in certain regions of the body. For example, the inguinal nodes are in the groin and the axillary nodes are in the armpits.

The **spleen** (see fig. 7.1) is located in the upper left abdominal cavity just beneath the diaphragm. The construction of the spleen is similar to a lymph node. Outer connective tissue divides the organ into lobules, which contain sinuses. In the spleen, however, the sinuses are filled with blood instead of lymph. The blood vessels of the spleen can expand, which enhances the carrying capacity of this organ to serve as a blood reservoir and make blood available in times of low pressure or when the body needs extra oxygen carrying capacity.

A spleen nodule contains red pulp and white pulp. Red pulp contains red blood cells, lymphocytes, and macrophages. The white pulp contains only lymphocytes and macrophages. Both types of pulp help to purify blood that passes through the spleen. If the spleen ruptures due to injury, it can be removed, and although its functions are duplicated by other organs, the individual is expected to be slightly more susceptible to infections and may have to take antibiotic therapy indefinitely.

The **thymus** (see fig. 7.1) is located along the trachea behind the sternum in the upper thoracic cavity. This gland varies in size, but it is larger in children than in adults and may disappear completely. Some believe this contributes to the increased incidence of cancer as we age.

The thymus is divided into lobules by connective tissue. The T lymphocytes mature in these lobules. Those in the interior (medulla) are more mature than those in the exterior (cortex) of a lobule. The thymus secretes thymosin, a molecule that is believed to be an inducing factor; that is, it causes pre-T lymphocytes to become T lymphocytes. Thymosin may also have other functions in immunity.

> The lymphoid organs have specific functions that assist immunity. White blood cells are made in the bone marrow; lymph is cleansed in lymph nodes; blood is cleansed in the spleen; and T lymphocytes mature in the thymus.

Nonspecific Defenses

Immunity is the ability of the body to defend itself against infectious agents, foreign cells, and even body cells that have gone astray, such as cancer cells. Natural immunity ordinarily keeps us free of diseases, including those caused by microbes and also cancer. There is evidence that among their various detrimental effects, pesticides suppress the immune system. Certainly, this must contribute to the ability of pesticides to cause cancer. The Ecology Focus for this chapter suggests that we cut down drastically on our use of pesticides and offers alternatives to their use.

Immunity includes nonspecific and specific defenses. The 3 nonspecific defenses—barriers to entry, the inflammatory reaction, and protective proteins—are useful against all types of infectious agents.

Barriers to Entry: From Skin to Resident Bacteria

Skin and the mucous membrane lining the respiratory and digestive tracts serve as mechanical barriers to entry by microbes. Oil gland secretions contain chemicals that weaken or kill bacteria on skin. The respiratory tract is lined by cells that sweep mucus and trapped particles up into the throat, where they can be swallowed. The stomach has an acidic pH, which inhibits the growth of many types of bacteria. The various bacteria that normally reside in the intestine and other organs, such as the vagina, prevent pathogens from taking up residence.

Inflammatory Reaction: Call to Arms

Whenever the skin is broken due to a minor injury, a series of events occurs that is known as the **inflammatory reaction.** The inflamed area has 4 symptoms: redness, pain, swelling, and heat. Figure 7.3 illustrates the participants in the inflammatory reaction. The mast cells, one type of participant, are derived from basophils (see fig. 7.2), which take up residence in the tissues.

When an injury occurs, a capillary and several tissue cells are apt to rupture and to release **bradykinin,** a molecule that initiates nerve impulses, resulting in the sensation of pain, and stimulates mast cells to release **histamine,** which together with bradykinin causes the capillary to dilate and become more permeable. The enlarged capillaries cause the skin to redden, and the increased permeability allows proteins and fluids to escape so swelling results. A rise in temperature reduces the number of invading microbes and increases phagocytosis by white blood cells.

Any break in the skin allows microbes to enter the body and triggers a migration of neutrophils and monocytes to the site of injury. Neutrophils and monocytes are amoeboid; they can change shape and squeeze through capillary walls to enter tissue fluid. Neutrophils phagocy-

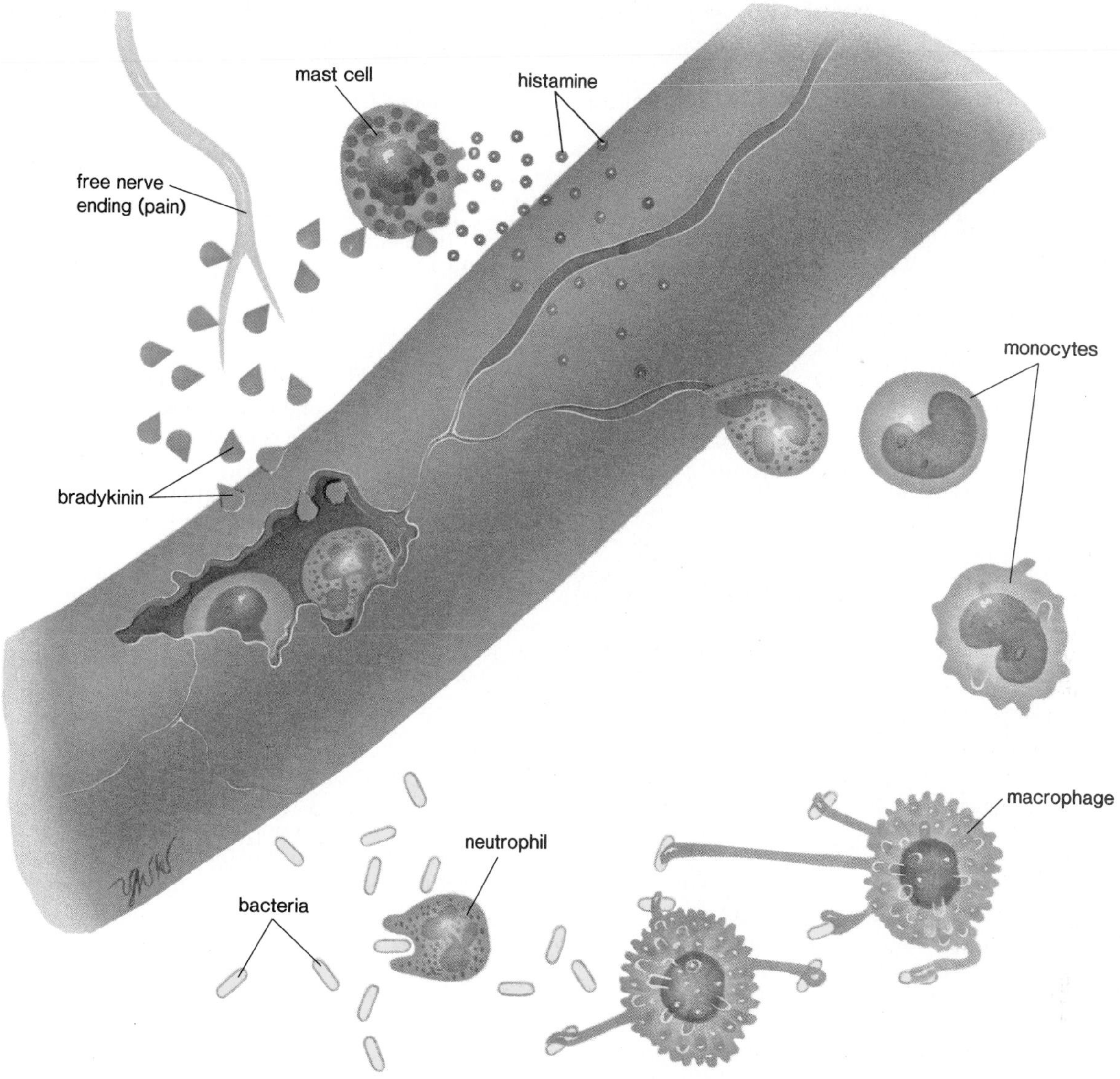

Figure 7.3

Inflammatory reaction. When a blood vessel is injured, mast cells release histamine, which dilates blood vessels, and bradykinin, which stimulates the pain nerve endings. Neutrophils and monocytes congregate at the injured site and squeeze through the capillary wall. The neutrophils begin to phagocytize bacteria. The monocytes become macrophages, large cells that are especially good at phagocytosis and also stimulate other white blood cells to action.

tize bacteria and when phagocytosis occurs, an endocytic vesicle forms. The engulfed bacteria are destroyed by hydrolytic enzymes when the vesicle combines with a lysosome (an organelle discussed in chapter 2).

Monocytes differentiate into **macrophages,** large phagocytic cells that are able to devour a hundred bacteria or viruses and still survive. Some tissues, particularly connective tissue, have resident macrophages, which routinely act as scavengers, devouring old blood cells, bits of dead tissue, and other debris. Macrophages also are capable of bringing about an explosive increase in the number of leukocytes by liberating a growth factor, which passes by way of blood to the red bone marrow, where it stimulates the production and the release of white blood cells, usually neutrophils.

As the infection is being overcome, some neutrophils die. These, along with dead tissue, cells, bacteria, and living white blood cells, form **pus,** a thick, yellowish fluid. Pus indicates that the body is trying to overcome the infection.

> The inflammatory reaction is a "call to arms"—it marshals phagocytic white blood cells to the site of bacterial invasion.

Ecology Focus

Pesticides: A Benefit Gone Wrong

A pesticide is any one of 55,000 chemical products used to kill insects, fungi, or rodents that interfere with human activities. The argument for pesticides is attractive. It is claimed they kill off disease-causing agents, increase yield and thus lower the cost of food, and work quickly with minimal risk. Over time it has been found that pesticides do not meet these claims. Instead, pests become resistant to pesticides, which kill off natural enemies in addition to the pest. Then the pest population explodes. At first DDT did a marvelous job killing off the mosquitos that carry malaria; now malaria is as big a problem as ever. In the meantime, DDT has accumulated in the tissues of wildlife and humans, causing possible harmful effects. The problem was made obvious when birds of prey became unable to reproduce due to weak egg shells. The use of DDT is now banned in the United States.

Crop losses due to insects are now greater percentage-wise than they were in the 1940s before modern pesticides became available. In the meantime, farmers are sold ever-more expensive pesticides to control pests that are increasing in resistance and number. This is called the pesticide treadmill. At least 20 insect species are now apparently resistant to all known insecticides; some 80 species of weeds are resistant to one or more herbicides.

Increasingly, we have discovered that pesticides are harmful to the environment and humans. Death due to pesticide poisoning is a worldwide problem that involves mostly children who mistakenly ingest a large quantity. The worst industrial accident in the world occurred when a nerve gas used to produce an insecticide in India was released into the air; 3,700 people were killed, and 30,000 were injured. It is impossible to estimate how many people are eventually affected by exposure to pesticides at work or play. Pesticides are mostly used in farming, but 20% are used each year on lawns, gardens, parks, golf courses, and cemeteries. We even spray them in our houses! Vietnam veterans claim any manner of health effects due to contact with a herbicide called Agent Orange used as a defoliant during the Vietnam War. (The use of this herbicide is now also banned in the United States.) The EPA says that pesticide residues on food pose a serious cancer risk to the U.S. population. Other possible effects are birth defects, disorders of the nervous system, and suppression of the immune system.

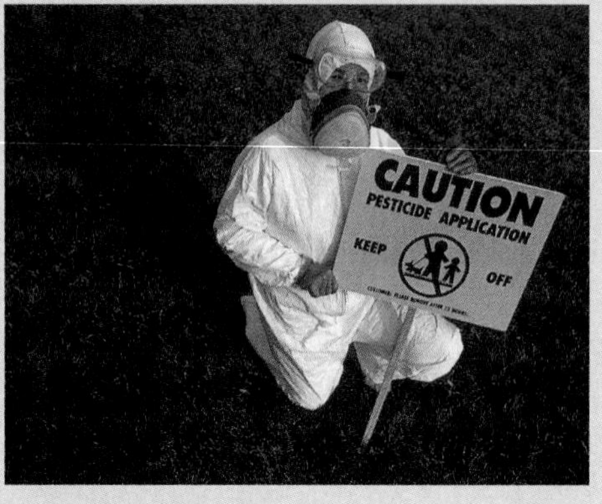

Figure 7A

Pesticide warning signs indicate that pesticides are harmful to our health.

There is an alternative to the addictive use of pesticides. Integrated pest management uses a diversified environment, mechanical and physical means, natural enemies, disruption of reproduction, and resistant plants to control rather than eradicate pest populations. Chemicals are used only as a last resort. Maintaining hedgerows, weedy patches, and certain trees can provide diverse habitats and food for predators and parasites that help control pests. The use of strip farming and crop rotation by farmers denies pests of a continuous food source. Natural enemies abound in the environment. When lacewings were released in cotton fields, they reduced the boll weevil population by 96% and increased cotton yield threefold. A predatory moth was used to reclaim 60 million acres in Australia overrun by the prickly pear cactus, and another type of moth is now used to control alligator weed in the southern United States. Similarly, in this country the *Chrysolina* beetle controls Klamath weed, except in shady places where a root boring beetle is more effective.

The use of sex attractants and sterile insects is also a type of biological control. In Sweden and Norway, scientists synthesized the sex pheromone of the *Ips* bark beetle, which attacks spruce trees. Almost 100,000 baited traps collected females normally attracted to males emitting this pheromone. Sterile males have also been used to reduce pest populations. The screwworm fly parasitizes cattle in the United States. Flies raised in a laboratory were made sterile by exposure to radiation. The entire Southeast was freed from this parasite when female flies mated with sterile males and then laid eggs that did not hatch.

In the past, only cross breeding could produce resistant plants. Now, genetic engineering, a new technique by which certain genes are introduced into organisms, including plants, may eventually make large-scale use of pesticides unnecessary (see chapter 19). Already, 50 types of genetically engineered plants that resist insects, viruses, or herbicides have entered small-scale field trials.

In general, biological control is a more sophisticated method of controlling pests than the use of pesticides. It requires an indepth knowledge of pests and/or their life cycles. Because it does not have an immediate effect on the pest population the effects of biological control may not be apparent to the farmer or gardener who benefits from it.

It is important for citizens to do all they can to promote biological control of pests instead of relying on pesticides, which are unhealthy for the environment and humans.

- Urge elected officials to support legislation to protect humans and the environment against the use of pesticides.
- Allow native plants to grow on all or most of the land to give natural predators a place to live.
- Cut down on the use of pesticides, herbicides, and fertilizers for the lawn, garden, and house.
- Use alternative methods such as cleanliness and good sanitation to keep household pests under control.
- Keep homes in good repair so that pests are not encouraged to live there.
- Dispose of pesticides in a safe manner.

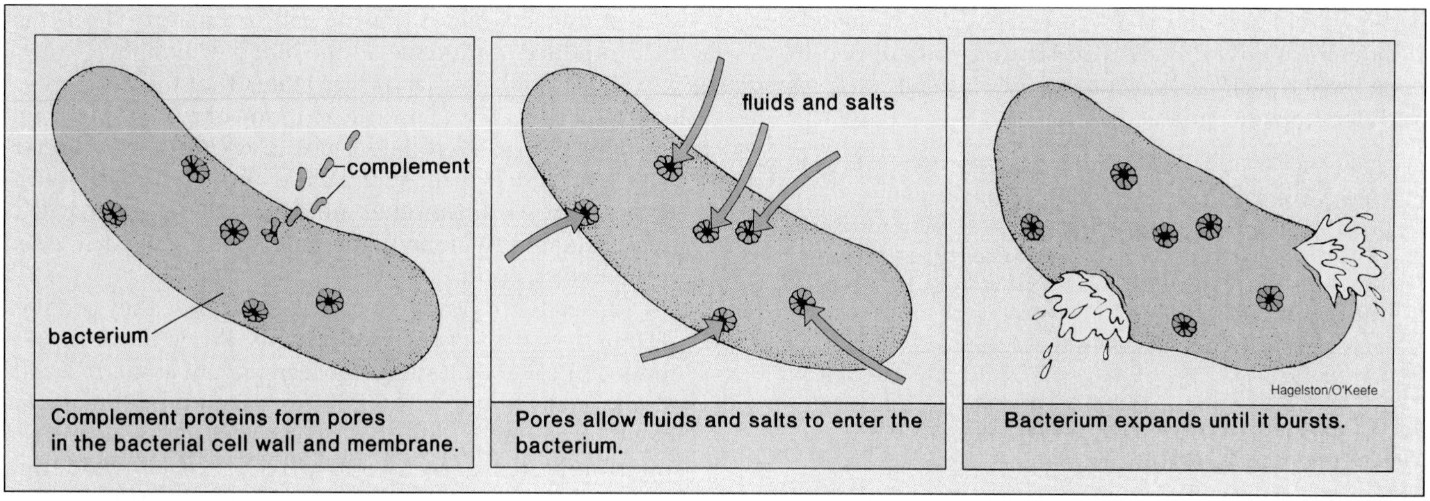
complement
bacterium
Complement proteins form pores in the bacterial cell wall and membrane.
fluids and salts
Pores allow fluids and salts to enter the bacterium.
Hagelston/O'Keefe
Bacterium expands until it bursts.

Figure 7.4

Action of the complement system against a bacterium. When complement proteins in the plasma are activated by an immune reaction, they form pores in bacterial cell walls and membranes, allowing fluids and salts to enter until the cell eventually bursts.

Protective Proteins: Plasma Proteins

The **complement system,** often simply called complement, is a number of plasma proteins designated by the letter *C* and a subscript. Once a complement protein is activated, it activates another protein, and the result is a set series of reactions. A limited amount of activated protein is needed because a cascade occurs: each protein in the series is capable of activating many proteins next in line.

Complement is activated when microbes enter the body. They are involved in and amplify the inflammatory response. Another series of reactions is complete when complement proteins form pores in bacterial cell walls and membranes. Potassium ions leave; fluids and salt enter the bacterial cell to the point that it bursts (fig. 7.4).

Complement also releases chemicals that attract phagocytes to the scene. It "complements" certain immune responses, and this accounts for its name. For example, some complement proteins bind to the surface of microbes already coated with antibodies; this ensures that the microbes will be phagocytized by a neutrophil or macrophage.

When viruses infect a tissue cell, the affected cell produces and secretes **interferon,** a type of lymphokine. This class of molecules is discussed later. Interferon binds to receptors of noninfected cells, and this action causes the cells to prepare for possible attack by producing substances that interfere with viral replication. Interferon is specific to the species; therefore, only human interferon can be used in humans. Although it once was a problem to collect enough interferon for clinical and research purposes, interferon is now made by recombinant DNA technology.

> Immunity includes these nonspecific defenses: barriers to entry, the inflammatory reaction, and protective proteins.

Specific Defenses

Sometimes, we are threatened by an invasion of microbes that cannot be successfully counteracted by a nonspecific defense mechanism. In such cases, it is necessary to rely on a specific defense against a particular antigen. An **antigen** is a protein (or polysaccharide) molecule that the body recognizes as nonself. Microbes have antigens, but antigens can also be part of a foreign cell or a cancer cell. Because we do not ordinarily become immune to our own cells, it is said that the immune system is able to tell self from nonself.

Immunity usually lasts for some time. For example, once we recover from the measles, we usually cannot be infected by the measles virus a second time. Immunity is primarily the result of the action of the **B lymphocytes** and the **T lymphocytes.** B (for bone marrow) lymphocytes mature in the bone marrow, and T (for thymus) lymphocytes mature in the thymus gland. B lymphocytes, also called B cells, give rise to plasma cells, which produce **antibodies,** proteins that are capable of combining with and neutralizing antigens. These antibodies are secreted into the blood and the lymph. In contrast, T lymphocytes, also called T cells, do not produce antibodies. Instead, certain T cells directly attack cells that bear antigens. Other T cells regulate the immune response.

Lymphocytes are capable of recognizing an antigen because they have receptor molecules on their surface. The shape of the receptors on any particular lymphocyte is complementary to a specific antigen. It is often said that the receptor and the antigen fit together like *a lock and a key.* It is estimated that during our lifetime, we encounter a million different antigens, so we need the same number of different lymphocytes for protection against those antigens. It is remarkable that diversification occurs to such an extent during the maturation process that there is a different lymphocyte type for each possible antigen. Despite this great diversity, none of the lymphocytes ordinarily attacks the body's own cells. It is believed that if by chance a lymphocyte arises that is equipped to respond to the body's own proteins, it is normally suppressed and develops no further.

> There are 2 types of lymphocytes. B cells produce and secrete antibodies, which combine with antigens. Certain T cells directly attack antigen-bearing cells, and others regulate the immune response.

B Cells: Make Plasma and Memory Cells

The receptor on a B cell is called a membrane-bound antibody because of its structure. When a B cell encounters a bacterial cell or a toxin bearing an appropriate antigen, the membrane-bound antibody binds to the antigen. This reaction causes the B cell to undergo clonal expansion; it divides and produces many plasma cells, which secrete antibodies against this antigen. A **plasma cell** is a mature B cell that mass produces antibodies in the blood or lymph.

The *clonal selection theory* states that the antigen selects which B cell will produce a clone of plasma cells (fig. 7.5). Notice that a B cell does not clone until its antigen is present and that it recognizes the antigen directly without assistance from another immune cell. However, B cells are stimulated to clone by helper T cells, as is discussed in the next section.

When B cells undergo clonal expansion, they produce plasma cells and also memory B cells. When the antigen is under control, the plasma cells decrease but memory B cells remain in the bloodstream. **Memory B cells** are the means by which active immunity is possible. If the same antigen enters the system again, memory B cells quickly divide and give rise to new antibody-producing plasma cells.

Defense by B cells is called **antibody-mediated immunity** because the various types of B cells produce antibodies. It is also called *humoral immunity* because these antibodies are present in the bloodstream. A *humor* is any fluid normally occurring in the body.

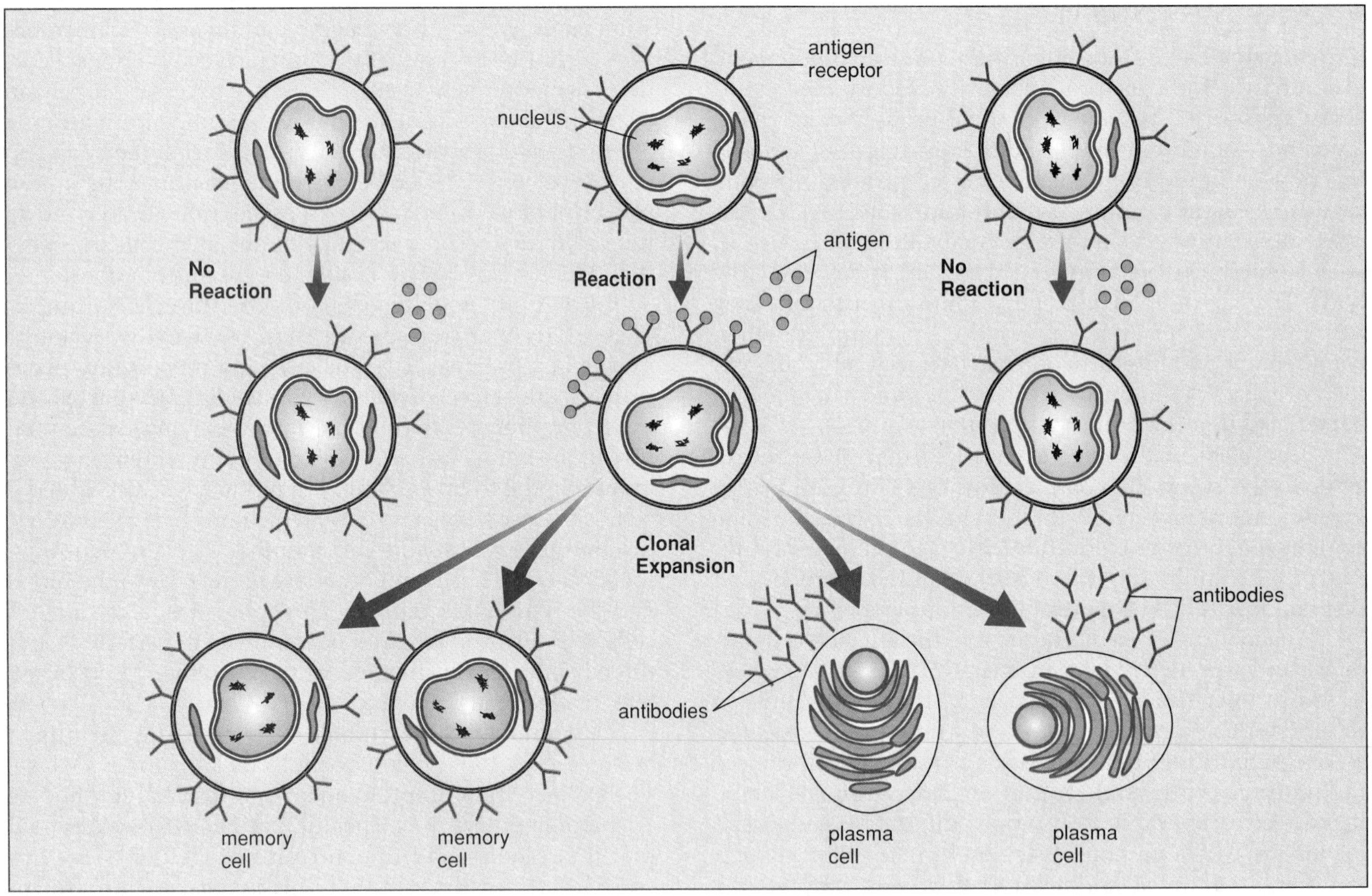

Figure 7.5

Clonal selection theory as it applies to B cells. An antigen activates the appropriate B cell, which undergoes clonal expansion if stimulated by a helper T cell. During the process, many plasma cells, which produce specific antibodies against this antigen, are produced. Memory cells, which retain the ability to secrete these antibodies, are produced also.

Characteristics of B Cells:

- Antibody-mediated immunity
- Produced and mature in bone marrow
- Direct recognition of antigen
- Clonal expansion produces antibody-secreting plasma cells as well as memory cells

Antibodies and Antigens Often Form Complexes

The most common type of antibody (IgG) is a Y-shaped protein molecule with 2 arms. Each arm has a "heavy" (long) chain and a "light" (short) chain of amino acids. These chains have *constant regions,* where the sequence of amino acids is set, and *variable regions,* where the sequence of amino acids varies (fig. 7.6). The constant regions are not identical among all the antibodies. Instead, they are the same within different classes of antibodies (table 7.1). The variable regions form an antigen-binding site, and their shape is specific to a particular antigen. The antigen combines with the antibody at the antigen-binding site in a lock-and-key manner.

The antigen-antibody reaction can take several forms, but quite often the antigen-antibody reaction produces complexes of antigens combined with antibodies. Such an antigen-antibody complex, sometimes called the immune complex, marks the antigen for destruction by other forces. For example, the complex may be engulfed by neutrophils or macrophages or it may activate complement. Complement makes microbes more susceptible to phagocytosis, as discussed.

Figure 7.6

Antigen-antibody reaction. **a.** An IgG antibody contains 2 heavy (long) amino acid chains and 2 light (short) amino acid chains arranged to give 2 variable regions, where a particular antigen is capable of binding with the antibody. **b.** Quite often the antigen-antibody reaction produces complexes of antigens combined with antibodies.

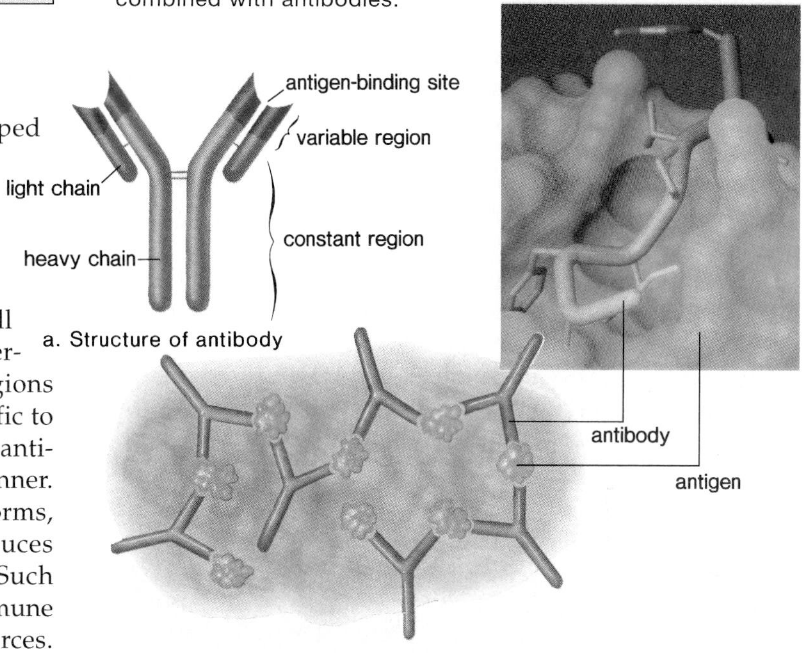

a. Structure of antibody

b. Antigen-antibody complex

Classes of Antibodies: Improving the Odds

There are 5 different classes of circulating antibodies (table 7.1). IgG antibodies are the major type in blood, and lesser amounts are also found in lymph and tissue fluid. IgG antibodies attack microbes and their toxins. A toxin is a specific chemical produced, for example by bacteria, that is poisonous to other living things. IgM antibodies are a cluster of 5 of the Y-shaped structures shown in figure 7.6*a.* These antibodies appear soon after an infection begins and disappear before it is over. IgA antibodies contain 2 Y-shaped structures and are the main type of antibody found in bodily secretions. They attack microbes and their toxins before they reach the bloodstream. IgD antibodies are identical to the membrane-bound antibodies on B cells. IgE antibodies are responsible for allergic reactions and are discussed later (p. 140).

An antigen combines with an antibody at the antigen-binding site in a lock-and-key manner. The reaction can produce antigen-antibody complexes, which contain several molecules of antibody and antigen.

Table 7.1

Antibodies

Classes	Presence	Function
IgG	Main antibody type in circulation	Attacks microbes* and bacterial toxins; enhances phagocytosis
IgM	Antibody type found in circulation, largest antibody	Activates complement; clumps cells
IgA	Main antibody type in secretions such as saliva and milk	Attacks microbes and bacterial toxins
IgD	Antibody type found as a membrane-bound receptor	Functions unknown
IgE	Antibody type found as membrane-bound receptor on basophils in blood and on mast cells in tissues	Responsible for allergic reactions

*Viruses and bacteria

T Cells: Cytotoxic, Helper, Memory, Suppressor

There are 4 different types of T cells: cytotoxic T cells, helper T cells, memory T cells, and suppressor T cells. Cytotoxic and helper T cells differentiate in the thymus. Memory T cells and suppressor T cells arise as discussed in this section. All 4 types look alike but can be distinguished by their functions.

Cytotoxic T cells sometimes are called "killer T cells." They attack and destroy antigen-bearing cells, such as virus-infected or cancer cells. Cytotoxic T cells have storage vacuoles containing perforin molecules. Perforin molecules perforate a cell membrane, forming a pore that allows water and salts to enter. The cell under attack then swells and eventually bursts (fig. 7.7). It often is said that T cells are responsible for **cell-mediated immunity,** characterized by destruction of antigen-bearing cells. Of all the T cells, only cytotoxic T cells are involved in this type of immunity.

Helper T cells regulate immunity by enhancing the response of other immune cells. When exposed to an antigen, they enlarge and secrete **lymphokines,** stimulatory molecules that cause helper T cells to clone and other immune cells to perform their functions. For example, lymphokines stimulate macrophages to phagocytize and B cells to become antibody-producing plasma cells. Because the HIV virus, which causes AIDS, attacks helper T cells, it inactivates the immune response. AIDS is discussed on pages 307–314.

When an activated helper T cell divides, the clone also contains **suppressor T cells** and **memory T cells.** Once there is a sufficient number of suppressor T cells, the immune response ceases. Following suppression, however, a population of memory T cells persists, perhaps for life. These cells are able to secrete lymphokines and to stimulate macrophages and B cells whenever the same antigen reenters the body. In this way they contribute to active immunity.

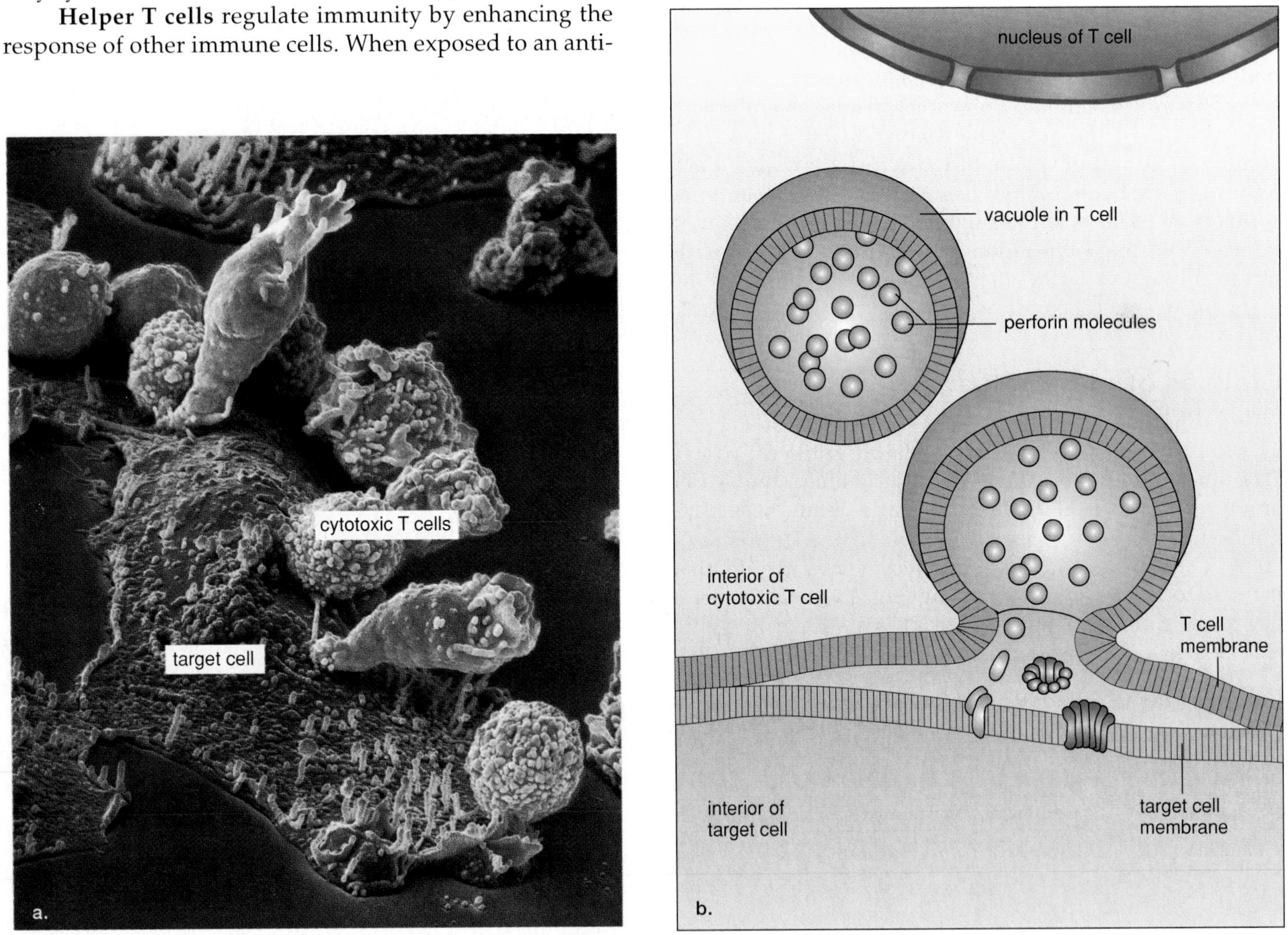

Figure 7.7

Cell-mediated immunity. **a.** The scanning electron micrograph shows cytotoxic T cells attacking and destroying a cancer cell. **b.** During the killing process, the vacuoles in a cytotoxic T cell fuse with the cell membrane and release units of the protein perforin. These units combine to form pores in the target cell membrane. Thereafter, fluid and salts enter so that the target cell eventually bursts.

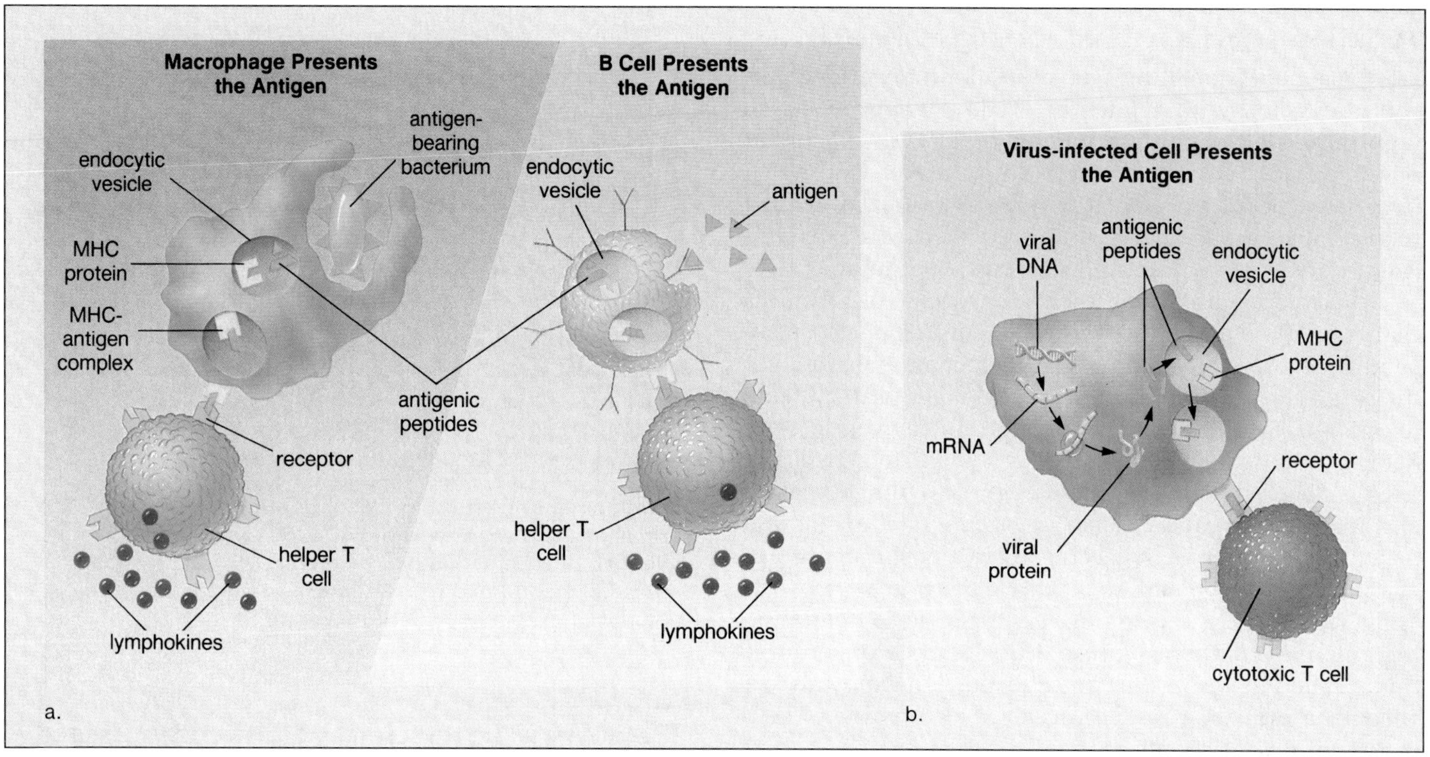

Figure 7.8

T cell activation. **a.** Either a macrophage or a B cell presents an antigen to a helper T cell. To accomplish this, the antigen has to be digested to peptides that are combined with an MHC protein. The complex is presented to the T cell. In return, the helper T cell produces and secretes lymphokines that stimulate T cells and other immune cells. **b.** Cells infected with a virus present one of the viral proteins along with an MHC protein to a cytotoxic T cell. This causes the cytotoxic T cell to attack and destroy any cell infected with the same virus (see fig. 7.7).

Activating Cytotoxic (Killer) and Helper T Cells

T cells have receptors just as B cells do. Unlike B cells, however, cytotoxic T cells and helper T cells are unable to recognize an antigen that simply is present in lymph or blood. Instead, the antigen must be presented to them by an *antigen-presenting cell* (*APC*). When an APC, usually a macrophage, engulfs a microbe, it is enclosed within an endocytic vesicle, where it is broken down to peptide fragments. These fragments are antigenic (have the properties of an antigen). The antigenic peptide fragment is linked to an **MHC** (major histocompatibility complex) **protein,** and together they are displayed at the cell membrane and presented to a T cell.

The importance of MHC proteins in cell membranes was first recognized when it was discovered that they contribute to the specificity of tissues and make it difficult to transplant tissue from one person to another. In other words, the donor and the recipient must be histo-(tissue) compatible (the same or nearly so) for a transplant to be successful without the administration of immunosuppressive drugs.

Figure 7.8 shows a macrophage presenting an antigen to a helper T cell. Once a helper T cell recognizes an antigen, it undergoes clonal expansion, producing suppressor T cells and memory T cells, which can also recognize this same antigen. Once a cytotoxic T cell recognizes an antigen, it attacks and destroys any cell that is infected with the same virus. In this way they contribute to active immunity.

Characteristics of T Cells:

- Cell-mediated immunity
- Produced in bone marrow, mature in thymus
- Antigen must be presented by macrophage
- Cytotoxic T cells search and destroy antigen-bearing cells
- Helper T cells secrete lymphokines and stimulate other immune cells

Induced Immunity

Induced immunity is immunity brought about artificially by medical intervention. There are 2 types of induced immunity: active and passive. In active immunity, the individual alone produces antibodies against an antigen and in passive immunity, the individual is given ready-made antibodies.

Active Immunity: Long-Lived

Active immunity sometimes develops naturally after a person is infected with a microbe. Today, however, active immunity is often induced when a person is well so that possible future infection will not take place. To prevent infections, people can be artificially immunized against them. One recommended immunization schedule for children is given in the Health Focus for this chapter.

Immunization involves the use of **vaccines,** substances that contain an antigen to which the immune system responds. Traditionally vaccines are the microbes themselves that have been treated so they are no longer virulent (able to cause disease). New methods of producing vaccines are being developed. For example, it is possible to genetically engineer bacteria to mass-produce a protein from microbes, and this protein can be used as vaccine. This method is being used to prepare a vaccine against malaria, a protozoan disease, and hepatitis B, a viral disease.

After a vaccine is given, it is possible to determine the amount of antibody present in a sample of serum—this is called the *antibody titer.* After the first exposure to a vaccine, a primary response occurs. For a period of several days, no antibodies are present; then, there is a slow rise in the titer, followed by first a plateau and then a gradual decline as the antibodies bind to the antigen or simply break down (fig. 7.9). After a second exposure, a secondary response is expected. The titer rises rapidly to a plateau level much greater than before. The second exposure is called a "booster" because it boosts the antibody titer to a high level. The high antibody titer now is expected to help prevent disease symptoms even if the individual is exposed to the disease-causing microbe. Immunological memory causes an individual to be actively immune.

Immunological Memory: A State of Readiness

Immunological memory is dependent upon the number of memory B and memory T cells capable of responding to a particular antigen. Memory B cells differ qualitatively from unprimed B cells, as they are prone to make IgG earlier and their receptors usually have a higher affinity for the antigen due to the selection process that occurred during the first exposure to the antigen. Both memory B cells and memory T cells can respond to lower doses of antigen. Good active immunity lasts as long as memory B and memory T cells are present in blood. Active immunity is usually long-lived.

> Active (long-lived) immunity can be induced by the use of vaccines when a person is well and in no immediate danger of contracting an infectious disease. Active immunity is dependent upon the presence of memory B cells and memory T cells in the body.

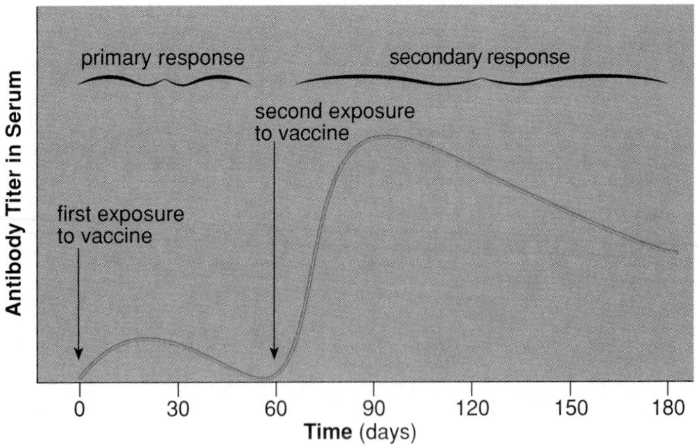

Figure 7.9

Development of active immunity due to immunization. The primary response, after the first exposure to a vaccine, is minimal, but the secondary response, which may occur after the second exposure, shows a dramatic rise in the amount of antibody present in serum.

Passive Immunity: Short-Lived

Passive immunity occurs when an individual is given ready-made antibodies (immunoglobulins) to combat a disease. Since these antibodies are not produced by the individual's B cells, passive immunity is short-lived. For example, newborn infants are passively immune to disease because antibodies have crossed the placenta from the mother's blood. These antibodies soon disappear, however, so that within a few months, infants become more susceptible to infections. Breast-feeding prolongs the passive immunity an infant receives from the mother because antibodies are present in the mother's milk.

Even though passive immunity does not last, it sometimes is used to prevent illness in a patient who has been unexpectedly exposed to an infectious disease. Usually, the patient receives a gamma globulin injection (serum that contains antibodies), perhaps taken from individuals who have recovered from the illness. In the past, horses were immunized, and serum was taken from them to provide the needed antibodies against such diseases as diphtheria, botulism, and tetanus. Occasionally, a patient who received these antibodies became ill because the serum contained proteins that the individual's immune system recognized as foreign. This was called serum sickness.

> Passive immunity is needed when an individual is in immediate danger of succumbing to an infectious disease. Passive immunity is short-lived because the antibodies are administered to and not made by the individual.

Health Focus

Immunization: The Great Protector

Vaccine	Months of Age				Years of Age
DTP (diphtheria, tetanus, whooping cough)	2	4	6	15★	4-6
OPV (oral polio vaccine)	2	4		15★	4-6
MMR (measles, mumps, rubella)				15†	4-6"
Hib (Haemophilus influenza, type b)	2	4	6‡	15§	
Td (tetanus and less diphtheria than DTP)					14-16#

★ Many experts recommend this dose of vaccine at 18 months.

† In some areas, this dose of MMR vaccine may be given at 12 months.

‡ This dose may not be required, depending on which Hib vaccine is used.

§ This dose may be given at 12 months, depending on which Hib vaccine is used.

" Some experts recommend that this dose of MMR vaccine be given at entry to middle or junior high school.

And every 10 years thereafter.

Source: U.S.Department of Health and Human Services, Parent's Guide to Childhood Immunization, revised May, 1991, page 26.

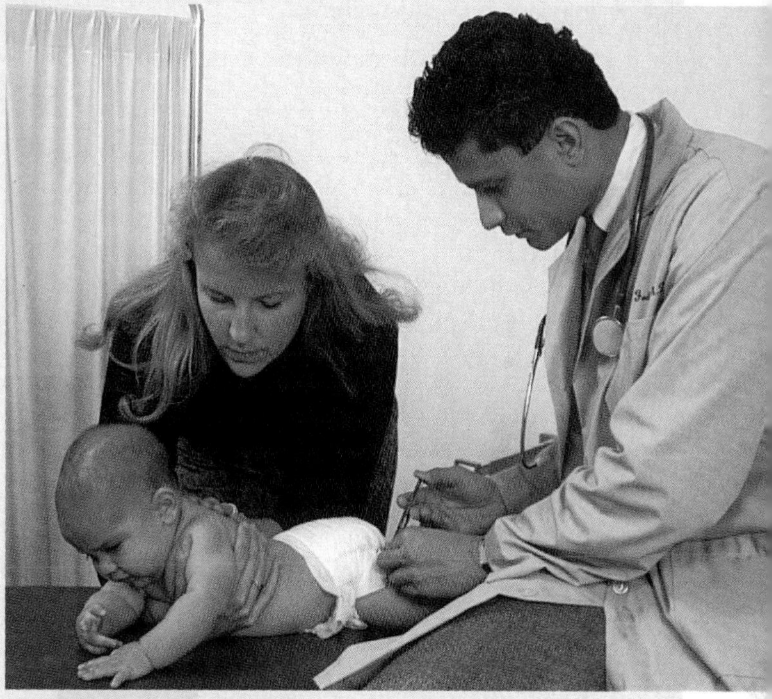

Figure 7B

Suggested immunization schedule for infants and young children. Children who are not immunized are subject to childhood diseases that can cause serious health consequences.

Immunization protects children and adults from diseases. The success of immunization is witnessed by the fact that the vaccination against smallpox is no longer required because the disease has been eradicated. However, parents today often fail to get their children immunized because they don't realize the importance of it or can't bear the expense. Therefore, we might read in the newspapers about an outbreak of measles at a college or hospital in the United States because adults were not immunized as children. A recommended immunization schedule for children is given in figure 7B, and the United States is now committed to the goal of immunizing all children against the common types of childhood diseases listed. Diphtheria, whooping cough, and an *Haemophilus influenzae* infection are all life-threatening respiratory diseases. Tetanus is characterized by muscular rigidity, including a locked jaw. These extremely serious infections are all caused by bacteria; the rest of the diseases listed are caused by viruses. Polio is a type of paralysis; measles and rubella, sometimes called German measles, are characterized by skin rashes; and mumps is characterized by enlarged parotid and other salivary glands.

Bacterial diseases are subject to cure by antibiotic therapy such as penicillin, streptomycin, tetracycline, and erythromycin. Even so, it is better to rely on immunization rather than antibiotic therapy if possible. Some patients are allergic to antibiotics, and the reaction can be fatal. Antibiotics not only kill off disease-causing bacteria, they also reduce the number of beneficial bacteria in the intestinal tract and elsewhere. These beneficial bacteria may have checked the spread of pathogens that now are free to multiply and to invade the body. This is why antibiotic therapy is often followed by a vaginal yeast infection in women. Antibiotic therapy also leads to strains that are resistant and therefore difficult to cure even with antibiotics. As mentioned previously, there are now resistant strains of bacteria that cause gonorrhea, a disease that has no vaccine.

Therefore, everyone should avail themselves of appropriate vaccinations. Preventing a disease by becoming actively immune to it is preferable to becoming ill and needing antibiotic therapy in order to be cured.

Lymphokines: Boost White Blood Cells

Lymphokines are being investigated as possible adjunct therapy for cancer and AIDS because they stimulate white blood cell formation and/or function. Both interferon and various other types of lymphokines called interleukins have been used as immunotherapeutical drugs, particularly to potentiate the ability of the individual's own T cells (and possibly B cells) to fight cancer.

Interferon, discussed previously on page 133, is a substance produced by leukocytes, fibroblasts, and probably most cells in response to a viral infection. When it is produced by T cells, interferon is called a lymphokine. Interferon still is being investigated as a possible cancer drug, but so far it has proven to be effective only in certain patients, and the exact reasons as yet cannot be discerned.

When and if cancer cells carry an altered protein on their cell surface, they should be attacked and destroyed by cytotoxic T cells. Whenever cancer does develop, it is possible that the cytotoxic T cells have not been activated. In that case, lymphokines might awaken the immune system and lead to the destruction of the cancer. In one technique being investigated, researchers first withdraw T cells from the patient and activate the cells by culturing them in the presence of an interleukin. The cells then are reinjected into the patient, who is given doses of interleukin to maintain the killer activity of the T cells.

Those who are actively engaged in interleukin research believe that interleukins soon will be used as adjuncts for vaccines, for the treatment of chronic infectious diseases, and perhaps for the treatment of cancer. Interleukin antagonists also may prove helpful in preventing skin and organ rejection, autoimmune diseases, and allergies.

> The interleukins and other lymphokines show some promise of potentiating the individual's own immune system.

Monoclonal Antibodies: Same Specificity

As previously discussed, every plasma cell derived from the same B cell secretes antibodies against a specific antigen. These are **monoclonal antibodies** because all of them are the same type (*mono*) and because they are produced by plasma cells derived from the same B cell (*clone*).

One method of producing monoclonal antibodies in vitro (in laboratory glassware) is depicted in figure 7.10. B lymphocytes are removed from the body (today, usually a mouse) and are exposed to a particular antigen. The activated B lymphocytes are fused with myeloma cells (malignant plasma cells that live and divide indefinitely). The fused cells are called hybridomas; *hybrid* because they result from the fusion of 2 different cells, and *oma* because one of the cells is a cancer cell.

At present, monoclonal antibodies are being used for quick and certain diagnosis of various conditions. For example, a particular hormone is present in the urine of a pregnant woman. A monoclonal antibody can be used to detect the hormone and so indicate that the woman is pregnant. Monoclonal antibodies also are used to identify infections. They are so accurate they can even sort out the different types of T cells in a blood sample. And because they can distinguish between cancer and normal tissue cells, they are used to carry radioactive isotopes or toxic drugs only to tumors so these can be selectively destroyed.

Monoclonal antibodies are considered to be a biotechnology product because the production process makes use of a living system to mass-produce the product.

> Monoclonal antibodies are produced in pure batches—they all react to just one type of molecule (antigen); therefore, they can distinguish one cell, or even one molecule, from another.

Immunological Side Effects and Illnesses

The immune system protects us from disease because it can tell self from nonself. Sometimes, however, the immune system is underprotective, as when an individual develops cancer, or is overprotective, as when individuals cannot tolerate certain types of blood.

Allergies: Overactive Immune System

Allergies are caused by an overactive immune system, which forms antibodies to substances that usually are not recognized as foreign substances. Unfortunately, allergies usually are accompanied by coldlike symptoms, or even at times, by severe systemic reactions, such as anaphylactic shock, which is a sudden drop in blood pressure.

Of the 5 varieties of antibodies—IgG, IgM, IgA, IgD, and IgE (see table 7.1)—IgE antibodies cause allergies. IgE antibodies are found in the bloodstream; but they, unlike the other types of antibodies, also reside in the membrane of *mast cells,* found in the tissues. As mentioned, mast cells are basophils that have left the bloodstream and taken up residence in the tissues. In any case, when the *allergen,* an antigen that provokes an allergic reaction, attaches to the IgE antibodies on mast cells, these cells release histamine and other substances, which cause mucus secretion and airway constriction. This results in the characteristic symptoms of allergy. On occasion, basophils and other white blood cells release these substances into the bloodstream. The increased capillary permeability that results from this can lead to fluid loss and shock.

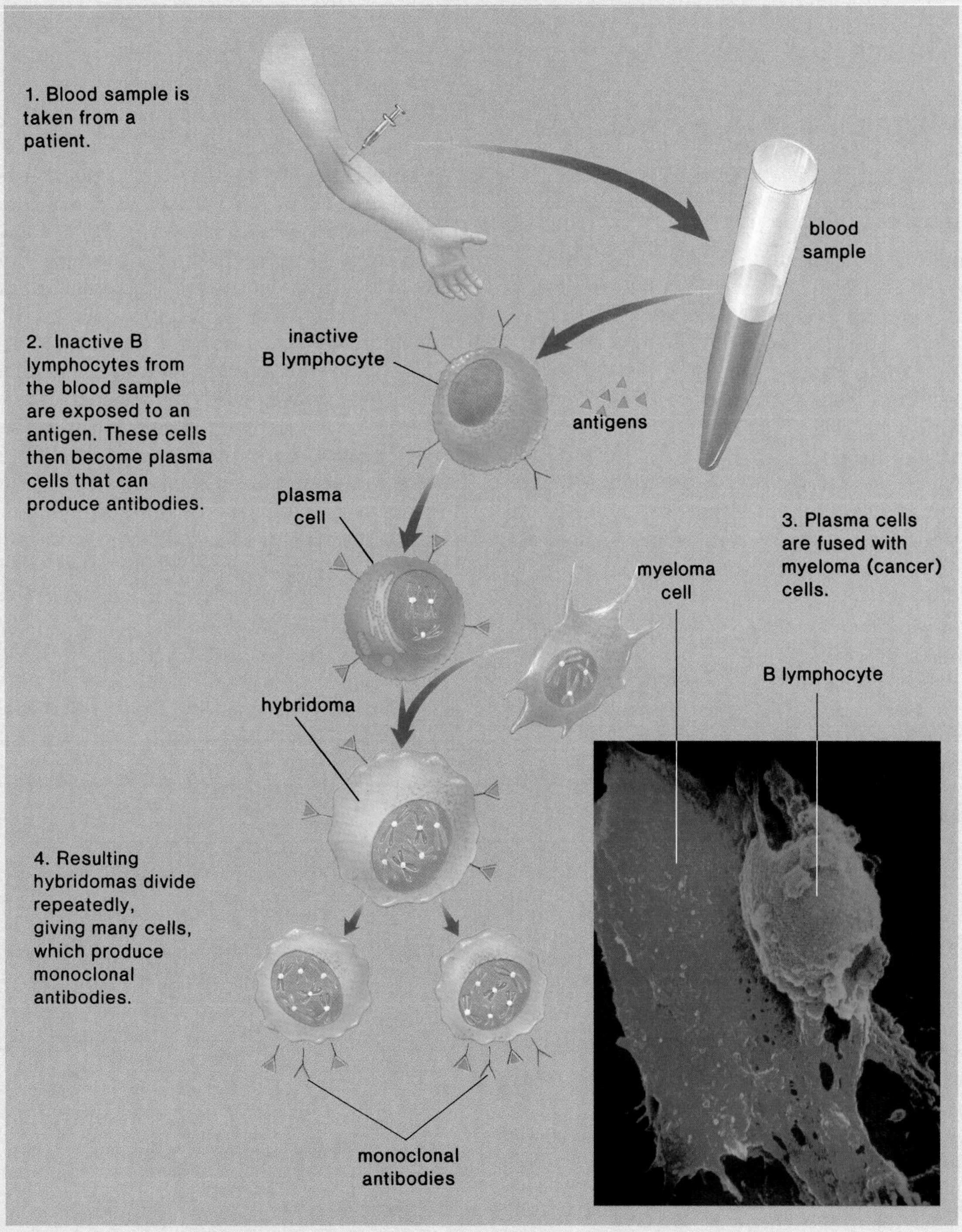

Figure 7.10

One possible method of producing human monoclonal antibodies.

Allergy shots sometimes prevent the onset of allergic symptoms. Injections of the allergen cause the body to build up high quantities of IgG antibodies, and these combine with allergens received from the environment before they have a chance to reach the IgE antibodies located in the membrane of mast cells.

> Histamine and other substances released by mast cells cause allergic symptoms.

Tissue Rejection: Foreign MHC Proteins

Certain organs, such as skin, the heart, and the kidneys, could be transplanted easily from one person to another if the body did not attempt to *reject* them. Rejection occurs because cytotoxic T cells bring about disintegration of foreign tissue in the body.

Organ rejection can be controlled by careful selection of the organ to be transplanted and the administration of immunosuppressive drugs. It is best if the transplanted organ has the same type of MHC proteins as those of the recipient, because cytotoxic T cells can recognize foreign MHC proteins. The immunosuppressive drug cyclosporine has been used for many years. A new drug, tacrolimus (formerly known as FK-506), shows some promise, especially in liver transplant patients. Both drugs, which act by inhibiting the production of interleukin-2, are known to adversely affect the kidneys as a toxic side effect.

> When an organ is rejected, the immune system is attacking cells that bear different MHC proteins from those of the individual.

Autoimmune Diseases: The Body Attacks Itself

Certain human illnesses are referred to as **autoimmune diseases** because they are due to an attack on tissues by the body's own antibodies and T cells. Exactly what causes autoimmune diseases is not known, but they seem to appear after the individual has recovered from an infection. Some bacteria have been observed to produce toxic products that can cause T cells to bind prematurely to macrophages. Perhaps at that time the T cell learns to recognize the body's own tissues. This might be the cause of at least some autoimmune diseases. In myasthenia gravis, neuromuscular junctions do not work properly and muscular weakness results. In multiple sclerosis (MS), the myelin sheath of nerve fibers is attacked, and this causes various neuromuscular disorders. A person with systemic lupus erythematosus (SLE) has various symptoms prior to death due to kidney damage. In rheumatoid arthritis, the joints are affected. It is suspected that heart damage following rheumatic fever and type 1 diabetes are also autoimmune illnesses. There are no cures for autoimmune diseases.

> Autoimmune diseases seem to be preceded by an infection that results in cytotoxic T cells attacking the body's own organs.

Summary

The lymphatic system consists of lymphatic vessels and lymphoid organs. The lymphatic vessels collect fat molecules at lacteals and excess tissue fluid and carry these to the cardiovascular system. Lymphocytes are produced and accumulate in the lymphoid organs (bone marrow, thymus, lymph nodes, and spleen).

Immunity involves nonspecific and specific defenses. Nonspecific defenses include barriers to entry, the inflammation reaction, and protective proteins. Specific defenses require 2 types of lymphocytes, both of which are produced in the bone marrow. B cells mature in the bone marrow, and T cells mature in the thymus.

B cells directly recognize an antigen and give rise to antibody-secreting plasma cells and memory B cells if stimulated to do so by helper T cells. Memory B cells

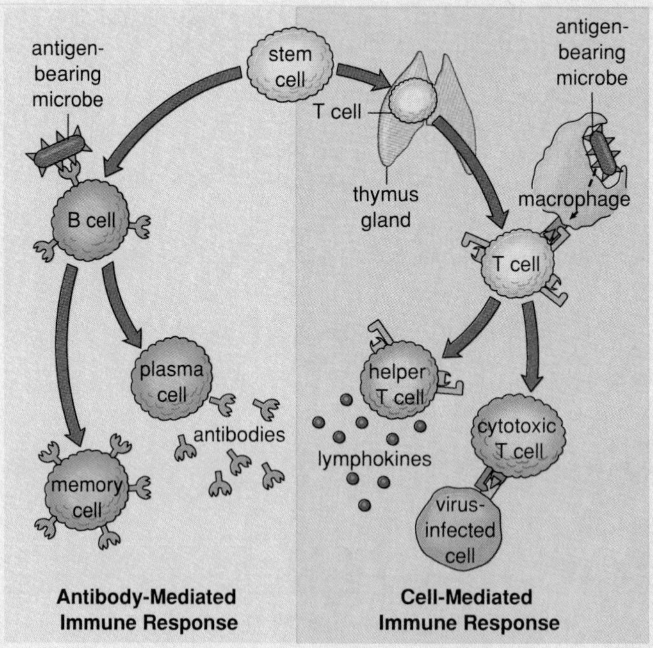

respond if the same antigen enters the body at a later date. For the T cell to recognize an antigen, the antigen must be presented by an APC, usually a macrophage. There are 4 types of T cells. Cytotoxic T cells kill cells on contact; helper T cells stimulate other immune cells and produce lymphokines; suppressor T cells suppress the immune response; and memory T cells remain in the body to provide long-lasting immunity.

Immunity can be induced in various ways. Vaccines are available to promote long-lived active immunity, and antibodies sometimes are available to provide an individual with short-lived passive immunity. Lymphokines, notably interferon and interleukins, are used in an attempt to promote the body's ability to recover from cancer and to treat AIDS.

Immunity has certain undesirable side effects. Allergies result when an overactive immune system forms antibodies to substances not normally recognized as foreign. Cytotoxic T cells attack transplanted organs, although immunosuppressive drugs are available. Autoimmune illnesses occur when antibodies and T cells attack the body's own tissues.

Study and Thought Questions

1. What is the lymphatic system, and what are its 3 functions? (p. 128)

   Critical Thinking Why would you expect edema when blood pressure rises rather than when it decreases?

2. Describe the structure and the function of the bone marrow, lymph nodes, the spleen, and the thymus. (pp. 129–130)

   Ethical Issue John Moore's spleen (removed for medical reasons) was used to develop the technique for mass-production of a stem cell growth factor. Researchers were handsomely rewarded, but John was not. Should he have been? Why or why not?

3. What are the body's nonspecific defense mechanisms? (p. 130)
4. Describe the inflammatory reaction, and give a role for each type of cell and molecule that participates in the reaction. (pp. 130–131)
5. B cells are responsible for which type of immunity? What is the clonal selection theory? (p. 134)
6. Describe the structure of an antibody, including the terms *variable regions* and *constant regions.* (p. 135)
7. Name the 4 types of T cells, and state their functions. (p. 136)
8. Explain the process by which a T cell is able to recognize an antigen. (p. 137)

   Critical Thinking Name 2 ways that immune cells communicate with one another.

9. How is active immunity achieved? How is passive immunity achieved? (p. 138)
10. What are lymphokines, and how are they used in immunotherapy? (p. 140)
11. How are monoclonal antibodies produced, and what are their applications? (p. 140)

    Critical Thinking Explain the term *monoclonal,* and tell why monoclonal antibodies are specific.

12. Discuss allergies, tissue rejection, and autoimmune diseases as they relate to the immune system. (pp. 141–142)

Writing Across the Curriculum

1. Argue both for and against the suggestion that the lymphatic system should be considered part of the blood circulation system.
2. There are more cells in the immune system than any other system. What does this tell us about the environment of organisms?
3. The immune system has various ways of attacking foreign antigens. Does this seem to be a duplication of effort or a necessity?

Objective Questions

1. Lymphatic vessels collect excess ____________ and return it to the _________________ veins.
2. The function of lymph nodes is to ______________ the lymph.
3. T lymphocytes mature in the _________________.
4. A stimulated B cell produces antibody-secreting ____________ cells and ____________ cells, which are ready to produce the same type of antibody at a later time.
5. B cells are responsible for _____________-mediated immunity.
6. For a T cell to recognize an antigen, it must be presented by an _______________ along with an MHC protein.
7. T cells produce ___________, which are stimulatory chemicals for all types of immune cells.
8. Cytotoxic T cells are responsible for ___________-mediated immunity.
9. Allergic reactions are associated with the release of __________ and other substances from mast cells.
10. The body recognizes foreign cells because they bear different ______________ proteins than the body's cells.
11. Immunization with ______________ brings about active immunity.
12. Hybridomas produce ______________ antibodies.
13. Give the function of the 4 types of T cells shown.

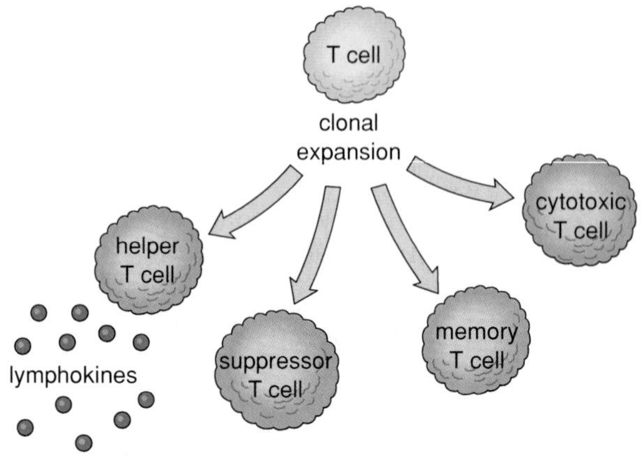

Selected Key Terms

antibody A protein produced in response to the presence of an antigen; each antibody combines with a specific antigen. 133

antigen (ant´ i-jen) A foreign substance, usually a protein, that stimulates the immune system to react, such as to produce antibodies. 133

autoimmune disease A disease that results when the immune system mistakenly attacks the body's own tissues. 142

B lymphocyte A lymphocyte that matures in the bone marrow and, when stimulated by the presence of a specific antigen, gives rise to antibody-producing plasma cells. 133

complement system A series of proteins in plasma that form a nonspecific defense mechanism against a microbe invasion; complements the antigen-antibody reaction. 133

cytotoxic T cell T lymphocyte that attacks and kills antigen-bearing cells; killer T cell. 136

helper T cell T lymphocyte that releases lymphokines and stimulates other immune cells to perform their respective functions. 136

inflammatory reaction A tissue response to injury that is characterized by redness, swelling, pain, and heat. 130

interferon (in″ ter-fēr´ on) A protein formed by a cell infected with a virus that can increase the resistance of other cells to the virus. 133

lymph Fluid having the same composition as tissue fluid and carried in lymphatic vessels. 129

lymphatic system A one-way vascular system that takes up excess fluid in the tissues, filters it, and transports it to cardiovascular veins in the shoulders. 128

lymphokine (lim´ fo-kīn) Molecule secreted by T lymphocytes that has the ability to affect the activity of all types of immune cells. 136

memory B cell Persistent population of B cells ready to produce antibodies specific to a particular antigen; accounts for the development of active immunity. 134

memory T cell Persistent population of T cells ready to recognize an antigen that previously invaded the body. 136

MHC protein A membrane protein that serves to identify the cells of a particular individual. 137

monoclonal antibody Antibody of one type produced by a hybridoma—a lymphocyte that has fused with a cancer cell. 140

plasma cell A differentiated B lymphocyte that is specialized to mass-produce antibodies. 134

pus Thick, yellowish fluid composed of dead phagocytes, dead tissue, and bacteria. 131

suppressor T cell T lymphocyte that suppresses certain other T and B lymphocytes from continuing to divide and perform their respective functions. 136

T lymphocyte A lymphocyte that matures in the thymus and occurs in 4 varieties, one of which kills antigen-bearing cells outright. 133

vaccine Antigens prepared in such a way that they can promote active immunity without causing disease. 138

Chapter 8

Respiratory System

Fish take their oxygen from the water. Why must we take our oxygen with us when we descend into the depths? For one thing, we are warm blooded and therefore require more oxygen per unit of time than do fish. Water contains 5–7 ppt (parts of oxygen per thousands parts of water), but air contains 210 ppt. If our oxygen demands were lower, could we breathe water? The answer seems to be yes, particularly if the water is cold. Scientists have observed that when frigid water splashes over the forehead and nose of whales and seals, nerves signal the brain to divert oxygen-rich blood from the limbs to the heart and brain. It can be reasoned that water actually entering the lungs would further cool the body and slow down the metabolic rate, thereby reducing the brain's oxygen requirements. It was not viewed as astonishing, then, when physicians were able to revive a toddler who fell into an ice-cold creek near Salt Lake City and was not found for 66 minutes. Once the child's temperature rose from 19° C to 25° C, her heart began beating and her pupils reacted to light.

Chapter Concepts

1 As air passes along the respiratory tract, it is filtered, warmed, and saturated with water before gas exchange takes place across a very extensive moist surface. 147

2 During inspiration, the pressure in the lungs decreases and then air comes rushing in. During expiration, increased pressure in the thoracic cavity causes air to leave the lungs. 151

3 During external respiration, the respiratory pigment hemoglobin combines with oxygen in the lungs, and during internal respiration, hemoglobin gives up oxygen to the cells. 154

4 Hemoglobin also aids in the transport of carbon dioxide from the tissues to the lungs. 156

5 The respiratory tract is especially subject to disease because it is exposed to infectious agents. Polluted air contributes to 2 major lung disorders—emphysema and cancer. 159

Chapter Outline

Breathing is more continuously necessary than eating. While it is possible to stop eating altogether for several days, it is not possible to remain alive for longer than several minutes without breathing. The normal breathing rate is about 14 to 20 times per minute. The more energy expended, the greater the breathing rate. The average young adult male utilizes about 250 ml of oxygen per minute in a basal, or restful, state. Exercise and the digestion of food raise the need for oxygen. The average amount of oxygen needed with mild exercise is 500 ml of oxygen per minute.

Breathing supplies the body with the oxygen (O_2) needed for aerobic cellular respiration, as indicated in the following equation:

$$38\ ADP + 36\ (P) \rightarrow 36\ ATP$$

$$C_6H_{12}O_6 + 6\ O_2 \longrightarrow 6\ H_2O + 6\ CO_2$$

As glucose is broken down to water and carbon dioxide, ATP molecules are formed.

Breathing not only draws oxygen into the body, it also pushes carbon dioxide out. It is only the first step of respiration, which can be said to include the following steps (fig. 8.1):

1. **Breathing:** entrance or exit of air into and out of the lungs;
2. **External and internal respiration:** External respiration is the movement of the gases oxygen (O_2) and carbon dioxide (CO_2) between air and blood; internal respiration is the movement of these gases between blood and tissue fluid.
3. **Aerobic cellular respiration:** production of ATP in cells.

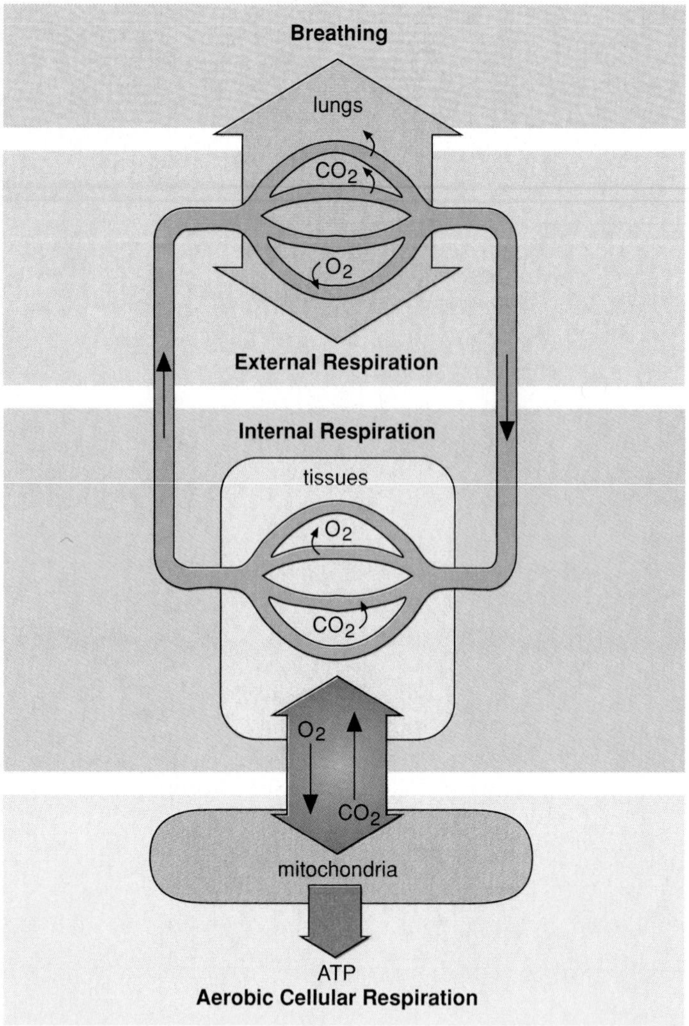

Figure 8.1

Respiration is divided into 4 components: breathing brings oxygen into the lungs; external respiration is the exchange of gases between lungs and blood; internal respiration is the exchange of gases between blood and tissues; and aerobic cellular respiration is the production of ATP in cells—an oxygen-requiring process.

Table 8.1

Path of Air

Structure	Description	Function
Nasal cavities	Hollow spaces in nose	Filter, warm, and moisten air
Pharynx	Chamber behind oral cavity and between nasal cavity and larynx	Connection to surrounding regions
Glottis	Opening into larynx	Passage of air into larynx
Larynx	Cartilaginous organ that contains vocal cords (voice box)	Sound production
Trachea	Flexible tube that connects larynx with bronchi (windpipe)	Passage of air to bronchi
Bronchi	Major divisions of trachea that enter lungs	Passage of air to each lung
Bronchioles	Branched tubes that lead from the bronchi to the alveoli	Passage of air to each alveolus
Lungs	Soft, cone-shaped organs that occupy a large portion of the thoracic cavity	Gas exchange

Respiratory Tract

During **inspiration** (breathing in) and **expiration** (breathing out), air is conducted toward or away from the lungs by a series of cavities, tubes, and openings, listed in order in table 8.1 and illustrated in figure 8.2.

As air moves in along the air passages, it is filtered, warmed, and moistened. Filtering is accomplished by coarse hairs and cilia in the region of the nostrils and by cilia alone in the rest of the nose and the trachea. In the nose, the hairs and the cilia act as a screening device. In the trachea, cilia beat upward, carrying mucus, dust, and occasional bits of food that "went down the wrong way" into the pharynx, where the accumulation can be swallowed or expectorated. The air is warmed by heat given off by the blood vessels lying close to the surface of the lining of the air passages, and it is moistened by the wet surface of these passages.

Conversely, as air moves out during expiration, it cools and loses its moisture. As the air cools, it deposits its moisture on the lining of the windpipe and the nose, and the nose may even drip as a result of this condensation. The air still retains so much moisture, however, that upon expiration on a cold day, it condenses and forms a small cloud.

> Air is warmed, filtered, and moistened as it moves from the nose toward the lungs.

Each portion of the respiratory tract has its own structure and function, as described in the sections that follow.

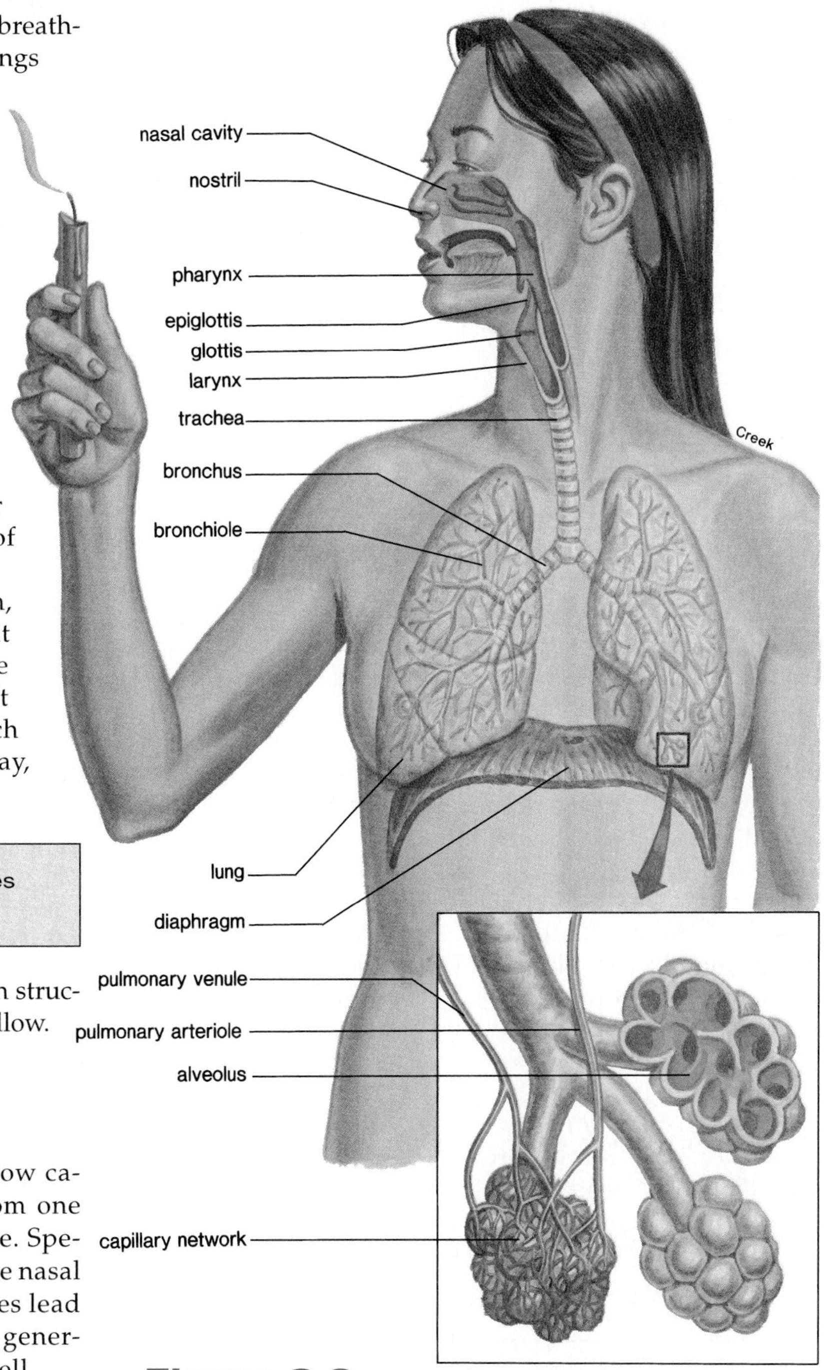

Figure 8.2

The human respiratory tract, with an enlargement showing the internal structure of a lung portion. Gas exchange occurs between the alveoli and blood within the surrounding capillary network. Notice that the pulmonary arteriole carries deoxygenated blood (colored blue) and the pulmonary venule carries oxygenated blood (colored red) (see p. 100).

The Nose: Two Cavities

The nose contains 2 *nasal cavities,* which are narrow canals with convoluted lateral walls separated from one another by a wall composed of bone and cartilage. Special ciliated cells in the narrow upper recesses of the nasal cavities (see fig. 12.3) act as odor receptors. Nerves lead from these cells to the brain, where the impulses generated by the odor receptors are interpreted as smell.

The tear (lacrimal) glands drain into the nasal cavities by way of tear ducts. For this reason, crying produces a runny nose. The nasal cavities also communicate with the cranial sinuses, air-filled, mucous membrane-lined spaces in the skull. If these membranes are inflamed due to a cold or an allergic reaction, mucus can accumulate in the sinuses, causing a sinus headache.

The nasal cavities empty into the nasopharynx, the upper portion of the pharynx. The *eustachian tubes* lead from the nasopharynx to the middle ears (see fig. 12.11).

> The nasal cavities, which receive air, open into the nasopharynx.

The Pharynx: Crossroads

Air taken in by either the nose or the mouth enters the *pharynx,* which is, essentially, the throat. In the pharynx, the air (trachea) and food (esophagus) passages temporarily join. The trachea, which lies in front of the esophagus, is normally open, allowing the passage of air, but the esophagus is normally closed and opens only when swallowing occurs. The larynx lies at the top of the trachea.

> Air from either the nose or the mouth enters the pharynx, as does food. The passage of air continues in the larynx and then the trachea proper.

The Larynx: Voice Box

The **larynx** can be imagined as a triangular box whose apex, the Adam's apple, is located at the front of the neck. At the top of the larynx is a variable-sized opening called the *glottis.* When food is swallowed, the glottis is covered by a flap of tissue called the *epiglottis* so that no food passes into the larynx. If, by chance, food or some other substance does enter the larynx, reflex coughing usually expels the substance. If this reflex is not sufficient, it may be necessary to resort to the Heimlich maneuver (fig. 8.3).

The **vocal cords** are mucous membrane folds supported by elastic ligaments, which occur at the edges of the glottis (fig. 8.4). These cords vibrate when air is expelled past them through the glottis, and their vibration produces sound. The high or low pitch of the voice depends upon the length, the thickness, and the degree of elasticity of the vocal cords and the tension at which they are held. The loudness, or intensity, of the voice depends upon the amplitude of the vibrations, or the degree to which vocal cords vibrate.

At the time of puberty, the growth of the larynx and the vocal cords is much more rapid and accentuated in the male than in the female, causing the male to have a more prominent Adam's apple and a deeper voice. The voice "breaks" in the young male due to his inability to control the longer vocal cords.

> The larynx is the voice box because it contains the vocal cords at the sides of the glottis, an opening covered by the epiglottis during swallowing.

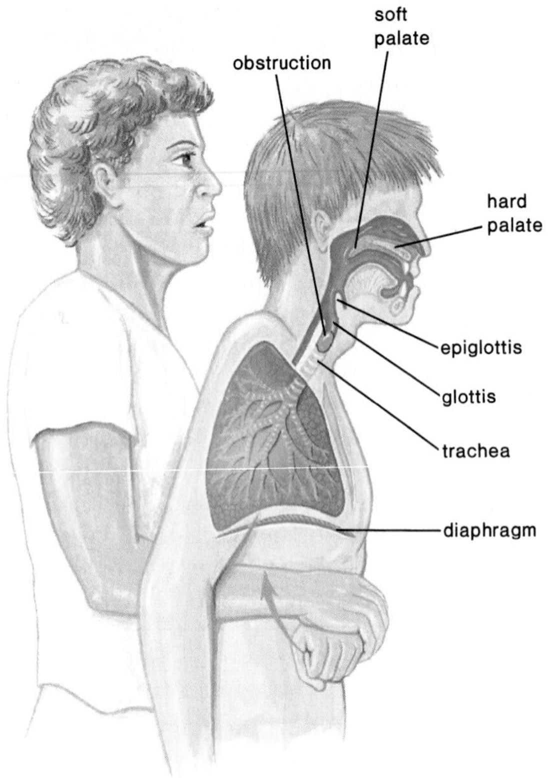

Figure 8.3

The Heimlich maneuver. More than 8 Americans choke to death each day on food lodged in the trachea. A simple process termed the abdominal thrust, or Heimlich maneuver, can save the life of a person who is choking. If the person is standing or sitting: (1) Stand behind the person or the person's chair, and wrap your arms around the choking person's waist; (2) grasp your fist with your other hand, and place the fist against the abdomen, slightly above the navel and below the rib cage; (3) press your fist into the abdomen with a quick upward thrust; (4) repeat several times if necessary. If the choking person is lying down: (1) Position this person so that the face is directed upward; (2) face the choking person, and kneel astride the hips; (3) with one of your hands on top of the other, place the heel of your bottom hand on the abdomen, slightly above the navel and below the rib cage; (4) press into the abdomen with a quick upward thrust; (5) repeat several times if necessary. If you are alone and choking, use anything that applies force just below your diaphragm. Press into a table or a sink, or use your own fist. The Heimlich maneuver should be used only when no air is being exchanged. If there is a partial rather than a complete airway obstruction, the Heimlich maneuver should *not* be used.

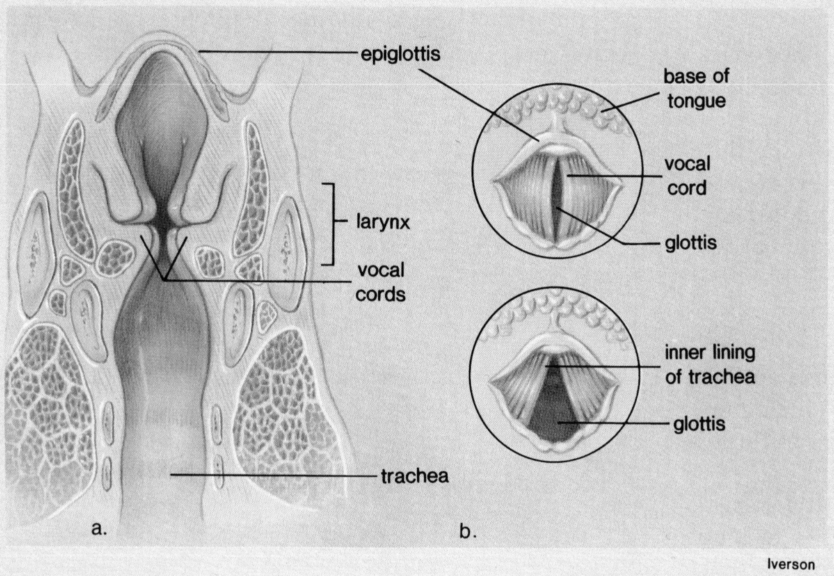

Figure 8.4

Placement of the vocal cords. The vocal cords are on either side of the glottis, an opening into the larynx. The epiglottis covers the glottis during swallowing. **a.** Longitudinal section of the larynx, showing the placement of the vocal cords. When air is expelled from the larynx, the cords vibrate. **b.** The glottis is narrow when we produce a high-pitched sound (*top*) and widens as the pitch deepens (*bottom*).

The Trachea: Windpipe

The **trachea** is a tube held open by C-shaped cartilaginous rings. Cilia that project from the epithelium of the trachea keep the lungs clean by sweeping mucus and debris toward the throat. Smoking is known to destroy the cilia, and consequently the soot in cigarette smoke collects in the lungs. Smoking is discussed more fully at the end of this chapter.

If the trachea is blocked because of illness or accidental swallowing of a foreign object, it is possible to insert a tube by way of an incision made in the trachea. This tube acts as an artificial air intake and exhaust duct. The operation is called a *tracheostomy.*

Bronchi: Air Tubes

The trachea divides into 2 *bronchi* (sing., **bronchus**), which lead respectively into the right and left lungs (fig. 8.5). The bronchi branch into a great number of smaller passages called **bronchioles.** The bronchi resemble the trachea in structure, but as the bronchiolar tubes divide and subdivide, their walls become thinner and the small rings of cartilage are no longer present. During an asthma attack, the bronchioles constrict even to the point of closing, and movement of air through the narrowed tubes may result in the characteristic wheezing. Each bronchiole terminates in an elongated space enclosed by a multitude of air pockets, or sacs, called *alveoli* (sing., **alveolus**) (see fig. 8.2). The alveoli make up the lungs.

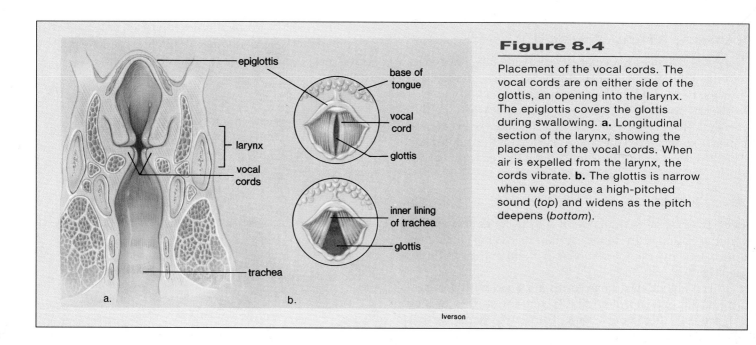

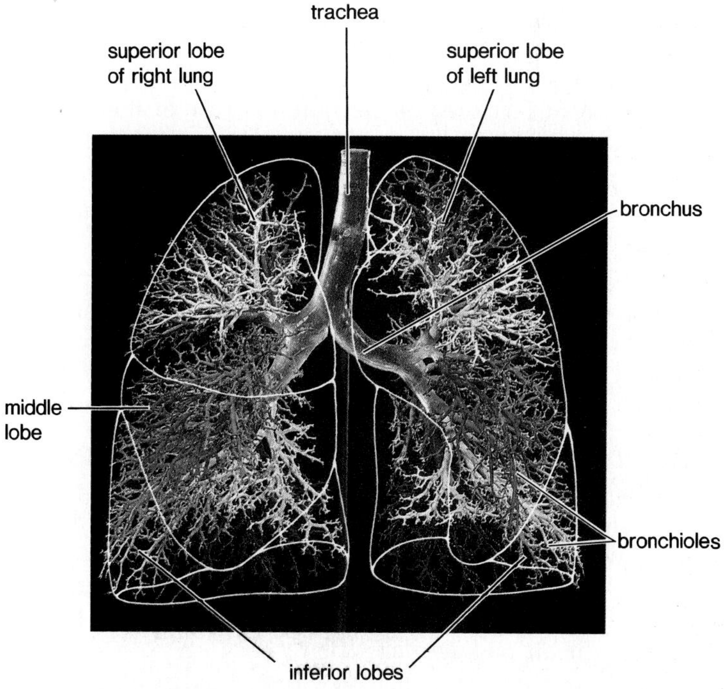

Figure 8.5

Airways to the lungs. The trachea divides into the bronchi, which give rise to the bronchioles. The bronchioles have many branches and terminate at the alveoli. Each bronchiole and its branches are similarly colored.

Lungs: Many Alveoli

The lungs are cone-shaped organs; the right lung has 3 lobes and the left lung has 2 lobes. A lobe is further divided into lobules, each of which has a bronchiole serving many alveoli. The lungs lie on either side of the heart in the thoracic (chest) cavity. The base of each lung is broad and concave so that it fits the convex surface of the diaphragm. The other surfaces of the lungs follow the contours of the ribs and the organs in the thoracic cavity.

Alveoli: 300 Million Air Sacs

Each alveolar sac is made up of simple squamous epithelium surrounded by blood capillaries. Gas exchange occurs between air in the alveoli and blood in the capillaries (see fig. 8.2).

The alveoli of human lungs are lined with a surfactant, a film of lipoprotein that lowers the surface tension and prevents them from closing. The lungs collapse in some newborn babies, especially premature infants, who lack this film. The condition, called *infant respiratory distress syndrome,* is now treatable by surfactant replacement therapy.

There are approximately 300 million alveoli, with a total cross-sectional area of 50–70 m^2. This is at least 40 times the surface area of the skin. Because of their many air spaces, the lungs are very light; normally, a piece of lung tissue dropped in a glass of water floats. Such a test can be used to determine if a deceased baby has taken a breath or not.

> Air moves from the trachea and the 2 bronchi, which are held open by cartilaginous rings, into the lungs. The lungs are composed of air sacs called alveoli.

The Mechanisms of Breathing

To understand **ventilation,** the manner in which air is drawn into and expelled from the lungs, it is necessary to remember first that when we are breathing, there is a continuous column of air from the pharynx to the alveoli of the lungs; that is, the air passages are open.

Secondly, the lungs lie within the sealed-off thoracic cavity. The **rib cage** forms the top and sides of the thoracic cavity. It contains the ribs, hinged to the vertebral column at the back and to the sternum (breastbone) at the front, and the intercostal muscles that lie between the ribs. The **diaphragm,** a dome-shaped horizontal muscle, forms the floor of the thoracic cavity.

The lungs are enclosed by the **pleural membranes**. An infection of the pleural membranes is called pleurisy. The outer pleural membrane adheres to the rib cage and the diaphragm, and the inner membrane is fused to the lungs. The 2 pleural layers lie very close to one another, separated only by a small amount of fluid. Normally, the intrapleural pressure is lower than atmospheric pressure by 4 mm Hg.

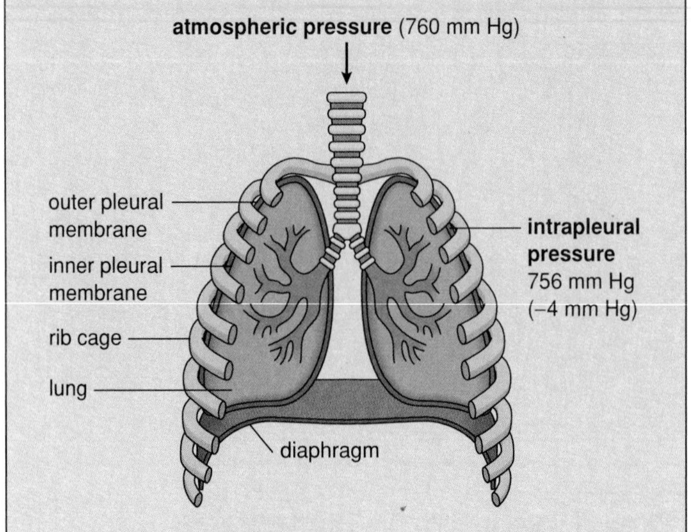

The importance of the reduced intrapleural pressure is demonstrated when, by design or accident, air enters the intrapleural space. The lungs collapse, and inspiration is impossible.

> The pleural membranes enclose the lungs and line the thoracic cavity. Intrapleural pressure is lower than atmospheric pressure.

Inspiration Is Active

The **respiratory center,** located in the medulla, consists of a group of neurons that exhibit an automatic rhythmic discharge that triggers inspiration. Carbon dioxide (CO_2) and hydrogen ions (H^+) are the primary stimuli that cause changes in the activity of this center. This center is not affected by low oxygen (O_2) levels. There also are chemoreceptors in the *carotid bodies,* located in the carotid arteries, and in the *aortic bodies,* located in the aorta, that respond primarily to hydrogen ion concentration [H^+] but also to the level of carbon dioxide and oxygen in blood. These bodies communicate with the respiratory center. When the level of carbon dioxide and hydrogen rise, the rate and depth of breathing increase.

> In humans, carbon dioxide and hydrogen ions are the primary stimuli regulating the breathing rate.

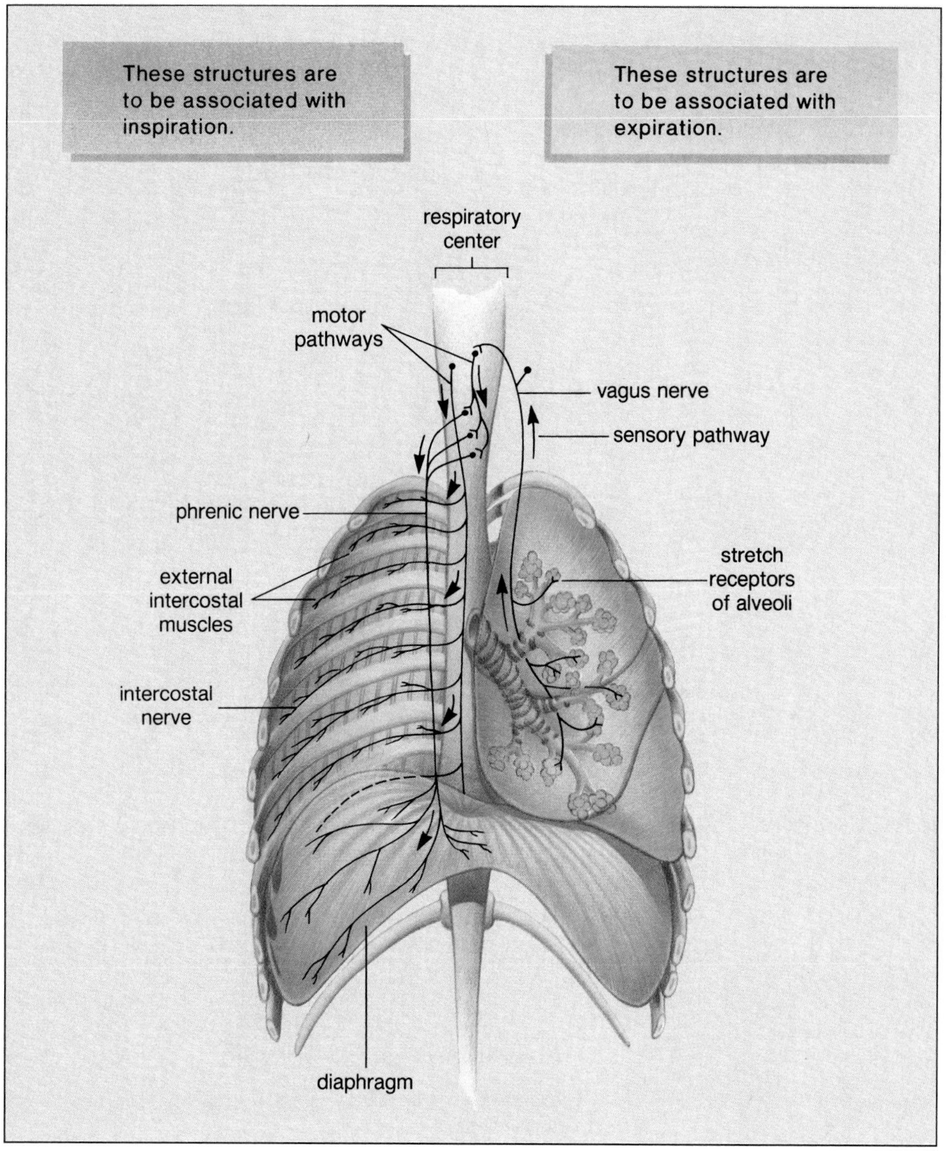

Figure 8.6

Nervous control of breathing. During inspiration, the respiratory center stimulates the intercostal (rib) muscles to contract via the intercostal nerves and the diaphragm to contract via the phrenic nerve. Should the lung volume increase above 1.5 liters, stretch receptors send inhibitory nerve impulses to the respiratory center via the vagus nerve. In any case, expiration occurs due to a lack of stimulation from the respiratory center to the diaphragm and intercostal muscles.

The respiratory center sends out nerve impulses by way of nerves to the diaphragm and the rib cage (fig. 8.6). In its relaxed state, the diaphragm is dome-shaped, but upon stimulation, it contracts and lowers. Also, the external intercostal muscles contract, and the rib cage moves upward and outward. Both of these contractions serve to increase the size of the thoracic cavity. As the thoracic cavity increases in size, the lungs expand. When the lungs expand, air pressure within the enlarged alveoli lowers and is immediately rebalanced by air rushing in through the nose or the mouth.

Inspiration is the active phase of breathing (fig. 8.7). During this time, the diaphragm and the rib muscles contract, intrapleural pressure decreases even more, the lungs expand, and air comes rushing in. Note that air comes in because the lungs already have opened up; air does not force the lungs open. This is why it is sometimes said that *humans breathe by negative pressure.* The creation of a partial vacuum sucks air into the lungs.

Expiration Is Usually Passive

When the respiratory center stops sending signals to the diaphragm and the rib cage, the diaphragm relaxes and it resumes its dome shape. The abdominal organs press up against the diaphragm, and the rib cage moves down and inward (fig. 8.7). Now, the elastic lungs recoil, and air is pushed out. The respiratory center acts rhythmically to bring about breathing at a normal rate and volume. If by chance we inhale more deeply, the lungs are expanded and the alveoli stretch. This stimulates stretch receptors in the alveolar walls, and they initiate inhibitory nerve impulses that travel from the inflated lungs to the respiratory center. This causes the respiratory center to stop sending out nerve impulses.

It is clear that while inspiration is the active phase of breathing, expiration is normally passive—the diaphragm and external intercostal muscles are relaxed when expiration occurs. In deeper and more rapid breathing, expiration can also be active. Contraction of internal intercostal muscles can force the rib cage to move downward and inward. Also, when the abdominal wall muscles are contracted, increased pressure helps to expel air.

During inspiration, due to nervous stimulation the diaphragm lowers and the rib cage lifts up and out. During expiration, due to a lack of nervous stimulation the diaphragm rises and the rib cage lowers.

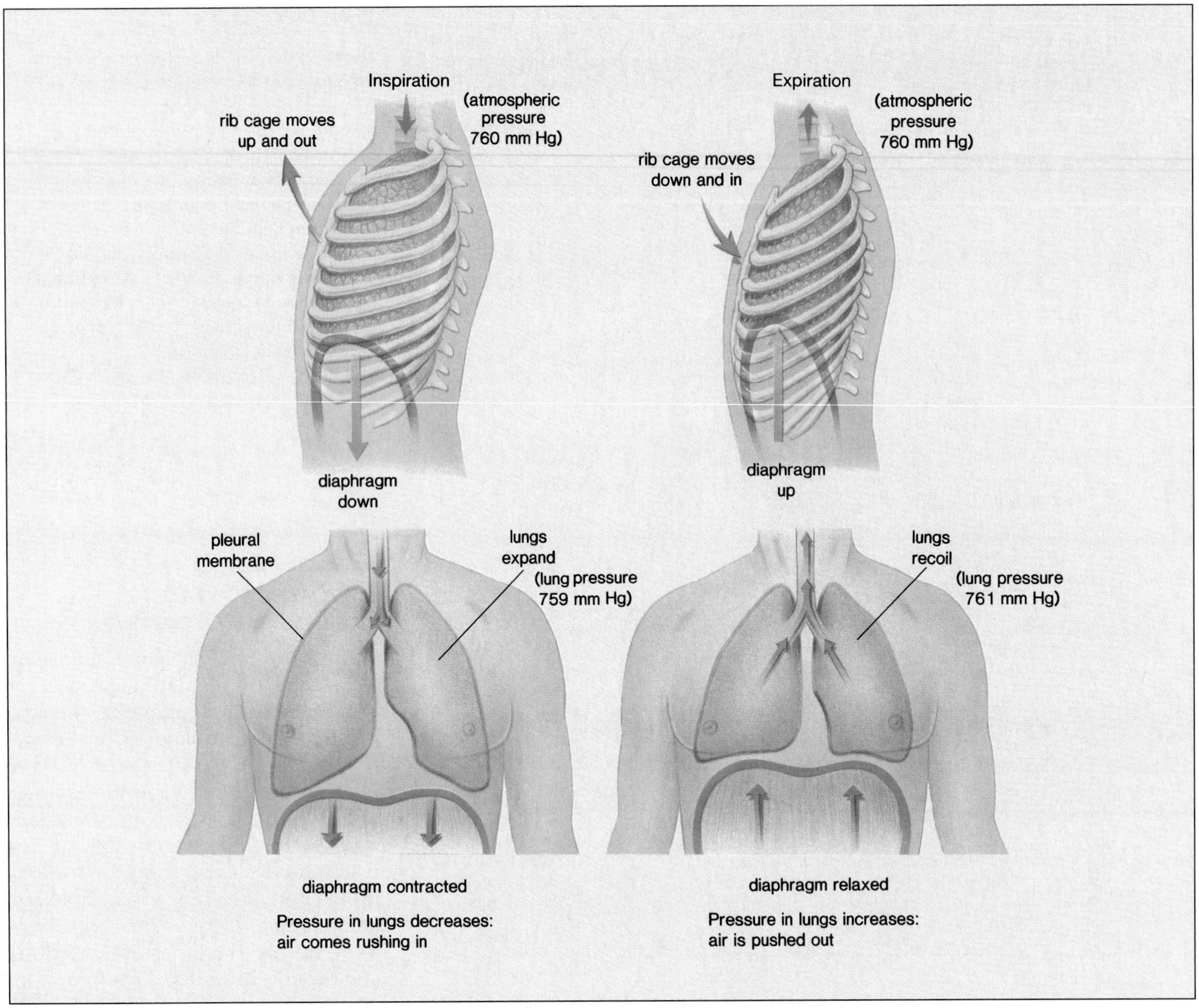

Figure 8.7

Inspiration versus expiration. During inspiration, the rib cage lifts up and out, the diaphragm lowers, the lungs expand, and air is drawn in. This sequence of events is only possible because the pressure within the intrapleural space, containing a thin film of fluid, is less than atmospheric pressure. During expiration, the rib cage lowers, the diaphragm rises, the lungs recoil, and air is forced out.

How Much Do Lungs Hold?

When we breathe, the amount of air moved in and out with each breath is called the **tidal volume.** Normally, the tidal volume is about 500 ml, but we can increase the amount inhaled and exhaled by deep breathing. The maximum volume of air that can be moved in and out during a single breath is called the **vital capacity** (fig. 8.8). First, we can increase inspiration by as much as 3,100 ml of air by forced inspiration. This is called the *inspiratory reserve volume.* Similarly, we can increase expiration by contracting the thoracic muscles. This is called the *expiratory reserve volume,* and it measures approximately 1,400 ml of air. Vital capacity is the sum of tidal, inspiratory reserve, and expiratory reserve volumes.

Note in figure 8.8 that even after very deep breathing, some air (about 1,000 ml) remains in the lungs; this is called the **residual volume.** This air is no longer useful for gas exchange purposes. In some lung diseases, such as emphysema (p. 159), the residual volume builds up because the individual has difficulty emptying the lungs. This means that the lungs tend to be filled with useless air, and as you can see from examining figure 8.8, the vital capacity is reduced.

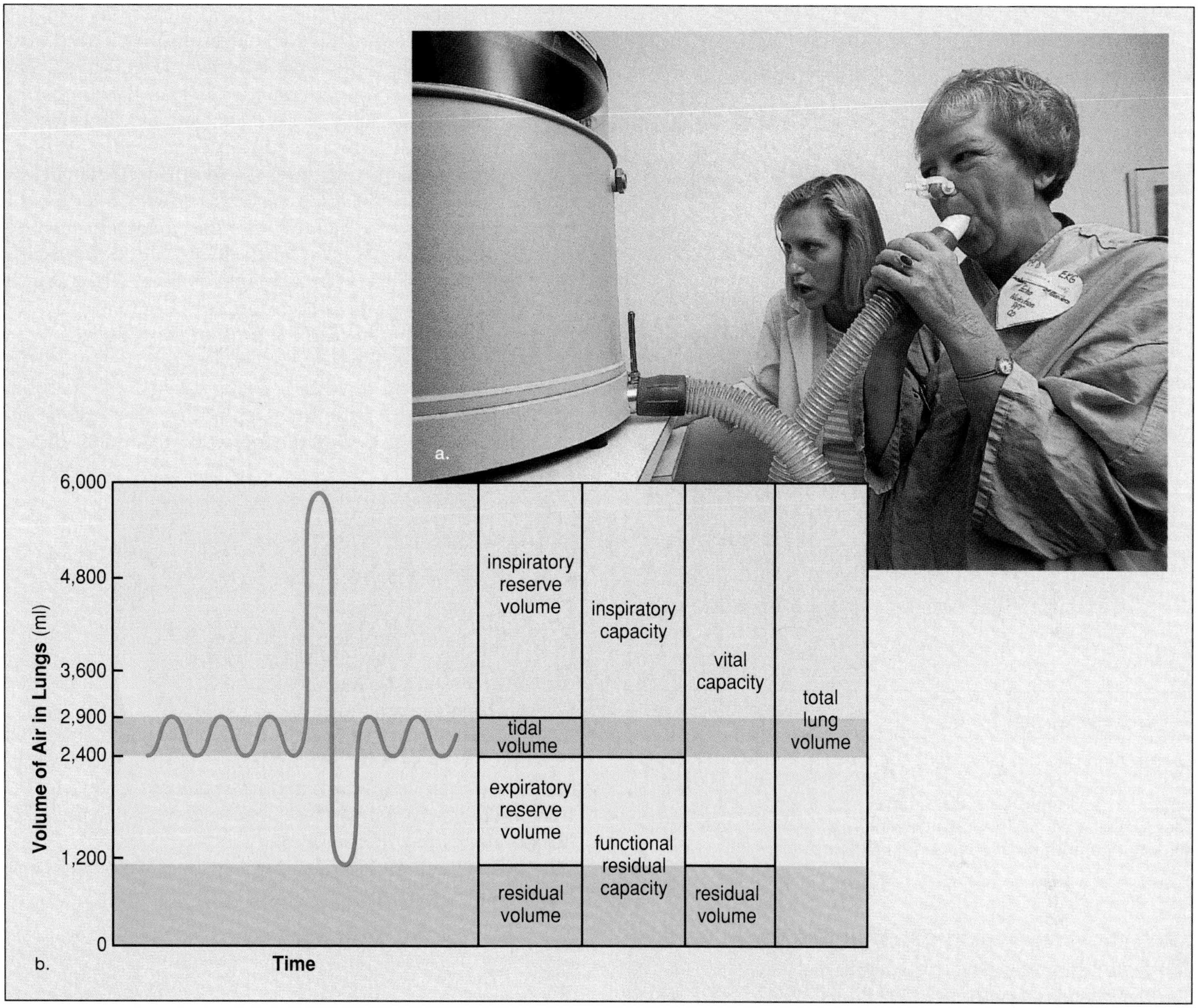

Figure 8.8

Vital capacity. **a.** This individual is using a spirometer, which measures the maximum amount of air that can be maximally inhaled and exhaled. When she inspires, a pen moves up, and when she expires, a pen moves down. **b.** The resulting pattern, such as the one shown here, is called a spirograph.

Dead Space in Airways

Some of the inspired air never reaches the lungs; instead it fills the conducting airways (fig. 8.9). These passages are not used for gas exchange and therefore are said to contain *dead space.* To ventilate the lungs, then, it is better to breathe slowly and deeply because this ensures that a greater percentage of the tidal volume reaches the lungs.

Breathing through a very long tube increases the amount of dead space beyond maximum inspiratory capacity. Thereafter, death will occur because the air inhaled never reaches the alveoli.

The volume of fresh air reaching the lungs can vary.

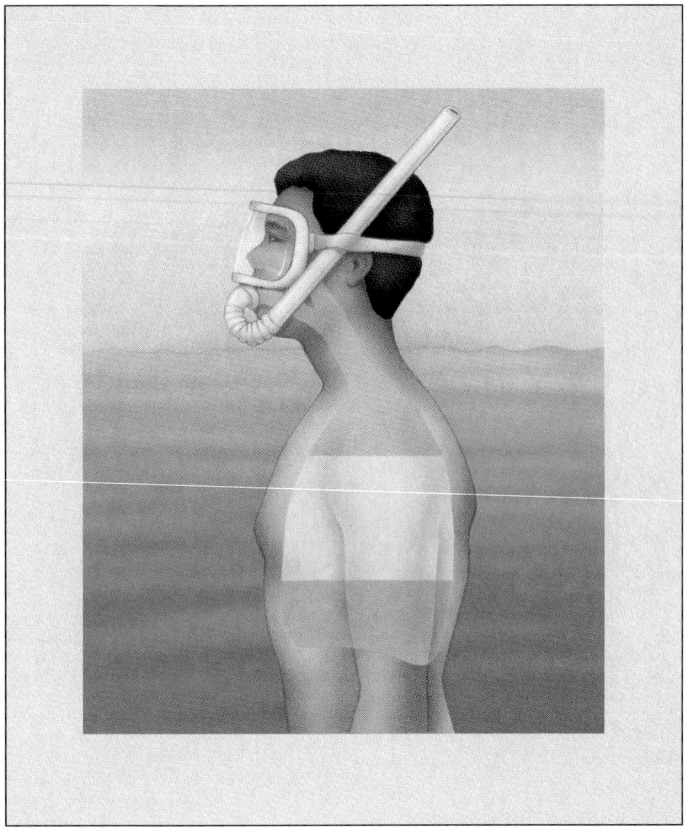

Figure 8.9

Distribution of air in the lungs. The air colored blue does not reach the alveoli immediately; therefore, this is called dead space. The air colored purple represents the amount of residual air that has not left the lungs. Only the air colored pink brings additional oxygen (O_2) for respiration.

External and Internal Respiration

Figure 8.10 shows both external respiration and internal respiration. The term *external respiration* refers to the exchange of gases between air in the alveoli and blood in the pulmonary capillaries. The term *internal respiration* refers to the exchange of gases between blood in systemic capillaries and tissue fluid.

External Respiration: Air Sacs and Blood

In the lungs, the walls of both an alveolus and a blood capillary consist of a thin, single layer of cells. Since neither wall offers resistance to the passage of gases, *diffusion* alone governs the exchange of oxygen (O_2) and carbon dioxide (CO_2) between alveolar air and blood. Active cellular absorption and secretion do not appear to play a role. Rather, the direction in which the gases move is determined by the pressure gradients between blood and inspired air.

Blood flowing into the lung capillaries has a higher concentration of carbon dioxide than atmospheric air. Therefore, *carbon dioxide diffuses out of blood into the alveoli.* The pressure pattern is the reverse for oxygen. Blood coming into the pulmonary capillaries is deoxygenated, and alveolar air is oxygenated; therefore, *oxygen diffuses into the capillary.* Breathing at high altitudes is less effective than at low altitudes because the air pressure is lower, making the concentration of oxygen (and other gases) lower than normal; therefore, less oxygen diffuses into blood. Breathing problems do not occur in airplanes because the cabin is pressurized to maintain an appropriate pressure. Emergency oxygen is available in case the pressure is reduced.

As blood enters the pulmonary capillaries (fig. 8.10), most of the carbon dioxide is being carried as bicarbonate ions (HCO_3^-). As the little remaining free carbon dioxide begins to diffuse out, the following reaction is driven to the right:

$$H^+ + HCO_3^- \longrightarrow H_2CO_3 \longrightarrow H_2O + CO_2\uparrow$$

bicarbonate ion

Up arrow indicates carbon dioxide is leaving the body.

The enzyme *carbonic anhydrase* (p. 157), present in red blood cells, speeds up the reaction. As the reaction proceeds, the respiratory pigment *hemoglobin* gives up the hydrogen ions (H^+) it has been carrying; HHb becomes Hb. Hb is called deoxyhemoglobin.

Now, hemoglobin more readily takes up oxygen and becomes oxyhemoglobin.

$$Hb + \downarrow O_2 \longrightarrow HbO_2$$

deoxyhemoglobin — oxyhemoglobin

"Down" arrow indicates that oxygen is entering the body.

It is remarkable that at the partial pressure[1] of oxygen in the lungs (PO_2 = about 100 mm Hg), hemoglobin is about 98% saturated (fig. 8.11). Hemoglobin takes up oxygen in increasing amounts as the PO_2 increases and likewise gives it up as the PO_2 decreases. The curve begins to level off at about 90 mm Hg. This means that hemoglobin easily retains oxygen in the lungs but tends to release it in the tissues. This effect is potentiated by the fact that hemoglobin takes up oxygen more readily in the cool temperature (fig. 8.11*a*) and neutral pH (fig. 8.11*b*) of

1. Air exerts pressure, and the amount of pressure each gas exerts in air is called its partial pressure, symbolized by a capital P.

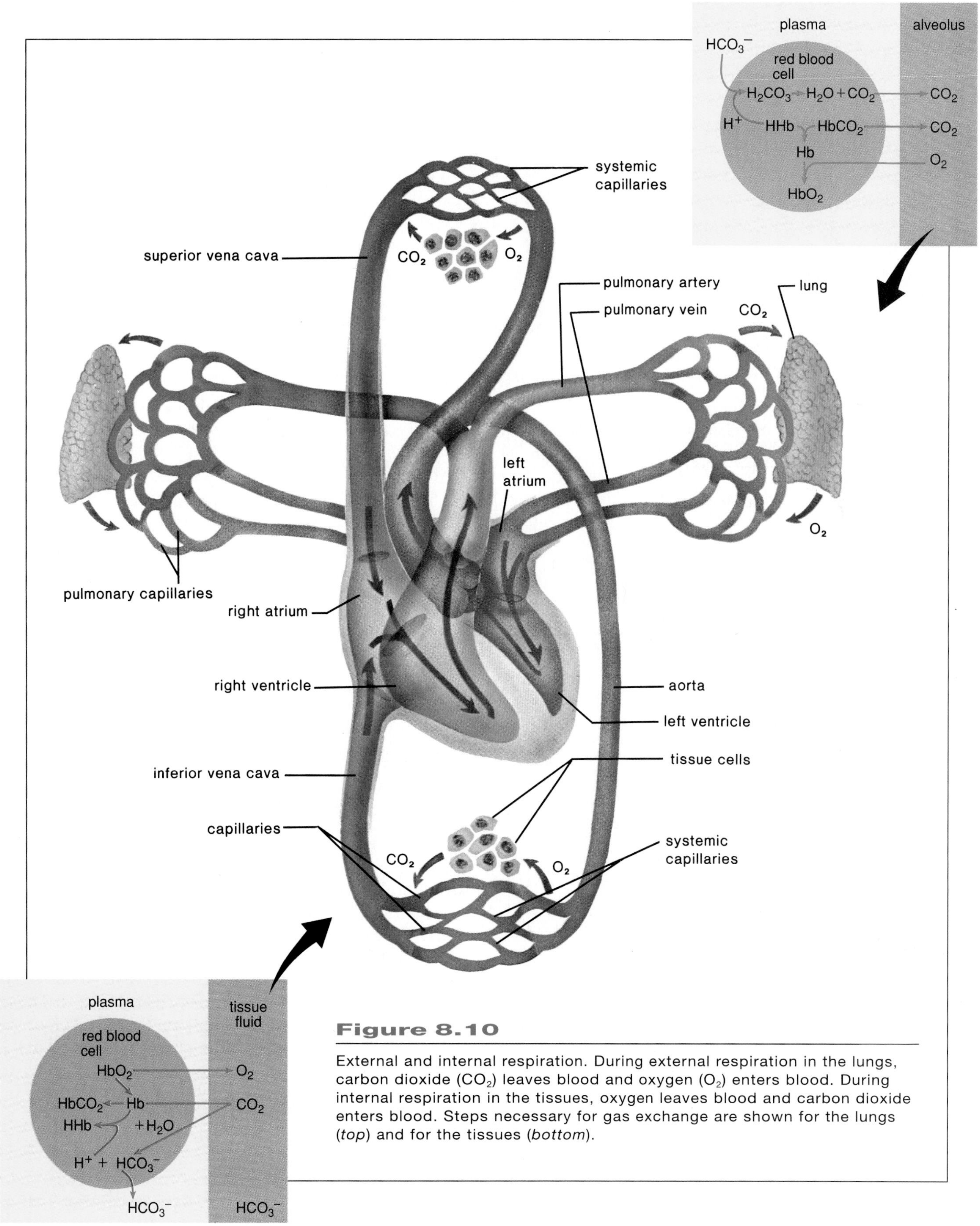

Figure 8.10

External and internal respiration. During external respiration in the lungs, carbon dioxide (CO_2) leaves blood and oxygen (O_2) enters blood. During internal respiration in the tissues, oxygen leaves blood and carbon dioxide enters blood. Steps necessary for gas exchange are shown for the lungs (*top*) and for the tissues (*bottom*).

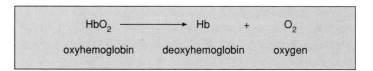

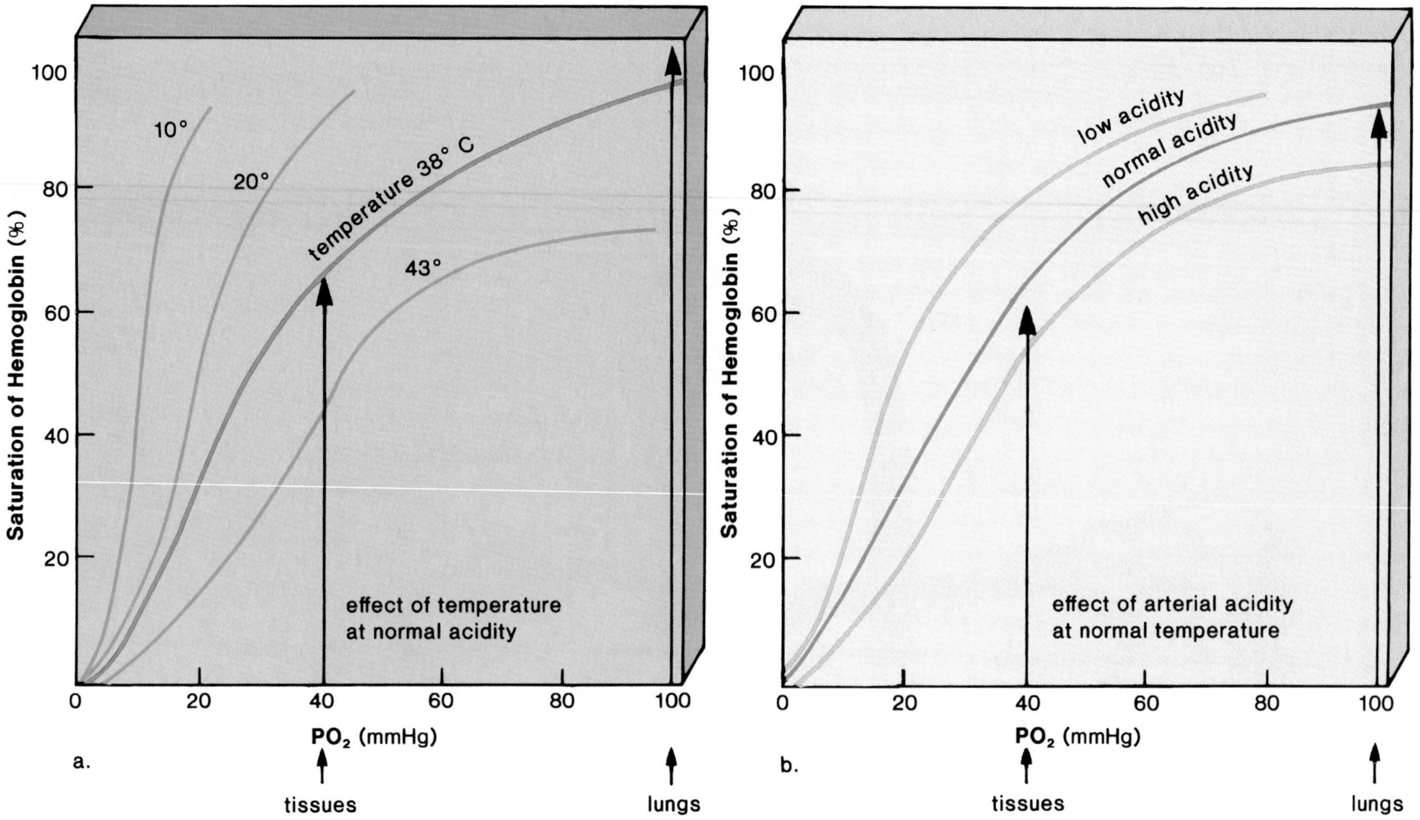

Figure 8.11

Effect of conditions on hemoglobin saturation. Hemoglobin becomes more saturated in the lungs because the partial pressure of oxygen (PO_2) increases. Oxygen take-up is also enhanced because **a.** temperature and **b.** acidity decrease in the lungs. The situation is the reverse in the tissues. Hemoglobin is about 60%–70% saturated in the tissues and about 98%–100% saturated in the lungs.

the lungs. On the other hand, it gives up oxygen more readily at the warmer temperature and more acidic pH of the tissues.[2]

> External respiration, the movement of oxygen (O_2) and carbon dioxide (CO_2) between air within alveoli and blood in pulmonary capillaries, is dependent on the process of diffusion.

Internal Respiration: Blood and Tissue Fluid

Blood that enters the systemic capillaries is bright red in color because red blood cells contain oxyhemoglobin. Oxyhemoglobin gives up oxygen, which diffuses out of blood into the tissues (see fig. 8.10).

$$HbO_2 \longrightarrow Hb + O_2$$

oxyhemoglobin — deoxyhemoglobin — oxygen

Oxygen diffuses out of blood into the tissues because the oxygen concentration of tissue fluid is low—the cells continuously use up oxygen in aerobic cellular respiration. *Carbon dioxide diffuses into blood from the tissues* because the carbon dioxide concentration of tissue fluid is high. Carbon dioxide, produced continuously by cells, collects in tissue fluid.

After carbon dioxide diffuses into blood, it enters the red blood cells, where a small amount is taken up by hemoglobin, forming **carbaminohemoglobin.** Most of the carbon dioxide combines with water forming carbonic acid (H_2CO_3), which dissociates to hydrogen ions (H^+) and bicarbonate ions (HCO_3^-). The increased concentration of carbon dioxide in the blood causes the reaction to proceed to the right.

2.	**pH**	**Temperature**
Lungs	7.40	37° C
Body	7.38	38° C

$$CO_2 + H_2O \rightleftharpoons H_2CO_3 \rightleftharpoons H^+ + HCO_3^-$$

carbon dioxide | water | carbonic acid | hydrogen ion | bicarbonate ion

The enzyme carbonic anhydrase, present in red blood cells, speeds up the reaction. The globin portion of hemoglobin combines with excess hydrogen ions produced by the reaction, and Hb becomes HHb called **reduced hemoglobin.** In this way, the pH of blood remains fairly constant. Bicarbonate ions diffuse out of red blood cells and are carried in the plasma. Blood that leaves the capillaries is deep purple in color because red blood cells contain reduced hemoglobin.

Internal respiration, the movement of oxygen and carbon dioxide between blood in the systemic capillaries and tissue fluid, is dependent on the process of diffusion.

Respiration and Health

We have seen that the entire respiratory tract has a warm, wet, mucous membrane lining, which is constantly exposed to environmental air. The quality of this air, as discussed in the Ecology Focus for this chapter, can affect our health.

Respiratory Tract Infections

Microbes frequently spread from one individual to another by way of the respiratory tract. Droplets from one single sneeze can be loaded with billions of bacteria or viruses. The mucous membranes are protected by mucus and by the constant beating of the cilia, but if the number of infective agents is large and/or our resistance is reduced, respiratory infections such as colds and influenza (flu) can result. Other more serious infections and disorders are discussed here.

Bronchitis: Acute and Chronic

Viral infections can spread from the nasal cavities to the sinuses (sinusitis), to the middle ears (otitis media), to the larynx (laryngitis), and to the bronchi (bronchitis). Acute bronchitis (fig. 8.12) usually is caused by a secondary bacterial infection of the bronchi, resulting in a heavy mucus discharge with much coughing. Acute bronchitis usually responds to antibiotic therapy. Chronic bronchitis, on the other hand, is not necessarily due to infection. It is often caused by constant irritation of the lining of the bronchi, which as a result undergo degenerative changes, including the loss of cilia and their normal cleansing action. There is frequent coughing, and the individual is more susceptible to respiratory infections. Chronic bronchitis is most often seen in cigarette smokers or those exposed to secondhand smoke or other types of polluted air.

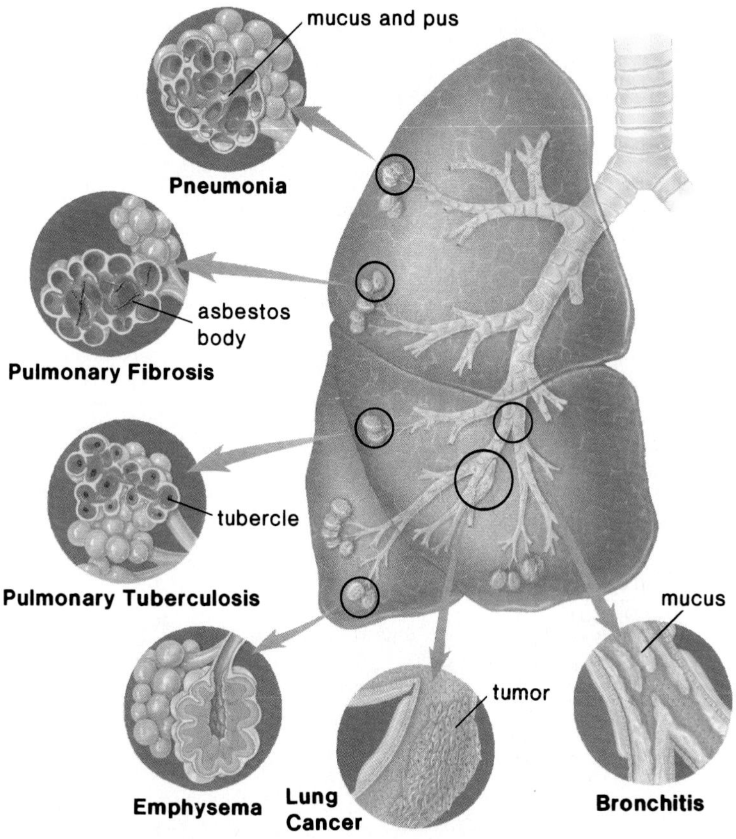

Figure 8.12

Common bronchial and pulmonary infectious diseases and disorders. Exposure to infectious microbes and/or polluted air, including cigarette and cigar smoke, causes the diseases and disorders shown here.

Strep Throat: Risks Rheumatic Fever

Strep throat is a very severe throat infection caused by the bacterium *Streptococcus pyogenes.* Swallowing may be difficult, and there is fever. Unlike a viral infection, strep throat should be treated with antibiotics. If not treated, it can lead to complications such as rheumatic fever, which can permanently damage the heart valves.

Lung Disorders

Pneumonia and tuberculosis are 2 serious infections of the lungs ordinarily controlled by antibiotics. Two other illnesses discussed, emphysema and lung cancer, are not due to infections; in most instances, they are due to cigarette smoking.

Ecology Focus

Photochemical Smog Can Kill

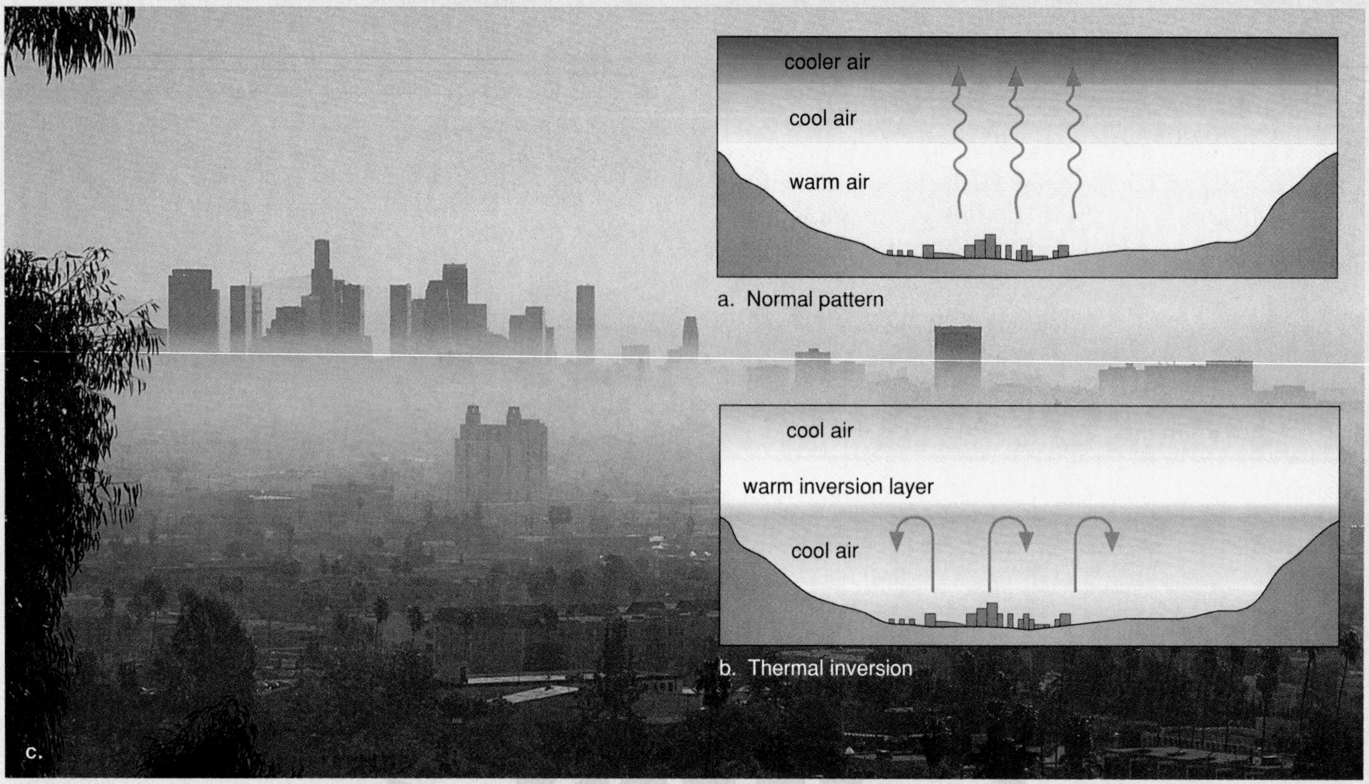

Figure 8A

Thermal inversion. **a.** Normally, pollutants escape into the atmosphere when warm air rises. **b.** During a thermal inversion, a layer of warm air (warm inversion layer) overlies and traps pollutants in cool air below. **c.** Los Angeles is particularly susceptible to thermal inversions, and this accounts for why this city is the "air pollution capital" of the United States.

Most industrialized cities have photochemical smog at least occasionally. Photochemical smog arises when primary pollutants react with one another under the influence of sunlight to form a more deadly combination of chemicals. For example, the primary pollutants nitrogen oxides (NO_x) and hydrocarbons (HC) react with one another in the presence of sunlight to produce nitrogen dioxide (NO_2), ozone (O_3), and PAN (peroxylacetyl nitrate). Ozone and PAN are commonly referred to as oxidants. Breathing oxidants affects the respiratory and nervous systems, resulting in respiratory distress, headache, and exhaustion.

Cities with warm, sunny climates, such as Los Angeles, Denver, and Salt Lake City in the United States, Sydney in Australia, Mexico City in Mexico, and Buenos Aires in Argentina, are particularly susceptible to photochemical smog because they are large and industrialized. Los Angeles contains 13.8 million people, 8.5 million cars, and thousands of factories. Cities surrounded by hills are particularly subject to thermal inversions, which aggravate the situation. Normally, warm air near the ground rises, so that pollutants are dispersed and carried away by air currents. But sometimes during a thermal inversion, smog gets trapped near the earth by a blanket of warm air (fig. 8A). This may occur when a cold front brings in cold air, which settles beneath a warm layer. The trapped pollutants cannot disperse and the results can be disastrous. In 1963, about 300 people died, and in 1966, about 168 people died in New York City when air pollutants accumulated over the city. Even worse were the events in London in 1962, when 700 people died, and in 1957, when 700 to 800 people died, due to the effects of air pollution.

Even though we have federal legislation to bring air pollution under control, more than half the people in the United States live in cities polluted by too much smog. We should place our emphasis on pollution prevention because, in the long run, prevention is usually easier and cheaper than pollution cleanup methods. Some prevention suggestions are as follows:

- Build more efficient automobiles or burn fuels that do not produce pollutants.
- Reduce the amount of waste to be incinerated by recycling materials.
- Reduce our energy use so that power plants need to provide less, and/or use renewable energy sources such as solar, wind, or water power.
- Require that industries meet clean air standards.

Pneumonia: Lobules Fill and Breathing Ceases

Most forms of pneumonia are caused by either a bacterium or a virus that has infected the lungs. AIDS patients are subject to a particularly rare form of pneumonia caused by the protozoan *Pneumocystis carinii.* Sometimes, pneumonia is localized in specific lobules of the lungs. These lobules become nonfunctional as they fill with mucus and pus. Obviously, the more lobules involved, the more serious the infection.

Pulmonary Tuberculosis: Past and Recent Threat

Pulmonary tuberculosis is caused by the tubercle bacillus, a type of bacterium. When a person has tuberculosis, the alveoli burst and are replaced by inelastic connective tissue. It is possible to tell if a person has ever been exposed to tuberculosis with a skin test in which a highly diluted extract of the bacilli is injected into the skin of the patient. A person who has never been in contact with the bacillus shows no reaction, but one who has developed immunity to the organism shows an area of inflammation that peaks in about 48 hours. If these bacilli invade the lung tissue, the cells build a protective capsule about the foreigners and isolate them from the rest of the body. This tiny capsule is called a *tubercle.* If the resistance of the body is high, the imprisoned organisms die, but if the resistance is low, the organisms eventually can be liberated. If a chest X ray detects active tubercles, the individual is put on appropriate drug therapy to ensure the localization of the disease and the eventual destruction of any live bacterial organisms.

Tuberculosis killed about 100,000 people in the United States each year before the middle of this century, when antibiotic therapy brought it largely under control. In recent years, however, the incidence of tuberculosis is on the rise, particularly among AIDS patients, the homeless, and the rural poor. Worse, the new strains are resistant to the usual antibiotic therapy. Therefore, some physicians would like to again use sanitoriums to quarantine patients needing treatment.

Emphysema: Bronchioles Collapse and Alveoli Burst

Emphysema refers to the destruction of lung tissue, with accompanying ballooning or inflation of the lungs due to trapped air. The trouble stems from the damage and the collapse of the bronchioles. When this occurs, the alveoli are cut off from renewed oxygen supply and the air within them is trapped. The trapped air very often causes the alveolar walls to rupture (fig. 8.12) and also a loss of elasticity makes breathing difficult. The victim is breathless and may have a cough. Since the surface area for gas exchange is reduced, not enough oxygen reaches the heart and the brain. Even so, the heart works furiously to force more blood through the lungs, which can lead to a heart condition. Lack of oxygen to the brain can make the person feel depressed, sluggish, and irritable.

Pulmonary Fibrosis: Inhaling Particles

Inhaling particles such as silica (sand), coal dust, and asbestos (fig. 8.12) can lead to pulmonary fibrosis, a condition in which fibrous connective tissue builds up in the lungs. Breathing capacity can be seriously impaired, and the development of cancer is common. Since asbestos has been used so widely as a fireproofing and insulating agent, unwarranted exposure has occurred. It is projected that 2 million deaths could be caused by asbestos exposure—mostly in the workplace—between 1990 and 2020.

Lung Cancer: Women Catch Up

Lung cancer used to be more prevalent in men than in women, but recently it has surpassed breast cancer as a cause of death in women. This can be linked to an increase in the number of women who smoke today. Autopsies on smokers have revealed the progressive steps by which the most common form of lung cancer develops. The first event appears to be thickening and callusing of the cells lining the bronchi. (Callusing occurs whenever cells are exposed to irritants.) Then there is a loss of cilia so that it is impossible to prevent dust and dirt from settling in the lungs. Following this, cells with atypical nuclei appear in the callused lining. A tumor consisting of disordered cells with atypical nuclei is considered to be cancer *in situ* (at one location). A final step occurs when some of these cells break loose and penetrate other tissues, a process called metastasis. Now the cancer has spread. The tumor may grow until the bronchus is blocked, cutting off the supply of air to that lung. The entire lung then collapses, the secretions trapped in the lung spaces become infected, and pneumonia or a lung abscess (localized area of pus) results. The only treatment that offers a possibility of cure is to remove a lobe or the lung completely before secondary growths have had time to form. This operation is called *pneumonectomy.*

> The incidence of lung cancer is over 20 times higher in individuals who smoke than in those who do not.

Current research indicates that *involuntary smoking,* simply breathing in air filled with cigarette smoke, can also cause lung cancer and other illnesses associated with smoking. The Health Focus for this chapter lists the various illnesses associated with smoking. If a person stops both voluntary and involuntary smoking and if the body tissues are not already cancerous, they usually return to normal over time.

The Most Often Asked Questions About Smoking, Tobacco, and Health,

Is there a safe way to smoke?

No. *All* cigarettes can cause damage and smoking even a small amount is dangerous. Cigarettes are perhaps the only legal product whose advertised and intended use—that is, smoking them—will hurt the body. Some people try to make smoking safer by smoking fewer cigarettes, but most smokers find this difficult. Some people think that switching from high tar/nicotine cigarettes to those with low tar/nicotine content makes smoking safer, but this does not always happen.... Even if smokers who switch to lower tar brands avoid these changes in smoking behavior, the health benefits from switching would be insignificant compared with the benefits of quitting altogether....

Does smoking cause cancer?

Yes, and not only lung cancer. Tobacco use is responsible for about 30% of all cancer deaths in the United States. Cigarette smoking causes about 87% of lung cancer deaths. Besides lung cancer, cigarette smoking is also a major cause of cancers of the mouth, larynx (voice box), and esophagus (swallowing tube). In addition, smoking increases the risk of cancer of the bladder, kidney, pancreas, stomach, and the uterine cervix.

What are the chances of being cured of lung cancer?

Very low; the 5-year survival rate is only 13%. Most forms of the disease start without producing any warning signs, so that it is rarely detected in the early stages when it is more likely to be cured. The past 15 years have brought little significant progress in the earlier diagnosis or treatment of lung cancer. Fortunately, lung cancer is a largely preventable disease. That is, by not smoking it can be prevented....

Do cigarettes cause other lung diseases?

Yes. Cigarette smoking causes other lung diseases which can be just as dangerous as lung cancer. It leads to chronic bronchitis—a disease where the airways produce excess mucus, which forces the smoker to cough frequently. Cigarette smoking is also the major cause of emphysema—a disease which slowly destroys a person's ability to breathe.... Chronic obstructive pulmonary disease (COPD), which includes chronic bronchitis and emphysema,kills about 81,000 people each year; cigarette smoking is responsible for 65,000 of these deaths.

Why do smokers have "smoker's cough"?

Cigarette smoke contains chemicals which irritate the air passages and lungs. When a smoker inhales these substances, the body tries to protect itself by coughing. The well-known "early morning" cough of smokers happens for a different reason. Normally, cilia (tiny hairlike formations which line the airways) beat outwards and "sweep" harmful material out of the lungs. Cigarette smoke, however, decreases this sweeping action, so some of the poisons in the smoke remain in the lungs.... Smokers are more likely to get pneumonia because damaged or destroyed cilia cannot protect the lungs from bacteria and viruses that float in the air....

If you smoke but don't inhale, is there any danger?

Yes. Wherever smoke touches living cells, it does harm. So, even if smokers don't inhale—including pipe and cigar smokers—they are at an increased risk for lip, mouth, and tongue cancer. Because it is virtually impossible to avoid inhaling tobacco smoke totally, these smokers also have an increased chance of getting lung cancer. Lung cancer is much more likely to occur in a person who has always smoked cigars or pipes than in a person who has never smoked at all.

Does cigarette smoking affect the heart?

Yes. Smoking cigarettes increases the risk of heart disease, which is America's number one killer. About 150,000 Americans die each year from heart attacks and other forms of heart disease caused by smoking. Smoking, high blood pressure, high blood cholesterol, and lack of exercise are all risk factors for heart disease. Smoking alone doubles the risk of heart disease. When a person smokes and has other risk factors, his chance of getting heart disease increases dramatically. For example, if smoking is combined with high blood pressure or high cholesterol, then the risk goes up four times. Put all three together—smoking, high blood pressure, and high cholesterol—and the risk goes up eight times. Smokers who have already had one heart attack are also more likely than nonsmokers to have another attack.

Is there any risk for pregnant women and their babies?

Pregnant women who smoke endanger the health and lives of their unborn babies. When a pregnant woman smokes, she really is smoking for two because the nicotine, carbon monoxide, and other dangerous chemicals in smoke enter the mother's bloodstream and then pass into the baby's body. Women who smoke during pregnancy risk having a miscarriage, or a premature or stillborn baby. Their babies are also more likely to be underweight, by an average of one-half pound....

Does smoking cause any special health problems for women?

Yes. Nonsmoking women who use oral contraceptives ("the Pill") double their chances of having a heart attack. However, when women use the Pill and smoke, they are 10 times more likely to suffer a heart attack than nonsmoking women who don't take the Pill. Women who smoke and use the Pill have an increased risk of stroke and blood clots in the legs as well.

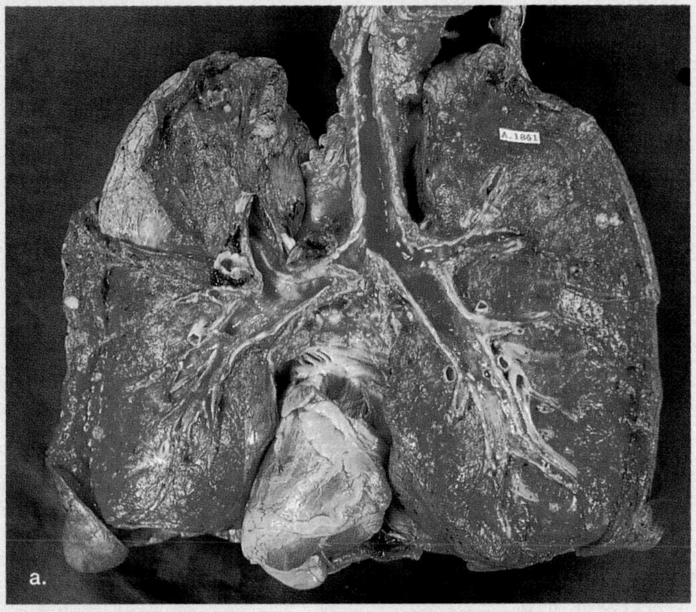

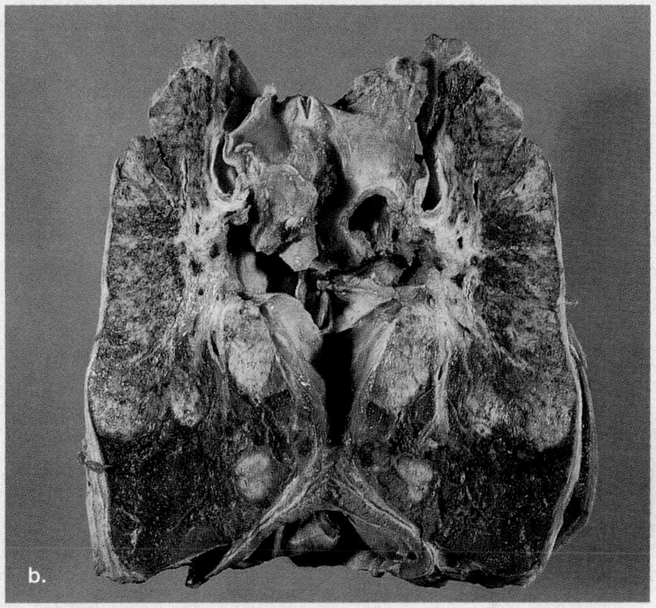

Figure 8B

Normal lung versus cancerous lung. **a.** Normal lung with heart in place. Notice the healthy red color. **b.** Lungs of a heavy smoker. Notice how black the lungs are except where cancerous tumors have formed.

Women who smoke also run the risk of having trouble getting pregnant; the more they smoke, the more likely it is that they will have difficulty. Some studies show that female smokers, especially the elderly, are at a higher risk for osteoporosis (a disease which weakens the bones and makes them more likely to break) than nonsmoking women. In addition, women who smoke increase their chances of getting cancer of the uterine cervix.

What are some of the short-term effects of smoking cigarettes?

Almost immediately, smoking can make it hard to breathe. Within a short time, it can also worsen asthma and allergies. Nicotine reaches the brain only seven seconds after taking a puff (faster than it takes heroin to reach the brain), where it produces a variety of effects....

Are there any other risks to the smoker?

Yes. There are *many* other risks. As we already mentioned briefly, smoking cigarettes causes stroke, which is the third leading cause of death in America. Smoking causes lung cancer, but if a person smokes and is exposed to radon or asbestos, the risk increases even more. Smokers are also more likely to have and die from stomach ulcers than nonsmokers. In addition, cigarettes can interact with medication the smoker is taking in unwanted ways—like preventing the drug from doing what it is supposed to do....

What are the dangers of passive smoking?

...Passive smoking causes lung cancer in healthy nonsmokers. Children whose parents smoke are more likely to suffer from pneumonia or bronchitis in the first two years of life than children who come from smoke-free households. Nonsmokers who are married to smokers have a 30% greater risk for developing lung cancer than nonsmokers married to other nonsmokers....

Are chewing tobacco and snuff safe alternatives to cigarette smoking?

No, they are not. Many people who use chewing tobacco or snuff believe it can't harm them because there is no smoke. Wrong. Smokeless tobacco contains nicotine, the same addicting drug found in cigarettes. Snuff dippers also take in an average of over 10 times more cancer-causing substances (called nitrosamines) than cigarette smokers. In fact, some brands of smokeless tobacco contain as much as *20,000* times the legal limit of nitrosamines which are permitted in certain foods and consumer products (such as bacon, beer, and baby bottle nipples). While not inhaled through the lungs, the juice from smokeless tobacco is absorbed through the lining of the mouth. There it can cause sores and white patches which often lead to cancer of the mouth....

Excerpted from "The Most Often Asked Questions About Smoking, Tobacco, and Health, and... The Answers," revised July 1993. © American Cancer Society, Inc., Atlanta, GA. Used with permission.

SUMMARY

During the process of breathing, air enters and exits the lungs by way of the respiratory tract, which consists of the nose, the nasopharynx, the pharynx, the larynx (which also contains the vocal cords), the trachea, the bronchi, and the bronchioles. The right and left lungs, located on either side of the heart, are covered by pleural membranes. The chest cavity is bounded by the rib cage and by the diaphragm. The bronchi, along with the pulmonary arteries and veins, penetrate the lungs. Thereafter, the bronchi divide into the bronchioles, which enter the alveoli, air sacs surrounded by extensive pulmonary capillaries.

Inspiration begins when the respiratory center in the medulla oblongata sends excitatory nerve impulses to the diaphragm and the muscles of the rib cage. As they contract, the diaphragm lowers and the rib cage moves upward and outward; the lungs expand, creating a partial vacuum, which causes air to rush in. The respiratory center now stops sending impulses to the diaphragm and rib cage. As the diaphragm relaxes, it resumes its dome shape, and as the rib cage retracts, air is pushed out of the lungs during expiration.

External respiration occurs when carbon dioxide (CO_2) leaves blood and oxygen (O_2) enters blood at the alveoli. Oxygen is transported to the tissues in combination with hemoglobin as oxyhemoglobin (HbO_2). Internal respiration occurs when oxygen leaves blood and carbon dioxide enters blood at the tissues. Carbon dioxide is mainly carried to the lungs within the plasma as the bicarbonate ion (HCO_3^-). Hemoglobin combines with hydrogen ions and becomes reduced (HHb).

There are a number of illnesses associated with the respiratory tract. In addition to colds and flu, the lungs may be infected by the more serious pneumonia and tuberculosis. Two illnesses that have been attributed to breathing polluted air are emphysema and lung cancer.

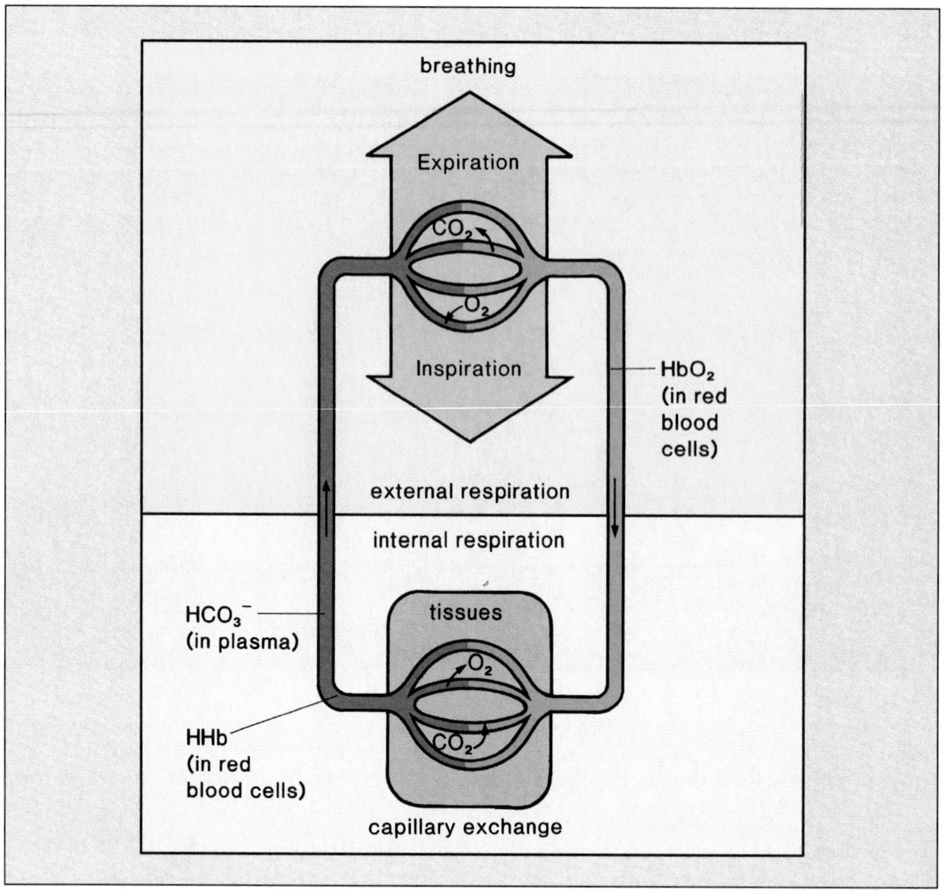

STUDY AND THOUGHT QUESTIONS

1. Name and explain the 3 steps of respiration. (p. 146)
2. List the parts of the respiratory tract. What are the special functions of the nasal cavity, the larynx, and the alveoli? (pp. 147–148, 150)

   Critical Thinking The larynx has no function in respiration. Should it be considered a part of the respiratory system?
3. What occurs during inspiration and expiration? How is breathing controlled? (p. 150)

   Critical Thinking Why is it better to give a nonbreathing person a mixture of oxygen and carbon dioxide rather than pure oxygen?
4. Why can't we breathe through a very long tube? (p. 153)
5. What physical process is believed to explain gas exchange? (p. 156)
6. What 2 equations are needed to explain external respiration? (p. 154)
7. How is hemoglobin remarkably suited to its job? (p. 157)

8. What 2 equations are needed to explain internal respiration? (pp. 156–157)
9. Name and discuss some respiratory tract infections. (p. 157)
10. What are emphysema and pulmonary fibrosis, and how do they affect a person's health? (p. 159)
11. By what steps is cancer believed to develop in the person who smokes? (p. 159)

Ethical Issue Do you believe smokers have rights or do you think smoking should be restricted to protect everyone's health? Give reasons for your answer.

Writing Across the Curriculum

1. If the human respiratory system were a flow-through system, it would have an opening both for intake and outtake. How would this affect various features of the respiratory system?
2. Fishes live in the water and expose their gills to the external environment. We live on land and have our lungs deep within the body. Explain how this is adequate.
3. Newborn infants have poorly developed chest muscles and therefore they rely primarily on movements of the diaphragm to ventilate their lungs. Explain the reason an infant's body appears to lengthen during inhalation and shorten during exhalation.

Objective Questions

1. In tracing the path of air, the ____________________ immediately follows the pharynx.
2. The lungs contain air sacs called _______________.
3. The breathing rate is primarily regulated by the amount of ___________ and ___________ in blood.
4. Air enters the lungs after they have _____________.
5. Carbon dioxide (CO_2) is carried in blood as the ________________ ion.
6. The hydrogen (H^+) ions given off when carbonic acid (H_2CO_3) dissociates are carried by ____________________.
7. Gas exchange is dependent on the physical process of ____________________.
8. Reduced hemoglobin becomes oxyhemoglobin in the __________________ of the body.
9. The most likely cause of emphysema and chronic bronchitis is ______________________.
10. Most cases of lung cancer actually begin in the ____________________.
11. Label the diagram (*right*) of the human respiratory tract.

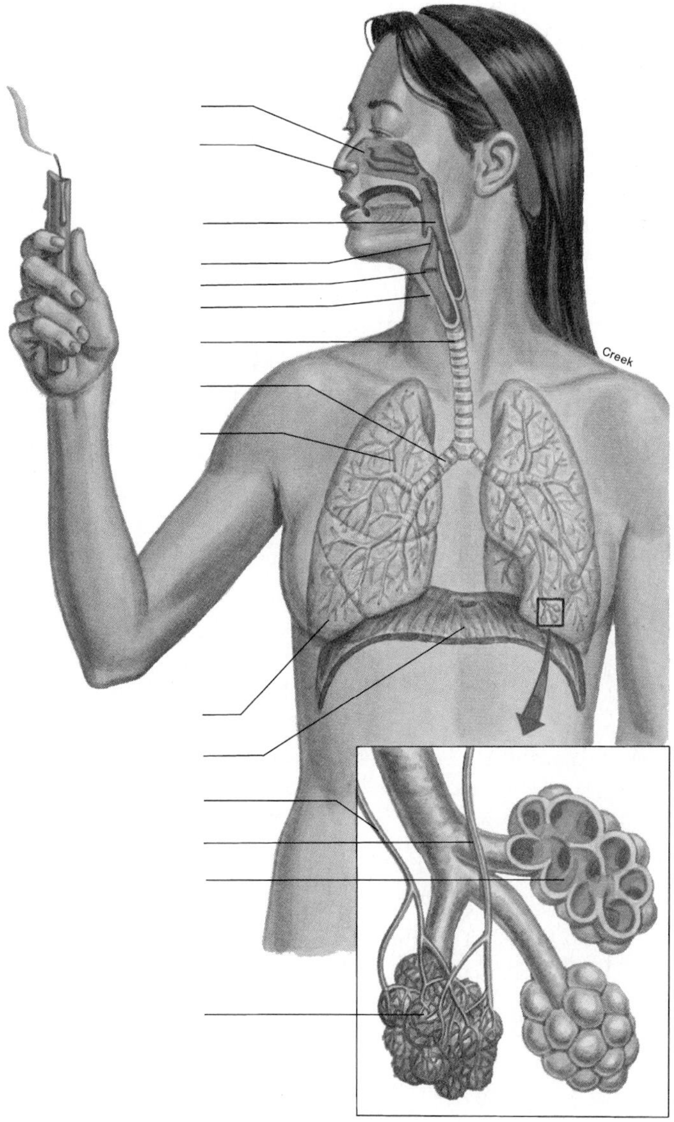

Selected Key Terms

alveolus (pl., alveoli) Air sac of a lung. 149

bronchiole (brong-kéōl) The smaller air passages in the lungs that eventually terminate at alveoli. 149

bronchus (brong-kus) (pl., bronchi) One of the 2 major divisions of the trachea leading to the lungs. 149

diaphragm A sheet of muscle that separates the thoracic cavity from the abdominal cavity in higher animals. 150

expiration The act of expelling air from the lungs; exhalation. 147

external respiration Exchange between blood and alveoli of carbon dioxide and oxygen. 146

inspiration The act of taking air into the lungs; inhalation. 147

internal respiration Exchange between blood and tissue fluid of oxygen and carbon dioxide. 146

larynx (lar´ingks) Cartilaginous organ located between pharynx and trachea, which contains the vocal cords; voice box. 148

respiratory center Group of nerve cells in the medulla oblongata that send out nerve impulses on a rhythmic basis, resulting in inspiration. 150

rib cage The top and sides of the thoracic cavity; contains ribs and intercostal muscles. 150

trachea (trá ke-ah) A tube that is supported by C-shaped cartilaginous rings: lies between the larynx and the bronchi; also called the windpipe. 149

ventilation Breathing; the process of moving air into and out of the lungs. 150

vocal cord Fold of tissue within the larynx; creates vocal sounds when it vibrates. 148

Chapter 9

Urinary System and Excretion

Everyone knows that if you cross a desert without a supply of fresh water, death is likely. Dehydration occurs because the body loses water and salts through sweating and urination. But what if you were lost at sea? Can you drink salt water? Unfortunately not. Humans have no special way to rid the body of excess salt, and the kidneys would have to excrete more liquid than was consumed to wash out the excess salt. If there is no fresh water available, don't drink at all and don't eat salty fish either!

Chapter Concepts

1 Excretion rids the body of unwanted substances, particularly the end products of metabolism. 166

2 Several organs assist in the process of excretion but only the kidneys produce urine. 166

3 The urinary system consists of organs that produce, store, and rid the body of urine. 167–168

4 Nephrons within the kidneys produce urine in several steps. 169–173

5 The kidneys rid the body of nitrogenous wastes, and regulate the pH and the salt/water balance of blood. 174–176

6 Malfunction of the kidneys causes illness and even death. 176

Chapter Outline

The composition of blood serving the tissues remains relatively constant due to the action of several organs. In previous chapters, we discussed how the digestive tract and the lungs add nutrients and oxygen to blood. In this chapter, we discuss how the organs of excretion and in particular the kidneys contribute to homeostasis by removing waste products of metabolism.

At this point, it is helpful to remember that the term *defecation,* and not *excretion,* is used to refer to the elimination of feces from the body. Undigested food and bacteria, which make up feces, have never passed through cell membranes and entered the body proper. **Excretion** rids the body of metabolic wastes, which come from the breakdown of substances that have actually entered the body.

Four Excretory Organs

The kidneys are the primary excretory organs, but other organs also function in excretion (fig. 9.1), including those described in the discussion that follows.

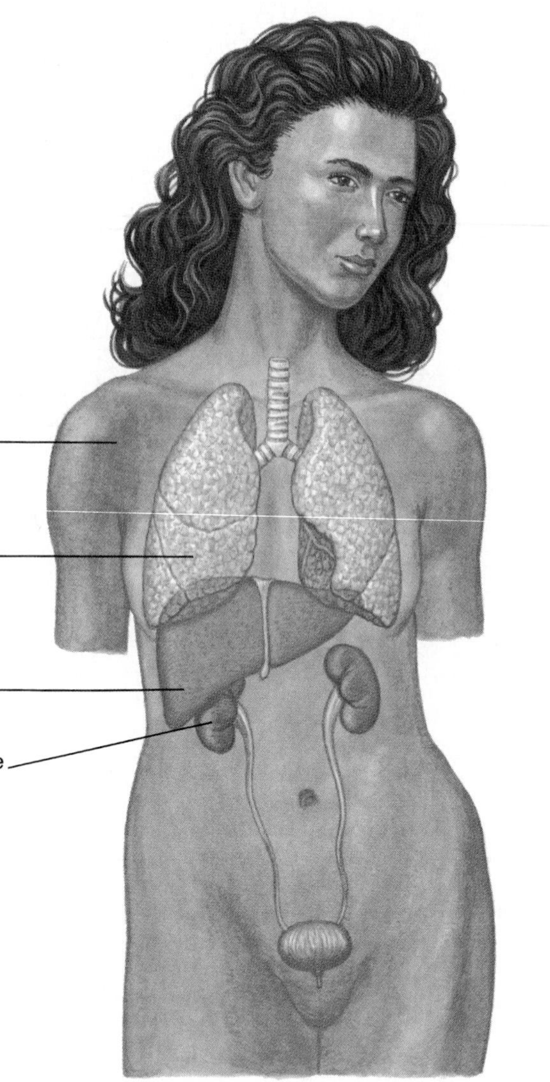
Skin excretes water and salts in perspiration.
Lungs excrete carbon dioxide as a gas.
Liver excretes bile pigments in bile.
Kidneys excrete nitrogenous wastes in urine.

Figure 9.1

The organs of excretion.

Skin Excretes Perspiration

The sweat glands in the skin (see fig. 3.8) excrete perspiration, which is a solution of water, salt, and some urea, an end product discussed following. In the dermis, a sweat gland is a coiled tubule, but then it straightens as it passes through and exits the epidermis. Although perspiration is a form of excretion, we perspire not so much to rid the body of waste as to cool it. The body cools because heat is lost as perspiration evaporates. Sweating keeps the body temperature within normal range during muscular exercise or when the outside temperature rises. In times of kidney failure, urea is excreted by the sweat glands and forms a so-called urea frost on the skin.

Liver Excretes Bile Pigments

The liver excretes bile pigments, which are incorporated into bile, a substance stored in the gallbladder before it passes into the small intestine by way of ducts. The yellow pigment found in urine, called urochrome, like bile pigments, is derived from the breakdown of heme, but this pigment is deposited in blood and is subsequently excreted by the kidneys.

Lungs Remove CO_2 and H_2O

The process of expiration (breathing out) not only removes carbon dioxide (CO_2) from the body, it also results in the loss of water (H_2O). The air we exhale contains moisture, as demonstrated by breathing onto a cool mirror.

Kidneys Produce Urine

The kidneys produce urine, which ordinarily contains the nitrogenous wastes and inorganic salts listed in table 9.1. Amino acids, nucleotides, and creatine phosphate all contain nitrogen, and their metabolism results in nitrogenous wastes.

Amino acid metabolism ends with **urea,** the primary nitrogenous end product of humans. Urea, which is reasonably soluble, is produced by the liver but excreted by the kidneys. The liver produces urea by a complicated series of reactions called the urea cycle. In this cycle, carrier molecules take up carbon dioxide and 2 molecules of ammonia to release urea:

$$H_2N - \overset{\overset{\displaystyle O}{||}}{C} - NH_2 \quad \text{urea}$$

Creatine phosphate, which serves as a reservoir of high-energy phosphate in muscles, results in the end product *creatinine.* The breakdown of nucleotides produces **uric acid,** which is rather insoluble. If too much uric acid is present in blood it precipitates out. Crystals of uric acid sometimes collect in the joints, producing a painful ailment called gout.

As stated, urine contains these nitrogenous end products (table 9.1). The kidneys produce urine by a process that is outlined later in the chapter.

> There are various organs that excrete metabolic wastes, but, primarily, the kidneys rid the body of urea.

Table 9.1

Composition of Urine

Water	95%
Solids	5%
Nitrogenous wastes	(per 1,500 ml of urine)
Urea	30 grams
Creatinine	1–2 grams
Uric acid	1 gram
Salts	25 grams
Positive ions	
Sodium	
Potassium	
Magnesium	
Ammonium	
Calcium	
Negative ions	
Chlorides	
Sulfates	
Phosphates	

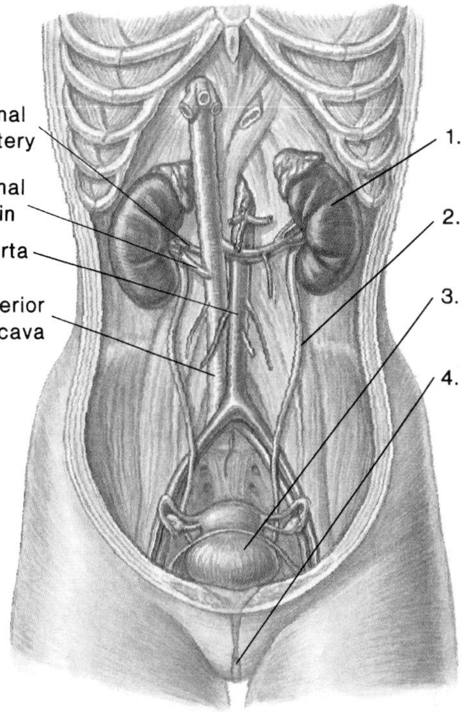
renal artery
renal vein
aorta
inferior vena cava
1. Kidneys produce urine.
2. Ureters transport urine.
3. Urinary bladder stores urine.
4. Urethra passes urine to outside.

Figure 9.2

The urinary system. Urine is found only within the kidneys, the ureters, the urinary bladder, and the urethra.

The Urinary System

The kidneys are a part of the urinary system, which includes the structures illustrated in figure 9.2. The organs of the urinary system contribute to homeostasis. They rid the body of nitrogenous wastes and help to keep the pH and salt/water balance of blood within a normal range.

The Path of Urine

Urine is made by the **kidneys,** bean-shaped, reddish brown organs, each about the size of a fist. One kidney is found on either side of the vertebral column, just below the diaphragm. The kidneys lie in depressions against the deep muscles of the back, beneath the membranous lining of the abdominal cavity, where they also receive some protection from the lower rib cage. Each is covered by a tough fibrous capsule of connective tissue overlaid by adipose tissue. The kidneys can be damaged by blows on the back; kidney punches are not allowed in boxing.

The **ureters** are muscular tubes that convey the urine from the kidneys toward the bladder by peristalsis. Urine enters the bladder by peristaltic contractions, in jets that occur at the rate of 5 per minute.

The **urinary bladder,** which can hold up to 600 ml of urine, is a hollow, muscular organ that gradually expands as urine enters. In the male, the urinary bladder lies in front of the rectum, the seminal vesicles, and the vas deferens. In the female, the urinary bladder is in front of the uterus and the upper vagina.

The **urethra,** which extends from the urinary bladder to an external opening, differs in length in the female and the male. In the female, the urethra lies ventral to the vagina and is only about 4 cm long. The short length of the female urethra invites bacterial invasion and explains why the female is more prone to urinary tract infections. The Health Focus for this chapter discusses this and suggests ways to prevent urinary tract infections. In the male, the urethra averages 20 cm when the penis is flaccid. As the urethra leaves the male urinary bladder, it is encircled by the prostate gland (see fig. 14.1). In older men, enlargement of the prostate gland can prevent urination, a condition that usually can be corrected surgically.

There is no connection between the genital (reproductive) and urinary systems in females, but there is a connection in males (compare figs. 14.1 and 14.5). In males, the urethra also carries sperm during ejaculation (p. 274). This double function does not alter the path of urine, and it is important to realize that urine is found only in those structures noted in figure 9.2.

Urination and the Nervous System

When the urinary bladder fills with urine, stretch receptors send nerve impulses to the spinal cord; nerve impulses leaving the cord then cause the urinary bladder to contract and the sphincters to relax so that urination is possible. In older children and adults, it is possible for the brain to control this reflex, delaying urination until a suitable time.

> Only the urinary system, consisting of the kidneys, the urinary bladder, the ureters, and the urethra, ever hold urine.

Health Focus

Urinary Tract Infections Require Attention

Females have more urinary tract infections than do males. Their urethral and anal openings are closer together, and the shorter urethra makes it easier for bacteria from the bowels to enter and start an infection. Although it is possible to have no outward signs of an infection, usually urination is painful and patients often describe a burning sensation. The urge to pass urine is frequent, but it may be difficult to start the stream. Chills with fever, nausea, and vomiting may be present.

Urinary tract infections can be confined to the urethra, in which case urethritis is present. If the bladder is involved it is called cystitis, and should the infection reach the kidneys, the person has pyelonephritis. The causative agent is usually *Escherichia coli* (*E. coli*), a normal resident of the large intestine. Since the infection is caused by a bacterium, it is curable by antibiotic therapy. The problem is, however, that reinfection is possible as soon as antibiotic therapy is finished.

It makes sense to try to prevent infection in the first place. These tips might help.

Drink lots of water. Try to drink from 2–2.5 liters of liquid a day. Try to avoid caffeinated drinks, which may be irritating. Cranberry juice is recommended because its acidity helps prevent infection.

Personal hygiene and the hygiene of any sex partners is terribly important. The expression *honeymoon cystitis* was coined because of the common association of urinary tract infections with sexual intercourse. Washing the genitals before having sex and being careful not to introduce bacteria from the anus into the urethra is recommended. Also, voiding immediately before and after sex will help to flush out any bacteria that are present. A diaphragm may press on the urethra and prevent adequate emptying of the bladder, and estrogen such as in birth-control pills can increase the risk of cystitis.

Females should wipe from the front to the back after using the toilet. Perfumed toilet paper and any other perfumed products that come in contact with the genitals may be irritating. Wearing loose clothing and cotton underwear discourages the growth of bacteria while tight clothing, such as jeans and panty hose, provides an environment for the growth of bacteria.

Personal hygiene is especially important too at the time of menstruation. Hands should be washed before and after changing napkins and/or tampons. Superabsorbent tampons are not best if they are changed infrequently, as this may encourage the growth of bacteria. Also, sexual intercourse may cause menstrual flow to enter the urethra.

A sex partner may have an asymptomatic urinary infection that causes a woman to become infected repeatedly. Testing and antibiotic therapy may, therefore, be in order. Keep in mind also that sexually transmitted diseases such as gonorrhea, chlamydia, or herpes can cause urinary tract infections to occur. All personal behaviors should be examined carefully, and suitable adjustments should be made to avoid urinary tract infections.

Kidneys: Have Three Regions

On the concave side of each kidney there is a depression where the renal blood vessels and the ureters enter (fig. 9.3). When a kidney is sliced lengthwise, it is possible to identify 3 regions. The **renal cortex** is an outer granulated layer that dips down in between a radially striated, or lined, inner layer called the **renal medulla.** The renal medulla contains cone-shaped tissue masses called *renal pyramids* (fig. 9.3). The **renal pelvis** is a central space, or cavity, that is continuous with the ureter.

Over One Million Nephrons

Microscopically, the kidney is composed of over one million **nephrons,** sometimes called renal or kidney tubules (figs. 9.3*c* and 9.4). Each nephron is made up of several parts. The closed end of the nephron is pushed in on itself to form a cuplike structure called **Bowman's capsule** (also called the glomerular capsule). The outer layer of Bowman's capsule is composed of squamous epithelial cells; the inner layer is composed of specialized cells that allow easy passage of molecules. Next, there is a **proximal** (meaning near the Bowman's capsule) **convoluted tubule,** lined by cells with many mitochondria and an inner brush border (tightly packed microvilli). Then simple squamous epithelium appears as the tube narrows and makes a U-turn. This portion of the tubule is called the **loop of Henle** (or the loop of the nephron). Then comes the **distal** (far from Bowman's capsule) **convoluted tubule,** lined by cells again with mitochondria but without a brush border. The distal convoluted tubules of several nephrons enter one **collecting duct.** A kidney contains many collecting ducts, which carry urine to the renal pelvis.

As shown in figure 9.4, Bowman's capsule and the convoluted tubules always lie within the renal cortex. The loop of Henle dips down into the renal medulla; a few nephrons have a very long loop of Henle, which penetrates deep into the renal medulla. Collecting ducts are also located in the renal medulla, and they give the renal pyramids their lined appearance.

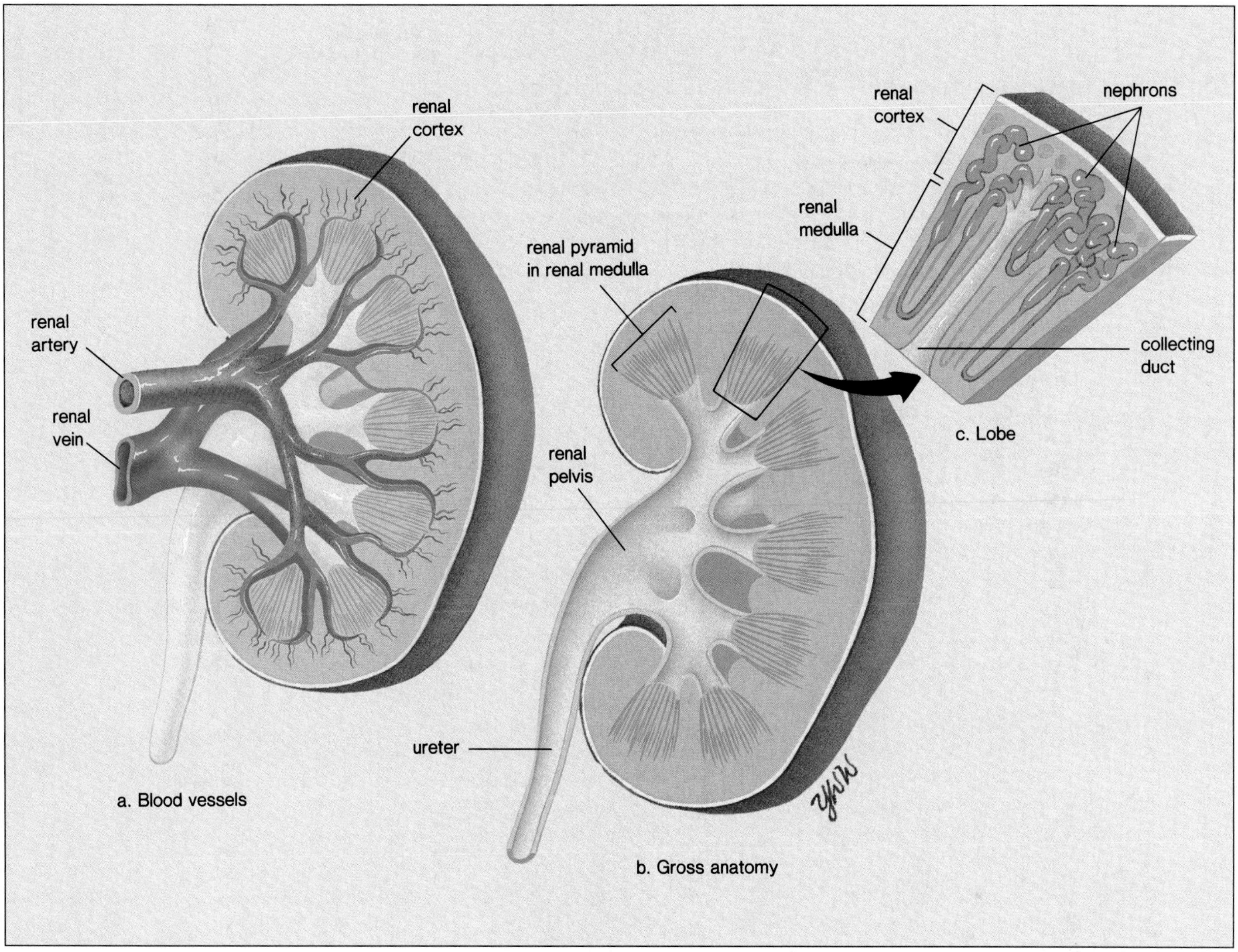

Figure 9.3

Gross anatomy of the kidney. **a.** A longitudinal section of the kidney showing the blood supply. Note that the renal artery divides to give smaller arteries. Smaller veins join to form the renal vein. **b.** The same section without the blood supply. Now it is easier to make out the renal cortex, the renal medulla, and the renal pelvis, which connects with the ureter. The renal pyramids make up the renal medulla. **c.** An enlargement of a lobe, showing the placement of nephrons in a renal pyramid.

Urine Formation

Each nephron has its own blood supply, including 2 capillary regions. The **glomerulus** is a tuft of capillaries inside Bowman's capsule, and the **peritubular capillaries** surround the rest of the nephron (fig. 9.4 and table 9.2). Urine formation requires the movement of molecules between these capillaries and the nephron. Three steps are involved: *pressure filtration, selective reabsorption,* and *tubular excretion.*

The pattern of blood flow about the nephron is critical to urine formation.

Table 9.2

Circulation About a Nephron

Name of Structure	Comment
Afferent arteriole	Brings arteriolar blood toward Bowman's capsule
Glomerulus	Capillary tuft enveloped by Bowman's capsule
Efferent arteriole	Takes arteriolar blood away from Bowman's capsule
Peritubular capillary network	Capillary bed that envelops the rest of the tubule
Venule	Takes venous blood away from the tubule

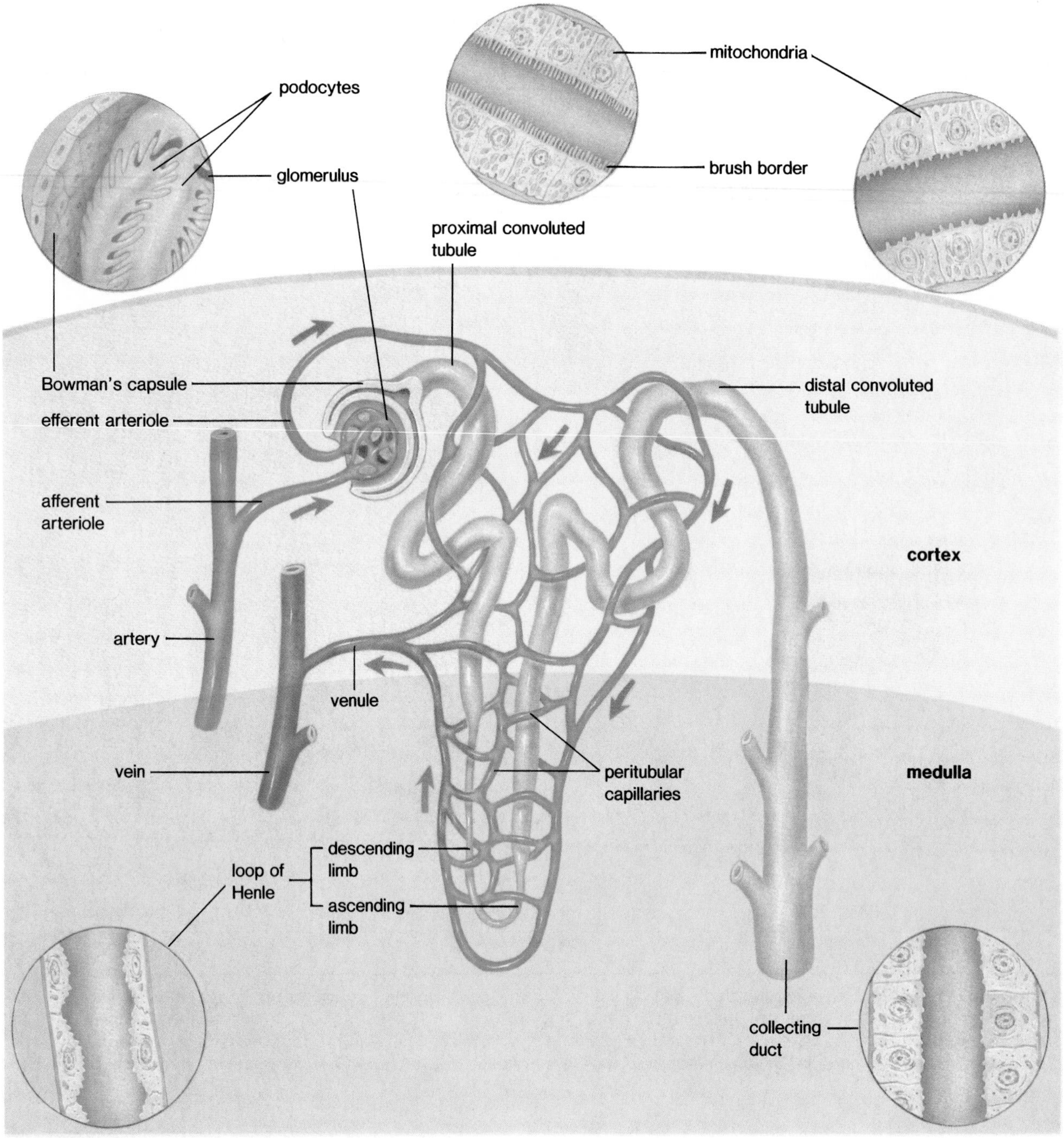

Figure 9.4

Nephron macroscopic and microscopic anatomy. A nephron is made up of Bowman's capsule, the proximal convoluted tubule, the loop of Henle, the distal convoluted tubule, and the collecting duct. The blowups show the types of tissue at these different locations. You can trace the path of blood about the nephron by following the arrows.

Pressure Filtration: Divides the Blood

Figure 9.5 gives a simple overview of urine formation. Whole blood, of course, enters the afferent arteriole and the glomerulus. Under the influence of glomerular blood pressure, which is usually about 60 mm Hg, water and small molecules move from the glomerulus to the inside of Bowman's capsule, across the thin walls of each. This is a **pressure filtration** process because large molecules and formed elements are unable to pass through these thin walls. In effect, then, blood that enters the glomerulus is divided into 2 portions: the filterable components and the nonfilterable components.

Filterable Blood Components	Nonfilterable Blood Components
Water	Formed elements (blood cells and platelets)
Nitrogenous wastes	Proteins
Nutrients	
Salts (ions)	

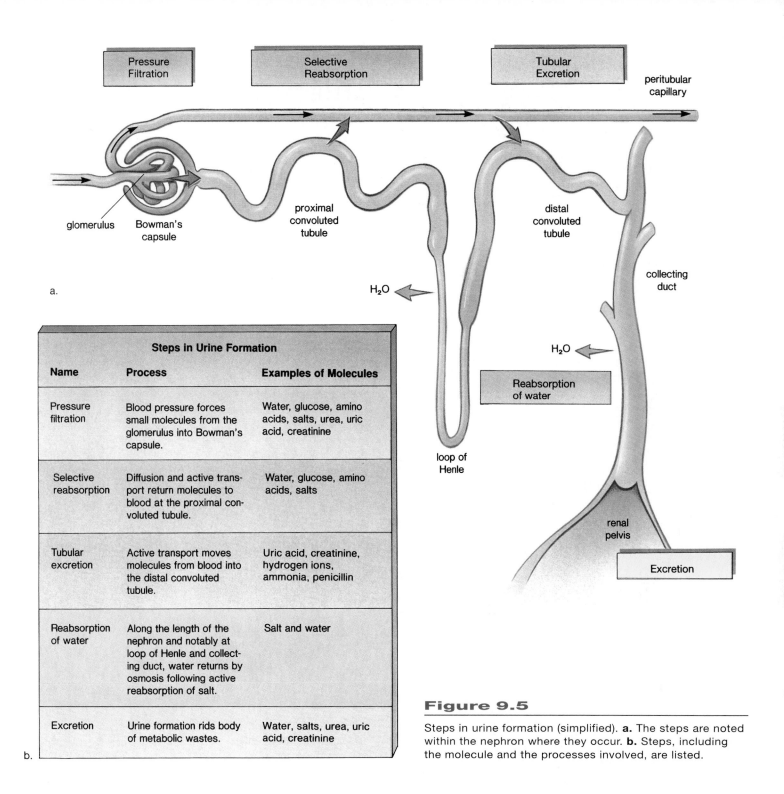

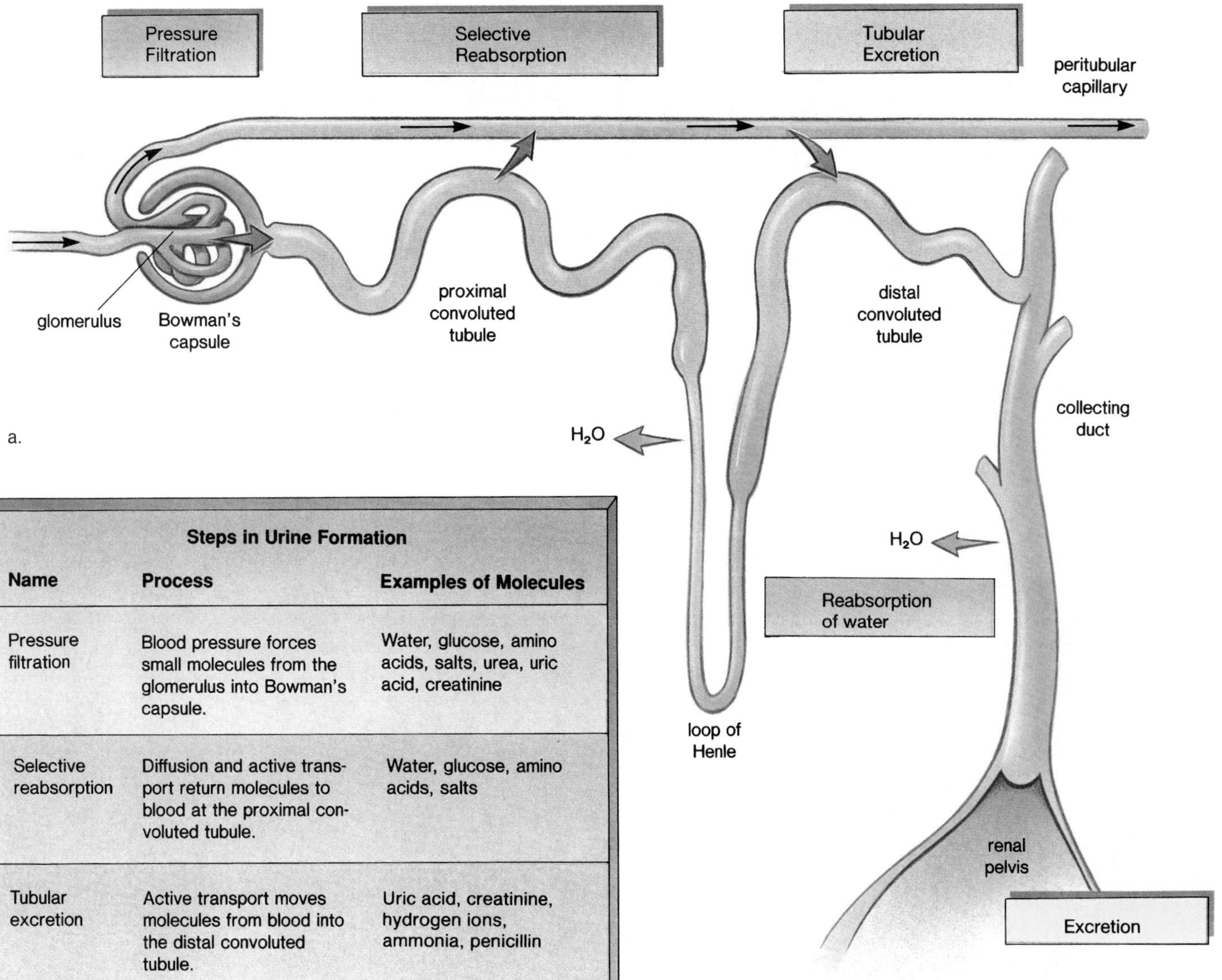

a.

Steps in Urine Formation

Name	Process	Examples of Molecules
Pressure filtration	Blood pressure forces small molecules from the glomerulus into Bowman's capsule.	Water, glucose, amino acids, salts, urea, uric acid, creatinine
Selective reabsorption	Diffusion and active transport return molecules to blood at the proximal convoluted tubule.	Water, glucose, amino acids, salts
Tubular excretion	Active transport moves molecules from blood into the distal convoluted tubule.	Uric acid, creatinine, hydrogen ions, ammonia, penicillin
Reabsorption of water	Along the length of the nephron and notably at loop of Henle and collecting duct, water returns by osmosis following active reabsorption of salt.	Salt and water
Excretion	Urine formation rids body of metabolic wastes.	Water, salts, urea, uric acid, creatinine

b.

Figure 9.5

Steps in urine formation (simplified). **a.** The steps are noted within the nephron where they occur. **b.** Steps, including the molecule and the processes involved, are listed.

The filterable components form the **glomerular filtrate,** which contains small dissolved molecules in approximately the same concentration as plasma. The filtrate stays inside Bowman's capsule, and the nonfilterable components leave the glomerulus by way of the efferent arteriole.

A consideration of the preceding filterable substances leads us to conclude that if the composition of urine were the same as that of the glomerular filtrate, the body would continually lose nutrients, water, and salts. Death from dehydration, starvation, and low blood pressure would quickly follow. Therefore, we can assume that the composition of the filtrate must be altered as this fluid passes through the remainder of the tubule.

> During pressure filtration, water, salts, nutrient molecules, and waste molecules move from the glomerulus to the inside of Bowman's capsule. The filtered substances are called the glomerular filtrate.

Selective Reabsorption: Active and Passive

Both passive and active reabsorption of molecules from the nephron to blood in the peritubular capillary occur as the filtrate moves along the *proximal convoluted tubule.*

Selective reabsorption is a process that returns useful products to the blood. However, we are particularly interested in the passive reabsorption of water (H_2O). Two factors aid this process: the nonfilterable proteins remain in blood, and salt is returned to blood. Following active reabsorption of sodium ions (Na^+), chloride ions (Cl^-) follow passively, as does water. Therefore, water moves to the area of greater solute concentration. This process occurs along the length of the nephron, until eventually nearly all water and sodium ions have been reabsorbed (table 9.3).

The cells lining the proximal convoluted tubule are anatomically adapted for *active reabsorption* (fig. 9.6). These cells have numerous microvilli, each about 1 µm in length, which increase the surface area for reabsorption. In addition, the cells contain numerous mitochondria, which produce the energy (ATP) necessary for active transport. Reabsorption by active transport is **selective reabsorption** because only molecules recognized by carrier molecules are actively reabsorbed. After passing through the tubule cells, the molecules enter blood in the peritubular capillary.

Glucose is an example of a molecule that ordinarily is completely reabsorbed because there is a plentiful supply of carrier molecules for it. However, every substance has a maximum rate of transport, and after all its carriers are in use, any excess in the filtrate will appear in the urine. For example, as reabsorbed levels of glucose approach

Table 9.3

Reabsorption from Nephron

Substance	Amount Filtered (per day)	Amount Excreted (per day)	Reabsorption (%)
Water, L	180	1.8	99.0
Sodium, g	630	3.2	99.5
Glucose, g	180	0.0	100.0
Urea, g	54	30.0	44.0

L= liters, g= grams

From A. J. Vander, et al. Human Physiology, *4th ed. © 1985 McGraw-Hill Publishing Company, New York, NY. Reprinted by permission of McGraw-Hill, Inc.*

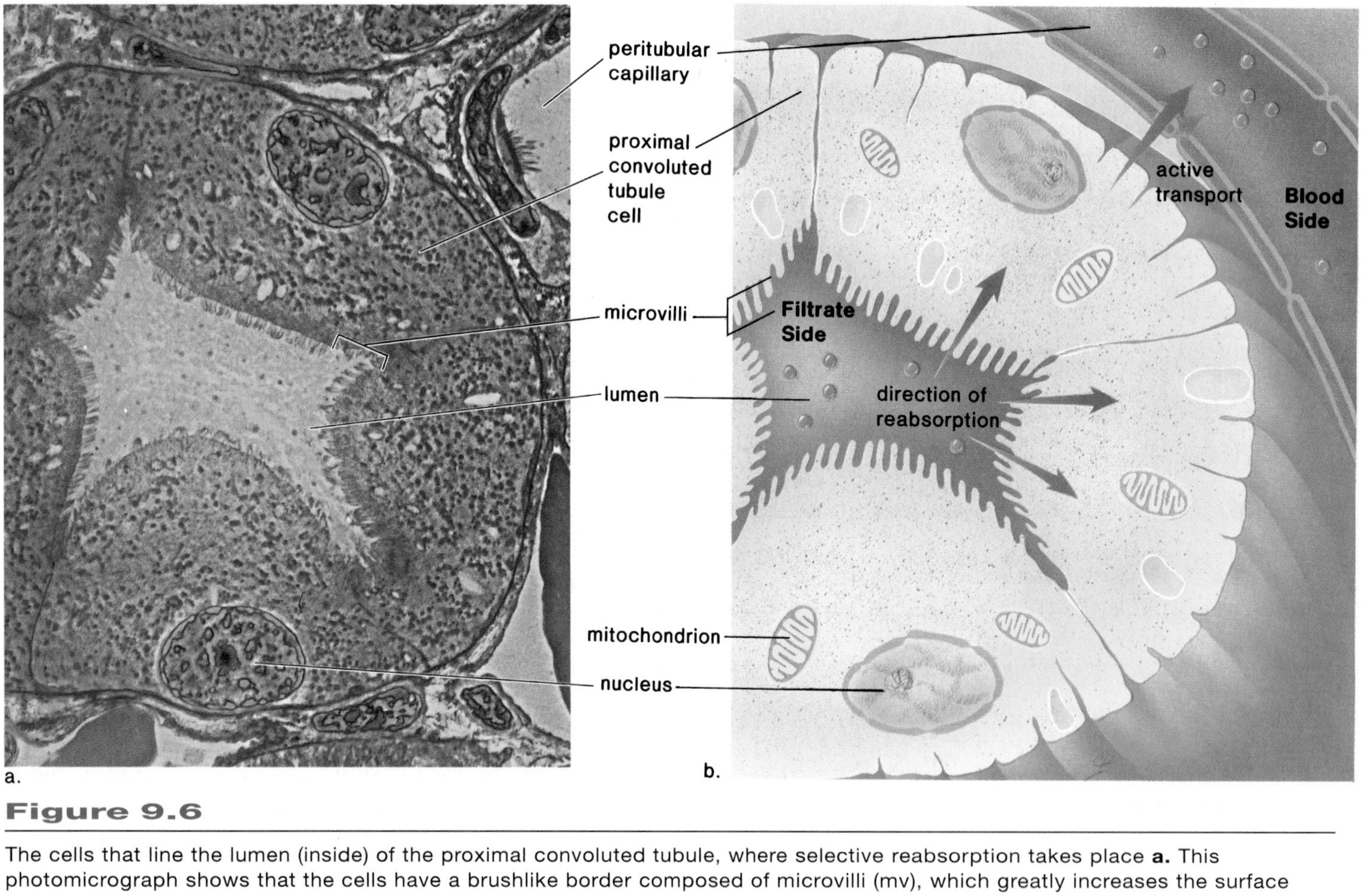

Figure 9.6

The cells that line the lumen (inside) of the proximal convoluted tubule, where selective reabsorption takes place **a.** This photomicrograph shows that the cells have a brushlike border composed of microvilli (mv), which greatly increases the surface area exposed to the lumen. The peritubular capillary surrounds the cells (nu = nucleus). **b.** Diagrammatic representation of **a.** shows that each cell has many mitochondria, which supply the energy needed for active transport, the process that moves molecules (green) from the lumen to the capillary.

the usual amount in plasma, the rest will appear in the urine. In diabetes mellitus, excess glucose occurs in the blood and then the filtrate because the liver and muscles fail to store glucose as glycogen.

We have seen that the filtrate that enters the proximal convoluted tubule is divided into 2 portions: the components that are reabsorbed from the tubule into blood and the components that are nonreabsorbed.

Reabsorbed Filtrate Components	Nonreabsorbed Filtrate Components
Most water	Some water
Nutrients	Much nitrogenous waste
Required salts (ions)	Excess salts (ions)

The substances that are not reabsorbed become the tubular fluid, which enters the loop of Henle.

During selective reabsorption, nutrient and salt molecules are actively reabsorbed from the proximal convoluted tubule into the peritubular capillary, and water follows passively.

Tubular Excretion: Second Addition

Tubular excretion[1] is a second way substances are removed from blood and added to tubular fluids. Hydrogen and ammonium ions, creatinine, and drugs such as penicillin move from blood in the peritubular capillaries into the distal convoluted tubule. The cells that line this portion of the tubule have numerous mitochondria because tubular excretion is an active process, which required ATP. In the end, urine contains substances that underwent pressure filtration and substances that underwent tubular excretion. However, pressure filtration is the more important of the 2 processes. Tubular excretion also controls blood pH by transporting hydrogen ions from blood into the tubule.

During tubular excretion, certain molecules are actively secreted from the peritubular capillary into the fluid of the distal convoluted tubule. These molecules are found in urine. Tubular excretion also regulates blood pH.

Reabsorbing Water

Water is reabsorbed along the whole length of the nephron, but the excretion of a hypertonic urine (one that is more concentrated than blood) is dependent upon the action of the loop of Henle and the collecting duct.

1. The term *tubular excretion* is used instead of tubular secretion because it better represents the end result of the process.

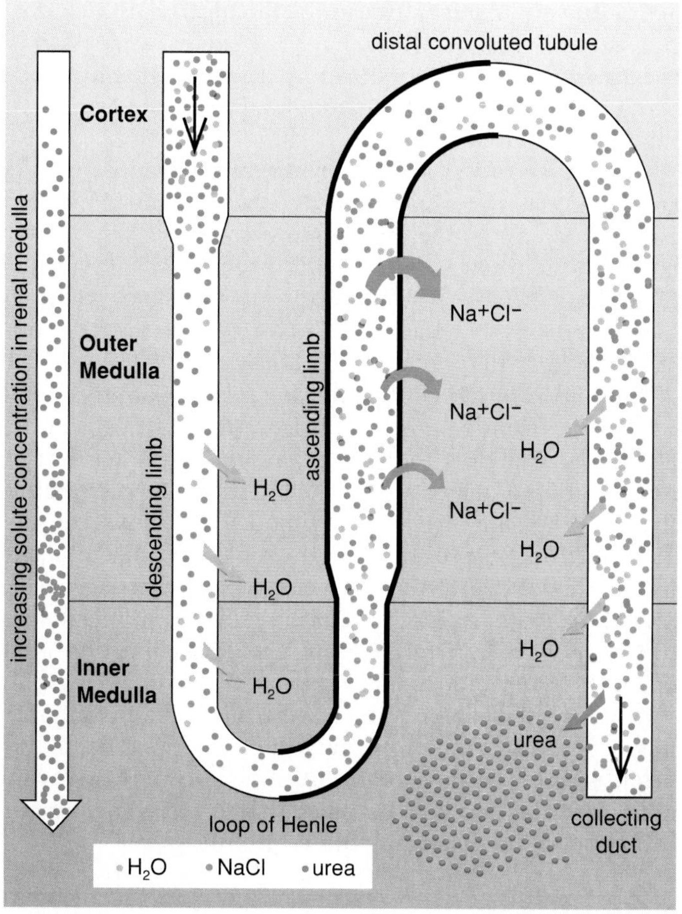

Figure 9.7

Reabsorption of water at the loop of Henle and the collecting duct. Salt (Na^+Cl^-) diffuses and is actively transported out of the ascending limb of the loop of Henle into the renal medulla; also, urea is believed to leak from the collecting duct and to enter the tissues of the renal medulla. This creates a hypertonic environment, which draws water out of the descending limb and the collecting duct. This water is returned to the circulatory system.

A long *loop of Henle,* which typically penetrates deep into the renal medulla, is made up of a *descending* (going down) limb and an *ascending* (going up) limb. Salt (Na^+Cl^-) passively diffuses out of the lower portion of the ascending limb, but the upper, thick portion of the limb actively transports salt out into the tissue of the outer renal medulla (fig. 9.7). Less and less salt is available for transport from the tubule as fluid moves up the thick portion of the ascending limb. In the end, there is an osmotic gradient within the tissues of the renal medulla: the concentration of salt is greater in the direction of the inner medulla. (Note that water cannot leave the ascending limb because the limb is impermeable to water.)

If you examine figure 9.7 carefully, you can see that the *innermost* portion of the inner medulla has the highest concentration of solutes. This cannot be due to salt because active transport of salt does not start until the thick portion of the ascending limb. Urea is believed to leak from the lower portion of the collecting duct, and it is this molecule that contributes to the high solute concentration of the inner medulla.

Because of the solute concentration gradient of the renal medulla, water leaves the descending limb of the loop of Henle along its length. This is a *countercurrent mechanism*—the increasing concentration of solute encounters the decreasing number of water molecules in the descending limb, ensuring that water continues to leave the descending limb from the top to the bottom.

Fluid entering a *collecting duct* comes from the distal convoluted tubule. This fluid is isotonic to the cells of the cortex. This means that to this point, the net effect of reabsorption of water and salt is the production of a fluid that has the same tonicity as blood. Now, however, the collecting duct passes through the renal medulla, which is increasingly hypertonic, as previously explained (fig. 9.7). Therefore, water diffuses out of the collecting duct into the renal medulla, and the urine within the collecting duct becomes hypertonic to blood plasma.

Urine, the composition of which is listed in table 9.1, now passes out of the collecting duct into the renal pelvis of the kidney. Urine contains all the molecules that were not reabsorbed as well as those that underwent tubular excretion at the distal convoluted tubule.

Water diffuses from both the descending limb of the loop of Henle and the collecting duct due to an increasingly hypertonic renal medulla. Urine (table 9.1) formation is complete.

Regulatory Functions of the Kidneys

The kidneys regulate the pH, the salt balance, and the volume of blood.

Maintaining Blood pH and Salt Balance

The kidneys help to maintain the pH level of blood around 7.4, and the whole nephron takes part in this process. The excretion of hydrogen ions (H^+) and ammonia (NH_3), together with the reabsorption of sodium ions (Na^+) and bicarbonate ions (HCO_3^-), is adjusted to keep the pH within normal bounds. If blood is acidic, hydrogen ions are excreted in combination with ammonia, while sodium ions and bicarbonate ions are reabsorbed. This restores the pH because $NaHCO_3$ is a base. If blood is basic, fewer hydrogen ions are excreted and fewer sodium ions and bicarbonate ions are reabsorbed.

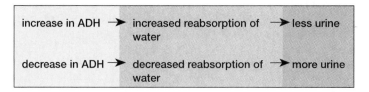

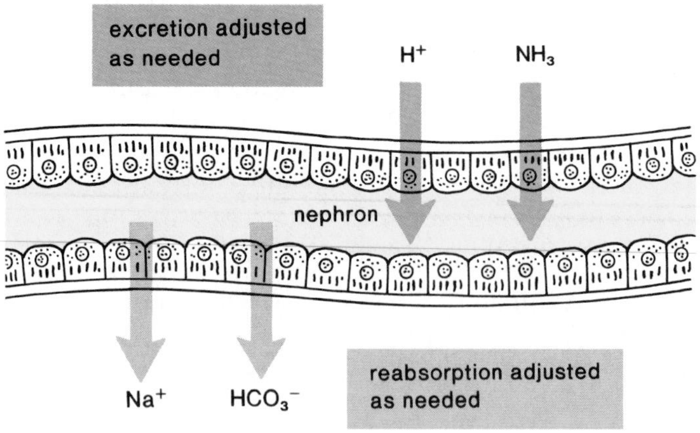

These examples also show that the kidneys regulate the salt balance in blood by controlling the excretion and the reabsorption of various ions. Sodium (Na^+) is an important ion in plasma that must be regulated, but the kidneys also excrete or reabsorb other ions, such as bicarbonate ions, potassium ions, and magnesium ions, as needed.

Maintaining Blood Volume

Maintenance of blood volume and salt balance is under the control of hormones. **Antidiuretic hormone (ADH),** secreted by the posterior pituitary, primarily maintains blood volume. ADH increases the permeability of the collecting duct so that more water can be reabsorbed into the blood. To understand the function of this hormone, consider its name. *Diuresis* means increased amount of urine, and *antidiuresis* means decreased amount of urine. When ADH is present, more water is reabsorbed into the blood and a decreased amount of urine results.

The secretion of this hormone is dependent on whether blood volume needs to be increased or decreased. When water is reabsorbed at the collecting duct, blood volume increases, and when water is not reabsorbed, blood volume decreases. In practical terms, if an individual does not drink much water on a certain day, the *posterior lobe of the pituitary* releases ADH, more water is reabsorbed, blood volume is maintained at a normal level, and consequently, there is less urine. On the other hand, if an individual drinks a large amount of water and does not perspire much, the posterior lobe of the pituitary does not release ADH, more water is excreted, blood volume is maintained at a normal level, and a greater amount of urine is formed.

increase in ADH → increased reabsorption of water → less urine

decrease in ADH → decreased reabsorption of water → more urine

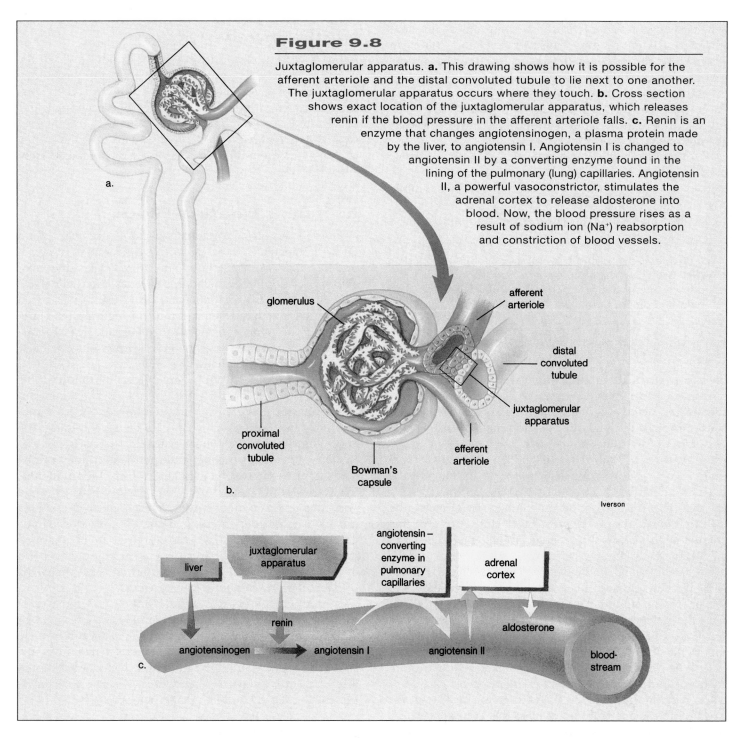

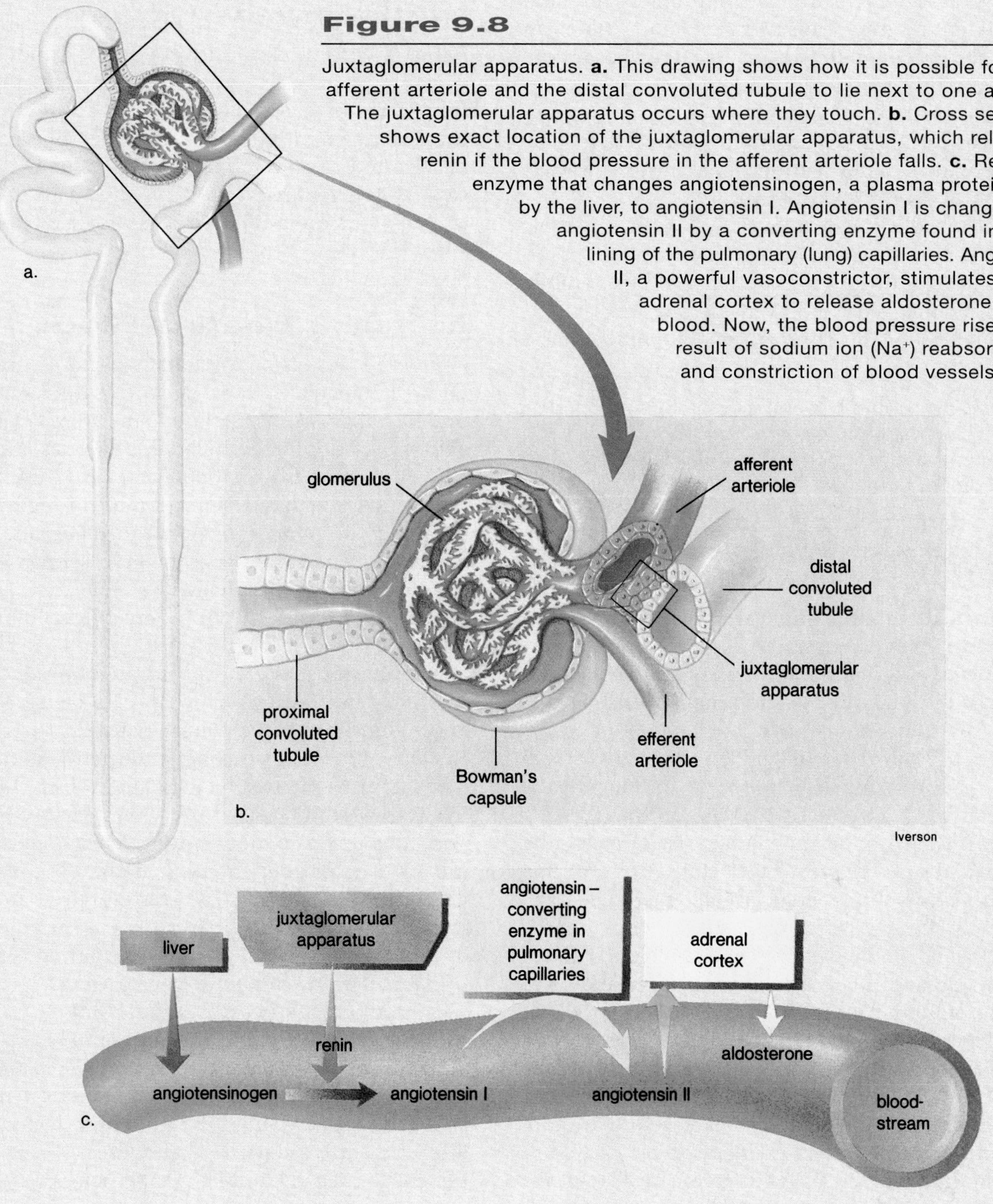

Figure 9.8

Juxtaglomerular apparatus. **a.** This drawing shows how it is possible for the afferent arteriole and the distal convoluted tubule to lie next to one another. The juxtaglomerular apparatus occurs where they touch. **b.** Cross section shows exact location of the juxtaglomerular apparatus, which releases renin if the blood pressure in the afferent arteriole falls. **c.** Renin is an enzyme that changes angiotensinogen, a plasma protein made by the liver, to angiotensin I. Angiotensin I is changed to angiotensin II by a converting enzyme found in the lining of the pulmonary (lung) capillaries. Angiotensin II, a powerful vasoconstrictor, stimulates the adrenal cortex to release aldosterone into blood. Now, the blood pressure rises as a result of sodium ion (Na^+) reabsorption and constriction of blood vessels.

Drinking alcohol causes diuresis because it inhibits the secretion of ADH. The dehydration that follows is believed to contribute to the symptoms of a "hangover." Caffeine blocks the action of ADH at the tubule level, therefore resulting in diuresis. Drugs called diuretics often are prescribed for high blood pressure, pulmonary edema, and congestive heart failure. The drugs cause salts and water to be excreted; therefore, they reduce blood volume and blood pressure. Concomitantly, any *edema* (p. 129) that is present is also reduced.

Aldosterone, secreted by the adrenal cortex, is a hormone that primarily maintains sodium ion (Na^+) and potassium ion (K^+) balance. It causes the reabsorption of sodium ions into the blood at the distal convoluted tubule and the excretion of potassium ions. The increase of sodium ions in blood causes water to be reabsorbed, leading to an increase in blood volume and blood pressure.

Blood pressure is constantly monitored by the *juxtaglomerular apparatus,* a region of contact between the afferent arteriole and the distal convoluted tubule (fig. 9.8).

When blood pressure is insufficient to promote pressure filtration, afferent arteriole cells secrete renin. *Renin* is an enzyme that changes angiotensinogen (a large plasma protein produced by the liver) into angiotensin I. Later, angiotensin I is converted to angiotensin II in the lungs by angiotensin-converting enzyme. Angiotensin II, a powerful vasoconstrictor, stimulates the adrenal cortex to release aldosterone and blood pressure rises.

The renin-angiotensin-aldosterone system always seems to be active in some patients with hypertension. In response to this possibility, a drug for hypertension that inhibits angiotensin-converting enzyme is available. The drug is called ACE (for angiotensin-converting enzyme) inhibitor.

> The kidneys contribute to homeostasis by excreting urea. They also maintain both the pH and the salt balance of blood and regulate the volume of blood, 3 very important functions.

Problems with Kidney Function

Because of the great importance of the kidneys to the maintenance of body fluid homeostasis, renal failure is a life-threatening event. There are many types of illnesses that cause progressive renal disease and renal failure.

Urinary tract infections are a fairly common occurrence, particularly in the female since the shorter female urethra invites bacterial invasion more than the longer male urethra. *Urethritis* is an infection of the urethra; *cystitis* involves the bladder, and, if the kidneys are infected, the infection is called *pyelonephritis.* The Health Focus for this chapter suggests ways to prevent urinary tract infections.

Glomerular damage sometimes leads to blockage of the glomeruli so that no fluid moves into the tubules, or damage can cause the glomeruli to become more permeable than usual. This is detected when a **urinalysis** is done. If the glomeruli are too permeable, albumin, white blood cells, or even red blood cells appear in the urine. A trace amount of protein in the urine is not a matter of concern, however.

When glomerular damage is so extensive that more than two-thirds of the nephrons are inoperative, waste substances accumulate in blood. This condition is called *uremia* because urea is one of the substances that accumulates. Although nitrogenous wastes can cause serious damage, the retention of water and salts (ions) is of even greater concern. The latter causes edema, fluid accumulation in the body tissues. Imbalance in the ionic composition of body fluids can even lead to loss of consciousness and to heart failure.

Replacing the Kidney

Patients with renal failure sometimes undergo a kidney transplant operation, during which a functioning kidney from a donor is received. As with all organ transplants, there is the possibility of organ rejection. Receiving a kidney from a close relative has the highest chance of success. The current one-year survival rate is 97% if the kidney is received from a relative and 90% if it is received from a nonrelative.

Dialysis: Treating Blood

Frequently, a satisfactory kidney donor is not available to allow a transplant, and it is necessary for the patient to undergo dialysis, utilizing either an artificial kidney machine or continuous ambulatory peritoneal (abdominal) dialysis (CAPD). *Dialysis* is defined as the diffusion of dissolved molecules through a semipermeable membrane. These molecules, of course, move across a membrane from the area of greater concentration to one of lesser concentration.

During hemodialysis (fig. 9.9), the patient's blood is passed through a semipermeable membranous tube, which is in contact with a balanced salt (dialysis) solution. Substances more concentrated in blood diffuse into the dialysis solution, which is also called the dialysate, and substances more concentrated in the dialysate diffuse into blood. Accordingly, the artificial kidney can be utilized either to extract substances from blood, including waste products or toxic chemicals and drugs, or to add substances to blood—for example, bicarbonate ions (HCO_3^-) if blood is acidic. In the course of a 6-hour hemodialysis, from 50–250 grams of urea can be removed from a patient, which greatly exceeds the amount excreted by normal kidneys. Therefore, a patient only needs to undergo treatment about twice a week.

In the case of CAPD, a fresh amount of dialysate is introduced directly into the abdominal cavity from a bag attached to a permanently implanted plastic tube. Waste and water molecules pass into the dialysate from the surrounding organs before the fluid is collected 4 or 8 hours later. The individual can go about his or her normal activities during CAPD, unlike during hemodialysis.

> Kidney transplants and hemodialysis are available corrective procedures for persons who have suffered renal failure.

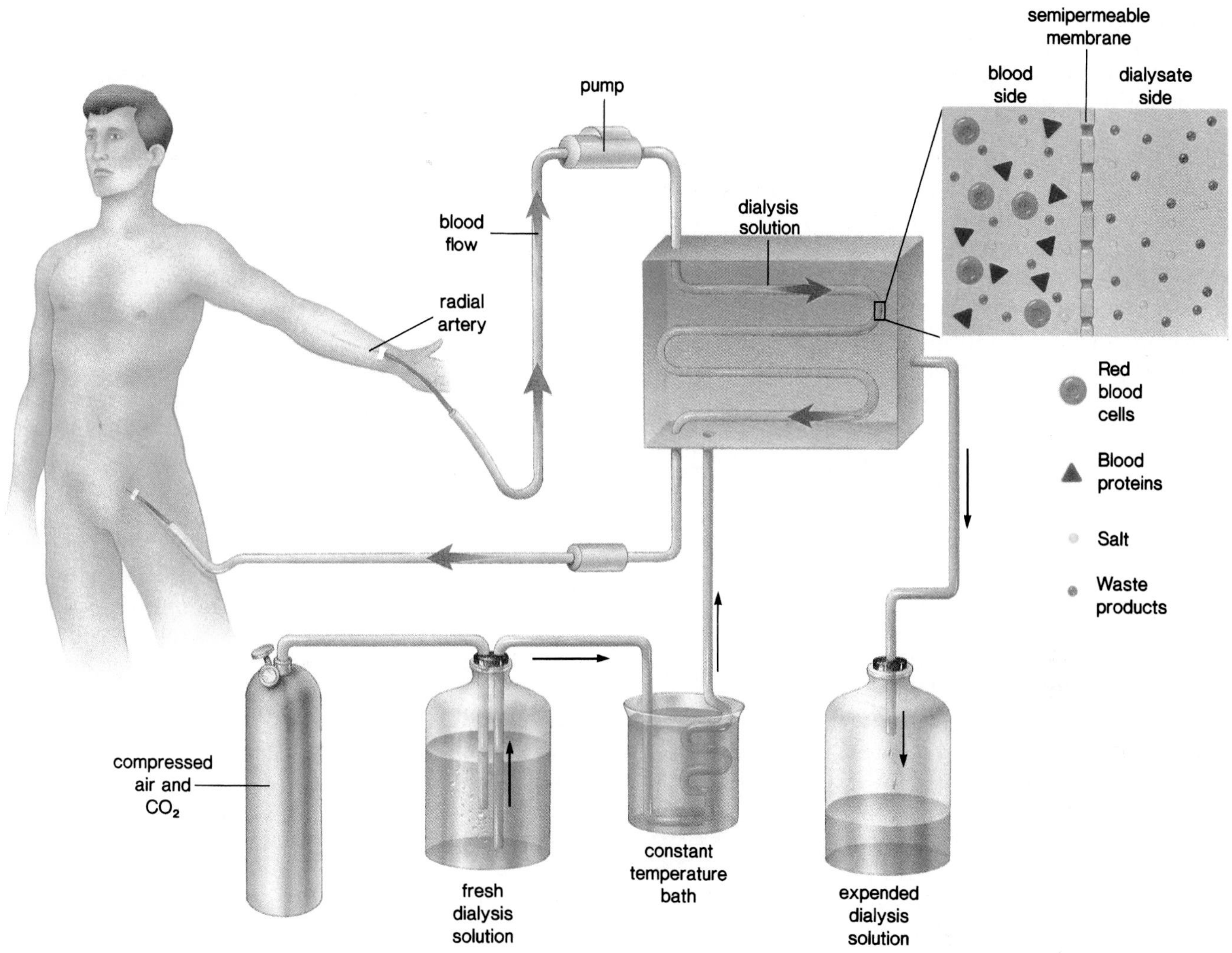

Figure 9.9

An artificial kidney machine. As the patient's blood circulates through dialysis tubing, it is exposed to a dialysis solution (dialysate). Wastes exit from blood into the solution because of a preestablished concentration gradient. In this way, blood is not only cleansed, its pH can also be adjusted.

SUMMARY

The end products of metabolism for the most part are nitrogenous wastes, such as urea, uric acid, and creatinine, all of which are primarily excreted by the kidneys. Other end products of metabolism, such as bile pigments and carbon dioxide, are excreted by the liver and lungs, respectively.

The kidneys are part of the urinary system, and they produce urine. Urine is composed primarily of nitrogenous end products and ions (salts) in water. Urine passes to the ureters, which take it to the bladder, where it may be stored for a time. From there it leaves the body by way of the urethra.

Macroscopically, the kidney is made up of 3 parts: a pelvis, medulla, and cortex. Microscopically, it is made up of over one million nephrons. The nephrons, which account for the kidney's macroscopic anatomy, have several parts: Bowman's capsule, proximal convoluted tubule, loop of Henle, distal convoluted tubule, and the collecting duct. The loop of Henle and collecting duct are primarily in the medulla, while the other portions are primarily in the cortex. Each nephron has its own blood supply:

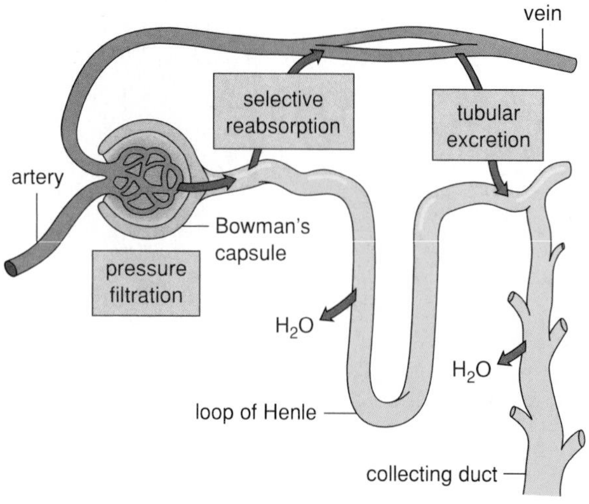

the afferent arteriole approaches Bowman's capsule and divides to become a capillary tuft called the glomerulus, which is enclosed by the capsule. The efferent arteriole leaves the capsule and immediately branches into a capillary bed, which is in close contact with all other parts of the tubule.

During the first step of urine formation, termed *pressure filtration*, small components of plasma pass into Bowman's capsule from the glomerulus, due to blood pressure. This filtrate of blood contains water, nutrients, nitrogenous wastes, and ions (salts). During the second step, termed *selective reabsorption*, nutrients and sodium are actively reabsorbed from the proximal convoluted tubule back into the blood. During the third step in urine formation, termed *tubular excretion*, a few types of substances are actively transported into the distal convoluted tubule. Much water is removed from the filtrate in the region of the loop of Henle and collecting duct (180 liters of glomerular filtrate to 1.8 liters of urine/day).

STUDY AND THOUGHT QUESTIONS

1. Name several excretory organs, as well as the substances they excrete. (p. 166)
2. State the path of urine and the function of each organ mentioned. (p. 167)
3. Name 3 nitrogenous end products, and explain how each is formed in the body. (pp. 166–167)
4. Describe the macroscopic anatomy of a kidney and the microscopic anatomy of a nephron. (p. 168)
5. Trace the path of blood about the nephron. (pp. 169–170)
6. Describe how urine is made by telling what happens at each part of a nephron. (pp. 170–174)

   Critical Thinking If there is a loss of proteins from blood into Bowman's capsule, would the filtration rate increase or decrease? (Assume constant blood pressure.) Explain.
7. Explain these terms: *pressure filtration, selective reabsorption,* and *tubular excretion.* (pp. 170–173)
8. What is the composition of urine? (p. 167)

   Critical Thinking If 99% of water is reabsorbed (table 9.3), how can urine be 95% water (table 9.1)?
9. Tell how pH is regulated and how hormonal effects on the kidneys regulate blood volume. (pp. 174–176)
10. Explain how the artificial kidney machine and CAPD work. (p. 176)

    Critical Thinking Why is more urea excreted during hemodialysis than during urine formation?

    Ethical Issue Should people feel free to refuse donation of a kidney to a relative? Why or why not?

WRITING ACROSS THE CURRICULUM

1. Would animals that live in fresh water or on land have well-developed loops of Henle? What is accomplished by having a well-developed loop of Henle?
2. Why is the term *filtration* used both in regard to a capillary in the tissues and to the glomerulus in the nephron? What differences are there between these 2 processes?
3. Physicians recognize that the presence of protein in the urine is the single most important indicator of kidney disease. How does kidney disease cause plasma proteins to be lost via the kidneys? What are the bodily consequences of losing plasma proteins in this way?

Objective Questions

1. The primary nitrogenous end product of humans is ___________________.
2. The liver excretes _________________ , which are derived from the breakdown of ________________.
3. Urine leaves the urinary bladder by way of the ___________________.
4. The capillary tuft inside Bowman's capsule is called the ______________________.
5. ___________________ is a substance that is found in the filtrate, is reabsorbed, and is still in urine.
6. ___________________ is a substance that is found in the filtrate, is minimally reabsorbed, and is concentrated in the urine.
7. Tubular excretion takes place at the ____________, a portion of the nephron.
8. Reabsorption of water from the collecting duct is regulated by the hormone ___________________.
9. In addition to excreting nitrogenous wastes, the kidneys adjust the __________ and __________ of blood.
10. Persons who have nonfunctioning kidneys often have their blood cleansed by __________________ machines.
11. Label this diagram of a nephron.

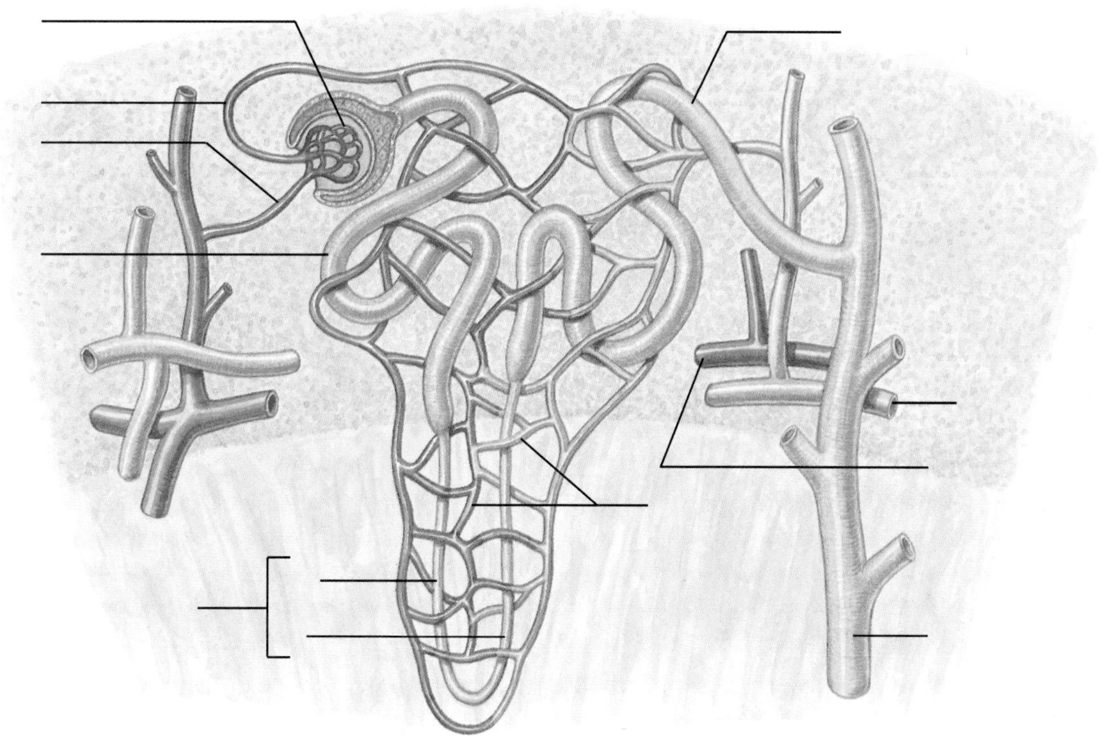

Selected Key Terms

antidiuretic hormone (ADH) (an″ tĭ-di″ u-ret′ ik hōr′ mōn) A hormone secreted by the posterior pituitary that controls the degree to which water is reabsorbed by the kidneys. 174

Bowman's capsule A double-walled cup that surrounds the glomerulus at the beginning of the nephron. 168

collecting duct A tube that receives urine from the distal convoluted tubules of several nephrons within a kidney. 168

distal convoluted tubule Highly coiled region of a nephron that is distant from Bowman's capsule, where tubular excretion takes place. 168

excretion Removal of metabolic wastes from the body. 166

glomerular filtrate (glo-mer′ u-lar fil′ trāt) The filtered portion of blood contained within Bowman's capsule. 171

glomerulus (glo-mer′ u-lus) A cluster; for example, the cluster of capillaries surrounded by Bowman's capsule in a nephron, where pressure filtration takes place. 169

kidney An organ in the urinary system that produces and excretes urine. 167

nephron (nef´ ron) The anatomical and functional unit of the kidney; kidney tubule. 168

peritubular capillary Capillary that surrounds a nephron and functions in reabsorption during urine formation. 169

pressure filtration The movement of small molecules from the glomerulus into Bowman's capsule due to the action of blood pressure. 170

proximal convoluted tubule Highly coiled region of a nephron near Bowman's capsule, where selective reabsorption takes place. 168

renal pelvis A hollow chamber in the kidney that lies inside the renal medulla and receives freshly prepared urine from the collecting ducts. 168

selective reabsorption The movement of nutrient molecules, as opposed to waste molecules, from the contents of the nephron into blood at the proximal convoluted tubule. 172

tubular excretion The movement of certain molecules from blood into the distal convoluted tubule so that they are added to urine. 173

urea (u-re´ ah) Primary nitrogenous waste of humans derived from amino acid breakdown. 166

ureter (u´ re´ ter) One of 2 tubes that take urine from the kidneys to the urinary bladder. 167

urethra (u´ re´ thrah) Tube that takes urine from the bladder to outside. 167

uric acid Waste product of nucleotide metabolism. 166

urinary bladder Organ where urine is stored before being discharged by way of the urethra. 167

Further Readings for Part Two

Berns, M. B. June 1991. Laser surgery. *Scientific American.*

Boon, T. March 1993. Teaching the immune system to fight cancer. *Scientific American.*

Cohen, I. R. April 1988. The self, the world and autoimmunity. *Scientific American.*

Creager, J. G. 1992. *Human anatomy and physiology.* 2d ed. Dubuque, Iowa: Wm. C. Brown Publishers.

Eigen, M. July 1993. Viral quasispecies. *Scientific American.*

Fischetti, V. A. February 1992. Streptococcal M protein. *Scientific American.*

Fox, S. I. 1990. *Human physiology.* 3d ed. Dubuque, Iowa: Wm. C. Brown Publishers.

Golde, D. W. December 1991. The stem cell. *Scientific American.*

Golub, E. S., and Green, D. R. 1991. *Immunology: A synthesis.* 2d ed. Sunderland, Mass.: Sinauer Associates.

Green, H. November 1991. Cultured cells for the treatment of disease. *Scientific American.*

Hales, D. 1992. *An invitation to health.* 5th ed. Redwood City, Calif.: Benjamin/Cummings Publishing.

Harken, A. July 1993. Surgical treatment of cardiac arrhythmias. *Scientific American.*

Hirshhorn, N., and Greenough, W. B., III. May 1991. Progress in oral rehydration therapy. *Scientific American.*

Hole, J. W., Jr. 1992. *Essentials of human anatomy and physiology.* 4th ed. Dubuque, Iowa: Wm. C. Brown Publishers.

Houston, C. S. October 1992. Mountain sickness. *Scientific American.*

Janeway, C. A., Jr. September 1993. How the immune system recognizes invaders. *Scientific American.*

Johnson, H. M., Russell, J. K., and Pontzer, C. H. April 1992. Superantigens in human disease. *Scientific American.*

Klein, J. December 1993. MHC polymorphism and human origins. *Scientific American.*

Lawn, R. M. June 1992. Lipoprotein(a) in heart disease. *Scientific American.*

Marieb, E. N. 1991. *Essentials of human anatomy and physiology.* 3d ed. Redwood City, Calif.: Benjamin/Cummings Publishing.

Marrack, P., and Kappler, J. W. September 1993. How the immune system recognizes the body. *Scientific American.*

Moberg, C. L., and Cohn, Z. A. May 1991. Rene Jules Dubos. *Scientific American.*

Nossal, G. J. V. September 1993. Life, death and the immune system. *Scientific American.*

Phillips, M. July 1992. Breath tests in medicine. *Scientific American.*

Rasmussen, H. October 1989. The cycling of calcium as an intracellular messenger. *Scientific American.*

Rennie, J. December 1990. The body against itself. *Scientific American.*

Rietschel, E. T., and Brade, H. August 1992. Bacterial endotoxins. *Scientific American.*

Schultz, J. S. August 1991. Biosensors. *Scientific American.*

Schwartz, R. August 1993. *T* cell anergy. *Scientific American.*

Scientific American. September 1993. Entire issue is devoted to the immune system.

Scrimshaw, N. S. October 1991. Iron deficiency. *Scientific American.*

Sharon, N., and Lis, H. January 1993. Carbohydrates in cell recognition. *Scientific American.*

Smith, K. A. March 1990. Interleukin-2. *Scientific American.*

Snyder, S. H., and Bredt, D. S. May 1992. Biological roles of nitric oxide. *Scientific American.*

Spence, A. P. 1990. *Basic human anatomy.* 3d ed. Redwood City, Calif.: Benjamin/Cummings Publishing.

vonBoehmer, H., and Kisielow, P. October 1991. How the immune system learns about self. *Scientific American.*

Weissman, G. January 1991. Aspirin. *Scientific American.*

Weissman, I. L., and Cooper, M. D. September 1993. How the immune system develops. *Scientific American.*

Young, J., and Cohn, Z. January 1988. How killer cells kill. *Scientific American.*

Zivin, J. A., and Choi, D. W. July 1991. Stroke therapy. *Scientific American.*

Part THREE

Integration and Coordination in Humans

The nervous and hormonal systems help maintain a relatively constant internal environment, known as homeostasis, by coordinating the functions of the body's other systems. The nervous system acts quickly but provides short-lived regulation, and the hormonal system acts more slowly but provides a more sustained regulation of body parts.

Organisms must also be able to react with the external environment to survive. The sense receptors are the organs that inform the organism about the outside environment. This information is then processed by the nervous system, and the individual responds through the muscular system. These attributes allow humans to be aware of the external environment and its possible incompatibility with the internal environment.

Chapter 10

Nervous System

Paraplegia, paralysis of the legs, and quadriplegia, paralysis of all 4 limbs, occur when the spinal cord is partially or wholly severed. Not until this decade did scientists believe there was hope for a cure. Now they have learned to coax nerve regrowth within the central nervous system. In hamsters with severed optic nerves, researchers not only got the optic nerves to regrow, they detected nerve signals in the visual center of the brain after flashing a light in front of the hamsters' eyes. They reported the work demonstrates that "one can make functional connections between neurons that are widely separated by injury."

Chapter Concepts

Chapter Outline

The nervous system tells us that we exist and, along with the muscles and sense organs, accounts for our distinctly animal characteristic of quick reaction to environmental stimuli. The nerve cell is called a **neuron,** and it is neurons that carry nerve impulses (messages) from sense organs to the central nervous system (brain and spinal cord), where they are interpreted and coordinated before other impulses go to the muscles or the glands, which then react.

Neuron Structure

All neurons have 3 parts: dendrite(s), cell body, and axon (fig. 10.1). A **dendrite** conducts nerve impulses toward the **cell body,** the part of a neuron that contains the nucleus. An **axon** conducts nerve impulses away from the cell body.

There are 3 types of neurons: sensory neurons, motor neurons, and interneurons (fig. 10.1). A **sensory neuron** takes a message from a **receptor** to the central nervous system (CNS) and typically has a long dendrite and a short axon. A **motor neuron** takes a message away from the CNS to an **effector,** a muscle or a gland, and has short dendrites and a long axon. Because motor neurons cause muscle fibers and glands to react, they are said to **innervate** these structures. Sometimes a sensory neuron is referred to as the *afferent neuron,* and the motor neuron is called the *efferent neuron.* These words, which are derived from Latin, mean *running to* and *running away from,* respectively. Obviously, they refer to the relationship of these neurons to the CNS.

An **interneuron** (also called association neuron or connector neuron), which is always found completely within the CNS, conveys messages between parts of the system. An interneuron has short dendrites and either a long or a short axon. Table 10.1 summarizes the 3 types of neurons, which are also illustrated in figure 10.1.

> Although all neurons have the same 3 parts, each is specialized in structure and in function. Specialization is dependent on the location of the neuron in relation to the CNS.

The dendrites and the axons of neurons sometimes are called fibers or processes. Most long fibers, whether dendrites or axons, are covered by tightly packed spirals of *Schwann cells* (neurolemmocytes) (fig. 10.2). Schwann cells are one of several types of neuroglial cells in the nervous system. Neuroglial cells service the neurons—they have supportive and nutritive functions. Schwann cells encircle a fiber, leaving gaps called the **nodes of Ranvier** (neurofibril nodes). Schwann cells wrap themselves around the axon

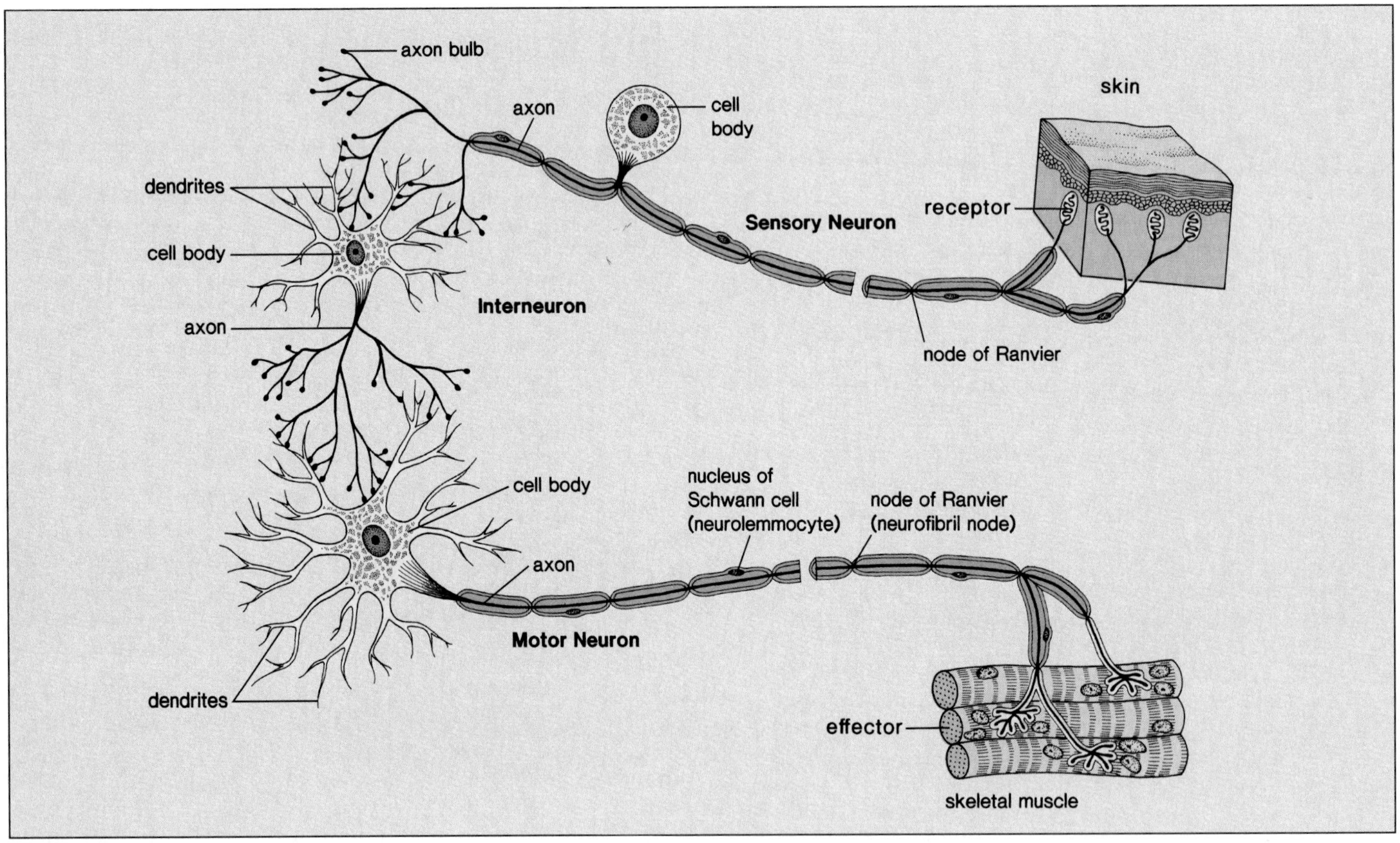

Figure 10.1

Types of neurons. A sensory neuron, an interneuron, and a motor neuron are drawn here to show their arrangement in the body. (The breaks indicate that the fibers are much longer than shown.) How does this arrangement correlate with the function of each neuron?

Table 10.1

Neurons

Neurons	Structure	Function
Sensory neuron (afferent)	Long dendrite, short axon	Carries nerve impulses (messages) from a receptor to the CNS
Motor neuron (efferent)	Short dendrite, long axon	Carries nerve impulses (messages) from the CNS to an effector
Interneuron	Short dendrites, long or short axon	Carries nerve impulses (messages) within the CNS

many times, and they lay down several layers of cellular membrane containing myelin, a lipid substance that is an excellent insulator. Myelin gives nerve fibers their white, glistening appearance. Because of the manner in which Schwann cells wrap themselves about nerve fibers, 2 sheaths are formed (fig. 10.2). The outermost sheath is called the **neurilemma,** or the cellular sheath, and the inner one is called the **myelin sheath.** (Scarlike patches begin to replace myelin in multiple sclerosis, and the motor difficulties result.) The neurilemma plays an important role in nerve regeneration. If a nerve fiber is accidently severed, the part distant from the nerve fiber degenerates, except for the neurilemma. The neurilemma then serves as a passageway for new axonal growth.

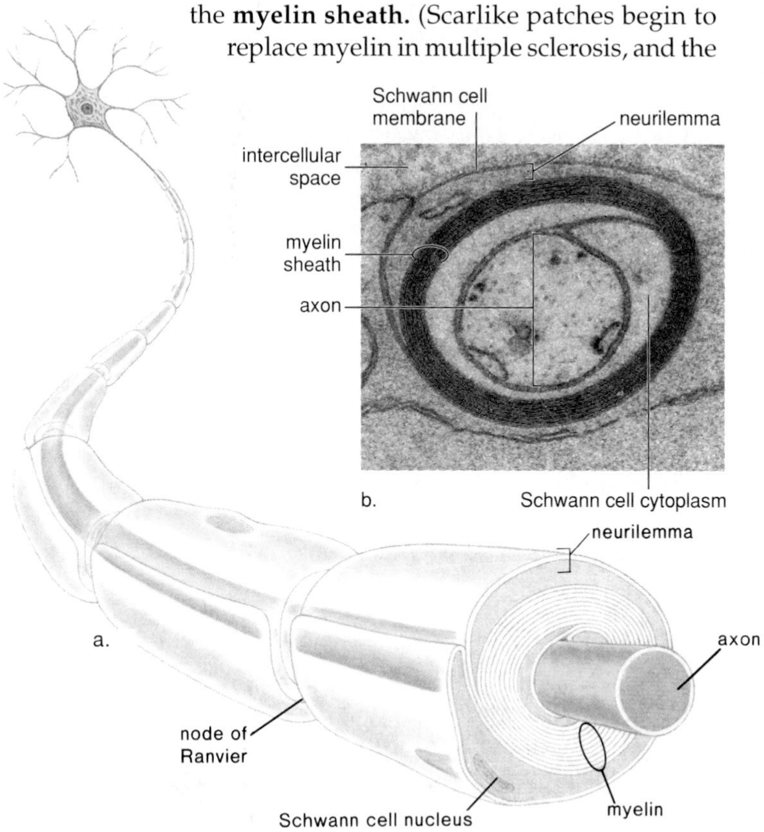

Figure 10.2

Neurilemma and myelin sheath. **a.** Axon of a motor neuron ending in a cross section of neurilemma and myelin sheath, which enclose the long fibers of all neurons. The myelin sheath is composed of many layers of Schwann cell membrane and has a white, glistening appearance in the body. **b.** Electron micrograph of a cross section of an axon surrounded by neurilemma and myelin sheath.

Nerve Impulse

A **nerve impulse** is the way a neuron transmits information. The nature of a nerve impulse has been studied using giant axons from the squid and an instrument called a voltmeter. Voltage is a measure of the electrical potential difference between 2 points, which in this case are the inside and the outside of the axon. The change in voltage is displayed on an *oscilloscope*, an instrument with a screen that shows a trace, or pattern, indicating a change in voltage with time.

In humans, the speed of conduction is quite rapid (200 meters per second) because the nerve impulse jumps from node of Ranvier to node of Ranvier. This is called saltatory (saltatory means jumping) conduction.

Resting Potential: Inside Is Negative

In the experimental setup shown in figure 10.3, an oscilloscope is wired to 2 electrodes, one inside and one outside a giant axon of the squid. The axon is essentially a membranous tube filled with axoplasm (cytoplasm of the axon). When the axon is not conducting an impulse, the oscilloscope records a *membrane potential* (potential difference across a membrane) equal to about –65 mV. This reading indicates that the inside of the neuron is negative compared to the outside. This is called the **resting potential** because the axon is not conducting an impulse.

The existence of this polarity (charge difference) can be correlated with a difference in ion distribution on either side of the axomembrane (cell membrane of the axon). As figure 10.4*a* shows, the concentration of sodium ions (Na^+) is greater outside the axon than inside and the concentration of potassium ions (K^+) is greater inside the axon than outside. The unequal distribution of these ions is due to the action of the sodium-potassium pump. This is an active transport system in the axomembrane that pumps sodium ions out of and potassium ions into the axon (p. 187). The work of the pump maintains the unequal distribution of sodium ions and potassium ions across the axomembrane.

The pump is always working because the membrane is somewhat permeable to these ions and they tend to diffuse toward their lesser concentration. Since the membrane is more permeable to potassium than to sodium, there are always more positive ions outside the axomembrane than inside; this accounts for the polarity recorded by the oscilloscope. There are also large, negatively charged proteins in the axoplasm, which are termed *immobile* in figure 10.4*a* because they are too large to cross the axomembrane.

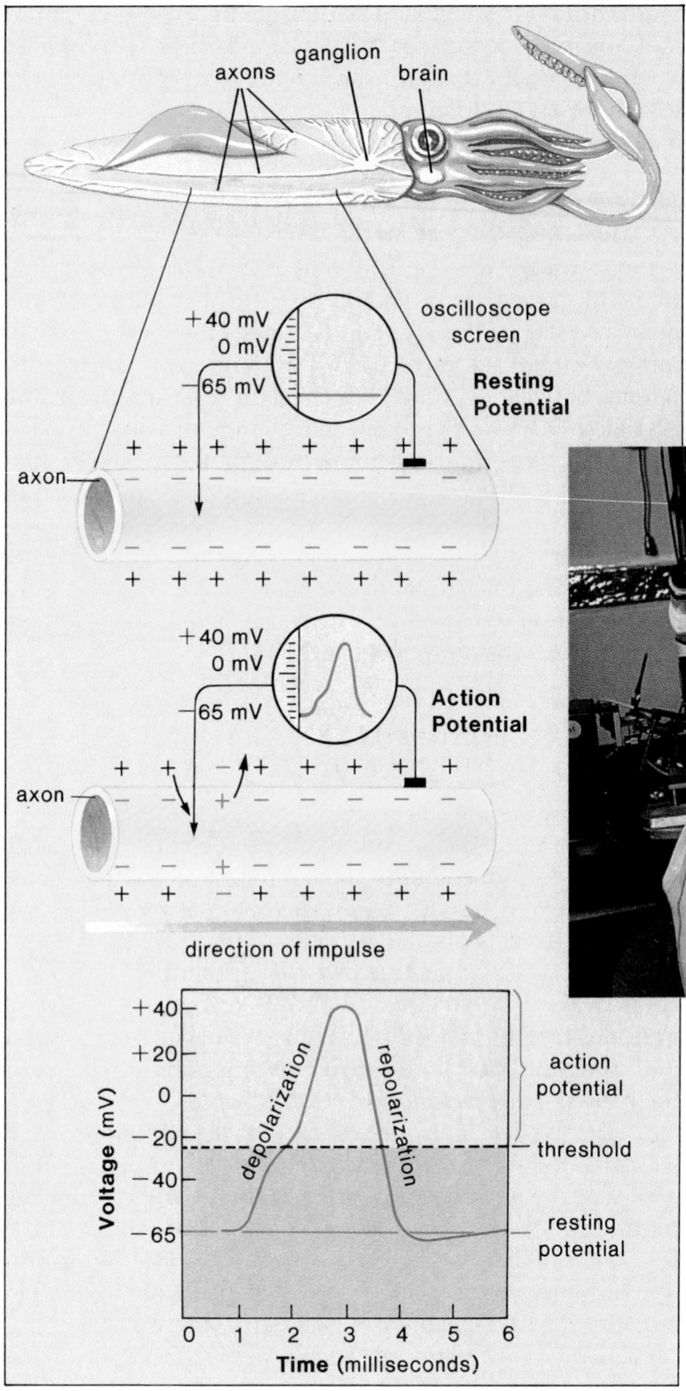

Figure 10.3

Study of nerve impulse requires the use of an oscilloscope and a nerve fiber, such as a giant squid axon. The squid axon is so large (about 1 mm in diameter) that a microelectrode can be inserted into it. When the axon is not conducting a nerve impulse, the electrode registers and the oscilloscope shows a resting potential of –65 mV. When an action potential is achieved, the axon conducts a nerve impulse and there is a rapid change in potential from –65 mV to +40 mV (called depolarization), followed by a return to –65 mV (called repolarization).

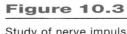

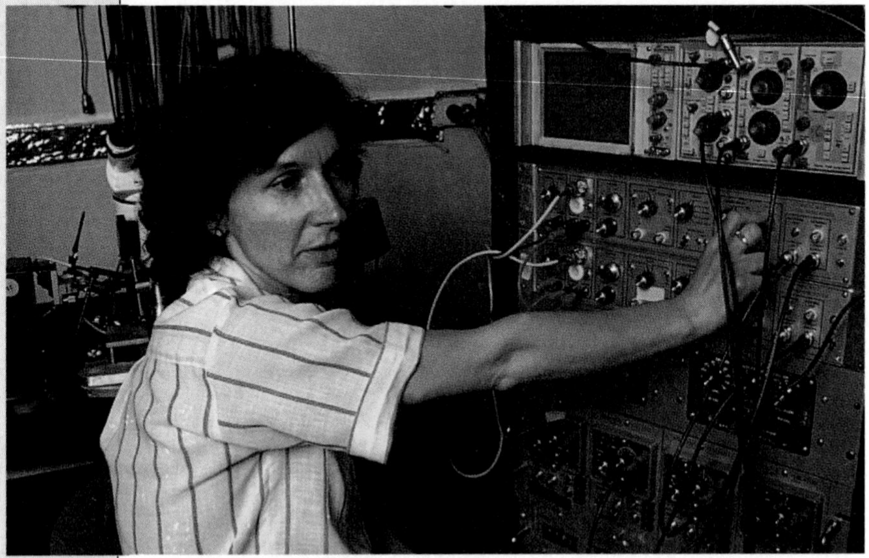

Action Potential: Upswing and Downswing

If the axon is stimulated to conduct a nerve impulse by an electric shock, a sudden difference in pH, or a pinch, then a pattern appears on the oscilloscope screen. This pattern, caused by rapid polarity changes and called the **action potential,** has an upswing and a downswing.

Sodium Gates Open

As the action potential swings up from –65 mV to +40 mV, sodium ions (Na^+) rapidly move across the axomembrane to the inside of the axon. The stimulation of the axon causes the gates of the sodium channels to open temporarily, allowing sodium to flow into the axon. This sudden permeability of the axomembrane causes the oscilloscope to record a *depolarization:* the charge inside of the fiber changes from negative to positive as sodium ions enter the interior (fig. 10.4*b*).

Potassium Gates Open

As potassium ions (K^+) rapidly move from the inside to the outside of the axon, the action potential swings down from +40 mV to at least –65 mV. The axomembrane is suddenly permeable to potassium because the potassium gates of the potassium channels temporarily open, allowing potassium ions to flow out of the axon. The oscilloscope records a *repolarization:* the inside of the axon resumes a negative charge (fig. 10.4*c*).

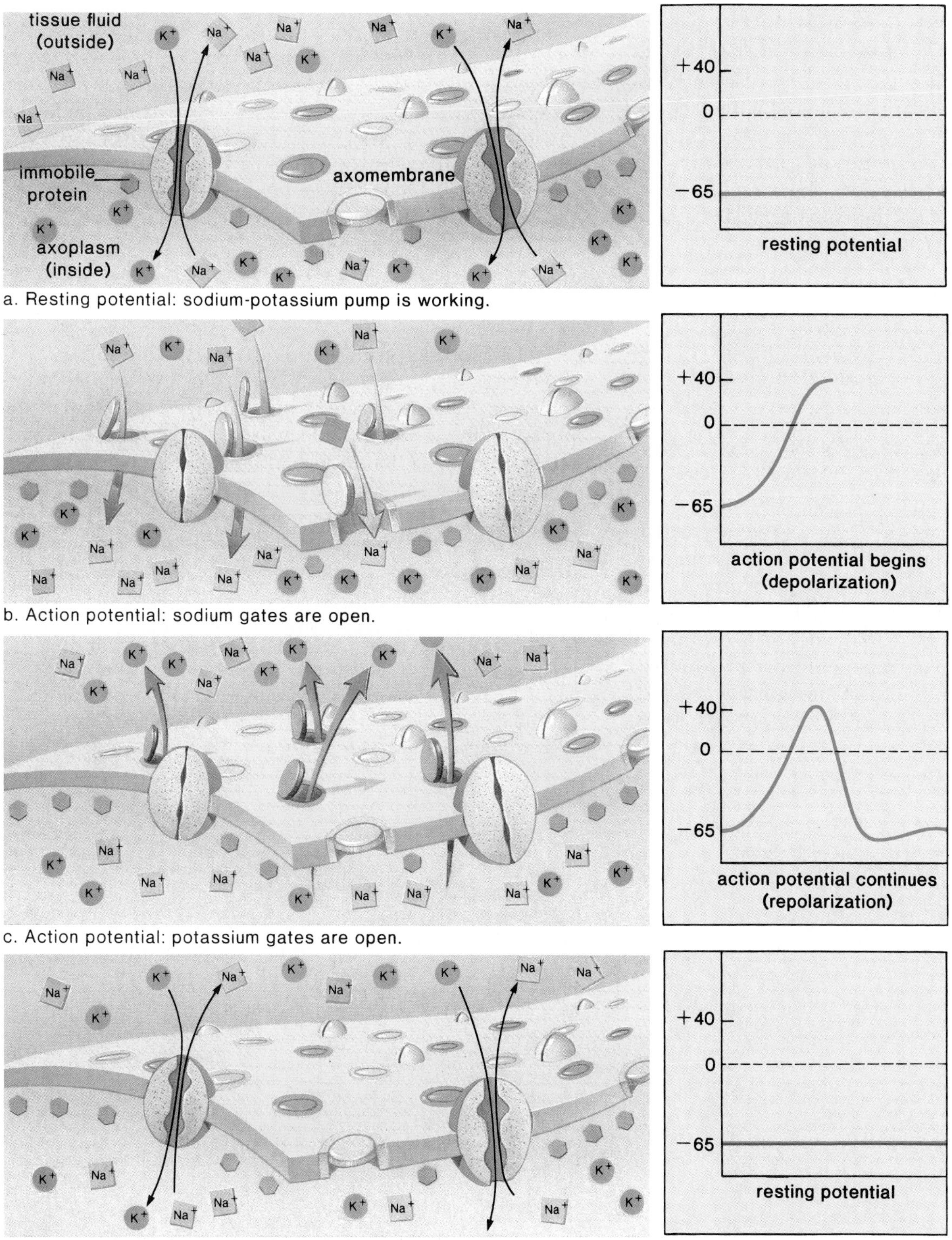

a. Resting potential: sodium-potassium pump is working.

When a neuron is not conducting a nerve impulse, the sodium (Na^+) and potassium (K^+) gates are closed. The sodium-potassium pump maintains the uneven distribution of these ions across the axomembrane. The oscilloscope registers a resting potential of −65 mV inside compared to outside.

b. Action potential: sodium gates are open.

An action potential begins when the sodium gates open and sodium ions move to the inside. The oscilloscope registers a depolarization as the axoplasm reaches +40 mV compared to tissue fluid.

c. Action potential: potassium gates are open.

Action potential continues as the sodium gates close and the potassium gates open, allowing potassium ions to move to the outside. The oscilloscope registers repolarization as the axoplasm again becomes −65 mV compared to tissue fluid.

d. Refractory period: sodium-potassium pump is working.

The oscilloscope registers −65 mV again, but the sodium-potassium pump is working to restore the original sodium and potassium ion distribution. The sodium and potassium gates are now closed but will open again in response to another stimulus.

Figure 10.4

Action potential and resting potential. The action potential is the result of an exchange of sodium ions (Na^+) and potassium ions (K^+), and it is shown (*right*) as a change in polarity by an oscilloscope. So few ions are exchanged for each action potential that it is possible for a nerve fiber to repeatedly conduct nerve impulses. Whenever the fiber rests, the sodium-potassium pump restores the original distribution of ions.

Refractory Period: Sodium-Potassium Pump Keeps Working

A fiber can conduct a volley of nerve impulses because only a small number of ions are exchanged with each impulse. When the fiber rests, however, there is a refractory period, during which the sodium-potassium pump continues to return sodium ions (Na^+) to the outside and potassium ions (K^+) to the inside of the axon (fig. 10.4*d*). During the absolute refractory period, a neuron is unable to conduct a nerve impulse.

> All neurons, whether sensory or motor, transmit the same type of nerve impulse—an electrochemical charge, which is propagated along the nerve fiber(s).

Transmission Across a Synapse

The mechanism by which an action potential passes from one neuron to another is not the same as the mechanism by which an action potential is conducted along a neuron. Every axon branches into many fine terminal branches, each of which is tipped by a small swelling, the axon bulb (fig. 10.5*a*). Each bulb lies very close to the dendrite (or the cell body) of another neuron. This region of close proximity is called a **synapse.** At a synapse the axomembrane is called the **presynaptic membrane,** and the membrane of the next neuron is called the **postsynaptic membrane.** The small gap between is the **synaptic cleft.**

Transmission of nerve impulses across a synaptic cleft is carried out by **neurotransmitters,** which are stored in synaptic vesicles (fig. 10.5*b*), before their release. When nerve impulses traveling along an axon reach a synaptic ending, the axomembrane becomes permeable to calcium ions (Ca^{++}). These ions then interact with microfilaments, causing the microfilaments to pull the synaptic vesicles to the presynaptic membrane. When the vesicles merge with this membrane, a neurotransmitter is discharged into the synaptic cleft. The neurotransmitter molecules diffuse across the cleft to the postsynaptic membrane, where they bind with a receptor in a lock-and-key manner (fig. 10.5*c*).

Depending on the type of neurotransmitter and/or the type of receptor, the response can be excitation or inhibition. If excitation occurs, the membrane potential of the postsynaptic membrane decreases, the sodium ion (Na^+) channels open at that locale, and the likelihood of the neuron firing (transmitting a nerve impulse) increases. If inhibition occurs, the membrane potential of the postsynaptic membrane increases as the inside becomes more negative, and the likelihood of a nerve impulse decreases.

> Transmission across a synapse is dependent on the release of neurotransmitters, which diffuse across a small space, the synaptic cleft separating neuron from neuron. The neurotransmitter changes the membrane potential of the next neuron.

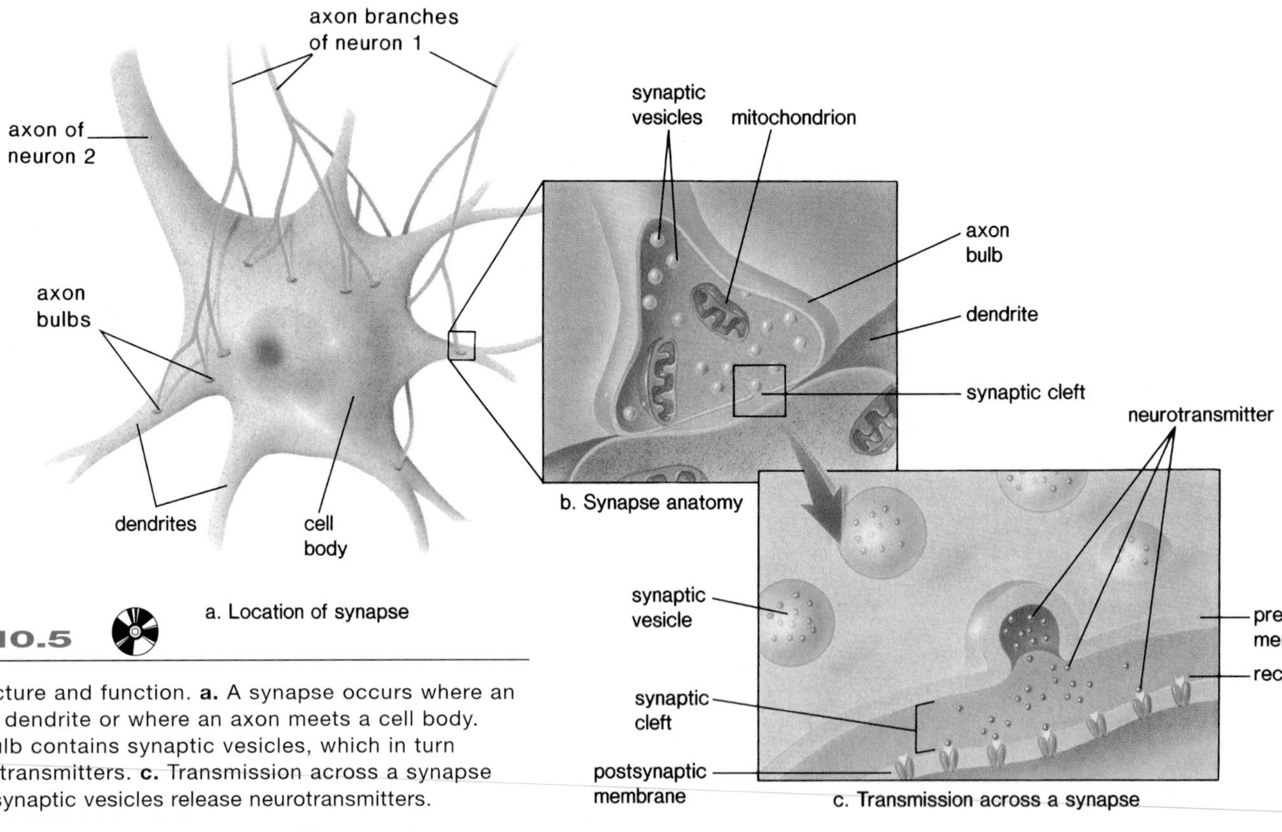

Figure 10.5

Synapse structure and function. **a.** A synapse occurs where an axon meets a dendrite or where an axon meets a cell body. **b.** An axon bulb contains synaptic vesicles, which in turn contain neurotransmitters. **c.** Transmission across a synapse occurs when synaptic vesicles release neurotransmitters.

Neurotransmitters: Quick Acting

Acetylcholine (ACh) and **norepinephrine (NE)** are well-known neurotransmitters active in both the **peripheral nervous system (PNS)** and the CNS. They are excitatory or inhibitory according to the type of receptor at the postsynaptic membrane.

Once a neurotransmitter has been released into a synaptic cleft, it has only a short time to act. In some synapses, the cleft contains enzymes that rapidly inactivate the neurotransmitter. For example, the enzyme **acetylcholinesterase (AChE),** or simply cholinesterase, breaks down acetylcholine. In other synapses, the synaptic ending rapidly absorbs the neurotransmitter, possibly for repackaging in synaptic vesicles or for chemical breakdown. The short existence of neurotransmitters in the synapse prevents continuous stimulation (or inhibition) of postsynaptic membranes.

> Neurotransmitters act for a short time and the body has specific means to break down or reprocess them.

Peripheral Nervous System

The peripheral nervous system (PNS) is made up of nerves (fig. 10.6). *Nerves* are structures that contain many long fibers—long dendrites and/or long axons (table 10.2). The cell bodies of neurons are found in the CNS—that is, brain and spinal cord—or in ganglia. Ganglia (sing., **ganglion**) are collections of cell bodies within the PNS. In the PNS,

Table 10.2

Nerves

Type	Structure	Function
Sensory nerve	Long dendrites of sensory neurons only	Carries message from receptors to CNS
Motor nerve	Long axons of motor neurons only	Carries message from CNS to effectors
Mixed nerve	Both long dendrites of sensory neurons and long axons of motor neurons	Carries message in dendrite to CNS and away from CNS in axons

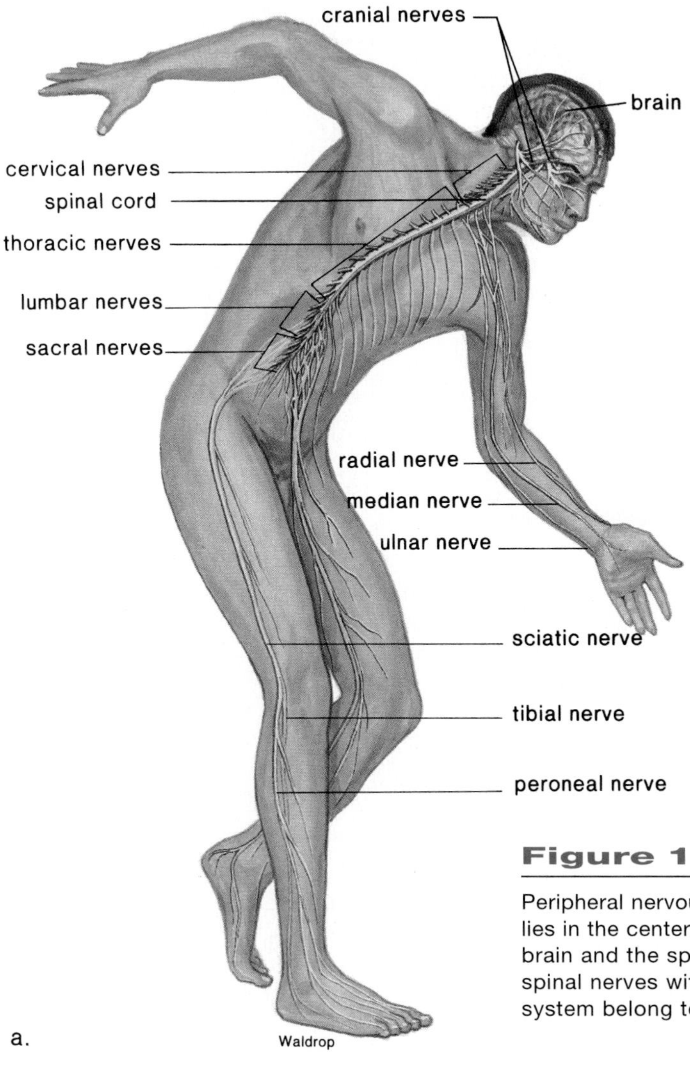

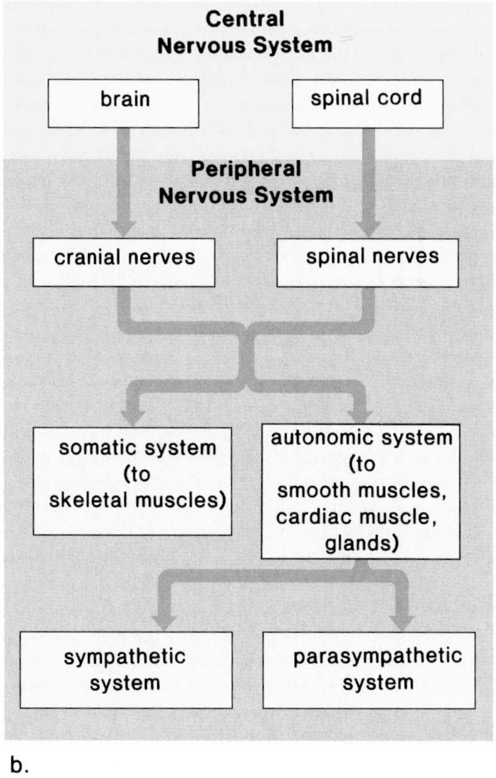

Figure 10.6

Peripheral nervous system (PNS) compared with central nervous system (CNS). **a.** The CNS lies in the center of the body, and the PNS lies to either side. **b.** The CNS consists of the brain and the spinal cord, and the PNS consists of the nerves. There are both cranial and spinal nerves within the somatic and autonomic systems. The nerves of the autonomic system belong to either the sympathetic system or the parasympathetic system.

the **somatic system** contains nerves that control skeletal muscles, skin, and joints, and the **autonomic system** contains nerves that control glands and the smooth muscles of the internal organs.

Humans have 12 pairs of **cranial nerves** attached to the brain. Some of these are sensory, some are motor, and others are mixed nerves. All cranial nerves, except the vagus, are concerned with the head, neck, and facial regions of the body; the vagus nerve has many branches serving the internal organs.

Humans have 31 pairs of **spinal nerves.** Each spinal nerve emerges from the spinal cord by 2 short branches, or roots, which lie within the vertebral column (fig. 10.7). The *dorsal root* contains the dendrites of sensory neurons, which conduct impulses to the cord. The *ventral root* contains the axons of motor neurons, which conduct impulses away from the cord. These 2 roots join just before a spinal nerve leaves the vertebral column. Therefore, all spinal nerves are mixed nerves that contain many sensory dendrites and motor axons. Each spinal nerve serves the particular region of the body in which it is located.

An individual nerve fiber obeys an all-or-none law, meaning that it fires maximally or it does not fire. A nerve does not obey the all-or-none law—a nerve can have a number of degrees of performance because it contains many fibers, any number of which can be carrying nerve impulses.

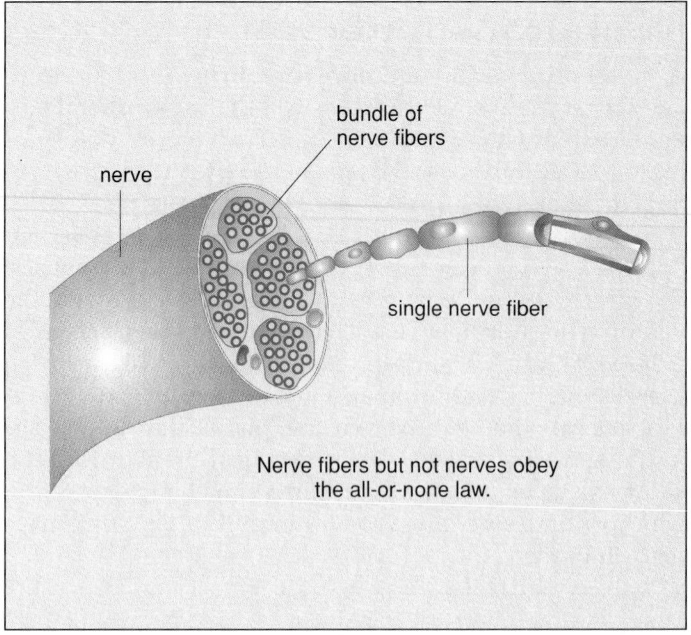

In the PNS, cranial nerves take impulses to and/or from the brain, and spinal nerves take impulses to and from the spinal cord.

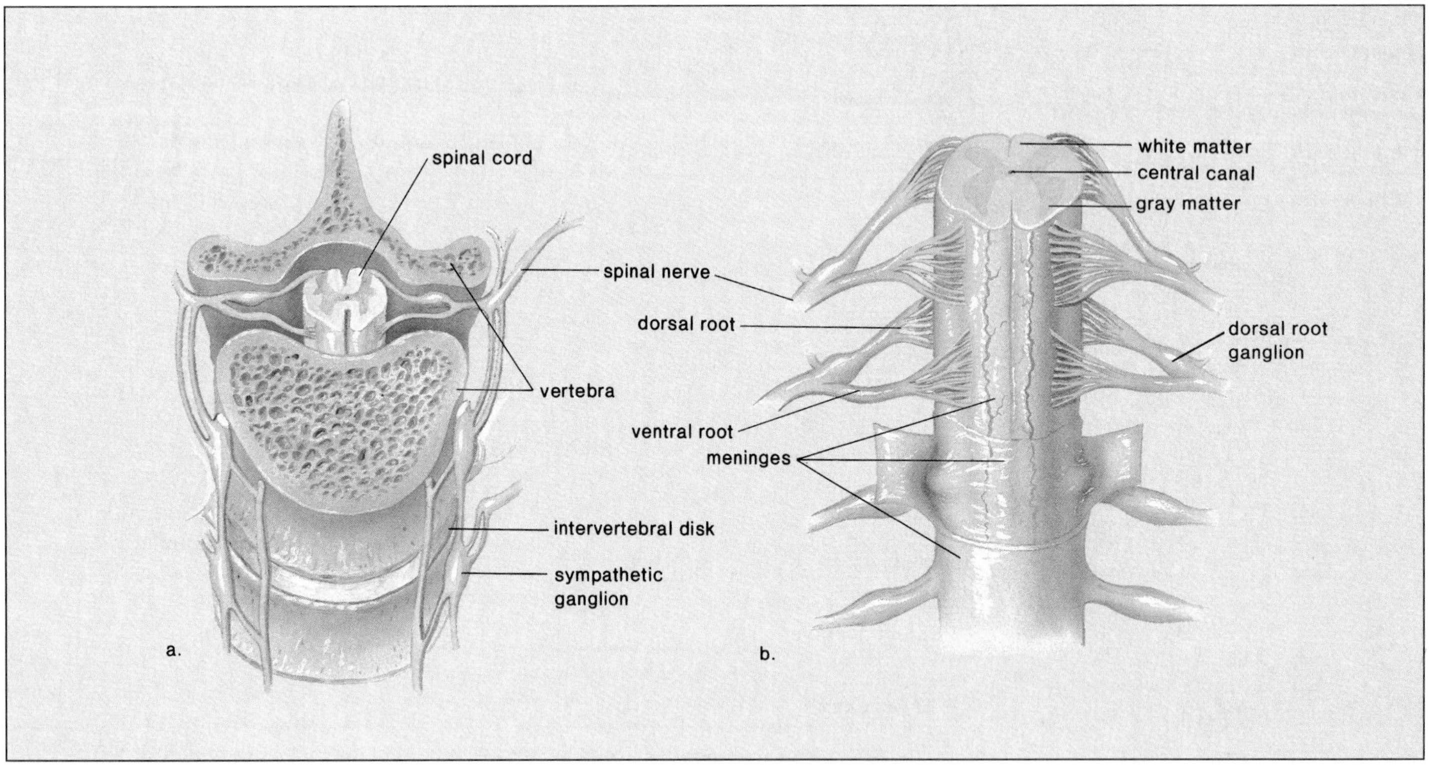

Figure 10.7

The anatomy of the spinal cord. **a.** Cross section of the spine, showing spinal nerves. The human body has 31 pairs of spinal nerves. **b.** This cross section of the spinal cord shows that a spinal nerve has a dorsal root and a ventral root. Also, the cord is protected by 3 layers of tissue called the meninges. Spinal meningitis is an infection of these layers.

Reflex Arc: A Functional Unit

Reflexes are automatic, involuntary responses to changes occurring inside or outside the body. *Receptors* receive environmental stimuli and then initiate nerve impulses. Muscles are *effectors*, which bring about a reaction to the stimulus. In the somatic system, outside stimuli often initiate a reflex. Some reflexes, such as blinking your eye, involve the brain, while others, such as withdrawing your hand from a hot object, do not necessarily involve the brain. Figure 10.8 illustrates the path of the second type of reflex. If you touch a very hot object, a receptor in the skin generates nerve impulses, which move along the dendrite of a *sensory neuron* toward the cell body and the CNS. The cell body of a sensory neuron is located in the **dorsal-root ganglion,** just outside the cord. From the cell body, the impulses travel along the axon of the sensory neuron and enter the cord by the dorsal root of a spinal nerve. The impulses then pass to many interneurons, one of which connects with a motor neuron. The short dendrites and the cell body of the *motor neuron* lead to the axon, which leaves the cord by way of the ventral root of a spinal nerve. The nerve impulses travel along the axon to *muscle fibers,* which then contract so that you withdraw your hand from the hot object. (See table 10.3 for a listing of these events.)

Table 10.3

Path of a Simple Reflex

1. Receptor (formulates message)*	Generates nerve impulses
2. Sensory neuron (takes message to CNS)	Impulses move along dendrite (spinal nerve)† and proceed to cell body (dorsal-root ganglia) and then go from cell body to axon (spinal cord)
3. Interneuron (passes message to motor neuron)	Impulses picked up by dendrites and pass through cell body to axon (spinal cord)
4. Motor neuron (takes message away from CNS)	Impulses travel through short dendrites and cell body (spinal cord) to axon (spinal nerve)
5. Effector (receives message)	Receives nerve impulses and reacts: glands secrete and muscles contract

** Phrases within parentheses state overall function.*
† Words within parentheses indicate location of structure.

Various other reactions usually accompany a reflex; you may look in the direction of the object, jump back, and utter appropriate exclamations. This whole series of responses is explained by the fact that the sensory neuron stimulates several interneurons, which take impulses

Figure 10.8

A reflex arc showing the path of a reflex. When a receptor in the skin is stimulated, nerve impulses (*see arrows*) move along a sensory neuron to the spinal cord. (Note that the cell body of a sensory neuron is in a ganglion outside the cord.) The nerve impulses are picked up by an interneuron, which lies completely within the cord, and pass to the dendrites and the cell body of a motor neuron that lies ventrally within the cord. The nerve impulses then move along the motor neuron to an effector, such as a muscle, which contracts. The brain receives information concerning sensory stimuli by way of other interneurons, with long fibers in tracts that run up and down the cord within the white matter.

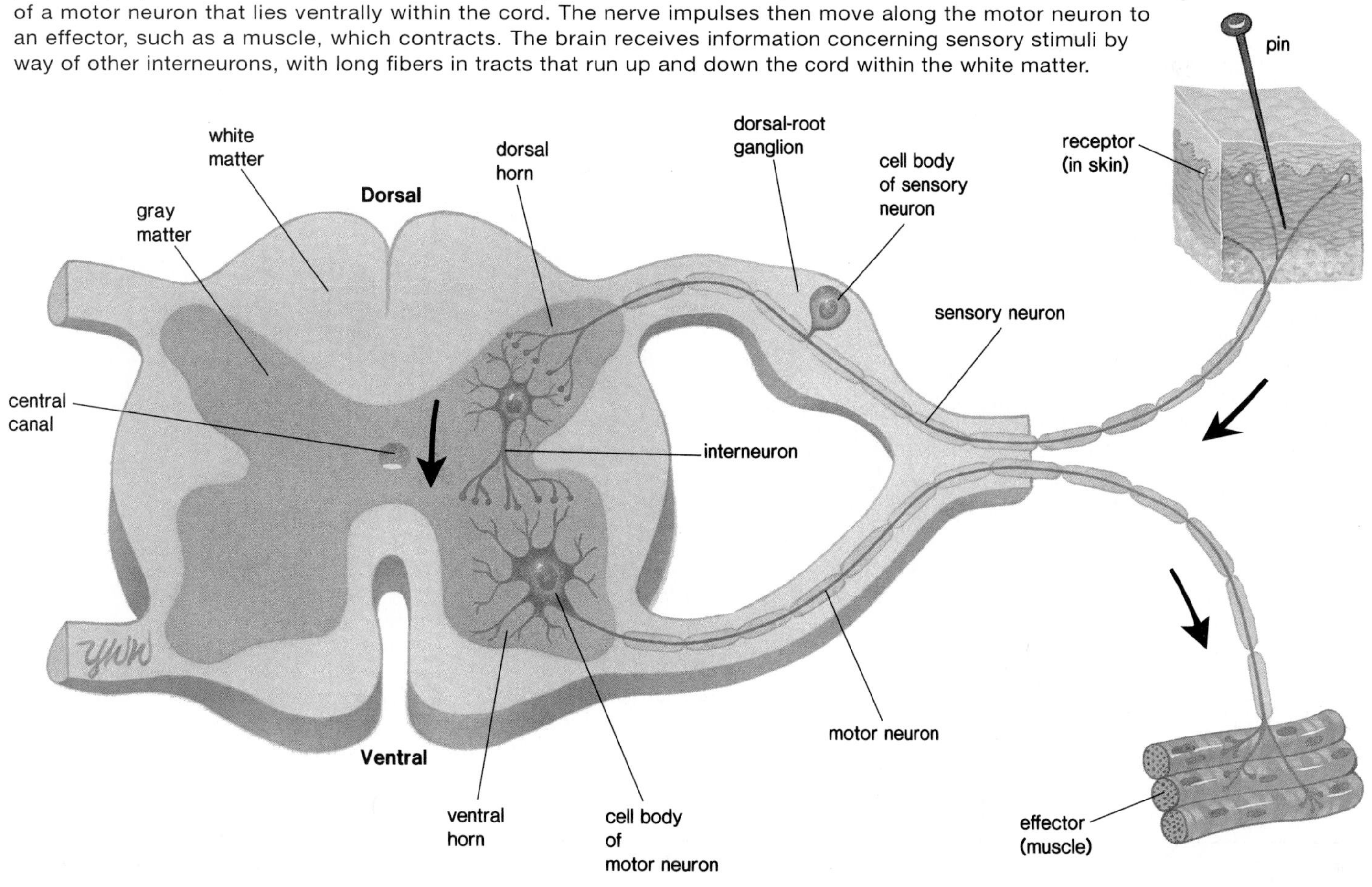

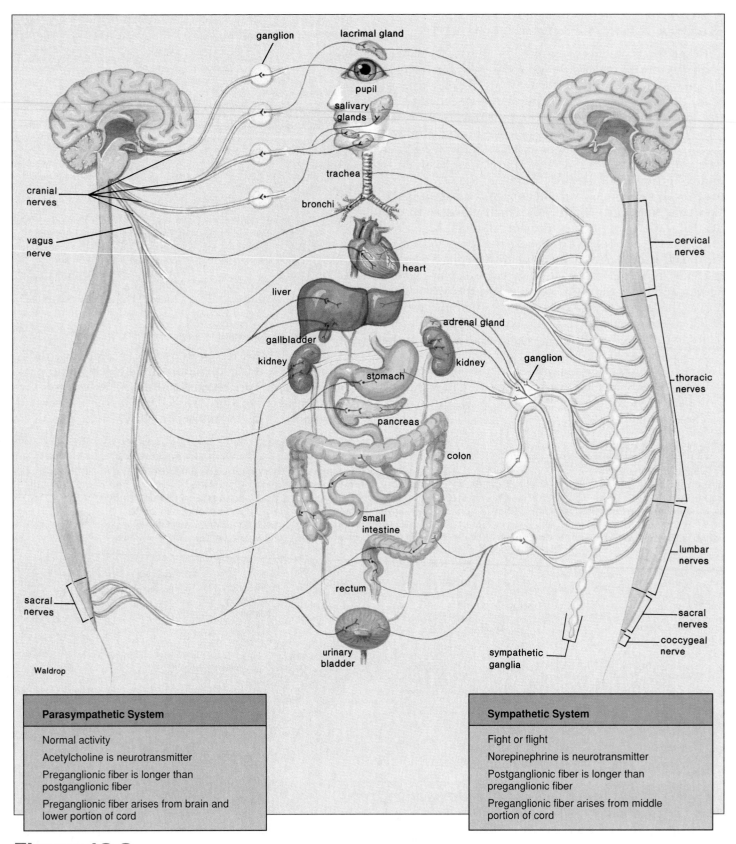

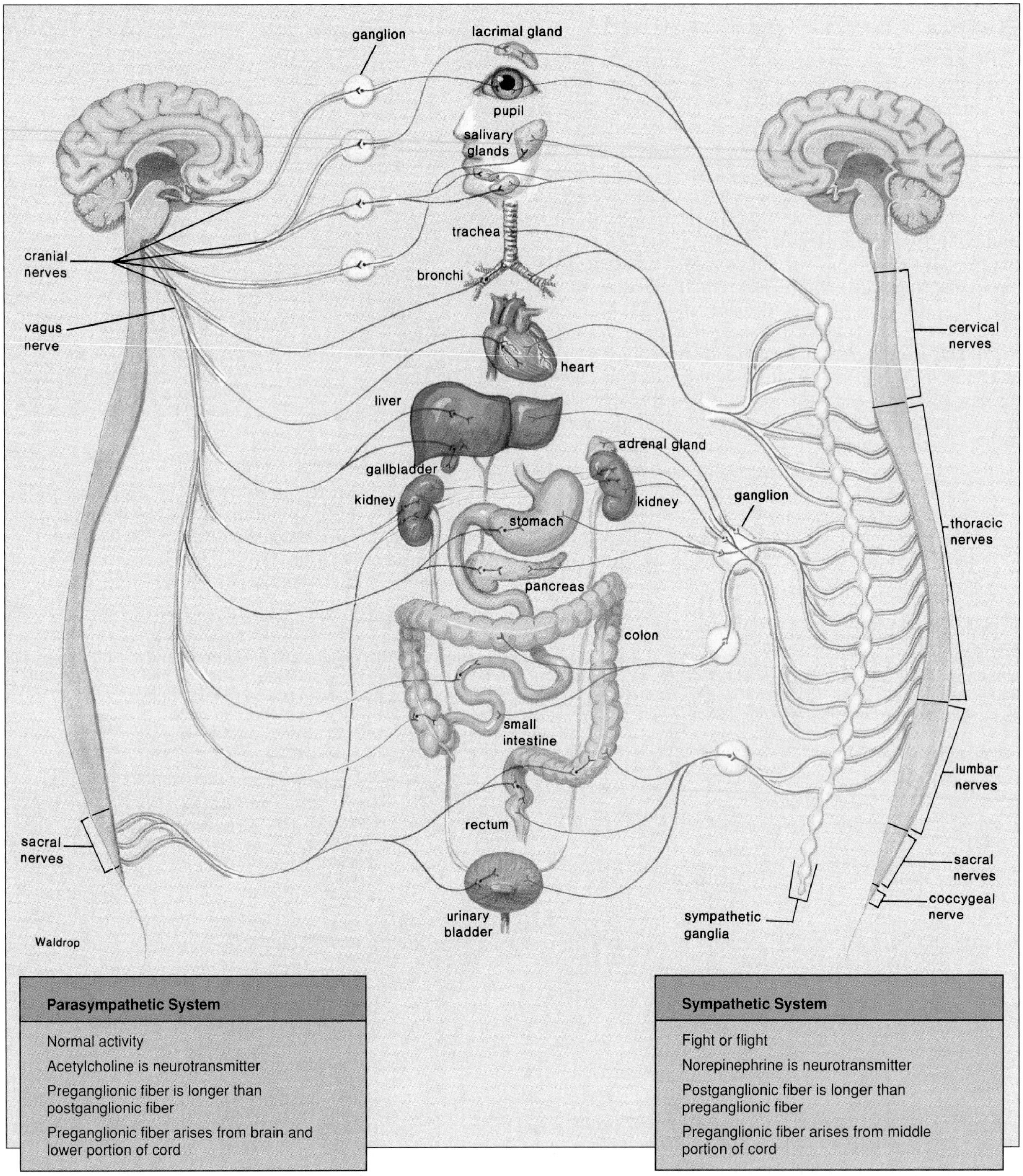

Parasympathetic System
Normal activity
Acetylcholine is neurotransmitter
Preganglionic fiber is longer than postganglionic fiber
Preganglionic fiber arises from brain and lower portion of cord

Sympathetic System
Fight or flight
Norepinephrine is neurotransmitter
Postganglionic fiber is longer than preganglionic fiber
Preganglionic fiber arises from middle portion of cord

Figure 10.9

Structure and function of the autonomic system. The sympathetic fibers arise from the thoracic and lumbar portions of the spinal cord; the parasympathetic fibers arise from the brain and the sacral portion of the spinal cord. Each system innervates the same organs but have contrary effects. For example, the sympathetic system speeds up the beat of the heart, while the parasympathetic system slows it down.

to all parts of the CNS, including the cerebrum, which in turn, makes you conscious of the stimulus and your reaction to it.

> The reflex arc is an important functional unit of the nervous system. It allows us to react to internal and external stimuli.

Autonomic System: For Internal Organs

The autonomic system (fig. 10.9), a part of the PNS, is made up of motor neurons that control the internal organs automatically and usually without need for conscious intervention. The sensory neurons that come from the internal organs allow us to feel internal pain. The cell bodies for these sensory neurons are in dorsal-root ganglia along with the cell bodies of somatic sensory neurons.

There are 2 divisions of the autonomic system: the sympathetic and parasympathetic systems. Both of these (1) function automatically and usually subconsciously in an involuntary manner; (2) innervate all internal organs; and (3) utilize 2 motor neurons and one ganglion for each impulse. The first of these 2 neurons has a cell body within the CNS and a *preganglionic fiber*. The second neuron has a cell body within the ganglion and a *postganglionic fiber.*

> The autonomic system controls the function of internal organs, without conscious control.

Sympathetic System: Fight or Flight

Most preganglionic fibers of the **sympathetic system** (fig. 10.9) arise from the middle, or *thoracic-lumbar*, portion of the spinal cord and almost immediately terminate in ganglia that lie near the cord. Therefore, in this system, the preganglionic fiber is short, but the postganglionic fiber that makes contact with an organ is long.

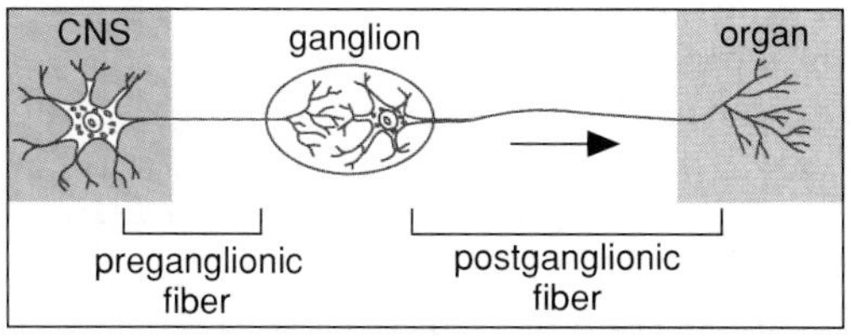

The sympathetic system is especially important during emergency situations and is associated with "fight or flight." If you need to fend off a foe or flee from danger, active muscles require a ready supply of glucose and oxygen. The sympathetic system accelerates the heartbeat, dilates the bronchi, and increases the breathing rate. On the other hand, the sympathetic system inhibits the digestive tract—digestion is not an immediate necessity if you are under attack. The neurotransmitter released by the postganglionic axon is primarily norepinephrine (NE), a chemical close in structure to epinephrine (adrenaline), a medicine used as a heart stimulant.

> The sympathetic system brings about those responses we associate with "fight or flight."

Parasympathetic System: Housekeeper

A few cranial nerves, including the vagus nerve, together with fibers that arise from the sacral (bottom) portion of the spinal cord, form the **parasympathetic system** (fig. 10.9). Therefore, this system often is referred to as the *craniosacral portion* of the autonomic system. In the parasympathetic system, the preganglionic fiber is long and the postganglionic fiber is short because the ganglia lie near or within the organ.

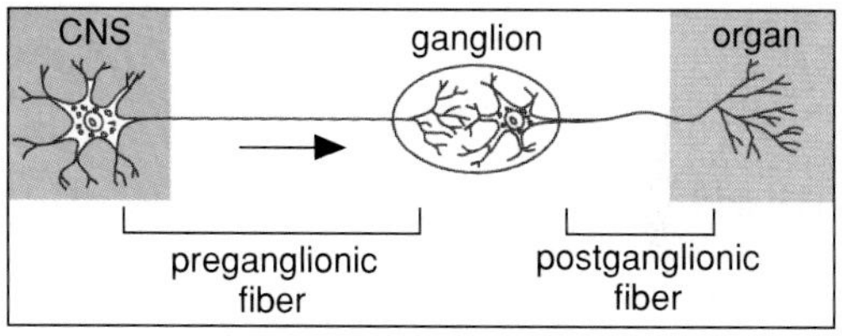

The parasympathetic system, sometimes called the "housekeeper system," promotes all the internal responses we associate with a relaxed state; for example, it causes the pupil of the eye to contract, promotes digestion of food, and retards the heartbeat. The neurotransmitter utilized by the parasympathetic system is acetylcholine (ACh).

> The parasympathetic system brings about the responses we associate with a relaxed state.

Central Nervous System

The **central nervous system (CNS)** consists of the spinal cord and the brain. As figures 10.7 and 10.10 illustrate, the CNS is protected by bone: the spinal cord is surrounded by vertebrae and the brain is enclosed by the skull. Also, both the spinal cord and the brain are wrapped in 3 protective membranes known as **meninges** (sing., meninx); meningitis is an infection of these coverings (see fig. 10.7*b*). The spaces between the meninges are filled with **cerebrospinal fluid,** which cushions and protects the CNS. Cerebrospinal fluid is contained within the *ventricles* of the brain, which are interconnecting spaces that produce and serve as a reservoir for cerebrospinal fluid, and in the **central canal** of the spinal cord. A small amount of this fluid sometimes is withdrawn for laboratory testing when a spinal tap (i.e., lumbar puncture) is performed.

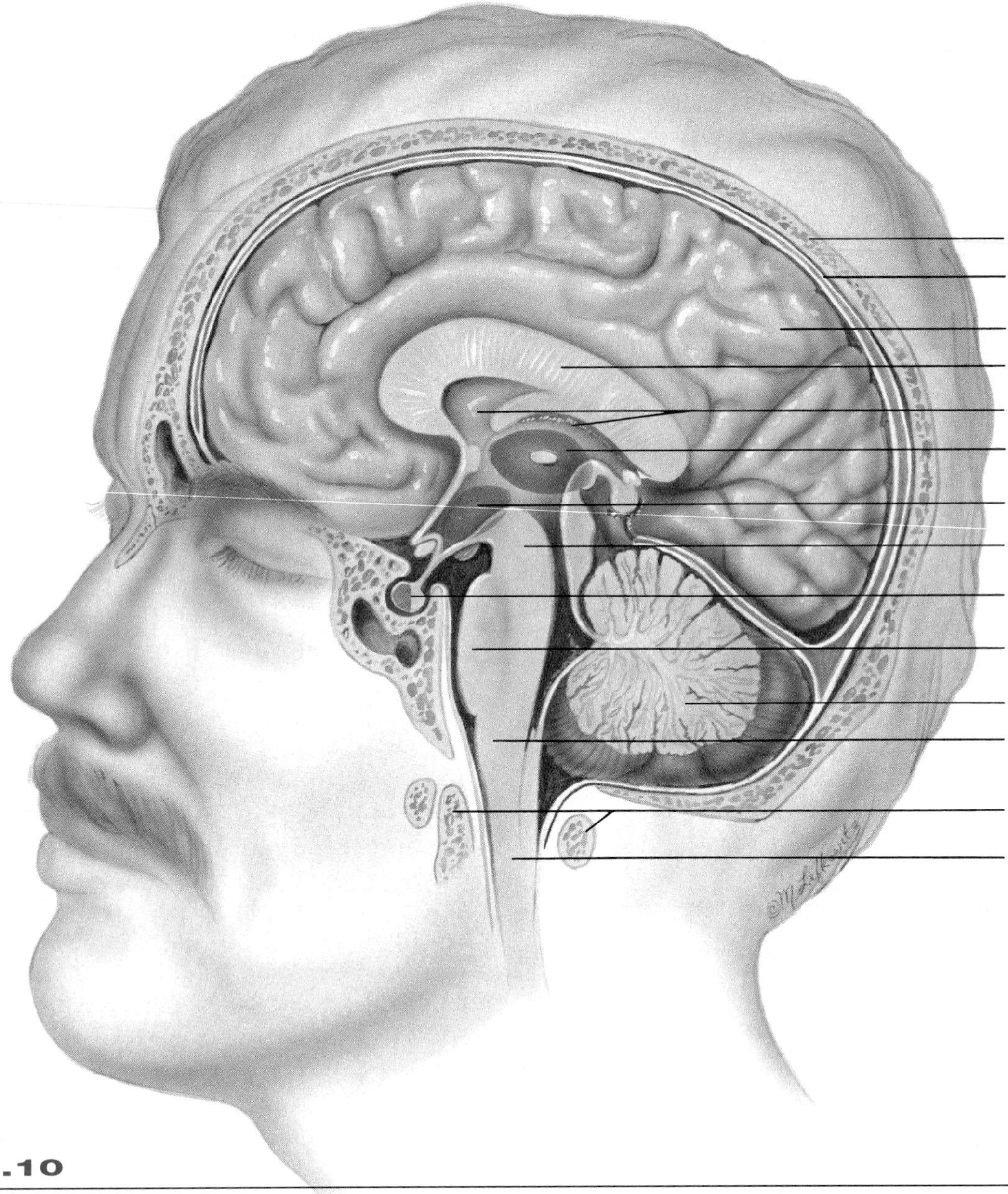

Figure 10.10

The human brain. Note how large the cerebrum is compared to the rest of the brain.

Spinal Cord: Two Main Functions

The spinal cord lies along the middorsal line of the body (see fig. 10.7). It has 2 main functions: (1) it is the center for many reflex actions, and (2) it provides a means of communication between the brain and the spinal nerves, which leave the spinal cord.

The path of a spinal reflex passes through the gray matter of the cord (see fig. 10.8). Unmyelinated cell bodies and short fibers give this area its gray color. In cross section, the gray matter looks like a butterfly or the letter H. The axons of sensory neurons are found in the dorsal regions (horns) of the gray matter, and the dendrites and cell bodies of motor neurons are found in the ventral regions (horns) of the gray matter. Short interneurons connect sensory neurons to motor neurons on the same and the opposite sides of the spinal cord.

The white matter of the spinal cord is found in between the regions of the gray matter (see fig. 10.8). Myelinated long fibers of interneurons that run together in bundles called *tracts* give white matter its color. These tracts connect the spinal cord to the brain. Dorsally, there are primarily ascending tracts taking information to the brain, and ventrally, there are primarily descending tracts carrying information from the brain. Because the tracts at one point cross over, the left side of the brain controls the right side of the body and the right side of the brain controls the left side of the body.

> The CNS lies in the midline of the body and consists of the brain and the spinal cord. Sensory information is received and motor control is initiated in the CNS.

Brain: Control Centers

The human brain is divided into these parts: medulla oblongata, cerebellum, pons, midbrain, hypothalamus, thalamus, and cerebrum (fig. 10.10). The brain has 4 cavities, called **ventricles:** 2 lateral ventricles, the third ventricle, and the fourth ventricle.

Brain Stem: Our Reflex Centers

The medulla oblongata, pons, and midbrain lie in a portion of the brain known as the *brain stem.* The **medulla oblongata** lies between the spinal cord and the pons and is anterior to the cerebellum. It contains a number of *vital centers* for regulating heartbeat, breathing, and vasoconstriction (blood pressure). It also contains the reflex centers for vomiting, coughing, sneezing, hiccoughing, and swallowing. The medulla contains tracts that ascend or descend between the spinal cord and the brain's higher centers.

As its name suggests, the **pons** is a "bridge"; it contains bundles of axons traveling between the cerebellum and the rest of the CNS. In addition, the pons functions with the medulla to regulate breathing rate and has reflex centers concerned with head movements in response to visual and auditory stimuli. Aside from acting as a relay station for tracts passing between the cerebrum and the spinal cord or cerebellum, the **midbrain** has reflex centers for visual, auditory, and tactile responses.

Diencephalon: The Third Ventricle

The hypothalamus and thalamus are in a portion of the brain known as the diencephalon, where the third ventricle is located. The hypothalamus forms the floor of the third ventricle. The **hypothalamus** maintains homeostasis, or the constancy of the internal environment, and contains centers for regulating hunger, sleep, thirst, body temperature, water balance, and blood pressure. The hypothalamus controls the pituitary gland and thereby serves as a link between the nervous and endocrine systems.

The thalamus is in the roof of the third ventricle. The **thalamus** is the last portion of the brain for sensory input before the cerebrum. It serves as a central relay station for sensory impulses traveling upward from other parts of the body and brain to the cerebrum. It receives all sensory impulses (except those associated with the sense of smell) and channels them to appropriate regions of the cortex for interpretation.

Cerebellum: Our Balance

The **cerebellum,** which lies below the posterior portion of the cerebrum, is separated from the brain stem by the fourth ventricle. The cerebellum has 2 portions that are joined by a narrow median portion. The surface of the cerebellum is gray matter, and the interior is largely white matter. The cerebellum functions in muscle coordination, integrating impulses received from higher centers to ensure that all of the skeletal muscles work together to produce smooth and graceful motions. The cerebellum is also responsible for maintaining normal muscle tone and transmitting impulses that maintain posture. It receives information from the inner ear indicating position of the body and then sends impulses to the muscles, whose contraction maintains or restores balance (fig. 10.11).

Figure 10.11

Rollerblading. This sport requires muscular coordination and balance, which are controlled by the cerebellum.

> The cerebellum controls balance and complex muscular movements.

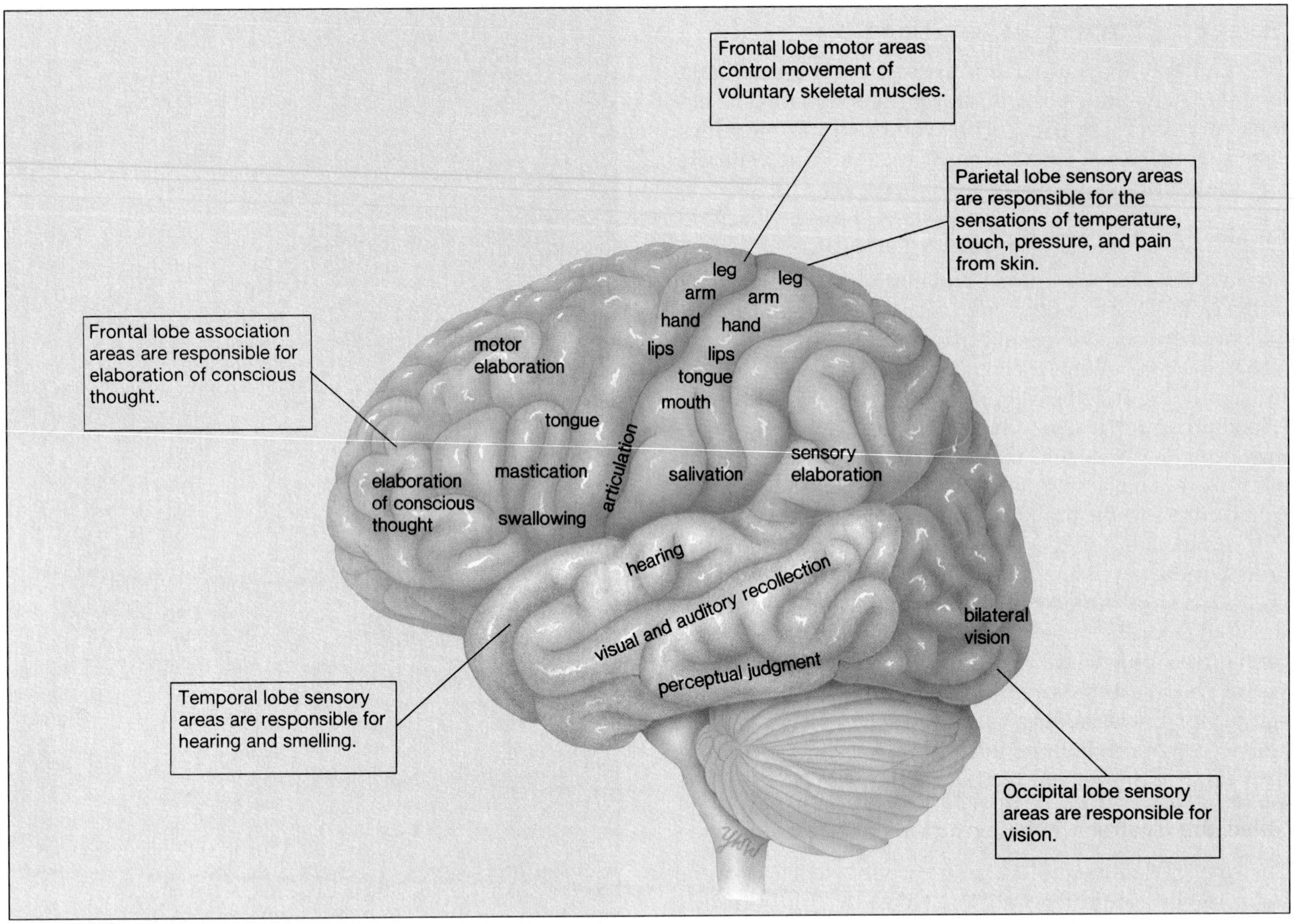
Frontal lobe motor areas control movement of voluntary skeletal muscles.
Parietal lobe sensory areas are responsible for the sensations of temperature, touch, pressure, and pain from skin.
Frontal lobe association areas are responsible for elaboration of conscious thought.
leg
arm
hand
lips
leg
arm
hand
lips
tongue
mouth
motor elaboration
tongue
articulation
mastication
swallowing
elaboration of conscious thought
salivation
sensory elaboration
hearing
visual and auditory recollection
perceptual judgment
bilateral vision
Temporal lobe sensory areas are responsible for hearing and smelling.
Occipital lobe sensory areas are responsible for vision.

Figure 10.12

The convoluted cortex of the cerebrum is divided into 4 surface lobes: frontal, parietal, temporal, and occipital. It is possible to map the cerebral cortex since each area has a particular function.

Cerebrum: Our Consciousness

The **cerebrum,** the foremost part of the brain, is the only area responsible for consciousness. It is the largest portion of the brain in humans. The outer layer of the cerebrum, called the *cerebral cortex,* is gray in color and contains cell bodies and unmyelinated short fibers. The cerebrum is divided into halves, known as the right and left **cerebral hemispheres.** Each hemisphere contains 4 surface lobes: **frontal, parietal, temporal,** and **occipital** (fig. 10.12). Little is known about the functions of a fifth lobe, the *insula,* which lies beneath the surface.

Certain areas of the cerebral cortex have been "mapped" in great detail. Physiologists have identified the *motor areas* of the frontal lobe, which initiate contraction of skeletal (voluntary) muscles, the *sensory areas* of the parietal lobe, which receive impulses from receptors, and the *association areas,* which receive information from the other lobes and integrate it into higher, more complex levels of consciousness. Association areas are concerned with intellect, artistic and creative abilities, learning, and memory.

Basal Ganglia

Basal ganglia, constituting the central gray matter of the cerebrum, lie between the white matter of the cerebrum and the thalamus of the diencephalon. The precise functions of the basal ganglia are not known, but they may have some control over voluntary muscle action because when they are diseased, Parkinson disease and St. Vitus' dance develop. Both of these are neuromuscular disturbances.

> Consciousness is the province of the cerebrum, the most developed portion of the human brain. The cerebrum is responsible for higher mental processes, including the interpretation of sensory input and the initiation of voluntary muscular movements.

There has been a great deal of testing to determine whether the right and left halves of the cerebrum serve different functions. These studies tend to suggest that the left half of the brain is the verbal (word) half and the right half of the brain is the visual (spatial relation) and artistic half. However, other test results indicate that such a strict dichotomy does not always exist between the 2 halves. In any case, the 2 cerebral hemispheres normally share information because they are connected by a horizontal tract called the **corpus callosum** (see fig. 10.10).

Severing the corpus callosum can control severe epileptic seizures. In the past, this severing meant the 2 halves of the brain no longer communicated; each half has its own memories and thoughts. Today, use of the laser permits more precise treatment without this side effect. *Epilepsy* is caused by a disturbance of normal communication between the brain stem and the cerebral cortex. In a grand mal epileptic seizure, the cerebrum is extremely excited due to reverberation of signals; the individual loses consciousness, even while convulsions are occurring. Finally, the neurons fatigue and the signals cease. Following an attack, the brain is so fatigued the person must sleep for a while.

EEG: Brain Wave Patterns

The electrical activity of the brain can be recorded in the form of an **electroencephalogram (EEG).** Electrodes are taped to different parts of the scalp, and an instrument called the electroencephalograph records the so-called brain waves (fig. 10.13).

When the subject is awake, 2 types of waves are usual: *alpha waves,* with a frequency of about 6–13 per second and a potential of about 45 mV, which predominate when the eyes are closed, and *beta waves,* with higher frequencies but lower voltage, which appear when the eyes are open.

During sleep the waves become larger, slower, and more erratic. At times, the eyes move back and forth rapidly, and this is called **REM** (rapid eye movement) **sleep.** When subjects are awakened during REM sleep, they always report that they had been dreaming. The significance of REM sleep is still being debated, but some studies indicate that it is needed for memory development.

The EEG is a diagnostic tool; for example, an irregular pattern can signify epilepsy or a brain tumor. A flat EEG signifies lack of electrical activity of the brain, or brain death; therefore, it can be used to determine the precise time of death.

Limbic System: Emotions and Memory

The **limbic system** involves portions of both the subconscious and the conscious brain (fig. 10.14). It lies just beneath the cerebral cortex and contains neural pathways that connect portions of the frontal lobes, the temporal lobes, the thalamus, and the hypothalamus. The basal nuclei are also a part of the limbic system.

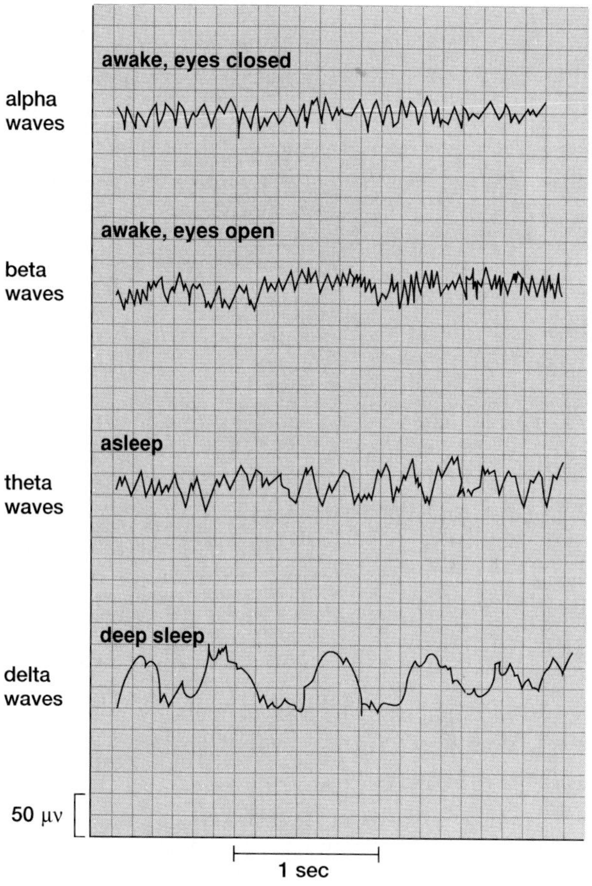

Figure 10.13

Encephalograms (EEGs), recordings of the electrical activity of the brain. The alpha waves, which appear when the subject is awake with eyes closed, are the most common. Second most common are the beta waves, recorded when the subject is awake with eyes open. Sleep has various stages, as indicated.

Stimulation of different areas of the limbic system causes the subject to experience pain, pleasure, rage, affection, sexual interest, fear, or sorrow. By causing pleasant or unpleasant feelings about experiences, the limbic system apparently guides the individual into behavior that is likely to increase the chance of survival.

The limbic system also is involved in the processes of learning and memory. Learning requires memory, but just what permits memory development is not definitely known. Experimentation with invertebrates such as slugs and snails indicates that learning is accompanied by an increase in the number of synapses in the brain, while forgetting involves a decrease in the number of synapses.

Experiments with monkeys have led to the conclusion that the limbic system is absolutely essential to both short-term and long-term memory. An example of short-term memory in humans is the ability to recall a telephone number long enough to dial it; an example of long-term memory

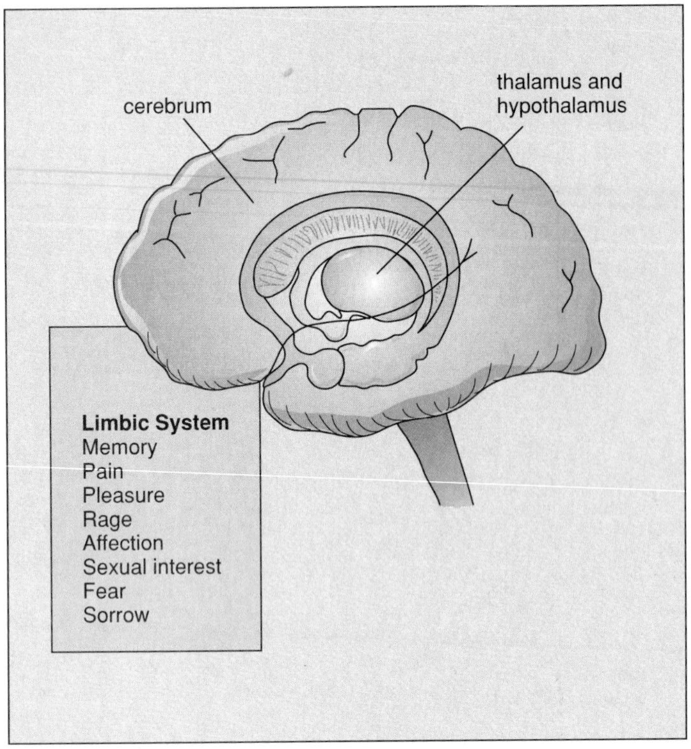

Figure 10.14

The limbic system. The limbic system, which includes portions of the cerebrum, the thalamus, and hypothalamus, is sometimes called the emotional brain because it seems to control the emotions listed.

is the ability to recall the events of the day. It is believed that at first, impulses move only within the limbic circuit, but eventually the basal nuclei transmit the neurotransmitter acetylcholine (ACh) to the sensory areas where memories are stored. The involvement of the limbic system certainly explains why emotionally charged events result in our most vivid memories. The fact that the limbic system communicates with the sensory areas for touch, smell, vision, hearing, and taste accounts for the ability of any particular sensory stimulus to awaken a complex memory.

> The limbic system is particularly involved in emotions and in memory and learning.

Alzheimer Disease

Alzheimer disease (AD) is a disorder characterized by a gradual loss of reason that begins with memory lapses and ends with an inability to perform any type of daily activity. Personality changes signal the onset of AD. People with AD forget the name of a neighbor who visits daily. With time, they have trouble traveling and cannot perform simple errands. People afflicted with AD become confused and tend to repeat the same question. Signs of mental disturbances eventually appear, and patients gradually become bedridden and die of a complication such as pneumonia.

The AD neuron has 2 abnormalities not seen in the normal neuron. Bundles of fibrous protein, called neurofibrillary tangles, surround the nucleus in the cell and protein-rich accumulations, called amyloid plaques, envelop the axon branches. These abnormal neurons are especially seen in the portions of the brain that are involved in reason and memory (frontal lobe and limbic system). A chemical test can check brain tissue for the presence of a protein called Alzheimer disease associated protein (ADAP), which is believed to be the protein contained in the neurofibrillary tangles. Once it is proven that ADAP is the protein involved in AD, it may become routine to test patients for this protein by obtaining a spinal tap of cerebrospinal fluid.

Drug Abuse

A wide variety of drugs can be used to alter the mood and/or emotional state (see appendix C). Drugs that affect the nervous system have 2 general effects: (1) they impact the limbic system, and (2) they either promote or decrease the action of a particular neurotransmitter (fig. 10.15). Stimulants can either enhance the action of an excitatory neurotransmitter or block the action of an inhibitory neurotransmitter. Depressants can either enhance the action of an inhibitory neurotransmitter or block the action of an excitatory neurotransmitter. Increasingly, researchers believe that dopamine, a neurotransmitter in the brain, primarily affects mood. Cocaine is known to potentiate the effects of dopamine by interfering with its uptake from synaptic clefts. Many new medications developed to counter drug addiction and mental illness affect the release, reception, or breakdown of dopamine.

Drug abuse is apparent when a person takes a drug at a dose level and under circumstances that increase the potential for a harmful effect. Drug abusers are apt to display either a psychological and/or a *physical dependence* on the drug. Dependence has developed when the person spends much time thinking about the drug or arranging to get it and often takes more of the drug than was intended. With physical dependence, formerly called an addiction to the drug, the person is *tolerant* to the drug—that is, must increase the amount of the drug to get the same effect and has *withdrawal symptoms* when he or she stops taking the drug.

> Drugs that affect the nervous system can cause physical dependence and withdrawal symptoms.

Alcohol, which is discussed in the Health Focus for this chapter, and these drugs are the ones most apt to be abused today:

Marijuana. The dried flowering tops, leaves, and stems of the Indian hemp plant *Cannabis sativa* contain and are covered by a resin that is rich in THC (tetrahydrocannabinol) (fig. 10.16). The names *cannabis* and *marijuana* apply to either the plant or THC.

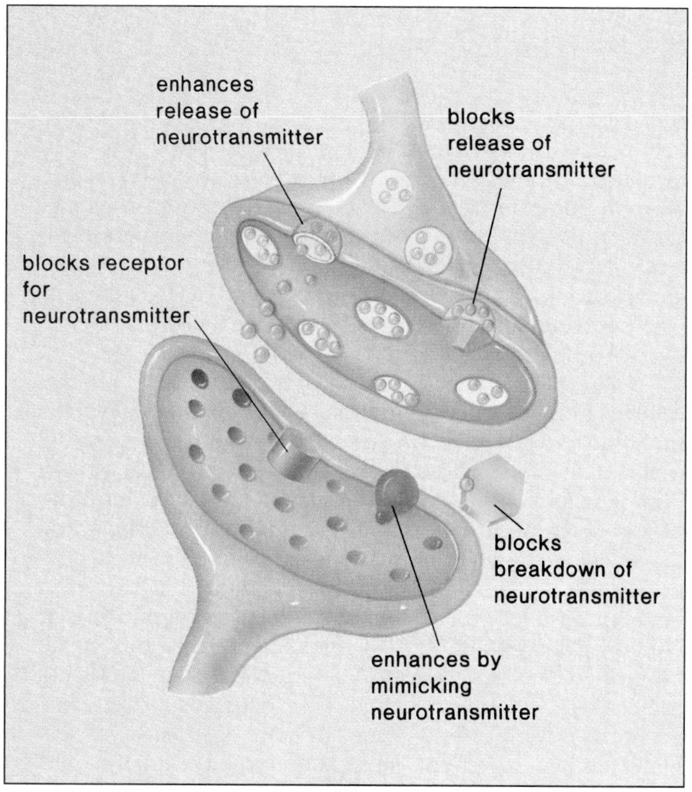

Figure 10.15

Drug action at synapses.

Figure 10.16

Cannabis sativa. This plant, which is used to make marijuana, often is smoked in the same manner as tobacco.

The effects of marijuana differ depending upon the strength and the amount consumed, the expertise of the user, and the setting in which it is taken. Usually, the user reports experiencing a mild euphoria along with alterations in vision and judgment, which result in distortions of space and time. Motor incoordination occurs as well as the inability to concentrate and to speak coherently.

Intermittent use of low-potency marijuana generally is not associated with obvious symptoms of toxicity, but heavy use can produce chronic intoxication. Intoxication is recognized by the presence of hallucinations, anxiety, depression, rapid flow of ideas, body image distortions, paranoid reactions, and similar psychotic symptoms. The terms *cannabis psychosis* and *cannabis delirium* refer to such reactions.

Marijuana is classified as a hallucinogen. It is possible that, like LSD (lysergic acid diethylamide), it has an effect on the action of serotonin, an excitatory neurotransmitter in the brain.

The use of marijuana does not seem to produce physical dependence, but a psychological dependence on the euphoric and sedative effects can develop. Craving or difficulty in stopping use also can occur as a part of regular heavy use.

Marijuana has been called a *gateway drug* because adolescents who have used marijuana also tend to try other drugs. For example, in a study of 100 cocaine abusers, 60% had smoked marijuana for more than 10 years.

Usually marijuana is smoked in a cigarette form called a joint. Since this allows toxic substances, including carcinogens, to enter the lungs, chronic respiratory disease and lung cancer are considered dangers of long-term, heavy use. Some researchers claim that marijuana use leads to long-term brain impairment. Others report that males and females suffer reproductive dysfunctions. *Fetal cannabis syndrome,* which resembles fetal alcohol syndrome, has been reported.

Some psychologists are very concerned about the use of marijuana among adolescents. Marijuana can be used to avoid dealing with the personal problems that often develop during this maturational phase.

> Although marijuana does not produce physical dependence, it does produce psychological dependence.

Cocaine. Cocaine is an alkaloid derived from the shrub *Erythroxylum cocoa.* Cocaine is sold in powder form and as *crack,* a more potent extract (fig. 10.17). Users often describe the feeling of euphoria that follows intake of the drug as a *rush.* Snorting (inhaling) produces this effect in a few minutes, injection, within 30 seconds, and smoking, in less than 10 seconds. Persons dependent upon the drug are, therefore, most likely to smoke cocaine. The rush only lasts a few seconds and then is replaced by a state of arousal, which

Health Focus

The Bad Effects of Alcohol

The negative effects of alcohol begin in the digestive tract. Alcohol increases stomach secretions, leading to gastritis, and interferes with absorption of nutrients by the small intestine. Alcohol requires no digestion and can easily cross cell membranes. Since it is a solvent of lipids, alcohol can destroy cell membranes and therefore cells themselves. About 20% of the alcohol consumed is absorbed through the walls of the stomach; the rest enters at the small intestine. Alcohol can reach the brain within a minute of entering the tract.

Alcohol is primarily metabolized in the liver, where it disrupts the normal workings of glycolysis and the Krebs cycle. The liver contains dehydrogenase enzymes, which carry out these reactions, reducing NAD in the process.

$$\text{alcohol} \xrightarrow{\text{NAD} \rightarrow \text{NADH}_2} \text{acetaldehyde} \xrightarrow{\text{NAD} \rightarrow \text{NADH}_2} \text{active acetate}$$

The supply of NAD in liver cells is used up by these reactions, and there is not enough free NAD left to run glycolysis and the Krebs cycle. The cell begins to ferment and lactic acid builds up. The pH of the blood decreases and becomes acidic.

Since the Krebs cycle is not working, excess active acetate cannot be broken down and it is converted to fat—the liver turns fatty. Fat accumulation, the first stage in liver deterioration, begins after only a single night of heavy drinking. If heavy drinking continues, fibrous scar tissue appears during a second stage of deterioration. If heavy drinking stops, the liver can still recover and become normal once again. If not, the final and irrevocable stage, cirrhosis of the liver, occurs: liver cells die, harden and turn orange (cirrhosis means orange).

It is customary to define heavy drinking in terms of the number of drinks. A "drink" is the amount of wine, beer, or hard liquor (whiskey, scotch, rum, or vodka) that contains 1/2 oz of pure ethanol:

3 to 4 oz of wine
8 to 12 oz beer
1 oz of hard liquor

It is presumed that an average-sized healthy man can consume 3 drinks a day and an average-sized healthy woman can consume 2 drinks a day without any long-term health effects. After that, liver damage is expected.

The surgeon general recommends that pregnant women drink no alcohol at all. Alcohol crosses the placenta freely and causes fetal alcohol syndrome, which is characterized by mental retardation and various physical defects.

Good nutrition is difficult when a person is a heavy drinker. Alcohol is energy intensive—the $NADH_2$ molecules that result from its breakdown can be used to produce ATP molecules. However, these calories are empty because they do not supply any amino acids, vitamins, and minerals as other energy sources do. Without adequate vitamins, red blood cells and white blood cells cannot be formed in the bone marrow. The immune system is depressed, and the chances of stomach, liver, lung, pancreas, colon, and tongue cancer are increased. Protein digestion and amino acid metabolism is so upset that even adequate protein intake will not prevent amino acid deficiencies. Muscles atrophy and weakness results. Fat deposits occur in the heart wall and hypertension develops. There is an increased risk of cardiac arrhythmias and stroke.

Most people drink because it gives them a feeling of euphoria. The effects of alcohol on the brain begin with the frontal lobe and progress to the brain stem as the blood alcohol concentration increases. At low blood concentrations (0.05%), alcohol affects only the frontal lobe; reasoning and judgment are impaired. Then at 0.10%, speech, vision, and coordination are also affected. Driving a car can now endanger the drinker and others on the road. At 0.15%, control of large muscles fails and the person begins to stagger and weave when walking. At high concentrations (0.30%), stupor and unconsciousness occur. If rapid consumption preceded unconsciousness, blood concentration can continue to rise. Breathing and heart rate centers can fail, and then coma and death follow.

Aside from health, alcohol abuse also damages social relationships and job efficiency and increases the possibility of legal difficulties. Table 10A lists some of the questions that can help to identify the alcohol-dependent person.

---

lasts from 5 minutes to 30 minutes. Then the user begins to feel restless, irritable, and depressed. To overcome these symptoms, the user is apt to take more of the drug, repeating the cycle until there is no more drug left. A binge of this sort can go on for days, after which the individual suffers a *crash.* During the binge period, the user is hyperactive and has little desire for food or sleep but has an increased sex drive. During the crash period, the user is fatigued, depressed, and irritable, has memory and concentration problems, and displays no interest in sex. Indeed, men are often impotent. Other drugs, such as marijuana, alcohol, or heroin, often are taken to ease the symptoms of the crash.

Cocaine affects the concentration of dopamine, a neurotransmitter associated with behavioral states. After release into a synapse, dopamine ordinarily is withdrawn into the presynaptic cell for recycling and reuse. Cocaine prevents the reuptake of dopamine by the presynaptic membrane; this causes an excess of dopamine in the synaptic cleft so that the user experiences the sensation of a rush. The epinephrine-like effects of dopamine account for the state of arousal that lasts for some minutes after the rush experience.

With continued cocaine use, the body begins to make less dopamine to compensate for a seemingly excess supply. The user, therefore, now experiences *tolerance, withdrawal*

Table 10A

Some Questions to Identify the Alcohol-Dependent Person

1. Do you occasionally drink heavily after a disappointment, a quarrel, or when the boss gives you a bad time?
2. When you are having trouble or feel under pressure, do you always drink more heavily than usual?
3. Have you noticed that you are able to handle more alcohol than you did when you were first drinking?
4. Did you ever wake up the "morning after" and discover that you could not remember part of the evening before, even though your friends tell you that you did not "pass out"?
5. When drinking with other people, do you try to have a few extra drinks when others will not know it?
6. Are there certain occasions when you feel uncomfortable if alcohol is not available?
7. Have you recently noticed that when you begin drinking you are in more of a hurry to get the first drink than you used to be?
8. Do you sometimes feel a little guilty about your drinking?
9. Are you secretly irritated when your family or friends discuss your drinking?
10. Have you recently noticed an increase in the frequency of your memory "blackouts"?
11. When you are sober, do you often regret things you have done or said while drinking?
12. Have you often failed to keep the promises you have made to yourself about controlling or cutting down on your drinking?
13. Do more people seem to be treating you unfairly without good reason?
14. Do you eat very little or irregularly when you are drinking?
15. Do you get terribly frightened after you have been drinking heavily?

From National Council on Alcoholism and Drug Dependence, Inc., New York, NY. Reprinted by permission.

symptoms, and an intense *craving* for cocaine. These are indications that the person is highly dependent upon the drug or, in other words, that cocaine is extremely addictive.

Overdosing on cocaine is a real possibility. The number of deaths from cocaine and the number of emergency-room admissions for drug reactions involving cocaine have increased greatly. High doses can cause seizures and cardiac and respiratory arrest.

Individuals who snort the drug can suffer damage to the nasal tissues and even perforation of the septum between the nostrils. Whether long-term cocaine abuse causes brain damage is not yet known, but this possibility is

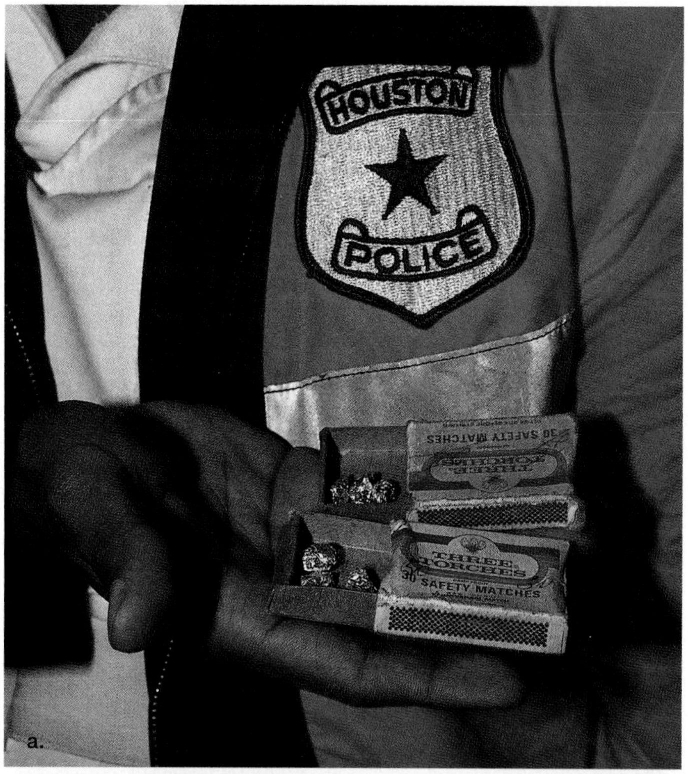

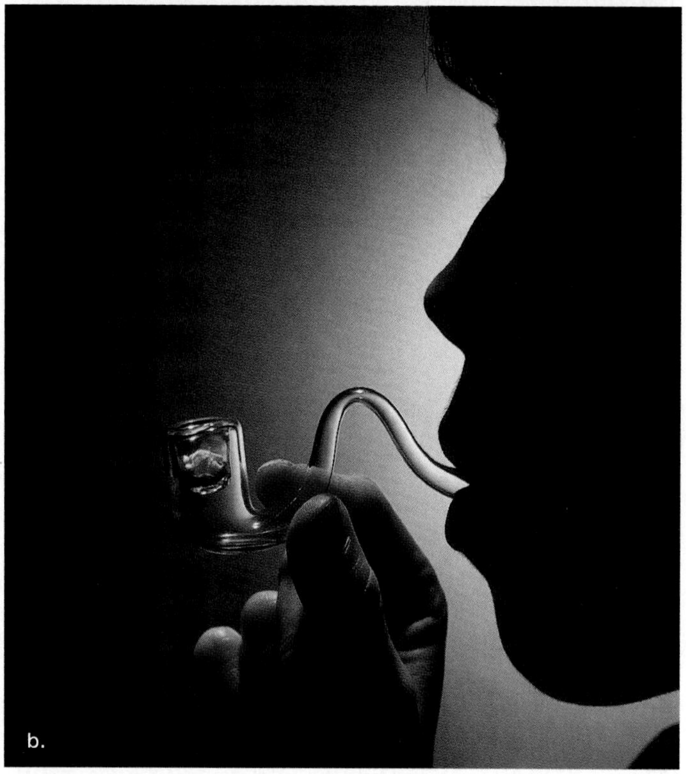

Figure 10.17

Cocaine use. **a.** Crack is the ready-to-smoke form of cocaine. It is a more potent and a more deadly form than the powder. **b.** Users often smoke crack in a glass water pipe. The high produced consists of a "rush" lasting a few seconds, followed by a few minutes of euphoria. Continuous use makes the user extremely dependent on the drug.

under investigation. It is known that babies born to addicts suffer withdrawal symptoms and may suffer neurological and developmental problems.

Heroin. Heroin is derived from morphine, an alkaloid of *opium.* Heroin usually is injected. After intravenous injection, the onset of action is noticeable within one minute and reaches its peak in 3–6 minutes. There is a feeling of euphoria along with relief of pain. Side effects can include nausea, vomiting, dysphoria, and respiratory and circulatory depression leading to death.

Heroin binds to receptors meant for the endorphins, the special neurotransmitters that kill pain and produce a feeling of tranquility. They are believed to alleviate pain by preventing the release of a neurotransmitter termed *substance P* from certain sensory neurons in the region of the spinal cord. When substance P is released, pain is felt, and when substance P is not released, pain is not felt. Endorphins and heroin also bind to receptors on neurons that travel from the spinal cord to the limbic system. Stimulation of these can cause a feeling of pleasure.

Individuals who inject heroin become physically dependent on the drug. With time, the body's production of endorphins decreases. *Tolerance* develops so that the user needs to take more of the drug just to prevent *withdrawal* symptoms. The euphoria originally experienced upon injection is no longer felt.

Heroin withdrawal symptoms include perspiration, dilation of pupils, tremors, restlessness, abdominal cramps, gooseflesh, defecation, vomiting, and increase in systolic pressure and respiratory rate. Those who are excessively dependent may experience convulsions, respiratory failure, and death. Infants born to women who are physically dependent also experience these withdrawal symptoms.

Cocaine and heroin produce a very strong physical dependence. An overdose of these drugs can cause death.

Methamphetamine (Ice). Methamphetamine is related to amphetamine, a well-known stimulant. Both methamphetamine and amphetamine have been drugs of abuse for some time, but a new form of methamphetamine known as "ice" is now used as an alternative to cocaine. Ice is a pure, crystalline hydrochloride salt that has the appearance of sheetlike crystals. Unlike cocaine, ice can be illegally produced in this country in laboratories and does not need to be imported.

Ice, like crack, will vaporize in a pipe, so it can be smoked, avoiding the complications of intravenous injections. After rapid absorption into the bloodstream, the drug moves quickly to the brain. It has the same stimulatory effect as cocaine, and subjects report they cannot distinguish between the 2 drugs after intravenous administration. Methamphetamine effects, however, persist for hours instead of a few seconds. Therefore, it is the preferred drug of abuse by many.

Designer Drugs. Designer drugs are analogues; that is, they are chemical compounds of controlled substances slightly altered in molecular structure. One such drug is MPPP (1-methyl-4-phenylprionoxy-piperidine), an analogue of the narcotic fentanyl. Even small doses of the drug are very toxic; MPPP already has caused many deaths on the West Coast.

Summary

The anatomical unit of the nervous system is the neuron, of which there are 3 types: sensory, motor, and interneuron. Each of these is made up of a cell body, an axon, and a dendrite(s). Axons and dendrites make up nerves that project from the brain and cord.

The nerve impulse is the same in all neurons. It simply consists of a change in permeability of the membrane so that sodium ions move to the inside of a neuron and potassium ions move to the outside. The movement of ions produces an electrochemical charge that can be recorded in terms of millivolts by an oscilloscope. Depolarization occurs with the initiation of the action potential; repolarization restores the resting potential. Transmission of the nerve impulse from one neuron to another takes place when

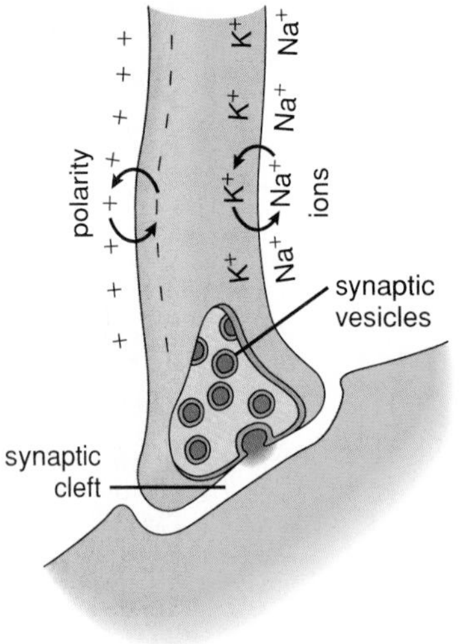

synaptic vesicles found at the ends of axons release a neurotransmitter that diffuses across a synapse to the postsynaptic membrane of the next neuron.

While the somatic division of the peripheral nervous system controls skeletal muscle, the autonomic division controls smooth muscle of the internal organs and glands. The autonomic system contains 2 parts: the sympathetic system, which is often associated with those reactions that occur during times of stress, and the parasympathetic system, which is often associated with those activities that occur during times of relaxation.

Cranial and spinal nerves are found in the peripheral nervous system. A nerve contains the long fibers of sensory and/or motor neurons. A simple reflex requires the use of neurons that make up a reflex arc. A sensory neuron conducts the nerve impulse from a receptor to an interneuron, which in turn transmits the impulse to a motor

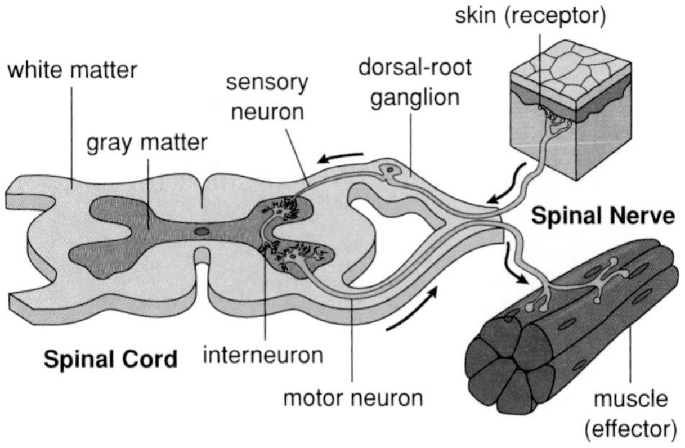

neuron, which conducts it to an effector. Reflexes are automatic, and some do not require involvement of the brain.

The central nervous system consists of the spinal cord and brain. The gray matter of the cord contains cell bodies; the white matter contains tracts that consist of the long axons of interneurons. These run from all parts of the cord, even up to the cerebrum. The brain integrates all nervous system activity and commands all voluntary activities. In the brain stem, the medulla oblongata has centers for visceral functions and the cerebellum coordinates muscle contractions. In the diencephalon, the hypothalamus, in particular, controls homeostasis; the thalamus specializes in sense reception; and the cerebrum gives us consciousness. The cerebrum, which alone is responsible for consciousness, can be mapped, and each lobe seems to have particular functions.

Neurological drugs, although quite varied, have been found to affect the limbic system by either promoting or preventing the action of neurotransmitters.

Study and Thought Questions

1. What are the 3 types of neurons? How are they similar, and how are they different? (p. 184)
2. What does the term *resting potential* mean, and how is it brought about? (p. 185) Describe the 2 parts of an action potential and the change that can be associated with each part. (p. 186)

   Critical Thinking If a neurotransmitter is inhibitory, would you expect a higher or lower voltage reading compared to –65 mV on the oscilloscope?
3. What is the sodium-potassium pump, and when is it active? (p. 188)
4. What is a neurotransmitter, where is it stored, how does it function, and how is it destroyed? (p. 188) Name 2 well-known neurotransmitters. (p. 189)

   Critical Thinking Why do nerve impulses travel only from axon to dendrite or cell body across a synapse in the body?
5. What are the 3 types of nerves, and how are they structurally different? functionally different? Distinguish between cranial and spinal nerves. (pp. 189–190)

   Ethical Issue Research utilizing animals has led to the possibility of curing paraplegia. Do you think animals should be used in medical research? Why or why not?
6. What are the 2 main divisions of the nervous system? Explain why these names are appropriate. (p. 189)
7. Trace the path of a reflex action after discussing the structure and the function of the spinal cord and the spinal nerve. (p. 191)
8. What is the autonomic system, and what are its 2 major divisions? Give several similarities and differences between these divisions. (p. 193)
9. Name the major parts of the brain, and give a function for each. (pp. 195–196)
10. Describe the EEG, and discuss its importance. (p. 197)
11. Describe the physiological effects and mode of action of alcohol, marijuana, cocaine, heroin, and methamphetamine. (pp. 198–202)

Writing Across the Curriculum

1. Skeletal muscle fibers are innervated only by somatic motor neurons. Speculate why 2 different motor neurons (one from the sympathetic system and the other from the parasympathetic system) are used for internal organs and glands.
2. The limbic system is sometimes called the animal brain. Considering the function of the limbic system, why might this term be appropriate? What purposes are served by the "primitive" brain beneath the cerebral cortex?
3. The nervous system is divided into the central nervous system (CNS) and the peripheral nervous system (PNS). In what way does this division seem awkward and artificial?

Objective Questions

1. A(n) ____________________ carries nerve impulses away from the cell body.
2. During the upswing of the action potential, ____________ ions are moving to the ____________ of the nerve fiber.
3. The space between the axon of one neuron and the dendrite of another is called the ________________.
4. ACh is broken down by the enzyme ____________ after ACh has altered the permeability of the postsynaptic membrane.
5. Motor nerves innervate ________________________.
6. The vagus nerve is a(n) ________________ nerve that controls ________________________.
7. In a reflex arc, only the neuron called the ____________________ is completely within the CNS.
8. The brain and the spinal cord are covered by protective layers called ________________________.
9. The ______________________ is the part of the brain that allows us to be conscious.
10. The ______________________ is the part of the brain responsible for coordination of body movements.
11. Label this diagram. See figure 10.1 in text.

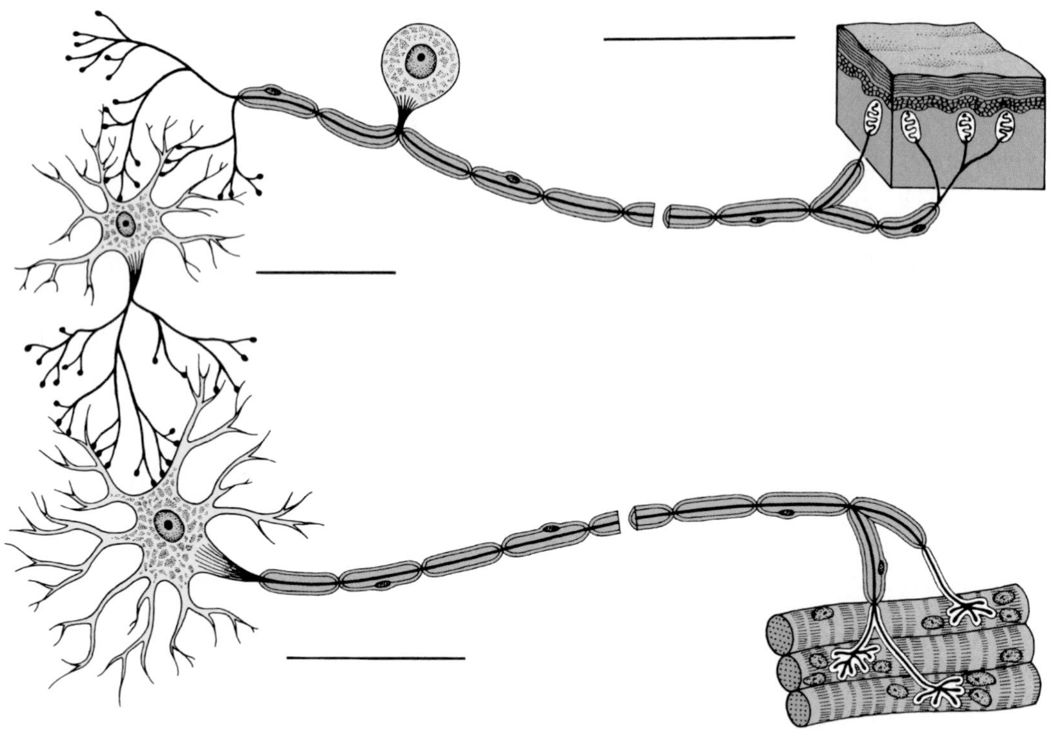

Selected Key Terms

autonomic system A branch of the peripheral nervous system that has control over the internal organs; consisting of the sympathetic and parasympathetic systems. 190

axon Fiber of a neuron that conducts nerve impulses away from the cell body. 184

cell body Portion of a neuron that contains a nucleus and from which the nerve fibers extend. 184

central nervous system (CNS) Portion of the nervous system containing the brain and spinal cord. 193

cerebral hemisphere (ser´ ĕ-bral hem´ĭ-sfer) One of the large, paired structures that together constitute the cerebrum of the brain. 196

cerebrospinal fluid A fluid found in the ventricles of the brain, in the central canal of the spinal cord, and in association with the meninges. 193

dendrite Fiber of a neuron, typically branched, that conducts nerve impulses toward the cell body. 184

effector A structure such as a muscle or a gland that allows an organism to respond to environmental stimuli. 184

ganglion (gang´ gle-on) A collection of neuron cell bodies within the peripheral nervous system. 189

innervate To activate an organ, muscle, or gland by motor neuron stimulation. 184

interneuron A neuron found within the central nervous system that takes nerve impulses from one portion of the system to another. 184

limbic system A portion of the brain concerned with memory and emotions. 197

motor neuron A neuron that takes nerve impulses from the central nervous system to the effectors. 184

myelin sheath (mi´ĕ-lin shēth) The Schwann cell membranes that cover long neuron fibers giving them a white, glistening appearance. 185

nerve impulse An electrochemical change due to increased membrane permeability that is propagated along a neuron from the dendrite to the axon following excitation. 185

neurotransmitter A chemical made at the ends of axons that is responsible for transmission across a synapse. 188

parasympathetic system That part of the autonomic system that usually promotes activities associated with a normal state. 193

peripheral nervous system (PNS) Nerves and ganglia that lie outside the central nervous system. 189

receptor A structure specialized to receive information from the environment and generate nerve impulses. 184

sensory neuron A neuron that takes nerve impulses to the central nervous system and typically has a long dendrite and a short axon; afferent neuron. 184

somatic system That portion of the peripheral nervous system containing motor neurons that control skeletal muscles. 190

sympathetic system That part of the autonomic system that usually promotes activities associated with emergency (fight or flight) situations. 193

synapse The region between 2 nerve cells where the nerve impulse is transmitted from one to the other, usually from axon to dendrite. 188

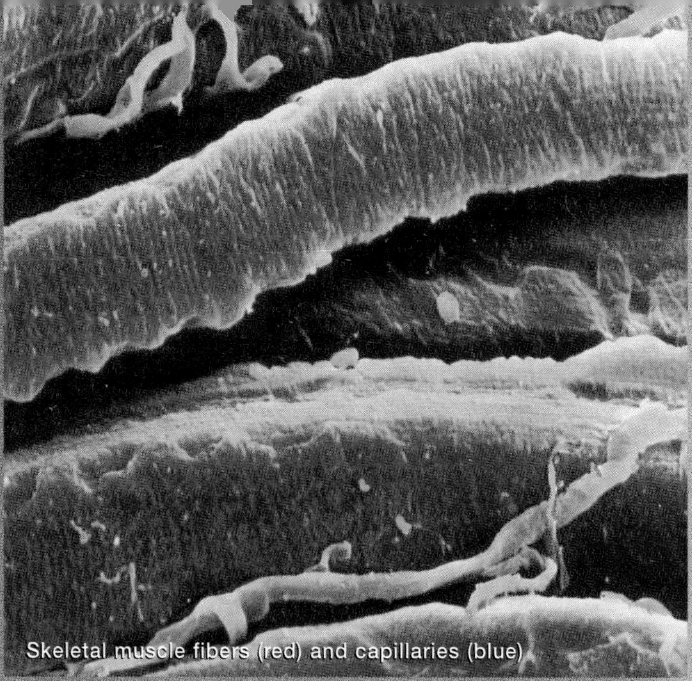

Skeletal muscle fibers (red) and capillaries (blue)

Chapter 11

Musculoskeletal System

Exercise causes a muscle to grow larger—there is an increase in the number and size of the protein filaments within each muscle cell and in the number of mitochondria. Strong muscles also contain more energy sources in the form of glycogen and fat, and they resist fatigue.

Apparently, however, you are born with a certain proportion of slow-twitching versus fast-twitching muscle cells (fibers), and exercise cannot drastically alter the proportion. The slow-twitching fibers found in dark meat are generously supplied with mitochondria and capillaries. They produce a steady pulling power so necessary to swimming, running, and skiing. The fast-twitching fibers found in light meat aren't as well supplied with blood vessels nor do they have as many mitochondria. They supply bursts of energy for sprinting, weight lifting, or swinging a golf club. No wonder, then, that athletes have their favorite sports at which they perform best.

Chapter Concepts

1 Bone, a rigid material, makes up the human skeleton. 209

2 The axial portion of the skeleton contains the skull and vertebral column, which are centrally placed. 209

3 The appendicular portion of the skeleton contains the rest—the girdles and the limbs. 212

4 The skeleton is jointed; the joints differ in movability. 213

5 Macroscopically, skeletal muscles work in antagonistic pairs and exhibit certain physiological characteristics. 216

6 Microscopically, muscle fiber contraction is dependent on filaments of both actin and myosin and a ready supply of calcium ions (Ca^{++}) and ATP. 221

Chapter Outline

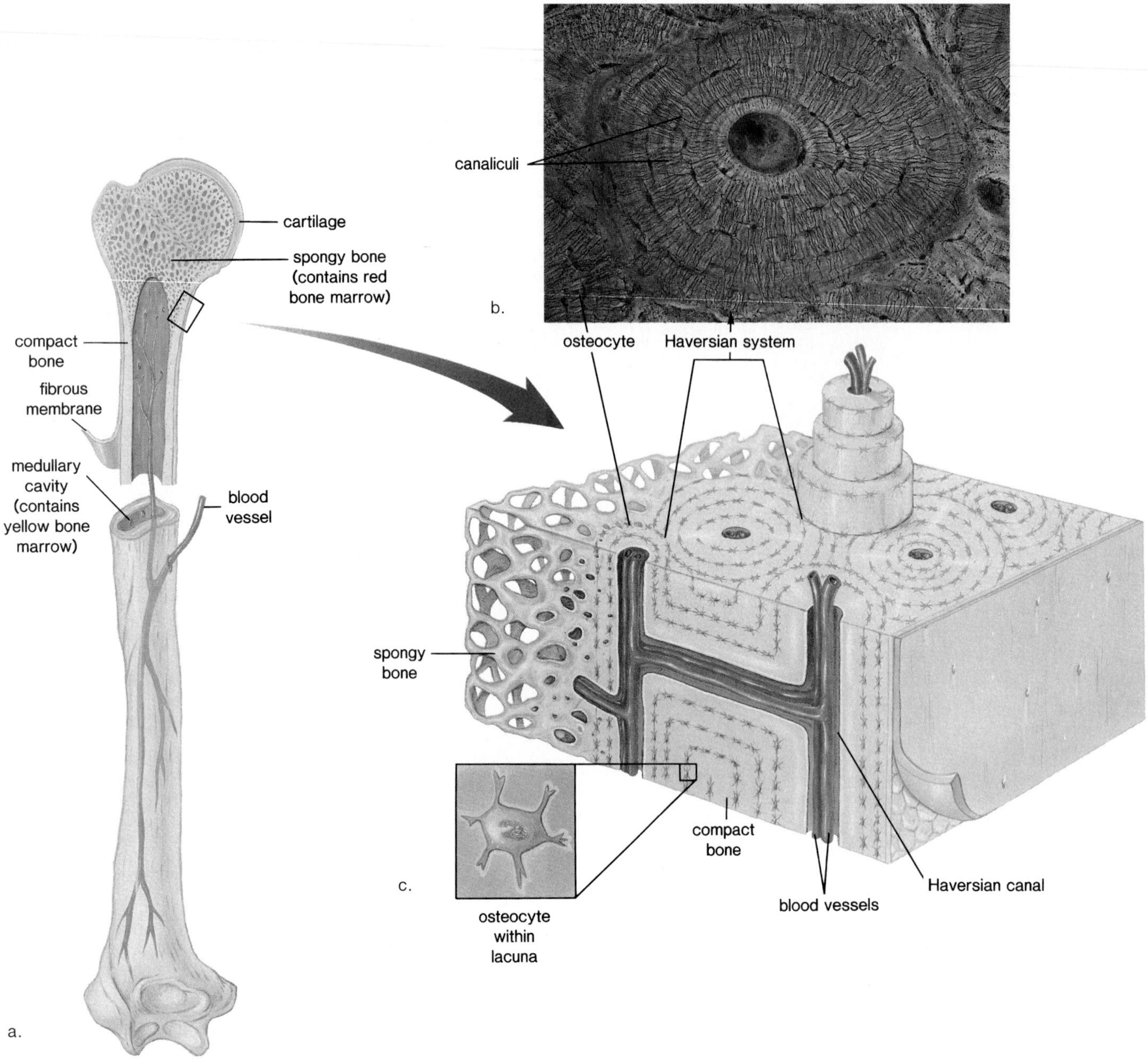

Figure 11.1

Anatomy of the long bone. **a.** A long bone is encased by fibrous membrane except where it is covered by cartilage at the ends. The central shaft is composed of compact bone, but the ends are spongy bone encased by a thin layer of compact bone. The spongy bone may contain red bone marrow. A central medullary cavity contains yellow bone marrow. Microscopic anatomy is shown in **b.** micrograph and **c.** art.

Bone Structure and Growth

Bone is a living tissue that continually renews itself in a manner to be described.

Bone Structure: Compact and Spongy

A long bone, such as the femur, illustrates principles of bone anatomy. When the bone is split open, as in figure 11.1, the longitudinal section shows that it is not solid but has a cavity called the medullary cavity bounded at the sides by compact bone and at the ends by spongy bone. Beyond the spongy bone there is a thin shell of compact bone and finally a layer of cartilage.

Compact bone, as discussed previously (p. 50), contains bone cells in tiny chambers called lacunae, arranged in concentric circles around Haversian canals. The Haversian canals contain blood vessels and nerves. The lacunae are separated by a matrix that contains protein fibers of collagen and mineral deposits, primarily of calcium and phosphorus salts.

Spongy bone contains numerous bony bars and plates separated by irregular spaces. Although lighter than compact bone, spongy bone is still designed for strength. Just as braces are used for support in buildings, the solid portions of spongy bone follow lines of stress. The spaces in spongy bone are often filled with *red bone marrow,* a specialized tissue that produces red blood cells. The cavity of a long bone usually contains *yellow bone marrow,* which is a fat-storage tissue.

> A long bone is designed for strength. It has a medullary cavity filled with yellow marrow and bounded by compact bone. The ends contain spongy bone and are covered by cartilage.

Bone Growth: Constant Renewal

Most of the bones of the skeleton are cartilaginous during prenatal development. Later, the cartilage is replaced by bone due to the action of bone-forming cells known as **osteoblasts.** (Once they are isolated in lacunae, the cells are called osteocytes.) At first, there is only a primary ossification center at the middle of a long bone, but later, secondary centers form at the ends of the bones. There remains a *cartilaginous disk* between the primary ossification center and each secondary center, which can increase in length. The rate of growth is controlled by hormones such as growth hormones and the sex hormones. Eventually, though, the disks disappear, and the bone stops growing as the individual attains adult height.

In the adult, bone is continually being broken down and built up again. Bone-absorbing cells, called **osteoclasts,** are derived from cells carried in the bloodstream. As they break down bone, they remove worn cells and deposit calcium in the blood. Apparently after a period of about 3 weeks, they disappear. The destruction caused by the work of osteoclasts is repaired by osteoblasts. As they form new bone, they take calcium from the blood. Eventually some of these cells get caught in the matrix they secrete and are converted to **osteocytes,** the cells found within Haversian systems (pp. 49–50).

> Bone is a living tissue, and it is always being rejuvenated.

Skeleton: Six Functions

The skeleton (fig. 11.2), notably the large, heavy bones of the legs, *supports the body* against the pull of gravity. The skeleton also *protects soft body parts.* For example, the skull forms a protective encasement for the brain, as does the rib cage for the heart and the lungs. Flat bones, such as those of the skull, the ribs, and the breastbone, *produce red blood cells* in both adults and children. All bones are *storage areas* for inorganic calcium and phosphorus salts. Bones also provide *sites for muscle attachment.* The long bones, particulary those of the legs and the arms, *permit flexible body movement.*

> The skeleton not only permits flexible movement, it also supports and protects the body, produces red blood cells, and serves as a storehouse for certain inorganic salts.

Axial Skeleton

The skeleton has 2 divisions: the axial skeleton and the appendicular skeleton. The **axial skeleton** lies in the midline of the body and consists of the skull, vertebral column, sternum, and ribs.

Skull: Bones of the Head and Face

The skull is formed by the cranium and the facial bones.

Cranium

The cranium protects the brain and is composed of 8 bones fitted tightly together in adults. In newborns, certain bones are not completely formed and instead are joined by membranous regions called *fontanels,* all of which usually close

Figure 11.2

Major bones (*right*) and skeletal muscles (*left*) of the human body. The axial skeleton, composed of the skull, the vertebral column, the sternum, and the ribs (red labels), lies in the midline; the rest of the bones belong to the appendicular skeleton (black leaders and labels).

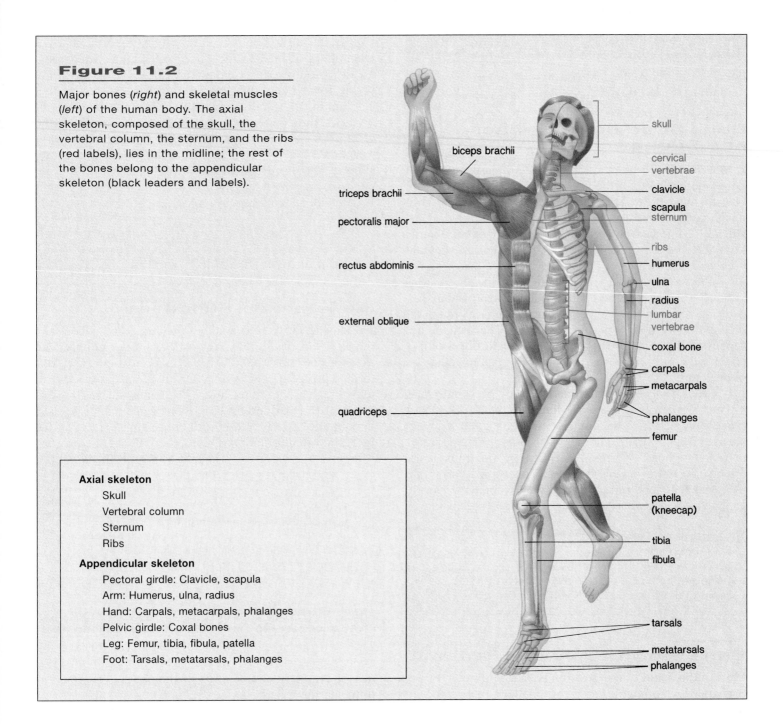

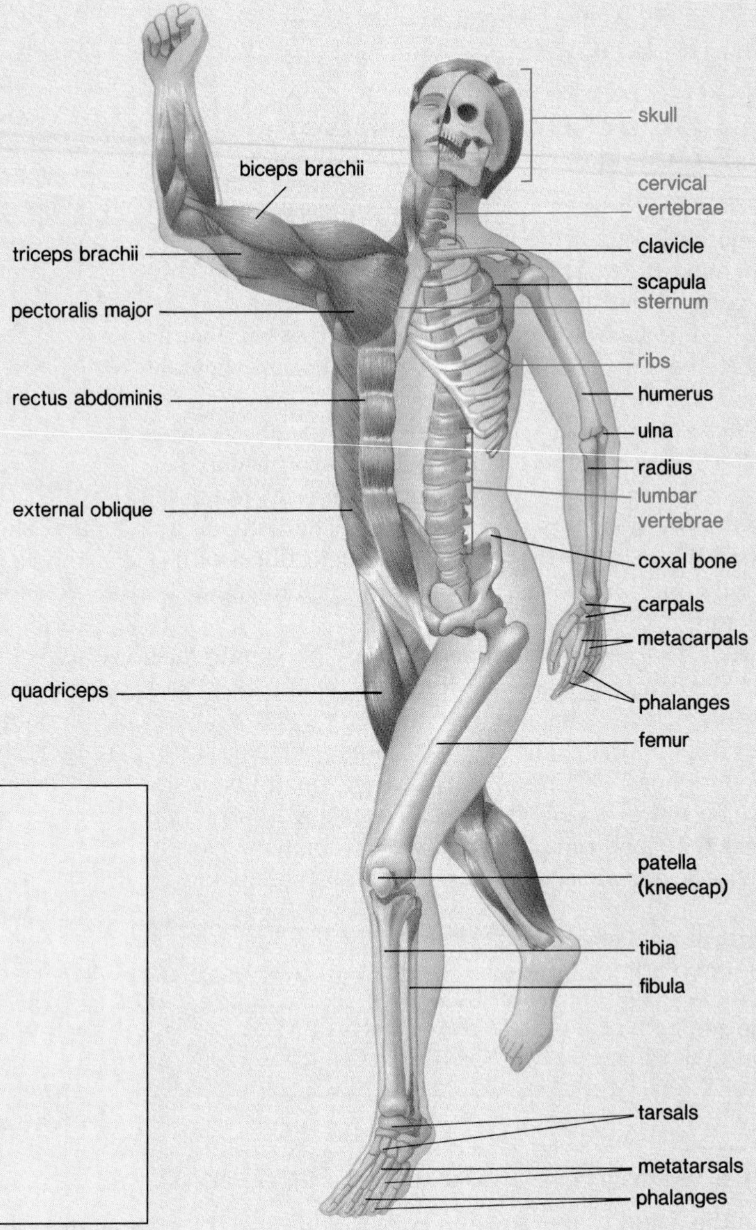

Axial skeleton
- Skull
- Vertebral column
- Sternum
- Ribs

Appendicular skeleton
- Pectoral girdle: Clavicle, scapula
- Arm: Humerus, ulna, radius
- Hand: Carpals, metacarpals, phalanges
- Pelvic girdle: Coxal bones
- Leg: Femur, tibia, fibula, patella
- Foot: Tarsals, metatarsals, phalanges

by the age of 16 months. The bones of the cranium contain the **sinuses,** air spaces lined by mucous membrane, which reduce the weight of the skull and give a resonant sound to the voice. Two sinuses called the mastoid sinuses drain into the middle ear. *Mastoiditis,* a condition that can lead to deafness, is an inflammation of these sinuses.

The major bones of the cranium have the same names as the lobes of the brain: frontal, parietal, temporal, and occipital. On the top of the cranium (fig. 11.3*a*), the *frontal bone* forms the forehead, the *parietal bones* extend to the sides, and the *occipital bone* curves to form the base of the skull. Here there is a large opening, the **foramen magnum** (fig. 11.3*b*), through which the spinal cord passes and becomes the brain stem. Below the much larger parietal bones, each *temporal bone* has an opening that leads to the middle ear. The *sphenoid bone* not only completes the sides of the skull, it also contributes to the floors and walls of the eye sockets. Likewise, the *ethmoid bone,* which lies in front of the sphenoid, is a part of the orbital wall and, in addition, is a component of the nasal septum.

> The cranium contains 8 bones; the frontal, 2 parietal, the occipital, 2 temporal, the sphenoid, and the ethmoid.

Figure 11.3

Skull. **a.** Lateral view. **b.** Inferior view.

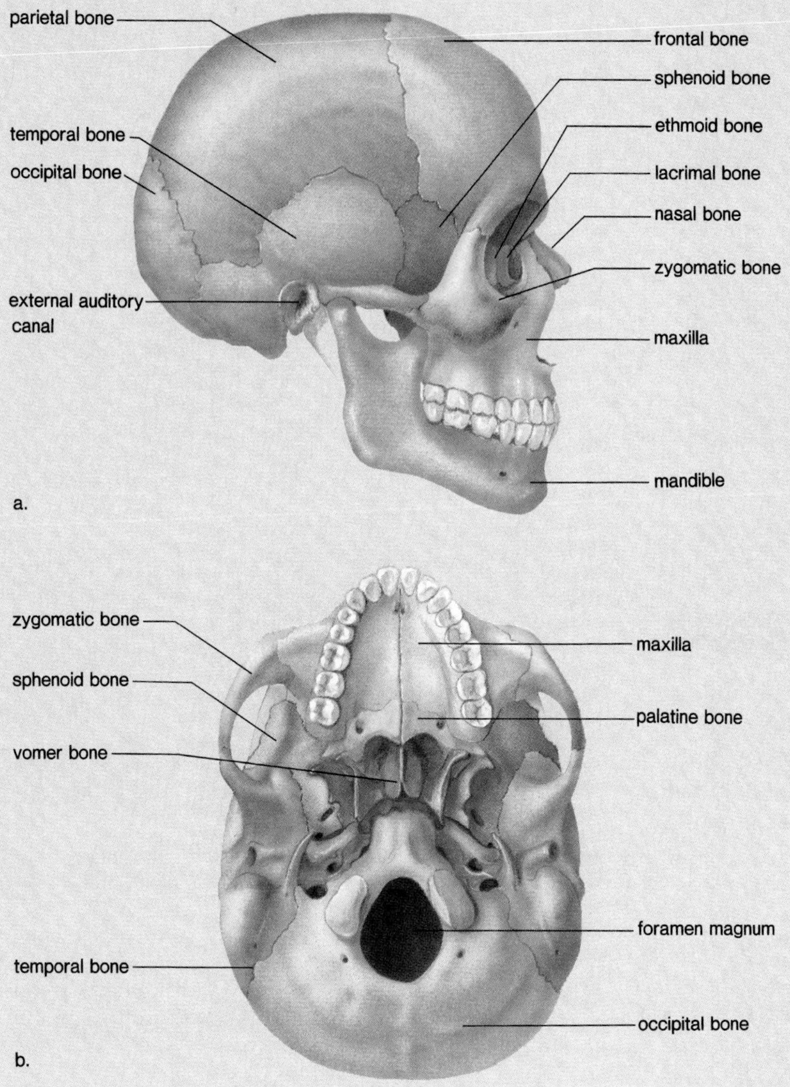

Figure 11.4

The vertebral column. The vertebrae are named according to their location in the column, which is flexible due to the intervertebral disks. Note the presence of the coccyx, also called the tailbone.

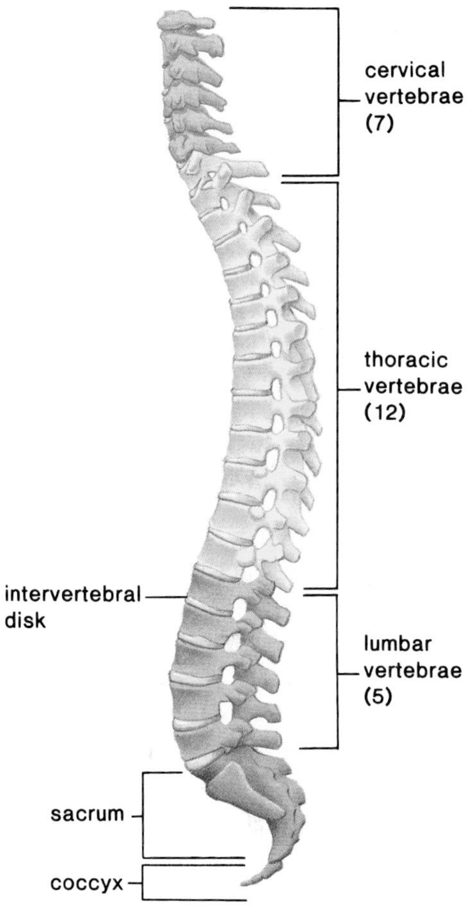

Facial Bones

The **mandible,** or lower jaw, is the only movable portion of the skull (fig. 11.3*a*), and its action permits us to chew our food. Tooth sockets are located on this bone and on the **maxillae** (maxillary bones), the upper jaw that also forms the anterior portion of the hard palate. The *palatine bones* make up the posterior portion of the hard palate and the floor of the nasal cavity. The *zygomatic bones* give us our cheekbone prominences, and the nasal bones form the bridge of the nose. Each thin, scalelike *lacrimal bone* lies between an ethmoid bone and a maxillary bone, and the thin, flat *vomer* joins with the perpendicular plate of the ethmoid to form the nasal septum.

> The facial bones include the mandible, 2 maxillae, 2 palatine, 2 zygomatic, 2 lacrimal, 2 nasal, and the vomer.

Vertebral Column: Many Vertebrae

The **vertebral column** extends from the skull to the pelvis. Normally, the vertebral column has 4 curvatures that provide more resiliency and strength than a straight column could. The various vertebrae are named according to their location in the vertebral column (fig. 11.4). When the vertebrae join, they form a canal through which the spinal cord passes. The *spinous processes* of the vertebrae can be felt as bony projections along the midline of the back.

There are intervertebral disks between the vertebrae that act as a kind of padding. They prevent the vertebrae from grinding against one another and absorb shock caused by movements such as running, jumping, and even walking. Unfortunately, these disks become weakened with age and can slip or even rupture. This causes pain when the damaged disk presses up against the spinal cord and/or spinal nerves. The body may heal itself, or the disk can be removed surgically. If the latter occurs, the vertebrae can be fused together, but this limits the flexibility of the body. The presence of the disks allows motion between the vertebrae so that we can bend forward, backward, and from side to side.

The vertebral column, directly or indirectly, serves as an anchor for all the other bones of the skeleton (see fig. 11.2). All of the 12 pairs of **ribs** connect directly to the thoracic vertebrae in the back, and all but 2 pairs connect either directly or indirectly via shafts of cartilage to the **sternum** (breastbone) in the front. The lower 2 pairs of ribs are called "floating ribs" because they do not attach to the sternum.

The vertebral column contains the vertebrae and serves as the backbone for the body. Disks between the vertebrae provide padding and account for flexibility of the column.

Appendicular Skeleton

The **appendicular skeleton** consists of the bones within the pectoral and pelvic girdles and the attached limbs (see fig. 11.2). The pectoral (shoulder) girdle and upper limbs (arms) are specialized for flexibility, but the pelvic girdle (hipbones) and lower limbs (legs) are specialized for strength.

Pectoral Girdle and Arm: Work Together

The components of the **pectoral girdle** are loosely linked together by ligaments (fig. 11.5). Each **clavicle** (collarbone) connects with the sternum in front and the **scapula** (shoulder blade) behind, but the scapula is freely movable and held in place only by muscles. This allows it to follow freely the movements of the arm. The single long bone in the upper arm, the **humerus,** has a smoothly rounded head that fits into a socket of the scapula. The socket, however, is very shallow and much smaller than the head. Although this means that the arm can move in almost any direction, there is little stability. Therefore, this is the joint that is most apt to dislocate. The opposite end of the humerus meets the 2 bones of the lower arm, the **ulna** and the **radius,** at the elbow. (The prominent bone in the elbow is the topmost part of the ulna.) When the arm is held so that the palm is turned frontward, the radius and ulna are about parallel to one another. When the arm is turned so that the palm is next to the body, the radius crosses in front of the ulna, a feature that contributes to the easy twisting motion of the forearm.

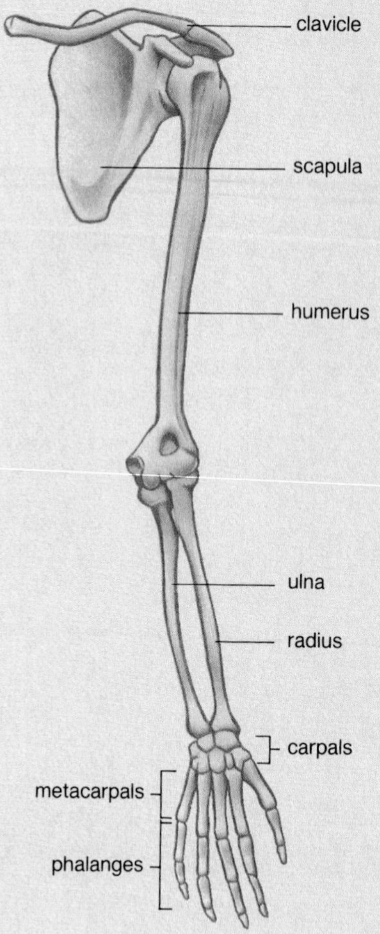

Figure 11.5

The bones of the pectoral girdle, the arm, and the hand. The humerus becomes the "funny bone" of the elbow. The sensation upon bumping it is due to the activation of a nerve that passes across its end.

The many bones of the hand increase its flexibility. The wrist has 8 **carpal** bones, which look like small pebbles. From these, 5 *metacarpal* bones fan out to form a framework for the palm. The metacarpal bone that leads to the thumb is placed in such a way that the thumb can reach out and touch the other digits. (**Digits** is a term that refers to either fingers or toes.) Beyond the metacarpals are the *phalanges,* the bones of the fingers and the thumb. The phalanges of the hand are long, slender, and lightweight.

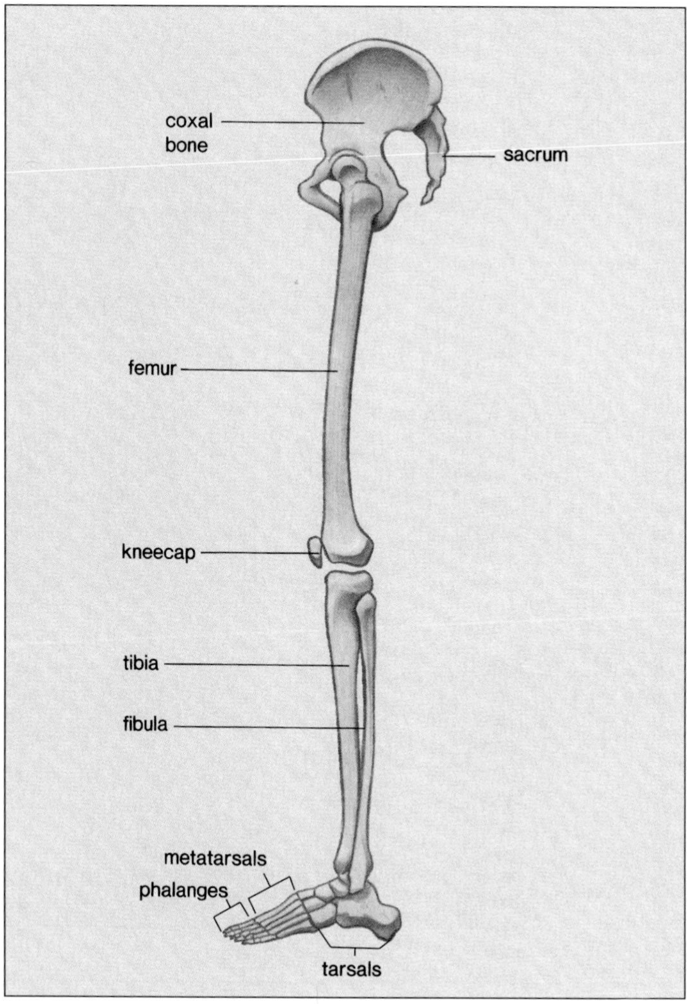

Figure 11.6

The bones of the pelvic girdle, the leg, and the foot. The femur is our strongest bone—it withstands a pressure of 540 kilograms per 2.5 cm^3 when we walk.

Pelvic Girdle and Leg: Work Together

The **pelvic girdle,** or pelvis (fig. 11.6), consists of 2 heavy, large *coxal* bones (hipbones). The coxal bones are anchored to the sacrum, and together these bones form a hollow cavity that is wider in females than males. The weight of the body is transmitted through the pelvis to the legs and then onto the ground. The largest bone in the body is the femur, or thighbone. Although the femur is a strong bone, it is doubtful that the femurs of a fairytale giant could support its weight. If a giant were 10 times taller than an ordinary human being, he would also be about 10 times wider and thicker, making him weigh about 1,000 times as much. This amount of weight would break even giant-size femurs.

In the lower leg, the larger of the 2 bones, the tibia (fig. 11.6), has a ridge we call the shin. Both of the bones of the lower leg have a prominence that contributes to the ankle—the tibia on the inside of the ankle and the *fibula* on the outside of the ankle. Although there are 7 tarsal bones in the ankle, only one receives the weight and passes it on to the heel and the ball of the foot. If you wear high-heeled shoes, the weight is thrown even further forward toward the front of your foot. The *metatarsal* bones form the arches of the foot. There is a longitudinal arch from the heel to the toes and a transverse arch across the foot. These provide a stable, springy base for the body. If the tissues that bind the metatarsals together become weakened, flatfeet are apt to result. The bones of the toes are called *phalanges,* just like those of the fingers, but in the foot the phalanges are stout and extremely sturdy.

> The appendicular skeleton contains the bones of the girdles and limbs.

Joints

Bones are joined at the joints, which are classified as fibrous, cartilaginous, and synovial. Fibrous joints such as those between the cranial bones are immovable. Cartilaginous joints such as those between the vertebrae are slightly movable. The vertebrae are also separated by disks, which increase their flexibility. The 2 hipbones are slightly movable because they are ventrally joined by cartilage. Owing to hormonal changes, this joint becomes more flexible during late pregnancy, and this allows the pelvis to expand during childbirth.

Synovial Joints: Freely Movable

Most joints are *freely movable* **synovial joints,** in which the 2 bones are separated by a cavity. *Ligaments* composed of fibrous connective tissue bind the 2 bones to one another, holding them in place as they form a capsule. In a "double-jointed" individual, the ligaments are unusually loose. The joint capsule is lined by synovial membrane, which produces *synovial fluid,* a lubricant for the joint.

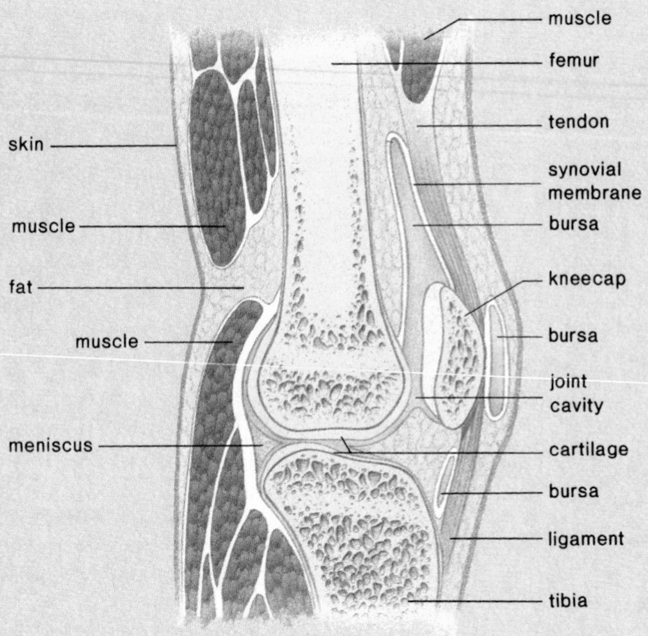

Figure 11.7

The knee joint, which is an example of a synovial joint. Notice the cavity between the bones, which is encased by ligaments and lined by synovial membrane. The kneecap protects the joint and keeps tendons in proper alignment.

The knee is an example of a synovial joint (fig. 11.7). In the knee, as in other freely movable joints, the bones are capped by cartilage, although there are also crescent-shaped pieces of cartilage between the bones called **menisci.** These give added stability, helping to support the weight placed on the knee joint. Unfortunately, athletes often suffer injury of the menisci, known as torn cartilage. The knee joint also contains 13 fluid-filled sacs called bursae, which ease friction between tendons and ligaments and between tendons and bones. Inflammation of the bursae is called bursitis. Tennis elbow is a form of bursitis.

There are different types of movable joints. The knee and elbow joints are *hinge joints* because, like a hinged door, they largely permit movement in one direction only. More movable are the ball-and-socket joints; for example, the ball of the femur fits into a socket on the hipbone. *Ball-and-socket joints* allow movement in all planes and even a rotational movement. The various movements of body parts at joints are depicted in figure 11.8.

Synovial joints are subject to *arthritis.* In rheumatoid arthritis, the synovial membrane becomes inflamed and thickens. Degenerative changes take place that make the joint almost immovable and painful to use. There is evidence that these effects are brought on by an autoimmune reaction. In old-age arthritis, or osteoarthritis, the cartilage at the ends of the bones disintegrates so that the 2 bones become rough and irregular. This type of arthritis is apt to affect the joints that have received the greatest use over the years.

There are 3 kinds of joints. Fibrous joints are immovable, cartilaginous joints are slightly movable, and synovial joints are freely movable.

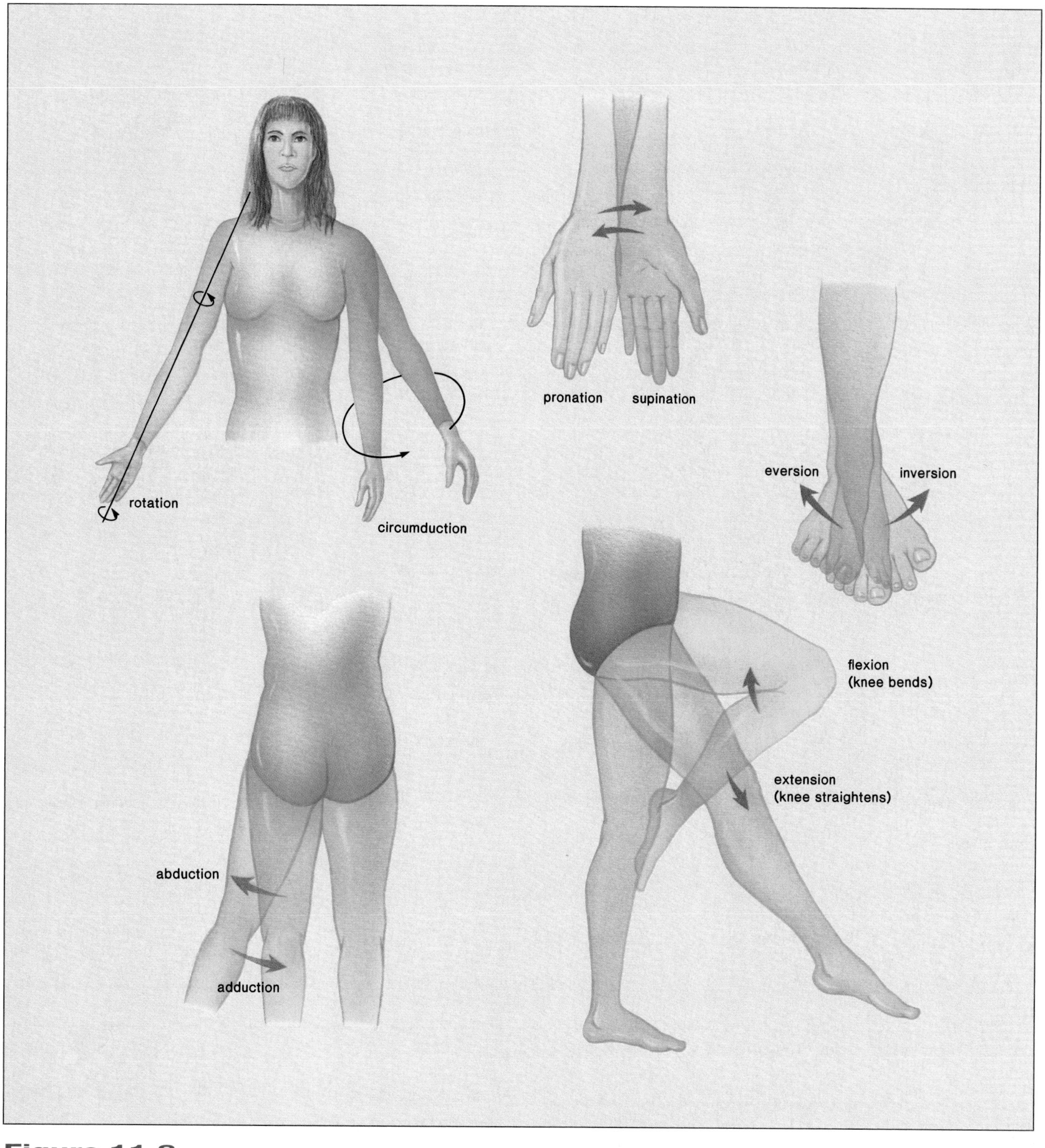

Figure 11.8

Movement of body parts in relation to joints. Muscles of the appendages are attached to bones across a joint. One of these bones remains steady while the other bone moves.

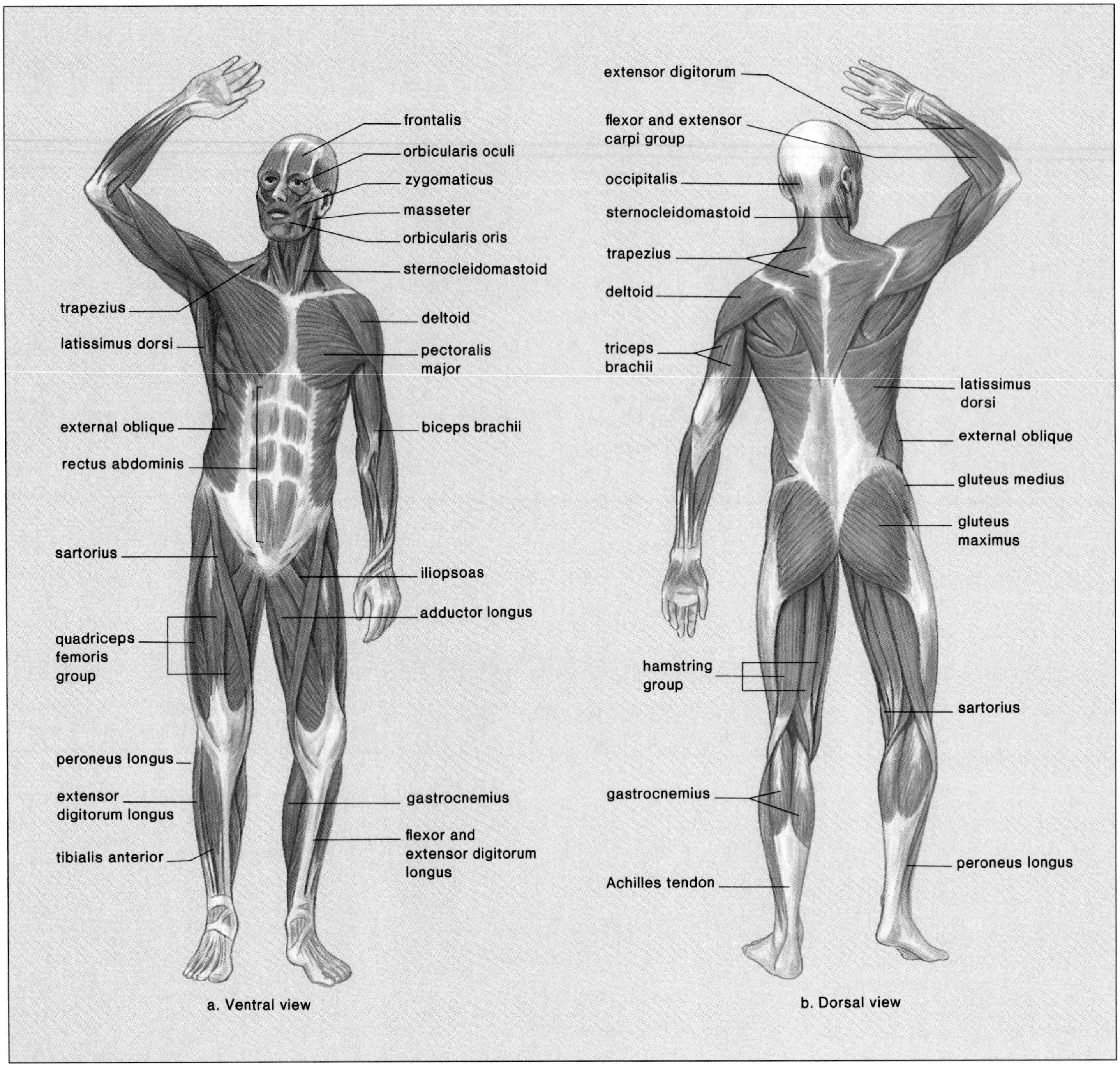

Figure 11.9

Superficial skeletal muscles. **a.** Ventral view. **b.** Dorsal view.

Skeletal Muscles: Macroscopic View

Muscles are effectors, which enable the organism to respond to a stimulus (p. 191). Skeletal muscles are attached to the skeleton, and their contractions account for voluntary movements (fig. 11.9). Involuntary muscles, both smooth and cardiac, were discussed on page 51.

Anatomy: Muscle Pairs

Muscles typically are attached to bone by *tendons* made of fibrous connective tissue. Tendons most often attach muscles to the far side of a joint so that the muscle extends across the joint (see fig. 11.7). When the central portion of the muscle, called the belly, contracts, one bone remains fairly stationary and the other one is moved. The **origin** of the

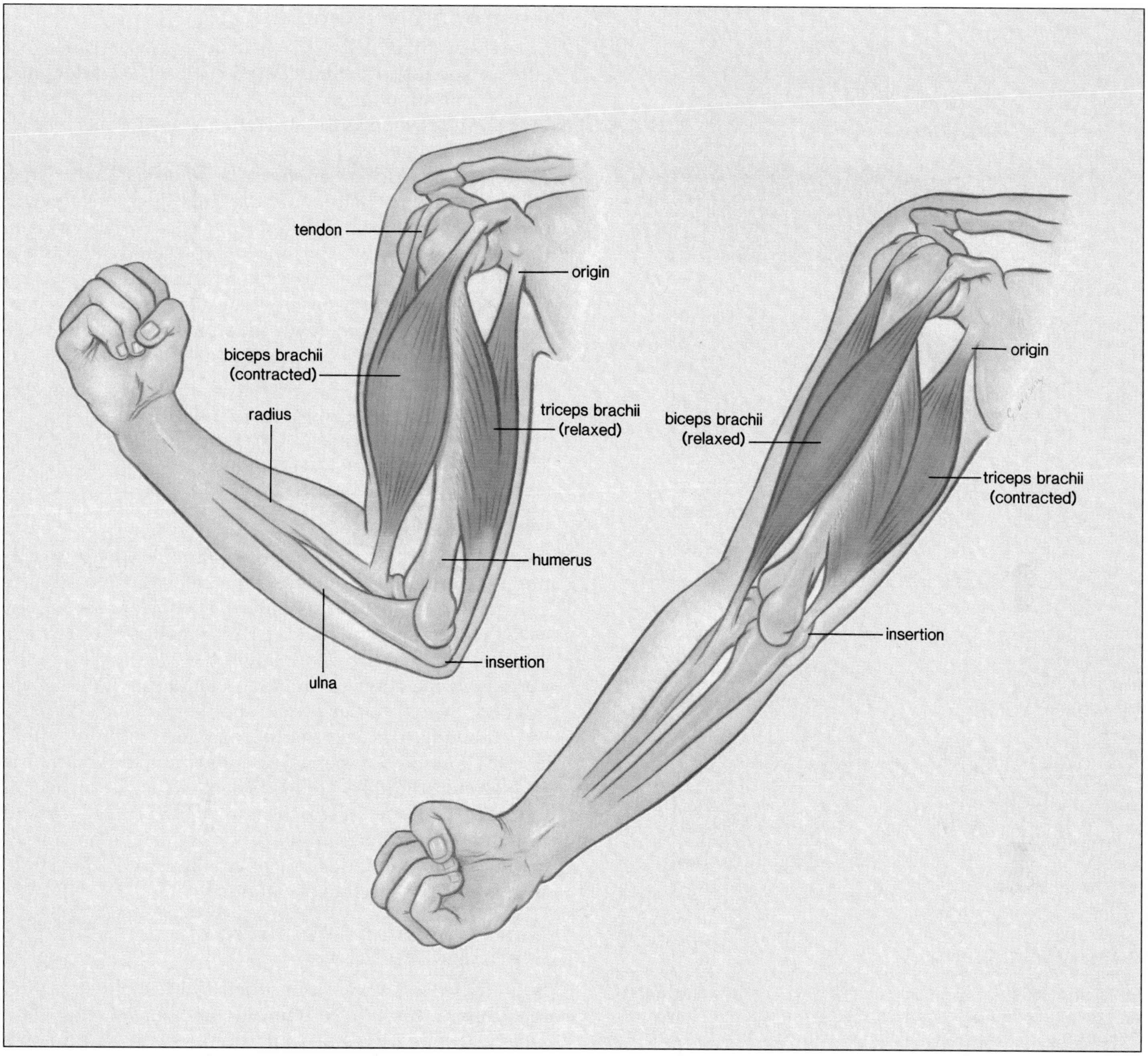

Figure 11.10

Attachment of skeletal muscles as exemplified by the biceps brachii and the triceps brachii. The origin of a muscle is fairly stationary, while the insertion moves. These muscles are antagonistic. When the biceps brachii contracts, the lower arm flexes, and when the triceps brachii contracts, the lower arm extends.

muscle is on the stationary bone, and the **insertion** of the muscle is on the bone that is moved.

When a muscle contracts, it shortens. Therefore, muscles can only pull; they cannot push. Because we need both to extend and to flex at a joint, muscles generally work in *antagonistic pairs.* For example, the biceps and the triceps are a pair of muscles that move the forearm up and down (fig. 11.10). When the biceps contracts, the lower arm flexes, and when the triceps contracts, the lower arm extends. The antagonistic pairs produce smooth, coordinated movements.

> Typically, muscles are attached across a joint. When a muscle contracts, the movable end (insertion) is pulled toward the fixed end (origin), and movement occurs at the joint. This system requires that muscles work in antagonistic pairs.

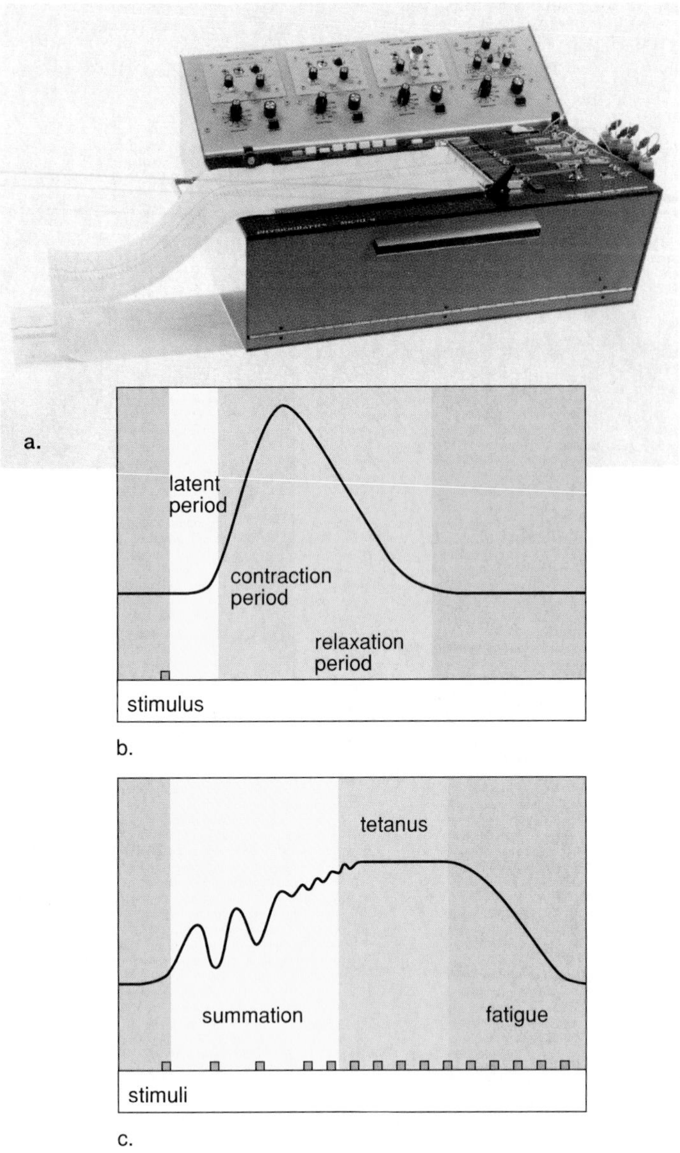

Figure 11.11

Physiology of skeletal muscle contraction. **a.** A physiograph® is an apparatus that can be used to record a myogram, a visual representation of the contraction of a muscle that has been dissected from an animal. **b.** Simple muscle twitch is composed of 3 periods: latent, contraction, and relaxation. **c.** Summation and tetanus. When a muscle is not allowed to relax completely between stimuli, the contraction gradually increases in intensity. The muscle becomes maximally contracted until it fatigues.

Physiology: When Whole Muscles Contract

It is possible to study the contraction of individual whole muscles in the laboratory. Customarily, a calf muscle removed from a frog is loosened at the tendon of Achilles and attached to a muscle lever. The muscle is stimulated and the mechanical force of contraction is transduced (changed) into an electrical current recorded by a *physiograph*® (fig. 11.11*a*). The resulting visual pattern is called a *myogram.*

Response to Stimulus

Like a nerve fiber, a *single* muscle fiber (muscle cell, p. 51) either responds to a stimulus and contracts or it does not. At first, the stimulus may be so weak that no contraction occurs, but as soon as the strength of the stimulus reaches the *threshold stimulus,* the muscle fiber contracts completely. Therefore, a muscle fiber obeys the *all-or-none law.*

Contrary to that of an individual fiber, the strength of contraction of a whole muscle can increase according to the degree of stimulus beyond the threshold stimulus. A whole muscle contains many fibers, and the degree of contraction is dependent on the total number of fibers contracting. The *maximum stimulus* is the one beyond which the degree of contraction does not increase.

> Although muscle fibers normally obey the all-or-none law, whole muscles do not. The degree of contraction is dependent on the total number of fibers contracting.

Muscle Twitch

If a muscle is placed in a physiograph® and is given a maximum stimulus, it contracts and then relaxes. This action—a single contraction that lasts only a fraction of a second—is called a muscle *twitch.* Figure 11.11*b* is a myogram of a twitch, which is customarily divided into the *latent period,* or the period of time between stimulation and initiation of contraction, the *contraction period,* and the *relaxation period.*

If a muscle is exposed to 2 maximum stimuli within a certain period of time, it responds to the first but not to the second stimulus. This is because it takes an instant following a contraction for the muscle fibers to recover and respond to the next stimulus. The very brief moment following stimulation, during which a muscle is unresponsive, is called the refractory period.

Summation and Tetanus: Blending Twitches

If a muscle is given a rapid series of threshold stimuli, it can respond to the next stimulus without relaxing completely. In this way, muscle tension *summates* until maximal sustained contraction, called **tetanus,** is achieved (fig. 11.11*c*). The myogram no longer shows individual twitches; rather, they are fused and blended completely into a straight line. Tetanus continues until the muscle fatigues due to depletion of energy reserves. *Fatigue* is apparent when a muscle relaxes even though stimulation continues.

Tetanic contractions occur whenever skeletal muscles are actively used. Ordinarily, however, only a portion of any particular muscle is involved—while some fibers are contracting, others are relaxing. Because of this, intact muscles rarely fatigue completely.

> Muscle twitch, summation, and tetanus are related to the frequency with which a muscle is stimulated.

Health Focus

Exercise, Exercise, Exercise

Exercise programs improve muscular strength, muscular endurance, and flexibility. Muscular strength is the force a muscle group (or muscle) can exert against a resistance in one maximal effort. Muscular endurance is judged by the ability of a muscle to contract repeatedly or sustain a contraction for an extended period. Flexibility is tested by observing the range of motion about a joint.

Exercise also improves cardiorespiratory endurance. The heart rate and capacity increase, and the air passages dilate so that the heart and lungs are able to support prolonged muscular activity. The blood level of high-density lipoprotein (HDL), the molecule that prevents the development of plaque in blood vessels, increases. Also, body composition, the proportion of protein to fat, changes favorably when you exercise.

Exercise seems to also help prevent certain kinds of cancer. Cancer prevention involves eating properly, not smoking, avoiding cancer-causing chemicals and radiation, undergoing appropriate medical screening tests, and knowing the early warning signs of cancer. However, studies show that people who exercise are less likely to develop colon, breast, cervical, uterine, and ovarian cancers.

Physical training with weights can improve muscular strength and endurance in all adults regardless of age. Even men and women in their eighties and nineties make substantial gains in muscle strength, which can help them lead more independent lives. Exercise also helps prevent osteoporosis, a condition in which the bones are weak and tend to break. Exercise promotes the activity of osteoblasts in young people as well as older people. The stronger the bones when a person is young, the less chance of osteoporosis as a person ages. Exercise helps prevent weight gain by increasing the level of activity but also because muscles metabolize faster than other tissues. As a person becomes more muscular, it is less likely that fat will accumulate.

Exercise relieves depression and enhances the mood. Some report exercise actually makes them feel more energetic, and after exercising, particularly in the late afternoon, they sleep better that night. Self-esteem rises not only because of improved appearance but due to other factors that are not well understood. It is known that vigorous exercise releases endorphins, hormonelike chemicals that are known to alleviate pain and provide a feeling of tranquility.

A sensible exercise program is one that provides all the benefits without the detriments of a too strenuous program. Overexertion can actually be harmful to the body and might result in sports injuries such as a bad back or knees. The programs suggested in table 11A are tailored according to age and, if followed, are beneficial.

Dr. Arthur Leon at the University of Minnesota performed a study involving 12,000 men, and the results showed that only moderate exercise is needed to lower the risk of a heart attack by one-third. In another study conducted by the Institute for Aerobics Research in Dallas, Texas, which included 10,000 men and more than 3,000 women, even a little exercise was found to lower the risk of death from circulatory diseases and cancer. Increasing daily activity by walking to the corner store instead of driving and by taking the stairs instead of the elevator can improve your health.

Table 11A

A Checklist for Staying Fit

Children, 7–12	Teenagers, 13–18	Adults, 19–55	Seniors, 55 and up
Vigorous activity 1–2 hours daily	Vigorous activity 3–5 times a week	Vigorous activity for 1/2 hour, 3 times a week	Moderate exercise 3 times a week
Free play	Build muscle with calisthenics	Exercise to prevent lower back pain: aerobics, stretching, yoga	Plan a daily walk
Build motor skills through team sports, dance, swimming	Plan aerobic exercise to control buildup of fat cells	Take active vacations: hike, bicycle, cross-country ski	Daily stretching exercise
Encourage more exercise outside of physical education classes	Pursue tennis, swimming, horseback riding—sports that can be enjoyed for a lifetime	Find exercise partners: join a running club, bicycle club, outing group	Learn a new sport or activity: golf, fishing, ballroom dancing
Initiate family outings: bowling, boating, camping, hiking	Continue team sports, dancing, hiking, swimming		Try low-impact aerobics
			Before undertaking new exercises, consult your doctor

Exercise: Variety of Benefits

A regular exercise program, such as the one described in the Health Focus for this chapter, has many benefits. Increased endurance and strength of muscles are 2 possible benefits. Endurance is measured by the length of time the muscle can work before fatiguing, and strength is the force a muscle (or a group of muscles) can exert against a resistance.

A regular exercise program brings about physiological changes that build endurance, such as increased stores of ATP in the muscles and increased tolerance to lactate buildup. Muscle strength increases as muscle enlargement occurs due to exercise. When a muscle enlarges, the number of muscle fibers does not usually increase, but the protein content of the muscle does. This happens because the contractile elements in muscles—called myofibrils—which contain the protein filaments actin and myosin, increase in number.

Aside from improved endurance and strength, an exercise program also helps many other organs of the body. Cardiac muscle enlarges, and the heart can work harder than before. The resting heart rate decreases. Lung and diffusion capacity increase. Body fat decreases, but bone density increases so that breakage is less likely. Blood cholesterol and fat levels decrease, as does blood pressure.

Skeletal Muscles: Microscopic View

A whole skeletal muscle is composed of a number of *muscle fibers* in bundles (fig. 11.12).

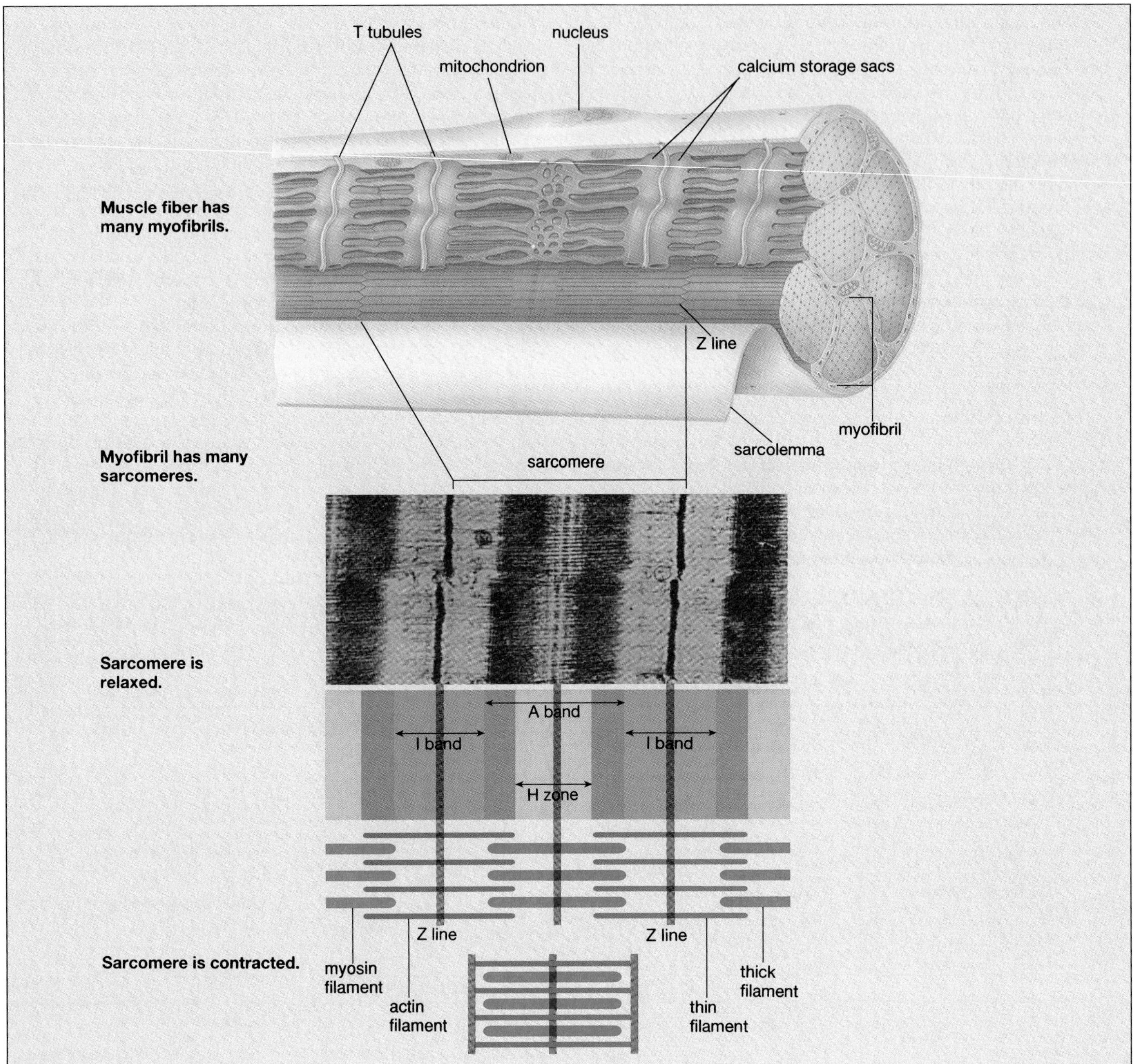

Figure 11.12

Skeletal muscle fiber structure and function. A muscle fiber contains many myofibrils divided into sarcomeres, which are contractile. A sarcomere contains actin (thin) filaments and myosin (thick) filaments. When a muscle fiber contracts, the thin filaments move toward the center so that the H zone gets smaller, to the point of disappearing.

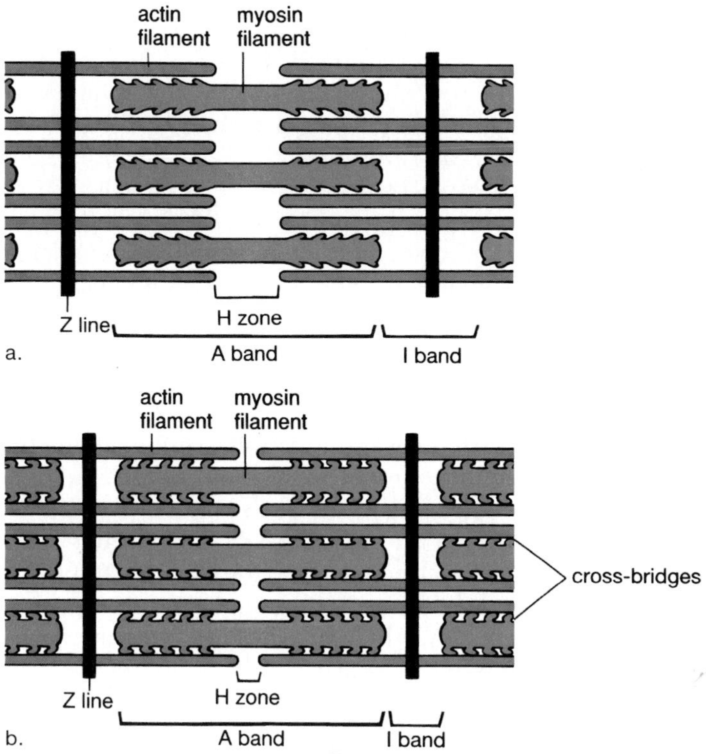

Figure 11.13

Sliding filament theory. **a.** Relaxed sarcomere. **b.** Contracted sarcomere. Note that during contraction, the I band and the H zone decrease in size. This indicates that the actin (thin) filaments slide past the myosin (thick) filaments. Even so, the myosin filaments do the work by pulling the actin filaments by means of cross-bridges.

Muscle Fiber: Structural Features

Each muscle fiber is a cell containing the usual cellular components but with special features. The cell membrane, called the **sarcolemma,** forms a *T* (for transverse) *system:* the *T tubules* penetrate, or dip down, into the cell so that they come into contact—but do not fuse—with expanded portions of the endoplasmic reticulum. These expanded portions, called calcium storage sacs, contain calcium ions (Ca^{++}), which are essential for muscle contraction. The endoplasmic reticulum encases hundreds and sometimes even thousands of **myofibrils,** which are the contractile portions of the fibers.

Myofibrils are cylindrical in shape and run the length of the muscle fiber. The light microscope shows that a myofibril has light and dark bands called *striations.* It is these bands that cause skeletal muscle to appear striated (see fig. 3.6*a*). The electron microscope shows that the striations of myofibrils are formed by the placement of protein filaments within contractile units called **sarcomeres.** A sarcomere contains 2 types of protein filaments. The thick filaments are made up of a protein called **myosin,** and the thin filaments are made up of a protein called **actin.** A sarcomere extends between 2 dark Z lines. The I band contains only actin filaments, and the H zone contains only myosin filaments.

Table 11.1

Muscle Contraction

Name	Function
Actin filaments	Slide past myosin, causing contraction
Ca^{++}	Needed for myosin to bind to actin
Myosin filaments	Pull actin filaments by means of cross-bridges; are enzymatic and split ATP
ATP	Supplies energy for muscle contraction

Physiology: Sliding Filaments

As a muscle fiber contracts, the sarcomeres within the myofibrils shorten. When a sarcomere shortens (fig. 11.13*b*), the actin (thin) filaments slide past the myosin (thick) filaments and approach one another. This causes the I band to shorten and the H zone to almost or completely disappear. The movement of actin filaments in relation to myosin filaments is called the **sliding filament theory** of muscle contraction. During the sliding process, the sarcomere shortens even though the filaments themselves remain the same length.

The participants in muscle contraction have the functions listed in table 11.1. Although it is the actin filaments that slide past the myosin filaments, it is the myosin filaments that do the work. In the presence of calcium ions (Ca^{++}), portions of a myosin filament called *cross-bridges* bend backward and attach to an actin filament (fig. 11.13). After attaching to the actin filament, the cross-bridges bend forward and the actin filament is pulled along. Now, ATP is broken down by myosin, and detachment occurs. Note, therefore, that myosin is not only a structural protein, it is also an ATPase enzyme. The cross-bridges attach and detach some 50 to 100 times as the thin filaments are pulled to the center of a sarcomere.

> The sliding filament theory states that actin filaments slide past myosin filaments because myosin has cross-bridges, which pull the actin filaments inward.

ATP provides the energy for muscle contraction to continue. To ensure a ready supply of ATP, muscle fibers contain **creatine phosphate** (phosphocreatine), a storage form of high-energy phosphate. Creatine phosphate does not directly participate in muscle contraction. Instead, it is used to regenerate ATP by the following reaction:

$$\text{creatine} \sim \text{P} + \text{ADP} \longrightarrow \text{ATP} + \text{creatine}$$

Oxygen Debt

When all of the creatine phosphate is depleted and no oxygen (O_2) is available for aerobic respiration, a muscle fiber can generate ATP using fermentation, an anaerobic process (p. 42). Fermentation, which is apt to occur during strenuous exercise, can supply ATP for only a short time because of lactate buildup. The buildup is noticeable when it produces muscle aches and fatigue upon exercising.

We all have had the experience of needing to continue deep breathing following strenuous exercise. This continued intake of oxygen is required to complete the metabolism of the lactate that has accumulated during exercise and represents an **oxygen debt,** which the body must pay to rid itself of lactate. The lactate is transported to the liver, where one-fifth of it is completely broken down to carbon dioxide (CO_2) and water (H_2O). The ATP gained by this respiration then is used to reconvert four-fifths of the lactate to glucose.

> Muscle contraction requires a ready supply of ATP. Creatine phosphate is used to generate ATP rapidly. If oxygen is in limited supply, fermentation produces ATP but results in oxygen debt.

Innervation: Muscle Activation

Muscles are innervated; that is, nerve impulses cause muscles to contract. A motor axon branches to several muscle fibers and collectively, this is called a motor unit. Each branch ends in several axon bulbs, which contain synaptic vesicles with the neurotransmitter acetylcholine (ACh).

The region where an axon bulb lies in close proximity to the sarcolemma of a muscle fiber is called a neuromuscular junction. A **neuromuscular junction** (fig. 11.14) has the same components as a synapse: a presynaptic membrane, a synaptic cleft, and a postsynaptic membrane. Only in this case the postsynaptic membrane is a portion of the sarcolemma of a muscle fiber. The sarcolemma, you will recall, forms T (for transverse) tubules, which dip down into the muscle fiber almost touching the calcium (Ca^{++}) storage sacs of the endoplasmic reticulum. The endoplasmic reticulum encases the myofibrils, which contain actin and myosin filaments (see fig. 11.12).

Nerve impulses travel down an axon, and it branches until nerve impulses reach the axon bulbs. Then the synaptic vesicles merge with the presynaptic membrane and ACh is released into the synaptic cleft. When ACh diffuses to and binds with receptor sites on the sarcolemma, the sarcolemma is depolarized. The result is **muscle action potentials,** which spread over the sarcolemma and down the T system to the region of calcium storage sacs. When a muscle action potential reaches a sac, calcium ions are released. They diffuse into the sarcoplasm, where they attach to the thin filaments. The attachment of calcium ions (Ca^{++}) changes the structure of the thin filaments so that myosin binding sites are exposed. Myosin cross-bridges reach out and attach to these binding sites, pulling the actin filaments toward the center of a sarcomere. The breakdown of ATP by myosin causes the cross-bridges to detach and then they reattach to different binding sites. In this way, the actin filaments are pulled along toward the center of a sarcomere and past the myosin filaments.

The sequence by which muscles are innervated can be stated as follows:

- Nerve impulses travel down an axon and it branches to axon bulbs.
- Synaptic vesicles merge with the presynaptic membrane of a neuromuscular junction.
- ACh is released and depolarizes the postsynaptic membrane (the sarcolemma of a muscle fiber).
- Muscle action potentials travel down the T system of a muscle fiber.
- Calcium ions (Ca^{++}) are discharged from storage sacs and attach to actin filaments exposing myosin binding sites.
- Myosin cross-bridges attach to and pull the actin filaments past the myosin filaments in each sarcomere within the myofibrils of a muscle fiber.

> A neuromuscular junction functions like a synapse except muscle action potentials cause calcium ions (Ca^{++}) to be released from calcium storage sacs, and thereafter muscle contraction occurs.

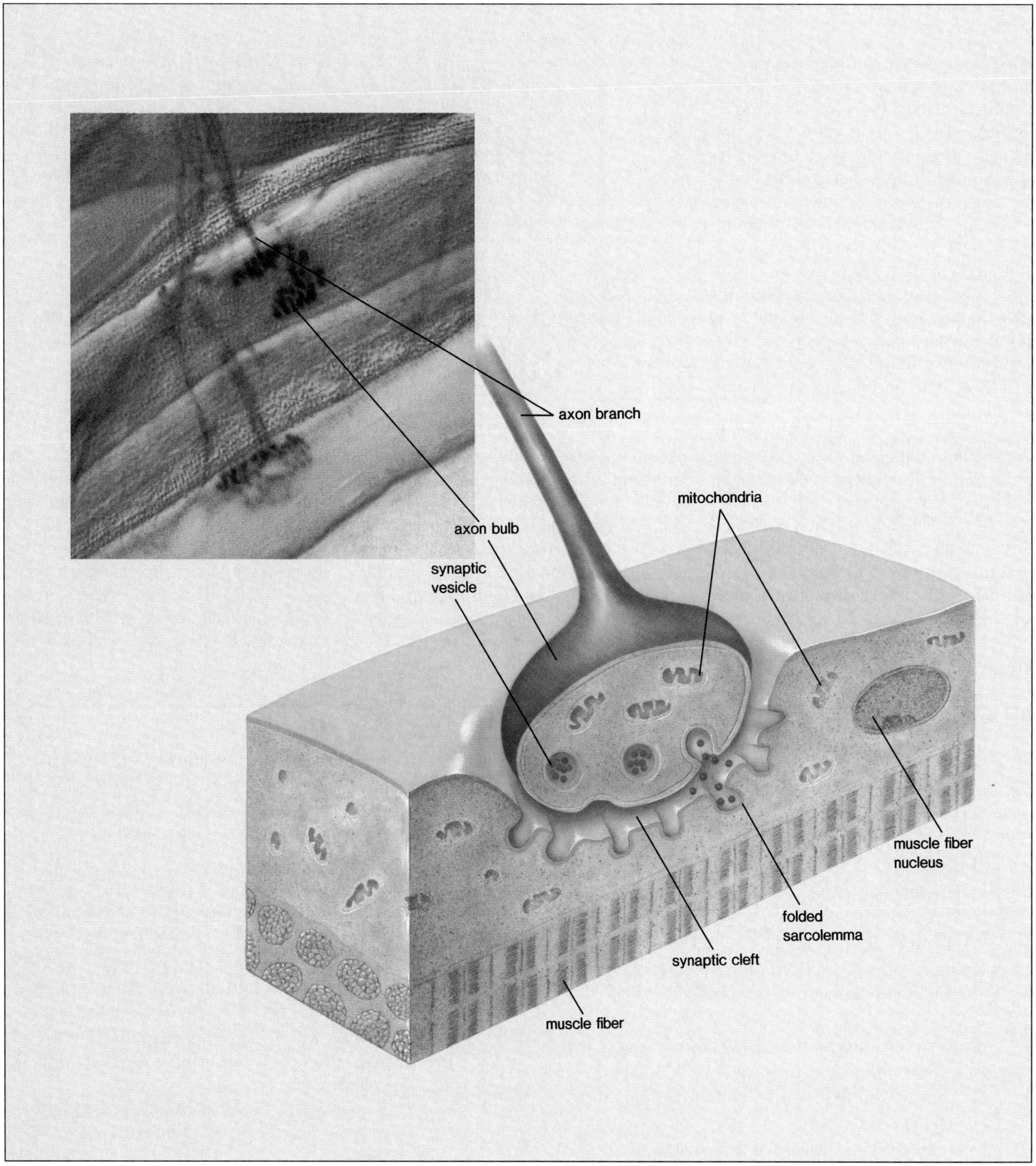

Figure 11.14

Neuromuscular junction. The branch of an axon terminating in an axon bulb approaches but does not touch a muscle fiber. A synaptic cleft separates the axon bulb from the sarcolemma of the muscle fiber. When nerve impulses travel down an axon and its branches, synaptic vesicles discharge a neurotransmitter, which diffuses across the synaptic cleft. A muscle action potential begins and is accompanied by muscle fiber contraction.

SUMMARY

Bones serve as deposits for inorganic salts, and some bones are sites for blood cell production. Bone is constantly being renewed: osteoclasts break down bone and osteoblasts build new bone. Osteocytes are found in the lacunae of Haversian systems. Bones are found in the skeleton, which aids movement while it also supports and protects the body. The skeleton is divided into 2 parts: (1) the axial skeleton, which is made up of the skull, the ribs, the sternum, and the vertebrae; and (2) the appendicular skeleton, which is composed of the girdles and their appendages. Joints are regions where bones are linked.

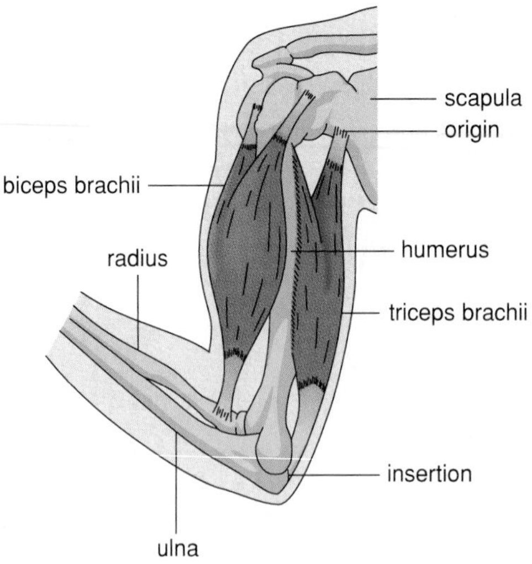

Whole skeletal muscles can only shorten when they contract; therefore, for a bone to be returned to its original position or the muscle to its original length, muscles must work in antagonistic pairs. For example, biceps brachii contraction raises the forearm and triceps brachii contraction lowers the forearm.

Muscle fibers obey the all-or-none law; it is possible to study a single contraction (muscle twitch) and sustained contraction (summation and tetanus) using a physiograph.

Muscle fibers are cells that contain myofibrils in addition to the usual cellular components. Longitudinally, myofibrils are divided into sarcomeres, where it is possible to note the arrangement of actin filaments and myosin filaments. When a sarcomere contracts, the actin filaments slide past the myosin filaments and the H zone all but disappears. Myosin has cross-bridges, which attach to and pull the actin filaments along. ATP breakdown by myosin is necessary for detachment to occur.

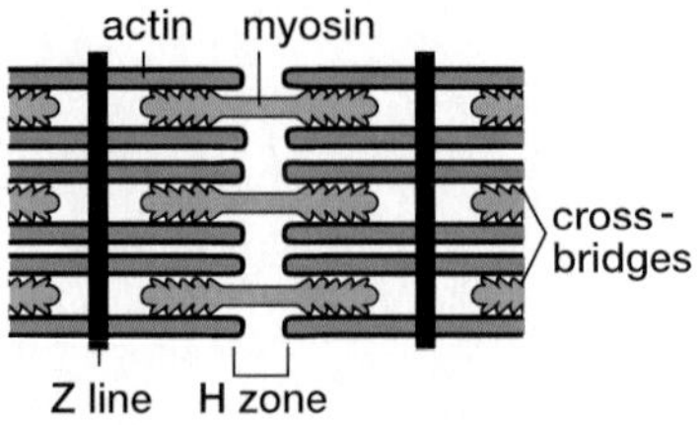

Innervation of a muscle fiber begins at a neuromuscular junction. Here, synaptic vesicles release ACh into the synaptic cleft. When the sarcolemma receives ACh, a muscle action potential moves down the T system to calcium storage sacs. When calcium ions (Ca^{++}) are released, contraction occurs. When calcium ions are actively transported back into the storage sacs, muscle relaxation occurs.

STUDY AND THOUGHT QUESTIONS

1. Describe the anatomy of a long bone. (pp. 208–209)
2. Distinguish between the axial and appendicular skeletons. (pp. 209–213)
3. List the bones that form the pectoral and pelvic girdles. (pp. 212–213)

   Critical Thinking The female pelvis is wider than the male pelvis. With what might this be associated?
4. How are joints classified? Describe the anatomy of a freely movable joint. (pp. 213–214)

   Critical Thinking Why would you expect persons with stronger muscles to have stronger bones?
5. Describe how muscles are attached to bones. What is accomplished by muscles acting in antagonistic pairs? (p. 216)
6. Describe the significance of threshold and maximum stimuli, muscle twitch, summation, and tetanic contraction. (p. 218)
7. Discuss the microscopic structural features of a muscle fiber and a sarcomere. What is the sliding filament theory? (pp. 220–221)
8. What is the role of creatine phosphate? (p. 221)

   Ethical Issue Do we in the United States place too much emphasis on physical prowess? Should more time and money be devoted to artistic and academic endeavors? Why or why not?
9. Discuss the availability and the specific role of ATP during muscle contraction. (p. 221) What is oxygen debt, and how is it repaid? (p. 222)
10. What causes a muscle action potential? How does the muscle action potential bring about sarcomere and muscle fiber contraction? (p. 222)

    Critical Thinking When we exercise, blood brings more oxygen to the muscles. What do muscles do with this oxygen?

Writing Across the Curriculum

1. Do nerves and muscles use ATP similarly? In general, what can you say about the use of ATP?
2. What is the value of having a jointed skeleton? Demonstrate this by discussing the different types of joints.
3. Muscles and bones vary in bulk. Relate the amount of bulk in the bones and muscles of the appendages to their role(s).

Objective Questions

1. The skull, the ribs, and the sternum are all in the ______________________ skeleton.
2. The vertebral column protects the ______________ cord.
3. The 2 bones of the lower arm are the ____________ and the _____________________.
4. Most joints are freely movable __________________ joints in which the 2 bones are separated by a cavity.
5. Muscles work in ____________________ pairs; the biceps brachii flexes and the triceps brachii extends the lower arm.
6. Maximal sustained contraction of a muscle is called ______________________.
7. In a muscle fiber, actin filaments and myosin filaments are found within ___________________, which are subdivided into units called ______________________.
8. The molecule ______________________ serves as an immediate source of high-energy phosphate for ATP production in muscle cells.
9. The juncture between axon ending and muscle cell sarcolemma is called a _________________ junction.
10. A muscle action potential causes ____________ ions to be released from storage sacs, which signals the muscle fiber to contract.
11. Label this diagram of a muscle fiber, using these terms: myofibril, mitochondrion, T tubules, sarcomere, sarcolemma, endoplasmic reticulum.

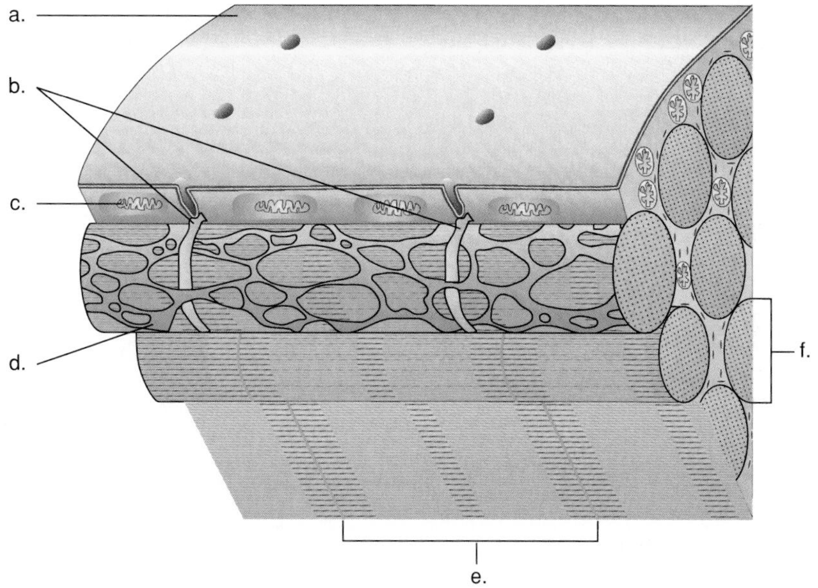

SELECTED KEY TERMS

actin One of 2 major proteins of muscle; makes up thin filaments in myofibrils of muscle fibers. *See* myosin. 221

appendicular skeleton Portion of the skeleton forming the upper extremities, the pectoral girdle, the lower extremities, and the pelvic girdle. 212

axial skeleton Portion of the skeleton that supports and protects the organs of the head, the neck, and the trunk. 209

creatine phosphate (kre´ ah-tin fos´ fāt) Compound unique to muscles that contains a high-energy phosphate bond. 221

insertion The end of a muscle that is attached to a movable bone. 217

muscle action potential An electrochemical change due to increased sarcolemma permeability that is propagated down the T system and results in muscle contraction. 222

myofibril (mi″ o-fi´ bril) The contractile portion of muscle fibers. 221

myosin One of 2 major proteins of muscle; makes up thick filaments in myofibrils and is capable of breaking down ATP. *See* actin. 221

neuromuscular junction The point of contact between a nerve cell and a muscle fiber. 222

origin End of a muscle that is attached to a relatively immovable bone. 216

osteocyte (os´te-o-sīt) A mature bone cell. 209

oxygen debt Oxygen that is needed to metabolize lactate, a compound that accumulates during vigorous exercise. 222

sarcomere (sar″ ko-mir) Structural and functional unit of a myofibril; contains actin and myosin filaments. 221

synovial joint (si-nó ve-al joint) A freely movable joint. 213

tetanus Sustained muscle contraction without relaxation. 218

Chapter 12

Senses

We view the world through our sense organs, and we can be fooled. The sense organs send information to be processed by the brain, and its interpretation is not always accurate. The boy and the man in the accompanying photo seem to be equally distant; therefore, our brain tells us that the boy is larger than the man. Actually the man's corner is farther away and has a higher ceiling than the boy's corner in this room,[1] which is specially prepared to fool the mind.

Chapter Concepts

1. Receptors are sensitive to environmental stimuli. Each receptor is especially sensitive to one type of stimulus, but they all initiate nerve impulses. 228
2. The skin contains receptors for touch, pressure, pain, and temperature (hot and cold). 228
3. The special sense organs (taste buds, nose, eyes, and ears) contain groupings of specialized receptors. 229
4. The microvilli of taste cells have receptors for the molecules in food. Certain receptors are responsible for bitter, sour, salty, or sweet tastes. 229
5. The receptors for sight contain visual pigments, which respond to light rays. 234
6. The receptors for hearing are hair cells, which respond to pressure waves. 243

Chapter Outline

[1] See diagram in appendix E, p. 490.

Receptors, present in sense organs, receive external and internal stimuli. Each type of receptor is especially sensitive to only one type of stimulus. They monitor changes in our external and internal environments. Receptors are the first component of a reflex arc, as described in chapter 10. When a receptor is stimulated, it generates nerve impulses that are transmitted to the spinal cord and/or the brain, but we are conscious of a sensation only if the impulses reach the cerebrum. The sensory portion of the cerebrum can be mapped according to the parts of the body and types of sensation realized (see fig. 10.12).

> Receptors only initiate nerve impulses, which are conducted to the cord and/or brain. Sensation is dependent upon the cerebrum.

Skin

The dermis of skin contains receptors for touch, pressure, pain, and temperature (hot and cold) (fig. 12.1). It is a mosaic of these tiny receptors, as you can determine by slowly passing a metal probe over your skin. At certain points, there is a feeling of pressure, and at others, there is a feeling of hot or cold (depending on the temperature of the probe). Certain parts of the skin contain more receptors for a particular sensation; for example, the fingertips have an abundance of touch receptors.

Adaptation occurs when the receptor becomes so accustomed to stimulation that it stops generating impulses, even though the stimulus is still present. The touch receptors adapt: soon after we put on an article of clothing, we are no longer aware of the feel of clothes against our skin.

> Specialized receptors in the human skin respond to touch, pressure, pain, and temperature.

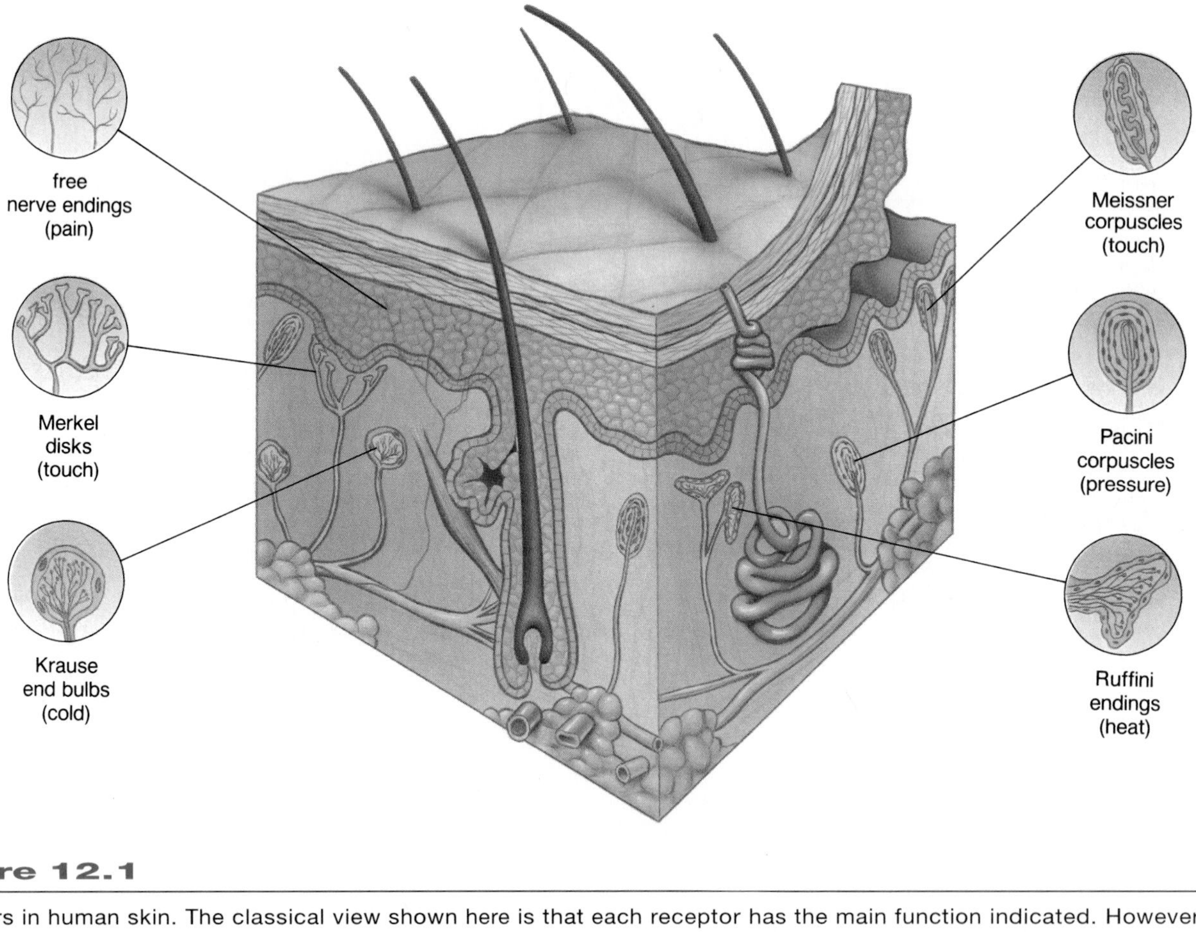

Figure 12.1

Receptors in human skin. The classical view shown here is that each receptor has the main function indicated. However, investigations in this century indicate that matters are not so clear-cut. For example, microscopic examination of the skin of the ear shows only free nerve endings (pain receptors), and yet the skin of the ear is sensitive to all sensations. Therefore, it appears that the receptors of the skin are somewhat but not completely specialized.

Chemoreceptors

Taste buds and the nose are *special sense organs.* Special sense organs contain groupings of specialized receptors (table 12.1). Taste and smell are called the *chemical senses* because the receptors for these senses are sensitive to chemical substances in the food we eat and the air we breathe. Therefore, they are called chemoreceptors.

Taste Buds: For Tasting

Taste buds are located primarily on the tongue (fig. 12.2). Many lie along the walls of the papillae, the small elevations on the tongue that are visible to the naked eye. Isolated ones are also present on the hard palate, the pharynx, and the epiglottis.

Taste buds are pockets of cells that extend through the tongue epithelium and open at a taste pore. Taste buds have supporting cells and a number of elongated taste cells that end in microvilli. Taste cells, which have associated sensory nerve fibers, are sensitive to certain chemicals. Nerve impulses most probably are generated when molecules bind to receptor sites found on the microvilli.

Table 12.1

Special Sense Organs

Sense Organ	Type of Receptor	Specific Receptor	Senses
Taste buds	Chemoreceptor	Taste cells	Taste
Nose	Chemoreceptor	Olfactory cells	Smell
Eye	Photoreceptor	Rods and cones in retina	Vision
Ear	Mechanoreceptor	Hair cells in utricle, saccule, and semicircular canals	Equilibrium
		Hair cells in organ of Corti	Hearing

It is believed that there are 4 types of tastes (bitter, sour, salty, sweet) and that taste buds for each are concentrated on the tongue in particular regions (fig. 12.2*a*). Sweet receptors are most plentiful near the tip of the tongue. Sour receptors occur primarily along the margins of the tongue. Salt receptors are most common on the tip and the upper front portion of the tongue. Bitter receptors are located toward the back of the tongue.

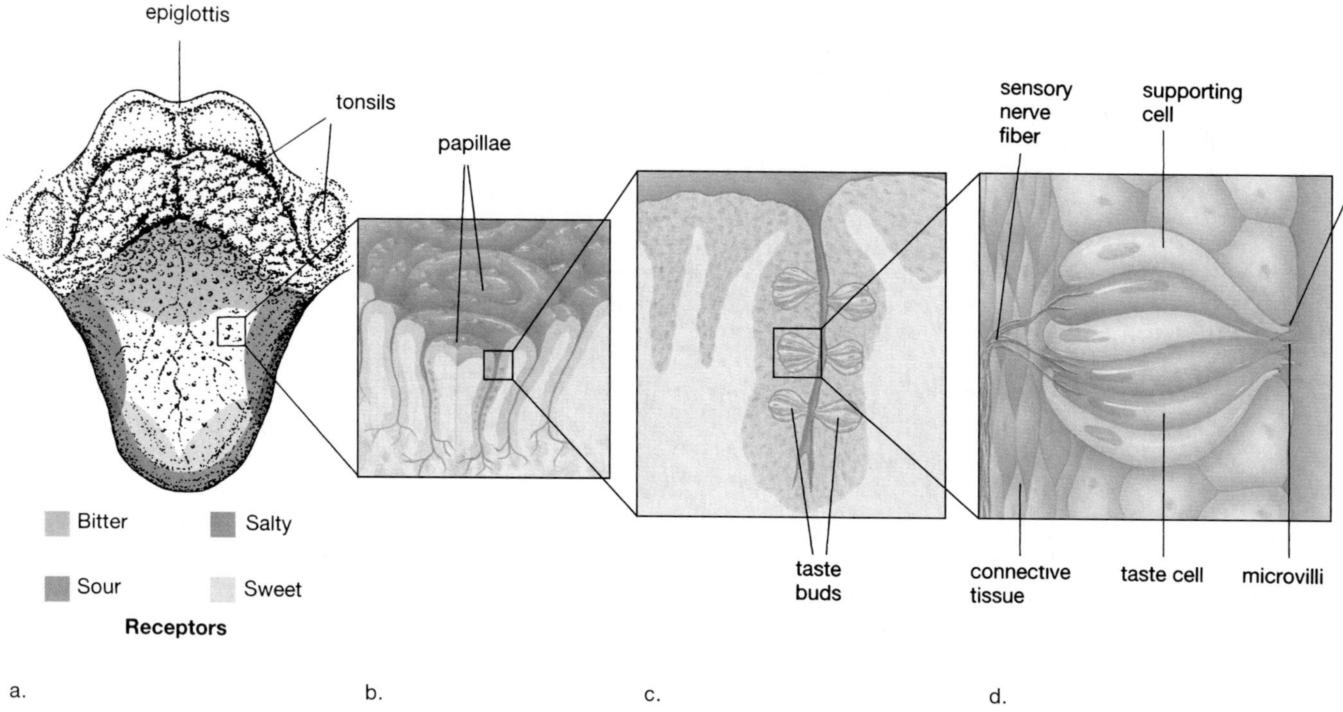

Figure 12.2

Taste buds. **a.** Elevations on the tongue are called papillae. The location of those containing taste buds responsive to sweet, sour, salt, and bitter is indicated. **b.** Enlargement of papillae. **c.** The taste buds occur along the walls of the papillae. **d.** Drawing shows the various cells that make up a taste bud. Taste cells in a bud end in microvilli that are sensitive to the chemicals exhibiting the tastes noted in **a.** When the chemicals combine with membrane-bound receptors, nerve impulses are generated.

The Nose: For Smelling

Our sense of smell is dependent on olfactory cells located high in the roof of the nasal cavity (fig. 12.3). Each cell ends in a tuft of about 5 cilia, which bear receptor sites for various chemicals. Research resulting in the stereochemical theory of smell suggests that different types of smell are related to the various shapes of molecules. When molecules combine with the receptor sites, nerve impulses are generated in the olfactory nerve fibers. Within the olfactory bulbs, the paired masses of gray matter beneath the frontal lobes of the cerebrum, an olfactory tract takes this sensory information to an olfactory area of the cerebrum.

The olfactory receptors, like touch and temperature receptors, adapt to outside stimuli. In other words, after a while, the presence of a particular chemical no longer causes the olfactory cells to generate nerve impulses, and we are no longer aware of a particular smell.

The sense of taste and the sense of smell supplement each other, creating a combined effect when interpreted by the cerebrum. For example, when you have a cold, you think that food has lost its taste, but actually you have lost the ability to sense its smell. This may work in reverse also. When you smell something, some of the molecules move from the nose down into the mouth region and stimulate the taste buds there. Therefore, part of what we refer to as smell may actually be taste.

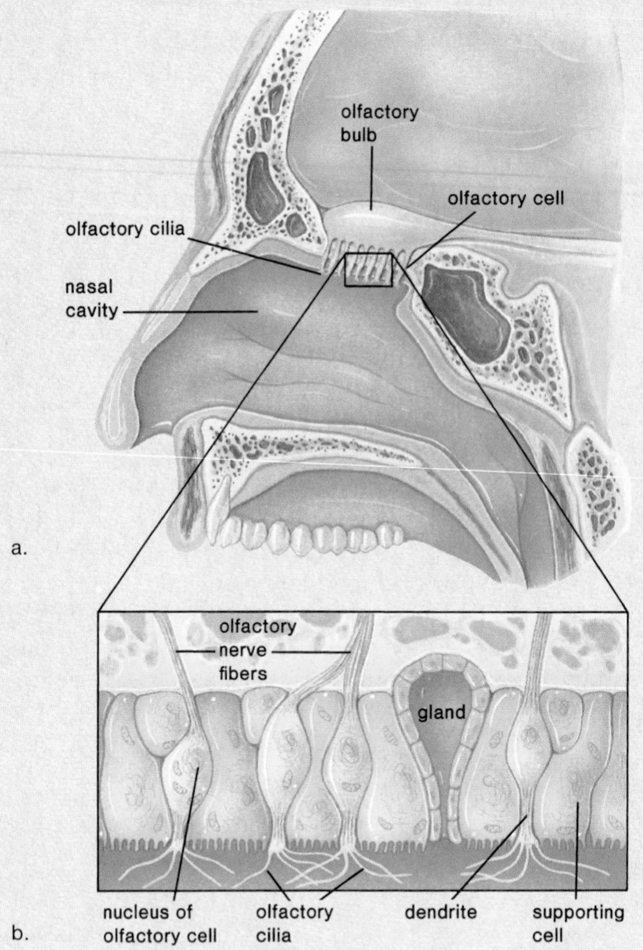

Figure 12.3

Olfactory cell location and anatomy. **a.** The olfactory area in humans is located high in the nasal cavity. **b.** Enlargement of the olfactory cells shows they are modified neurons located between supporting cells. When olfactory cells are stimulated by chemicals, olfactory nerve fibers conduct nerve impulses to the olfactory bulb. An olfactory tract within the bulb takes the nerve impulses to the brain.

> The receptors for taste (taste cells) and the receptors for smell (olfactory cells) work together to give us our sense of taste and our sense of smell.

Eye

The anatomy and physiology of the eye is complex, and we will consider each topic separately.

How the Eye Looks

The eyeball, an elongated sphere about 2.5 cm in diameter, has 3 layers, or coats: the sclera, the choroid, and the retina (fig. 12.4 and table 12.2). The outer **sclera** is a white, fibrous, protective layer except for the transparent **cornea,** the window of the eye. The middle, thin, dark-brown layer, the **choroid,** contains many blood vessels and absorbs stray light rays. Toward the front, the choroid thickens and forms the ring-shaped ciliary body. The **ciliary body** contains the *ciliary muscle,* which controls the shape of the lens for near and far vision. Finally, the choroid becomes a thin, circular, muscular, and pigmented diaphragm, the **iris,** which regulates the size of the **pupil,** a hole in the center through which light enters the eyeball. The **lens,** attached to the ciliary body by ligaments, divides the cavity of the eye into 2 smaller cavities. A viscous, gelatinous material, the **vitreous humor,** fills the posterior cavity behind the lens. The anterior cavity between the cornea and the lens is filled with an alkaline, watery solution secreted by the ciliary body and called the **aqueous humor.**

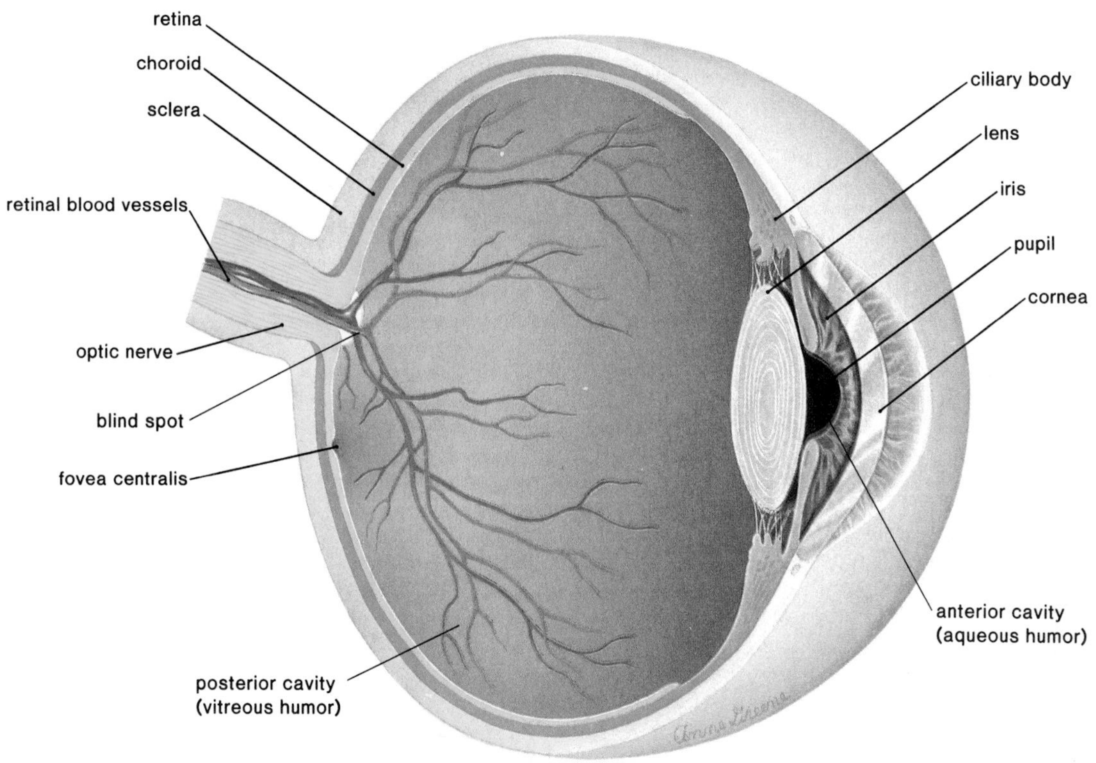

Figure 12.4

Anatomy of the human eye. Notice that the sclera becomes the cornea and the choroid becomes the ciliary body and the iris. The retina contains the receptors for vision; the fovea centralis is the region where vision is most acute. A blind spot occurs where the optic nerve leaves the retina. There are no receptors for light in this location.

Table 12.2

Function of Parts of the Eye

Part	Function
Lens	Refracts and focuses light rays
Iris	Regulates light entrance
Pupil	Admits light
Choroid	Absorbs stray light
Sclera	Protects eyeball
Cornea	Refracts light rays
Humors	Refract light rays
Ciliary body	Holds lens in place, accommodation
Retina	Contains receptors for sight
Rods	Make black-and-white vision possible
Cones	Make color vision possible
Optic nerve	Transmits impulse to brain
Fovea centralis	Makes acute vision possible

A small amount of aqueous humor is continually produced each day. Normally, it leaves the anterior cavity by way of tiny ducts located where the iris meets the cornea. When a person has glaucoma, these drainage ducts are blocked, and aqueous humor builds up. If glaucoma is not treated, the resulting pressure compresses the arteries that serve the nerve fibers of the retina, where the receptors for sight are located. The nerve fibers begin to die due to lack of nutrients, and the person becomes partially blind. Over time, total blindness can result.

Retina: Three Layers of Cells

The inner layer of the eye, the **retina,** has 3 layers of cells (fig. 12.5). The layer closest to the choroid contains the sense receptors for vision, the **rods** and the **cones;** the middle layer contains bipolar cells; and the innermost layer contains ganglionic cells, whose axons become the **optic nerve.** Only the rods and the cones contain light-sensitive pigments, and therefore light must penetrate to the back of the retina before nerve impulses are generated.

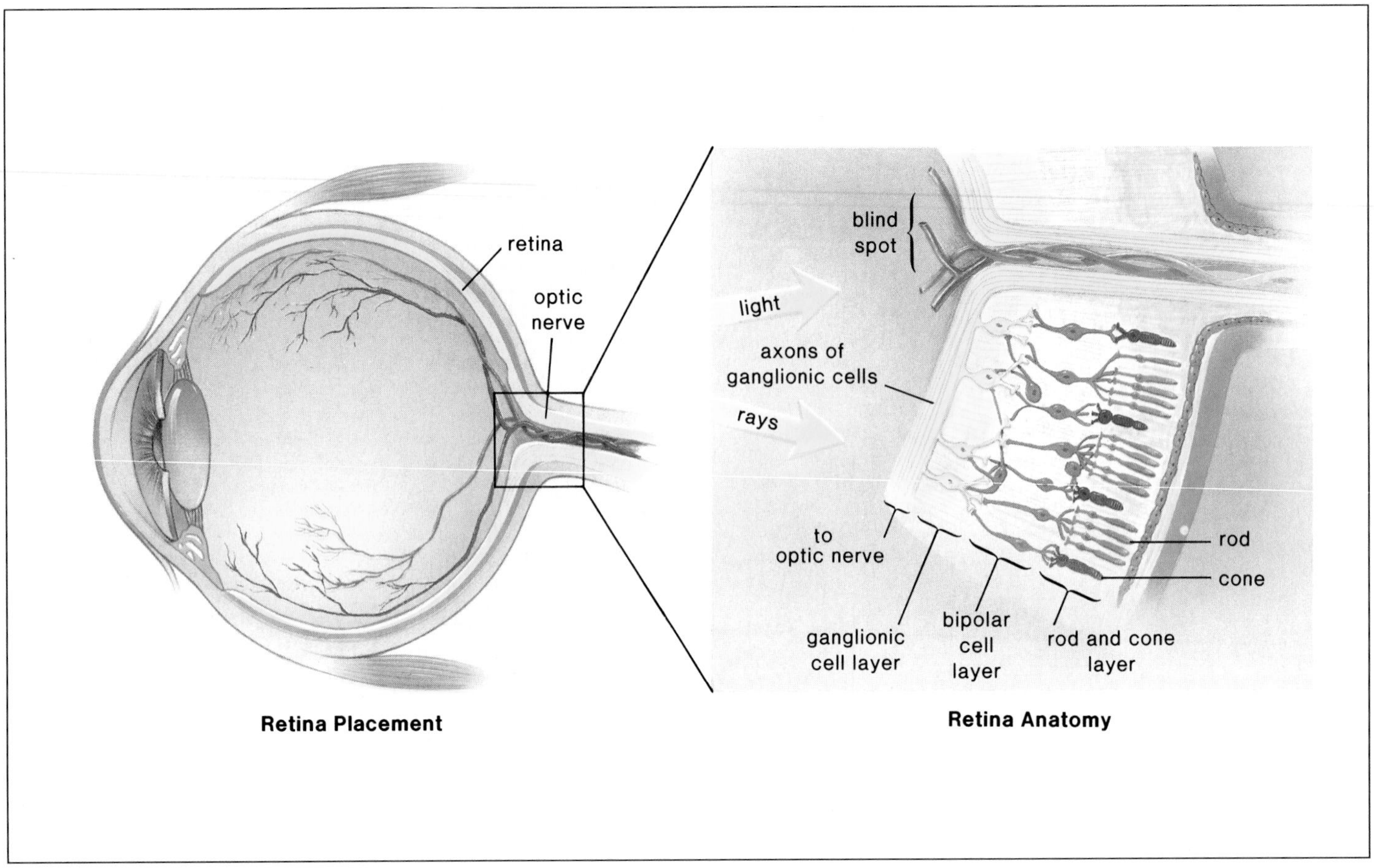

Figure 12.5

Anatomy of the retina. The retina is the inner layer of the eye. Rods and cones are located at the back of the retina, followed by the bipolar cells and the ganglionic cells, whose fibers become the optic nerve. (Notice that rods share bipolar cells but cones do not. Cones, therefore, distinguish more detail.) The optic nerve carries impulses to the occipital lobe of the cerebrum.

Nerve impulses initiated by the rods and the cones are passed to the bipolar cells, which in turn pass them to the ganglionic cells. The axons of the ganglionic cells pass in front of the retina, forming the optic nerve, which turns to pierce the layers of the eye. Notice in figure 12.5 that there are many more rods and cones than ganglionic cells. In fact, the retina has as many as 150 million rods but only one million ganglionic cells and optic nerve fibers. This means that there is considerable mixing of messages and a certain amount of integration before nerve impulses are sent to the thalamus and then on to the occipital lobe of the cerebrum. There are no rods or cones where the optic nerve passes through the retina; therefore, this is a **blind spot,** where vision is impossible.

The retina contains a very special region called the **fovea centralis** (see fig. 12.4), an oval, yellowish area with a depression where there are only cones. Vision is most acute in the fovea centralis.

> The eye has 3 layers: the outer sclera, the middle choroid, and the inner retina. Only the retina contains receptors for light energy.

Focusing: Bending Light

When we look at an object, light rays are **focused** on the retina (fig. 12.6*a*). In this way, an *image* of the object appears on the retina. The image on the retina occurs when

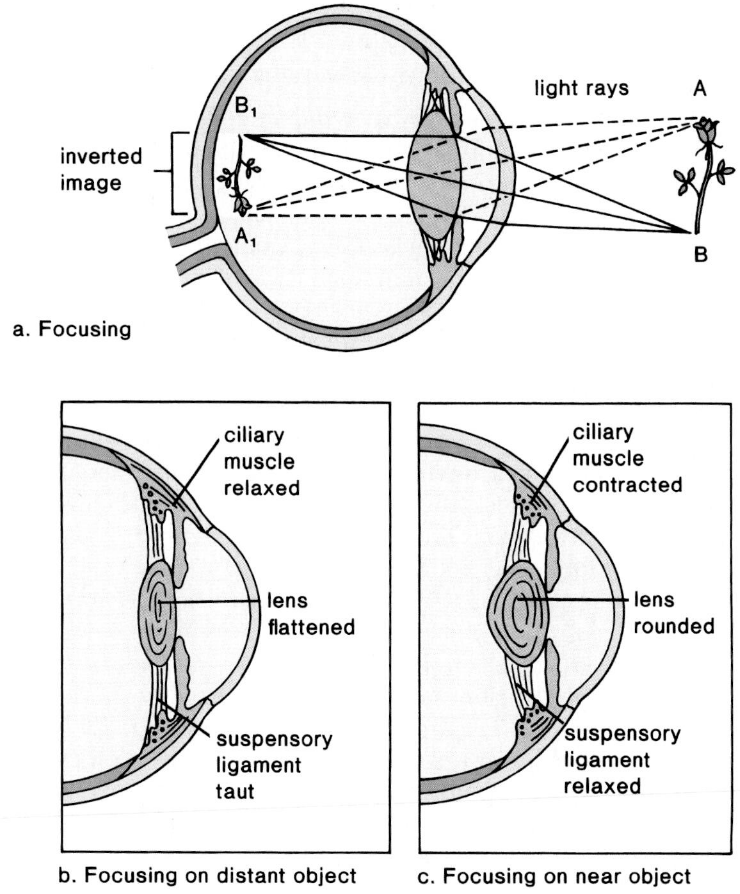

Figure 12.6

Focusing. **a.** Light rays from each point on an object are bent by the cornea and the lens in such a way that they are directed to a single point after emerging from the lens. By this process, an inverted image of the object forms on the retina. **b.** When focusing on a distant object, the lens is flat because the ciliary muscle is relaxed and the suspensory ligament is taut. **c.** When focusing on a near object, the lens accommodates: it rounds up because the ciliary muscle contracts, causing the suspensory ligament to relax.

the rods and the cones in a particular region are excited. Obviously, the image is much smaller than the object. To produce this small image, light rays must be bent (refracted) and brought into focus. They are initially bent as they pass through the cornea, and further bending occurs as the rays pass through the lens and the humors.

Light rays are reflected from an object in all directions. For distant objects, only nearly parallel rays enter the eye, and the cornea alone is needed for focusing. For close objects, many of the rays are at sharp angles to one another, and additional focusing is required. The lens provides this additional focusing power as **accommodation** occurs: the lens remains flat when we view distant objects, but it rounds up when we view near objects.

The shape of the lens is controlled by the ciliary muscle within the ciliary body. When we view a distant object, the ciliary muscle is relaxed, causing the suspensory ligaments attached to the ciliary body to be taut; therefore, the lens remains relatively flat (fig. 12.6*b*). When we view a near object, the ciliary muscle contracts, releasing the tension on the suspensory ligaments, and the lens rounds up due to its natural elasticity (fig. 12.6*c*). Because close work requires contraction of the ciliary muscle, it very often causes eyestrain.

Usually after the age of 40, the lens loses some of its elasticity and is unable to accommodate. Corrective lenses are then usually necessary, as is discussed on page 236. Also with aging or possibly exposure to the sun (see Health

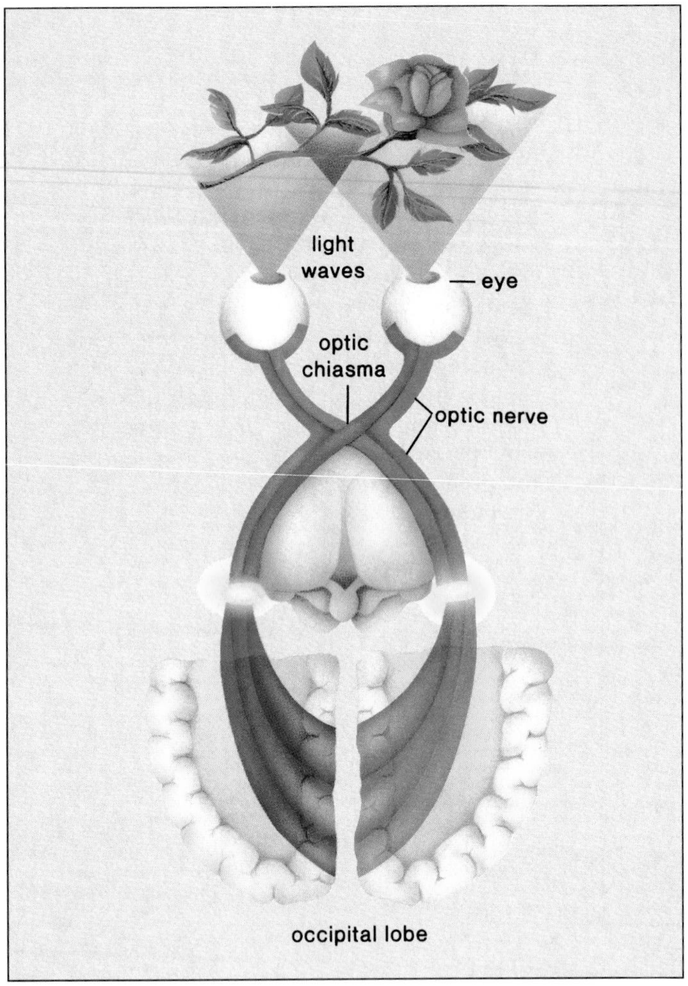

Figure 12.7

Optic chiasma. Both eyes "see" the entire object, but data from the right half of each retina go to the right occipital lobe of the cerebrum, and data from the left half of the retina go to the left occipital lobe because of the optic chiasma. When the data are combined, the brain "sees" the entire object in depth.

Focus), the lens is subject to *cataracts;* the lens can become opaque and therefore incapable of transmitting rays of light. Recent research suggests that cataracts develop when crystallin proteins contained by special cells within the interior oxidize, causing the three-dimensional shape to change. If so, researchers believe that eventually they may be able to find ways to restore the normal configuration of crystallin so that cataracts can be treated medically instead of surgically.

For the present, however, surgery is the only viable treatment for cataracts. First, a surgeon opens the eye near the rim of the cornea. Zonulysin, an enzyme, may be used to digest away the ligaments holding the lens in place. Most surgeons then use a cryoprobe, which freezes the lens for easy removal. An intraocular lens attached to the iris can then be implanted so that the patient does not need to wear thick glasses or contact lenses.

> The lens, assisted by the cornea and the humors, focuses images on the retina.

The Upside Down Image

The image formed on the retina is inverted (fig. 12.6*a*), and it is thought that perhaps this image is righted in the brain by experience. In one experiment, scientists wore glasses that inverted the field of vision. At first, they had difficulty adjusting to the placement of the objects, but they soon became accustomed to their inverted world. Experiments such as these suggest that if we see the world upside down, the brain learns to see it right side up.

Stereoscopic Vision

We can see well with either eye alone, but the 2 eyes functioning together provide us with **stereoscopic vision.** Normally, the 2 eyes are directed by the eye muscles toward the same object, and therefore the object is focused on corresponding points of the 2 retinas. Each eye, however, sends its own data to the brain about the placement of the object because each forms an image from a slightly different angle. These data are pooled to produce depth perception by a 2-step process. First, because the optic nerves cross at the optic chiasma (fig. 12.7), one-half of the brain receives information from both eyes about the same part of an object. Later, the 2 halves of the brain communicate to arrive at a three-dimensional interpretation of the whole object.

> The anatomy and the physiology of the brain allow us to see the world right side up and in 3 dimensions.

Seeing: Is Chemical

In *dim light,* the iris causes the pupils to enlarge so that more rays of light can enter the eyes. The 150 million rods located in the periphery, or sides, of the eyes are sensitive enough to be stimulated by this faint light. The rods do not detect fine detail or color, so at night, for example, all objects appear to be blurred and a shade of gray. Rods do detect even the slightest motion, however, because of their abundance and position in the eyes.

The rods contain **rhodopsin,** a molecule that contains the protein opsin and the pigment retinal (fig. 12.8). When light strikes rhodopsin, it breaks down to its components and this generates nerve impulses. The more rhodopsin present in the rods, the more sensitive the eyes are to dim light. Therefore, during the time required for adjustment to dim light, when it is difficult to see, rhodopsin is being formed in the rods. Retinal is a derivative of vitamin A, which is abundant in carrots, so the suggestion that eating carrots helps vision is not without foundation.

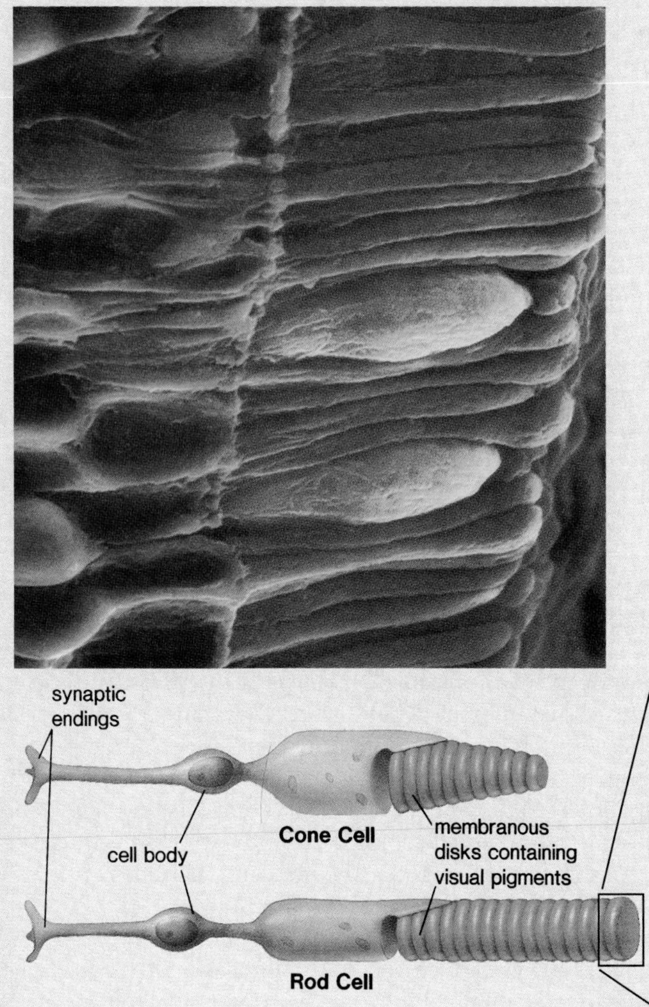

Figure 12.8

Structure and function of rods and cones. The outer segment of rods and cones is composed of stacks of membranous disks, which contain visual pigments. In rods, the membrane of each disk contains rhodopsin, a complex molecule containing the protein opsin and the pigment retinal. When retinal absorbs energy, it changes shape and disengages from opsin. This leads to nerve impulses in bipolar cells (see fig. 12.5).

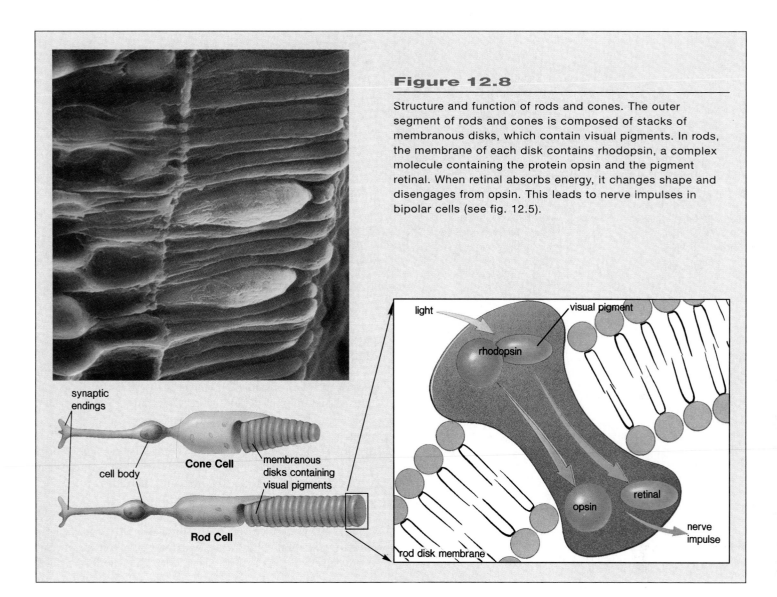

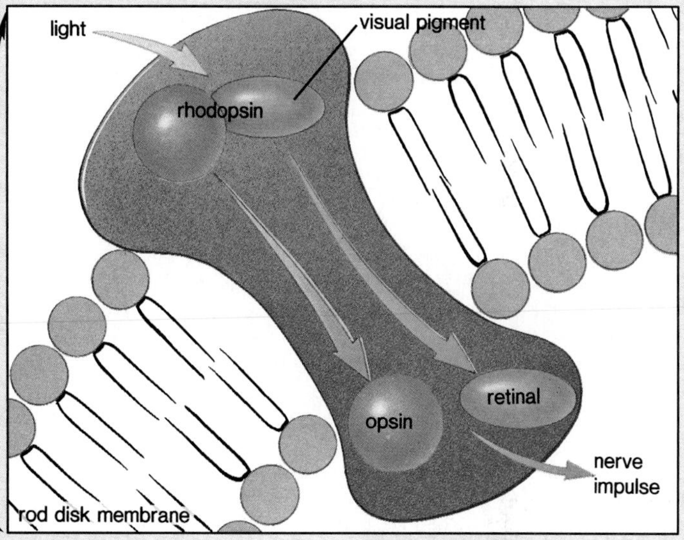

The cones, located primarily in the fovea centralis, detect the fine detail and color of an object. To perceive depth, as well as to see color, we turn our eyes so that reflected light from the object strikes the fovea centralis. Color vision has been shown to depend on the 3 kinds of cones that contain pigments sensitive to blue, green, or red light. The nerve impulses generated from one type of cone not only stimulate certain cells in the visual cortex of the brain, they also inhibit the reception of impulses from other types of cones. For example, when we see red, certain cells in the brain are prohibited from receiving impulses from green cones. Similarly, impulses sent through blue cones tend to oppose the combination of signals sent by red and green cones—which together produce yellow. This process assists integration and enables the brain to tell the location of various colors in the environment.

Complete color blindness is extremely rare. In most instances, a particular type of cone is lacking or deficient in number. The lack of red or green cones is the most common, affecting about 5%–8% of human males. If the eye lacks red cones, the green colors are accentuated, and vice versa (fig. 12.9).

The sense receptors for sight are the rods and the cones. The rods are responsible for vision in dim light, and the cones are responsible for vision in bright light and for color vision. When either is stimulated, nerve impulses are transmitted in the optic nerve to the brain.

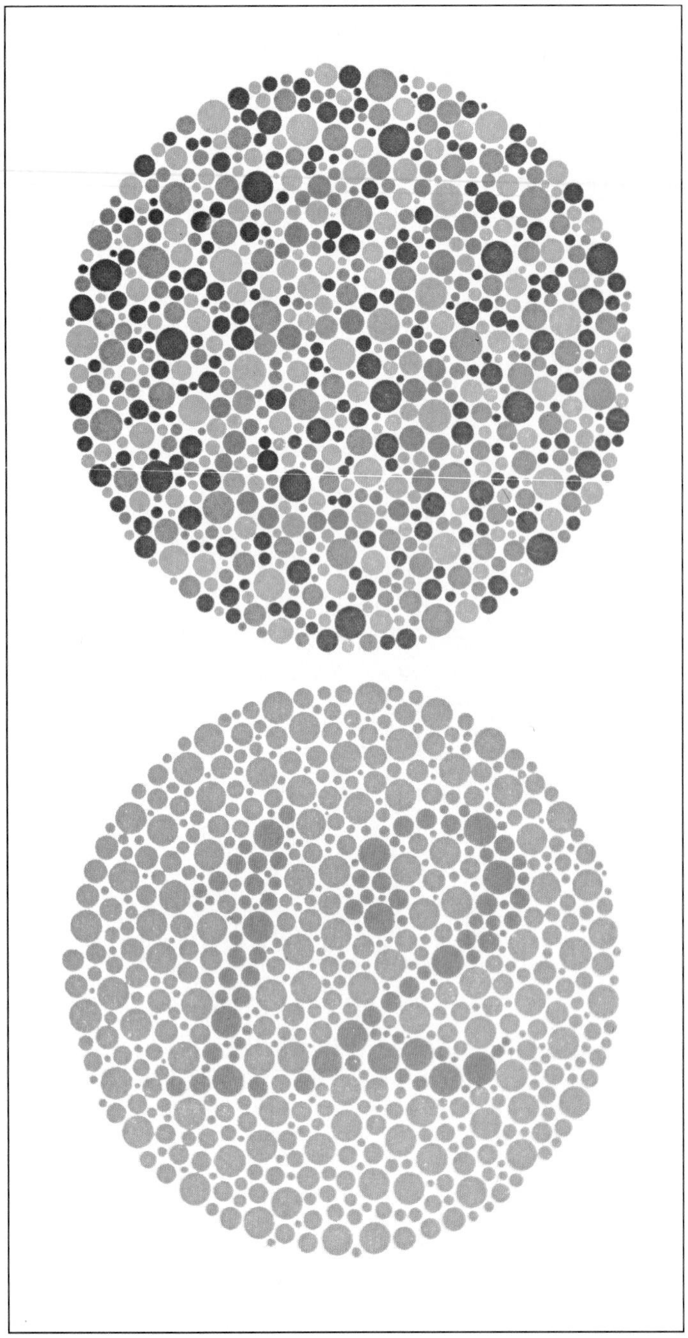

Figure 12.9

Test plates for color blindness. When looking at the top plate, a person with normal color vision sees the number 8, and when looking at the bottom plate, a person with normal color vision sees the number 12. The most common form of color blindness involves an inability to distinguish reds from greens.

These plates are reproduced from Ishihara's Tests for Colour Blindness published by Kanehara & Co., Ltd., Tokyo, Japan, but tests for color blindness cannot be conducted with this material. For accurate testing, the original plate should be used.

Corrective Lenses

The majority of people can see what is designated as a size 20 letter 20 feet away and so are said to have 20/20 vision. Persons who can see near objects but cannot see the letters from this distance are said to be *nearsighted.* Nearsighted people can see near objects better than they can see objects at a distance. These individuals often have an elongated eyeball, and when they attempt to look at a distant object, the image is brought to focus in front of the retina (fig. 12.10). They can see near objects because they can adjust the lens to allow the image to focus on the retina, but to see distant objects, these people must wear concave lenses, which diverge the light rays so that the image can be focused on the retina. There is a new treatment for nearsightedness called radial keratotomy, or radial K. From 4 to 8 cuts are made in the cornea so that they radiate out from the center like spokes in a wheel. When the cuts heal, the cornea is flattened. Although some patients are satisfied with the result, others complain of glare and varying visual acuity.

Persons who can easily see the optometrist's chart but cannot see near objects well are *farsighted;* these individuals can see distant objects better than they can see close objects. They often have a shortened eyeball, and when they try to see near objects, the image is focused behind the retina. When the object is distant, the lens can compensate for the short eyeball, but when the object is near, these persons must wear a convex lens to increase the bending of light rays so that the image can be focused on the retina.

When the cornea or lens is uneven, the image is fuzzy. The light rays cannot be evenly focused on the retina. This condition, called *astigmatism,* can be corrected by an unevenly ground lens to compensate for the uneven cornea.

As We Age

With normal aging, the lens loses some of its ability to change shape and focus on near objects. Because nearsighted individuals still have difficulty seeing objects clearly in the distance, they must wear bifocals, which means that the upper part of the lens is for distant vision and the remainder is for near vision.

The shape of the eyeball determines the need for corrective lenses; the inability of the lens to accommodate as we age also requires corrective lenses for near vision.

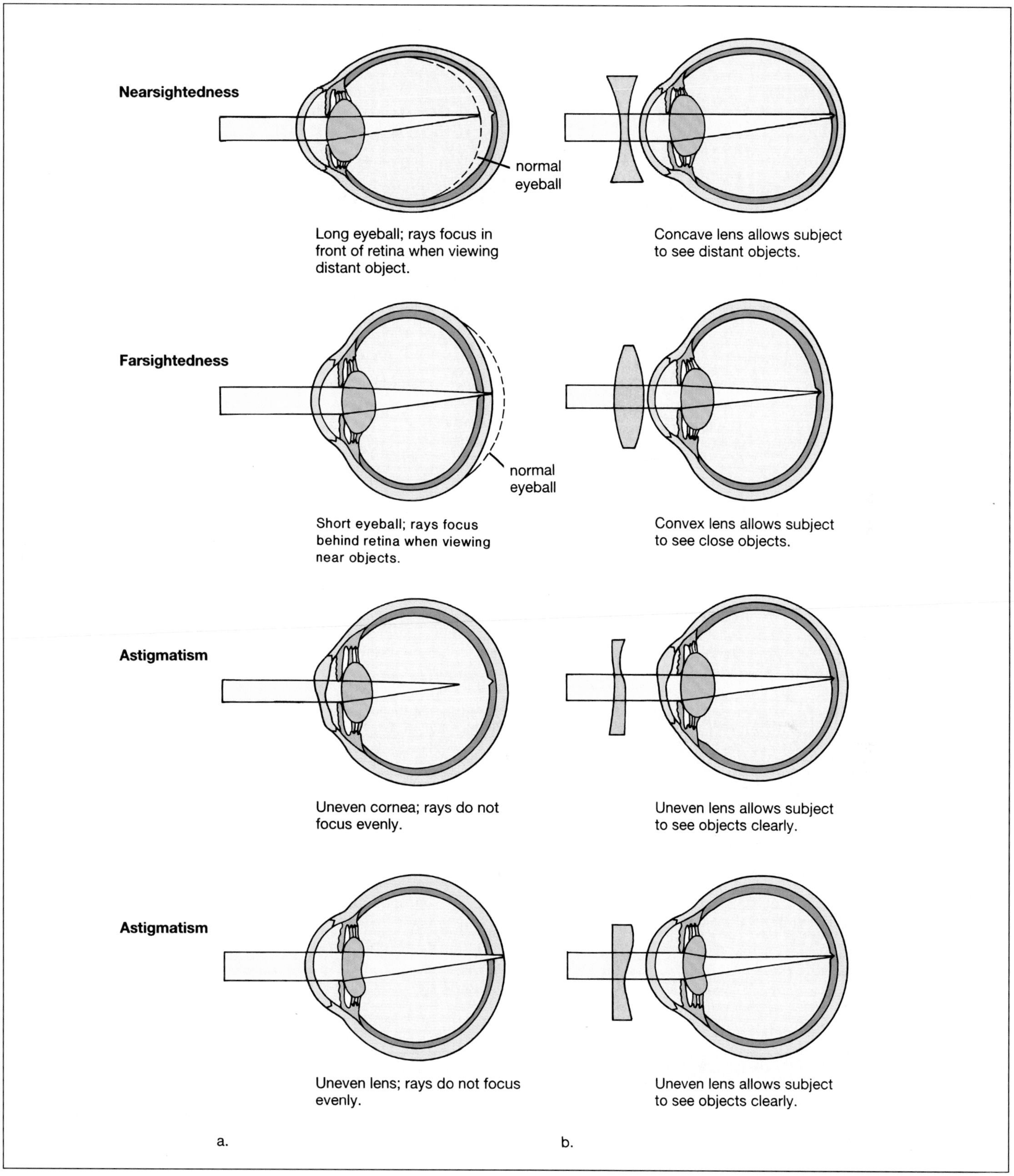
Nearsightedness
normal eyeball
Long eyeball; rays focus in front of retina when viewing distant object.
Concave lens allows subject to see distant objects.
Farsightedness
normal eyeball
Short eyeball; rays focus behind retina when viewing near objects.
Convex lens allows subject to see close objects.
Astigmatism
Uneven cornea; rays do not focus evenly.
Uneven lens allows subject to see objects clearly.
Astigmatism
Uneven lens; rays do not focus evenly.
Uneven lens allows subject to see objects clearly.
a.
b.

Figure 12.10

Common abnormalities of the eye, with possible corrective lenses. **a.** The cornea and the lens function in bringing light rays (*lines*) to focus, but sometimes they are unable to compensate for the shape of the eyeball or for an uneven cornea. **b.** In these instances corrective lenses can allow the individual to see normally.

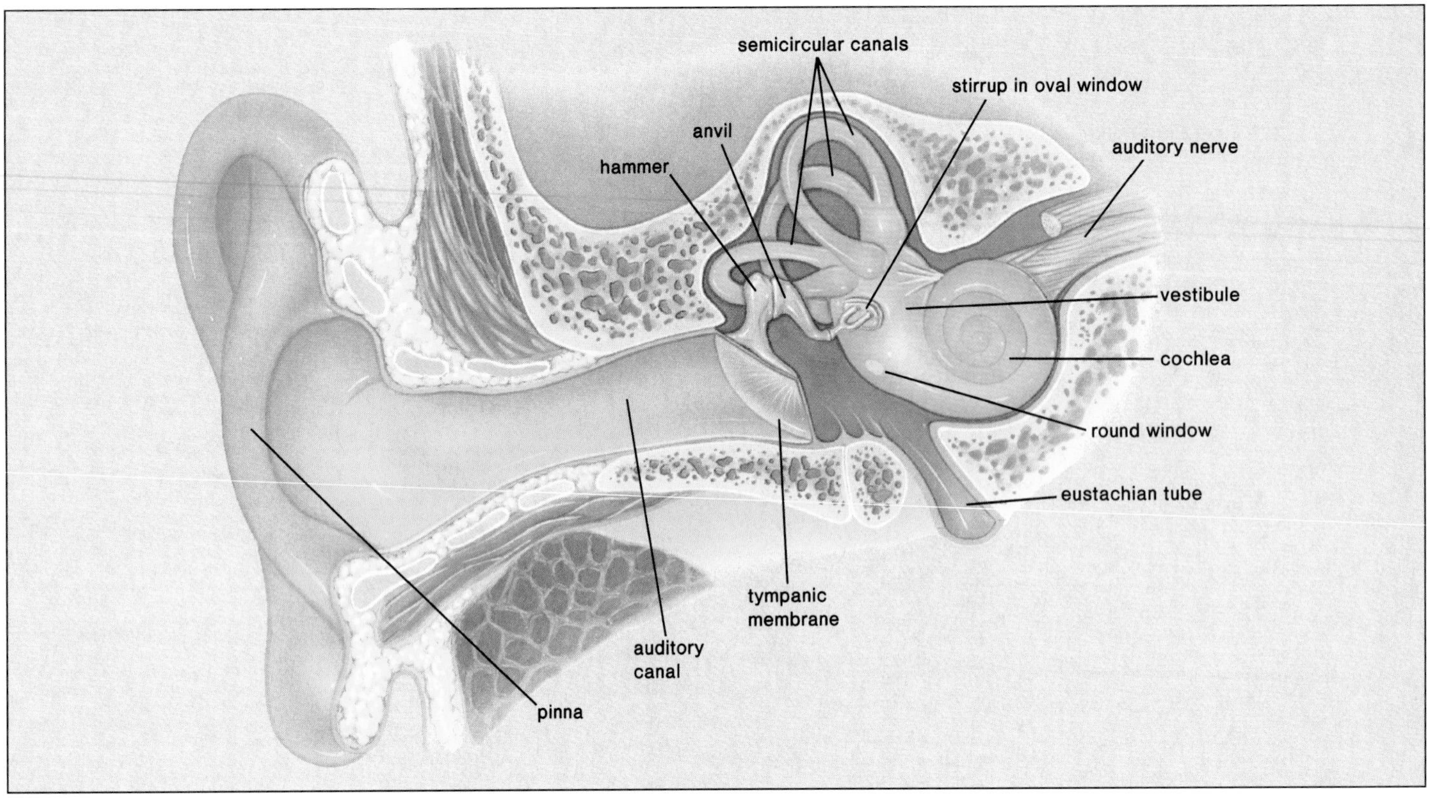

Figure 12.11

Anatomy of the human ear. In the middle ear, the hammer (malleus), the anvil (incus), and the stirrup (stapes) amplify sound waves. Otosclerosis is a condition in which the stirrup becomes attached to the inner ear and is unable to carry out its normal function. It can be replaced by a plastic piston, and thereafter the individual hears normally because sound waves are transmitted as usual to the cochlea, which contains the receptors for hearing.

Table 12.3

The Ear

	Outer Ear	Middle Ear	Inner Ear	
			Cochlea	***Sacs and Semicircular Canals***
Function	Directs sound waves to tympanic membrane	Picks up and amplifies sound waves	Hearing	Maintains equilibrium
Anatomy	Pinna Auditory canal	Tympanic membrane Ossicles	Organ of Corti; auditory nerve	Saccule and utricle Semicircular canals
Medium	Air	Air (eustachian tube)	Fluid	Fluid

Path of vibration: Sound waves—vibration of tympanic membrane—vibration of hammer, anvil, and stirrup—vibration of oval window—fluid pressure waves in canals of inner ear lead to stimulation of hair cells—bulging of round window.

Ear

The ear accomplishes 2 sensory functions: equilibrium (balance) and hearing. The receptors for both of these are located in the inner ear and consist of hair cells with cilia that respond to mechanical stimulation. Each hair cell has from 30 to 150 cilia. When the cilia of any particular hair cell are displaced in a certain direction, the cell generates nerve impulses, which are sent along a cranial nerve to the brain.

How the Ear Appears

Figure 12.11 is a drawing of the ear, and table 12.3 lists the parts of the ear. The ear has 3 divisions: outer, middle, and inner. The **outer ear** consists of the **pinna** (external flap) and the **auditory canal.** The opening of the auditory canal is lined with fine hairs and sweat glands. Modified sweat glands are located in the upper wall of the canal; they secrete earwax, a substance that helps to guard the ear against the entrance of foreign materials, such as air pollutants.

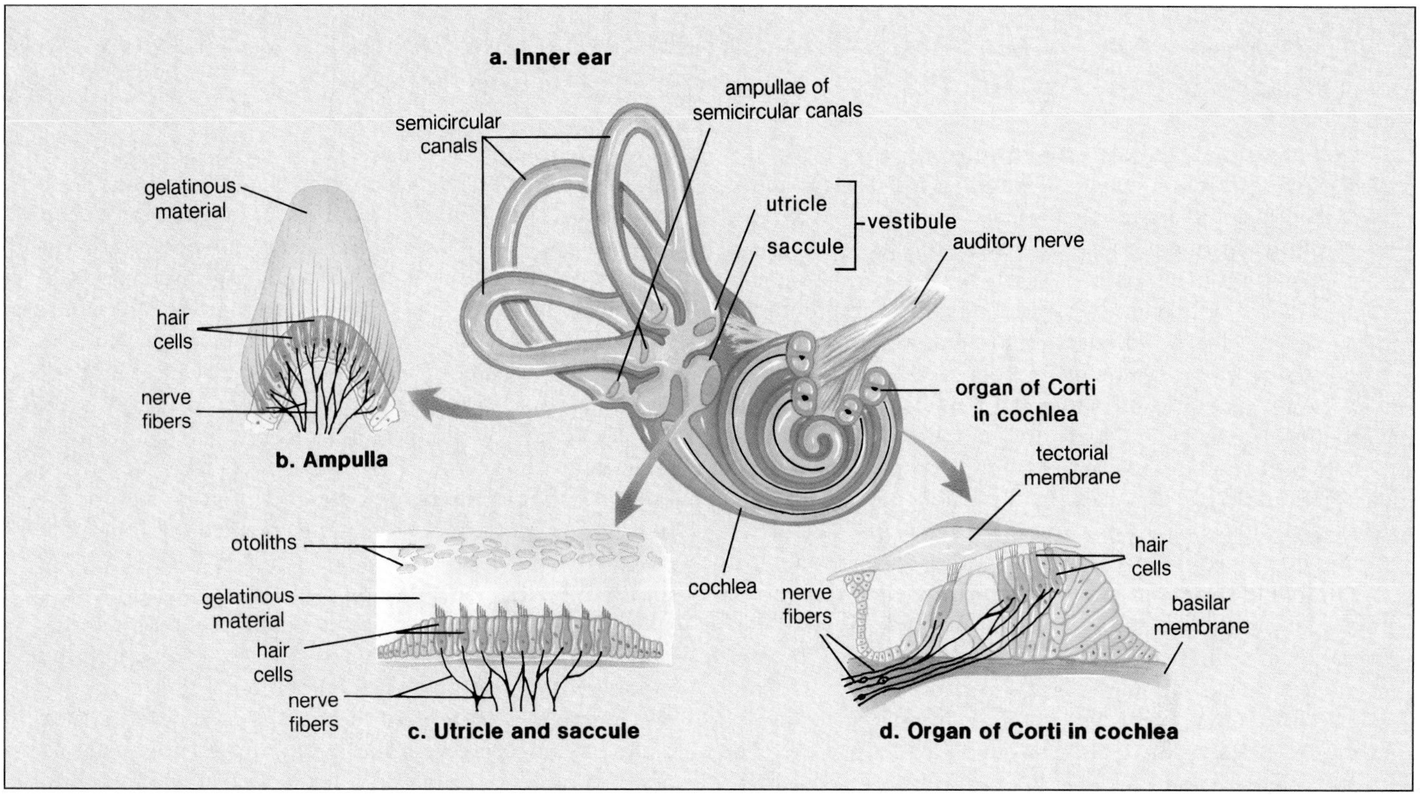

Figure 12.12

Anatomy of the inner ear. **a.** The inner ear contains the semicircular canals, the utricle and the saccule within a vestibule, and the cochlea. The cochlea has been cut to show the location of the organ of Corti. **b.** An ampulla at the base of each semicircular canal contains the receptors (hair cells) for dynamic equilibrium. **c.** The utricle and the saccule are small sacs that contain the receptors (hair cells) for static equilibrium. **d.** The receptors for hearing (hair cells) are in the organ of Corti.

The **middle ear** begins at the **tympanic membrane** (eardrum) and ends at a bony wall containing 2 small openings covered by membranes. These openings are called the **oval** and **round windows.** Three small bones are found between the tympanic membrane and the oval window. Collectively called the **ossicles,** individually they are the **hammer** (malleus), the **anvil** (incus), and the **stirrup** (stapes) because their shapes resemble these objects (fig. 12.11). The hammer adheres to the tympanic membrane, and the stirrup touches the oval window. The posterior wall has an opening that leads to mastoid sinuses of the skull.

A **eustachian tube,** which extends from each middle ear to the nasopharynx, permits equalization of air pressure. Chewing gum, yawning, and swallowing in elevators and airplanes help to move air through the eustachian tubes upon ascent and descent.

Whereas the outer ear and the middle ear contain air, the **inner ear** is filled with fluid. The inner ear (fig. 12.12*a*), anatomically speaking, has 3 areas: the first and second, the semicircular canals and the vestibule, are concerned with equilibrium; the third, the cochlea, is concerned with hearing.

The **semicircular canals** are arranged so that there is one in each dimension of space. The base of each of the 3 canals, called the **ampulla** (fig. 12.12*b*), is slightly enlarged. Little hair cells whose cilia are inserted into a gelatinous material are found within the ampullae.

A vestibule, or chamber, lies between the semicircular canals and the cochlea. It contains 2 small membranous sacs called the **utricle** and the **saccule** (fig. 12.12*c*). Both of these sacs contain little hair cells, whose cilia protrude into a gelatinous material. Calcium carbonate ($CaCO_3$) granules, or **otoliths,** rest on this material.

The **cochlea** resembles the shell of a snail because it spirals. Three canals are located within the tubular cochlea: the vestibular canal, the **cochlear canal,** and the tympanic canal. Along the length of the basilar membrane, which forms the lower wall of the cochlear canal, are little hair cells whose cilia come into contact with another membrane, called the **tectorial membrane.** The hair cells of the cochlear canal plus the tectorial membrane are called the **organ of Corti** (fig. 12.12*d*). The nerve impulses this organ sends to the brain stem are eventually relayed to the temporal lobe of the cerebrum, where they are interpreted as sound.

> The ear has 3 major divisions: outer ear, middle ear, and inner ear. The outer ear contains the auditory canal; the middle ear contains the ossicles; and the inner ear contains the semicircular canals, a vestibule, and the cochlea.

Health Focus

Protecting Vision and Hearing

The eye is subject to both injuries and disorders. Although flying objects sometimes penetrate the cornea and damage the iris, lens, or retina, careless use of contact lenses is the most common cause of injuries to the eye. Injuries cause only 4% of all cases of blindness; the most frequent causes are retinal disorders, glaucoma, and cataracts, in that order. Retinal disorders are varied. In diabetic retinopathy, which blinds many people between the ages of 20 and 74, capillaries to the retina burst and blood spills into the vitreous fluid. Careful regulation of blood glucose levels in these patients may be preventative. In macular degeneration, the cones are destroyed because thickened choroid vessels no longer function as they should. Glaucoma occurs when the drainage system of the eyes fails, so that fluid builds up and destroys nerve fibers responsible for peripheral vision. Eye doctors always check for glaucoma, but it is advisable to be aware of the disorder in case it comes on quickly. Those who have experienced acute glaucoma report that the eyeball feels as heavy as a stone. In cataracts, cloudy spots on the lens of the eye eventually pervade the whole lens. The milky yellow-white lens scatters incoming light and blocks vision.

Accumulating evidence suggests that both macular degeneration and cataracts, which tend to occur in the elderly, are caused by long-term exposure to the ultraviolet rays of the sun. It is recommended, therefore, that everyone, especially those who live in sunny climates or work outdoors, wear sunglasses that absorb ultraviolet light. Large lenses worn close to the eyes offer further protection. The Sunglass Association of America has devised the following system for categorizing sunglasses:

- Cosmetic lenses absorb at least 70% of UV-B and 20% UV-A and 60% of visible light. Such lenses are worn for comfort rather than protection.
- General purpose lenses absorb at least 95% of UV-B and 60% of UV-A and 60%–92% of visible light. They are good for outdoor activities in temperate regions.
- Special purpose lenses block at least 99% of UV-B and 60% UV-A, and 20%–97% of visible light. They are good for bright sun combined with sand, snow, or water.

Studies also suggest that age-associated hearing loss can be prevented if ears are protected from loud noises starting even during infancy. Hospitals are now aware of the problem and are taking steps to make sure neonatal intensive care units and nurseries are as quiet as possible.

In today's society, exposure to the types of noises listed in figure 12A is a common occurrence. Noise is measured in decibels, and any noise above a level of 80 decibels could result in damage to the hair cells of the organ of Corti. Eventually the cilia and then the hair cells disappear completely. If listening to city traffic for extended periods can damage hearing, it stands to reason that frequent attendance at rock concerts or constantly playing a stereo loudly is also damaging to hearing. The first hint of danger could be temporary hearing loss, a "full" feeling in the ears, muffled hearing, or tinnitus (e.g., ringing in the ears). If you have any of these symptoms, modify your listening habits immediately to prevent further damage. If exposure to noise is unavoidable, specially designed noise-reduction ear muffs are available, and it is also possible to purchase ear plugs made from a compressible, spongelike material at the drug store or sporting good store. These ear plugs are not the same as those worn for swimming, and they should not be used interchangeably.

Aside from loud music, noisy indoor or outdoor equipment such as a rug-cleaning machine or a chain saw are also troublesome. Even motorcycles and recreational vehicles such as snowmobiles and motocross bikes can contribute to a gradual loss of hearing. Exposure to intense sounds of short duration, such as a burst of gunfire, can result in an immediate hearing loss. Hunters may have a significant hearing reduction in the ear opposite to the shoulder where the gun is carried. The butt of the rifle offers some protection to the ear nearest the gun when it is shot.

Finally, people need to be aware that some medicines are ototoxic. Anticancer drugs, most notably cisplatin, and certain antibiotics (e.g., streptomycin, kanamycin, gentamicin) make ears especially susceptible to a hearing loss. Anyone taking such medications needs to be careful to protect the ears from any loud noises.

Type of Noise	Sound Level (decibels)	Effect
"Boom car" Jet engine Shotgun Rock concert	over 125	Beyond threshold of pain; potential for hearing loss high
Discotheque, "boom box," thunderclap	over 120	Hearing loss likely
Chain saw, pneumatic drill, jackhammer, symphony orchestra Snowmobile Garbage truck, cement mixer	100–200	Regular exposure of more than 1 minute risks permanent hearing loss
Farm tractor Newspaper press Subway, motorcycle	90–100	15 minutes of umprotected exposure potentially harmful
Lawn mower, food blender	85–90	Continous daily exposure for more than 8 hours can cause hearing damage
Diesel truck Average city traffic noise	80–85	Annoying; constant exposure may cause hearing damage

a.

Source: National institute on Deafness and Other Communication Disorders, National Institutes of Health, January 1990.

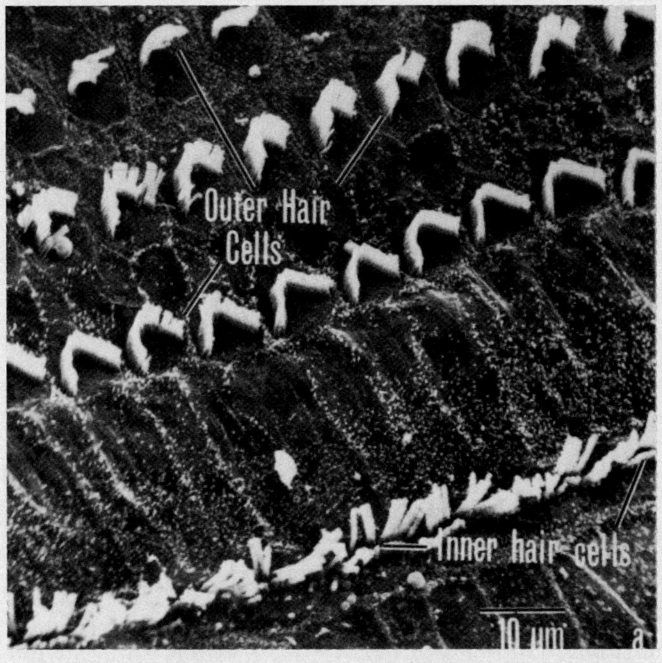

b. Normal organ of Corti

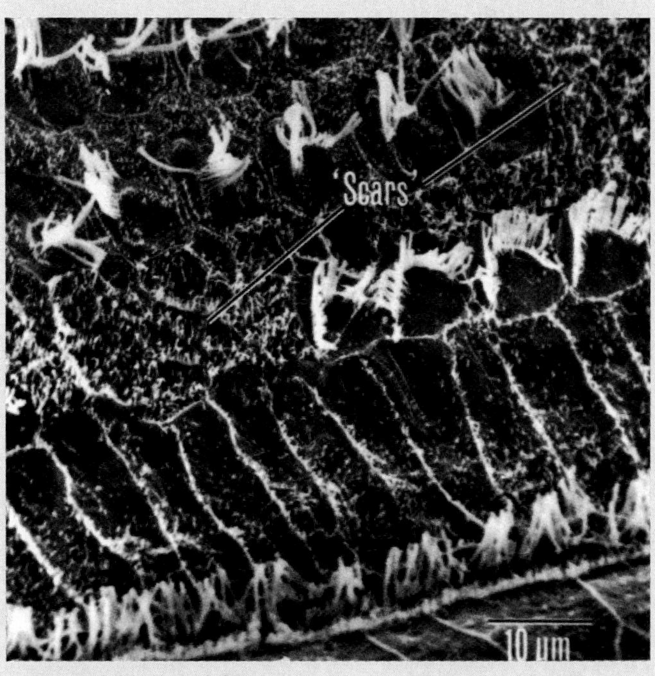

c. Damaged organ of Corti

Figure 12A

a. The higher the decibel reading, the more likely a noise will damage hearing. **b.** Normal versus **c.** damaged hair cells in the organ of Corti of a guinea pig. This damage occurred after 24-hour exposure to a noise level typical of rock music.

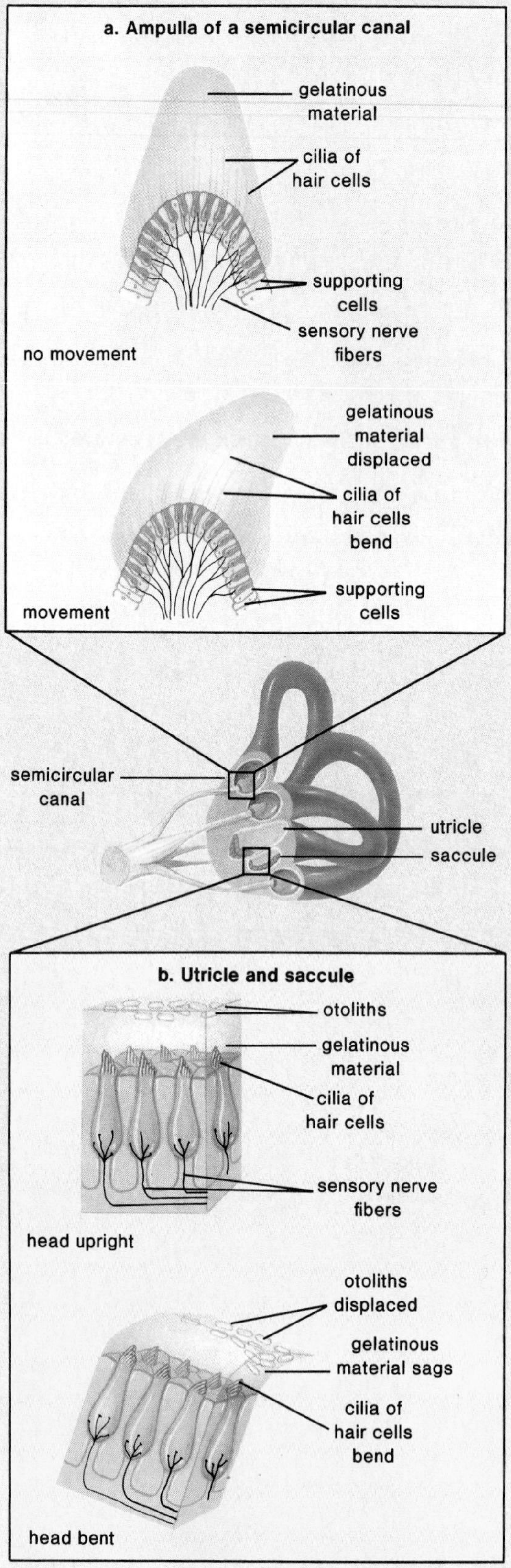

Figure 12.13

Receptors for balance. **a.** The ampullae of the semicircular canals contain hair cells with cilia embedded in a gelatinous material. When the head rotates, the material is displaced, and bending cilia initiate nerve impulses in sensory nerve fibers. This permits dynamic equilibrium. **b.** The utricle and the saccule contain hair cells with cilia embedded in a gelatinous material. When the head bends, otoliths are displaced, causing the gelatinous material to sag and the cilia to bend. The bending initiates nerve impulses in sensory nerve fibers. This permits static equilibrium.

Balance: Two Kinds

The sense of balance has been subdivided into 2 senses: *dynamic equilibrium,* requiring a knowledge of angular and/or rotational movement, and *static equilibrium,* requiring a knowledge of movement in one plane, either vertical or horizontal.

Dynamic equilibrium is required when the body is moving. At that time, the fluid within the semicircular canals flows over and displaces the gelatinous material within the ampullae (fig. 12.13*a*). This causes the cilia of the hair cells to bend, which initiates nerve impulses that travel to the brain. Continuous movement of the fluid in the semicircular canals causes one form of motion sickness.

When the body is still, the otoliths in the utricle and the saccule rest on the gelatinous material above the hair cells (fig. 12.13*b*). Static equilibrium is required when the body moves horizontally or vertically. At that time, the otoliths are displaced and the gelatinous material sags, bending the cilia of the hair cells beneath. The hair cells then generate nerve impulses that travel to the brain.

> Movement of fluid within the semicircular canals contributes to the sense of dynamic equilibrium. Movement of the otoliths within the utricle and the saccule is important for static equilibrium.

Hearing: Rubbing Cilia

The process of hearing begins when sound waves enter the auditory canal (see fig. 12.11). Just as ripples travel across the surface of a pond, sound travels by the successive vibrations of molecules. Ordinarily, sound waves do not carry much energy, but when a large number of waves strike the tympanic membrane (eardrum), it moves back and forth (vibrates) ever so slightly. The hammer then takes the pressure from the inner surface of the tympanic membrane and passes it by way of the anvil to the stirrup in such a way that the pressure is multiplied about 20 times as it moves from the tympanic membrane to the stirrup. The stirrup strikes the oval window, causing it to vibrate, and in this way, the pressure is passed to the fluid within the cochlea (fig. 12.14).

Figure 12.14

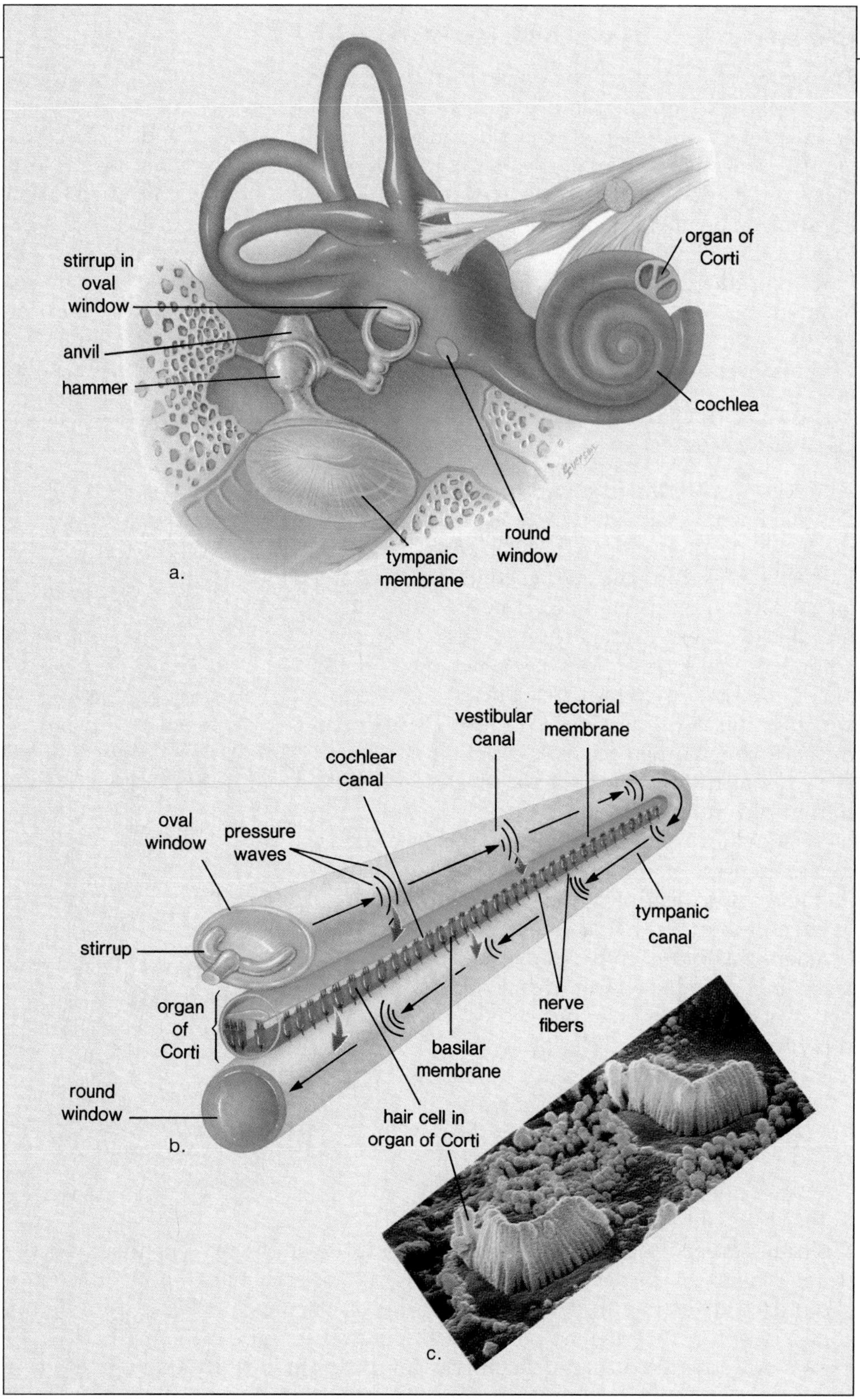

Receptors for hearing. **a.** The organ of Corti is located within the cochlea. **b.** In the unwound cochlea, note that the organ of Corti consists of hair cells resting on the basilar membrane with the tectorial membrane above. The arrows represent the pressure waves that move from the oval window to the round window due to the motion of the stirrup. These pressure waves cause the basilar membrane to vibrate and the cilia of at least a portion of the 16,000 hair cells to bend against the tectorial membrane. The generated nerve impulses result in hearing. **c.** The micrograph shows the hair cells in the organ of Corti.

If the cochlea is unwound, as shown in figure 12.14*b*, you can see that the vestibular canal connects with the tympanic canal and that pressure waves move from one canal to the other toward the round window, a membrane that bulges to absorb the pressure. As a result of the movement of the fluid within the cochlea, the basilar membrane moves up and down, and the cilia of the hair cells rub against the tectorial membrane. This bending of the cilia initiates nerve impulses, which pass by way of the **auditory nerve** to the temporal lobe of the cerebrum, where the impulses are interpreted as a sound.

The organ of Corti is narrow at its base but widens as it approaches the tip of the cochlear canal. Each part of the organ is sensitive to different wave frequencies, or pitch. Near the tip, the organ of Corti responds to low pitches, such as a tuba, and near the base, it responds to higher pitches, such as a bell or a whistle. The neurons from each region along the length of the cochlea lead to slightly different areas in the brain. The pitch sensation we experience depends upon which of these areas of the brain is stimulated.

Volume is a function of the amplitude of sound waves. Loud noises cause the fluid of the cochlea to vibrate to a greater degree, and this, in turn, causes the basilar membrane to move up and down to a greater extent. The resulting increased stimulation is interpreted by the brain as volume. It is believed that the tone of a sound is an interpretation of the brain based on the distribution of hair cells stimulated.

> The sense receptors for sound are hair cells on the basilar membrane (the organ of Corti). When the basilar membrane vibrates, the cilia of the delicate hair cells touch the tectorial membrane, initiating nerve impulses that are transmitted in the auditory nerve to the brain.

Deafness: Two Major Types

There are 2 major types of deafness: *conduction deafness* and *nerve deafness.* Conduction deafness can be due to a congenital defect, such as that which occurs when a pregnant woman contracts German measles during the first trimester of pregnancy. (For this reason every female should be immunized against rubella before the childbearing years.) Conduction deafness can also be due to infections that have caused the ossicles to fuse, restricting their ability to magnify sound waves. Because respiratory infections can spread to the ear by way of the eustachian tubes, every cold and ear infection should be taken seriously.

Nerve deafness most often occurs when cilia on the sense receptors within the cochlea have worn away. Since this can happen with normal aging, old people are more likely than younger persons to have trouble hearing. However, as discussed in the Health Focus in this chapter, nerve deafness also occurs when people listen to loud music amplified to or above 130 decibels. Because the usual types of hearing aids are not helpful for nerve deafness, it is wise to avoid subjecting the ears to any type of continuous loud noise. Costly cochlear implants, which stimulate the auditory nerve directly, are available, but those who have these electronic devices report that the speech they hear is like that of a robot.

Summary

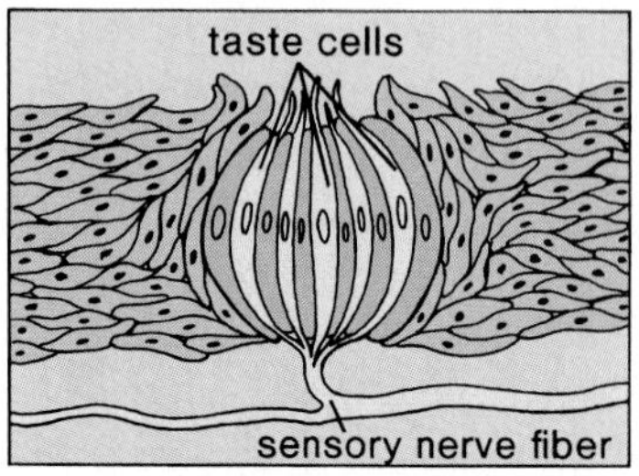

Taste Bud

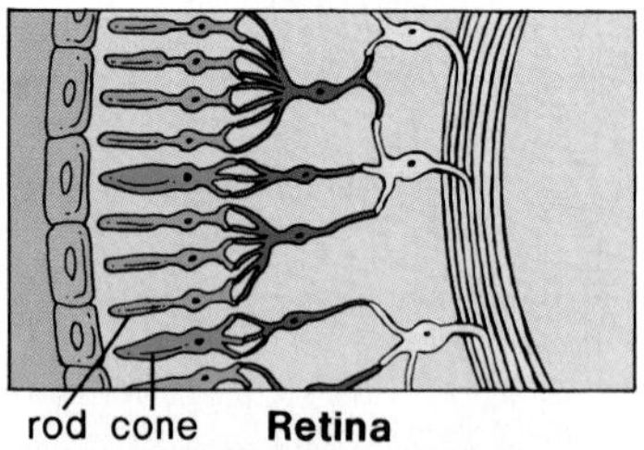

Retina

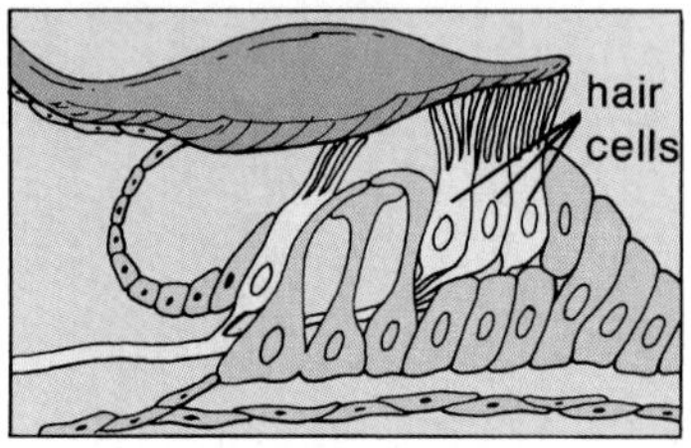

Organ of Corti

All receptors are the first part of a reflex arc. Each type is sensitive to a particular kind of stimulus in the external or internal environment. When stimulation occurs, receptors initiate nerve impulses that are transmitted to the spinal cord and/or brain. Only when nerve impulses reach the cerebrum are we conscious of sensation. The cerebrum can be mapped according to the type of sensation felt and the localization of sensation.

The skin contains receptors for touch, pressure, pain, and temperature (hot and cold). Taste and smell are due to chemoreceptors that are stimulated by chemicals in the environment. The taste buds contain cells in contact with nerve fibers, while olfactory cells are specialized nerve fibers. The former end in microvilli and the latter end in cilia.

Vision is dependent on the eye, the optic nerve, and the visual cortex of the cerebrum. The eye has 3 layers. The outer layer, the sclera, can be seen as the white of the eye; it also becomes the transparent bulge in the front of the eye called the cornea. The rods, receptors for vision in dim light, and the cones, receptors that depend on bright light and provide color and detailed vision, are located in the retina, the inner layer of the eyeball. The cornea, the humors, and especially the lens bring the light rays to focus on the retina. To see a near object, accommodation occurs as the lens rounds up. Due to the optic chiasma, both sides of the brain must function together to give three-dimensional vision.

The ear also contains receptors for our sense of balance. Dynamic equilibrium is dependent on the stimulation of hair cells within the ampullae of the semicircular canals. Static equilibrium relies on the stimulation of hair cells by otoliths within the utricle and the saccule.

Hearing is a specialized sense dependent on the ear, the auditory nerve, and the auditory cortex of the cerebrum. The ear is divided into 3 parts: outer, middle, and inner. The outer ear consists of the pinna and the auditory canal, which direct sound waves to the middle ear. The middle ear begins with the tympanic membrane and contains the ossicles (hammer, anvil, and stirrup). The hammer is attached to the tympanic membrane, and the stirrup is attached to the oval window, which is covered by membrane. The inner ear contains the cochlea, the semicircular canals, plus the utricle and saccule. The outer and middle portions of the ear simply convey and magnify the sound waves that strike the oval window. Its vibrations set up pressure waves within the cochlea, which contains the organ of Corti, consisting of hair cells with the tectorial membrane above. When the cilia of the hair cells strike this membrane, nerve impulses are initiated that finally result in hearing.

Study and Thought Questions

1. List and discuss the receptors of the skin. (p. 228)
2. Discuss the chemoreceptors. (pp. 229–230)
3. Describe the anatomy of the eye (p. 230), and explain focusing and accommodation. (pp. 232–233)

   Critical Thinking Devise a categorization for the parts of the eye listed in table 12.2. Justify your system.
4. Describe sight in dim light. What chemical reaction is responsible for vision in dim light? (p. 234) Discuss color vision. (p. 235)
5. Relate the need for corrective lenses to 3 possible eye shapes. Discuss bifocals. (p. 236)
6. Describe the anatomy of the ear (pp. 238–239) and how we hear. (pp. 242–243)
7. Describe the role of the utricle, the saccule, and the semicircular canals in balance. (p. 242)

   Critical Thinking Both balance and hearing utilize mechanical stimulation and hair cells. Why does one result in the sense of balance and the other in the sense of hearing?
8. Discuss the 2 causes of deafness, including why younger people frequently suffer a hearing loss. (p. 244)

   Ethical Issue What should be done, if anything, to protect teenagers who play music so loudly that their hearing is endangered? Explain your answer.

Writing Across the Curriculum

1. In general, list the steps, from sense organ to brain, by which sensation occurs. Discuss the weaknesses and strengths of depending on such a method for knowledge of the outside world.
2. Question 1 assumes that the brain is responsible for sensation. Based on figure 12.5, why might there be some integration of data in the eye itself? Does this fact lend strength or weakness to the way by which we see?
3. What do the retina and organ of Corti have in common?

Objective Questions

1. Vision, hearing, taste, and smell do not occur unless nerve impulses reach the proper portion of the _____.
2. Taste cells and olfactory cells are _______________ because they are sensitive to chemicals in the air and in food.
3. The receptors for vision, the ___________ and the _________, are located in the ____________, the inner layer of the eye.
4. The cones give us ___________ vision and work best in ___________ light.
5. Vision in dim light is dependent on the presence of ___________ in the rods.
6. The lens ________________ for viewing near objects.
7. People who are nearsighted cannot see objects that are ____________. A ___________ lens restores this ability.
8. The ossicles are the __________, the _____________, and the ____________________.
9. The semicircular canals are involved in our sense of ______________________________.
10. The organ of Corti is located in the ___________ canal of the __________.
11. Label this diagram of the eye.

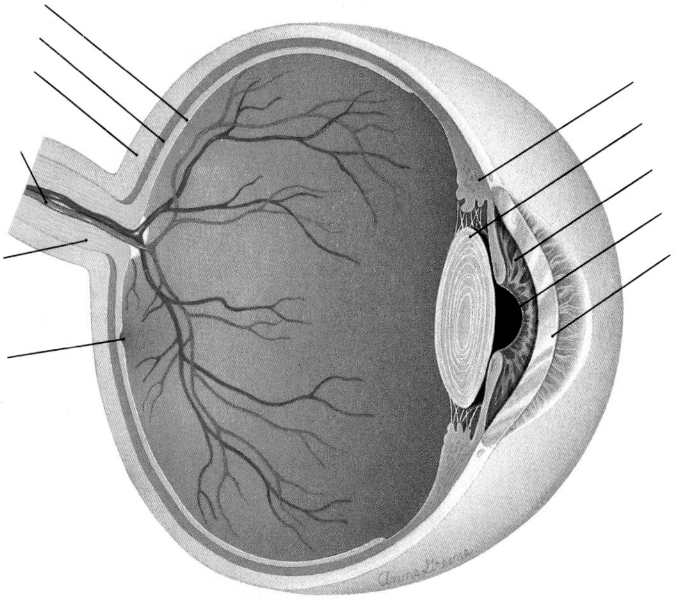

Selected Key Terms

accommodation Lens adjustment to see near objects. 233

choroid (ko´roid) The vascular, pigmented middle layer of the eyeball. 230

ciliary body (sil´e-er″e bod´e) Structure associated with the choroid layer that contains ciliary muscle, which controls the shape of the lens of the eye. 230

cochlea (kok´le-ah) That portion of the inner ear that resembles a snail's shell and contains the organ of Corti, the sense organ for hearing. 239

cone Bright-light receptor in the retina of the eye that detects color and provides visual acuity. 231

fovea centralis (fo´ve-ah sen-tral´is) Region of the retina consisting of densely packed cones that is responsible for the greatest visual acuity. 232

lens A clear membranelike structure found in the eye behind the iris; brings objects into focus. 230

organ of Corti A portion of the inner ear that contains the receptors for hearing. 239

ossicle One of the small bones of the middle ear—hammer, anvil, stirrup. 239

otolith (o-to-lith) Calcium carbonate granule associated with ciliated cells in the utricle and the saccule. 239

retina (ret´ĭ nah) The innermost layer of the eyeball, which contains the rods and the cones. 231

rhodopsin (ro-dop´sin) Visual purple, a pigment found in the rods. 234

rod Dim-light receptor in the retina of the eye that detects motion but no color. 231

saccule (sak´ūl) A saclike cavity that makes up part of the membranous labyrinth of the inner ear; contains receptors for static equilibrium. 239

sclera (skle´rah) White, fibrous, outer layer of the eyeball. 230

semicircular canal Tubular structure within the inner ear that contains the receptors responsible for the sense of dynamic equilibrium. 239

tympanic membrane (tim-pan´ik mem´brān) Located between the outer and middle ear and receives sound waves; the eardrum. 239

utricle (u´tre-k´l) Saclike cavity that makes up part of the membranous labyrinth of the inner ear; contains receptors for static equilibrium. 239

Chapter 13

Endocrine System

The first pregnancy test was developed in 1928 after it was discovered that the urine of a pregnant woman contains a hormone—the hormone is secreted in such excess that it spills over into the urine. When the urine of a pregnant woman was injected into a rabbit, characteristic changes in the animal's reproductive organs signaled pregnancy. When the urine was injected into a female frog, it caused the female frog to ovulate and produce eggs; when injected into a male frog, it caused the male frog to eject sperm. The basis of a pregnancy test is the same today except that antibodies to the hormone are added to a blood or urine sample. An observable antibody-antigen reaction (called agglutination) indicates that the woman is pregnant.

Chapter Concepts

1 Chemical messengers controlled by negative feedback systems bring about coordination of body parts. 248

2 The hypothalamus, a part of the brain, controls the function of the pituitary gland, which in turn controls several other glands. 249

3 The thyroid (which speeds metabolism) and parathyroid glands (which control blood calcium levels) are entirely separate in structure and function. 254–256

4 The adrenal medulla (which is active during stress), and the adrenal cortex (which helps us recover from stress) are parts of the adrenal gland. 256–258

5 The pancreas secretes hormones that control the blood glucose level. 259

6 Three types of environmental signals are identified; hormones are one of those types. 264

Chapter Outline

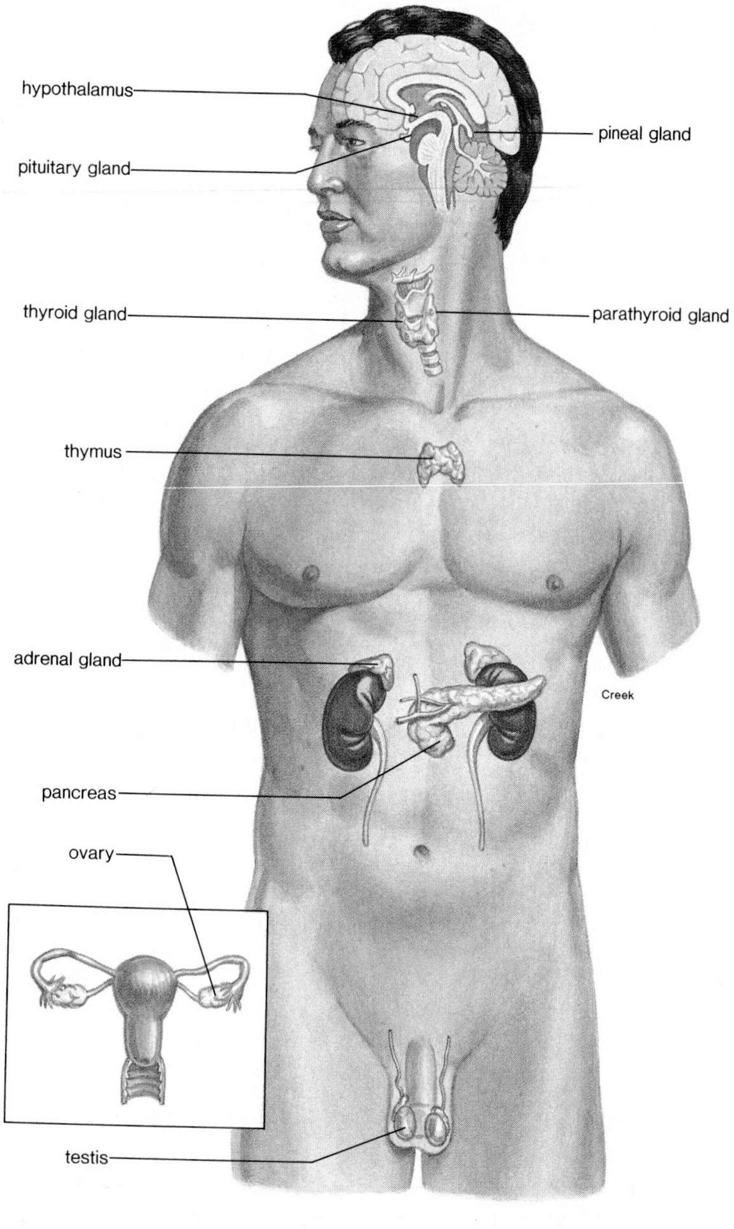

Figure 13.1

Anatomical location of major endocrine glands in the body.

Along with the nervous system, hormones coordinate the functioning of body parts. Their presence or absence affects our metabolism, our appearance, and our behavior. **Hormones** are produced by *endocrine glands,* which secrete their products internally, placing them directly in the blood (fig. 13.1). Since these glands do not have ducts for the transport of their secretions, they are sometimes called ductless glands. All hormones are carried throughout the body by the blood, but each one affects only a specific body part or parts, appropriately termed *target organs.*

> Endocrine glands secrete hormones into the bloodstream for transport to target organs.

How Hormones Work

Hormones are substances that fall into 2 basic categories: (1) peptides (used here to include amino acid, polypeptide, and protein hormones) and (2) steroid hormones. Steroids are complex rings of carbon and hydrogen atoms. The difference between steroids is due to the atoms attached to these rings. Steroid hormones are produced by the adrenal cortex, the ovaries, and the testes. All of the other glands produce peptide hormones (table 13.1).

Peptide Hormones: Activate Existing Enzymes

Peptide hormones bind to cell-membrane receptors (fig. 13.2*a*). The hormone-receptor complex then activates an enzyme that produces cyclic adenosine monophosphate (cAMP). cAMP is a compound made from ATP, but it contains only one phosphate group, which is attached to adenosine at 2 locations. The cAMP now activates the enzymes of the cell to carry out their normal functions. Notice that the peptide hormone never enters the cell. Therefore, these hormones sometimes are called the *first messenger,* while cAMP, which sets the metabolic machinery in motion, is called the *second messenger.*

Steroid Hormones: Synthesis of New Enzymes

Steroid hormones do not bind to cell-surface receptors; they can enter the cell freely because they are relatively small, lipid-soluble molecules. Once inside, steroid hormones bind to receptors in the cytoplasm (fig. 13.2*b*). The hormone-receptor complex then enters the nucleus, where it binds with chromatin at a location that promotes activation of particular genes. Protein synthesis follows. In this manner, steroid hormones lead to protein synthesis. Steroid hormones act more slowly than peptides because it takes more time to synthesize new proteins than to activate enzymes already present in the cell.

> Hormones are chemical messengers that influence the metabolism of the cell either directly or indirectly, depending on the hormone type.

Negative Feedback: Reduces Stimulus

In a self-regulating negative feedback mechanism, an adaptive response dampens or even cancels the stimulus that brought about the response (see fig. 3.12). The function of an endocrine gland is usually controlled by negative feedback. An endocrine gland can be sensitive

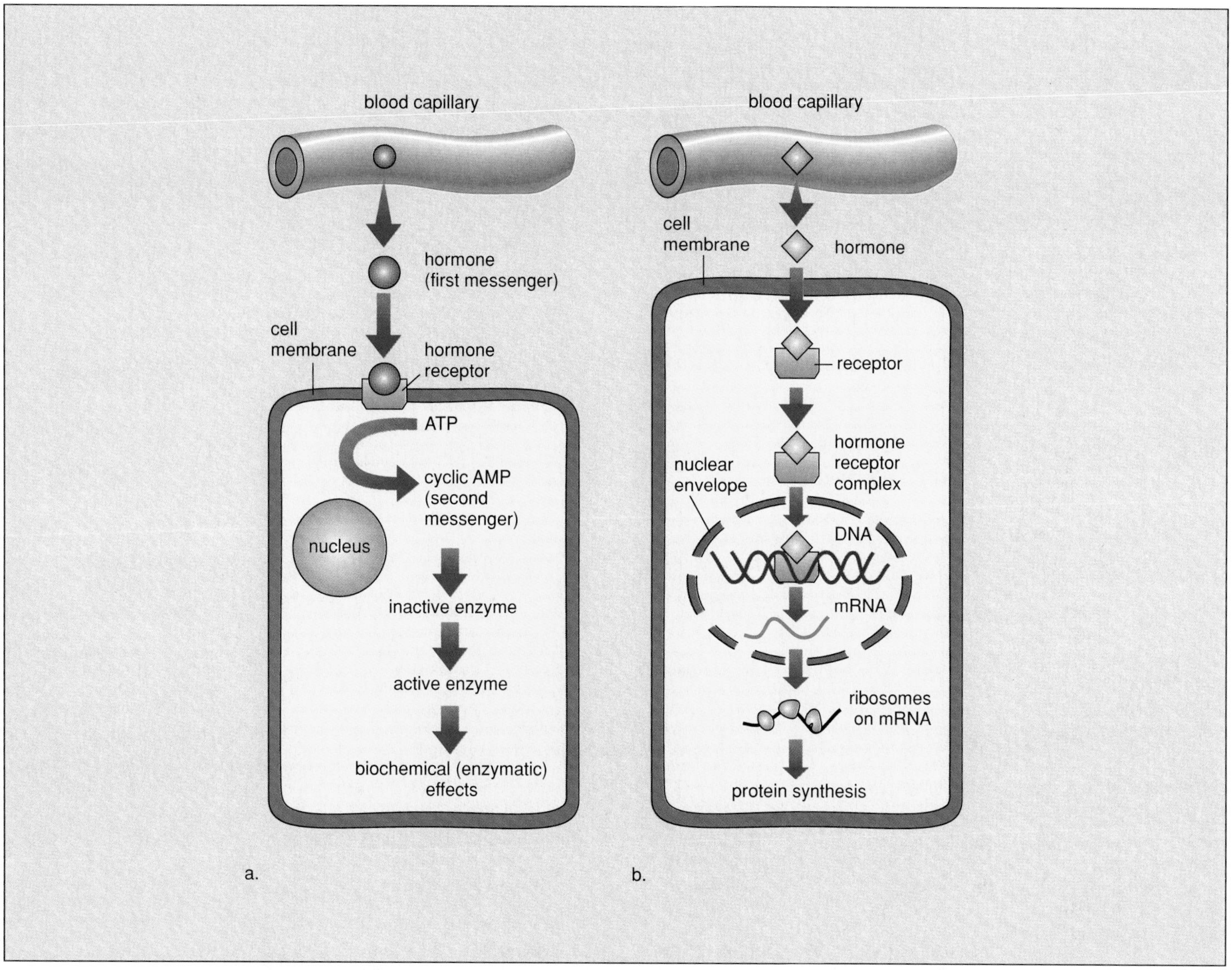

Figure 13.2

Cellular activity of hormones. **a.** Peptide hormones combine with receptors located on the cell membrane. This promotes the production of cyclic AMP, which in turn leads to activation of a particular enzyme. **b.** Steroid hormones pass through the cell membrane to combine with receptors; the complex activates certain genes, leading to protein synthesis.

to either the condition it is regulating or to the blood level of the hormone it is producing. For example, when blood is concentrated, the hypothalamus produces a hormone that causes blood dilution. The hypothalamus then stops producing the hormone. On the other hand, the pituitary gland produces a hormone that stimulates the thyroid gland. When the blood level of a hormone produced by the thyroid rises, the pituitary gland no longer stimulates the thyroid gland.

Hypothalamus and Pituitary Gland

The *hypothalamus,* located beneath the thalamus in the third ventricle of the brain, regulates the internal environment. For example, it helps to control heart rate, body temperature, and water balance, as well as the activity of the **pituitary gland.** The pituitary gland (only about 1 cm in diameter) lies just below the hypothalamus (see fig. 13.1) and is divided into 2 portions: the **posterior pituitary** and the **anterior pituitary.**

Table 13.1

The Principal Endocrine Glands and Their Hormones

Endocrine Gland	Hormone Released	Target Tissues/ Organ	Chief Function of Hormone	Disorders (too much/too little)
Hypothalamus	Hypothalamic-releasing and release-inhibiting hormones	Anterior pituitary	Regulate anterior pituitary hormones	*See* Anterior pituitary
Posterior pituitary (storage of hypothalamic hormones)	Antidiuretic hormone (ADH, vasopressin)	Kidneys	Stimulates water reabsorption by kidneys	Diverse*/diabetes insipidus
	Oxytocin	Uterus, mammary glands	Stimulates uterine muscle contraction and release of milk by mammary glands	
Anterior pituitary	Growth hormone (GH, somatotropin)	Soft tissues, bones	Stimulates cell division, protein synthesis, and bone growth	Giantism, acromegaly/ dwarfism
	Prolactin (PRL)	Mammary glands	Stimulates milk production and secretion	
	Melanocyte-stimulating hormone (MSH)	Melanocytes in skin	Regulates skin color in lower vertebrates; unknown function in humans	
	Thyroid-stimulating hormone (TSH)	Thyroid	Stimulates thyroid	*See* Thyroid
	Adrenocorticotropic hormone (ACTH)	Adrenal cortex	Stimulates adrenal cortex	*See* Adrenal cortex
	Gonadotropic hormones	Gonads	Controls gamete and sex hormone production	*See* Testes and Ovaries
Thyroid	Thyroxin	All tissues	Increases metabolic rate; helps to regulate growth and development	Exophthalmic goiter/ simple goiter, myxedema, cretinism
	Calcitonin	Bones, kidneys, intestine	Lowers blood calcium level	Tetany/weak bones
Parathyroids	Parathyroid hormone (PTH)	Bones, kidneys, intestine	Raises blood calcium level	Weak bones/tetany
Adrenal medulla	Epinephrine and norepinephrine	Cardiac and other muscles	Stimulate fight or flight reactions; raise blood glucose level	
Adrenal cortex	Glucocorticoids (e.g., cortisol)	All tissues	Raise blood glucose level; stimulate breakdown of protein	
	Mineralocorticoids (e.g., aldosterone)	Kidneys	Stimulate kidneys to reabsorb sodium and to excrete potassium	Cushing syndrome/ Addison disease
	Sex hormones	Sex organs, skin, muscles, bones	Stimulate development of secondary sex characteristics (particularly in male)	
Pancreas	Insulin	Liver, muscles, adipose tissue	Lowers blood glucose level; promotes formation of glycogen, proteins, and fats	Shock/diabetes mellitus
	Glucagon	Liver, muscles, adipose tissue	Raises blood glucose level; promotes breakdown of glycogen, proteins, and fats	
Gonads Testes	Androgens (testosterone)	Sex organs, skin, muscles, bones	Stimulate spermatogenesis; develop and maintain secondary male sex characteristics	Diverse/feminization
Ovaries	Estrogen and progesterone	Sex organs, skin, muscles, bones	Stimulate growth of uterine lining; develop and maintain secondary female sex characteristics	Diverse/masculinization
Thymus	Thymosins	T lymphocytes	Stimulate maturation of T lymphocytes	
Pineal gland	Melatonin	Circadian rhythms	Involved in circadian and circannual rhythms; possibly involved in maturation of sex organs	

*The word diverse in this table means that the symptoms have not been described as a syndrome in the medical literature.

Posterior Pituitary: Stores Two Hormones

The posterior pituitary is connected to the hypothalamus by means of a stalklike structure. Neurons in the hypothalamus, called *neurosecretory cells,* respond to neurotransmitters and produce the hormones that are stored in and released from the posterior pituitary (fig. 13.3). One hormone, called ADH, which was discussed in chapter 9, promotes the reabsorption of water from the collecting duct, a portion of the nephron. When osmoreceptors in the hypothalamus determine that blood is too concentrated, ADH is released into the bloodstream from the axon endings in the posterior pituitary. Once blood is diluted, the hormone is no longer released. This is an example of control by negative feedback (fig. 13.4).

The inability to produce ADH causes **diabetes insipidus** (watery urine), in which a person produces copious amounts of urine with a resultant loss of salts from blood. This condition can be corrected with the administration of ADH.

Oxytocin is another hormone that is made in the hypothalamus and stored in the posterior pituitary. Oxytocin causes the uterus to contract and is used to artificially induce labor. It also stimulates the release of milk from the mother's mammary glands when her baby is nursing.

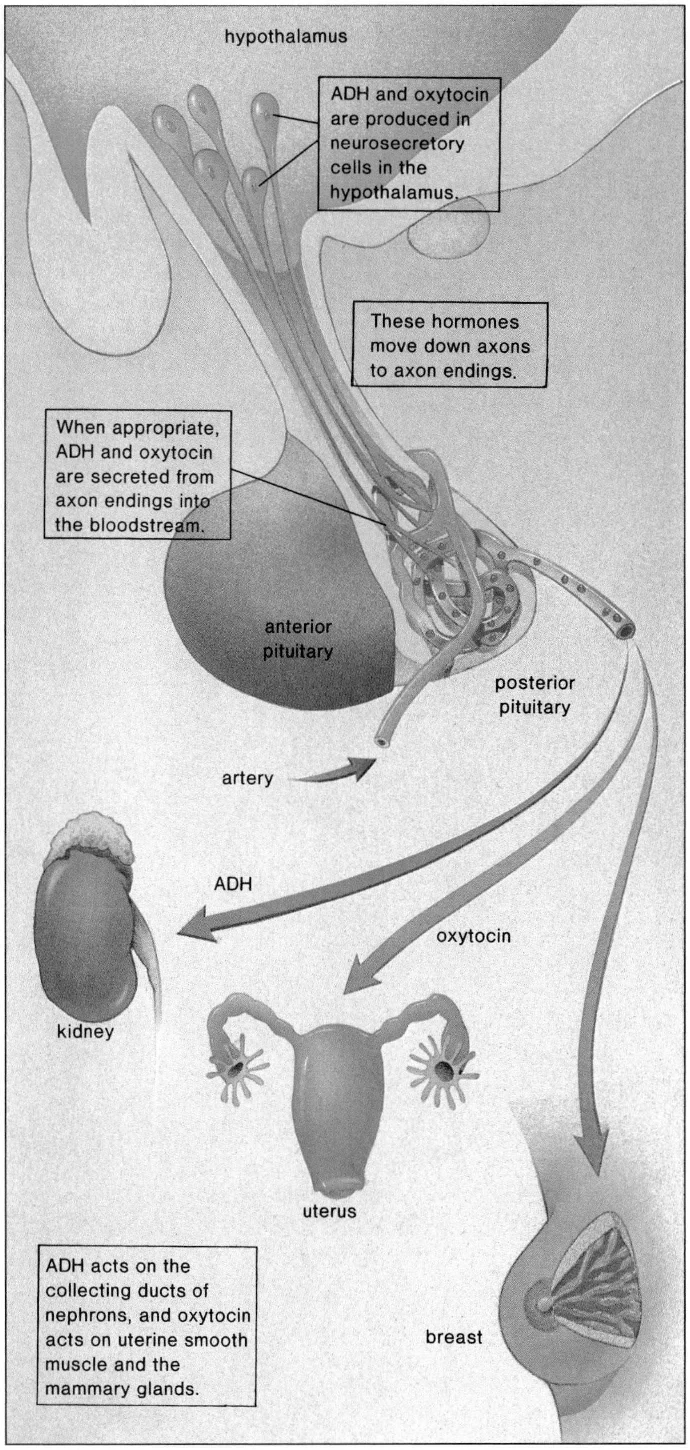

hypothalamus
ADH and oxytocin are produced in neurosecretory cells in the hypothalamus.
These hormones move down axons to axon endings.
When appropriate, ADH and oxytocin are secreted from axon endings into the bloodstream.
anterior pituitary
posterior pituitary
artery
ADH
kidney
oxytocin
uterus
breast
ADH acts on the collecting ducts of nephrons, and oxytocin acts on uterine smooth muscle and the mammary glands.

Figure 13.3

The hypothalamus produces 2 hormones, ADH and oxytocin, which are stored in and secreted by the posterior pituitary.

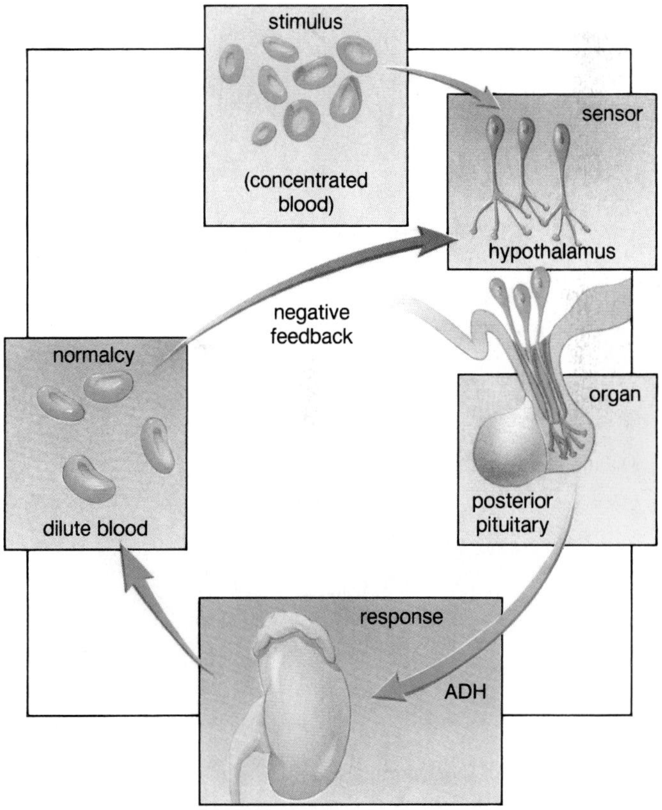

stimulus
(concentrated blood)
sensor
hypothalamus
negative feedback
normalcy
dilute blood
organ
posterior pituitary
response
ADH

Figure 13.4

Regulation of ADH secretion. When the blood is concentrated, neurosecretory cells in the hypothalamus produce ADH, which is released by the posterior pituitary. ADH causes the kidneys to reabsorb water. Once blood is diluted, ADH is no longer secreted.

It is appropriate to note that the neurosecretory cells in the hypothalamus provide an example of a way the nervous system and the endocrine system are joined. This topic is discussed again later.

> The posterior pituitary stores and releases 2 hormones, ADH and oxytocin, both of which are produced by neurosecretory cells in the hypothalamus.

Anterior Pituitary: Master Gland

The hypothalamus controls the anterior pituitary by producing *hypothalamic-releasing* and *release-inhibiting hormones.* For example, there is a thyroid-releasing hormone (TRH) and a thyroid-release-inhibiting hormone (TRIH).

Releasing and release-inhibiting hormones are transported from the hypothalamus to the anterior pituitary by way of a portal system connecting the 2 organs (fig. 13.5).

> The anterior pituitary is controlled by the hypothalamic-releasing hormones and hypothalamic-release-inhibiting hormones. These are produced in neurosecretory cells in the hypothalamus and pass to the anterior pituitary by way of a portal system.

Three of these hormones have a direct effect on the body. **Growth hormone (GH),** or somatotropin, dramatically affects physical appearance since it determines the height of the individual (fig. 13.6). If little or no GH is secreted by the anterior pituitary during childhood, the per-

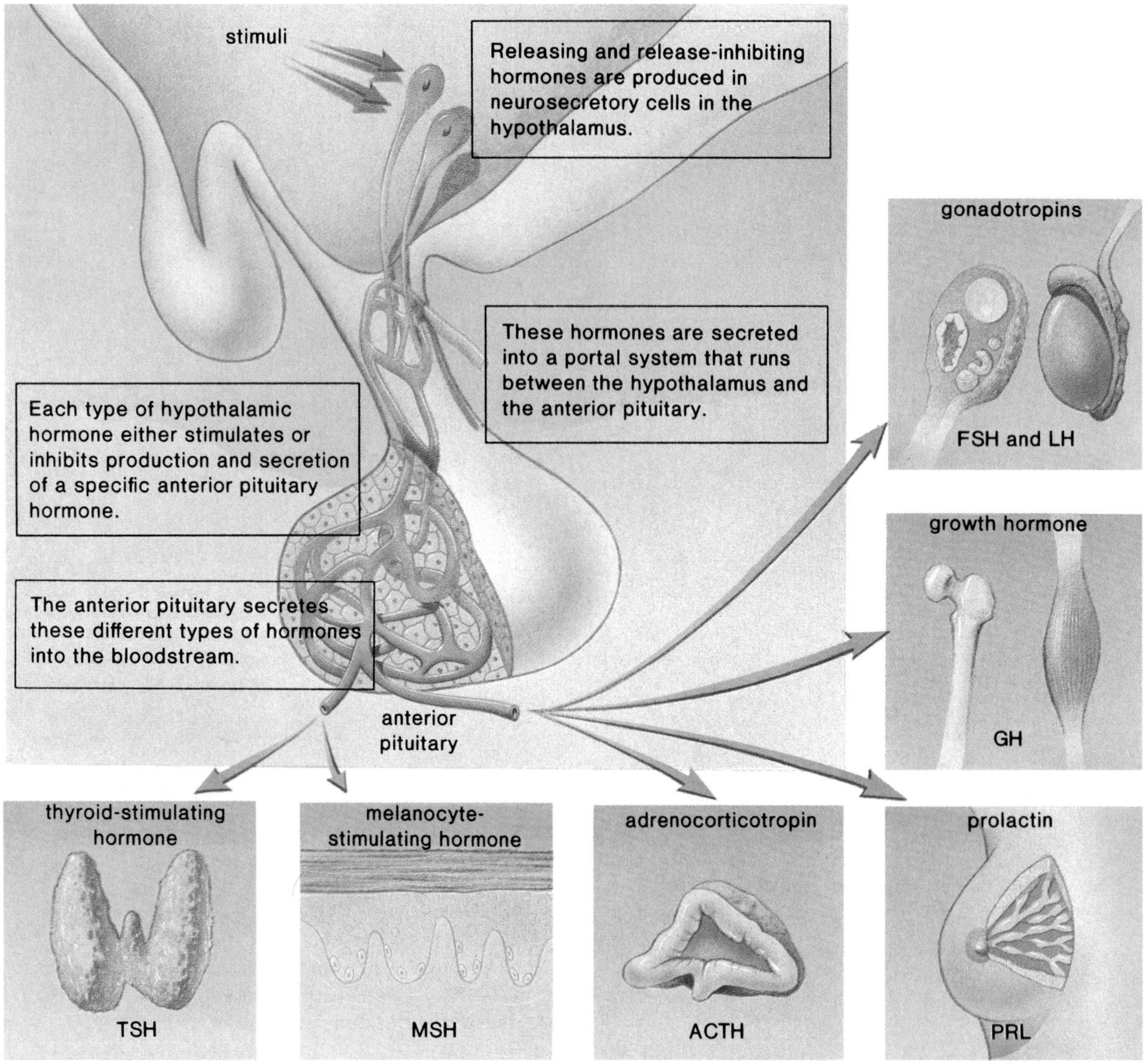
stimuli
Releasing and release-inhibiting hormones are produced in neurosecretory cells in the hypothalamus.
These hormones are secreted into a portal system that runs between the hypothalamus and the anterior pituitary.
Each type of hypothalamic hormone either stimulates or inhibits production and secretion of a specific anterior pituitary hormone.
The anterior pituitary secretes these different types of hormones into the bloodstream.
anterior pituitary
gonadotropins
FSH and LH
growth hormone
GH
thyroid-stimulating hormone
TSH
melanocyte-stimulating hormone
MSH
adrenocorticotropin
ACTH
prolactin
PRL

Figure 13.5

Hypothalamus and anterior pituitary. The hypothalamus controls the secretions of the anterior pituitary, and the anterior pituitary controls the secretions of these other endocrine glands.

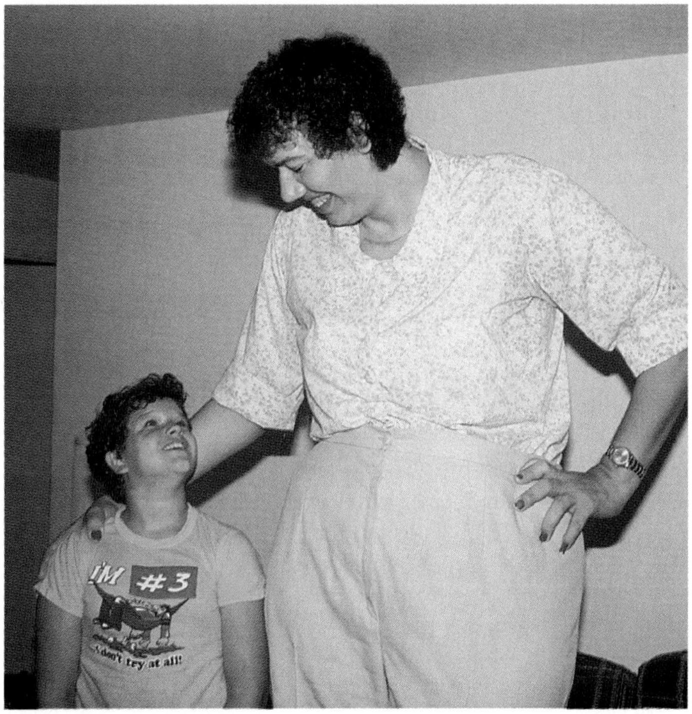

Figure 13.6

Giantism. Sandy Allen is one of the world's tallest women due to a higher than usual amount of GH, produced by the anterior pituitary.

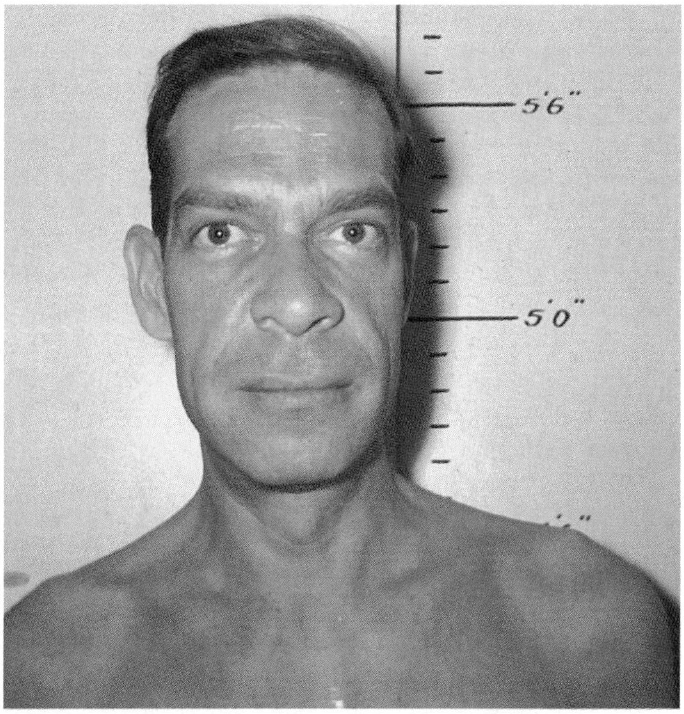

Figure 13.7

Acromegaly. This condition is caused by overproduction of GH in the adult. It is characterized by an enlargement of the bones in the face, the fingers, and the toes of an adult.

son will probably become a pituitary dwarf—although of perfect proportions, the individual will be quite small in stature. If too much GH is secreted, the person will probably become a giant. Giants usually have poor health, primarily because GH has a secondary effect on blood sugar level, promoting an illness called diabetes mellitus, discussed later.

GH promotes cell division, protein synthesis, and bone growth. It stimulates the transport of amino acids into cells and increases the activity of ribosomes, both of which are essential to protein synthesis. In bones, it promotes growth of the cartilaginous disks and causes osteoblasts to form bone (p. 209). Evidence suggests that the effects on cartilage and bone may actually be due to hormones called somatomedins, released by the liver. GH causes the liver to release somatomedins.

If the production of GH increases in an adult after full height has been attained, only certain bones respond. These are the bones of the jaw, the eyebrow ridges, the nose, the fingers, and the toes. When these begin to grow, the person takes on a slightly grotesque look, with enlargement of facial features and huge fingers and toes, a condition called **acromegaly** (fig. 13.7).

Prolactin (PRL) is produced in quantity only after childbirth. It causes the mammary glands in the breasts to develop and to produce milk.

Melanocyte-stimulating hormone (MSH) regulates skin color in lower vertebrates, but its function in humans is obscure. However, it is derived from a molecule that is also the precursor for adrenocorticotropic hormone (ACTH).

> GH and PRL are 2 hormones produced by the anterior pituitary. GH influences the height of children, and overproduction brings about a condition called acromegaly in adults. PRL promotes milk production after childbirth.

Other Hormones Produced by the Anterior Pituitary

The anterior pituitary sometimes is called the *master gland* because it controls the secretion of certain other endocrine glands (see fig. 13.5). As indicated in table 13.1, the anterior pituitary secretes the following hormones, which have an effect on other glands:

1. **Thyroid-stimulating hormone (TSH),** a hormone that stimulates the thyroid;
2. **Adrenocorticotropic hormone (ACTH),** a hormone that stimulates the adrenal cortex;
3. **Gonadotropic hormones**, which stimulate the gonads—the testes in males and the ovaries in females.

TSH causes the thyroid to produce thyroxin; ACTH causes the adrenal cortex to produce cortisol; and gonadoptropic hormones cause the gonads to secrete sex hormones. Notice that it is now possible to indicate a 3-tiered relationship between the hypothalamus, the anterior pituitary, and the other endocrine glands. The hypothalamus produces hormones, which control the anterior pituitary, and the anterior pituitary produces hormones that control the thyroid, the adrenal cortex, and the gonads. Figure 13.8 illustrates the negative feedback mechanism that controls the activity of all these glands.

The hypothalamus, the anterior pituitary, and the other endocrine glands controlled by the anterior pituitary are all involved in a self-regulating negative feedback system.

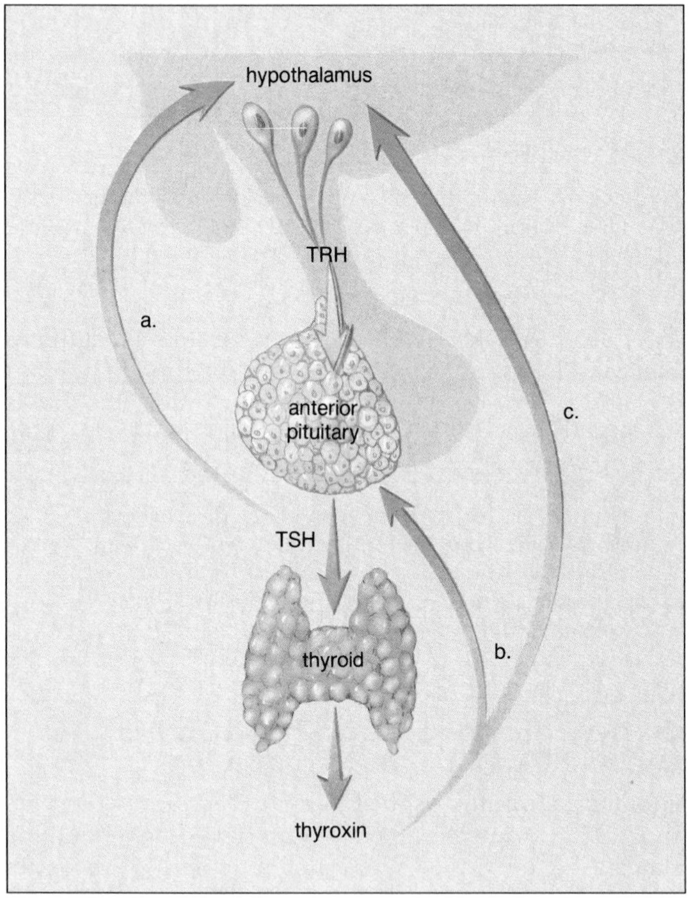

Figure 13.8

The hypothalamus-pituitary-thyroid control relationship. TRH (thyroid-releasing hormone) stimulates the anterior pituitary, and TSH (thyroid-stimulating hormone) stimulates the thyroid to secrete thyroxin. The level of thyroxin in the body is negatively controlled in 3 ways: **a.** The level of TSH exerts feedback control over the hypothalamus. **b.** The level of thyroxin exerts feedback control over the anterior pituitary. **c.** The level of thyroxin exerts feedback control over the hypothalamus. In this way, thyroxin controls its own secretion. Cortisol and sex hormone levels are controlled similarly.

Thyroid and Parathyroid Glands

The thyroid gland is located in the neck and is attached to the trachea just below the larynx (see fig. 13.1). The parathyroid glands are embedded in the posterior surface of the thyroid gland.

Thyroid Gland: Speeds up Metabolism

The **thyroid gland** is composed of a large number of follicles filled with thyroglobulin, the storage form of thyroxin. The production of both of these requires iodine. Iodine is actively transported into the thyroid gland, where the concentration can become as much as 25 times that of blood. If iodine is lacking in the diet, the thyroid gland enlarges, producing a goiter (fig. 13.9). In the United States, salt is iodized for this reason.

The cause of goiter becomes clear if we refer to figure 13.8. When there is a low level of thyroxin in blood, a condition called hypothyroidism, the anterior pituitary is stimulated to produce TSH. TSH causes the thyroid to increase in size so that enough **thyroxin** usually is produced. In this case, enlargement continues because enough thyroxin is never produced. An enlarged thyroid that produces some thyroxin is called a **simple goiter.**

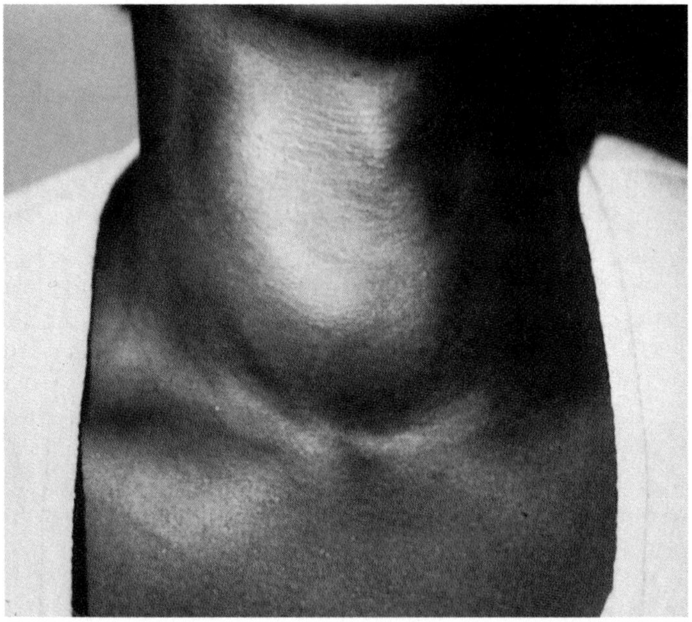

Figure 13.9

Simple goiter. An enlarged thyroid gland often is caused by a lack of iodine in the diet. Without iodine, the thyroid is unable to produce thyroxin, and continued anterior pituitary stimulation causes the gland to enlarge.

Thyroxin does not have a target organ; instead, it stimulates most of the cells of the body to metabolize at a faster rate. The number of respiratory enzymes in the cell increases, as does oxygen (O_2) uptake.

If the thyroid fails to develop properly, a condition called **cretinism** results. Individuals with this condition are short and stocky, and have had extreme hypothyroidism since infancy and/or childhood (fig. 13.10). Thyroxin therapy can initiate growth, but unless treatment is begun within the first 2 months of life, mental retardation results. The occurrence of hypothyroidism in adults produces the condition known as **myxedema** (fig. 13.11), which is characterized by lethargy, weight gain, hair loss, slowed pulse rate, decreased body temperature, and thickness and puffiness of the skin. The administration of adequate doses of thyroxin restores normal function and appearance.

In the case of hyperthyroidism (too much thyroxin), the thyroid gland is enlarged and overactive, causing a goiter to form and the eyes to protrude because of edema in the tissues of the eye sockets and swelling of muscles that move the eyes. This type of goiter is called **exophthalmic goiter** or Graves' disease (fig. 13.12). The individual usually becomes hyperactive, nervous, irritable, and suffers from insomnia. Removal or destruction of a portion of the thyroid by means of radioactive iodine sometimes is effective in curing the condition.

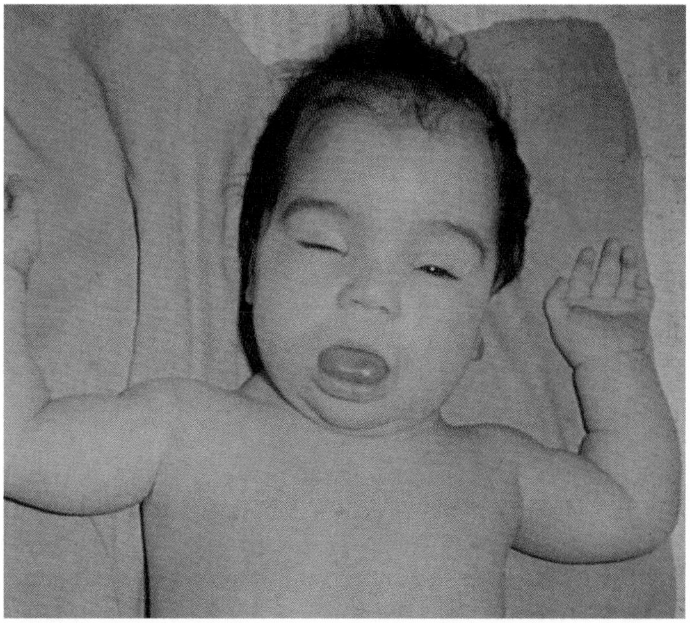

Figure 13.10

Cretinism. Individuals with this condition have suffered from thyroxin insufficiency since birth or early childhood. Skeletal growth is usually inhibited to a greater extent than soft tissue growth; therefore, the child appears short and stocky. Sometimes, the tongue becomes so large that it obstructs swallowing and breathing. Mental retardation also occurs.

In addition to thyroxin, the thyroid gland produces the hormone **calcitonin.** This hormone helps to regulate the calcium level in blood and opposes the action of parathyroid hormone. The interaction of these 2 hormones is discussed following.

> The anterior pituitary produces TSH, a hormone that promotes the production of thyroxin by the thyroid, a gland subject to goiters. Thyroxin speeds up metabolism, and an undersecretion can affect the body as a whole, as exemplified by cretinism and myxedema.

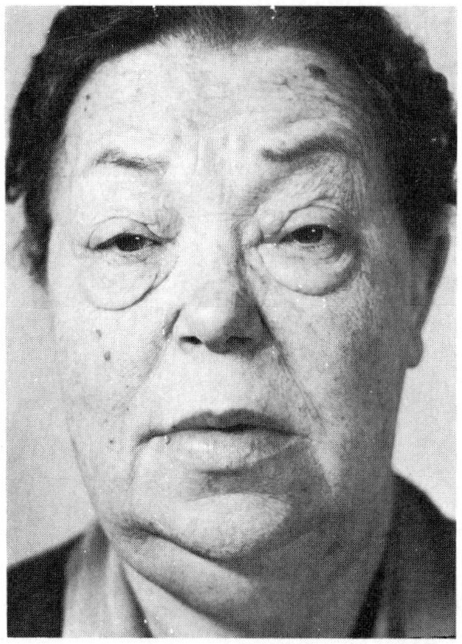

Figure 13.11

Myxedema. This condition is caused by thyroxin insufficiency in the older adult. An unusual type of edema leads to swelling of the face and bagginess under the eyes.

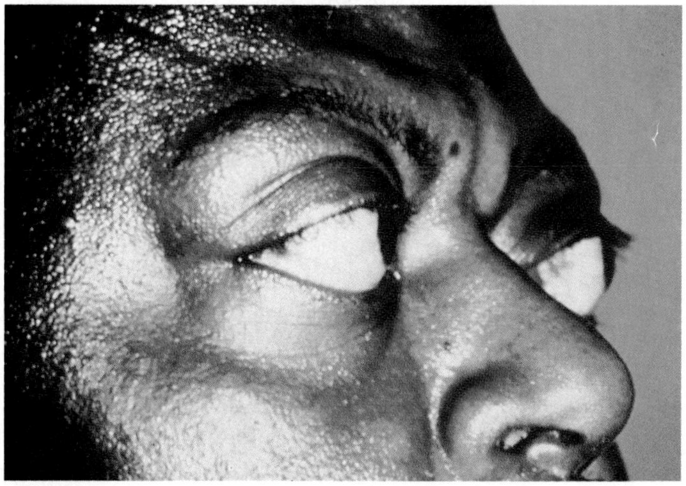

Figure 13.12

Exophthalmic goiter (Graves' disease). In hyperthyroidism, the eyes protrude because of edema in the tissues of the eye sockets.

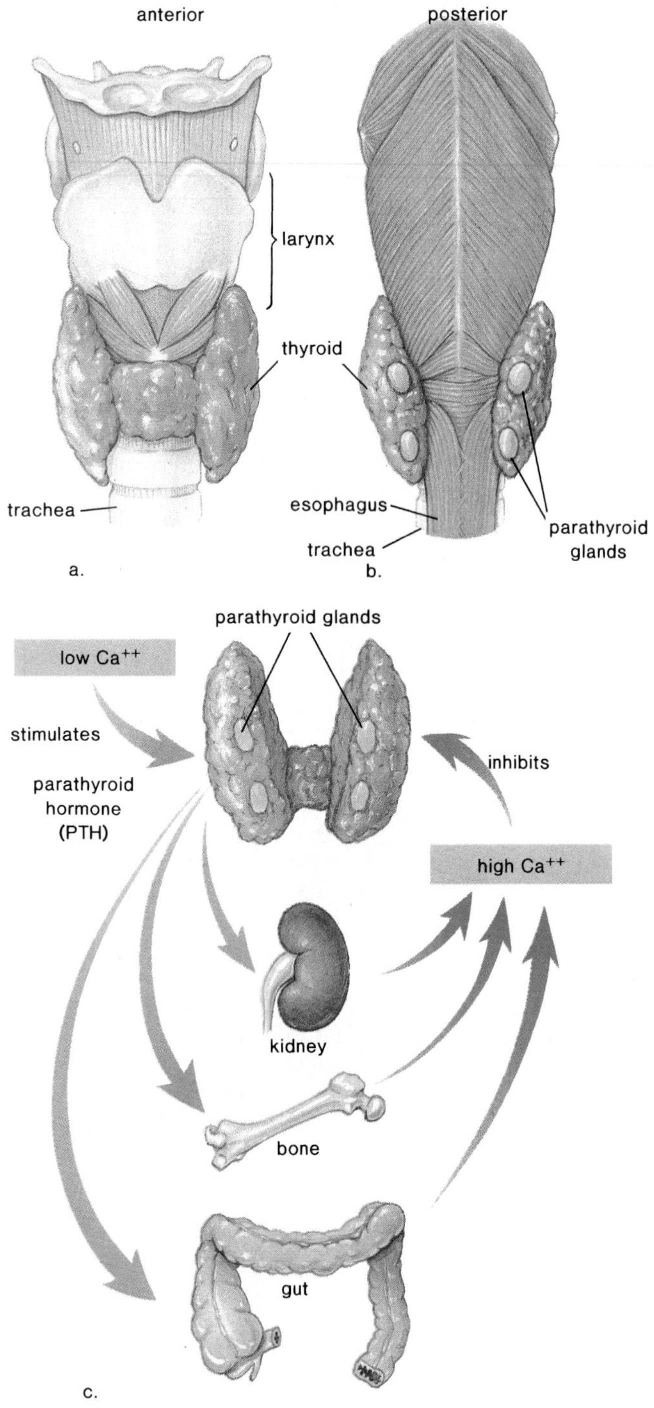

Figure 13.13

Thyroid and parathyroid glands. **a.** The thyroid gland is located in the neck in front of the trachea. **b.** The 4 parathyroid glands are embedded in the posterior surface of the thyroid gland. Yet, the parathyroid and thyroid glands have no anatomical or physiological connection with one another. **c.** Regulation of parathyroid hormone (PTH) secretion. A low blood calcium level causes the parathyroids to secrete PTH, which causes the kidneys and the intestine to retain calcium and osteoclasts to break down bone. The end result is a higher blood calcium level. A high blood calcium level inhibits secretion of PTH.

Parathyroid Glands: Regulate Calcium

The **parathyroid glands** are embedded in the posterior surface of the thyroid gland, as shown in figure 13.13*b*. Under the influence of **parathyroid hormone (PTH)**, the blood calcium level rises and the phosphate level lowers. The hormone stimulates the absorption of calcium from the digestive tract, the retention of calcium by the kidneys, and the demineralization of bone. In other words, PTH promotes the activity of *osteoclasts,* the bone-absorbing cells. Although this also raises the level of phosphate in blood, PTH acts on the kidneys to excrete phosphate in the urine.

If insufficient PTH is produced, the blood calcium level lowers, resulting in **tetany.** In tetany, the body shakes from continuous muscle contraction. The effect really is brought about by increased excitability of the nerves, which initiate nerve impulses spontaneously and repeatedly. Calcium plays an important role in both nervous conduction and muscle contraction.

The level of PTH secretion is controlled by a negative feedback mechanism involving calcium (fig. 13.13*c*). When the calcium level rises, PTH secretion is inhibited, and when the calcium level lowers, PTH secretion is stimulated.

As mentioned previously, the thyroid secretes calcitonin, which also influences blood calcium level. Although calcitonin has the opposite effect of PTH, particularly on the bones, its action is not believed to be as significant. Still, the 2 hormones function together to regulate the blood calcium level.

> PTH maintains a high blood level of calcium by promoting its absorption in the digestive tract, its reabsorption by the kidneys, and the demineralization of bone. These actions are opposed by calcitonin, produced by the thyroid.

Adrenal Glands

The **adrenal glands,** as their name implies (*ad* means *near; renal* means *kidney*), lie atop the kidneys (see fig. 13.1). Each consists of an inner portion, called the *medulla,* and an outer portion, called the *cortex.* These portions, like the anterior pituitary and the posterior pituitary, have no functional connection with one another.

Adrenal Medulla: Fight or Flight

The *adrenal medulla* secretes *norepinephrine* and epinephrine under conditions of stress. They bring about all the responses we associate with the "fight or flight" reaction: the blood glucose level and the metabolic rate increase, as do breathing and the heart rate. The blood vessels in the intestine constrict, and those in the muscles dilate. This increased circulation to the muscles causes them to have more stamina than usual. In times of emergency, the sympathetic nervous system *initiates* these responses, but they are maintained by secretions from the adrenal medulla. People can perform feats of strength such as lifting up a car to free a child.

The adrenal medulla is innervated by only one set of sympathetic nerve fibers. Recall from chapter 10 that usually there are pre- and postganglionic nerve fibers for each organ stimulated. In this instance, what happened to the postganglionic neurons? It appears that the adrenal medulla may have evolved from a modification of the postganglionic neurons. Like the neurosecretory neurons in the hypothalamus, these neurons also secrete hormones into the bloodstream.

> The adrenal medulla releases norepinephrine and epinephrine into the bloodstream. These hormones help us and other animals to cope with situations that threaten survival.

Adrenal Cortex: Keeps Things Steady

Although the adrenal medulla can be removed with no ill effects, the adrenal cortex is absolutely necessary to life. The 2 major classes of hormones made by the adrenal cortex are the *glucocorticoids* and the *mineralocorticoids*. The *adrenal cortex* also secretes a small amount of male sex hormone and an even smaller amount of female sex hormone. All of these hormones are steroids.

Of the various glucocorticoids, the hormone responsible for the greatest amount of activity is **cortisol.** Cortisol promotes the hydrolysis of muscle protein to amino acids, which enter blood. This leads to a higher blood glucose level when the liver converts these amino acids to glucose. Cortisol also favors metabolism of fatty acids rather than carbohydrate. In opposition to insulin, therefore, cortisol raises the blood glucose level. Cortisol also counteracts the inflammatory response, which leads to the pain and the swelling of joints in arthritis and bursitis. The administration of cortisol aids these conditions because it reduces inflammation.

The secretion of cortisol by the adrenal cortex is under the control of the anterior pituitary hormone ACTH (adrenocorticotropic hormone). Using the same kind of system shown in figure 13.8, the hypothalamus produces a cortisol-releasing hormone (CRH) that stimulates the anterior pituitary to release ACTH. ACTH, in turn, stimulates the adrenal cortex to secrete cortisol, which regulates its own synthesis by negative feedback on both CRH and ACTH synthesis.

The secretion of mineralocorticoids, the most significant of which is **aldosterone,** is not under the control of the anterior pituitary. Aldosterone regulates the blood sodium and potassium levels. Its primary target organ is the kidney, where it promotes renal absorption of sodium and renal excretion of potassium. The blood sodium level is particularly important to the maintenance of blood pressure because its concentration indirectly regulates the secretion of aldosterone (fig. 13.14). When the blood sodium level is low, the kidneys secrete renin. Renin is an enzyme that converts the plasma protein angiotensinogen to angiotensin I, which is changed to angiotensin II by a converting enzyme found in the lungs. Angiotensin II stimulates the adrenal cortex to release aldosterone (see fig. 9.8). The effect of this system, called the renin-angiotensin-aldosterone system, is to raise the blood pressure in 2 ways. First, angiotensin II constricts the arteries directly, and secondly, aldosterone causes the kidneys to reabsorb sodium. When the blood sodium level is high, water is reabsorbed, and blood volume and pressure are maintained.

The heart produces a hormone that acts contrary to aldosterone. This hormone is called the atrial natriuretic hormone because (1) it is produced by the atria of the heart and (2) it causes natriuresis, the excretion of sodium. Once sodium is excreted, so is water; therefore, blood volume and blood pressure decrease.

> Cortisol, which raises the blood glucose level, and aldosterone, which raises the blood sodium level, are 2 hormones secreted by the adrenal cortex.

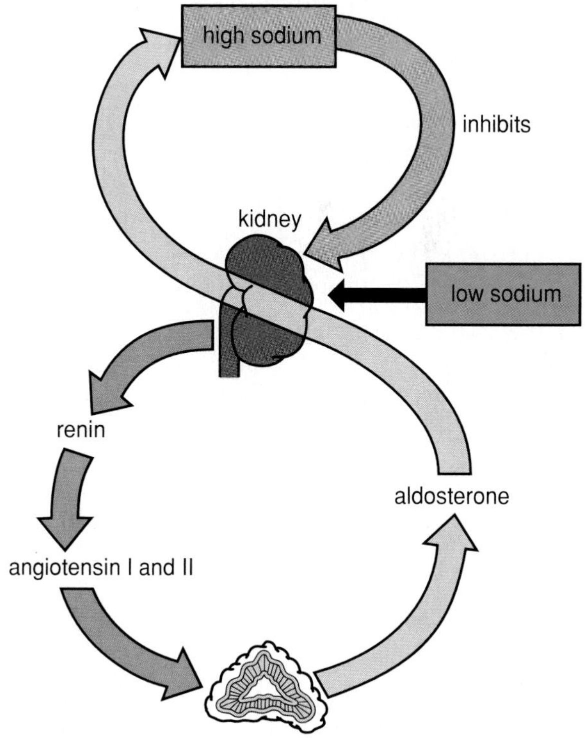

Figure 13.14

Renin-angiotensin-aldosterone system. If the blood sodium level is low, the kidneys secrete renin. Thereafter, there are increased levels of angiotensin I and II in the blood. Angiotensin II stimulates the adrenal cortex to release aldosterone. Aldosterone promotes reabsorption of sodium by the kidneys. When the blood sodium level rises, the kidneys stop secreting renin.

The adrenal cortex produces a small amount of both male and female sex hormones. In males, the cortex is a source of female sex hormones, and in females, it is a source of male hormones. A tumor in the adrenal cortex can cause the production of a large amount of sex hormones, which can lead to feminization in males and masculinization in females.

Disorders: Bronze Skin and Moon Face

When the level of adrenal cortex hormones is low, Addison disease results. When the level of adrenal cortex hormones in the body is high, Cushing syndrome results.

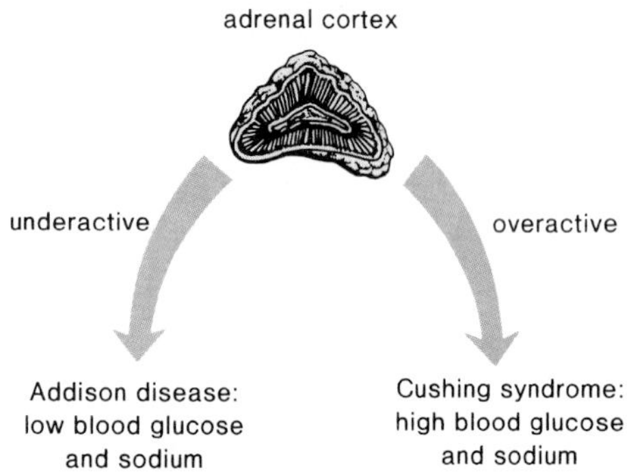

Because of the lack of cortisol, a person with **Addison disease** is unable to maintain the blood glucose level, tissue repair is suppressed, and there is a high susceptibility to any kind of bodily stress. Even a mild infection can cause death. Due to the lack of aldosterone, the blood sodium level is low, and the person experiences low blood pressure, along with low blood pH, called acidosis. In addition, the individual's skin has a peculiar bronze cast (fig. 13.15).

In **Cushing syndrome,** a high level of cortisol causes a tendency toward diabetes mellitus, a decrease in muscle protein, and an increase in subcutaneous fat. Because of these effects, the person usually develops thin arms and legs and an enlarged trunk. Due to the high level of sodium, blood is basic and the individual is hypertensive and has edema of the face, which gives it a moon shape (fig. 13.16).

Addison disease is due to adrenal cortex hyposecretion, and Cushing syndrome is due to adrenal cortex hypersecretion.

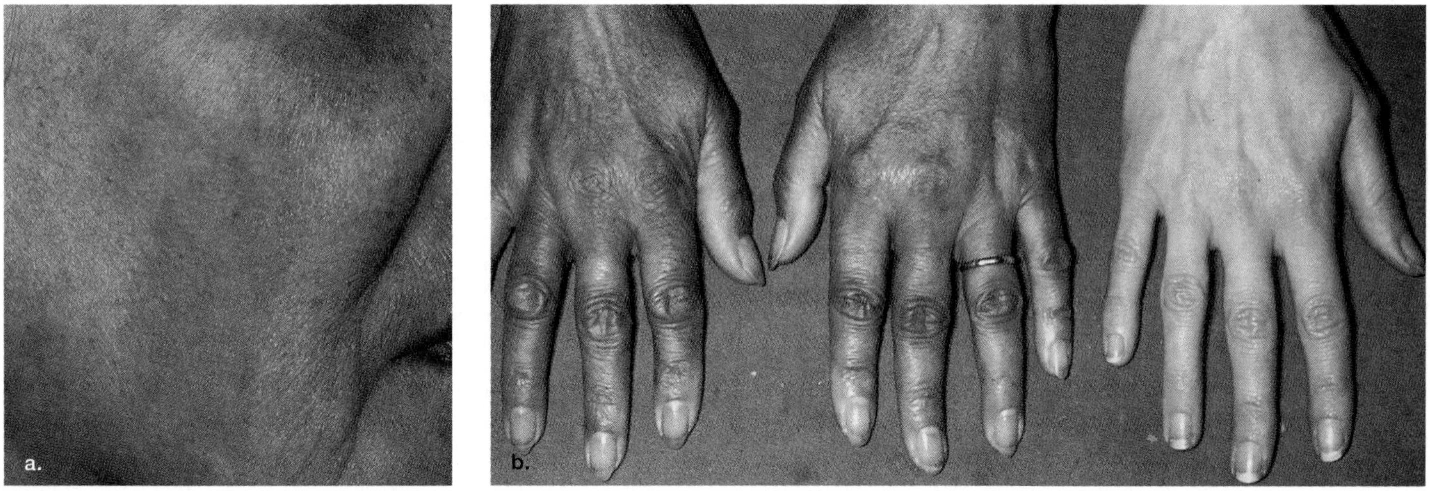

Figure 13.15

Addison disease. This condition is characterized by a peculiar bronzing of the skin. Note the color of **(a.)** the face and **(b.)** the hands compared to the hand of a normal individual.

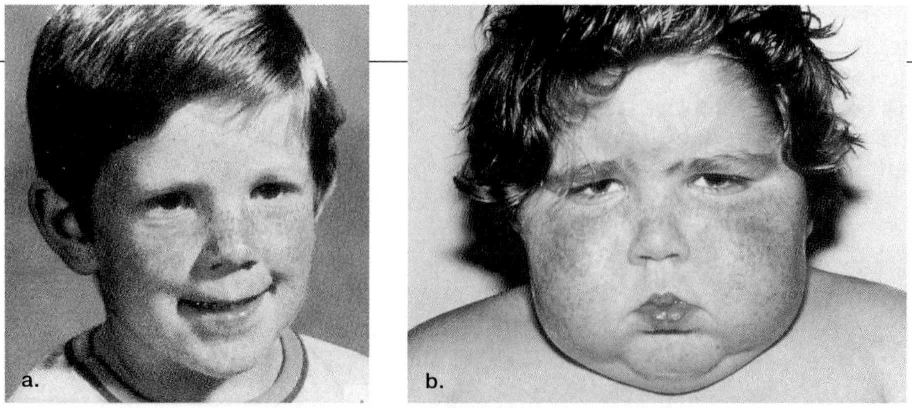

Figure 13.16

Cushing syndrome. This condition results from hypersecretion of glucocorticoid hormone by a tumor of the zona fasciculata of the adrenal cortex. **a.** First diagnosed with Cushing syndrome. **b.** Four months later.

The Pancreas

The **pancreas** is a long organ that lies transversely in the abdomen between the kidneys and near the duodenum of the small intestine (fig. 13.17). It is composed of 2 types of tissue—exocrine, which produces and secretes *digestive* juices that go by way of ducts to the small intestine, and endocrine, called the **pancreatic islets (of Langerhans)**, which produces and secretes the hormones **insulin** and **glucagon** directly into the blood.

All the cells of the body use glucose as an energy source; to preserve the health of the body, it is important that the glucose concentration remain within normal limits. Insulin is secreted when there is a high level of glucose in blood, which usually occurs just after eating. Insulin has 3 different actions: (1) it stimulates liver, fat, and muscle cells to take up and metabolize glucose; (2) it stimulates the liver and the muscles to store glucose as glycogen; and (3) it promotes the buildup of fats and proteins and inhibits their use as an energy source. Therefore, insulin is a hormone that promotes storage of nutrients so that they are on hand during leaner times. It also helps to lower the blood glucose level.

Glucagon is secreted from the pancreas in between eating, and its effects are opposite to those of insulin. Glucagon stimulates the breakdown of stored nutrients and causes the blood glucose level to rise (fig. 13.18).

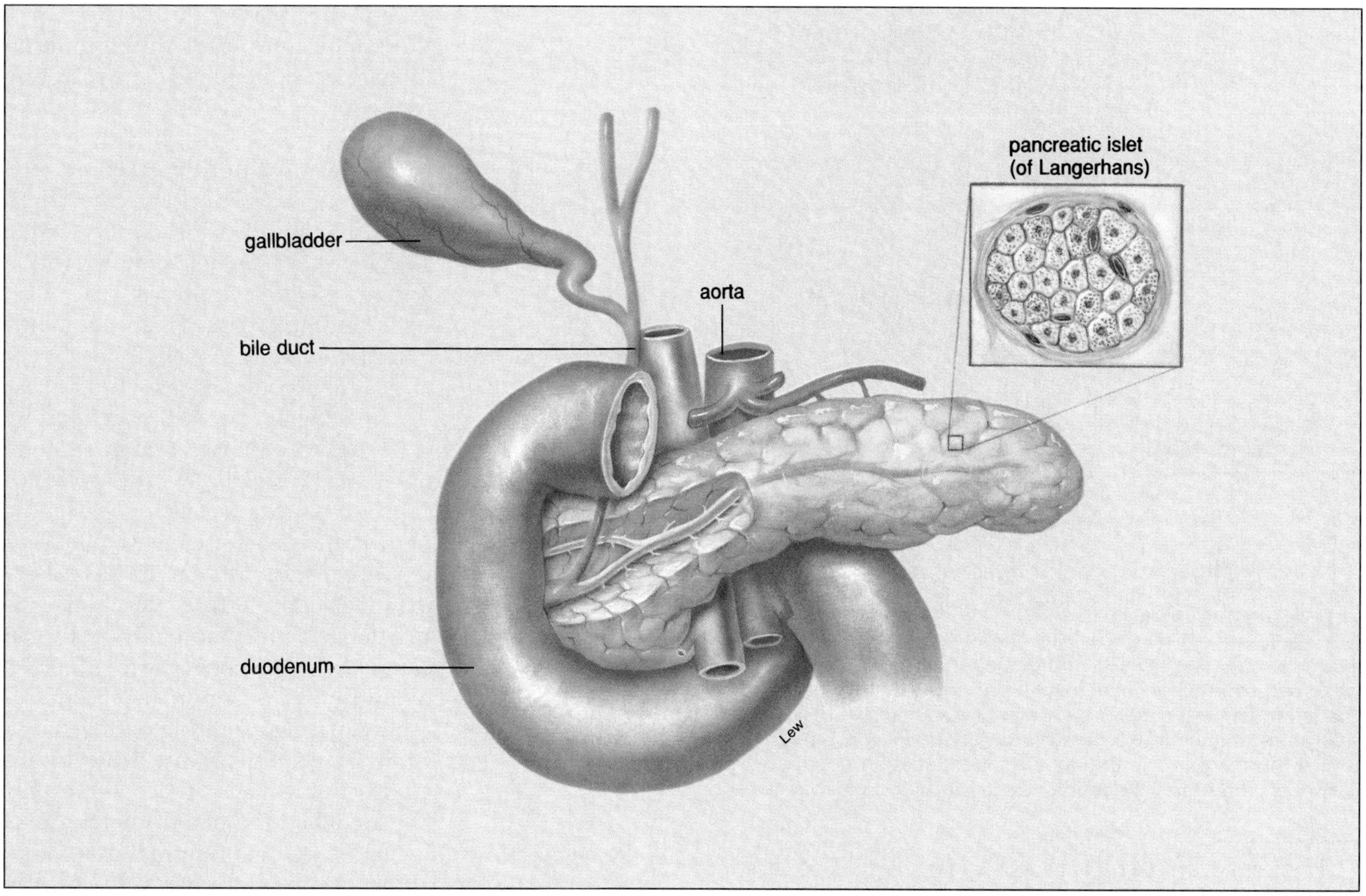

Figure 13.17

Gross and microscopic anatomy of the pancreas. The pancreas lies in the abdomen; extending from the C-shaped curvature of the duodenum to the left. As an exocrine gland, it secretes digestive enzymes that enter the duodenum by way of a duct utilized also by bile. As an endocrine gland, the pancreatic islets (of Langerhans) secrete insulin and glucagon into the bloodstream.

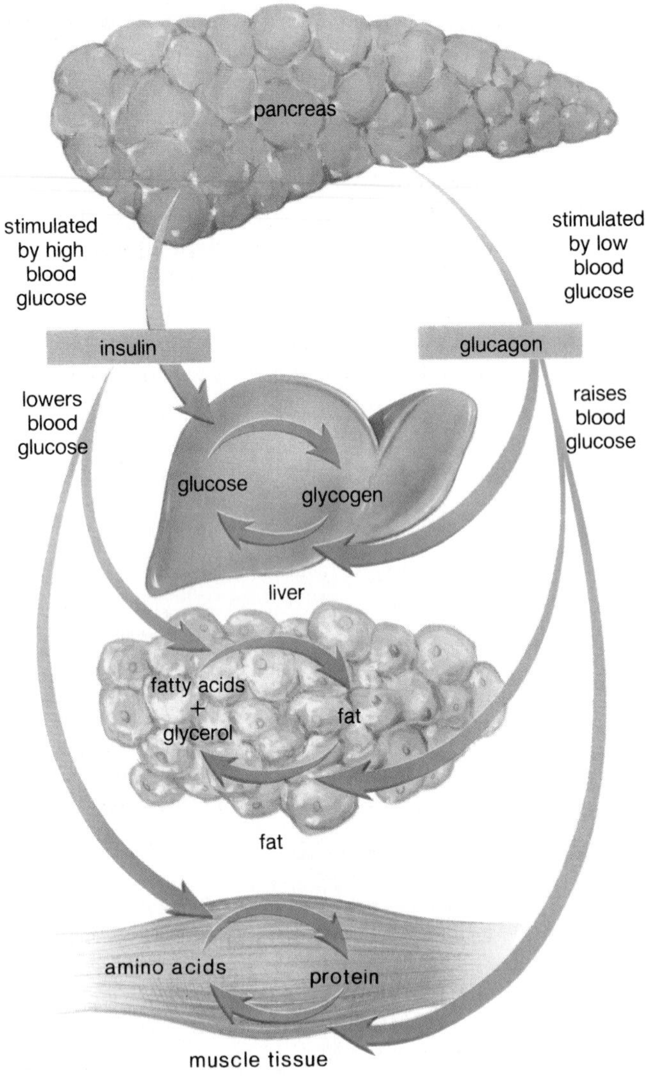

Figure 13.18

Contrary effects of insulin and glucagon. When blood glucose level is high, the pancreas secretes insulin. Insulin promotes the storage of glucose as glycogen and the synthesis of proteins and fats as opposed to their use as energy sources. Therefore, insulin lowers the blood glucose level. When the blood glucose level is low, the pancreas secretes glucagon. Glucagon acts in opposition to insulin in all respects; therefore, glucagon raises the blood glucose level.

Diabetes Mellitus: Insulin Lack and Insensitive Cells

The symptoms of **diabetes mellitus** include the following:

Sugar in the urine
Frequent, copious urination
Abnormal thirst
Rapid weight loss
General weakness
Drowsiness and fatigue
Itching of the genitals and the skin
Visual disturbances, blurring
Skin disorders, such as boils, carbuncles, and infection.

Many of these symptoms develop because sugar is not being metabolized by the cells. The liver fails to store glucose as glycogen, and all the cells fail to utilize glucose as an energy source. This means that the blood glucose level rises very high after eating, causing glucose to be excreted in the urine. More water than usual is therefore excreted so that the diabetic is extremely thirsty.

Since carbohydrate is not being metabolized, the body turns to the breakdown of protein and fat for energy. Unfortunately, the breakdown of these molecules leads to the buildup of ketones in the blood. The resulting reduction in blood volume and acidosis (acid blood) can eventually lead to coma and death of the diabetic. The symptoms of hyperglycemia (high blood sugar) develop slowly (table 13.2), and there is time for intervention and reversal of symptoms.

There are 2 types of diabetes. In *type I (insulin-dependent) diabetes,* the pancreas is not producing insulin. The condition is believed to be brought on by exposure to an environmental agent, most likely a virus, whose presence causes cytotoxic T cells to destroy the pancreatic islets (of Langerhans). As a result, the individual must have daily insulin injections. These injections control the diabetic symptoms but still can cause inconveniences since either an overdose of insulin or the absence of regular eating can bring on the symptoms of hypoglycemia (low blood sugar) (table 13.2). These symptoms appear when the blood glucose level falls below normal levels. Since the brain requires a constant supply of sugar, unconsciousness can result. The cure is quite simple: an immediate source of sugar, such as a sugar cube or fruit juice, can very quickly counteract hypoglycemia.

Obviously, insulin injections are not the same as a fully functioning pancreas that responds on demand to a high glucose level by supplying insulin. For this reason, some doctors advocate an islet transplant for type I diabetes.

Of the 12 million people who now have diabetes in the United States, at least 10 million have *type II (insulin-independent) diabetes*. This type of diabetes usually occurs in people of any age who are obese and inactive. The pancreas

Table 13.2

Symptoms of Hyperglycemia and Hypoglycemia

Hyperglycemia (High Blood Sugar)	Hypoglycemia (Low Blood Sugar)
Slow, gradual onset	Sudden onset
Dry, hot skin	Perspiration, pale skin
No dizziness	Dizziness
No palpitation	Heart palpitation
No hunger	Hunger
Excessive urination	Normal urination
Excessive thirst	Normal thirst
Deep, labored breathing	Shallow breathing
Fruity breath odor	Normal breath odor
Large amounts of urinary sugar	Urinary sugar absent or slight
Ketones in urine	No ketones in urine
Drowsiness and great lethargy leading to stupor	Confusion, disorientation, strange behavior

From Henry Dolger, M.D. and Bernard Seeman. How to Live with Diabetes. *Copyright © 1972, 1965, 1958 W.W. Norton & Company, Inc., New York, NY. Reprinted by permission of the publisher.*

produces insulin, but the cells do not respond to it. At first, cells lack the receptors necessary to detect the presence of insulin, and later, the organs and tissues listed in figure 13.18 are even incapable of taking up glucose. If type II diabetes is untreated, the results can be as serious as type I diabetes. (Diabetics are prone to blindness, kidney disease, and circulatory disorders, including strokes. Pregnancy carries an increased risk of diabetic coma, and the child of a diabetic is somewhat more likely to be stillborn or to die shortly after birth.) It is important, therefore, to prevent or to at least control type II diabetes. The best defense is a low-fat diet and regular exercise. If this fails, oral drugs that make the cells more sensitive to the effects of insulin or that stimulate the pancreas to make more of it are available.

> Diabetes mellitus is caused by the lack of insulin or the insensitivity of cells to insulin. Insulin lowers blood glucose levels by causing the cells to take up glucose and the liver to convert it to glycogen.

Testes and Ovaries

The sex organs are the testes in the male and the ovaries in the female. The *testes* are located in the scrotum, and the ovaries are located in the abdominal cavity. As is discussed in chapter 14, the testes produce the *androgens* (e.g., *testosterone*), which are the male sex hormones, and the *ovaries* produce estrogen and progesterone, the female sex hormones. The hypothalamus and pituitary gland control the hormonal secretions of these organs in the same manner as described for the thyroid gland in figure 13.8.

The male sex hormone, testosterone, has many functions. It is essential for the normal development and functioning of the sex organs in males. It is also necessary for the maturation of sperm.

Greatly increased testosterone secretion at the time of puberty stimulates the growth of the penis and the testes. Testosterone also brings about and maintains the secondary sex characteristics in males that develop at the time of puberty. Testosterone causes growth of a beard, axillary (underarm) hair, and pubic hair. It prompts the larynx and the vocal cords to enlarge, causing the voice to change. It is responsible for the muscular strength of males, and this is the reason some athletes take supplemental amounts of **anabolic steroids,** which are either testosterone or related chemicals. The contraindications of taking anabolic steroids are discussed in the Health Focus for this chapter. Testosterone also causes oil and sweat glands in the skin to secrete; therefore, it is largely responsible for acne and body odor. Another side effect of testosterone activity is baldness. Genes for baldness probably are inherited by both sexes, but baldness is seen more often in males because of the presence of testosterone.

Testosterone is believed to be largely responsible for the sex drive. It may even contribute to the supposed aggressiveness of males.

The female sex hormones, *estrogen* and *progesterone,* have many effects on the body. In particular, estrogen secreted at the time of puberty stimulates the growth of the uterus and the vagina. Estrogen is necessary for egg maturation and is largely responsible for the secondary sex characteristics in females. For example, it is responsible for female body hair and fat distribution. In general, females have a more rounded appearance than males because of a greater accumulation of fat beneath the skin. Also, the pelvic girdle enlarges in females so that the pelvic cavity has a larger relative size compared to males; this means that females have wider hips. Both estrogen and progesterone also are required for breast development and regulation of the uterine cycle, which includes monthly menstruation (discharge of blood from the uterus).

> The androgens, primarily testosterone, are the male sex hormones produced by the testes; estrogen and progesterone are the female sex hormones produced by the ovaries. The sex hormones maintain the sex organs and the secondary sex characteristics.

Health Focus

Dangers of Anabolic Steroids

Anabolic steroids are synthetic forms of the male sex hormone testosterone. These drugs were developed in the 1930s for medical reasons. They prevent muscular atrophy in patients with debilitating illness, they speed recovery in surgery and burn patients, and they are helpful in rare forms of anemia and breast cancer.

Trainers may have been the first to acquire anabolic steroids for weight lifters, body builders, and other athletes such as professional football players. When taken in large doses (10 to 100 times the amount prescribed by doctors for illnesses), anabolic steroids do promote larger muscles when the abuser also exercises. Theoretically, they function in the manner described in figure 13.2*b*—the end result in muscle cells is increased amounts of the proteins actin and myosin. Every once in a while steroid abuse makes the news because an Olympic winner tests positive for the drug and must relinquish a medal. Steroid use has been outlawed by the International Olympic Committee.

The Federal Drug Administration bans the entry of most steroids, but they are brought into the United States illegally and sold through the mail or in gyms and health clubs. According to federal officials, 1 million to 3 million Americans now take anabolic steroids; of great concern is their increased use by teenagers wishing to build bulk quickly. Some attribute this to society's emphasis on physical appearance and the need of insecure youngsters to feel better about how they look.

Physicians, teachers, and parents are quite alarmed about anabolic steroid drug abuse. It's even predicted that 2 or 3 months of high-dosage use in a youngster can cause death in the thirties and forties. The many and varied harmful effects of anabolic steroids on the body are listed in figure 13A. Unfortunately, these drugs also increase aggression and make a person feel invincible. One abuser even had his friend videotape him as he drove his car at 40 miles an hour into a tree!

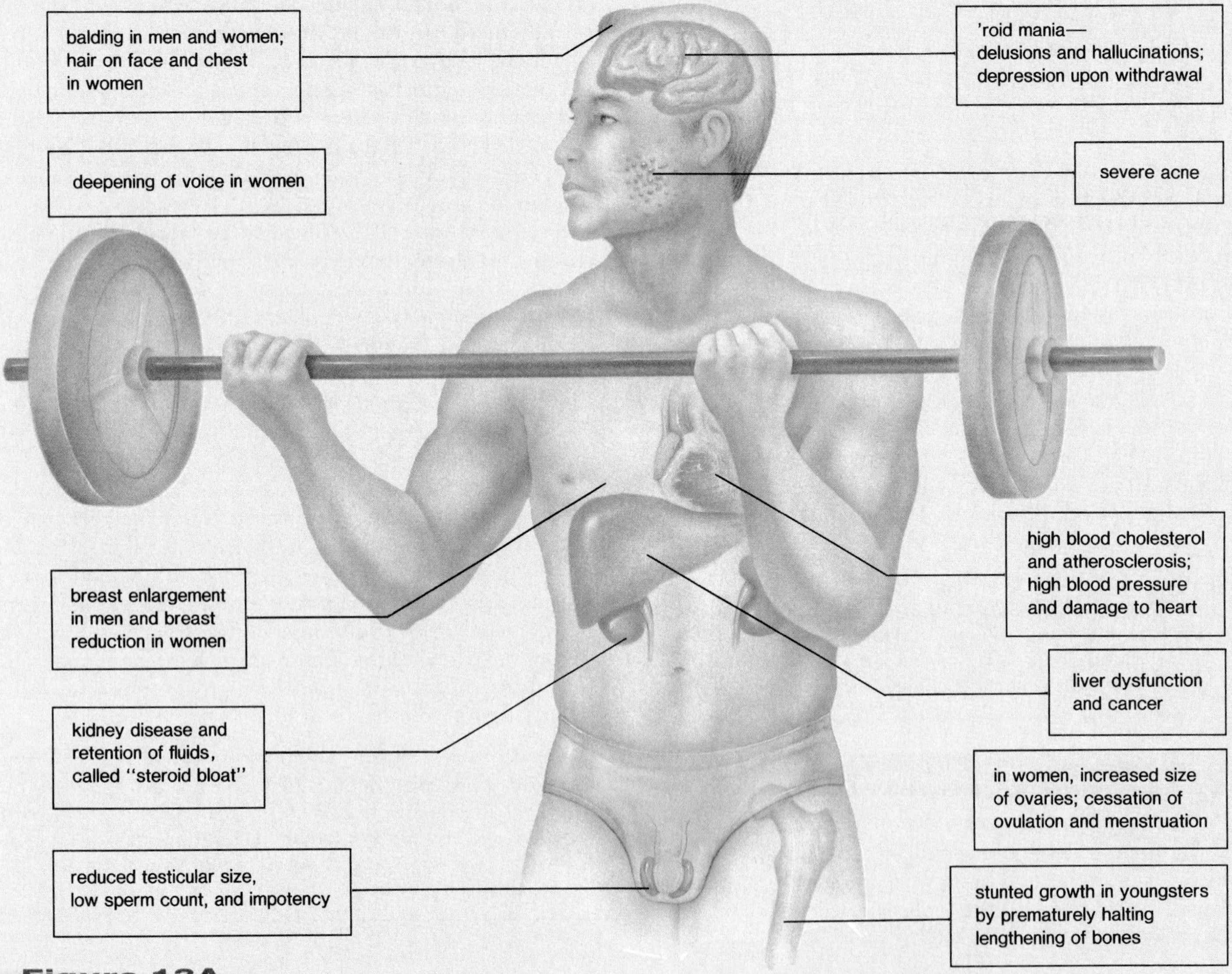

balding in men and women; hair on face and chest in women
'roid mania—delusions and hallucinations; depression upon withdrawal
deepening of voice in women
severe acne
high blood cholesterol and atherosclerosis; high blood pressure and damage to heart
breast enlargement in men and breast reduction in women
liver dysfunction and cancer
kidney disease and retention of fluids, called "steroid bloat"
in women, increased size of ovaries; cessation of ovulation and menstruation
reduced testicular size, low sperm count, and impotency
stunted growth in youngsters by prematurely halting lengthening of bones

Figure 13A

The effects of anabolic steroid use.

Other Endocrine Glands

There are some other glands in the body that produce hormones, and we will discuss some of these here.

Thymus: Most Active in Children

The *thymus* is a lobular gland that lies in the upper thoracic cavity (see fig. 13.1). This organ reaches its largest size and is most active during childhood. With aging, the organ gets smaller and becomes fatty. Certain lymphocytes that originate in the bone marrow and then pass through the thymus are transformed into T cells (p. 130). The thymus produces various hormones called *thymosins,* which aid the differentiation of T cells and may stimulate immune cells in general. There is hope that these hormones can be used in conjunction with lymphokine therapy to restore or to stimulate T cell function in patients suffering from AIDS or cancer.

Pineal Gland: Hormone at Night

The **pineal gland** produces the hormone called *melatonin,* primarily at night. In fishes and amphibians, the pineal gland is located near the surface of the body and is a "third eye," which receives light rays directly. In mammals, the pineal gland is located in the third ventricle of the brain and cannot receive direct light signals. However, it does receive nerve impulses from the eyes by way of the optic tract.

The pineal gland and melatonin are involved in daily cycles called **circadian rhythms.** Normally we grow sleepy at night when melatonin levels are high and awaken once daylight returns and melatonin levels are low. Shift work is usually troublesome because it upsets this normal daily rhythm. Similarly, travel to another time zone, as when going to Europe from the United States, results in jet lag because the body is still producing melatonin according to the old schedule. Some people even have Seasonal Affective Disorder (SAD); they become depressed and have an uncontrollable desire to sleep with the onset of winter. Giving melatonin makes their symptoms worse, but exposure to a bright light improves them.

Many animals go through a yearly cycle that includes enlargement of reproductive organs during the summer when melatonin levels are low. Mating occurs in the fall and young are born in the spring. It is of interest that children with a brain tumor that destroys the pineal gland experience early puberty. It's possible that the pineal gland is also involved in human sexual development.

Nontraditional Sources

Even organs that are not usually considered to be endocrine glands have been found to secrete hormones. The heart produces *atrial natriuretic hormone,* which helps to regulate the sodium and water balance of the body. It lowers blood pressure by promoting renal excretion of sodium and water, and it also inhibits the release of renin and the hormones aldosterone and ADH. Atrial natriuretic hormone is a peptide that is released not only by the atria but also by the aortic arch, ventricles, lungs, and pituitary gland in response to increases in blood pressure. The stomach and the small intestine produce the peptide hormones discussed in chapter 4.

A number of different types of organs and cells produce peptide *growth factors,* which stimulate cell division and mitosis. They are like hormones in that they act on cell types with specific receptors to receive them. Some, including lymphokines (p. 140) and blood cell growth factors, are released into blood, others diffuse to nearby cells. Other growth factors are described in the following listing:

Platelet-derived growth factor is released from platelets and from many other cell types. It helps in wound healing and causes an increase in the number of fibroblasts, smooth muscle cells, and certain cells of the nervous system.

Epidermal growth factor and *nerve growth factor* stimulate the cells indicated by their names as well as many others. These growth factors are also important in wound healing.

Tumor angiogenesis factor stimulates the formation of capillary networks and is released by tumor cells. One treatment for cancer is to prevent the activity of this growth factor.

Prostaglandins (PG) are another class of chemical messengers that also are produced and act locally. They are derived from fatty acids stored in cell membranes as phospholipids. When a cell is stimulated by reception of a hormone or even by trauma, a series of synthetic reactions takes place in the cell membrane, and PG is first released into the cytoplasm and then secreted from the cell. There are many different types of prostaglandins produced by many different tissues. In the uterus, certain prostaglandins cause muscles to contract; therefore, they are implicated in the pain and discomfort of menstruation in some women. (Antiprostaglandin therapy is useful in these cases.) On the other hand, certain prostaglandins are used to treat ulcers because they reduce gastric secretion, to treat hypertension because they lower blood pressure, and to prevent thrombosis because they inhibit platelet aggregation. Because the different prostaglandins can have contrary effects, it has been very difficult to standardize their use, and in most instances, prostaglandin therapy is still considered experimental.

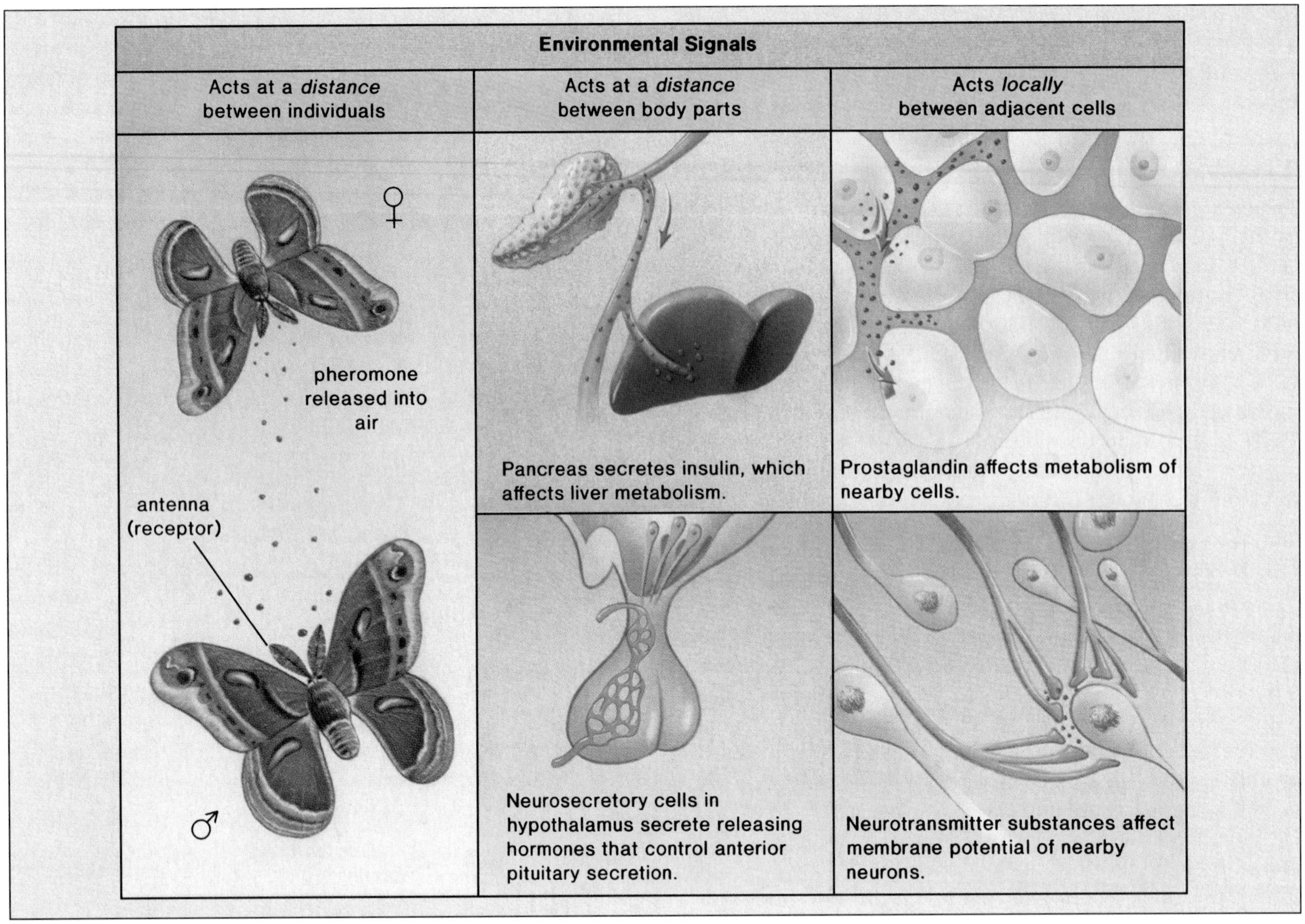
Environmental Signals
Acts at a *distance* between individuals
Acts at a *distance* between body parts
Acts *locally* between adjacent cells
♀
pheromone released into air
antenna (receptor)
♂
Pancreas secretes insulin, which affects liver metabolism.
Prostaglandin affects metabolism of nearby cells.
Neurosecretory cells in hypothalamus secrete releasing hormones that control anterior pituitary secretion.
Neurotransmitter substances affect membrane potential of nearby neurons.

Figure 13.19

The 3 categories of environmental signals. Pheromones are chemical messengers that act at a distance between individuals. Endocrine hormones and neurosecretions typically are carried in the bloodstream and act at a distance within the body of a single organism. Some chemical messengers have local effects only; they pass between cells that are adjacent to one another. This, of course, includes neurotransmitters.

Environmental Signals

In this chapter, we concentrated on the functions of the human endocrine glands and their hormonal secretions. We already know that hormones are only one type of chemical messenger, or environmental signal, between cells. In fact, the concept of the chemical messenger has now been broadened to include at least the following 3 categories of environmental signals (fig. 13.19).

Environmental signals that act at a distance between individuals. Many organisms release chemical messengers called **pheromones** into the air or in externally deposited body fluids. These are intended to be messages for other members of the species. For example, ants lay down a pheromone trail to direct other ants to food, and the female silkworm moth releases bombykol, a sex attractant that is received by male moth antennae even several kilometers away. This chemical is so potent that it has been estimated that only 40 out of 40,000 receptors on the male antennae need to be activated for the male to respond. Mammals, too, release pheromones; for example, the urine of dogs serves as a territorial marker. Some studies are being conducted to determine if humans have pheromones, and other studies have suggested that humans respond to pheromones. For example, investigators have observed that women who live in close quarters tend to have coinciding menstrual cycles. They reason that this might be caused by a pheromone, but they don't know what the pheromone might be or how it might exert its effect.

Environmental signals that act at a distance between body parts. This category includes the endocrine secretions, which traditionally have been called hormones. It also includes the secretions of the neurosecretory cells in the hypothalamus—the production and action of ADH and oxytocin illustrate the close relationship between the nervous system and the endocrine system. Neurosecretory cells produce these hormones, which are released when these cells receive nerve impulses. As another example of the overlap between the nervous and endocrine systems, consider that endorphins on occasion travel in the bloodstream, but they act on nerve cells to alter their membrane potential. Also, norepinephrine is secreted by the adrenal medulla but is also a neurotransmitter in the sympathetic nervous system.

Environmental signals that act locally between adjacent cells. Neurotransmitters belong in this category, as do substances that are sometimes called local hormones. For example, when the skin is cut, histamine, released by mast cells, promotes the inflammatory response.

> Today, hormones are categorized as one type of environmental signal. There are environmental signals that work at a distance between individuals (e.g., pheromones), at a distance between body parts (e.g., hormones), and locally between adjacent cells (e.g., neurotransmitters).

Summary

Hormones are chemical messengers having a metabolic effect on cells. Peptide hormones combine with receptor sites in the cell membrane. This leads to the formation of cAMP, which activates an enzyme within the cell. Steroid hormones combine with receptor molecules in the cytoplasm of the cell, and the hormone-receptor complex moves into the nucleus to combine with an activate DNA. Protein synthesis and new enzyme formation follows.

The hypothalamus produces the hormones ADH and oxytocin, which are released by the posterior pituitary. The hypothalamus also produces releasing hormones and release-inhibiting hormones, which control the production of hormones by the anterior pituitary. In addition to GH and PRL, which affect the body directly, the anterior pituitary secretes hormones that control other endocrine glands: ACTH stimulates the adrenal cortex to release glucocorticoids (cortisol) and mineralocorticoids (aldosterone); gonadotropic hormones (FSH and LH) stimulate the gonads to release the sex hormones; TSH stimulates the thyroid to release thyroxin. The secretion of hormones is controlled by a negative feedback mechanism; for example, the blood level of thyroxin controls the secretion of TRH from the hypothalamus and TSH from the anterior pituitary.

Endocrine glands not controlled by the anterior pituitary are the parathyroid glands, the adrenal medulla, and the pancreas. The parathyroids are embedded in the thyroid gland. They control the level of calcium and phosphate in the blood. Blood calcium level is maintained by promoting the absorption of calcium by the intestine, reabsorption by the kidneys, and demineralization of bone. These actions are opposed by calcitonin, which is produced by the thyroid. The adrenal medulla produces norepinephrine and epinephrine. These hormones prepare the body for fight or flight. The pancreas secretes insulin.

The most common illness due to hormonal imbalance is diabetes mellitus. This condition occurs when the pancreatic islets (of Langerhans) within the pancreas fail to produce insulin. Insulin promotes the conversion of glucose to glycogen, thereby lowering blood glucose levels. Without the production of insulin, the blood glucose level rises, and some glucose spills over into the urine. The real problem in diabetes mellitus, however, is acidosis, which may cause the death of the diabetic if therapy is not begun.

There are 3 categories of environmental signals: those that act at a distance between individuals (pheromones); those that act at a distance within the individual (traditional endocrine hormones and secretions of neurosecretory cells); and local messengers (such as prostaglandins and neurotransmitters). Since there is great overlap between these categories, perhaps the definition of a hormone now should be expanded to include all of them.

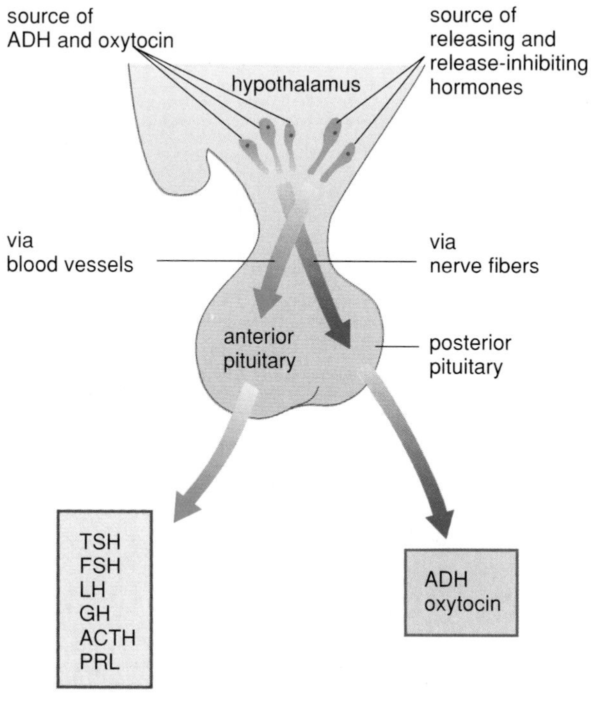

The hypothalamus and hormones produced by the pituitary

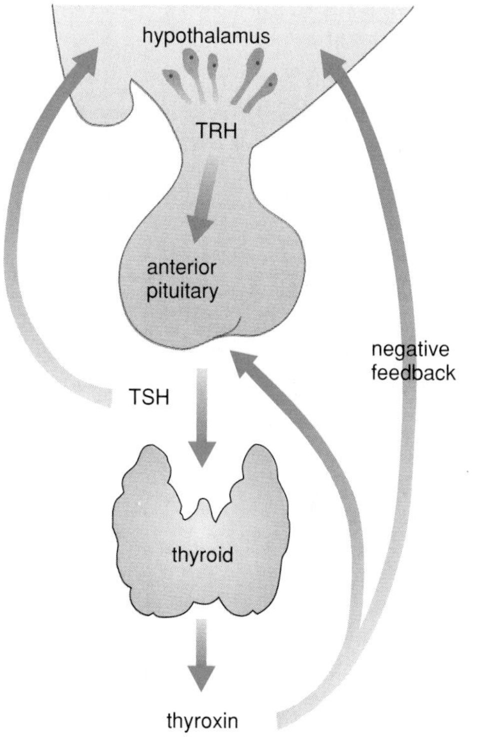

How a negative feedback system works

Study and Thought Questions

1. What is a hormone and how do hormones work? (p. 248)
2. Categorize endocrine hormones according to their chemical makeup. (p. 248)
3. Explain the concept of negative feedback mechanisms, and give an example involving ADH. (p. 251)
4. How does the hypothalamus control the posterior pituitary? The anterior pituitary? (pp. 251–252)
5. Discuss 2 hormones secreted by the anterior pituitary that have an effect on the body proper rather than on other glands. What medical conditions, if any, are associated with each hormone? (pp. 252–253)
6. Discuss the hormones secreted by the anterior pituitary that affect other endocrine glands. (pp. 253–254)
7. Explain the reason the anterior pituitary can be called the master gland. Give an example of the 3-tiered relationship between the hypothalamus, the anterior pituitary, and other endocrine glands. (pp. 253–254)

   Critical Thinking Explain why the birth-control pill, which contains estrogen and progesterone, prevents a woman from ovulating.
8. For each of the following endocrine glands, give the anatomical location, and name the hormone(s) secreted, the effect of the hormone(s), and the medical illnesses associated with each hormone: posterior pituitary, thyroid, parathyroids, adrenal cortex, adrenal medulla, pancreas. (pp. 251–261)

   Critical Thinking What hormones would you expect to be especially active during starvation? Why?

   Ethical Issue Goiters are fairly common in poorer countries but not in the United States. Are we obligated to improve the health of people in other countries? Explain you answer and suggest what should be done.
9. List and discuss the hormones produced by the testes and ovaries, the thymus, and the pineal gland. List and discuss the hormones produced by nontraditional sources. (pp. 261–263)

   Ethical Issue What pressures cause people to take steroids? How can such presures be alleviated?
10. Categorize environmental signals into 3 groups and give examples of each group. (pp. 264–265)

Writing Across the Curriculum

1. Use figure 13.8 to explain how control by negative feedback results in a fluctuation of hormonal blood levels about a mean.
2. Name several hormones that control the concentration of blood glucose, and show that this is not an unnecessary duplication of effects.
3. Argue that it is incorrect to speak of the nervous system versus the endocrine system and that instead the 2 systems should be considered one system.

Objective Questions

1. The hypothalamus produces the hormones ________________ and ________________, released by the posterior pituitary.
2. ________________ secreted by the hypothalamus control the anterior pituitary.
3. Generally, hormone production is self-regulated by a ________________ mechanism.
4. Growth hormone (GH) is produced by the ________________ pituitary.
5. Simple goiter occurs when the thyroid produces ____________ (too much or too little) ____________.
6. ACTH, produced by the anterior pituitary, stimulates the adrenal ________________.
7. An overproductive adrenal cortex results in the condition called ________________.
8. Parathyroid hormone increases the level of ________________ in blood.
9. Type I diabetes mellitus is due to a malfunctioning ______________, while type II diabetes mellitus is due to limited uptake of insulin by ______________.
10. Prostaglandins are not carried in ________________, as are hormones secreted by the endocrine glands.
11. Complete the following table:

Acronym	Name of Hormone	Secreted by	Function
TSH			
ACTH			
PRL			
GH			
ADH			
PTH			

Selected Key Terms

acromegaly (ak″ro-meg′ah-le) A condition resulting from an increase in GH production after adult height has been achieved. 253

anterior pituitary The portion of the pituitary gland that produces 6 types of hormones and is controlled by hypothalamic-releasing and release-inhibiting hormones. 249

cretinism (kre′tin-izm) A condition resulting from improper development of the thyroid in an infant. 255

diabetes mellitus (di″ ah-bĕt ēz me-li′tus) Condition characterized by a high blood glucose level and the appearance of glucose in the urine, due to a deficiency of insulin production or glucose uptake by cells. 251

exophthalmic goiter (ek″sof-thal′mik goi′ter) An enlargement of the thyroid gland accompanied by an abnormal protrusion of the eyes. 255

myxedema (mik″ să de-mah) A condition resulting from a deficiency of thyroid hormone in an adult. 255

pancreatic islets (of Langerhans) (lahng′ər-hanz) Distinctive group of cells within the pancreas that secretes insulin and glucagon. 259

pheromone (fer′o-mōn) A chemical substance secreted by one organism that influences the behavior of another. 264

posterior pituitary Back lobe of the pituitary gland that stores and secretes ADH and oxytocin produced by the hypothalamus. 249

simple goiter Condition in which an enlarged thyroid produces low levels of thyroxin. 254

FURTHER READINGS FOR PART THREE

Alkon, D. L. July 1989. Memory storage and neural systems. *Scientific American.*

Barlow, R. B., Jr. April 1990. What the brain tells the eye. *Scientific American.*

Berns, M. B. June 1991. Laser surgery. *Scientific American.*

Changeux, J. P. November 1993. Chemical signaling in the brain. *Scientific American.*

Creager, J. G. 1992. *Human anatomy and physiology.* 2d ed. Dubuque, Iowa: Wm. C. Brown Publishers.

Deutsch, D. September 1992. Paradoxes of musical pitch. *Scientific American.*

DeVries, P. J. October 1992. Singing caterpillars, ants and symbiosis. *Scientific American.*

Fischbach, G. D. September 1992. The developing brain. *Scientific American.*

Fox, S. I. 1990. *Human physiology.* 3d ed. Dubuque, Iowa: Wm. C. Brown Publishers.

Freeman, W. J. February 1991. The physiology of perception. *Scientific American.*

Hinton, G. E. September 1992. How neural networks learn from experience. *Scientific American.*

Hinton, G. E. October 1993. Simulating brain damage. *Scientific American.*

Hole, J. W., Jr. 1992. *Essentials of human anatomy and physiology.* 4th ed. Dubuque, Iowa: Wm. C. Brown Publishers.

Holloway, M. March 1991. Rx for addiction. *Scientific American.*

Kalil, R. E. December 1989. Synapse formation in the developing brain. *Scientific American.*

Kandel, E. R. and Hawkins, R. D. September 1992. The biological basis of learning and individuality. *Scientific American.*

Konishi, M. April 1993. Listening with two ears. *Scientific American.*

Koretz, J. F., and Handelman, G. H. July 1988. How the human eye focuses. *Scientific American.*

Mahowald, M. A., and Mead, C. May 1991. The silicon retina. *Scientific American.*

Marieb, E. N. 1991. *Essentials of human anatomy and physiology.* 3d ed. Redwood City, Calif.: Benjamin/Cummings Publishing.

Melzack, R. April 1992. Phantom limbs. *Scientific American.*

Ramachandran, V. S. May 1992. Blind spots. *Scientific American.*

Sataloff, R. T. December 1992. The human voice. *Scientific American.*

Schatz, C. J. September 1992. The developing brain. *Scientific American.*

Scientific American. September 1992. Mind and brain. Special Issue.

Selkoe, D. J. November 1991. Amyloid protein and Alzheimer's disease. *Scientific American.*

Spence, A. P. 1990. *Basic human anatomy.* 3d ed. Redwood City, Calif.: Benjamin/Cummings Publishing.

Tuomanen, E. February 1993. Breaching the blood-brain barrier. *Scientific American.*

Zeki, S. September 1992. The visual image in mind and brain. *Scientific American.*

Part FOUR

Human Reproduction

Human beings have 2 sexes, male and female. The anatomy of each sex functions to produce sex cells that join prior to the development of a new individual. The embryo develops into a fetus within the body of the female, and birth occurs when there is a reasonable chance for independent existence. The steps of human development can be outlined from the fertilized egg to the birth of a child.

We are in the midst of a sexual revolution. We have the freedom to engage in varied sexual practices and to reproduce by alternative methods of conception, such as in vitro fertilization. With freedom comes a responsibility to be familiar with the biology of reproduction and health-related issues, not only for ourselves but for our potential offspring.

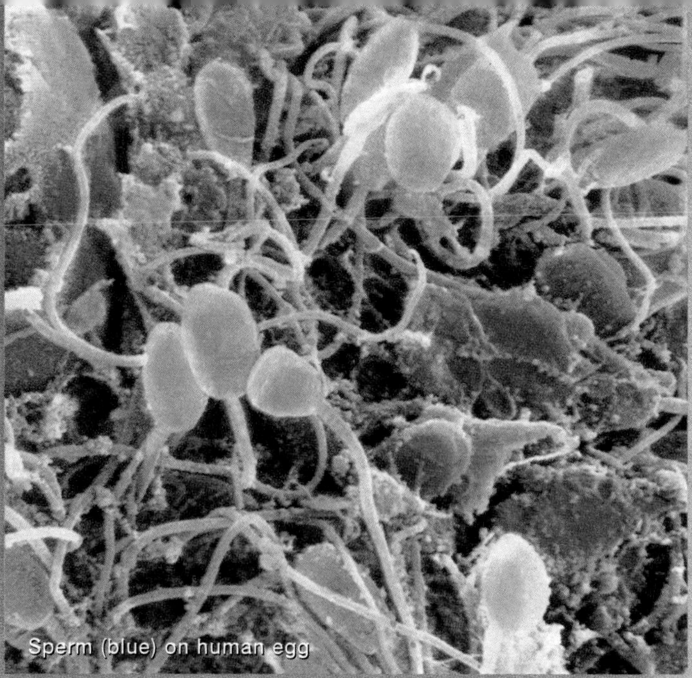

Chapter 14

Reproductive System

Organisms that reproduce in the water deposit their eggs and sperm in the water; the aquatic environment protects them from drying out. But organisms that reproduce on the land need a mechanism to protect the gametes and developing zygote from the drying effects of the air. In humans the egg is fertilized and the zygote undergoes development within the body of the female. The sperm pass from the male within seminal fluid, which if exposed would indeed dry out and be useless. Sexual intercourse prevents this. During sexual intercourse sperm are deposited in the female's vagina, which is lubricated by secretions. The human sex act is an adaptation to the land environment.

Chapter Concepts

1. The male reproductive system is designed for the continuous production of a large number of sperm within a fluid medium. 272
2. The female reproductive system is designed for the monthly production of an egg and preparation of the uterus to house the developing fetus. 277
3. Hormones control the reproductive process and the sex characteristics of the individual. 276, 279
4. Birth-control measures vary in effectiveness from those that are very effective to those that are minimally effective. 284
5. There are alternative methods of reproduction today, including in vitro fertilization followed by introduction to the uterus. 289

Chapter Outline

Sexual reproduction usually involves 2 types of gametes (sex cells). The sperm swim to the stationary egg, a much larger cell that contributes cytoplasm and organelles to the zygote. There are a large number of sperm deposited to ensure that a few find the egg. After puberty, the human male continually produces sperm, which are temporarily stored before being released.

Male Reproductive System

Figure 14.1 shows the reproductive system of the male, and table 14.1 lists the anatomical parts of this system and their functions.

Table 14.1

Male Reproductive System

ORGAN	FUNCTION
Testis	Produces sperm and sex hormones
Epididymis	Stores sperm as they mature
Vas deferens	Conducts and stores sperm
Seminal vesicle	Contributes to seminal fluid
Prostate gland	Contributes to seminal fluid
Urethra	Conducts sperm
Cowper's gland (bulbourethral gland)	Contributes to seminal fluid
Penis	Serves as organ of sexual intercourse

Testes: Millions of Sperm Daily

The testes (sing., **testis**) lie outside the abdominal cavity of the male within the **scrotum.** The testes first begin to develop inside the abdominal cavity but descend into the scrotal sacs during the last 2 months of fetal development. If by chance the testes do not descend and the male is not treated or operated on to place the testes in the scrotum, *sterility*—the inability to produce offspring—usually results. This is because the internal temperature of the body (37°C) is too high to produce viable sperm; the temperature in the scrotum is about 34° C. Wearing tight clothing can in-

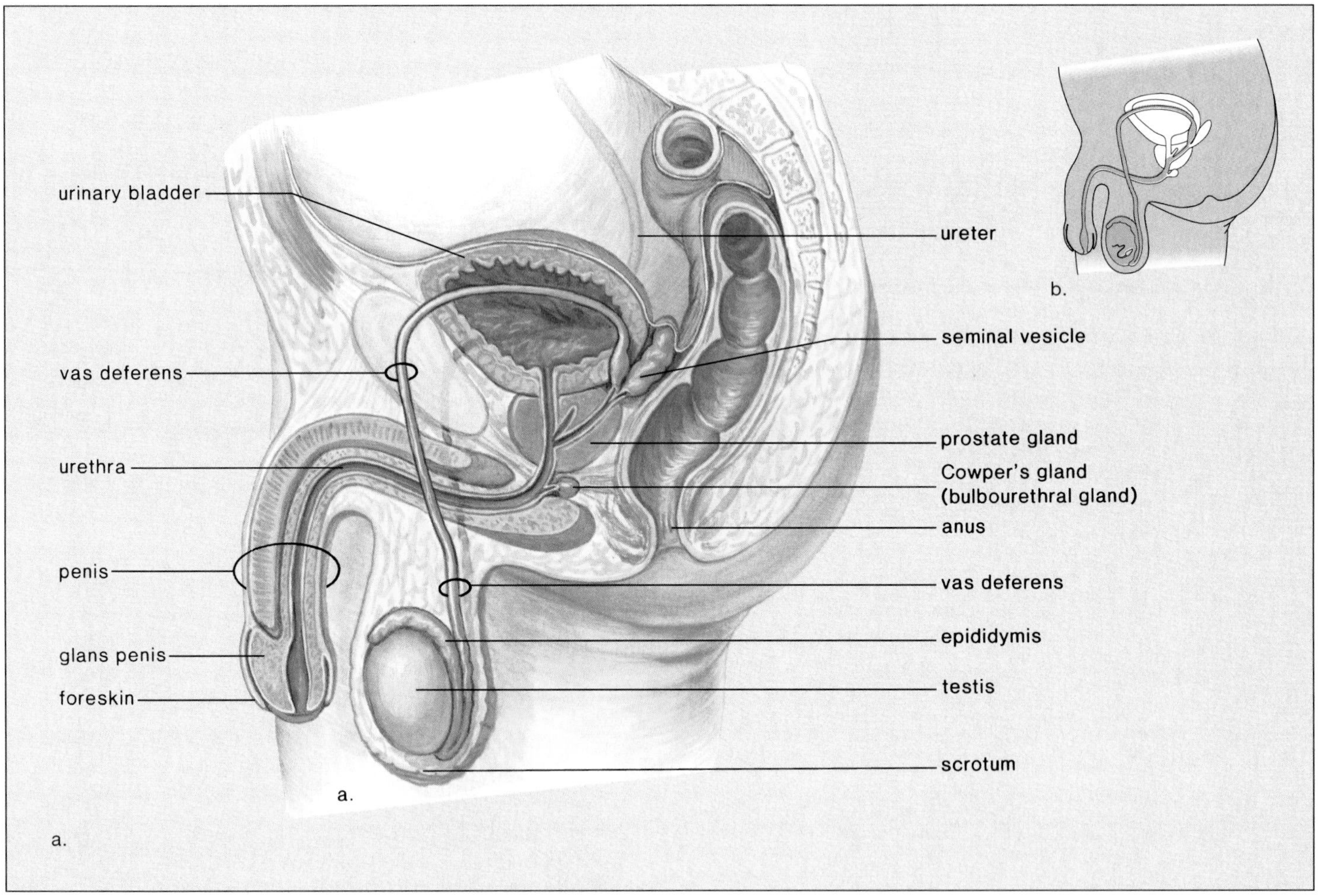

Figure 14.1

Side view of the male reproductive system. **a.** The testes produce sperm. The seminal vesicles, the prostate gland, and the Cowper's gland provide a fluid medium. Notice that the penis in this drawing is not circumcised—the foreskin is present. **b.** The path of sperm in the male genital tract.

crease this temperature and reduce sperm production. When the body is cold, the testes are normally held closer to the body, and this provides an optimum temperature.

Fibrous connective tissue forms the wall of a testis and divides it into lobules (fig. 14.2). Each lobule contains 1 to 3 tightly coiled **seminiferous tubules,** each of which is approximately 70 cm long when uncoiled. A microscopic cross section through a tubule shows that it is packed with cells undergoing spermatogenesis. These cells are derived from undifferentiated cells called spermatogonia (sing., spermatogonium), which lie just inside the outer wall of a tubule and divide mitotically, always producing new spermatogonia. Some newly formed spermatogonia move away from the outer wall to increase in size and become primary spermatocytes, which undergo meiosis, a type of cell division described in chapter 17. Primary spermatocytes, with 46 chromosomes, divide to give 2 secondary spermatocytes, each with 23 double-stranded chromosomes. Secondary spermatocytes divide to produce 4 spermatids, also with 23 chromosomes, but single-stranded. Spermatids then

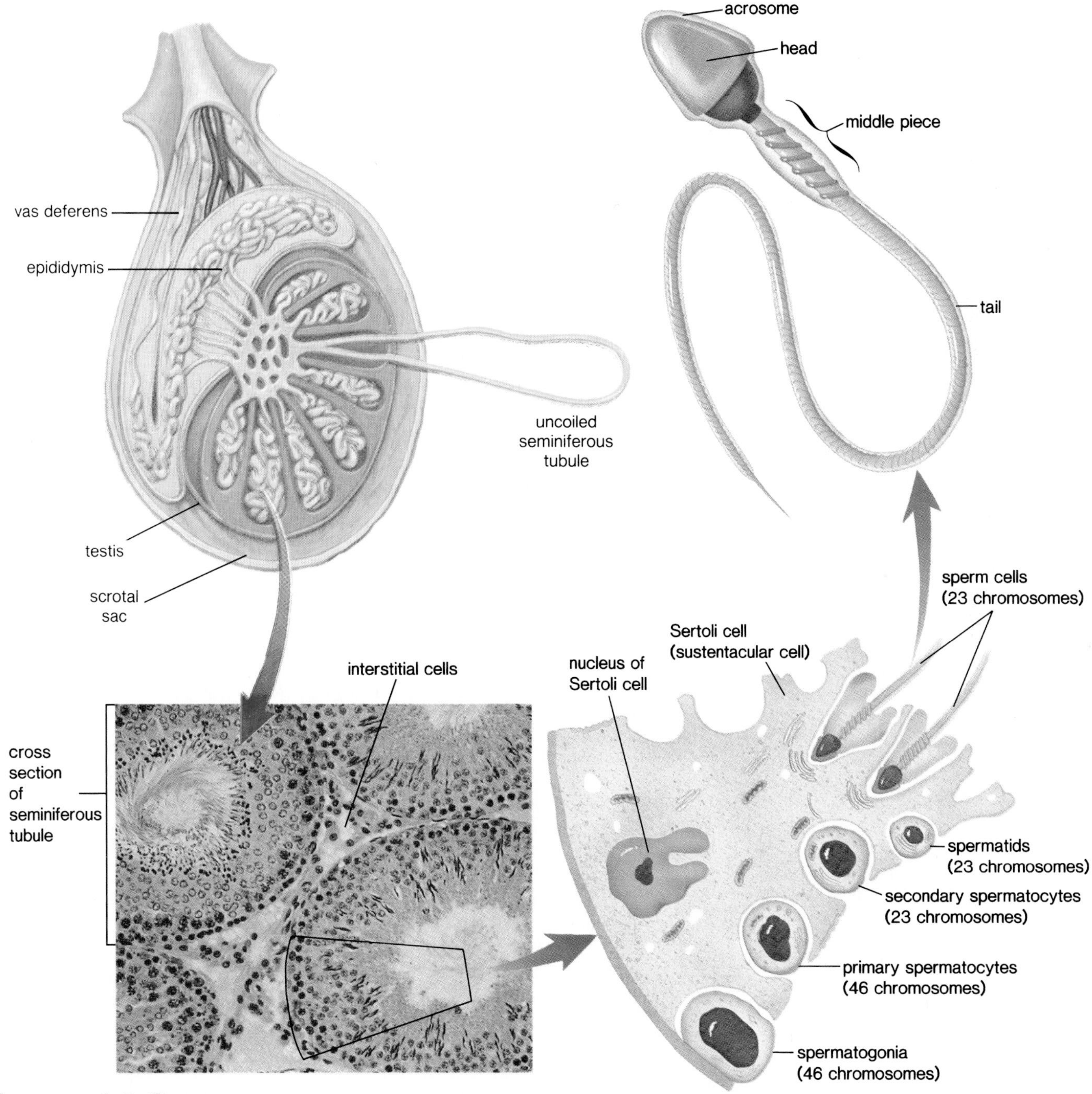

Figure 14.2

Testis and sperm. The lobules of a testis contain seminiferous tubules, where spermatogenesis occurs. A sperm has a head, a middle piece, and a tail. The nucleus is in the head, capped by the enzyme-containing acrosome.

differentiate into sperm (spermatozoa). The process of sperm production from spermatogonia to mature sperm cells takes approximately 48 days. Also present in the tubules are the *Sertoli cells,* which support, nourish, and regulate the spermatogenic cells.

The mature **sperm** has 3 distinct parts (fig. 14.2): a head, a middle piece, and a tail. The *tail* is a flagellum (see fig. 2.11); the *middle piece* contains energy-producing mitochondria; and the *head* contains the 23 chromosomes within a nucleus. Adhering to the nucleus is a specialized lysosome called the **acrosome,** which contains enzymes that facilitate penetration of the egg. The human egg is surrounded by several layers of cells and a mucoprotein substance called the zona pellucida; the acrosomal enzymes help a sperm to digest its way into an egg.

> In males, spermatogenesis occurs within the seminiferous tubules of the testes. Sperm have an acrosome-capped head, where 23 chromosomes reside in the nucleus, a mitochondria-containing middle piece, and a flagellum for a tail.

The **interstitial cells** lie between the seminiferous tubules (fig. 14.2). These cells produce the male sex hormones, called the androgens. The most important of the androgens is **testosterone,** whose functions are discussed on page 276.

Genital Tract: Testes to Glans Penis

Sperm are produced in the testes, but they mature and are stored in an **epididymis** (see fig. 14.1), a tightly coiled tubule about 5–6 meters in length. An epididymis lies just outside each testis. Each epididymis joins with a **vas** (ductus) **deferens,** which ascends through the *inguinal canal* and enters the abdomen, where it curves around the urinary bladder and empties into the urethra (see fig. 14.1). Sperm are also stored in the first part of a vas deferens. They pass from each vas deferens into the urethra only when ejaculation (p. 275) is imminent.

The urethra is located in the **penis**, the organ of sexual intercourse in males (fig. 14.3). The penis is a long shaft with an enlarged cone-shaped tip called the glans penis. At birth, the glans penis is covered by a layer of skin called the **foreskin,** or prepuce. Sometime near puberty, small glands located in the foreskin and glans begin to produce an oily secretion. This secretion, along with dead skin cells, forms a cheesy substance known as smegma. Before this time, no special cleansing method is needed to wash away smegma, but after puberty, the foreskin can be retracted to do so. **Circumcision** is the surgical removal of the foreskin, usually soon after birth.

When the male is sexually aroused, the penis becomes erect and ready for intercourse. **Erection** is achieved because blood sinuses within the erectile tissue of the penis fill with blood. Parasympathetic impulses dilate the arteries of the penis, while the veins are passively compressed so that blood flows into the erectile tissue under pressure and does not drain away. **Impotency** is the condition in which erection cannot be achieved. There are medical and surgical remedies for impotency.

> Sperm mature in an epididymis and are stored in a vas deferens before entering the urethra within the penis, just prior to ejaculation.

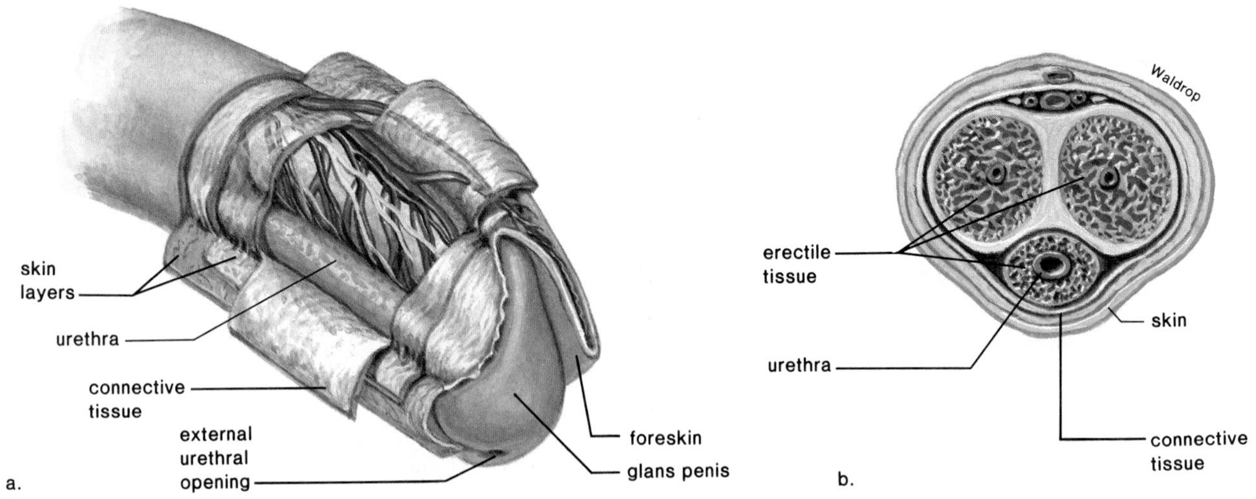

Figure 14.3

Penis anatomy. **a.** Beneath the skin and the connective tissue lies the urethra, surrounded by erectile tissue. This tissue expands to form the glans penis, which in uncircumcised males is partially covered by the foreskin. **b.** Two other columns of erectile tissue in the penis are located dorsally.

Male Orgasm: Upon Ejaculation

At the time of ejaculation, sperm leave the penis in a fluid called **seminal fluid** (also called semen). Three types of accessory glands add secretions to seminal fluid—the **seminal vesicles,** the prostate gland, and the Cowper's (bulbourethral) glands. The seminal vesicles lie at the base of the bladder, and each has a duct that joins with a vas deferens. The **prostate gland** is a single doughnut-shaped gland that surrounds the upper portion of the urethra just below the bladder. In older men, the prostate can enlarge and squeeze off the urethra, making urination painful and difficult. This condition can be treated medically or surgically. **Cowper's glands** are pea-sized organs that lie posterior to the prostate on either side of the urethra.

Each component of seminal fluid seems to have a particular function. Sperm are more viable in a basic solution, and seminal fluid, which is milky in appearance, has a slightly basic pH (about pH 7.5). Swimming sperm require energy, and seminal fluid contains the sugar fructose, which presumably serves as an energy source. Seminal fluid also contains prostaglandins, chemicals that cause the uterus in the female to contract. Some investigators now believe that uterine contraction is necessary to help propel the sperm toward the egg.

Once seminal fluid is in the urethra, rhythmic muscle contractions expel it from the penis in spurts. During ejaculation, a sphincter closes off the bladder so that no urine enters the urethra. (Notice that at different times the urethra carries either urine or seminal fluid but not both simultaneously.)

The contractions that expel seminal fluid from the penis are a part of male **orgasm**, the physiological and psychological sensations that occur at the climax of sexual stimulation. The psychological sensation of pleasure is centered in the brain, but the physiological reactions involve the genital (reproductive) organs and associated muscles, as well as the entire body. Marked muscular tension is followed by contraction and relaxation.

Following ejaculation and/or loss of sexual arousal, the penis returns to its normal flaccid state. The male typically experiences a refractory period, during which stimulation does not bring about an erection. The length of the refractory period increases with age.

There may be as many as 400–600 million sperm in the 3.5 ml of semen expelled during ejaculation. The sperm count can be much lower than this, however, and fertilization still can occur. Infertile males release fewer than 35–50 million per ejaculation and many (more than 25%) may be abnormal in appearance.

> The accessory glands (seminal vesicles, prostate gland, and Cowper's gland) add secretions to seminal fluid, which leaves the penis during ejaculation.

Regulating Male Hormone Levels

The hypothalamus has ultimate control of the testes' sexual functions because it secretes gonadotropic-releasing hormone (GnRH), which stimulates the anterior pituitary to release the gonadotropic hormones. Two gonadotropic hormones, **FSH (follicle-stimulating hormone)** and **LH (luteinizing hormone),** are named for their function in females but exist in both sexes, stimulating the appropriate gonads in each. FSH promotes spermatogenesis in the seminiferous tubules, and LH promotes the production of testosterone in the interstitial cells. LH in males is sometimes given the name interstitial cell-stimulating hormone (ICSH).

The hormones mentioned are involved in a negative feedback mechanism (fig. 14.4), which maintains the production of testosterone at a fairly constant level. For example, when the blood testosterone level rises, it causes

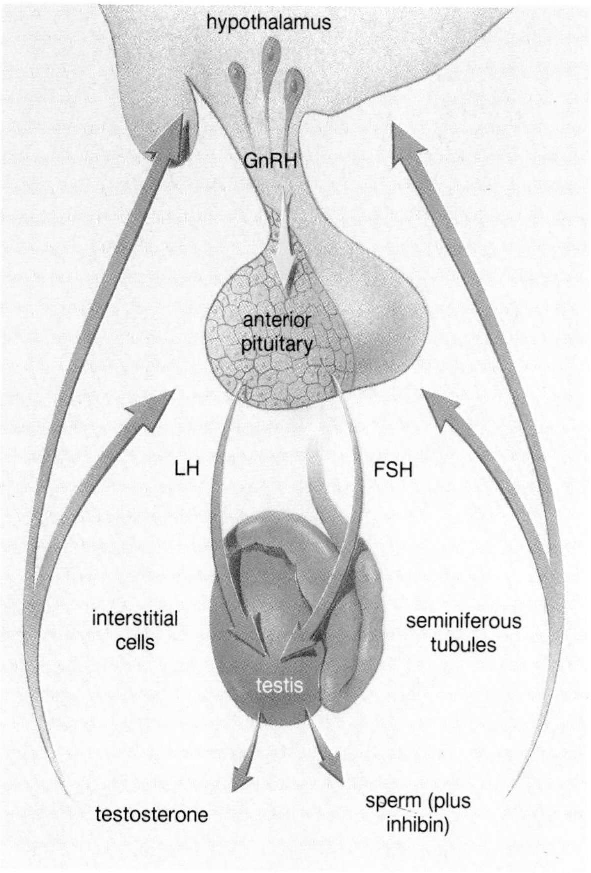

Figure 14.4

The hypothalamus-pituitary-testes control relationship. GnRH (gonadotropic-releasing hormone) stimulates the anterior pituitary to secrete the gonadotropic hormones FSH and LH. FSH stimulates the testes to produce sperm, and LH stimulates the testes to produce testosterone. Testosterone and inhibin exert negative feedback control over the hypothalamus and the anterior pituitary, and this ultimately regulates the blood testosterone level.

the anterior pituitary to decrease its secretion of LH. As the testosterone level begins to fall, the anterior pituitary increases its secretion of LH, and stimulation of the interstitial cells reoccurs. It should be emphasized that only minor fluctuations of testosterone level occur in the male and that the feedback mechanism in this case acts to maintain a normal testosterone level. It had long been suspected that the seminiferous tubules produce a hormone that blocks FSH secretion. This substance, termed *inhibin,* was isolated a few years ago.

Testosterone: The Main Male Sex Hormone

The male sex hormone, testosterone, has many functions. It is essential for normal development and function of the primary sex organs, those structures we have just discussed. It is also necessary for the production of sperm. FSH causes spermatogenic cells to take up testosterone, and it is testosterone that promotes their activity.

Greatly increased testosterone secretion at the time of puberty stimulates maturation of the penis and the testes. Testosterone also brings about and maintains the secondary sex characteristics in males, which develop at the time of puberty. Testosterone causes growth of a beard, axillary (underarm) hair, and pubic hair. It prompts the larynx and the vocal cords to enlarge, causing the voice to change. It is responsible for the greater muscular strength of males, and this is the reason some athletes take a supplemental *anabolic steroid,* which is either testosterone or a related chemical. The contraindications of anabolic steroid use are discussed in the Health Focus on page 262. Testosterone also causes oil and sweat glands in the skin to secrete; therefore, it contributes to acne and body odor. A side effect of testosterone activity is baldness. Genes for baldness are probably inherited by both sexes, but baldness is seen more often in males because of the presence of testosterone. This makes baldness a sex-influenced trait.

Testosterone is believed to be largely responsible for the sex drive. It may even contribute to the supposed aggressiveness of males.

> In males, FSH promotes spermatogenesis in the seminiferous tubules and LH promotes testosterone production by the interstitial cells. Testosterone stimulates growth of the male genitals during puberty and is necessary for maturation of sperm and development of the secondary sex characteristics.

Female Reproductive System

Figure 14.5 illustrates the female reproductive system, and table 14.2 lists the anatomical parts of this system and their functions.

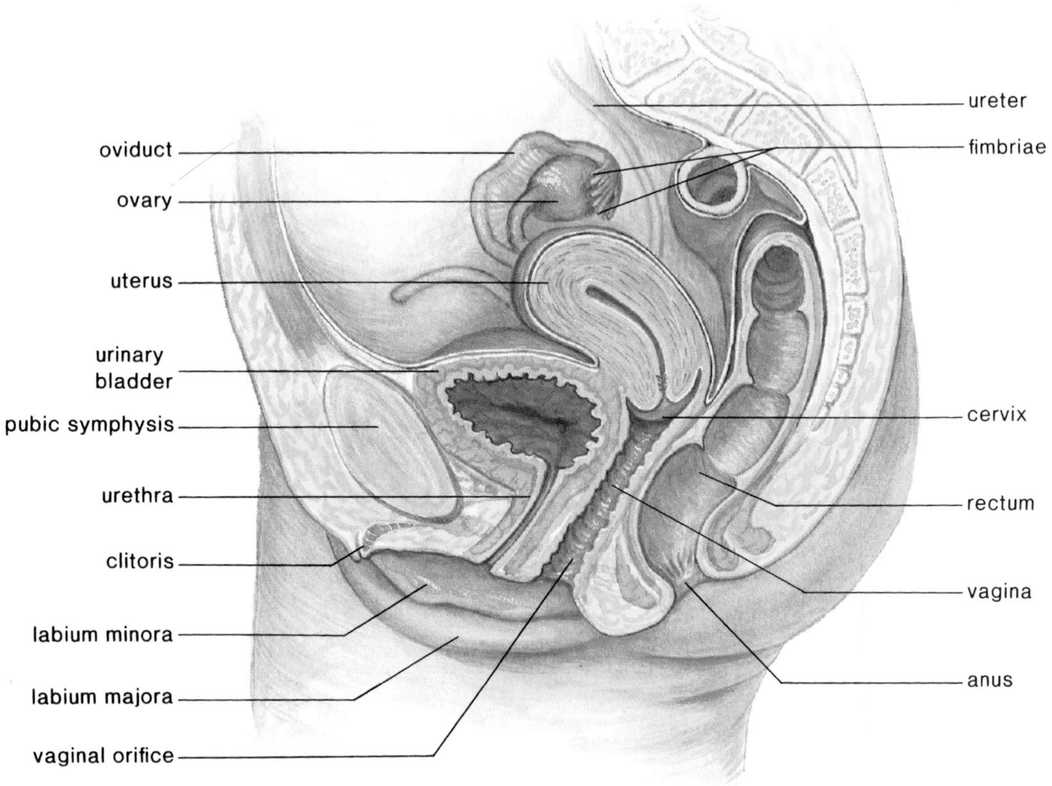

Figure 14.5

Side view of the female reproductive system. The ovaries normally only release one egg a month; fertilization occurs in the oviduct; and development occurs in the uterus. The vagina is the birth canal and the organ of sexual intercourse.

Table 14.2

Female Reproductive System

ORGAN	FUNCTION
Ovary	Produces eggs and sex hormones
Oviduct (fallopian or uterine tube)	Conducts eggs toward uterus
Uterus (womb)	Houses developing fetus
Vagina	Receives penis during sexual intercourse and serves as birth canal

Ovaries: An Egg a Month

The **ovaries** lie in shallow depressions, one on each side of the upper pelvic cavity. A longitudinal section through an ovary shows that it is made up of an outer cortex and an inner medulla. There are many saclike structures called **follicles** in the cortex, and each one contains an immature egg called an oocyte. A female is born with as many as 2 million follicles, but the number is reduced to 300,000 to 400,000 by the time of puberty. Only a small number of follicles (about 400) ever mature because the ovaries normally only release one egg per month during a female's reproductive years.

As the follicle undergoes maturation, it develops from a primary follicle to a secondary follicle to a **Graafian** (or vesicular) **follicle** (fig. 14.6). Oogenesis takes place in follicles. In a primary follicle, the primary oocyte divides meiotically into 2 cells, each having 23 duplicated chromosomes (see fig. 17.16). One of these cells, termed the *secondary oocyte,* receives almost all the cytoplasm. The other is a polar body, which disintegrates. A secondary follicle contains the secondary oocyte pushed to one side of a fluid-filled cavity. In a Graafian follicle, pressure within the fluid-filled cavity increases to the point that the follicle wall balloons out on the surface of the ovary and bursts, releasing the

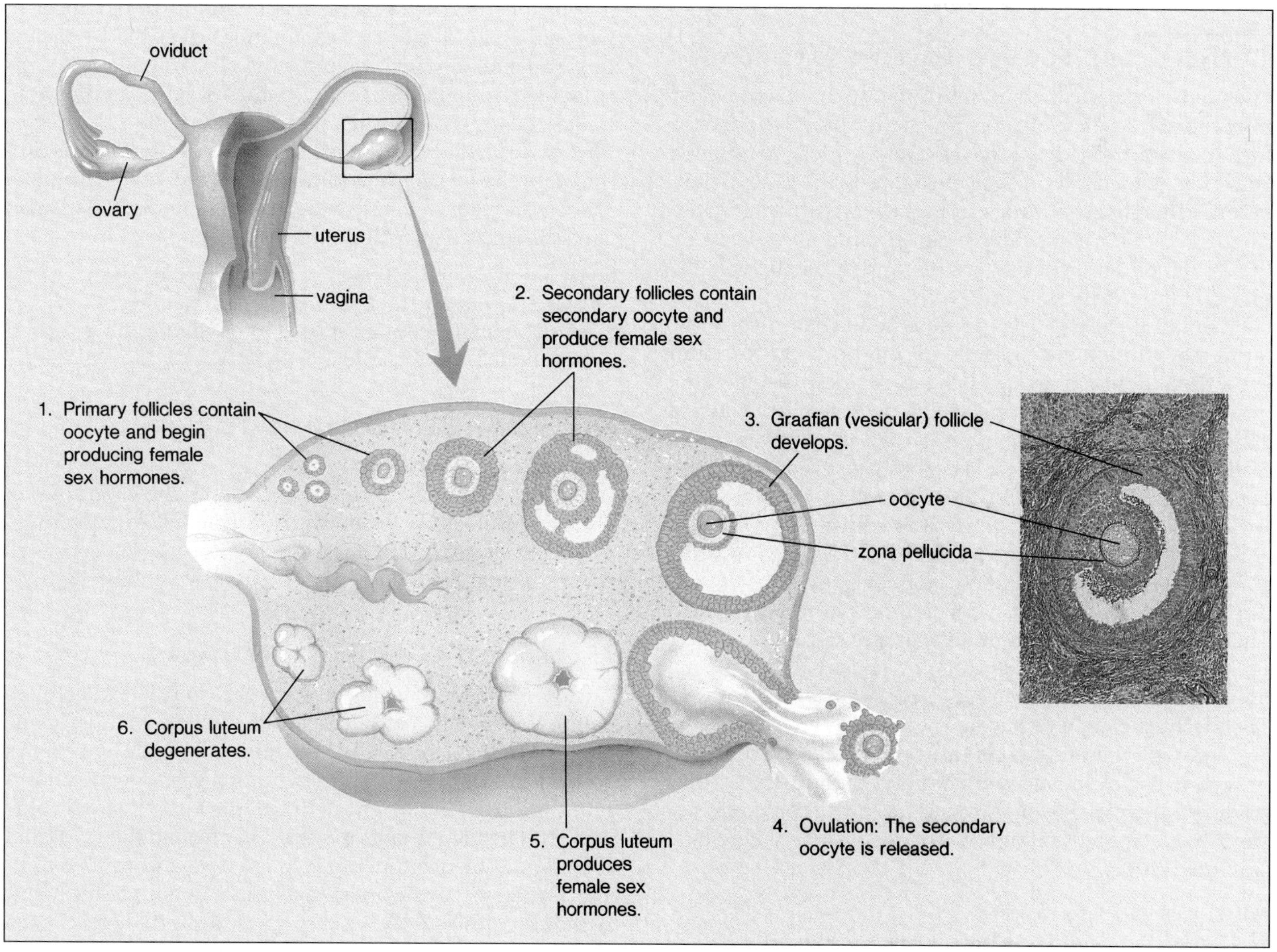
oviduct
ovary
uterus
vagina
2. Secondary follicles contain secondary oocyte and produce female sex hormones.
1. Primary follicles contain oocyte and begin producing female sex hormones.
3. Graafian (vesicular) follicle develops.
oocyte
zona pellucida
6. Corpus luteum degenerates.
5. Corpus luteum produces female sex hormones.
4. Ovulation: The secondary oocyte is released.

Figure 14.6

Anatomy of ovary and follicle. As a follicle matures, the oocyte enlarges and is surrounded by layers of follicular cells and fluid. Eventually, ovulation occurs, the mature follicle ruptures, and the secondary oocyte is released. A single follicle actually goes through all stages in one place within the ovary.

secondary oocyte (often called an egg for convenience) surrounded by a clear membrane, the *zona pellucida,* and follicular cells. The rupturing of a follicle is referred to as **ovulation.** Once a follicle has lost its egg, it develops into a **corpus luteum,** a hormone-secreting structure. If pregnancy does not occur, the corpus luteum begins to degenerate after about 10 days. If pregnancy does occur, the corpus luteum persists for 3–6 months. The follicle and the corpus luteum secrete the female sex hormones estrogen and progesterone, as discussed on page 279.

> In females, oogenesis occurs within the ovaries, where one follicle reaches maturity each month. This follicle balloons out of the ovary and bursts to release the egg. The ruptured follicle develops into a corpus luteum. The follicle and the corpus luteum produce the female sex hormones estrogen and progesterone.

Oviducts: Tubes to the Uterus

The oviducts, also called uterine or fallopian tubes, extend from the uterus to the ovaries. The oviducts are not attached to the ovaries; instead, they have fingerlike projections called **fimbriae,** which sweep over the ovary at the time of ovulation. When the egg bursts from the ovary during ovulation (fig. 14.6), it is usually swept up into an oviduct by the combined action of the fimbriae and the beating of cilia that line the oviducts.

Because the egg must cross a small space before entering an oviduct, it is possible for the egg to get lost and enter the abdominal cavity. Such eggs usually disintegrate, but in some rare cases, they have been fertilized in the abdominal cavity and have implanted themselves in the wall of an abdominal organ. Very rarely, such embryos have come to term, and the child is delivered by surgery.

Once in the oviduct, muscular contractions and the action of cilia of epithelial cells propel the egg slowly toward the uterus. Fertilization, the completion of oogenesis, and zygote formation normally occur in an oviduct. The developing embryo normally arrives at the uterus after several days and then *implants,* or embeds, itself in the uterine lining, which has been prepared to receive it. Occasionally, the embryo becomes embedded in the wall of an oviduct, where it begins to develop. These "tubular" pregnancies cannot succeed because the oviducts are not anatomically capable of allowing full development to occur. An *ectopic pregnancy* is one that begins anywhere outside the uterus.

The Uterus: House for Fetus

The **uterus** is a thick-walled, muscular organ about the size and the shape of an inverted pear. Normally, it lies above and is tipped over the urinary bladder. The oviducts join the uterus anteriorly, while posteriorly, the **cervix** enters into the vagina at nearly a right angle. A small opening in the cervix leads to the lumen of the vagina.

Development of the fetus normally takes place in the uterus. This organ, sometimes called the womb, is approximately 5 cm wide in its usual state but is capable of stretching to over 30 cm to accommodate the growing baby. The lining of the uterus, called the **endometrium,** participates in the formation of the placenta (p. 318), which supplies nutrients needed for fetal development. The endometrium has 2 layers: a basal layer and an inner functional layer. In the nonpregnant female, the functional layer of the endometrium varies in thickness according to a monthly reproductive cycle, called the uterine cycle (p. 280).

Cancer of the cervix is a common form of cancer in women. Early detection is sometimes possible by means of a **Pap smear,** which requires the removal of a few cells from the region of the cervix for microscopic examination. If the cells are cancerous, a *hysterectomy* may be recommended. A hysterectomy is the removal of the uterus. Removal of the ovaries in addition to the uterus is termed an *ovariohysterectomy.* Because the vagina remains, the woman can still engage in sexual intercourse.

> The egg enters the oviducts, which lead to the uterus, where implantation and development occur.

Vagina: Stretching Tube

The **vagina** is a tube that makes a 45° angle with the small of the back. The mucous membrane lining the vagina lies in folds, which extend as the wall stretches. This capacity to extend is especially important when the vagina serves as the birth canal, and it can also facilitate sexual intercourse.

External Genitals: Vulva

The external genitals of the female are known collectively as the **vulva** (fig. 14.7). The vulva includes 2 large, hair-covered folds of skin called the **labia majora.** They extend backward from the *mons pubis,* a fatty prominence underlying the pubic hair. The **labia minora** are 2 small folds lying just inside the labia majora. They extend forward from the vaginal orifice (opening) to encircle and form a foreskin for the *clitoris,* an organ that is homologous to the penis. Although quite small, the clitoris has a shaft of erectile tissue and is capped by a pea-shaped glans. The glans clitoris also has sense receptors, which allow it to function as a sexually sensitive organ.

The *vestibule,* a cleft between the labia minora, contains the orifices of the urethra and the vagina. The vagina may be partially closed by a ring of tissue called the hymen. The hymen is ordinarily ruptured by initial sexual intercourse; however, it can also be disrupted by other types of physical activities. If the hymen persists after sexual intercourse, it can be surgically ruptured.

Notice that the urinary and reproductive systems in the female are entirely separate. For example, the urethra carries only urine, and the vagina serves only as the birth canal and the organ for sexual intercourse.

> The vagina, the organ of sexual intercourse in females, opens into the vestibule, the location of female external genitals.

Sexual response of females has correlations with the sexual response of males. The clitoris is believed to be an especially sensitive organ for initiating sexual sensations. It is possible for the clitoris to become slightly erect as its erectile tissues engorge with blood, but vasocongestion is more obvious in the labia minora, which expand and deepen in color. Erectile tissue within the vaginal wall also expands with blood, and the added pressure in these blood vessels causes small droplets of fluid to squeeze through the vessel walls and to lubricate the vagina.

Release from muscular tension occurs in females during orgasm, especially in the region of the vulva and vagina but also throughout the entire body. Increased uterine motility may assist the transport of sperm toward the oviducts. Since female orgasm is not signaled by ejaculation, there is a wide range in normalcy of sexual response.

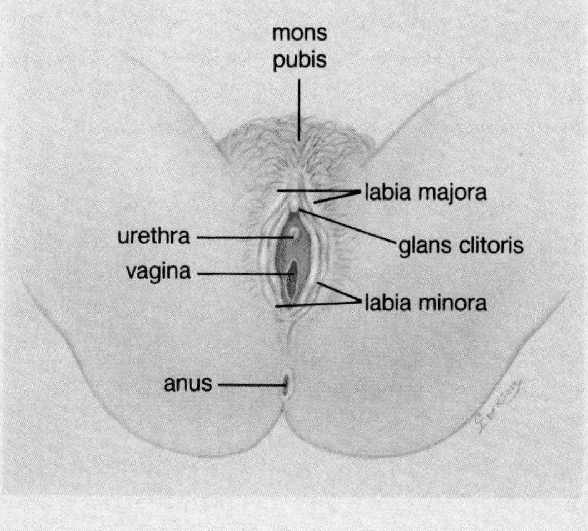

Figure 14.7

External genitals of the female. At birth, the opening of the vagina is partially blocked by a membrane called the hymen. Physical activities and sexual intercourse disrupt the hymen.

Female Hormone Levels

The following glands and hormones are involved in regulating female hormone levels.

Hypothalamus: secretes *GnRH* (gonadotropic-releasing hormone)

Anterior pituitary: secretes *FSH* (follicle-stimulating hormone) and *LH* (luteinizing hormone), the gonadotropic hormones

Ovaries: secrete **estrogen** and **progesterone,** the female sex hormones

The female sex hormones, estrogen and progesterone, have many effects on the body. In particular, estrogen secreted at the time of puberty stimulates the growth of the uterus and the vagina. Estrogen is necessary for egg maturation and is largely responsible for the secondary sex characteristics in females. For example, it is responsible for the onset of the uterine cycle, as well as female body hair and fat distribution. In general, females have a more rounded appearance than males because of a greater accumulation of fat beneath the skin. Also, the pelvic girdle enlarges in females so that the pelvic cavity has a larger relative size compared to males; this means that females have wider hips. Both estrogen and progesterone are also required for breast development.

Ovarian Cycle: FSH and LH

The **ovarian cycle** lasts an average of 28 days but may vary widely in individuals. For simplicity's sake, it is convenient to emphasize that during the first half of a 28-day cycle (days 1–13, see table 14.3), FSH secreted by the anterior pituitary promotes the development of a follicle in the ovary and that this follicle secretes estrogen. As the blood estrogen level rises, it exerts negative feedback control over the anterior pituitary secretion of FSH so that this follicular phase ends (fig. 14.8). The end of the follicular phase is marked by ovulation on the fourteenth day of the 28-day cycle. Similarly, it can be emphasized that during the last half of the ovarian cycle (days 15–28, see table 14.3), anterior pituitary production of LH promotes the development of a corpus luteum, which secretes progesterone. As the blood progesterone level rises, it exerts negative feedback control over anterior pituitary secretion of LH so that the corpus luteum degenerates. As the luteal phase ends, menstruation occurs.

Uterine Cycle: Estrogen and Progesterone

The effect that the female sex hormones have on the endometrium of the uterus causes the uterus to undergo a cyclical series of events known as the **uterine cycle** (see table 14.3). Cycles that last 28 days are divided as follows.

The hypothalamus produces GnRH (gonadotropic-releasing hormone).

GnRH stimulates the anterior pituitary to produce FSH (follicle-stimulating hormone) and LH (luteinizing hormone).

FSH stimulates the follicle to produce estrogen and LH stimulates the corpus luteum to produce progesterone.

Estrogen and progesterone affect the sex organs (e.g., uterus) and the secondary sex characteristics and exert feedback control over the hypothalamus and the anterior pituitary.

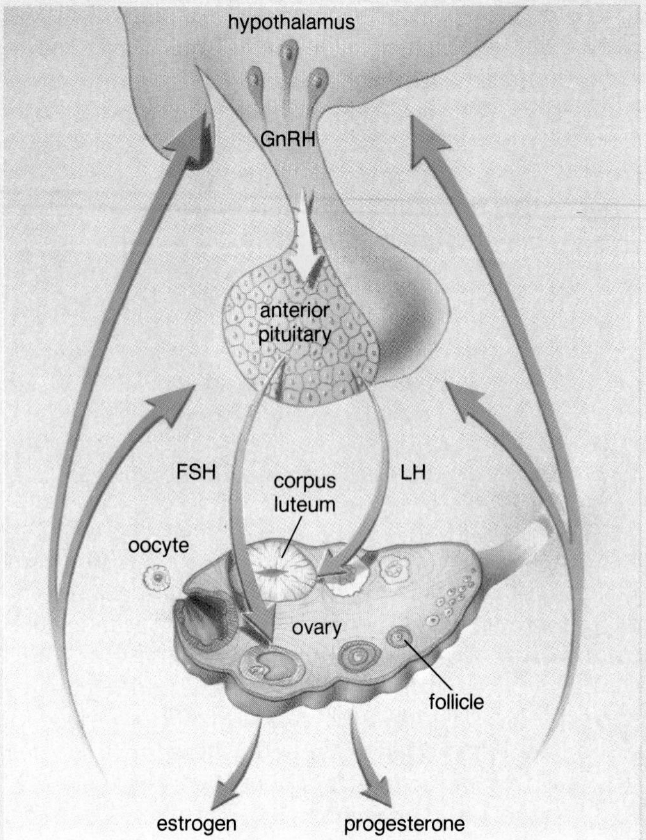

Figure 14.8

The hypothalamus-pituitary-ovary control relationship.

Table 14.3
Ovarian and Uterine Cycles (simplified)

Ovarian Cycle	Events	Uterine Cycle	Events
Follicular phase (days 1–13)	FSH stimulates ovary	Menstruation (days 1–5)	Endometrium breaks down
	Follicle matures	Proliferative phase (days 6–13)	Endometrium rebuilds
	Estrogen is released		
Ovulation (day 14*)			
Luteal phase (days 15–28)	LH stimulates ovary	Secretory phase (days 15–28)	Endometrium thickens and glands are secretory
	Corpus luteum develops		
	Progesterone is released		

**Assuming 28-day cycle.*

During *days 1–5,* there is a low level of female sex hormones in the body, causing the uterine lining to disintegrate and its blood vessels to rupture. A flow of blood, known as the *menses,* passes out of the vagina during **menstruation,** also known as the menstrual period. If menstruation is painful, the problem could be endometriosis, as discussed in the Health Focus for this chapter.

During *days 6–13,* increased production of estrogen by an ovarian follicle causes the endometrium to thicken, becoming vascular and glandular. This series of events is called the *proliferative phase* of the uterine cycle.

Ovulation usually occurs on the fourteenth day of the 28-day cycle.

Health Focus

Endometriosis

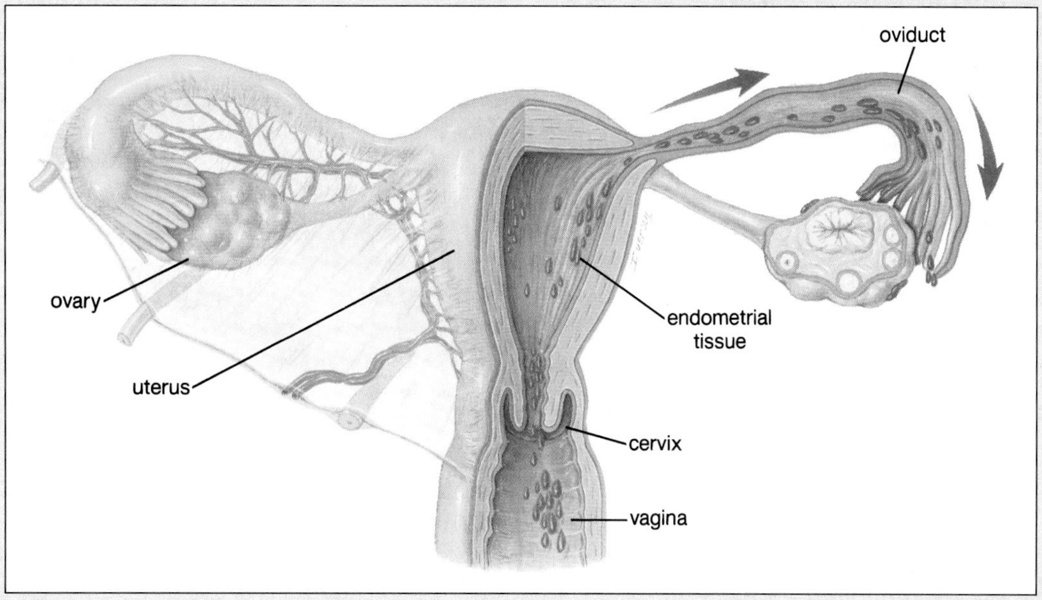

Figure 14A

Endometriosis. It is speculated that endometriosis is caused by a backward menstrual flow as represented by the arrows in this drawing. This allows endometrial cells to enter the abdominal cavity, where they take up residence and respond to the monthly cyclic changes in hormonal levels, including those that result in menstruation.

In the female reproductive tract, there is a small space between an ovary and the oviduct that leads to the uterus. The lining of the uterus, called the endometrium, loses its outer layer during menstruation and sometimes a portion of the menstrual discharge is carried backwards up the oviduct and into the abdominal cavity instead of being discharged through the opening in the cervix. There, endometrial tissue can become attached to and implanted in various organs such as the ovaries; the wall of the vagina, bowel, or bladder; or even on the nerves that serve the lower back or legs. This painful condition is called endometriosis, and affects 1%–3% of women of reproductive age.

Women with uterine cycles of less than 27 days and with a menstrual flow lasting longer than one week have an increased chance of endometriosis. Women who have taken the birth-control pill for a long time or have had several pregnancies, have a decreased chance of endometriosis.

When a woman has endometriosis, the displaced endometrial tissue reacts as if it were still in the uterus—thickening, becoming secretory, and then breaking down. The discomfort of menstruation is then felt in other organs of the abdominal cavity, resulting in pain. An area of endometriosis can degenerate and become a scar. Scars that hold two organs together are called adhesions, which can distort organs and lead to infertility.

Only direct observation of the abdominal organs can confirm endometriosis. First, a half-inch incision is made near the navel and the gas carbon dioxide is injected into the abdominal cavity to separate the organs. Then, a laparoscope (lit telescope) is inserted into the abdomen, allowing the physician to see the organs. Patches of endometriosis show up as purple, blue, or red spots, and there may be dark brown cysts filled with blood. When these rupture, there is a great deal of pain. In the severest cases, there is scarring and the formation of adhesions and abnormal masses around the pelvic organs. A second incision, usually at the pubic hairline, allows the insertion of other instruments that can be used to remove endometrial implants.

Further treatment can take one of two courses. The drug nafarelin can be administered as a nasal spray, which acts through hormonal controls to stop the production of estrogen and, therefore, to stop the uterine cycle. Or, the ovaries can be removed, stopping the uterine cycle. In either case, the woman will suffer symptoms of menopause that can be relieved in the first instance by withdrawing the drug nafarelin, or in the second instance by giving the woman estrogen in doses that do not reactivate the uterine cycle.

During *days 15–28,* increased production of progesterone by the corpus luteum causes the endometrium to double in thickness and the uterine glands to mature, producing a thick, mucoid secretion. This phase is called the *secretory phase* of the uterine cycle. The endometrium now is prepared to receive the developing embryo, but if pregnancy does not occur, the corpus luteum degenerates and the low level of sex hormones in the female body causes the uterine lining to break down. This is evidenced by the menstrual discharge that begins at this time. Even while menstruation is occurring, the anterior pituitary begins to increase its production of FSH and a new follicle begins to mature. Table 14.3 indicates how the ovarian cycle controls the uterine cycle. Changes in the blood hormone levels involved in the ovarian and uterine cycles are shown in figure 14.9.

The ovaries and the uterus undergo cycles that are ultimately under the control of the anterior pituitary and hypothalamus. During the follicular phase of the ovarian cycle, FSH causes the endometrium to proliferate. During the luteal phase of the ovarian cycle, LH causes the endometrium to become secretory.

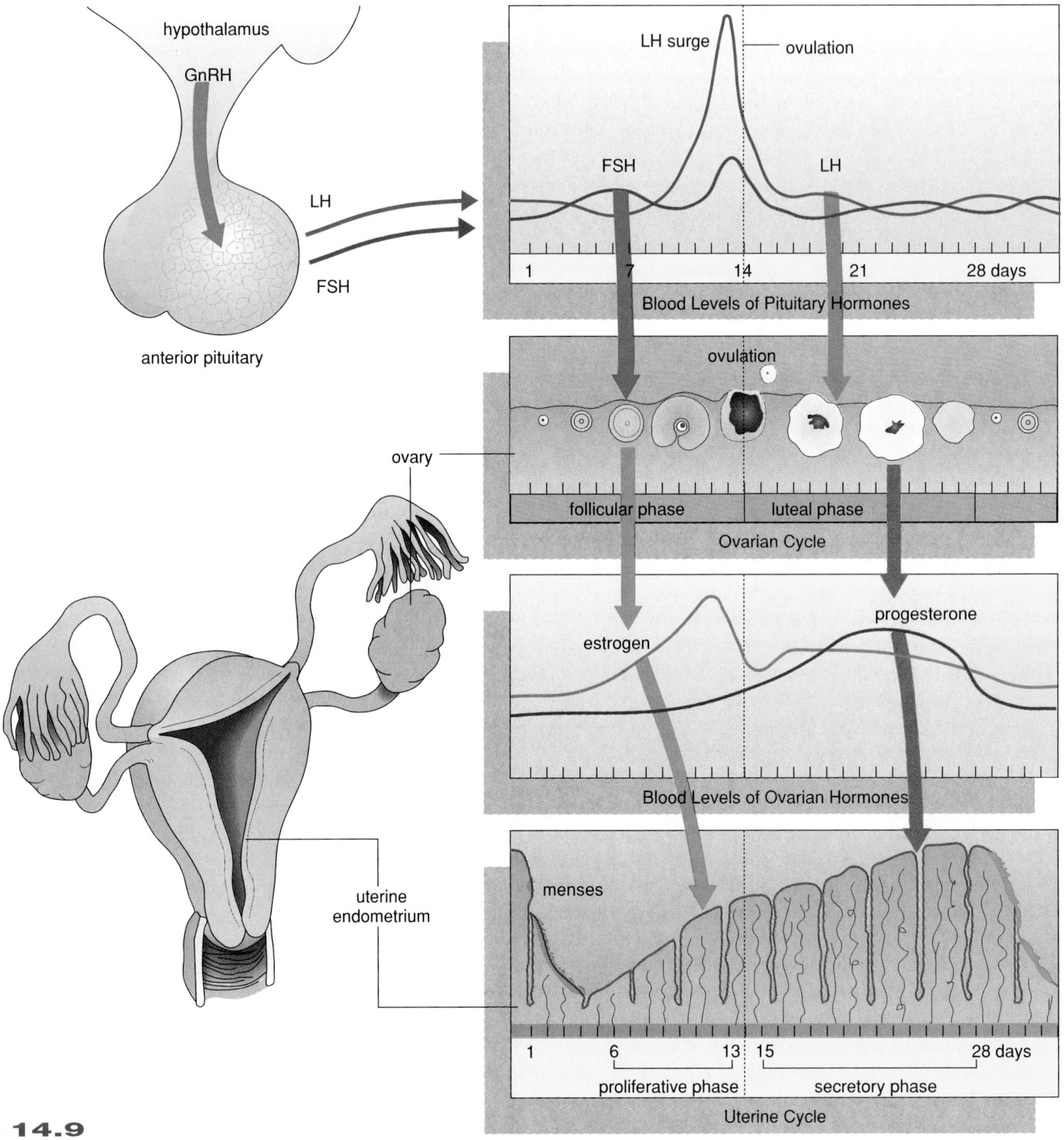

Figure 14.9

Blood hormone levels associated with the ovarian and uterine cycles. During the follicular phase, FSH released by the anterior pituitary promotes the maturation of a follicle in the ovary. The ovarian follicle produces increasing estrogen levels, which causes the endometrium to thicken. After ovulation and during the luteal phase, LH promotes the development of the corpus luteum. This structure produces increasing progesterone levels, which causes the endometrial lining to become secretory. Menstruation begins when progesterone production declines to a low level.

Pregnancy: Upon Implantation

Pregnancy occurs when the developing embryo embeds itself in the endometrium several days after fertilization. During **implantation,** a membrane surrounding the embryo produces a gonadotropic hormone called **human chorionic gonadotropic hormone (HCG)**, which prevents degeneration of the corpus luteum and instead causes it to secrete even larger quantities of progesterone. Progesterone (*pro* means *for; gestation* means *pregnancy*) also inhibits the motility of the uterus and together with estrogen prepares the breasts for lactation. The corpus luteum may be maintained for as long as 6 months, even after the placenta is fully developed.

The *placenta* (see fig. 16.4) originates from both maternal and fetal tissue and is the region of exchange of molecules between fetal and maternal blood, although there is no mixing of the 2 types of blood. After its formation, the placenta continues to produce HCG. It also begins to produce progesterone and estrogen, which have 2 effects: they shut down the anterior pituitary so that no new follicles mature, and they maintain the lining of the uterus so that the corpus luteum is not needed. There is no menstruation during the 9 months of pregnancy.

Menopause: Ovaries Don't Respond

Menopause, the period in a woman's life during which the ovarian and uterine cycles cease, is likely to occur between ages 45 and 55. The ovaries are no longer responsive to the gonadotropic hormones produced by the anterior pituitary, and the ovaries no longer secrete estrogen or progesterone. At the onset of menopause, the uterine cycle becomes irregular, but as long as menstruation occurs, it is still possible for a woman to conceive. Therefore, a woman usually is not considered to have completed menopause until there has been no menstruation for a year.

The hormonal changes during menopause often produce physical symptoms, such as "hot flashes," which are caused by circulatory irregularities, dizziness, headaches, insomnia, sleepiness, and depression. Again, there is great variation among women, and any of these symptoms may be absent altogether.

Women sometimes report an increased sex drive following menopause. It has been suggested that this may be due to androgen production by the adrenal cortex.

> During pregnancy, there is no menstruation because the placenta produces HCG and the hormones estrogen and progesterone, which shut down the pituitary. After menopause, FSH and LH are still produced by the anterior pituitary, but the ovaries are unable to respond.

Control of Reproduction

The use of birth-control (contraceptive) methods decreases the probability of pregnancy but does not, except for the condom, offer any protection against contracting a sexually transmitted disease such as AIDS. A common way to discuss pregnancy rate is to indicate the number of pregnancies expected per 100 women per year. For example, it is expected that 80 out of 100 young women, or 80%, who are regularly engaging in unprotected intercourse will be pregnant within a year. Another way to discuss birth-control methods is to indicate their effectiveness, in which case the emphasis is placed on the number of women who will not get pregnant. For example, with the least effective method given in figure 14.10, we expect that 70 out of 100, or 70%, sexually active women will not get pregnant, and 30 women will get pregnant within a year. The very best and surest method of birth control is total abstinence.

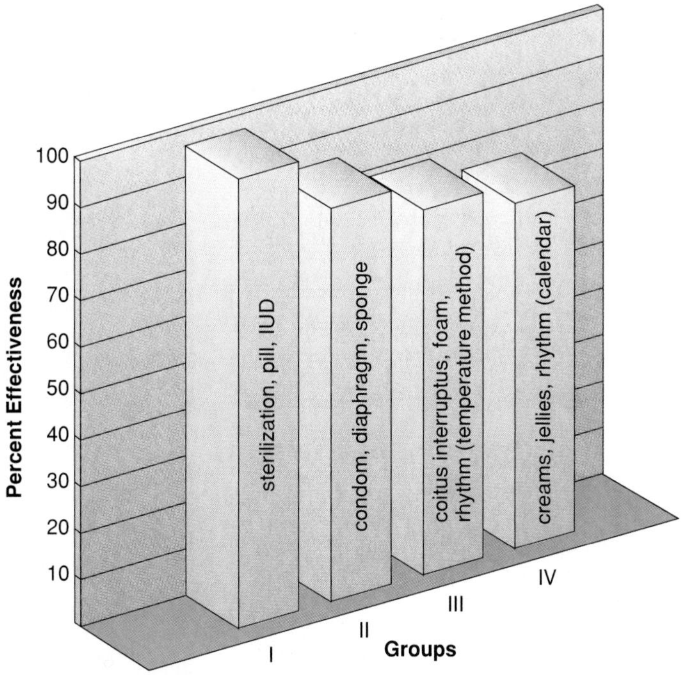

Figure 14.10

Effectiveness (the percentage of women who are not expected to be pregnant within one year) of various birth-control measures. Sterilization and the birth-control pill offer the best protection, while creams, jellies, and the rhythm method offer the least protection from the occurrence of pregnancy. This graph assumes that users are properly and faithfully using the various means of birth control.

Source: Data from Alan F. Guttmacher, Pregnancy, Birth, and Family Planning, *1973, New American Library, New York, NY.*

Birth Control: Variety of Methods

For the sake of discussion, the birth-control methods we will discuss are grouped according to effectiveness. Those methods in Group I are most effective; those in Group V are least effective in preventing pregnancy.

Group I

Sterilization is a surgical procedure that renders the individual incapable of reproduction. Sterilization operations do not affect the secondary sex characteristics nor sexual performance.

In the male, a **vasectomy** consists of cutting and sealing the vas deferens on each side so that the sperm are unable to reach the seminal fluid that is ejected at the time of orgasm. The sperm are then largely reabsorbed. Following this operation, which can be done in a doctor's office, the amount of ejaculate remains normal because sperm account for only about 1% of the volume of semen. Also, there is no effect on the secondary sex characteristics since testosterone continues to be produced by the testes.

In the female, **tubal ligation** consists of cutting and sealing the oviducts. Pregnancy rarely occurs because the passage of the egg through the oviducts has been blocked. Whereas major abdominal surgery was formerly required for a tubal ligation, today there are simpler procedures. Using a method called *laparoscopy,* which requires only 2 small incisions, the surgeon inserts a small, lit telescope to view the oviducts and a small surgical blade to sever them. An even newer method called hysteroscopy uses a telescope within the uterus to seal the tubes by means of an electric current.

Although recently developed microsurgical methods allow either a vas deferens or oviduct to be rejoined, it is still wise to view a vasectomy or tubal ligation as permanent. Even following successful reconnection, fertility is usually reduced by about 50%.

Oral contraception (*birth-control pill*) (fig. 14.11*a*) usually involves taking a combination of estrogen and progesterone for 21 days of a 28-day cycle (beginning at the end of menstruation). The estrogen and progesterone in the birth-control pill effectively shut down the pituitary production of both FSH and LH so that no follicle begins to develop in

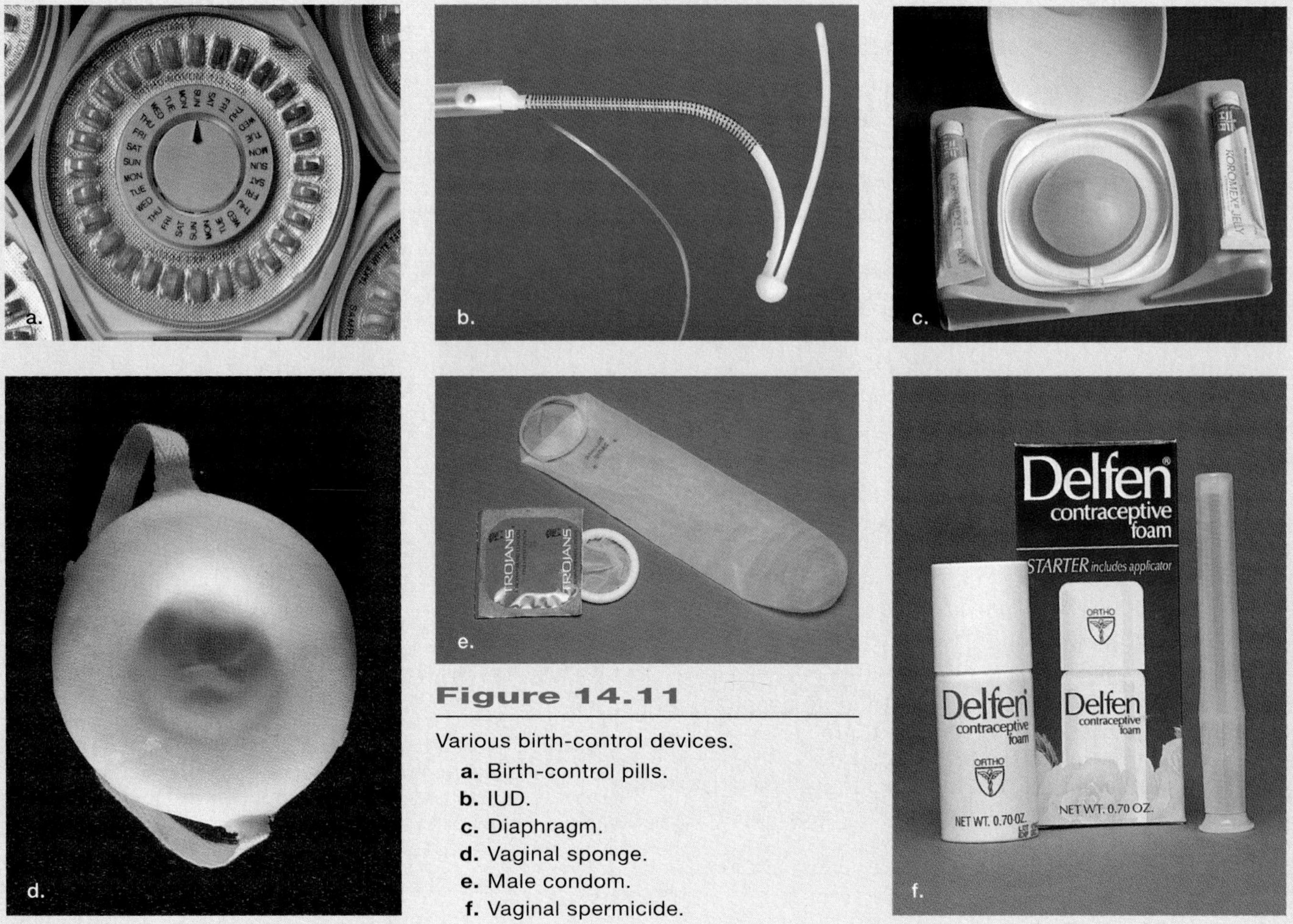

Figure 14.11

Various birth-control devices.
- **a.** Birth-control pills.
- **b.** IUD.
- **c.** Diaphragm.
- **d.** Vaginal sponge.
- **e.** Male condom.
- **f.** Vaginal spermicide.

the ovary; and since ovulation does not occur, pregnancy cannot take place. Both beneficial and adverse side effects have been linked to the birth-control pill. Women report relief of menstrual discomforts and acne. They also report several minor adverse side effects such as nausea and vomiting. Less common complaints are weight gain, headaches, and chloasma (areas of darkened skin on the face). One serious side effect of the birth-control pill is increased incidence of thromboembolism—almost exclusively in women who are over 35 and who smoke. Since there are possible side effects, those taking the birth-control pill should see a physician regularly.

The *Norplant system* and *Depo-Provera injections* are 2 new means of birth control that utilize a synthetic progesterone to prevent ovulation by disrupting the ovarian cycle. The Norplant system consists of 6 match-sized time-release capsules surgically inserted under the skin of a woman's upper arm. The effectiveness of this system may last 5 years. A Depo-Provera injection must be given every 3 months.

An *IUD* (intrauterine device) (fig. 14.11*b*) is a small piece of molded plastic that is inserted into the uterus by a physician. Two types of IUDs are now available: the copper type have a copper wire wrapped around the stem and the progesterone-releasing type have progesterone embedded in the plastic. IUDs are believed to alter the environment of the uterus and oviducts so that fertilization probably does not occur—but if it should occur, implantation cannot take place. With proper patient selection, side effects and complications of these 2 types of IUDs are rare. The best candidates for this use of birth control are women who have had at least one child, are of middle-to-older reproductive age, and have a stable relationship with a partner who does not have a sexually transmitted disease.

Group II

The *diaphragm* (fig. 14.11*c*) is a soft rubber or plastic cup with a flexible rim that lodges behind the pubic bone and fits over the cervix. Each woman must be properly fitted by a physician, and the diaphragm can be inserted into the vagina 2 hours at most before sexual relations. It must also be used with a spermicidal jelly or cream and should be left in place for at least 6 hours after sexual relations. If intercourse is repeated during this time, more jelly or cream should be inserted by means of a plastic insertion tube.

The *cervical cap* is a minidiaphragm that also must be fitted by a physician and should be used with a spermicide. It is made of natural rubber or plastic and fits over the cervix like a thimble. The cap must be inserted before intercourse and remain in place for 8 hours afterward. Unlike the diaphragm, the cervical cap is effective even if left in place for several days.

A *vaginal sponge* (fig. 14.11*d*) is permeated with spermicide and shaped to fit the cervix. Unlike the diaphragm and cervical cap, the sponge need not be fitted by a physician since one size fits everyone. It is effective immediately after placement in the vagina and remains effective for 24 hours. Like the other means of birth control in this category, the sponge is about 85% effective in preventing pregnancy.

A male *condom* (fig. 14.11*e*) is a thin skin (lambskin) or plastic sheath (latex) that fits over the erect penis.The ejaculate is trapped inside the sheath and, thus, does not enter the vagina. When used in conjunction with a spermicidal foam, cream, or jelly, the protection is better than with the condom alone. Today it is possible to purchase condoms that are already lubricated with a spermicide. The condom is generally recognized as giving protection against sexually transmitted diseases such as those discussed in the following chapter.

A female condom is now available. The closed end of a large plastic tube has a flexible ring that fits onto the cervix. The open end of the tube has a ring that covers the external genitals. The female condom also offers some protection against sexually transmitted diseases.

Group III

It is possible for the male to withdraw the penis just before ejaculation so that the semen is deposited away from the vaginal area. This method of birth control, called *coitus interruptus,* has a relatively high failure rate because a few drops of seminal fluid may escape from the penis before ejaculation takes place. Even a small amount of semen can contain numerous sperm.

Spermicidal jellies, creams, and foams (fig. 14.11*f*) contain sperm-killing ingredients and may be inserted into the vagina with an applicator up to 30 minutes before each occurrence of intercourse. Foams are considered the most effective of this group of contraceptives. When used alone, these are not highly effective means of birth control for those who have frequent intercourse. When used with a condom, they offer added protection against sexually transmitted disease; nonoxynol-9, a common ingredient, is a viral inhibitor giving some protection against AIDS.

Group IV

Natural family planning, formerly called the *rhythm method* of birth control, is based on the realization that a woman ovulates only once a month and that the egg and sperm are viable for a limited number of days. If the woman has a consistent 28-day cycle, then the period of "safe" days can be determined, as in figure 14.12. This method of birth control is not very effective because the days of ovulation can vary from month to month, and the viability of the egg and sperm varies perhaps monthly but certainly from person to person.

A more reliable way to practice natural family planning is to await the day of ovulation each month and then wait 3 more days before engaging in intercourse. The day of ovulation can be more accurately determined by noting the body temperature early each morning (body temperature rises at ovulation), by taking the pH of the vagina each day (near the day of ovulation the vagina becomes more alkaline), or by noting the consistency of the mucus at the cervix (at ovulation the mucus is thinner and more watery). Physicians can instruct women how to do these procedures.

Figure 14.12

Natural family planning. The calendar shows the "safe" (green) and "unsafe" (purple) days for intercourse. This calendar is *only* appropriate for women with regular 28-day cycles. Few women have regular cycles month after month.

Only medically recognized methods of birth control such as those discussed here should be used. Douching is of little value and position of intercourse will not prevent pregnancy at all. In fact, the proximate location of the penis (at the time of ejaculation) near but not in the vagina has been known to result in pregnancy.

> Numerous birth-control methods and devices are available for those who wish to prevent pregnancy. The more effective methods are sterilization, the birth-control pill, the Norplant system and Depo-Provera injections, and the IUD. A condom used with a spermicidal jelly or foam is also effective. The less effective methods are spermicidal foam or jelly alone, coitus interruptus, and natural family planning. (See also table 14.4.)

Searching for Other Means of Birth Control

Investigators have long searched for a "male pill." Analogues of gonadotropic-releasing hormone have been used to prevent the hypothalamus from stimulating the anterior pituitary. Or, inhibin has been used to prevent the anterior pituitary (see fig. 14.4) from producing FSH. Testosterone and/or related chemicals have been used to inhibit spermatogenesis in males, but there are usually feminizing side effects because an excess of testosterone is changed to estrogen by the body.

Morning-after Pills

There are morning-after regimens available by prescription that, depending on when a woman begins medication, either prevent fertilization altogether or stop the fertilized egg from ever implanting.

Table 14.4

Common Birth-Control Methods

Name	Procedure	Methodology	Effectiveness	Risk
Vasectomy	Vas deferens is cut and tied	No sperm in seminal fluid	Almost 100%	Irreversible sterility
Tubal ligation	Oviducts are cut and tied	No eggs in oviduct	Almost 100%	Irreversible sterility
Oral contraception (birth-control pill)	Hormone medication taken daily	Anterior pituitary does not release FSH and LH	Almost 100%	Thromboembolism, especially in smokers
Depo-Provera	Four injections of progesterone-like steroid a year	Anterior pituitary does not release FSH and LH	About 99%	Breast cancer? Osteoporosis?
Norplant	Tubes of progestin (form of progesterone) are implanted under skin	Anterior pituitary does not release FSH and LH	More than 90%	Presently none known
IUD	Plastic coil is inserted into uterus by physician	Prevents implantation	More than 90%	Infection (PID)
Vaginal sponge	Sponge permeated with spermicide is inserted in vagina	Kills sperm on contact	About 90%	Presently none known
Diaphragm	Plastic cup is inserted into vagina to cover cervix before intercourse	Blocks entrance of sperm to uterus	With jelly, about 90%	Presently none known
Cervical cap	Rubber cup is held by suction over cervix	Delivers spermicide near cervix	Almost 85%	Cancer of cervix?
Male condom	Latex sheath is fitted over erect penis at time of intercourse	Traps sperm and prevents STDs	About 85%	Presently none known
Female condom	Plastic tubing is fitted inside vagina	Blocks entrance of sperm to uterus and prevents STDs	About 85%	Presently none known
Coitus interruptus (withdrawal)	Male withdraws penis before ejaculation	Prevents sperm from entering vagina	75%	Presently none known
Jellies, creams, foams	Contain spermicidal chemicals that are inserted before intercourse	Kill a large number of sperm	About 75%	Presently none known
Natural family planning	Day of ovulation is determined by record keeping; various methods of testing	Intercourse avoided on certain days of the month	About 70%	Presently none known
Douche	Vagina and uterus are cleansed after intercourse	Washes out sperm	Less than 70%	Presently none known

These regimens involve taking pills containing synthetic progesterone and/or estrogen in a manner prescribed by a physician. Many women do not realize that this method of birth control is available, and yet it is estimated that use of these regimens could greatly reduce the number of unintended pregnancies. Effective treatment sometimes causes nausea and vomiting, which can be severe.

RU-486 is a pill presently used in France and Britain that is now being considered for use in this country. RU-486 causes the loss of an implanted embryo by blocking the progesterone receptors of the cells in the uterine lining. Without functioning receptors for progesterone, the uterine lining sloughs off, carrying the embryo with it. When taken in conjunction with a prostaglandin to induce uterine contractions, RU-486 is 95% effective. It is possible that some day the medication might be used by women who are experiencing delayed menstruation without knowing if they are actually pregnant.

> There are numerous well-known birth-control methods and devices available to those who wish to prevent pregnancy. Their effectiveness varies. In addition, new methods are expected to be developed.

Infertility: One Out of Four Couples

Sometimes, couples do not need to prevent pregnancy; conception or fertilization does not occur despite frequent intercourse. The American Medical Association estimates that 15% of all couples in this country are unable to have any children and therefore are properly termed *sterile*; another 10% have fewer children than they wish and therefore are termed *infertile.* The latter assumes that the couple has been unsuccessfully trying to become pregnant for at least one year.

What Causes Infertility?

The 2 major causes of infertility in females are blocked oviducts, possibly due to pelvic inflammatory disease (PID), discussed in chapter 15, and failure to ovulate due to low body weight. *Endometriosis,* discussed in the Health Focus for this chapter also contributes to infertility.

Sometimes the causes of infertility can be corrected surgically and/or medically so that couples can become parents (fig. 14.13). If no obstruction is apparent and body weight is normal, it is possible to give females a substance rich in FSH and LH extracted from the urine of postmenopausal women. This treatment causes multiple ovulations and sometimes multiple pregnancies.

The most frequent cause of infertility in males is low sperm count and/or a large proportion of abnormal sperm. Disease, radiation, high testes temperature, and manufactured chemicals can contribute to this condition. The Ecology Focus for this chapter discusses the effects of manufactured chemicals on fertility.

When reproduction does not occur in the usual manner, many couples adopt a child. Others sometimes first try one of the alternative reproductive methods discussed in the following paragraphs.

Alternative Reproduction: How to Have Five Parents

If all the alternative methods discussed are considered, it is possible to imagine a baby with 5 parents: (1) sperm donor, (2) egg donor, (3) surrogate mother, and (4) and (5) adoptive mother and father.

Artificial Insemination by Donor (AID). During **artificial insemination,** sperm are placed in the vagina by a physician. Sometimes a woman is artificially inseminated by her partner's sperm. Artificial insemination is especially helpful if her mate has a low sperm count—the sperm can be collected over a period of time and concentrated so that the sperm count is sufficient to result in fertilization. Often, however, a woman is inseminated by sperm acquired from an anonymous donor. At times, a mixture of partner and donor sperm is used.

Figure 14.13

Mother and child. Sometimes couples utilize alternative methods of reproduction to experience the joys of parenthood.

Ecology Focus

Fertility and the Environment

The last half century has seen an explosion in the number of manufactured chemicals. Worldwide, about 70,000 chemicals are in everyday use and about 1,000 new ones are added each year. Most of us are well aware that some of these chemicals are immunosuppressive and contribute to the development of cancer (see Ecology Focus, chapter 7). Now it appears that they may have more subtle effects, such as affecting human fertility.

Researchers at the Rigshospitalet, a major hospital in Copenhagen, tell us that sperm counts have declined significantly worldwide in the past 50 years and that the drop may be linked to exposure to certain chemicals such as DDT, TCDD, the most toxic form of dioxin, and PCBs (polychlorinated biphenyls), all of which have estrogen-like activity. In lab experiments, a single oral dose of TCDD fed to pregnant mother rats causes abnormal sperm counts and other reproductive malfunctions in male offspring. They note that besides a decline in human sperm counts, there has been a rise in testicular cancer. Such a combination of effects is known to occur when male fetuses are exposed to too high a ratio of female to male hormones in the womb. They speculate that the chemicals mentioned first accumulate in the mother's body and then are passed to the fetus by way of the placenta. The result is reduced fertility in male offspring. In another laboratory study, dioxin fed to young female rats was found to increase the chance of endometriosis several years later.

Exposure to chemicals in the workplace can also reduce fertility. Nitrous oxide is a gas that is used to kill pain in dental offices. Laboratory studies have shown that nitrous oxide can prevent conception in female rats. Recently, researchers decided to do a statistical test on 459 dental assistants who were exposed to nitrous oxide on the job. After controlling other factors known to prevent conception, they found a 41% drop in fertility when assistants were exposed to high levels of nitrous oxide. It is possible that the gas blocks secretion of gonadotropin-releasing hormone by the hypothalamus and therefore prevents ovulation.

The public needs to be aware that manufactured chemicals can be harmful to health and then decide if the benefit is worth the risk. If deemed appropriate, government can be pressured to use stricter regulations regarding the production of old and new chemicals.

A variation of AID is intrauterine insemination (IUI). IUI involves hormonal stimulation of the ovaries, followed by placement of the donor's sperm in the uterus rather than in the vagina.

In Vitro Fertilization (IVF). In the case of **in vitro fertilization (IVF),** hormonal stimulation of the ovaries is followed by laparoscopy. In this procedure, an aspiratory tube is used to retrieve preovulatory eggs (see fig. 14.6). Alternately, it is possible to pierce the vaginal wall with a needle and to guide it, using ultrasound, to the ovaries, where the needle is used to retrieve the eggs. This method is called transvaginal retrieval.

Concentrated sperm from the male are placed in a solution that approximates the conditions of the female genital tract. When the eggs are introduced, fertilization occurs. The resultant zygotes (fertilized eggs) begin development, and after about 2–4 days, the embryos are introduced into the uterus of the woman, who is now in the secretory phase of her uterine cycle. If implantation is successful, development is normal and continues to term.

Gamete Intrafallopian Transfer (GIFT). Gamete intrafallopian transfer (GIFT) was devised as a means to overcome the low success rate (15%–20%) of in vitro fertilization. The method is exactly the same as in vitro fertilization, except the eggs and the sperm are placed in the oviducts immediately after they have been brought together. This procedure is helpful to couples whose eggs and sperm never make it to the oviducts; sometimes the egg gets lost between the ovary and the oviducts, and sometimes the sperm never reach the oviducts. GIFT has an advantage in that it is a one-step procedure for the woman—the eggs are removed and are reintroduced all in the same time period. For this reason, it is less expensive—approximately $1,500 compared to $3,000 and up for in vitro fertilization.

Surrogate Mothers. A surrogate mother is a woman who has a baby for someone else, often for pay. Other individuals contribute sperm and/or eggs to the fertilization process in such cases. Later, the child is adopted by the gamete donors.

> Some couples are infertile due to various physical abnormalities. When corrective medical procedures fail, today it is possible to consider an alternative method of reproduction to be a parent.

Summary

In males, spermatogenesis occurs within the seminiferous tubules of the testes, which also produce testosterone within the interstitial cells. A sperm has a head, capped by an acrosome, where 23 chromosomes reside in the nucleus; a mitochondria-containing middle piece; and a tail, which is a flagellum. Sperm mature in the epididymis, and they are stored in the vas deferens before entering the urethra just prior to ejaculation.

The accessory glands (seminal vesicles, prostate gland, and Cowper's gland) produce seminal fluid, which leaves the penis during ejaculation. The penis, which has a foreskin that may be removed by circumcision, must become erect to be placed in the vagina of a female. Erection occurs when blood sinuses within erectile tissue fill with blood. If sexual stimulation is sufficient, ejaculation follows an erection, and this is an obvious sign of male orgasm. As many as 400 million sperm may be ejaculated.

Hormonal regulation in the male maintains testosterone at a fairly constant level. The hypothalamus produces GnRH, a hormone that stimulates the anterior pituitary to produce FSH and LH, which are present in both sexes. In males, FSH promotes spermatogenesis and LH promotes testosterone production. Testosterone stimulates growth of the male genitals during puberty and is necessary for maturation of sperm and development of the secondary sex characteristics—those features that we associate with the male body aside from the genitals.

In females, oogenesis occurs within the ovaries where one follicle reaches maturity each month. This follicle balloons out of the ovary and bursts to release the egg. The ruptured follicle develops into a corpus luteum. The follicle and corpus luteum produce the female sex hormones estrogen and progesterone.

The egg must cross a small space to enter the oviducts, which conduct it toward the uterus. Fertilization usually takes place within the oviducts. If fertilization occurs, the embryo embeds itself in the uterine lining and the corpus luteum is maintained. If fertilization does not take place, the corpus luteum degenerates. The vagina, the copulatory organ in females, opens into the vestibule where the urethra also opens. The vestibule is bounded by the labia minora, which come together at the clitoris, a highly sensitive organ. Outside the labia minora are the labia majora. The entire region of the external genitalia is called the vulva.

The female hormones, estrogen and progesterone, affect the female genitals, promote development of the egg, and maintain the secondary sex characteristics.

Hormone fluctuation in the female results in ovarian and uterine cycles. During the first half of the ovarian cycle, FSH from the anterior pituitary causes maturation of the follicle, which secretes estrogen. After ovulation, and during the second half of the cycle, LH from the anterior pituitary converts the follicle into the corpus luteum, which produces progesterone. Estrogen and progesterone regulate the menstrual cycle. During the first few days of the uterine cycle, a low level of hormones causes the endometrium to break down as menstruation occurs (days 1–5). As estrogen begins to be produced by the follicle, the endometrium begins to build (proliferation phase, days 6–13). Ovulation occurs on the fourteenth day of a 28-day cycle. As progesterone is produced by the corpus luteum, the endometrium thickens and becomes secretory (secretory phase, days 14–28).

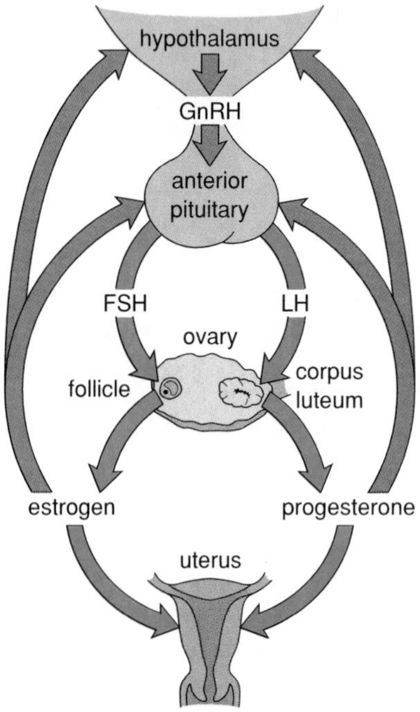

Numerous birth-control methods and devices are available for those who wish to prevent pregnancy. Infertile couples are increasingly resorting to alternative methods of reproduction.

Study and Thought Questions

1. Discuss the anatomy and the physiology of the testes. (pp. 272–273) Describe the structure of sperm. (pp. 273–274)
2. Give the path of sperm from the testes until they leave the penis. (p. 274)
3. Which glands produce seminal fluid? (p. 275)
4. Discuss the anatomy and the physiology of the penis. (p. 274) Describe ejaculation. (p. 275)
5. Discuss hormonal regulation in the male. Name 3 functions for testosterone. (pp. 275–276)

Critical Thinking Using figure 14.4 as a guide, hypothesize why anabolic steroids shrink the size of the testes.

6. Discuss the anatomy and the physiology of the ovaries. (p. 277) Describe ovulation. (pp. 277–278)
7. Give the path of the egg. Where do fertilization and implantation normally occur? Name 2 functions of the vagina. (p. 278)
8. Describe the external genitals in females. (p. 278)
9. Compare male and female orgasm. (p. 275, p. 279)

10. Name 4 functions of the female sex hormones. (p. 279)
11. Give the events of the uterine cycle, and relate them to the ovarian cycle. (pp. 279–280) In what way is menstruation prevented if pregnancy occurs? (p. 283)

    Critical Thinking Postmenopausal women have an increased level of FSH and LH in the blood. Why?
12. Discuss the various means of birth control and their relative effectiveness. (pp. 284–285)

    Ethical Issue Should society try to prevent teenage pregnancy? Explain your answer and suggest what should be done.
13. Discuss some of the methods employed to permit infertile couples to have children. (pp. 288–289)

    Ethical Issue In vitro fertilization sometimes results in embryos that are never utilized. Do you think such embryos have rights? Why or why not? If they do have rights, what can be done to protect these rights?

Writing Across the Curriculum

1. As a continuation of the introduction to this chapter, state specifically the anatomical and physiological means by which humans are adapted to reproduce on land.
2. All organisms have a life cycle that includes a reproductive strategy. For example, some insects spend much of their lives as larvae, undergo metamorphosis into winged forms, reproduce, and die all within a single season. They do not spend any time at all caring for their offspring and instead produce a large number of offspring, of which a few may survive. Outline and discuss the human life strategy. Include a reference to culture, mentioned on page 3 of the text.
3. Many people try each month to prevent pregnancy, while others try to achieve pregnancy. Are the methods available for use by both groups satisfactory? Can you suggest alternatives?

Objective Questions

1. In tracing the path of sperm, the structure that follows the epididymis is the __________.
2. The prostate gland, the Cowper's glands, and the __________ all contribute to seminal fluid.
3. The main male sex hormone is ____________________.
4. An erection is caused by the entrance of ______________ into sinuses within the penis.
5. In the female reproductive system, the uterus lies between the oviducts and the ____________________.
6. The vagina is the organ of ____________________ and the ____________________.
7. In the ovarian cycle, once each month a (n) ________________ produces an egg. In the uterine cycle, the ______________ lining of the uterus is prepared to receive the embryo.
8. The female sex hormones are ___________ and ___________.
9. In vitro fertilization occurs in __________________.
10. Label this diagram of the male reproductive system, and trace the path of sperm.

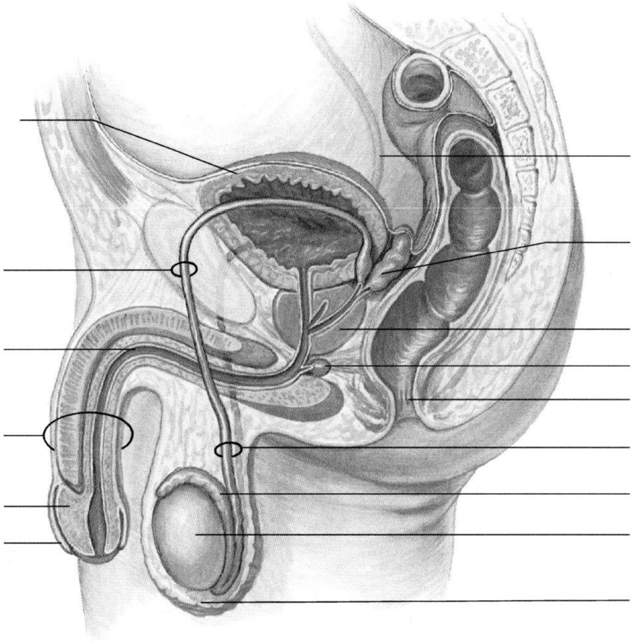

Selected Key Terms

endometrium (en″do-me′tre-um) The lining of the uterus, which becomes thickened and vascular during the uterine cycle. 278

erection Condition of the penis when it is turgid and erect, instead of being flaccid or lacking turgidity. 274

estrogen Female sex hormone, which, along with progesterone, maintains the primary sex organs and stimulates development of the female secondary sex characteristics. 279

Graafian follicle (graf′e-an fol′ĭk′l) Mature follicle within the ovaries, which contains a developing egg. 277

implantation The attachment and penetration of the embryo into the lining of the uterus (endometrium). 283

interstitial cell Hormone-secreting cell located between the seminiferous tubules of the testes. 274

menopause Termination of the ovarian and uterine cycles in older women. 283

menstruation Loss of blood and tissue from the uterus at the end of a uterine cycle. 280

ovarian cycle Monthly occurring changes in the ovary that determine the level of sex hormones in the blood. 279

ovary The female gonad, the organ that produces eggs, estrogen, and progesterone. 277

penis External organ in males through which the urethra passes and that serves as the organ of sexual intercourse. 274

progesterone Female sex hormone secreted by the corpus luteum of the ovary and by the placenta. 279

prostate gland Gland located around the male urethra below the urinary bladder; adds secretions to seminal fluid. 275

seminal fluid The sperm-containing secretion of males; also called semen. 275

seminiferous tubule (sem″ ĭ-nif′er-us tu-būl) Highly coiled duct within the male testis, which produces and transports sperm. 273

testis The male gonad, the organ that produces sperm and testosterone. 272

testosterone The main male sex hormone responsible for development of primary and secondary sex characteristics in males. 274

uterine cycle Monthly occurring changes in the characteristics of the uterine lining (endometrium). 279

uterus The womb, the organ located in the female pelvis where the fetus develops. 278

vagina Organ that leads from uterus to vestibule and serves as the birth canal and organ of sexual intercourse in females. 278

vas deferens Tube that leads from the epididymis to the urethra in males. 274

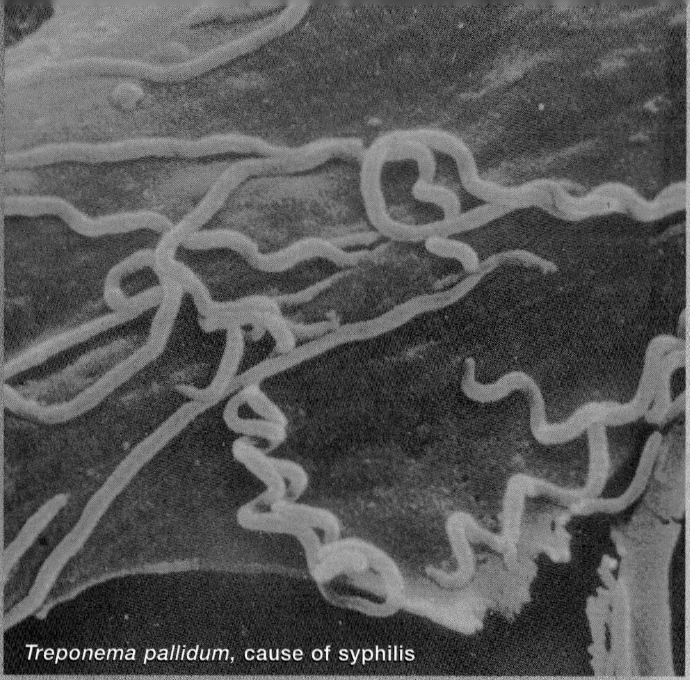
Treponema pallidum, cause of syphilis

Chapter 15

Sexually Transmitted Diseases

Can you imagine that a sexually transmitted disease could have altered the course of history? In the early part of the sixteenth century, Henry VIII contracted syphilis just before he married Catherine of Aragon. She bore him 4 sons, but all had congenital syphilis and were stillborn or fatally malformed. He blamed her for this tragedy and sought an annulment from the Catholic Church. When it was denied, he broke with the Church so that he could divorce Catherine and take another wife. England has been a Protestant country since that time.

Chapter Concepts

Chapter Outline

Viral in Origin

Sexually transmitted diseases (STDs) are contagious diseases caused by microorganisms that are passed from one human to another by sexual contact. Viruses cause numerous diseases in humans (table 15.1), including AIDS, herpes, and genital warts, 3 sexually transmitted diseases of great concern today. Since viruses are not cellular, they are incapable of independent reproduction and reproduce only inside a living cell. For this reason, they are called *obligate parasites.* A **parasite** requires a **host** organism to function properly, complete its life cycle, and reproduce. In the laboratory, viruses are maintained by injecting them into live chick embryos (fig. 15.1). Viruses are tiny particles that always have at least 2 parts: an outer coat of protein called a capsid and an inner core of nucleic acid, either DNA or RNA (fig. 15.2). Many viruses also have an outer envelope that contains lipid as well as protein molecules. The lipid molecules are derived from the host's membrane, but the protein is unique to the virus.

Most viruses are extremely specific. Not only do they prefer a particular type of organism, such as humans, they also prefer a particular tissue type. This specificity is due to the ability of the virus to combine with a particular molecular configuration, such as a receptor on the cell surface. Within a half hour, the virus, or simply the nucleic acid core depending on the virus, has entered the cell. Most often the viral genes, whether they are DNA or RNA viruses, immediately take over the genetic control, including the metabolic machinery, of the cell—then the virus undergoes reproduction. These are the steps required for a DNA virus to reproduce.

Table 15.1

Infectious Diseases Caused by Viruses

Respiratory Tract	**Nervous System**
Common colds	Encephalitis
Flu*	Polio*
Viral pneumonia	Rabies*
Skin Reactions	**Liver**
Measles*	Yellow fever*
German measles*	Hepatitis A, B, and C*
Chicken pox*	**Other**
Shingles	Mumps*
Warts	Cancer
Sexually Transmitted	
AIDS	
Herpes	
Genital warts	

**Vaccines available. Yellow fever, rabies, and flu vaccines are only given if the situation requires them. Smallpox vaccinations are no longer required.*

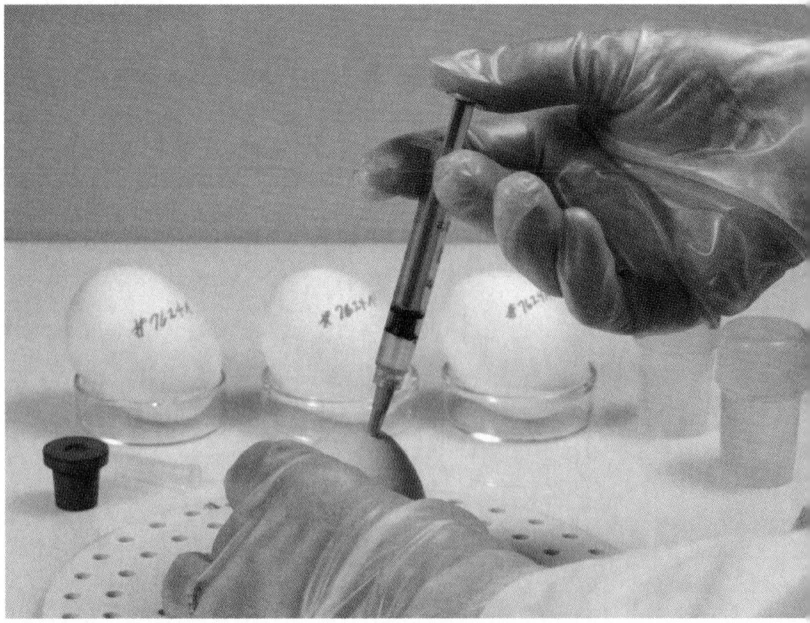

Figure 15.1

Inoculation of live chick eggs with virus particles. A virus only reproduces inside a living cell, not because it uses the cell for nutrition but because it takes over the metabolic machinery of the cell.

1. Viral DNA replicates repeatedly, utilizing the nucleotides within the host. Multiple copies of viral DNA result.
2. Viral DNA is transcribed into mRNA, which undergoes translation. Multiple copies of coat protein result.
3. Assemblage occurs. Viral DNA is packaged inside a coat protein. If the virus has an envelope, it is formed at the cell membrane just before the virus leaves the cell (fig. 15.2*a*).

There are a few types of viruses that do not immediately undergo reproduction; instead, viral DNA becomes integrated into the host DNA. RNA viruses that do this are called **retroviruses** because they have a special enzyme called *reverse transcriptase,* through which their RNA is transcribed into cDNA (a DNA copy of the RNA gene), which becomes incorporated into host DNA. Sometimes the virus remains *latent* (does not appear), and during this time the viral DNA is replicated along with host DNA. Eventually, viral reproduction may occur (fig. 15.2*a*). Certain environmental factors, such as ultraviolet radiation, can cause a latent virus to undergo reproduction. At this time, symptoms of the disease become evident.

> Viruses reproduce only inside host cells. Some viruses have the ability to incorporate their DNA within host DNA. Retroviruses are RNA viruses that are able to perform reverse transcription.

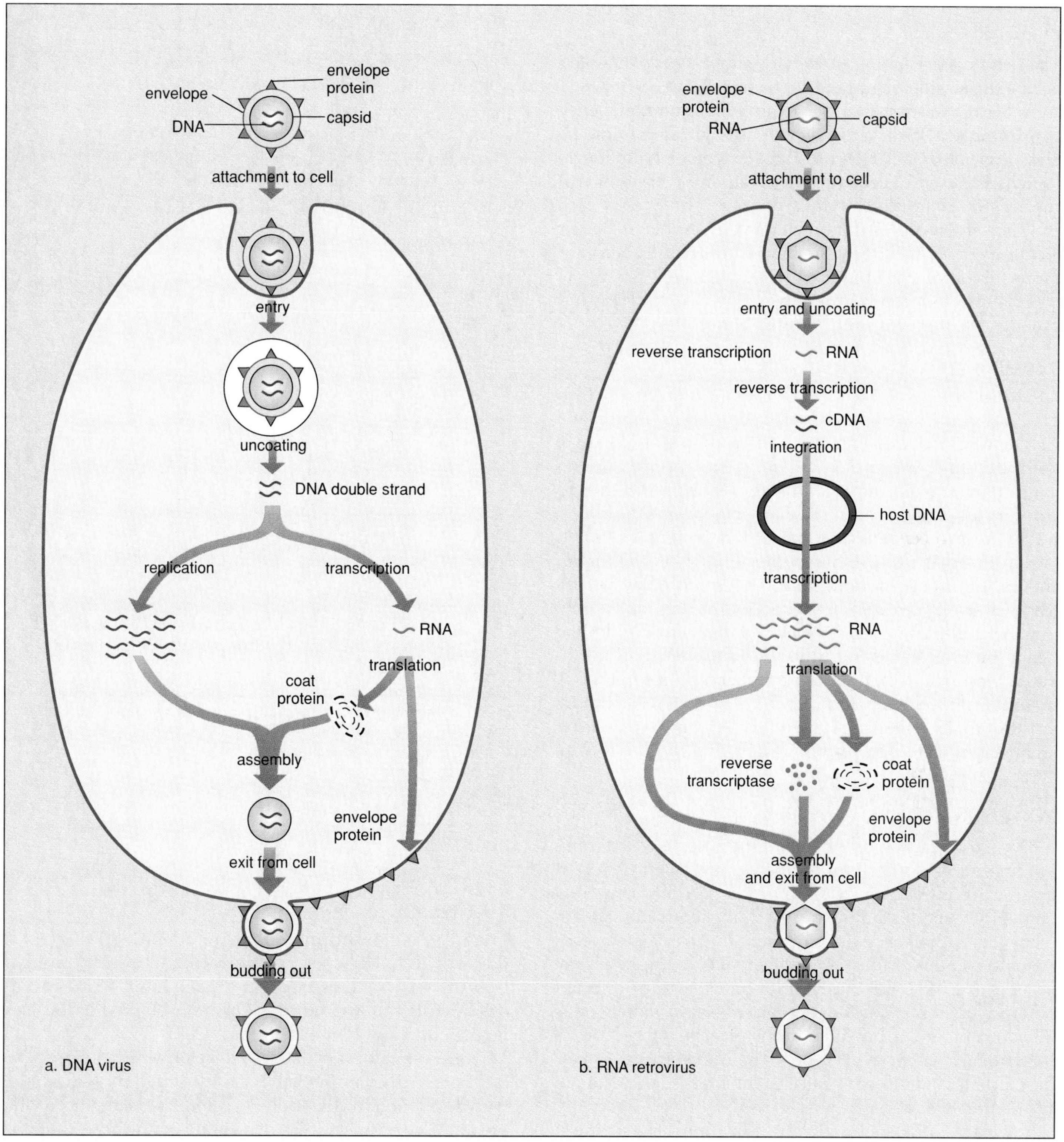

Figure 15.2

Life cycles of animal viruses. **a.** DNA virus. After entering by endocytosis, the virus becomes uncoated. The DNA then codes for proteins, some of which are capsid (coat) proteins and some of which are envelope proteins. Assembly follows replication of the DNA. When the virus exits by budding, it is enclosed by an envelope made up of host cell membrane lipids and viral envelope proteins of viral and host origin. **b.** RNA retrovirus. The life cycle includes steps not seen in **a.** The RNA genes are transcribed to cDNA (DNA copied off of RNA), which is integrated into the host DNA. Transcription produces many copies of the RNA genes, which direct the synthesis of 3 types of proteins: the enzyme reverse transcriptase, the capsid (coat) protein, and the envelope protein. Again, the virus buds from the host cell.

AIDS

The brief discussion of AIDS here may be supplemented by the discussion beginning on page 307.

The organism that causes **acquired immunodeficiency syndrome (AIDS)** is a virus called **human immunodeficiency virus (HIV).** HIV attacks the type of lymphocyte known as helper T cells (p. 136). Helper T cells, you will recall, stimulate the activities of B lymphocytes, which produce antibodies. After an HIV infection sets in, helper T cells begin to decline in number and the person becomes more susceptible to other types of infections.

Figure 15.3 gives the number of new AIDS cases, and other STDs, per year in the United States.

Symptoms

AIDS has 3 stages of infection. During the first stage, which may last about a year, the individual is an asymptomatic carrier. The AIDS blood test (an antibody test) is positive; the individual can pass on the infection, yet there are no symptoms. During the second stage, called AIDS-related complex (ARC), which may last about 6–8 years, the lymph nodes are swollen and there may also be weight loss, night sweats, fatigue, fever, and diarrhea. Infections like thrush (white sores on the tongue and in the mouth) and herpes (discussed on page 297) reoccur. Finally, the person may develop full-blown AIDS, especially characterized by the development of an opportunistic disease such as an unusual type of pneumonia, skin cancer, and also nervous disorders. Opportunistic diseases are ones that occur only in individuals who have little or no capability of fighting an infection. The AIDS patient usually dies about 7–9 years after infection.

Transmission

AIDS is transmitted by infected blood, semen, and vaginal secretions. In the United States, the 2 main affected groups are intravenous (IV) drug users and homosexual men. However, the number of HIV infections in male homosexuals appears to have peaked; in 1988, male homosexuals accounted for 61% of U.S. AIDS cases; now they account for 58%. In contrast, the percentage of infected heterosexuals is rising—every one in 5 new infections is now a woman. The potential for heterosexual infections is of extreme concern to everyone. The AIDS virus can cross the placenta, and presently infected infants account for about 1% of all AIDS cases.

Health officials emphasize that unprotected intercourse with multiple partners or a single infected partner increases the chance of transmission. The use of a latex condom reduces the risk, but the very best preventive measure at this time is a long-term, mutually monogamous relationship with a sexual partner who is free of the disease. Casual contact with someone who is infected, such as when shaking hands, eating at the same table, or swimming in the same pool, is not a mode of transmission.

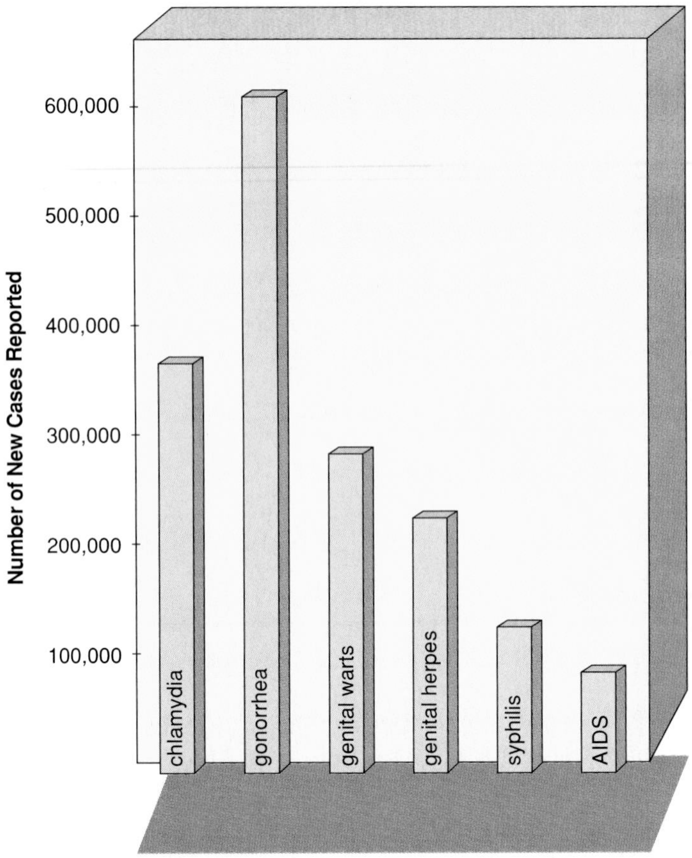

Figure 15.3

U.S. statistics for the most common sexually transmitted diseases show that chlamydia, gonorrhea, and genital warts are all much more common than herpes, syphilis, and AIDS. Chlamydia, gonorrhea, and syphilis can be cured with antibiotic therapy.

Sources: Data from Centers for Disease Control Surveillance Report for Chlamydia, Gonorrhea, Genital Warts, Genital Herpes, and Syphilis, Initial Visits 1991; *and* HIV/AIDS Surveillance Report, *July 1993, for cases reported June 1992–June 1993.*

Treatment

The 3 drugs now approved for treatment of AIDS (AZT, ddI, and ddC) all work by interfering with viral replication in cells. As discussed in this chapter's Health Focus, medical treatment of AIDS and other viral diseases is difficult.

Investigators are trying to develop a vaccine for AIDS. A whole-virus vaccine is being developed by Jonas Salk, who developed the first polio vaccine. There are those who are opposed to a whole-virus vaccine on the basis that it is too dangerous. Various subunit vaccines, all utilizing a different protein from the capsid or envelope, are under trial. In the end, these proteins have to be attached to a carrier that can be processed by a macrophage. HIV attaches to T lymphocytes at CD4 molecules. Antibodies against CD4 molecules have been made with the hope that they will attach to T4 cells in the body so that the way is blocked for viruses to be able to gain entry.

Genital Herpes

The different herpes viruses are large DNA viruses that cause various illnesses. Chicken pox, shingles, and mononucleosis, among other ailments, are due to herpes viruses. **Genital herpes** is caused by **herpes simplex virus** (fig. 15.4), of which there are 2 types: type 1 usually causes cold sores and fever blisters, while type 2 more often causes genital herpes. Cross-over infections do occur, however.

Transmission and Symptoms

Genital herpes is one of the more prevalent sexually transmitted diseases today; an estimated 40 million persons (17% of the U.S. population) have it, with an estimated 230,000 new cases appearing each year. Sometimes there are no symptoms. Or the individual may experience a tingling or itching sensation before blisters appear on the genitals (within 2–20 days). Once the blisters rupture, they leave painful ulcers that may take as long as 3 weeks or as little as 5 days to heal. These symptoms may be accompanied by fever, pain on urination, and swollen lymph nodes.

After ulcers heal, the disease is only dormant, and blisters can reoccur repeatedly at variable intervals. Sunlight, sex, menstruation, and stress seem to cause the symptoms of genital herpes to reoccur. While the virus is latent, it resides in nerve cells near the brain and spinal cord. Herpes occasionally infects the eye, causing an eye infection that can lead to blindness, and can cause central nervous system infections. Type 2 formerly was thought to cause a form of cervical cancer, but this is no longer believed to be the case.

Infection of the newborn can occur if the child comes in contact with a lesion in the birth canal. In 1–3 weeks, the infant is gravely ill and can become blind, have neurological disorders including brain damage, or die. Birth by cesarean section prevents these occurrences.

Treatment

Presently there is no cure for herpes. The drugs vidarabine and acyclovir disrupt viral reproduction. The ointment form of acyclovir relieves initial symptoms, but the oral form prevents the occurrence of outbreaks. Research is being conducted in an attempt to develop a vaccine.

Genital Warts

Genital warts are caused by a *human papillomavirus (HPV)*, which is a cuboidal DNA virus that reproduces in the nuclei of skin cells. Plantar warts and common warts are also caused by HPVs.

Transmission and Symptoms

Some HPVs are sexually transmitted. Quite often, carriers do not have any sign of warts, although flat lesions may be present. When present, the warts commonly are seen on the penis and foreskin of males and near the vaginal opening in females. If the warts are removed, they may reoccur.

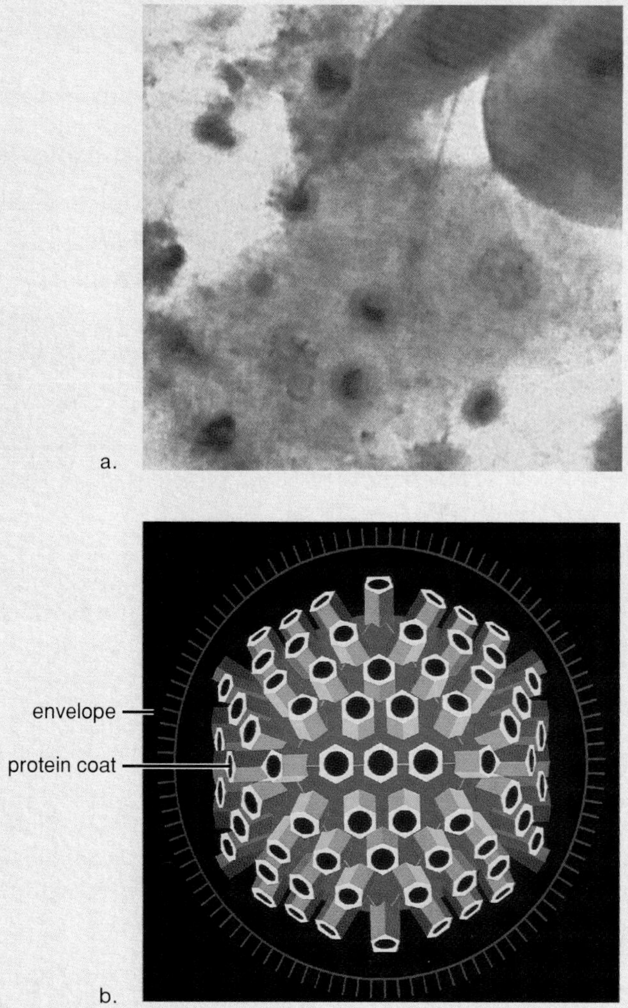

Figure 15.4

a. Cell infected with herpes virus. **b.** Enlarged model of herpes virus.

HPVs, rather than genital herpes, now are associated with cancer of the cervix, as well as tumors of the vulva, the vagina, the anus, and the penis. Some researchers believe that the viruses are involved in 90%–95% of all cases of cancer of the cervix. Teenagers with multiple sex partners seem to be particularly susceptible to HPV infections. More cases of cancer of the cervix are being seen among this age group.

Treatment

Presently, there is no cure for an HPV infection, but it can be treated effectively by surgery, freezing, application of an acid, or laser burning, depending on severity. A suitable medication to treat genital warts before cancer occurs is being sought, and efforts also are underway to develop a vaccine. But even after treatment, the virus can be transmitted.

Health Focus

STDs and Medical Treatment

Human beings are subject to many infectious diseases, caused by bacteria and viruses (see tables 15.1 and 15.2). Vaccines are available for many infections, particularly childhood diseases, but there are none available for the sexually transmitted diseases discussed in this chapter. There is a vaccine for hepatitis A, an infection of the liver that can also be sexually transmitted. Sexually transmitted diseases caused by bacteria can sometimes be cured with antibiotics.

An antibiotic is a medicine that kills bacteria and not human cells because it interferes with metabolic pathways found only in the bacteria. Antibiotics were introduced in the 1940s, and for several decades they worked so well that it appeared that infectious diseases had been brought under control. However, we now know that a bacterial strain can mutate and become resistant to a particular antibiotic. Worse yet, bacteria can swap bits of DNA and in this way resistance can be passed to other strains of bacteria. This is one reason why it is necessary to continue to develop new antibiotics to which resistance has not yet occurred. Antibiotic-resistant diseases can spread worldwide because the human population is large and motile. For example, Chinese Muslims visiting Mecca passed on a virulent strain of meningitis to Africans also there, who brought it home with them. Customers of Southeast Asian brothels carry penicillin-resistant gonorrhea and AIDS to all corners of the world.

Figure 15A

Sexually transmitted diseases among the poor who crowd together encourage the spread of TB. Adequate health care is required to prevent an epidemic among the general populace.

We must all try to prevent antibiotic-resistant diseases from arising in the first place. Antibiotics are overprescribed, and this practice should be stopped because every time a friendly bacteria is killed, it leaves more room for a resistant one. Many people take antibiotics for colds and flu caused by viruses, even though antibiotics are not effective against viruses. If you have an infection, don't take an antibiotic unless so directed by a physician. If you do need antibiotic therapy, be sure to complete the prescribed therapy, otherwise the bacteria remaining can become resistant to the prescribed antibiotic.

Also, it is not a wise practice to feed antibiotics to livestock. In the past decade, this practice has led to a doubling of antibiotic resistance in salmonella—an organism that causes severe diarrhea. In a study of farm families, the same resistance genes were found in the intestinal bacteria of both animals and human caretakers.

Antiviral drugs have only recently been developed. Viruses lack most enzymes and instead utilize the metabolic machinery of the host cell. Rarely has it been possible to find a drug that successfully interferes with viral reproduction without also interfering with host metabolism. The antiviral drugs, acyclovir (ACV) and vidarabine, are helpful against genital herpes; however, they do not cure herpes. In general, physicians are disappointed about the effectiveness of antiviral drugs developed to treat AIDS.

Possible drug resistance to antibiotics and the lack of effective antiviral drugs makes medical treatment of STDs troublesome. HIV attacks the immune system directly, but all STDs weaken the immune system somewhat. An alarming new development has been the upsurge of tuberculosis in persons with compromised immune systems. Tuberculosis is a chronic lung infection caused by the bacterium *Mycobacterium tuberculosis,* often called the tubercle bacillus (TB). If the immune system is healthy, the bacilli are walled off within tubercles that can sometimes be seen by X ray. More often a skin test is used to detect infection. If the immune system is weakened by an infection, particularly with HIV, advanced age, alcoholism, malnutrition, or diabetes, the bacilli may escape and resume growth. Each cough from an infected person discharges hordes of bacteria into the air.

Thanks to effective antibiotic therapy, the incidence of TB was brought under control in the 1940s. The situation remained stable until the 1980s, when the incidence began to rise. The year 1990 saw the largest single increase since records have been kept. The resurgence of tuberculosis in the United States is largely due to the spread of HIV. TB can pass from an HIV-infected person to a healthy person more easily than HIV. Sexual contact is not needed—only close contact under crowded conditions (fig. 15A). Two-thirds of TB cases occur among underprivileged groups such as the homeless, who often do not complete the prescribed therapy. This has contributed to the occurrence of antibiotic-resistant TB strains that can spread to the general populace.

It is obvious that we must do all we can to prevent HIV and TB infections. Adopting the behaviors suggested at the end of the AIDS supplement (p. 314) can help protect persons from AIDS. To combat a possible TB epidemic, public health programs need to be adequately funded; centralized diagnostic centers that utilize faster means of diagnosis and treatment need to be established; and new drugs need to be developed. Just as with HIV, research is needed for the development of an effective TB vaccine.

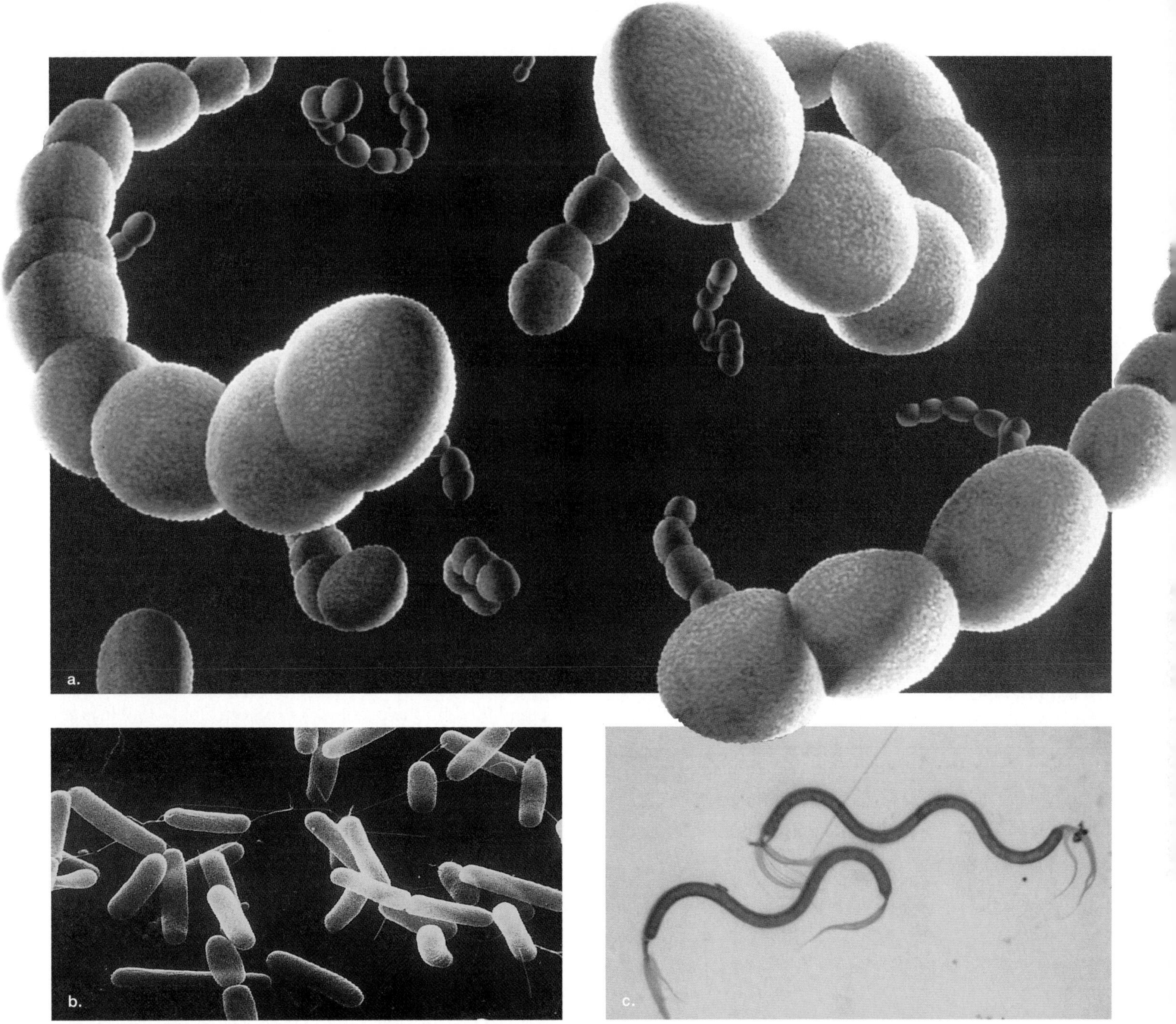

Figure 15.5

Scanning electron micrographs of bacteria. **a.** Spherical-shaped bacteria. **b.** Rod-shaped bacteria. **c.** Spiral-shaped bacteria that use flagella for locomotion. See figure 15.6 for a generalized drawing of a bacterium.

Bacterial in Origin

Although **bacteria** are generally larger than viruses, they are still microscopic. As a result, it is not always obvious that they are abundant in the air, water, and soil and on most objects. It has even been suggested that the combined weight of all bacteria would exceed that of any other type of organism on earth. Bacteria occur in 3 basic shapes (fig. 15.5): round, or spherical (coccus); rod (bacillus); and spiral (a curved shape called a spirillum). Some bacteria can locomote by means of flagella.

The bacterial cell is termed a *prokaryotic* (meaning, before the nucleus) cell to distinguish it from a *eukaryotic* (meaning, true nucleus) cell. Human cells are eukaryotic and contain numerous organelles in addition to a nucleus. Prokaryotic cells lack these organelles, except for ribosomes, but still they are functioning cells. They do have DNA; it is just not contained within a nuclear envelope, and although they have no mitochondria, they do have respiratory enzymes located in the cytoplasm. Notice in figure 15.6 that the bacterial cell is surrounded by a cell wall in addition to a cell membrane. Some bacteria are also surrounded by

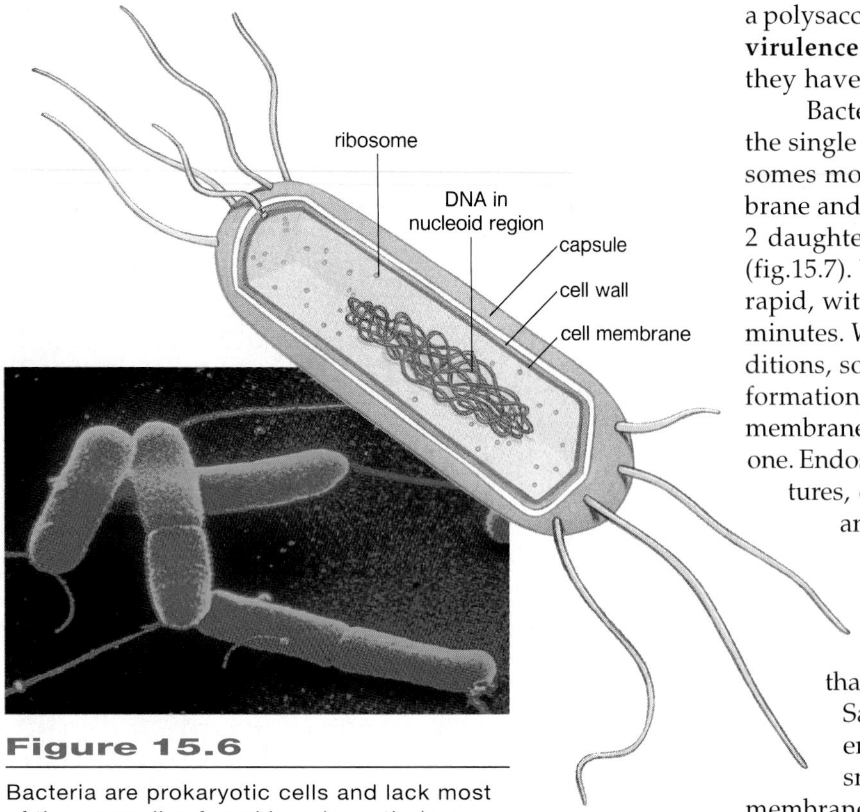

Figure 15.6

Bacteria are prokaryotic cells and lack most of the organelles found in eukaryotic (e.g., human) cells.

a polysaccharide or polypeptide capsule that enhances their **virulence** (i.e., ability to cause disease). If they are motile, they have flagella (sing., flagellum).

Bacteria reproduce sexually by **binary fission.** First, the single chromosome duplicates, and then the 2 chromosomes move apart into separate areas. Next the cell membrane and cell wall grow inward and partition the cell into 2 daughter cells, each of which has its own chromosome (fig.15.7). Under favorable conditions, growth may be very rapid, with cell division occurring as often as every 12–15 minutes. When faced with unfavorable environmental conditions, some bacteria can form **endospores.** During spore formation, the cell shrinks, rounds up within the former cell membrane, and secretes a new, thicker wall inside the old one. Endospores are amazingly resistant to extreme temperatures, drying out, and harsh chemicals, including acids and bases. When conditions are again suitable for growth, the spore absorbs water, breaks out of the inner shell, and becomes a typical bacterial cell capable of now reproducing.

Most bacteria are free-living **saprotrophs** that perform many useful services in the environment. Saprotrophs send out digestive enzymes into the environment to break down macromolecules into small molecules that can be absorbed across the cell membrane. Most bacteria are aerobic and require a constant supply of oxygen as we do, but a few are anaerobic, even being killed by the presence of oxygen. Table 15.2 lists the

Table 15.2

Some Infectious Diseases Caused by Bacteria

Respiratory Tract	**Nervous System**
Strep throat	Tetanus*
Pneumonia	Botulism
Whooping cough*	Meningitis
Tuberculosis*	**Digestive Tract**
Skin Reactions	Food poisoning
Staph (pimples and boils)	Dysentery
Other	Cholera*
Gas gangrene* (wound infections)	**Sexually Transmitted**
Diphtheria*	Gonorrhea
Typhoid fever*	Chlamydia
	Syphilis

**Vaccines are available. Tuberculosis vaccine is not used in this country. Typhoid fever, cholera, and gas gangrene vaccines are given if the situation requires it. Others are routinely given.*

Figure 15.7

Reproduction in bacteria. The single chromosome is attached to the cell membrane, where it is replicating. As the cell membrane and cell wall lengthen, the 2 chromosomes separate. Once fission has taken place, each bacterium has its own chromosome.

human diseases caused by bacteria; only a few serious illnesses are caused by anaerobic bacteria, such as botulism, gas gangrene, and tetanus. These bacteria and others produce toxins, chemicals that seriously interfere with the normal functioning of the body. Sometimes just a bacterial toxin is used to make a vaccine if a toxin alone invokes an antibody response.

Bacteria have long been used by humans to produce various products commercially. Chemicals, such as ethyl alcohol, acetic acid, butyl alcohol, and acetone, are produced by bacteria. Bacterial action is involved in the production of butter, cheese, sauerkraut, rubber, cotton, silk, coffee, wine, and cocoa. By means of gene splicing, bacteria are now used to produce human insulin and interferon, as well as other types of proteins (p. 390). Even certain antibiotics are produced by bacteria.

General cleanliness is the first step toward preventing the spread of infectious bacteria. Disinfectants and antiseptics also help reduce the number of infectious bacteria. **Sterilization,** a process that kills all living things, even endospores, is used whenever all bacteria must be killed. Sterilization can be achieved by use of an autoclave (fig. 15.8), a container that admits steam under pressure. If bacteria do invade our bodies and cause an infection, antibiotic therapy is often helpful.

> Bacteria are prokaryotic cells capable of independent existence. Most are free-living, but a few cause human diseases that can often be cured by antibiotic therapy.

An **antibiotic** is a chemical that selectively kills bacteria when it is taken into the body as a medicine. Most antibiotics are produced naturally by soil microorganisms. Penicillin is made by the fungus *Penicillium;* streptomycin, tetracycline, and erythromycin are all produced by the bacterium *Streptomyces.* Sulfa, an analog of a bacterial growth factor, can be produced in the laboratory.

Antibiotics are metabolic inhibitors specific for bacterial enzymes. This means that they poison bacterial enzymes without harming host enzymes. Penicillin blocks the synthesis of the bacterial cell wall; streptomycin, tetracycline, and erythromycin block protein synthesis; and sulfa prevents the production of a coenzyme.

There are problems associated with antibiotic therapy. Some individuals are allergic to antibiotics, and the reaction may even be fatal. Antibiotics not only kill off disease-causing bacteria, they also reduce the number of beneficial bacteria in the intestinal tract. The use of antibiotics sometimes prevents natural immunity from occurring, leading to the necessity for reoccurring antibiotic therapy. Most important, perhaps, is the growing resistance of certain strains of bacteria to a particular antibiotic. Penicillin and tetracycline, long used to cure gonorrhea, now have a failure rate of more than 20% against certain strains of *Gonococcus.* Antibiotics should only be administered when absolutely necessary, and the prescribed therapy should be finished. Otherwise, resistant strains of bacteria can completely replace present strains and antibiotic therapy will no longer be effective. These problems are also discussed in the Health Focus for this chapter.

Gonorrhea

Gonorrhea is caused by the bacterium *Neisseria gonorrhoeae,* which is a diplococcus, meaning that generally the 2 spherical cells stay together (fig. 15.9).

Figure 15.8

Sterilization by autoclaving. Hospital employees close the door and operate a large sterilizer. Sterilization of instruments and other materials permits surgical procedures to be done with reduced fear of subsequent infection. During autoclaving, steam under pressure kills bacterial cells and endospores.

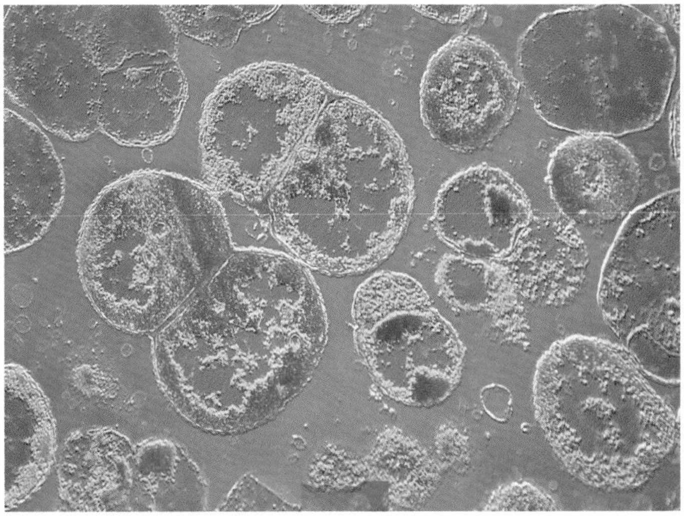

Figure 15.9

Gonorrheal bacteria (*Neisseria gonorrhoeae*) in male urethral discharge. Notice that these bacteria occur in pairs; for this reason, they are called diplococci.

Symptoms

The diagnosis of gonorrhea in the male is not difficult as long as he displays typical symptoms (as many as 20% of males may be asymptomatic). The patient complains of pain on urination and has a thick, greenish yellow urethral discharge 3–5 days after contact. In the female, the bacteria may first settle within the urethra or near the cervix, from which they may spread to the oviducts, causing **pelvic inflammatory disease (PID).** PID from gonorrhea is especially apt to occur in the female using an IUD (intrauterine device) as a birth-control measure (p. 285). As the inflamed tubes heal, they may become partially or completely blocked by scar tissue. As a result, the female is sterile or, at best, subject to ectopic pregnancy. Similarly, there may be inflammation in untreated males followed by scarring of the vas deferens. Unfortunately, 60%–80% of females are asymptomatic until they develop severe pains in the abdominal region due to PID. PID affects about one million women a year in the United States.

Homosexual males develop gonorrhea proctitis, or infection of the anus, with symptoms including anal pain and blood or pus in the feces. Oral sex can cause infection of the throat and the tonsils. Gonorrhea also can spread to other parts of the body, causing heart damage or arthritis. If, by chance, the person touches infected genitals and then his or her eyes, a severe eye infection can result (fig. 15.10).

Eye infection leading to blindness can occur as a baby passes through the birth canal. Because of this, all newborn infants receive eyedrops containing antibacterial agents, such as silver nitrate, tetracycline, or penicillin, as a protective measure.

Transmission and Treatment

Gonococci live for a very short time outside the body; therefore, most infections are spread by intimate contact, usually sexual intercourse. A female has a 50%–60% risk of contracting the disease after a single exposure to an infected male, whereas a male has a 20% risk after exposure to an infected female.

Blood tests for gonorrhea are being developed, but in the meantime, it is necessary to diagnose the condition by microscopically examining the discharge of males or by growing a culture of the bacterium from both males and females to positively identify the organism (fig. 15.11). Because there is no blood test, it is very difficult to recognize asymptomatic carriers, who are capable of passing on the condition without realizing it. If the infection is diagnosed, however, gonorrhea can be treated using antibiotics. Penicillin and tetracycline resistant strains do occur, however. There is no vaccine for gonorrhea, and immunity does not seem possible; therefore, it is possible to contract the disease many times over.

> Gonorrhea is one of the oldest known and most common of the sexually transmitted diseases. Untreated, an infection can cause PID and sterility in either sex. Appropriate antibiotic therapy usually cures the condition.

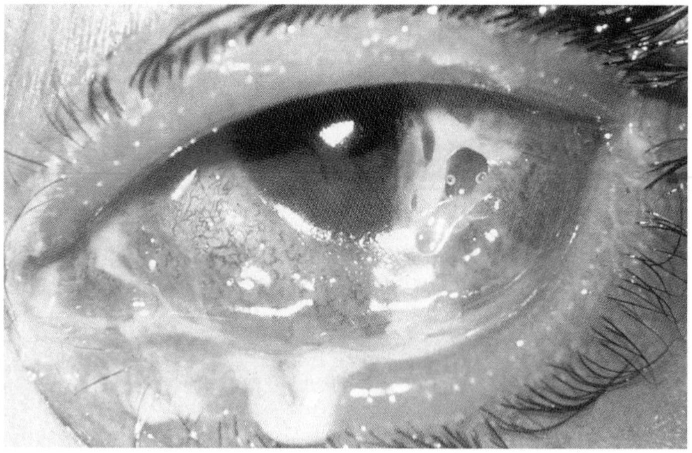

Figure 15.10

Gonorrhea infection of the eye is possible whenever the bacterium comes in contact with the eyes. This can happen when the newborn passes through the birth canal. Manual transfer from the genitals to the eyes is also possible.

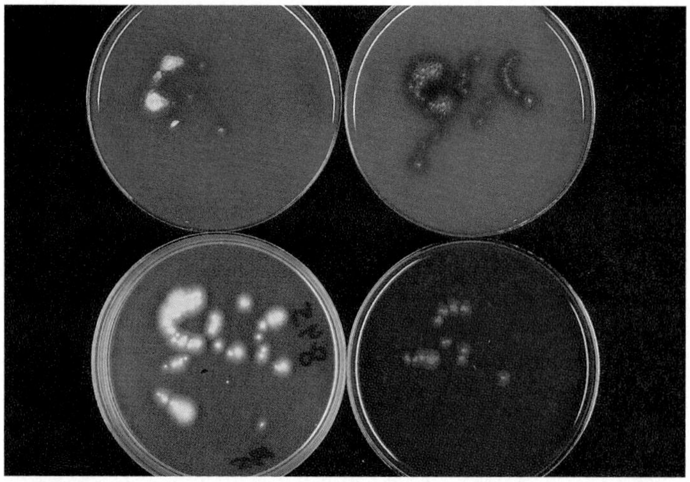

Figure 15.11

Culture plates with bacterial colonies. Visual examination and biochemical tests allow medical personnel to determine the type of bacteria, such as *N. gonorrhoeae*, growing on culture plates.

Chlamydia

Chlamydia is named for the tiny bacterium that causes it (*Chlamydia trachomatis*). For years, chlamydiae were considered to be more closely related to viruses than to bacteria, but today it is known that they are prokaryotic cells. Even so, they are obligate parasites due to their inability to produce ATP molecules. After a cell phagocytizes them, they develop inside the phagocytic vacuole, which eventually bursts and liberates many new infective chlamydiae.

New chlamydial infections occur at an even faster rate than gonorrheal infections. They are the most common cause of **nongonococcal urethritis (NGU).** About 8–21 days after infection, men experience a mild burning sensation on urination and a mucoid discharge. Women may have a vaginal discharge along with the symptoms of a urinary tract infection. Unfortunately, a physician mistakenly may diagnose a gonorrheal or urinary infection and prescribe the wrong type of antibiotic, or the person may never seek medical help. In either case, the infection can cause PID and sterility or ectopic pregnancy.

If a newborn comes in contact with chlamydia during delivery, inflammation of the eyes or pneumonia can result. Some believe that chlamydial infections increase the possibility of premature and stillborn births.

Detection and Treatment

New and faster laboratory tests are now available for chlamydia detection. Their expense sometimes prevents public clinics from using them, however. It's been suggested that these criteria could help physicians decide which women should be tested: no more than 24 years old; having had a new sex partner within the preceding 2 months; having a cervical discharge; bleeding during parts of the vaginal exam; and using a nonbarrier method of contraception. Some doctors, however, are routinely prescribing additional antibiotics appropriate to treating chlamydia for anyone who has gonorrhea because 40% of females and 20% of males with gonorrhea also have chlamydia.

As with AIDS, condoms serve as a protection against both gonorrheal and chlamydia infections. The concomitant use of a spermicide containing nonoxynol-9 gives added protection.

> PID and sterility are common effects of a chlamydia infection in the female. This condition often accompanies gonorrheal infection.

Syphilis

Syphilis is caused by a bacterium called *Treponema pallidum*, an actively motile, corkscrewlike organism that is classified as a spirochete. Because this bacterium is difficult to stain, it shows up only when viewed with a dark-field microscope. Syphilis is less common than gonorrhea (fig. 15.12), but it is the more serious of the 2 infections.

Syphilis has 3 stages, which can be separated by latent stages in which the bacteria are not multiplying again. During the primary stage, a hard chancre (ulcerated sore with hard edges) indicates the site of infection (fig. 15.12*a*). The chancre can go unnoticed, especially since it usually heals spontaneously, leaving little scarring. During the secondary stage, proof that bacteria have invaded and spread throughout the body is evident when the individual breaks out in a rash (fig. 15.12*b*).

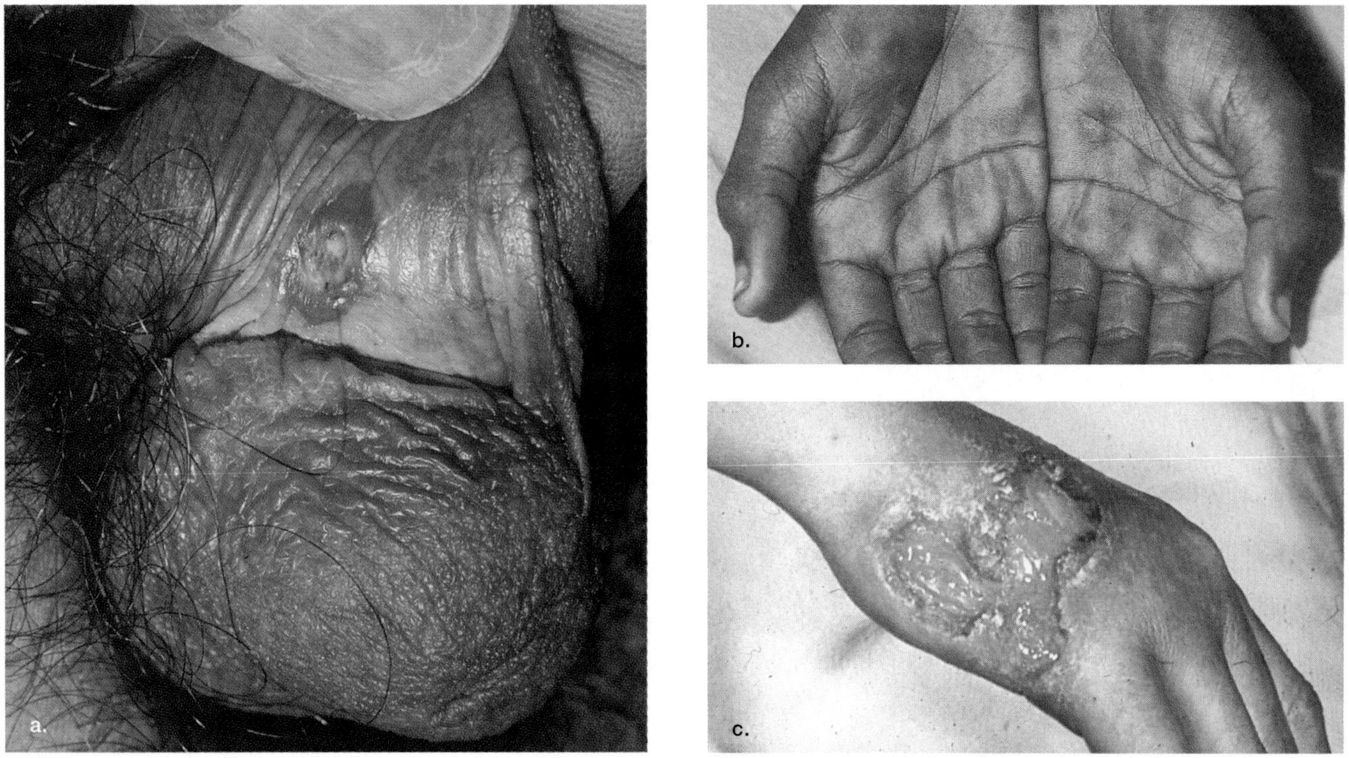

Figure 15.12

The 3 stages of syphilis. **a.** The first stage is a hard chancre where the bacterium enters the body. **b.** The second stage is a body rash that occurs even on the palms of the hands and soles of the feet. **c.** In the tertiary stage, gummas may appear on skin or internal organs.

Curiously, the rash does not itch and is seen even on the palms of the hands and the soles of the feet. There can be hair loss and infectious gray patches on the mucous membranes, including the mouth. These symptoms disappear of their own accord.

During a tertiary stage, which lasts until the patient dies, syphilis may affect the cardiovascular system, and weakened arterial walls (aneurysms) are seen, particularly in the aorta. In other instances, the disease may affect the nervous system. An infected person may become mentally impaired, become blind, walk with a shuffle, or show signs of insanity, for example. **Gummas,** large destructive ulcers (fig. 15.12*c*), may develop on the skin or within the internal organs.

Congenital syphilis is caused by syphilitic bacteria crossing the placenta. The child is stillborn, or blind, with many other possible anatomical malformations.

Transmission and Treatment

The syphilis bacterium is not present in the environment; it is only present within human beings. Only close intimate contact such as sexual intercourse transmits the condition from one person to another.

Diagnosis of syphilis can be made by dark-field microscopic examination of fluids from lesions for the bacterium, which is actively motile and has a corkscrew shape. There are also blood tests that are not positive until at least 6 weeks after initial infection.

Syphilis is a very devastating disease. Control of syphilis depends on prompt and adequate treatment of all new cases; therefore, it is very important for all sexual contacts to be traced so that they can be treated. Use of condoms can prevent exposure. As with other STDs, the incidence of syphilis is rising in most parts of the world. One hundred thousand cases are reported annually in the United States, and the highest incidence of these cases is among persons 29–39 years of age.

Other Diseases

Vaginitis

Two other types of organisms are of interest: protozoans and fungi. **Protozoa** are eukaryotes that usually exist as single cells. Each type of protozoan has its own mode of locomotion. There are some that move by extensions of the cytoplasm, called pseudopodia, some that move by cilia, and some that have flagella. Protozoa are most often found in an aquatic environment such as freshwater ponds, and the ocean simply teems with them. All protozoa require an outside source of nutrients, but only the parasitic ones take this nourishment from their host.

Most **fungi** are nonpathogenic saprotrophs, as are bacteria, but a few are parasitic. Usually the body of a fungus is made up of filaments called hyphae, but yeasts are an exception since they are single cells. Almost everyone is familiar with yeasts. They are used to make bread rise and to produce wine, beer, and whiskey because of their ability to ferment.

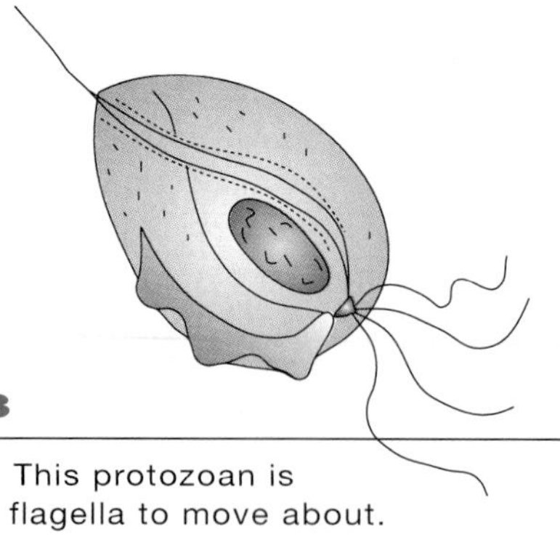

Figure 15.13

Trichomonas vaginalis. This protozoan is pear-shaped and uses flagella to move about.

Females very often have vaginitis, or infection of the vagina, which is caused by the flagellated protozoan *Trichomonas vaginalis* (fig. 15.13) or the yeast *Candida albicans.* The protozoan infection causes a frothy white or yellow foul-smelling discharge accompanied by itching, and the yeast infection causes a thick, white, curdy discharge, also accompanied by itching. *Trichomonas* is most often acquired through sexual intercourse, and the asymptomatic male is usually the reservoir of infection. *Candida albicans,* however, is a normal organism found in the vagina; its growth simply increases beyond normal under certain circumstances. For example, women taking birth-control pills are sometimes prone to yeast infections. Also, the indiscriminate use of antibiotics can alter the normal balance of organisms in the vagina so that a yeast infection flares up.

Pubic Lice (Crabs)

Small lice, animals that look like crabs under low magnification, infect the pubic hair, underarm hair, and even occasionally the eyebrows of humans. The females lay their eggs around the base of the hair, and these eggs hatch within a few days to produce a larger number of animals that suck blood from their host and cause severe itching, particularly at night.

The parasitic crab louse *Phthirus pubis* can be contracted by direct contact with an infected person or by contact with his or her clothing or bedding (fig. 15.14). In contrast to most types of sexually transmitted diseases, self-treatment that does not require shaving is possible. The lice can be killed by gamma benzene hexachloride, marketed in shampoo, lotion, or cream under the trade name Kwell; this treatment does not require shaving the pubic region.

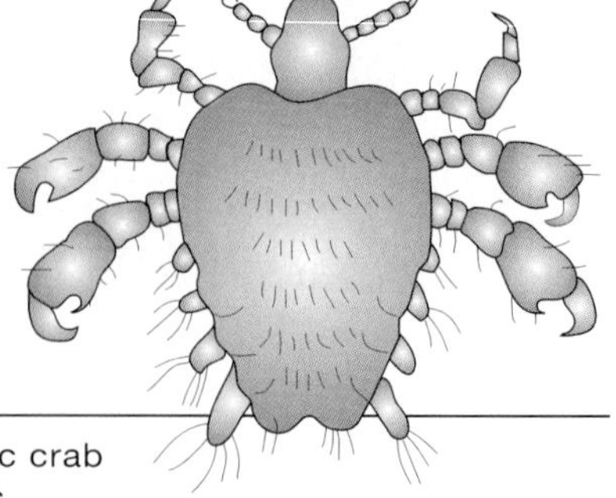

Figure 15.14

Phthirus pubis, the parasitic crab louse that infects the pubic hair of humans.

SUMMARY

Human beings are subject to sexually transmitted diseases (STDs). Some of these are viral in origin. Viruses are tiny particles that always have an outer coat of protein and an inner core of nucleic acid that may be DNA or RNA. DNA viruses are apt to undergo reproduction inside a cell immediately, while RNA viruses may be retroviruses that incorporate cDNA into host DNA for some time before reproduction. HIV is the cause of AIDS, a disease of the immune system that has 3 stages: asymptomatic carrier, AIDS-related complex, and full-blown AIDS. Because the immune system is compromised, an individual with AIDS tends to have opportunistic diseases—ones that do not occur in persons with normal immunity. Genital herpes is a disease that causes painful blisters on the genitals. Genital warts is a disease characterized by warts on the penis and foreskin in males and near the vaginal opening in females. At the present time, there are no methods of immunization or cure for any of these 3 STDs.

Some STDs are bacterial in origin. Bacteria are prokaryotic cells that reproduce by binary fission and may form endospores. Most bacteria are aerobic saprophytes that perform useful services for humans, but a few cause disease. Gonorrhea may not cause symptoms, particulary in females. In males, there may be painful urination and a thick, greenish yellow discharge. Chlamydia also produces symptoms of a urinary tract infection. Both gonorrhea and chlamydia can result in PID, leading to sterility. Syphilis is a systematic disease that should be cured in its early stages before deterioration of the nervous system and cardiovascular system possibly take place.

Other STDs are caused by a protozoan (vaginitis), a fungus (yeast infection), or insect (crabs).

STUDY AND THOUGHT QUESTIONS

1. Describe the structure and life cycle of viruses, including those that reproduce immediately and those that undergo a latent period. (pp. 294–295)
2. What are the symptoms for the 3 stages leading up to and including full-blown AIDS? (p. 296)
3. Among which groups of society is AIDS now increasing most rapidly? How might transmission be prevented? (p. 296)

   Ethical Issue Should condoms be dispensed in schools to prevent the spread of STDs? At what age should condoms become freely available?
4. Give the cause and the expected yearly incidence in the number of cases of herpes, gonorrhea, chlamydia, and syphilis. (pp. 296–297, 301–303)

   Critical Thinking Figure 15.3 indicates that chlamydia and gonorrhea are the most common STDs, yet they can be cured with antibiotic therapy. Discuss this apparent paradox.
5. Describe the progressive symptoms of a herpes infection. (p. 297)
6. List the 3 shapes of bacteria, and describe the structure of a prokaryotic cell. (pp. 299–300)
7. Describe the symptoms of a gonorrheal infection in the male and in the female. What is PID, and how does it affect reproduction? (p. 302)
8. Describe the 3 stages of syphilis. (pp. 303–304)

   Critical Thinking For what reason is syphilis a more dangerous disease than gonorrhea?
9. How does the newborn acquire an infection of AIDS, herpes, gonorrhea, chlamydia, or syphilis? What effects do these infections have on infants? (pp. 296, 297, 302, 303, 304)
10. List other common sexually transmitted diseases, and describe the associated symptoms. (p. 304)

WRITING ACROSS THE CURRICULUM

1. Parasites reproduce inside living hosts, and they all must have a means of transmission from host to host. Explain the necessity for this stage in the organism's life cycle.
2. Reexamine figure 15.2, and list all the steps in the life cycle of a retrovirus where a drug (medication) might be designed to interfere with the cycle. Discuss in specific terms.
3. From your knowledge of the human body, discuss the mode of entry by parasites in general. Give specific examples that may have been mentioned in previous chapters of the text.

Objective Questions

1. All viruses have an inner core of ________________ and a coat of ________________.
2. AIDS is caused by a type of RNA virus known as a ________________.
3. Although it is a sexually transmitted disease, intravenous drug users who share needles get AIDS because the virus lives in ________________ cells.
4. Herpes simplex virus type 1 causes ______________, and type 2 causes ________________.
5. Bacterial cells have one of 3 shapes: ______________, ______________, and ______________.
6. Females are often asymptomatic for gonorrhea until they develop ______________.
7. The use of a ______________ can help reduce the risk of acquiring an STD.
8. In the tertiary stage of syphilis, there may be large sores called ________________.
9. These 3 sexually transmitted diseases are usually curable by antibiotic therapy: ________________, ________________, and ________________.
10. Women who take birth-control pills are more likely to acquire a ________________ infection.
11. Identify the STD by these symptoms:
    a. ________________ greenish yellow discharge
    b. ________________ swollen lymph glands and night sweats
    c. ________________ chancre, followed by nonitching rash

Selected Key Terms

acquired immunodeficiency syndrome (AIDS) A disease caused by HIV and transmitted via body fluids; characterized by failure of the immune system. 296

bacterium Unicellular organism that is prokaryotic—its single cell lacks the complexity of a eukaryotic cell. 299

binary fission Reproduction by division into 2 equal parts by a process that does not involve a mitotic spindle. 300

chlamydia (klah-mid´e-ah) A tiny bacterium that causes a sexually transmitted disease particularly characterized by urethritis. 302

endospore A resistant body formed by bacteria when environmental conditions worsen. 300

fungus A eukaryote, usually composed of strands called hyphae, that is usually saprophytic; for example, mushroom and mold. 304

gonorrhea (gon″o-re´ah) Contagious sexually transmitted disease caused by bacteria and leading to inflammation of the urogenital tract. 301

gummas (gum´ahz) Large unpleasant sores that may occur during the tertiary stage of syphilis. 304

herpes simplex virus A virus of which type 1 causes cold sores and type 2 causes genital herpes. 297

host An organism on or in which another organism lives. 294

human immunodeficiency virus (HIV) Virus responsible for AIDS. 296

parasite An organism that resides on or within another organism and does harm to this organism. 294

pelvic inflammatory disease (PID) A disease state of the reproductive organs caused by a sexually transmitted organism. 302

retrovirus Virus that contains only RNA and carries out RNA to DNA transcription, called reverse transcription. 294

saprotroph (sap´ro-trōf) A heterotroph such as a bacterium or a fungus that externally digests dead organic matter before absorbing the products. 300

sterilization The absence of living organisms due to exposure to environmental conditions that are unfavorable to sustain life. 301

syphilis (sif´ĭ-lis) Chronic, contagious sexually transmitted disease caused by a bacterium that is a spirochete. 303

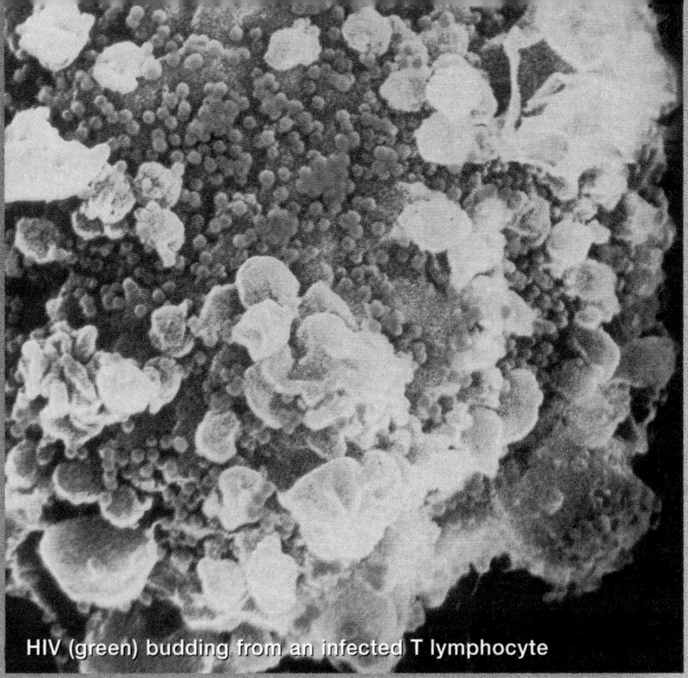

HIV (green) budding from an infected T lymphocyte

AIDS Supplement

The photo shows HIV budding from a T lymphocyte, the primary host for these viruses. No wonder the immune system falters in a person with AIDS. The very cells that orchestrate the immune response are under viral attack. Will medical science give us a cure for AIDS? Can any drug that interferes with viral replication also spare bone marrow stem cells? Seems unlikely. Will there instead be a vaccine? B lymphocytes would prepare a supply of antibodies to disarm HIV before the viruses have a chance to enter T cells. It would be a feat. There might be something about the virus—mode of transmission, or the course of the disease—that will make any type of vaccine ineffective. The burden, it seems, is on the individual. We all must come to realize the importance of our T lymphocytes and take measures to protect them from possible destruction by HIV. The Surgeon General's recommendations on page 314 should be faithfully followed.

Supplement Concepts

1 AIDS is a disease that originated about 30 or 40 years ago. 308

2 AIDS is a sexually transmitted disease that can also be spread by contaminated needles or by receiving contaminated blood. 308,309

3 An HIV infection passes through several stages before full-blown AIDS appears. In the end, the affected individual succumbs to disease(s) the immune system is unable to combat. 310–312

4 Several drugs and vaccines to cure and/or prevent an HIV infection are being investigated. 312–314

5 In the meantime, everyone should, if necessary, modify their behavior to prevent an HIV infection. 314

Supplement Outline

AIDS (acquired immunodeficiency syndrome) is caused by a group of related retroviruses known as HIV (human immunodeficiency viruses). The full name of AIDS can be explained in this way: *acquired* means that the condition is caught rather than inherited; *immune deficiency* means that the virus attacks the immune system so there is greater susceptibility to certain opportunistic infections and cancer; and *syndrome* means that some fairly typical infections and cancers usually occur in the infected person.

Origin of AIDS

The origin of the AIDS virus has not yet been determined. It has been suggested that HIV originated in Africa and then spread to Europe and the United States. Even today, there are monkeys in Africa infected with immunosuppressive viruses that could have mutated to HIV after humans ate monkey meat.

Most likely, HIV entered the United States on numerous occasions as early as the 1950s. Presently, the first documented case is a 15-year-old male who died in Missouri in 1969 with skin lesions now known to be characteristic of an AIDS-related cancer. Doctors froze some of his tissues because they could not identify the cause of death. Recently, these tissues were examined and found to be infected with HIV. Researchers also want to test the preserved tissue samples of a 49-year-old Haitian who died in New York during 1959 of a type of pneumonia now known to be AIDS related.

British scientists were able to show by examining preserved tissues that a Manchester seaman most probably died in 1959 of AIDS. This may be one of the first cases of AIDS because scientists believe the immunosuppressive monkey virus may have evolved into HIV during the late 1950s.

During the 1960s, it was the custom to list leukemia as the cause of death in immunodeficient patients. Most likely some of these people actually died of AIDS. Since HIV is not extremely infectious, it took several decades for the number of AIDS cases to increase to the point that AIDS became recognizable as a specific and separate disease. The name AIDS was coined in 1982, and HIV was found to be the cause of AIDS in 1983–84.

Since that time much has been learned about HIV, but no effective treatment has yet been found. The number of AIDS cases still continues to increase exponentially. The first 100,000 cases in the United States were reported over an 8-year period, and the second 100,000 occurred in only 2 years.

Prevalence of AIDS Today

Since it takes several years for full-blown AIDS to appear, the number of persons with full-blown AIDS represents only the "tip of the iceberg" when we consider the total number of persons infected with HIV. In the United States, 250,000 people have been diagnosed with AIDS, but about one million are believed to be infected with the virus. Worldwide, more than 3 million people have developed AIDS, and most of these individuals have already died. Many more people are infected with the virus but have not yet developed the symptoms of full-blown AIDS (fig. A1). Estimates vary, but as many as 20 million worldwide may now be infected, and there could be as many as 100 million individuals infected by the end of the century. A new infection is believed to occur every 15 seconds, the majority between heterosexuals.

AIDS is not distributed equally throughout the world nor in the United States. Considering our discussion thus far, it is not surprising that most people with AIDS live in Africa (69%). Europe and the United States account for another 22%, and the disease is now beginning to take hold in India and Southeast Asia, which has the fastest infection rate. New York City, San Francisco, and Los Angeles have about one-third of the U.S. cases. But AIDS is spreading to other U.S. cities and rural areas as well. Similarly, African Americans and Hispanics have a higher proportionate number of cases compared to Caucasians, but actually HIV poses a threat to all sexually active adults regardless of ethnicity and sexual orientation.

Transmission of AIDS

In Africa the majority of people infected are heterosexuals; in the United States the majority infected are homosexuals (fig. A2). Behavior plays a role in transmission because the virus spreads more easily within a group of sexually active persons who frequently change partners. A large proportion of prostitutes in Africa and now Southeast Asia are infected, and they spread the virus to their various clients. In this country AIDS was first seen among those male homosexuals who frequently sought new partners and therefore passed the virus from one to the other. The number of HIV infections in male homosexuals now appears to have peaked as many adopt a more conservative life-style. In 1988 male homosexuals accounted for 61% of U.S. AIDS cases; now they account for 58%. In contrast, the percentage of heterosexuals infected is rising—every one in 5 new infections is now a woman. The magnitude of the epidemic is potentially much greater among heterosexuals because they represent the majority of the population. One unhappy side effect to female infection is the fact

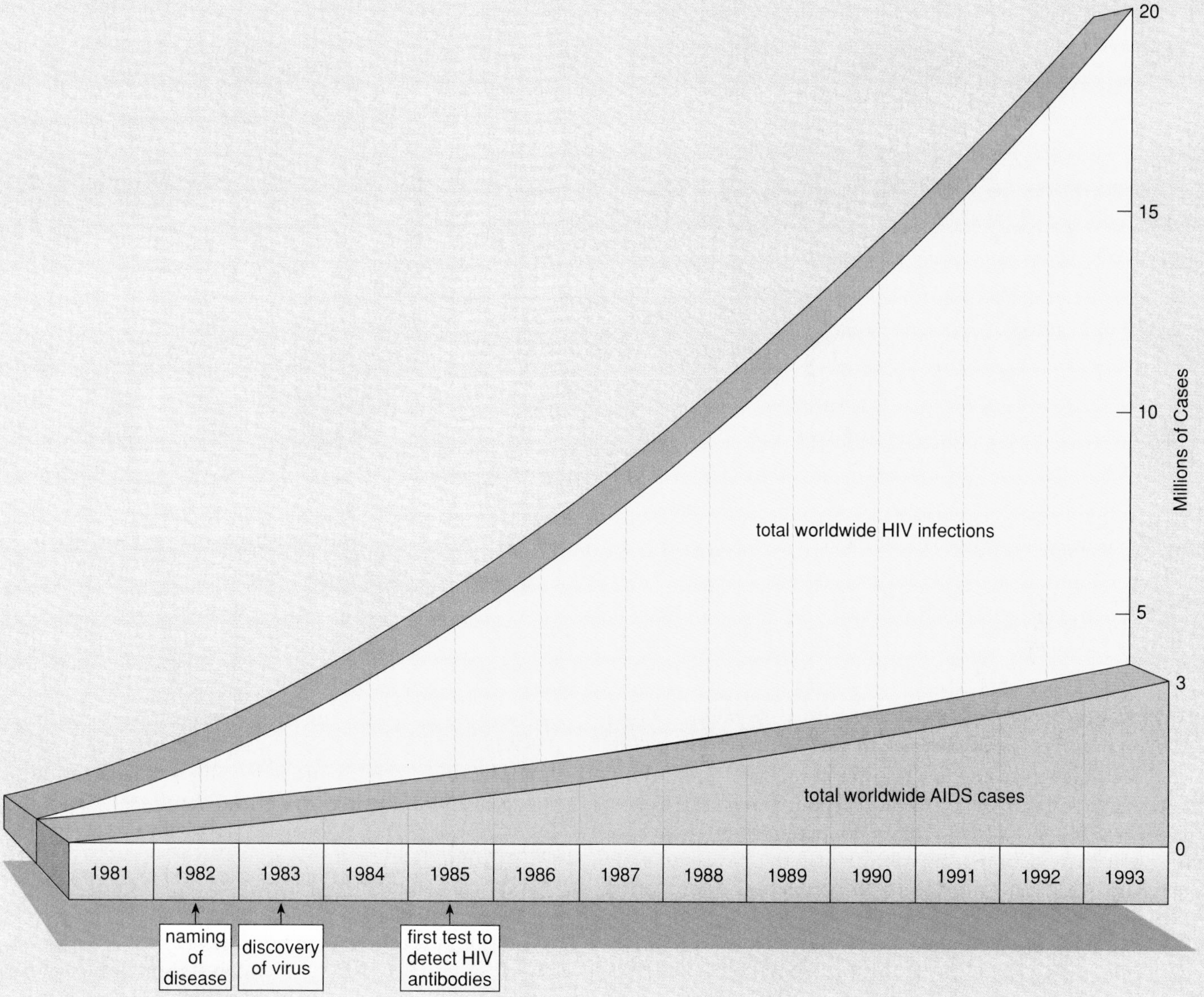

Figure A1

Estimated number of AIDS cases (worldwide) compared to number of people infected with HIV from the year 1981 to the year 1993. It can take several years before those infected show the symptoms of full-blown AIDS.

that the virus and infected lymphocytes can pass to a fetus via the placenta or to an infant via mother's milk. Presently, infected infants account for about 1% of all AIDS cases.

HIV comes in various strains. It could be that the first strain to enter the United States only spread by the more invasive anal sex act or by blood-to-blood contact. When intravenous drug users share contaminated needles, the infection spreads directly from one to the other. About 29% of total AIDS cases are intravenous drug users (fig. A2). A small number of persons (about 3% of total cases) have contracted AIDS by receiving infected blood during a blood transfusion. Particularly hemophiliacs, who require blood transfusions and/or products, have been known to get AIDS because of tainted blood. Since 1985, blood donations in the United States are tested for contamination with HIV. For various reasons, the testing process is not absolutely perfect and, therefore, a slight possibility of getting AIDS from a blood transfusion remains. Persons who are about to undergo an operation can predonate their own blood and/or have noninfected friends do so.

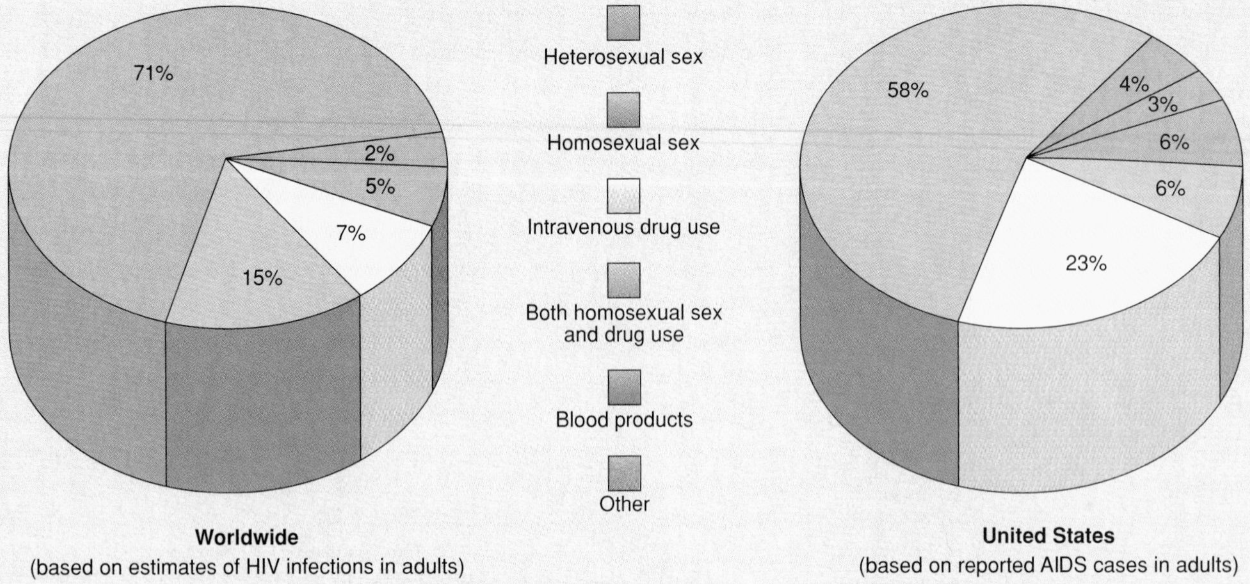

Figure A2

The distribution of AIDS among adults worldwide and in the United States.

Sources: Worldwide data from Harvard School of Public Health; United States data from U.S. Department of Health and Human Services; in Encyclopedia Britannica Medical and Health Annual 1993.

Symptoms of AIDS

A person has AIDS after HIV enters the blood and infects helper T lymphocytes (p. 133) of a type known as T4 cells. Therefore, it is possible to relate the progression of an HIV infection to a decline in the number of T4 cells, as is done in figure A3. This graph is based on actual data from a particular individual. For the sake of our discussion, figure A3 is divided into 3 stages: asymptomatic carrier, AIDS-related complex (ARC), and full-blown AIDS.

Asymptomatic Carrier

Usually people do not have any symptoms at all after initial infection. A few (1%–2%) do have mononucleosis-like symptoms that may include fever, chills, aches, swollen lymph nodes, and an itchy rash. These symptoms disappear, however, and there are no other symptoms for quite some time.

During the asymptomatic-carrier stage, the individual exhibits no symptoms and yet is highly infectious. According to figure A4, an immune system response then occurs. This means that antibodies are present in blood. The standard HIV blood test screens for the presence of antibodies and not for the presence of HIV itself. An individual is clearly infectious before the HIV test becomes positive.

A small percentage of false-negative and false-positive test results do occur with the HIV-antibody test. A positive result can be verified with a more expensive test that has a higher accuracy. Some people are against routine testing for HIV antibodies because it could subject those who test positive to unfair social discrimination; however, it is possible that routine testing could also help prevent the spread of AIDS. Without testing, people (including most young people), do not know they are positive because they have no symptoms during this initial stage.

AIDS-Related Complex (ARC)

Several months to several years after infection, the individual may begin to show some symptoms. The most common symptom of ARC is swollen lymph nodes in the neck, armpits, or groin that persist for 3 months or more. Notice that at this time little HIV can be detected in the blood, making it seem as if the infection is under control (fig. A4). The swollen lymph nodes may be a clue that the virus is now concentrated in the lymph nodes.

Other symptoms that indicate an HIV infection are severe fatigue not related to exercise or drug use; unexplained persistent or recurrent fevers, often with night sweats; persistent cough not associated with smoking, a cold, or the flu; and persistent diarrhea. Also possible are signs of nervous system impairment, including loss of memory, inability to think clearly, loss of judgment, and/or depression.

When the individual develops non-life-threatening but recurrent infections, it is a signal that full-blown AIDS will occur shortly. One possible infection is thrush, a fungal infection that is identified by the presence of white spots and ulcers on the tongue and inside the mouth. The fungus may also spread to the vagina, resulting in a chronic infection there. Another frequent infection is herpes simplex, with painful and persistent sores on the skin surrounding the anus, the genital area, and/or the mouth.

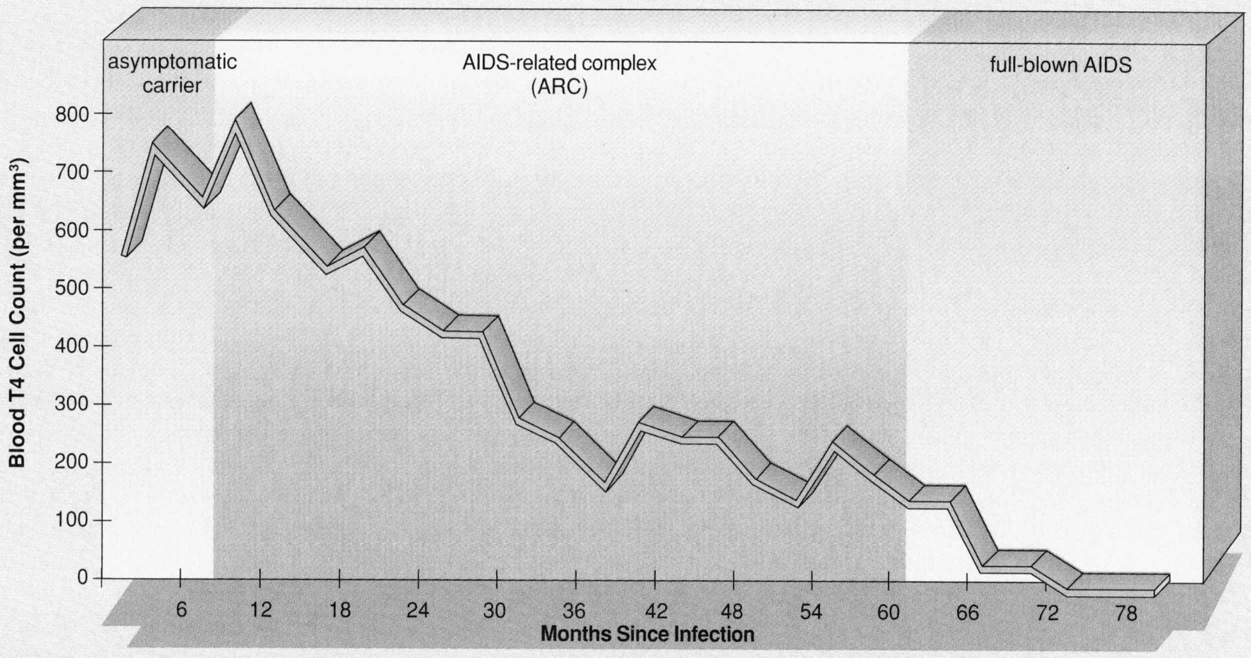

Figure A3

Blood T4 cell count in a young man whose HIV infection followed a typical course. After about 3 months, the T4 cell count increases as the body attempts to fight the infection. During this time, the individual is an asymptomatic carrier. Once the T4 cell count begins to decline, the individual develops the symptoms of ARC. Finally, full-blown AIDS is diagnosed when the patient has one or more of the opportunistic infections. By this time, the blood T4 cell count is below 100 per mm^3 and the individual soon dies.

Source: Data from R. R. Redfield and D. S. Burke, "HIV Infection: The Clinical Picture" in Scientific American, *October, 1988.*

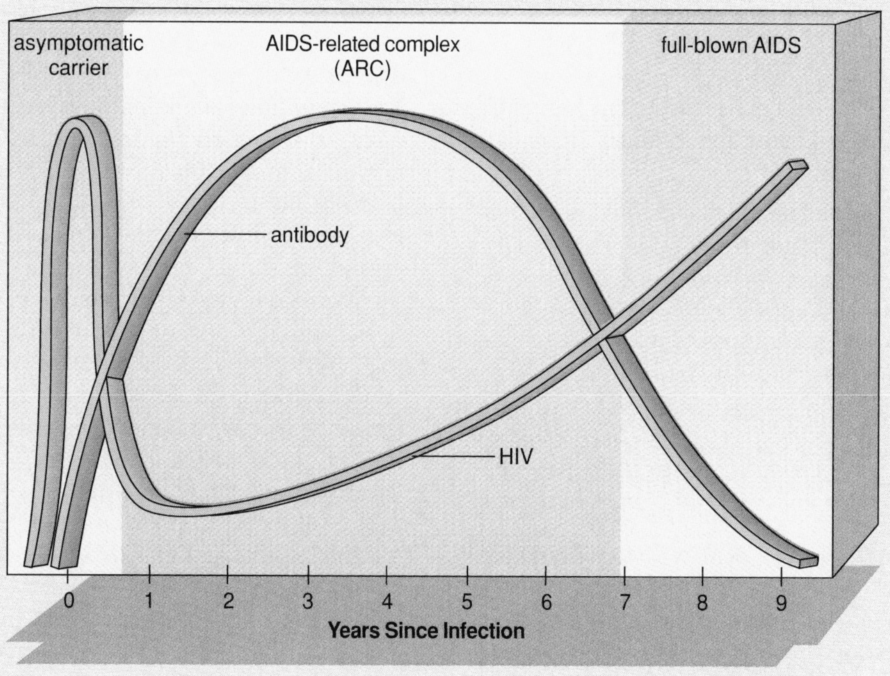

Figure A4

Antibody titer versus HIV in the blood. During the time that a person is an asymptomatic carrier, the blood concentration of HIV is high. Then, for reasons that are still unknown, there is little HIV in the blood despite the fact that the blood T4 cell count continues to decline (see fig. A3). Once a person has full-blown AIDS, the amount of HIV in the blood rises.

Source: Data from R. R. Redfield and D. S. Burke, "HIV Infection: The Clinical Picture" in Scientific American, *October, 1988.*

Full-Blown AIDS

The majority of people with ARC eventually develop full-blown AIDS. What causes the transition from ARC to full-blown AIDS is not known, but it is observed that at this time a blood count of 200 T4 cells per mm^3 is characteristic. At this time, too, the lymph nodes have degenerated and lost many of their specialized cells so vital to normal immunity.

In this final stage of an HIV infection, the AIDS patient, who is now suffering from "slim disease" (as AIDS is called in Africa)—severe weight loss and weakness due to persistent diarrhea and coughing—most likely succumbs to one of the so-called opportunistic infections. These infections are called opportunistic because they are caused by microbes that cannot ordinarily start an infection, but they have the opportunity in AIDS patients because of the severely impaired immune system. Some of the opportunistic infections are the following:

Pneumocystis carinii pneumonia. The lungs become useless as they fill with fluid and debris due to an infection with this organism. There is not a single documented case of *P. carinii* pneumonia in a person with normal immunity.

Toxoplasmic encephalitis is caused by a one-cell parasite that lives in cats and other animals as well as humans. Many persons harbor a latent infection in the brain or muscle, but in AIDS patients the infection leads to loss of brain cells, seizures, weakness, or decreased sensation on one side of the body.

Kaposi's sarcoma is an unusual cancer of blood vessels, which gives rise to reddish purple, coin-size spots and lesions on the skin.

Mycobacterium tuberculosis. This bacterial infection, usually of the lungs, is seen more often as an infection of lymph nodes and other organs in patients with AIDS. Of special concern, tuberculosis is spreading into the general populace and is multidrug resistant (see Health Focus, chapter 15).

Invasive cervical cancer. This cancer of the cervix spreads to nearby tissues. This condition has been added to the list because the incidence of AIDS has now increased in women.

Drugs have been developed to deal with opportunistic diseases in AIDS patients. These drugs help people with AIDS lead a fairly normal life for some months, but eventually patients are repeatedly hospitalized due to weight loss, constant fatigue, and multiple infections. Death usually follows in 2–4 years.

Treatment for AIDS

Much has been learned about the structure of HIV and its reproductive cycle. Treatment is based on this knowledge.

HIV has an outer envelope consisting of a lipid bilayer with embedded proteins (fig. A5). Some of these proteins, called spikes because they jut out of the envelope, consist of 4 molecules each of gp120 and gp41. Gp stands for glycoproteins (a protein linked to sugars, p. 31). Underneath the envelope are 2 layers of protein; the one containing p17 is a matrix, but the other containing p24 is the capsid that surrounds the virus's RNA genes and various viral enzymes. One of these enzymes is reverse transcriptase, the enzyme that produces a DNA copy of the viral RNA.

Figure A5 shows the reproductive cycle of the HIV virus. The capsid and nucleic acid core enter a T4 cell after the spike protein gp120 attaches to a CD4 molecule on the cell's surface. This is a lock-and-key reaction; only immune cells having a CD4 receptor molecule allow the virus to freely enter. Following entry, reverse transcriptase makes a DNA copy of the viral RNA, which enters the nucleus and integrates itself into a host chromosome. HIV is then called a provirus. Various regulatory proteins, including those coded for by *tat, nef,* and *rev* genes, are needed for full transcription of the provirus. After the various components of the virus are produced, final refinements are made by a viral enzyme called protease, and then assembly takes place. Copies of the virus bud from the cell; the envelope consists of host cell membrane plus the viral spikes.

The 3 drugs now approved for treatment of AIDS (AZT, ddI, and ddC[1]) all work by interfering with reverse transcription. Each is an analog (chemically altered form) of a nucleotide; when reverse transcriptase incorporates the analog instead of the normal nucleotide, transcription stops. Unfortunately, HIV genes mutate so rapidly (due to errors in replication) that a new form of the enzyme eventually appears that ignores the analogs and chooses the correct nucleotides instead. In other words, these drugs are effective in the patient for only a limited time.

Other types of AIDS drugs are planned or in clinical trial; each one interferes with a particular protein necessary to the reproductive cycle of the virus. Drugs are planned that act against the *tat* and *nef* regulatory proteins and against protease, for example.

Many investigators are working on a vaccine for AIDS. As discussed on page 138, vaccines are traditionally made by treating microbes chemically, thereby weakening them so that they can be injected into the body without causing disease. Jonas Salk, who developed the first polio vaccine, has taken this approach with the AIDS virus, and clinical

1. AZT (azidothymidine), ddI (dideoxyinosine), ddC (dideoxycytidine).

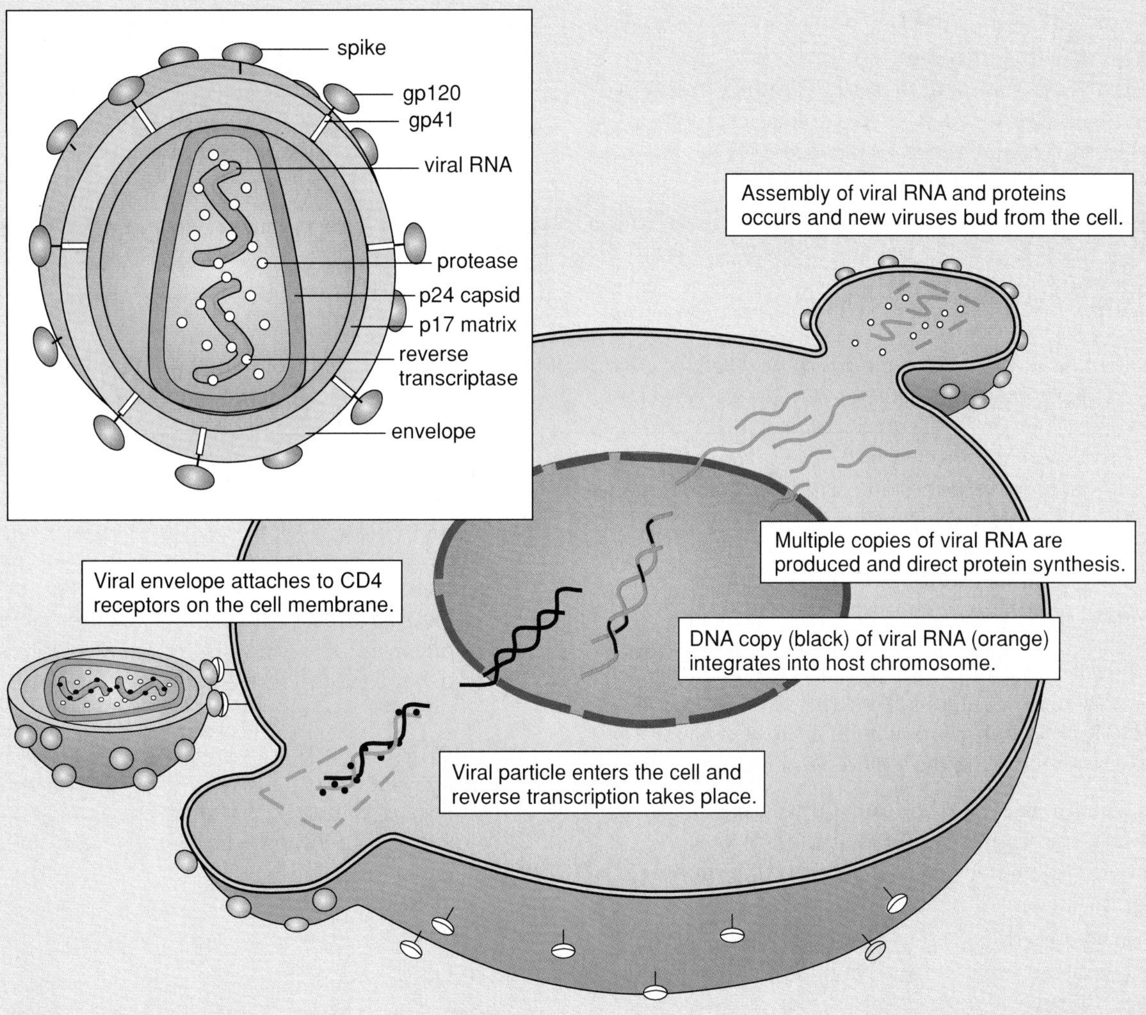
spike
gp120
gp41
viral RNA
protease
p24 capsid
p17 matrix
reverse transcriptase
envelope
Assembly of viral RNA and proteins occurs and new viruses bud from the cell.
Multiple copies of viral RNA are produced and direct protein synthesis.
Viral envelope attaches to CD4 receptors on the cell membrane.
DNA copy (black) of viral RNA (orange) integrates into host chromosome.
Viral particle enters the cell and reverse transcription takes place.

Figure A5

Anatomy of HIV and its reproductive cycle. HIV is a retrovirus with an outer envelope containing proteins, some of which are spikes consisting of gp120 and gp41 proteins. The core of the virus contains RNA genes and copies of reverse transcriptase, the enzyme that makes a DNA copy of RNA. During the reproductive cycle, the DNA copy first becomes integrated into a host chromosome and then is transcribed back into viral RNA.

trials are underway. Another group of investigators has developed a SIV (simian immunodeficiency virus) vaccine. This virus causes AIDS in monkeys, but it is well known that the smallpox vaccine, which is effective in humans, actually utilizes the cowpox virus and not smallpox virus. There are those who are opposed to a whole virus vaccine on the basis that it is too dangerous. There is no treatment if the vaccine by chance should cause the disease instead of preventing it.

The mutability of the virus is another factor hindering the development of a vaccine. The spike proteins gp120 and gp41 can be produced by genetically engineered bacteria and possibly used as vaccines; however, their composition changes too rapidly to make them good candidates. Still, these subunit vaccines are under trial as well as another, which uses gp160, a precursor to these 2 proteins. More success is expected with the matrix protein p17 because it is more stable than the spike proteins. To make a vaccine, p17 proteins have to be attached to a carrier that can be processed by a macrophage, because macrophages present the antigen to T cells, which then stimulate B cells to produce antibodies. A liposome, an artificial membranous sac, is one possible carrier.

New and novel ideas are being tried. Disabled cowpox viruses have been genetically engineered to carry the HIV genes that code for spike proteins, and these doctored viruses are being injected in AIDS patients. It is expected that the genes will be expressed and the viruses will produce spike proteins that will stimulate the immune system. Also, antibodies against CD4 molecules have been made with the hope that these antibodies, when injected into the body, will attach to T4 cells. Then the viruses will be unable to gain entry.

The various vaccines are at different stages in their development. None are close to completion of clinical trials, and therefore a vaccine is not expected before 1995.

Health Focus

AIDS Prevention

At the present time, there is no stopping the AIDS epidemic except by changing the behavior of human beings. The number of new HIV infections and the number of new AIDS cases will continue to increase dramatically in the near future. However, each person can take responsibility for himself or herself by following the guidelines given here.

AIDS is transmitted by sexual contact or by sharing contaminated needles or blood. Shaking hands, hugging, social kissing, coughing or sneezing, and swimming in the same pool will not transmit the AIDS virus. You cannot get AIDS from inanimate objects such as toilets, doorknobs, telephones, office machines, or household furniture.

The following behaviors will help prevent the spread of AIDS and decrease the projected number of new cases:

1. Do not use alcohol or drugs in a way that may prevent you from being able to control your behavior. Especially do not take up the habit of injecting drugs into veins.
2. If you are already a drug user and cannot stop your behavior, then always use a new sterile needle for injection or one that has been cleaned in bleach.
3. Either abstain from sexual intercourse or develop a long-term monogamous (always the same partner) sexual relationship with a partner who is free of HIV and is not an intravenous drug user.
4. If you do not know for certain that your partner has been free of HIV for the past 5 years, always use a latex condom during sexual intercourse. Be sure to follow the directions supplied by the manufacturer. Use of a spermicide containing nonoxynol-9 in addition to the condom can offer further protection because nonoxynol-9 also kills the virus and virus-infected lymphocytes.
5. Refrain from multiple sex partners, especially with homosexual or bisexual men or intravenous drug users of either sex. The risk of contracting AIDS is greater in persons who already have another STD.

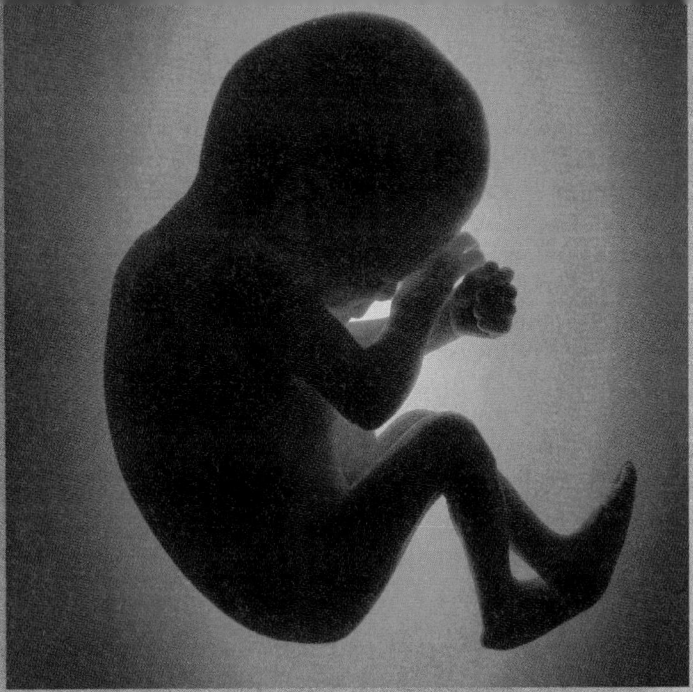

Chapter 16

Development and Aging

How do you define environment? The human embryo develops within the uterus of its mother. No doubt this is the embryo's environment, but is the mother's environment also the embryo's environment? In many ways it is. If the mother is X-rayed, has AIDS, or takes a harmful drug, the embryo can be irrevocably harmed. What about psychological assaults on the mother? Can exposure to stressful situations, for example, be harmful to the embryo and later to the fetus? Perhaps so. One study shows that during and after World War II, birth defects in German children were markedly increased. It has even been suggested that personality, too, is affected by prenatal environmental factors. Perhaps we are even psychologically affected by whether the womb is large or small or whether it contains another developing embryo.

Chapter Concepts

Chapter Outline

The male and female reproductive systems were discussed in a previous chapter. The male continually produces sperm in the testes, and the female produces one egg a month in the ovaries. The release of the egg, called *ovulation,* occurs on day 14 of an ovarian cycle that lasts 28 days. The hormones produced by the ovary control the uterine cycle, which consists of menstruation, a proliferative phase during which the endometrium (uterine lining) builds up, and a secretory phase during which the endometrium becomes secretory. Ovulation marks the end of the proliferative phase and the beginning of the secretory phase. Following ovulation, the egg enters the oviduct.

If intercourse is accompanied by ejaculation, some 200 to 600 million sperm are normally deposited at the rear of the vagina in the region of the cervix (see fig. 14.5). The semigelatinous seminal fluid protects the sperm from the acid of the vagina for several minutes, after which they are killed unless they have managed to enter the uterus. Whether or not the sperm enter depends in part on the consistency of the cervical mucus. Three to 4 days prior to ovulation and on the day of ovulation, the mucus is watery, and the sperm can penetrate it easily. During the other days of the uterine cycle, the mucus is thicker and has a sticky consistency and the sperm can rarely penetrate it.

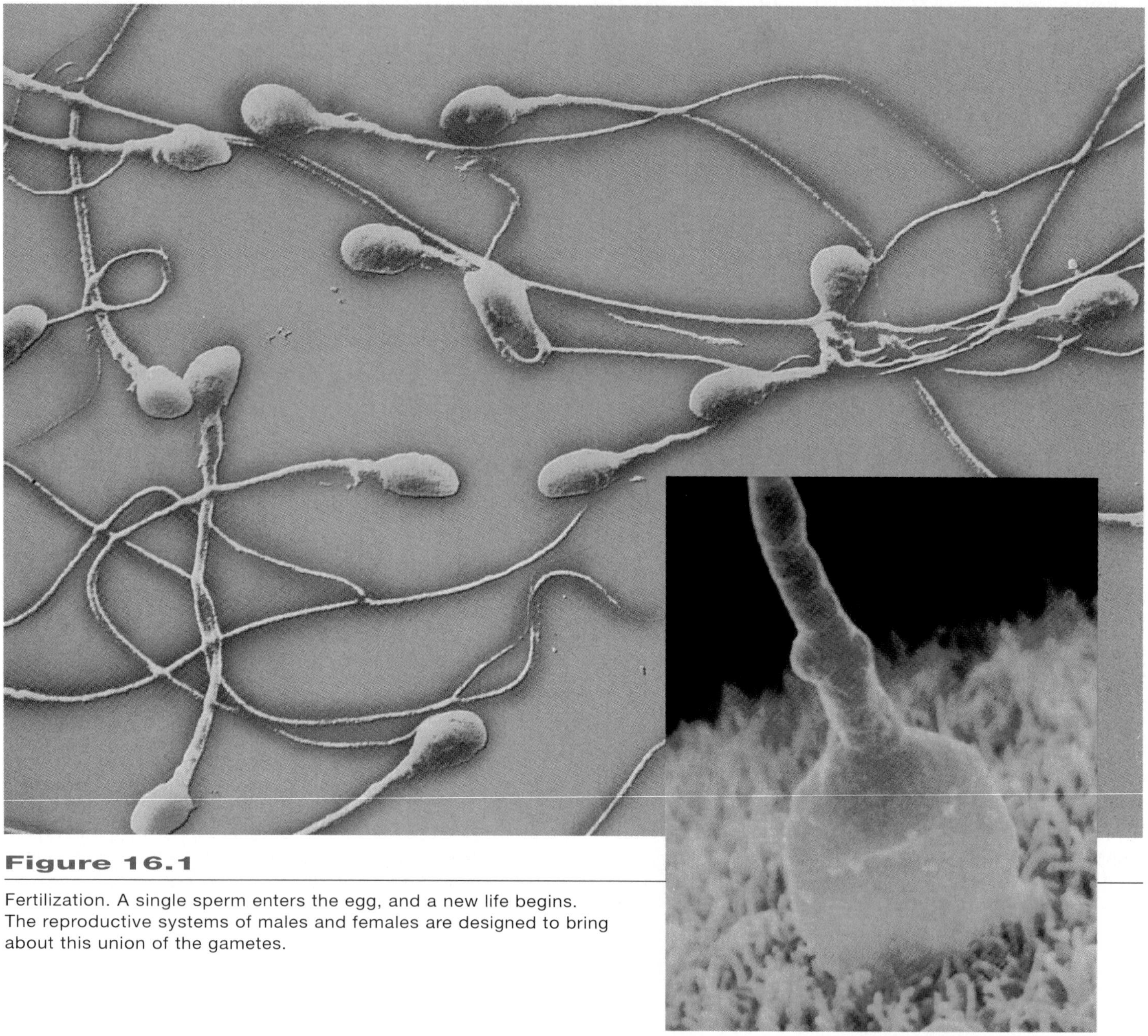

Figure 16.1

Fertilization. A single sperm enters the egg, and a new life begins. The reproductive systems of males and females are designed to bring about this union of the gametes.

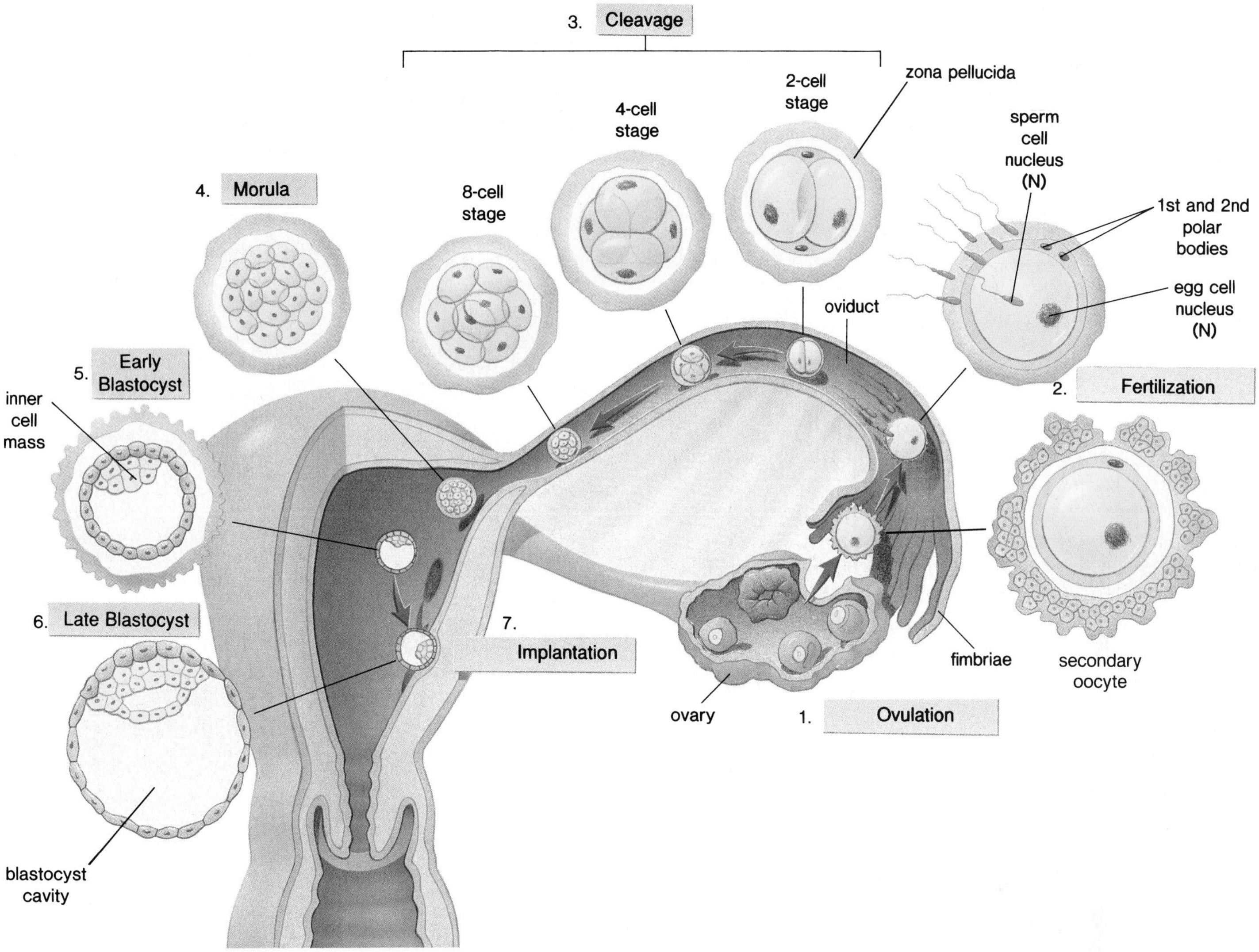

Figure 16.2

Human development before implantation. Structures and events proceed counterclockwise. At ovulation, the secondary oocyte leaves the ovary. A single sperm penetrates the zona pellucida and fertilization occurs in the oviduct. As the zygote moves along the oviduct, it undergoes cleavage to produce an embryo that implants itself in the uterine lining.

Development

Development encompasses the time from fertilization to birth (parturition). **Fertilization** (fig. 16.1) normally occurs in the upper third of the oviduct, and a small percentage of sperm (perhaps 100,000) usually arrive there within 5–30 minutes. It is believed that uterine and oviduct contractions transport the sperm and that prostaglandins within seminal fluid promote these contractions. Swimming action by sperm seems to be only needed when they are in the vicinity of the egg.

Fertilization: Sperm Meets Egg

When sperm make contact with the cells surrounding the egg (the corona radiata), acrosome enzymes are released that allow a sperm to enter the egg (see fig. 14.2). Only one sperm actually penetrates an elastic envelope (zona pellucida) and enters the egg; then the egg undergoes changes that prevent any more sperm from entering. The sperm nucleus moves through the cytoplasm and fuses with the egg nucleus. Fertilization is now complete.

> During fertilization, a single sperm enters an egg. The sperm nucleus fuses with the egg nucleus, and development begins immediately.

The fertilized egg, more properly called the **zygote,** begins dividing. The developing embryo travels very slowly down the oviduct to the uterus, where after 2–3 days it implants itself in the prepared uterine lining (fig. 16.2). Upon implantation, **conception** has occurred and the female is pregnant. If implantation takes place, the uterine lining is maintained, because cells surrounding the embryo begin to produce a hormone called HCG (human chorionic gonadotropin), which prevents degeneration of the corpus luteum (p. 283) and instead causes it to secrete even larger quantities of progesterone. This hormone causes the uterine lining to be maintained, and menstruation does not occur.

Pregnancy

In humans, the **gestation period,** or length of pregnancy, is approximately 9 months. It is customary to calculate the time of birth by adding 280 days to the start of the last menstruation because this date is usually known, whereas the day of fertilization is usually unknown. Because the time of birth is influenced by so many variables, only about 5% of babies actually arrive on the predicted date.

Pregnancy tests, which are readily available in hospitals, clinics, and now even drug and grocery stores, are based on the fact that HCG is present in the blood and the urine of a pregnant woman. Before the advent of monoclonal antibodies, only a hospital blood test using radioactive material was available to detect pregnancy before the first missed menstrual period. Now there is a monoclonal antibody (p. 140) test for the detection of pregnancy 10 days after conception. This test can be done on a urine sample in a doctor's office and the results are available within the hour.

The physical signs that often prompt a woman to have a pregnancy test are cessation of menstruation, increased frequency of urination, morning sickness, and increase in the size and the fullness of the breasts, as well as darkening of the areolae (see fig. 16.14).

Extraembryonic Membranes

One of the major events in early development is the establishment of the extraembryonic membranes (fig. 16.3). The term **extraembryonic membranes** is apt, because these membranes extend out beyond the embryo. One of the membranes, the **amnion,** provides a fluid environment for the developing embryo and fetus. It is a remarkable fact that all animals, even land-dwelling humans, develop in water. Amniocentesis is a process by which amniotic fluid and the fetal cells floating in it are withdrawn for examination. One authority describes the functions of amniotic fluid in this way:

> The colorless amniotic fluid by which the fetus is surrounded serves many purposes. It prevents the walls of the uterus from cramping the fetus and allows it unhampered growth and movement. It encompasses the fetus with a fluid of constant temperature which is a marvelous insulator against cold and heat. Above all, it acts as an excellent shock absorber. A blow on the mother's abdomen merely jolts the fetus, and it floats away.[1]

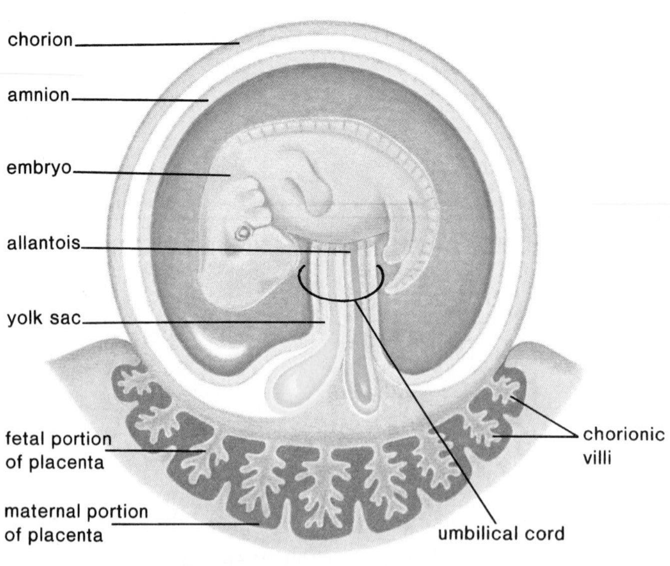

Figure 16.3

The extraembryonic membranes. The chorion and amnion surround the embryo. The 2 other extraembryonic membranes, the yolk sac and allantois, contribute to the umbilical cord.

The **yolk sac** is another extraembryonic membrane. Yolk is a nutrient material utilized by other animal embryos—the yellow of a chick's egg is yolk, for example. In humans, the yolk sac is the first site of red blood cell formation. Part of this membrane becomes incorporated into the umbilical cord. Another extraembryonic membrane, the **allantois,** contributes to the circulatory system: its blood vessels become umbilical blood vessels that transport fetal blood to and from the placenta. The **chorion,** the outer extraembryonic membrane, becomes part of the **placenta** (fig. 16.4), where the fetal blood exchanges molecules with maternal blood.

Placenta: For Exchange

The placenta begins formation once the embryo is implanted fully. Treelike extensions of the chorion called **chorionic villi** project into the maternal tissues. Later, these disappear in all areas except where the placenta develops. By the tenth week, the placenta is formed fully and begins to produce progesterone and estrogen (fig. 16.5). These hormones have 2 effects: due to their negative feedback effect on the hypothalamus and the anterior pituitary, they prevent any new follicles from maturing, and they maintain the lining of the uterus—the corpus luteum is not now needed. There is no menstruation during pregnancy.

1. A. F. Guttmacher. *Pregnancy, Birth and Family Planning* (New York: New American Library, 1974), p. 74.

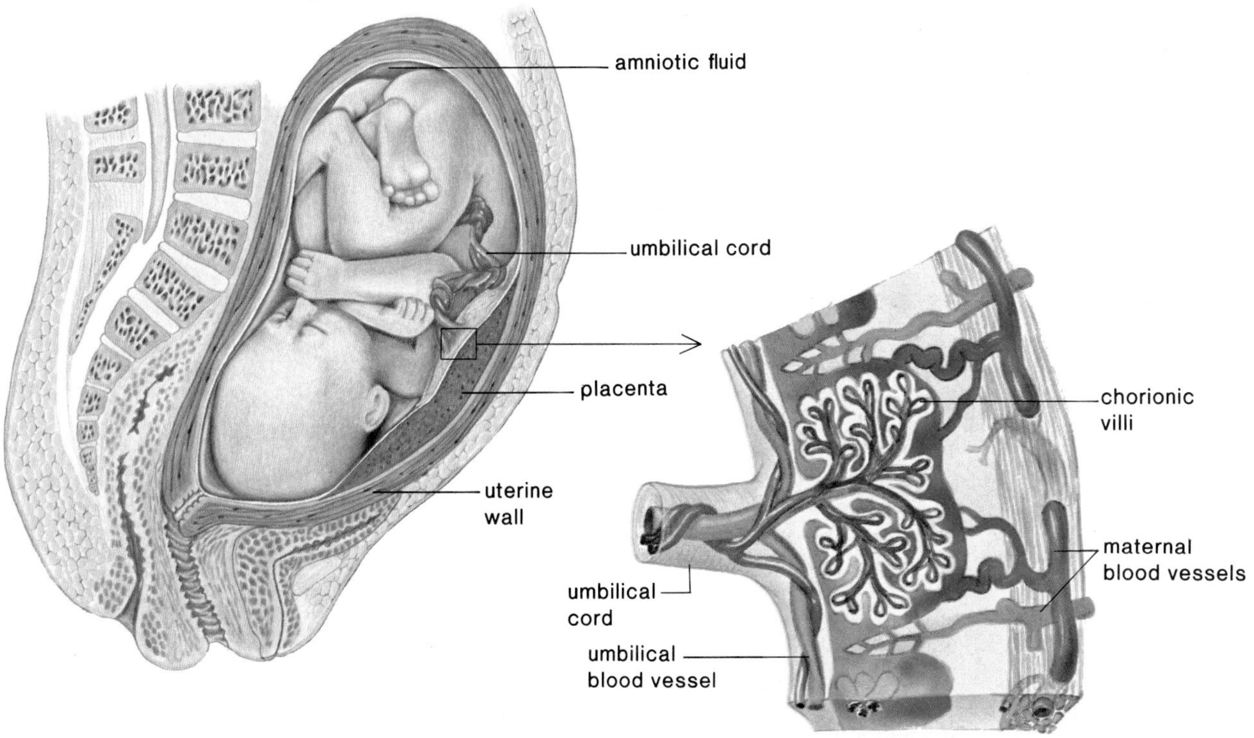

Figure 16.4

Anatomy of the placenta in a fetus at 6–7 months. The placenta is composed of both fetal and maternal tissues. Chorionic villi penetrate the uterine lining and are surrounded by maternal blood. Exchange of molecules between fetal and maternal blood takes place across the walls of the chorionic villi. Oxygen and nutrient molecules enter the fetal blood; carbon dioxide and urea exit from fetal blood.

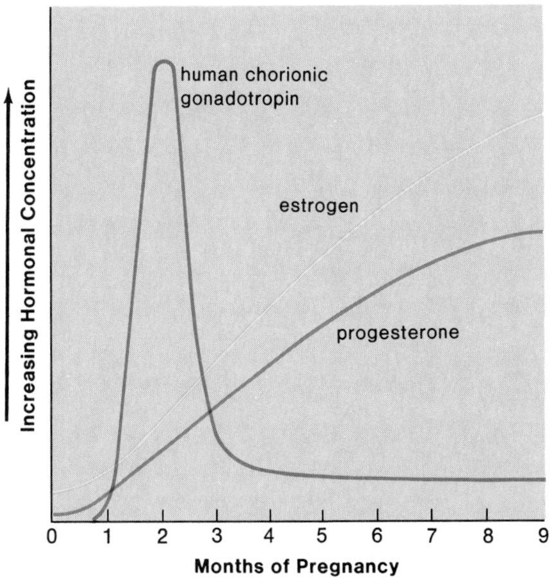

Figure 16.5

Hormones during pregnancy. Human chorionic gonadotropin is secreted by the chorion during the first 3 months of pregnancy. This maintains the corpus luteum, which continues to secrete estrogen and progesterone. At about 5 weeks of pregnancy, the placenta begins to secrete estrogen and progesterone in increasing amounts as the corpus luteum degenerates.

The placenta has a fetal side contributed by the chorion and a maternal side consisting of uterine tissues. Notice in figure 16.4 how the chorionic villi are surrounded by maternal blood; yet, the blood of the mother and the fetus never mix since exchange always takes place across cell membranes. Carbon dioxide and other wastes move from the fetal side to the maternal side, and nutrients and oxygen move from the maternal side to the fetal side of the placenta. The **umbilical cord** stretches between the placenta and the fetus. Although it may seem that the umbilical cord travels from the placenta to the intestine, actually it simply is taking fetal blood to and from the placenta. The umbilical cord is the lifeline of the fetus because it contains the umbilical arteries and vein, which transport waste molecules (carbon dioxide and urea) to the placenta for disposal and take oxygen and nutrient molecules from the placenta to the rest of the fetal circulatory system.

Harmful chemicals also can cross the placenta, and this is of particular concern during the embryonic period, when various structures are first forming. Each organ or part seems to have a sensitive period during which a substance can alter its normal function. The first Health Focus for this chapter concerns the origination of birth defects and explains ways to detect genetic defects before birth.

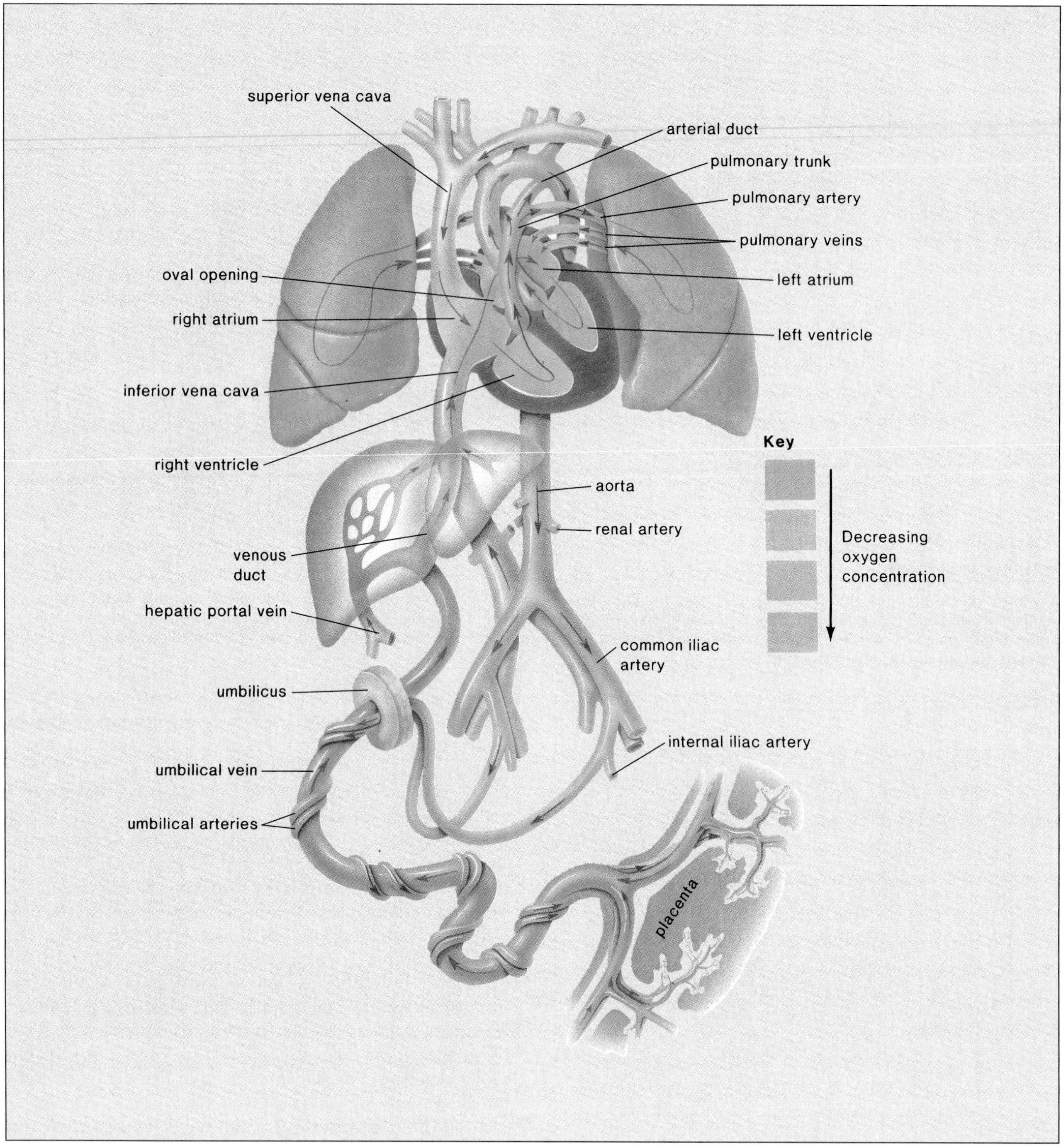

Figure 16.6

Fetal circulation. The umbilical arteries carry deoxygenated blood to the placenta, where gas exchange occurs. The umbilical vein carries oxygenated blood to the inferior vena cava via the venous duct. The inferior and superior vena cava enter the right atrium. From there mixed blood enters the left atrium by way of the oval opening or continues to the right ventricle. From the right ventricle blood enters the pulmonary trunk and then the aorta via the arterial duct. In this way blood is diverted from the lung. Notice, too, that the blood in the aorta is mixed blood with less oxygen.

Fetal Circulation

As figure 16.6 shows, the fetus has 4 circulatory features that are not present in adult circulation.

1. **Oval opening,** or *foramen ovale,* an opening between the 2 atria. This opening is covered by a flap of tissue that acts as a valve.
2. **Arterial duct,** or *ductus arteriosus,* a connection between the pulmonary artery and the aorta.
3. **Umbilical arteries and vein,** vessels that travel to and from the placenta, leaving waste and receiving nutrients.
4. **Venous duct,** or *ductus venosus,* a connection between the umbilical vein and the inferior vena cava.

All of these features can be related to the fact that the fetus does not use its lungs for gas exchange; it receives oxygen and nutrients from the mother's blood by way of the placenta.

To trace the path of blood in the fetus, begin with the right atrium (fig. 16.6). From the right atrium, the blood may pass directly into the left atrium by way of the oval opening or it may pass through the atrioventricular valve into the right ventricle. From the right ventricle, the blood goes into the pulmonary artery, but because of the arterial duct, most of the blood then passes into the aorta. Therefore, by whatever route the blood takes, most of the blood reaches the aorta instead of the lungs.

Blood within the aorta travels to the various branches, including the iliac arteries, which connect to the umbilical arteries leading to the placenta. Exchange between maternal blood and fetal blood takes place at the placenta. It is interesting to note that the blood in the umbilical arteries, which travels to the placenta, is low in oxygen, but the blood in the umbilical vein, which travels from the placenta, is high in oxygen. The umbilical vein enters the venous duct, which passes directly through the liver. The venous duct then joins with the inferior vena cava, a vessel that contains deoxygenated blood. The vena cava returns this "mixed blood" to the heart.

The most common of all cardiac defects in the newborn is the persistence of the oval opening. With the tying of the cord and the expansion of the lungs, blood enters the lungs in quantity. Return of this blood to the left side of the heart usually causes a flap to cover the opening. Incomplete closure occurs in nearly one out of 4 individuals, but even so, passage of the blood from the right atrium to the left atrium rarely occurs because either the opening is small or it closes when the atria contract. In a small number of cases, the passage of impure blood from the right side to the left side of the heart is sufficient to cause a "blue baby." Such a condition now can be corrected by open heart surgery.

The arterial duct closes because endothelial cells divide and block off the duct. Remains of the arterial duct and parts of the umbilical arteries and vein later are transformed into connective tissue.

> Fetal circulation shunts blood away from the lungs, toward and away from the placenta within the umbilical blood vessels located within the umbilical cord. Exchange of substances between fetal blood and maternal blood takes place at the placenta, which forms from the chorion, an extraembryonic membrane, and uterine tissue.

Embryonic Development: Getting Started

Human development can be divided into **embryonic development** (the second week through the eighth week) and **fetal development** (the third through ninth months). During embryonic development, all the major organs form, and during fetal development there is a refinement of these structures. The fetus, not the embryo, is recognizable as a human being.

Development includes these processes:

Cleavage Immediately after fertilization, the zygote begins to divide so that at first there are 2, then 4, 8, 16, and 32 cells, and so forth. Increase in size does not accompany these divisions.

Growth During embryonic development, cell division is accompanied by an increase in size of the daughter cells, and *growth* (in the true sense of the term) takes place.

Morphogenesis *Morphogenesis* refers to the shaping of the embryo and is first evident when certain cells are seen to move, or migrate, in relation to other cells. By these movements, the embryo begins to assume various shapes.

Differentiation When cells take on a specific structure and function, *differentiation* occurs. The first system to become visibly differentiated is the nervous system.

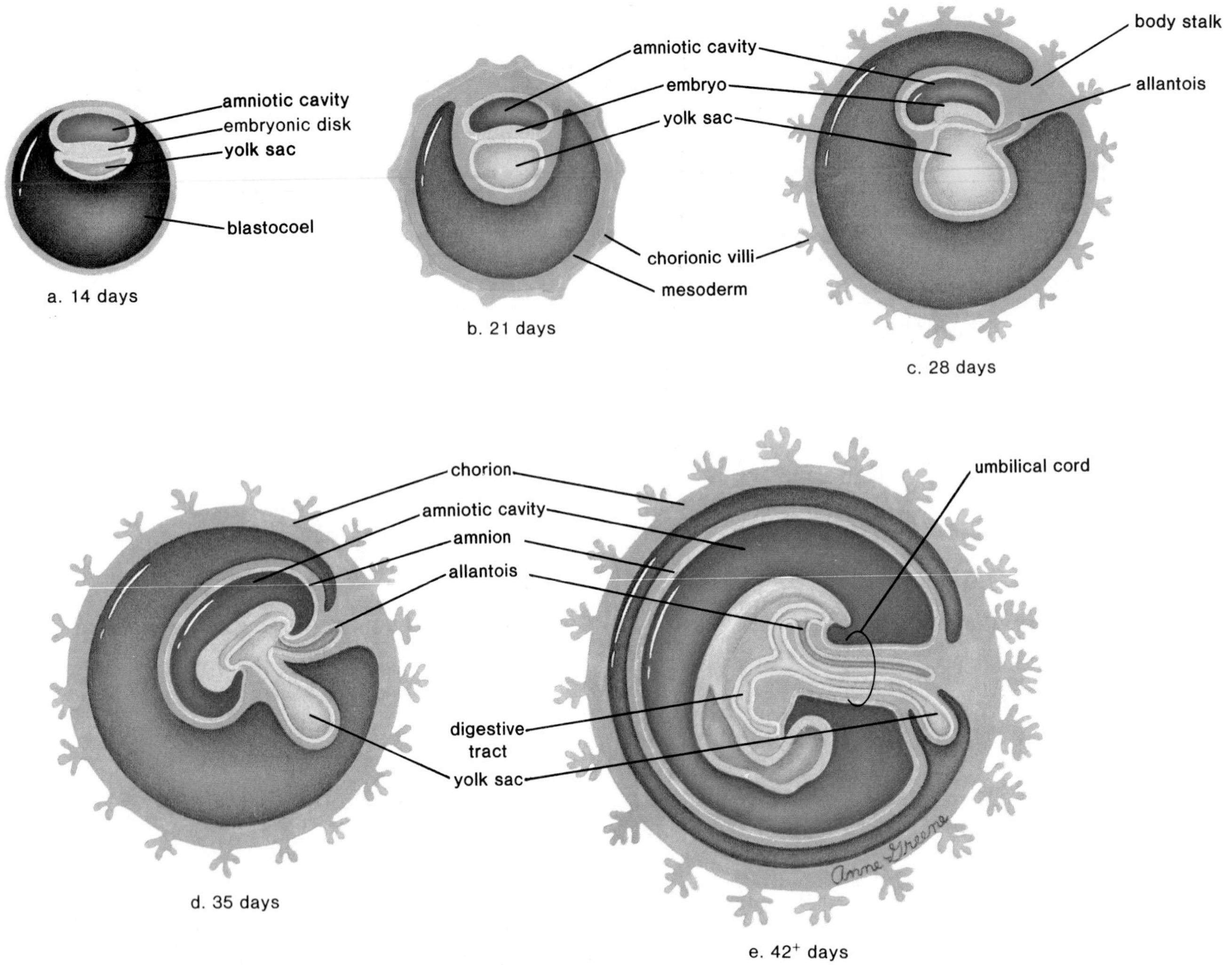

Figure 16.7

Embryonic development. **a.** At first there are only tissues present in the embryo. The amniotic cavity is above the embryo, and the yolk sac is below. **b.**The chorion is developing villi so important to exchange between mother and child. **c.** The allantois and yolk sac are 2 more extraembryonic membranes. **d.** These extraembryonic membranes are positioned inside the body stalk as it becomes the umbilical cord. **e.** At 42 days, the embryo has a head region and a tail region. The umbilical cord takes blood vessels between the embryo and the chorion (placenta).

First Week

Immediately after fertilization, the zygote divides repeatedly as it passes down the oviduct to the uterus. A **morula** is a compact ball of embryonic cells that becomes a **blastocyst.** The many cells of the blastocyst arrange themselves so that there is an inner cell mass surrounded by a layer of cells that becomes the chorion (see fig. 16.2). The early appearance of the chorion emphasizes the complete dependence of the developing embryo on this extraembryonic membrane. The inner cell mass is the **embryo.**

Once it has arrived in the uterus on or about the seventh day, the blastocyst begins to implant itself in the uterine lining (see fig. 16.2).

Second Week

Implantation is completed and embryonic development begins during the second week. With implantation, pregnancy has now taken place, and the placenta begins formation. The ever-growing number of cells is now arranged in tissues; the amniotic cavity is seen above the embryo and the yolk sac is below (fig. 16.7*a*).

Third Week

Another extraembryonic membrane, the allantois, makes its appearance briefly, but later it and the yolk sac become part of the umbilical cord as it forms (fig. 16.7*c*–*d*). Organs are already developed, including the spinal cord and heart. The nervous system and circulatory system have begun to develop.

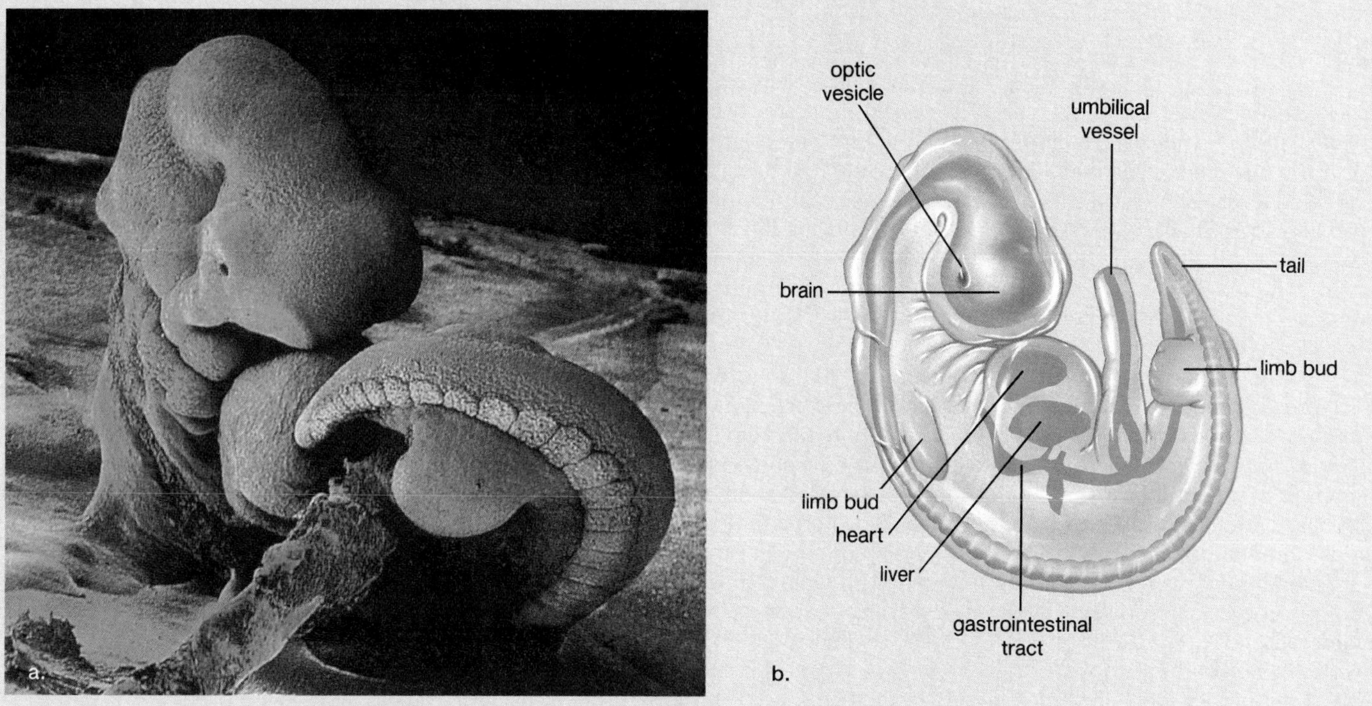

Figure 16.8

Human embryo at beginning of fifth week. **a.** Scanning electron micrograph. **b.** The embryo is curled so that the head touches the heart. The organs of the gastrointestinal tract are forming. The bones in the tail will regress and become those of the coccyx. The arms and legs will develop from the bulges that are called limb buds.

Fourth Week

By the end of the first month, the chorionic villi project into the uterine wall, and the placenta is producing enough (HCG) to maintain the corpus luteum and the uterine lining. Although the chorionic villi enlarge at one location only, they may eventually spread over 50% of the uterine endometrium (see fig. 16.4).

The embryo has a nonhuman appearance largely due to the presence of a tail, but also because the arms and legs, which begin as limb buds, resemble paddles. The head is much larger than the rest of the embryo, and the whole embryo bends under its weight (fig. 16.8). The eyes, ears, and nose are just appearing. The enlarged heart beats and the bulging liver takes over the production of blood cells for blood, which will carry nutrients to the developing organs and wastes from the developing organs.

Second Month

At the end of 2 months, the embryo's tail has disappeared, and the arms and legs are more developed with fingers and toes apparent (fig. 16.9). The head is very large, the nose is flat, the eyes are far apart, and the ears are distinctively present. Internally, all major organs have appeared. Embryonic development is now finished; table 16.1 outlines the main events.

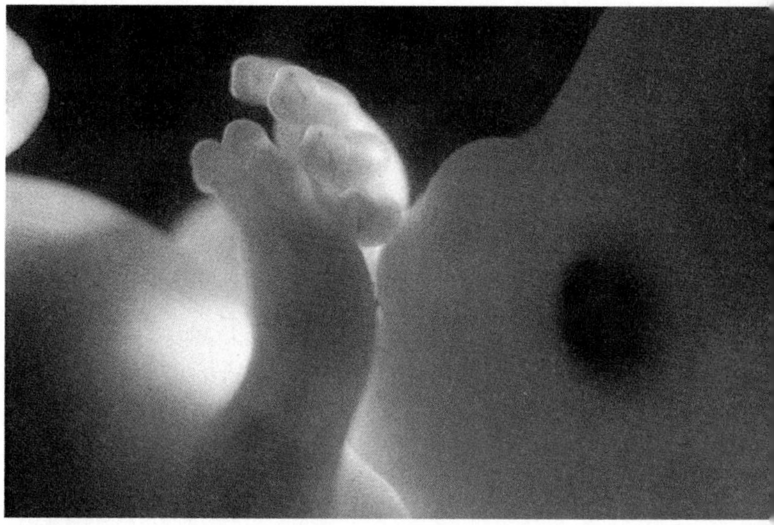

Figure 16.9

Eight-week-old embryo. The embryonic period is over, and from now on a more human appearance takes shape.

> At the end of the embryonic period all organ systems are established and there is a mature and functioning placenta. The embryo is only about 38 mm (1 1/2 in) long.

Health Focus

Preventing Birth Defects

It is believed that at least one in 16 newborns has a birth defect, either minor or serious, and the actual percentage may be even higher. Most likely, only 20% of all birth defects are due to heredity. Those that are hereditary can sometimes be detected before birth. **Amniocentesis** allows the fetus to be tested for abnormalities of development; chorionic villi sampling allows the embryo to be tested; and a new method has been developed for screening eggs to be used for in vitro fertilization (fig. 16A).

Treatment of the fetus in the womb is a rapidly developing area of medical expertise. Biochemical defects can sometimes be treated by giving the mother appropriate medicines. For example, if a baby is unable to use biotin efficiently, the mother can take these substances in doses large enough to prevent any untoward effects in the child. Structural defects can sometimes be corrected by intrauterine surgery. For example, if the fetus has water on the brain or is unable to pass urine, tubes that temporarily allow the fluid to pass out into the amniotic fluid can be inserted while the fetus is still in the womb. Physicians are hopeful that eventually all sorts of structural defects can be corrected by lifting the fetus from the womb long enough for corrective surgery to be performed.

It is recommended that all females take everyday precautions to protect any future and/or presently developing embryos and fetuses from defects that are not due to heredity. X-ray diagnostic therapy should be avoided during pregnancy because X rays cause mutations in the developing embryo or fetus. Children born to women who received X-ray treatment are apt to have birth defects and/or to develop leukemia later on. Toxic chemicals, such as pesticides and many organic industrial chemicals, are also mutagenic. Cigarette smoke not only contains carbon monoxide but also some of these very same fetotoxic chemicals. Babies born to smokers are often underweight and subject to convulsions.

Pregnant Rh^- women should receive an Rh immunoglobulin injection to prevent the production of Rh antibodies. These antibodies can cause nervous system and heart defects.

Sometimes, birth defects are caused by microbes. Females can be immunized before the childbearing years for rubella (German measles), which in particular causes such birth defects as deafness. Unfortunately, immunization for sexually transmitted diseases is not possible. The AIDS virus can cross the placenta, and over 1,500 babies who con-

Figure 16A

Three methods for genetic defect testing before birth. **a.** Amniocentesis is usually performed from the fifteenth to the seventeenth week of pregnancy. A long needle is passed through the abdominal wall to withdraw a small amount of amniotic fluid, along with fetal cells. Since there are only a few cells in the amniotic fluid, testing may be delayed as long as 4 weeks until cell culture produces enough cells for testing purposes. About 40 tests are available for different defects.

b. Chorionic villi sampling is usually performed from the eighth to the twelfth week of pregnancy. The doctor inserts a long, thin tube through the vagina into the uterus. With the help of ultrasound, which gives a picture of the uterine contents, the tube is placed between the lining of the uterus and the chorion. Then a sampling of the chorionic villi cells is suctioned. Chromosome analysis and biochemical tests for several different genetic defects can be done immediately on these cells.

c. Screening eggs for genetic defects is a new technique. Preovulatory eggs are removed by aspiration after a telescope with a fiberoptic illuminator, called a laparoscope, is inserted into the abdominal cavity through a small incision in the region of the navel. The prior administration of FSH ensures that several eggs are available for retrieval and screening. Only the chromosomes within the first polar body are tested because if the woman is heterozygous for a genetic defect and it is found in the polar body, then the egg must be normal. Normal eggs undergo in vitro fertilization and are placed in the prepared uterus. At present, only one in 10 attempts results in a birth, but it is known ahead of time that the child will be normal for the genetic traits tested.

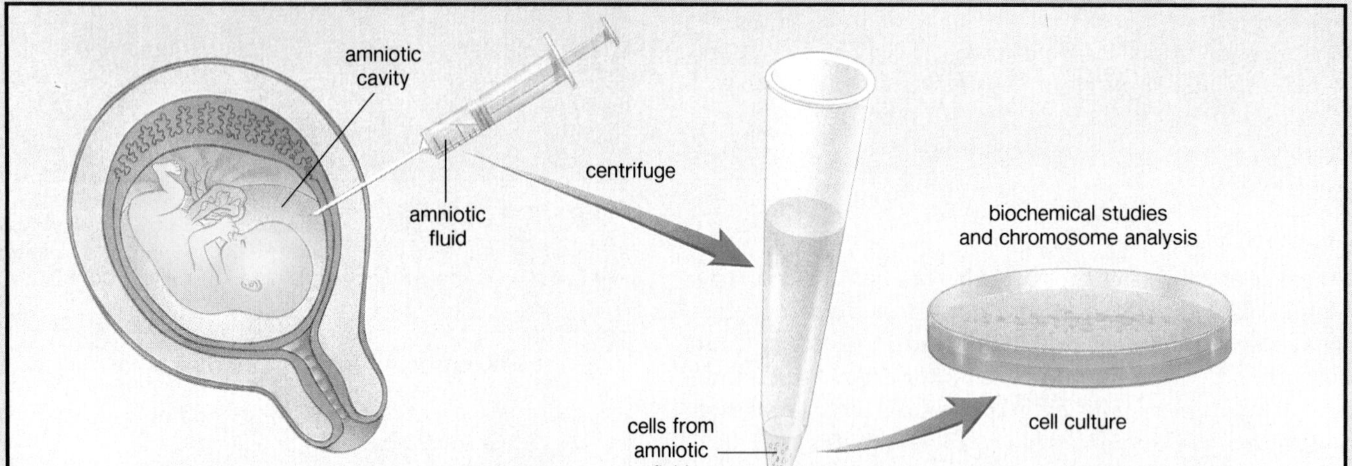

a. Amniocentesis

tracted AIDS while in their mother's womb are now mentally retarded. When a mother has herpes, gonorrhea, or chlamydia, newborns can become infected as they pass through the birth canal. Blindness and other physical and mental defects may develop. Birth by cesarean section could prevent these occurrences.

Pregnant women should not take any type of drug. Certainly illegal drugs, such as marijuana, cocaine, and heroin, should be completely avoided. "Cocaine babies" now make up 60% of drug-affected babies. Severe fluctuations in blood pressure accompany the use of cocaine; these temporarily deprive the developing brain of oxygen. Cocaine babies have visual problems, lack coordination, and are mentally retarded. The drugs aspirin, caffeine (present in coffee, tea, and cola), and alcohol should be severely limited. It is not unusual for babies of drug addicts and alcoholics to display withdrawal symptoms and to have various abnormalities. Babies born to women who have about 45 drinks a month and as many as 5 drinks on one occasion are apt to have fetal alcohol syndrome (FAS). These babies have decreased weight, height, and head size, with malformation of the head and face. Mental retardation is common in FAS infants.

Medications can also cause problems. When the synthetic hormone DES was given to pregnant women to prevent miscarriage, their daughters showed various abnormalities of the reproductive organs and an increased tendency toward cervical cancer. Other sex hormones, including birth-control pills, can possibly cause abnormal fetal development, including abnormalities of the sex organs. The tranquilizer thalidomide is well known for having caused deformities of the arms and legs in children born to women who took the drug. Therefore, a woman has to be very careful about taking medications while pregnant.

Now that physicians and laypeople are aware of the various ways in which birth defects can be prevented, it is hoped that the incidence of birth defects will decrease in the future.

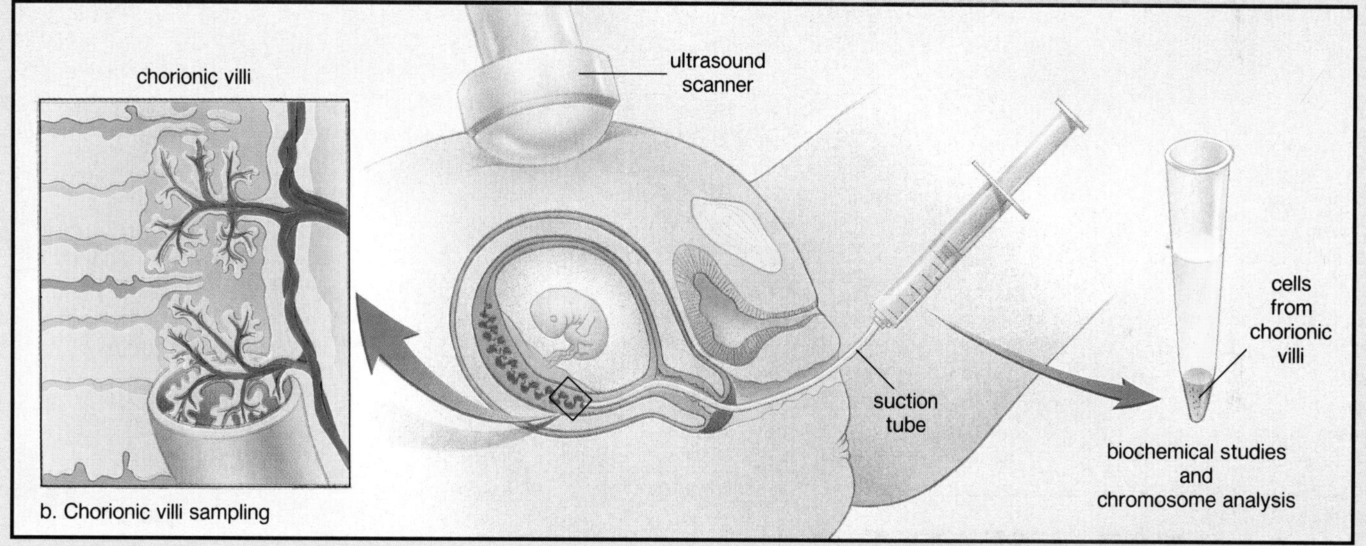

b. Chorionic villi sampling

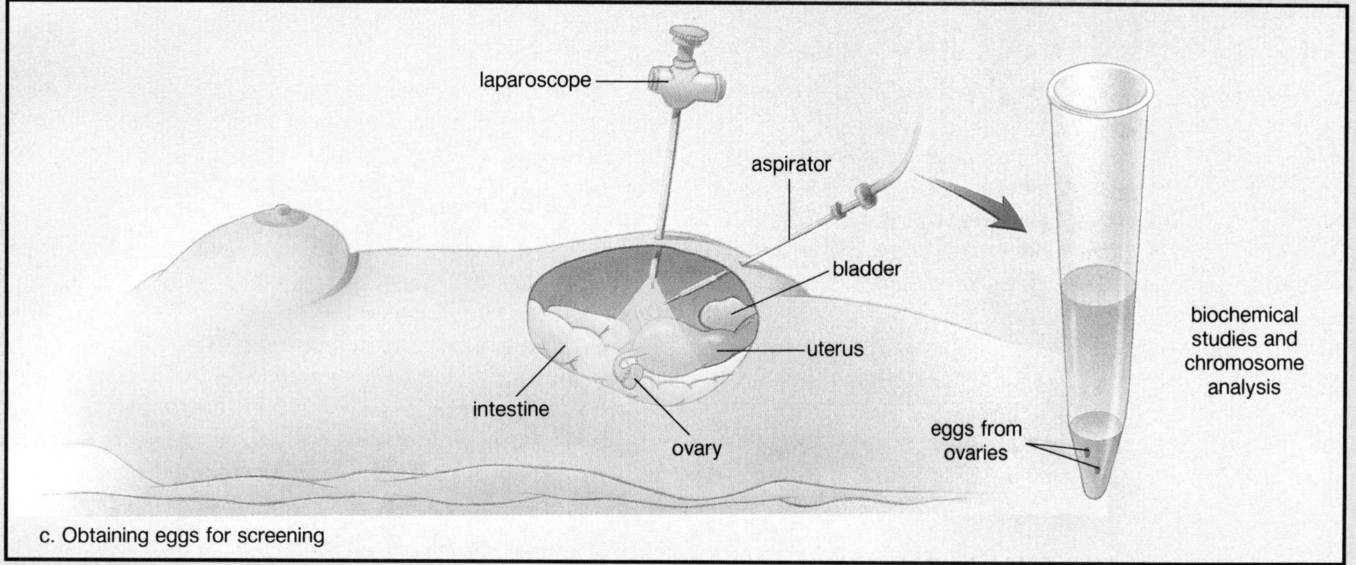

c. Obtaining eggs for screening

Table 16.1

Human Development

Time	Events for Mother	Events for Baby
Embryonic Development		
First week	Ovulation occurs.	Fertilization occurs. Cell division begins and continues. Chorion appears.
Second week	Symptoms of early pregnancy (nausea, breast swelling and tenderness, fatigue) are present.	Implantation. Amnion and yolk sac appear. Embryo has tissues. Placenta begins to form.
Third week	First missed menstruation. Blood pregnancy test is positive.	Nervous system begins. Allantois and blood vessels are present. Placenta is well formed.
Fourth week	Urine pregnancy test is positive.	Limb buds form. Heart is noticeable and beating. Nervous system is prominent. Embryo has tail. Other systems form.
Fifth week	Uterus is the size of hen's egg. Frequent need to urinate due to pressure of growing uterus on bladder.	Embryo is curved. Head is large. Limb buds show divisions. Nose, eyes, and ears are noticeable.
Sixth week	Uterus is the size of an orange.	Fingers and toes are present. Cartilaginous skeleton.
Two months	Uterus can be felt above the pubic bone.	All systems are developing. Bone is replacing cartilage. Refinement of facial features. 38 mm (1 1/2 in).
Fetal Development		
Third month	Uterus is the size of a grapefruit.	Possible to distinguish sex. Fingernails appear.
Fourth month	Fetal movement is felt by those who have been pregnant before.	Skeleton visible. Hair begins to appear. 150 mm (6 in), 170 g (6 oz).
Fifth month	Fetal movement is felt by those who have not been pregnant before. Uterus reaches up to level of umbilicus and pregnancy is obvious.	Protective cheesy coating begins to be deposited. Heartbeat can be heard.
Sixth month	Doctor can tell where baby's head, back, and limbs are. Breasts have enlarged, nipples and areolae are darkly pigmented, and colostrum is produced.	Body is covered with fine hair. Skin is wrinkled and red.
Seventh month	Uterus reaches halfway between umbilicus and rib cage.	Testes descend into scrotum. Eyes are open. 300 mm (12 in), 1,350 g (3 lb).
Eighth month	Weight gain is averaging about a pound a week. Difficulty in standing and walking because center of gravity is thrown forward.	Body hair begins to disappear. Subcutaneous fat begins to be deposited.
Ninth month	Uterus is up to rib cage, causing shortness of breath and heartburn. Sleeping becomes difficult.	Ready for birth. 530 mm (20 1/2 in), 3,400 g (7 1/2 lb).

Fetal Development: Finishing Up

Fetal development extends from the third to the ninth month as shown in table 16.1.

Third and Fourth Months

At the beginning of the third month (fig. 16.10), head growth begins to slow down as the rest of the body increases in length. Epidermal refinements, such as eyelashes, eyebrows, hair on the head, fingernails, and nipples, appear.

Cartilage is replaced by *bone* as ossification centers appear in most of the bones. Cartilage remains at the ends of the long bones, and ossification is not complete until the age of 18 or 20 years. The skull has 6 large *fontanels* (membranous areas), which permit a certain amount of flexibility as the head passes through the birth canal and allow rapid growth of the brain during infancy. The fontanels disappear by 2 years of age.

Sometime during the third month, it is possible to distinguish males from females. Once the testes differentiate, they produce androgens, the male sex hormones. The androgens, especially testosterone, stimulate the growth of the male external genitals. In the absence of androgens, female genitals form. The ovaries do not produce estrogen because there is plenty of it circulating in the mother's bloodstream.

At this time, the testes or ovaries are located within the abdominal cavity. Later, in the last trimester of male fetal development, the testes descend into the scrotal sacs of the scrotum. Sometimes the testes fail to descend, in which case an operation can be performed to place them in their proper location.

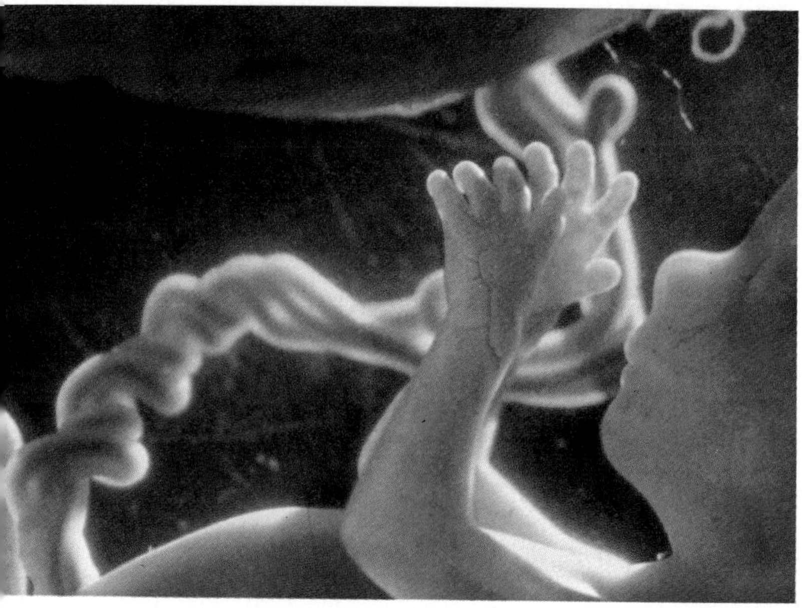

Figure 16.10

The 3- to 4-month-old fetus looks human. Face, hands, and fingers are well defined.

During the fourth month, the fetal heartbeat is loud enough to be heard when a physician applies a stethoscope to the mother's abdomen. By the end of this month, the fetus is less than 150 mm (6 in) in length and weighs a little more than 170 grams (6 oz).

> During the third and fourth months, the skeleton is becoming ossified. The sex of the individual is now distinguishable.

Fifth through Seventh Months

During the fifth through seventh months (fig. 16.11), the mother begins to feel movement. At first, there is only a fluttering sensation, but as the fetal legs grow and develop, kicks and jabs are felt. The fetus is in the fetal position with the head bent down and in contact with the flexed knees.

The wrinkled, translucent, pink-colored skin is covered by a fine down called **lanugo.** The lanugo is coated with a white, greasy, cheeselike substance called **vernix caseosa,** which probably protects the delicate skin from the amniotic fluid. During these months the eyelids open fully.

At the end of this period, the fetus is almost 300 mm (12 in), and the weight has increased to almost 1,350 grams (3 lb). It is possible that if born now, the baby will survive.

Eighth and Ninth Months

As the end of development approaches, the fetus usually rotates so that the head is pointed toward the cervix. However, if the fetus does not turn, then the likelihood of a breech birth (rump first) may require a cesarean section. It is very difficult for the cervix to expand enough to accommodate this form of birth, and asphyxiation of the baby is more likely to occur.

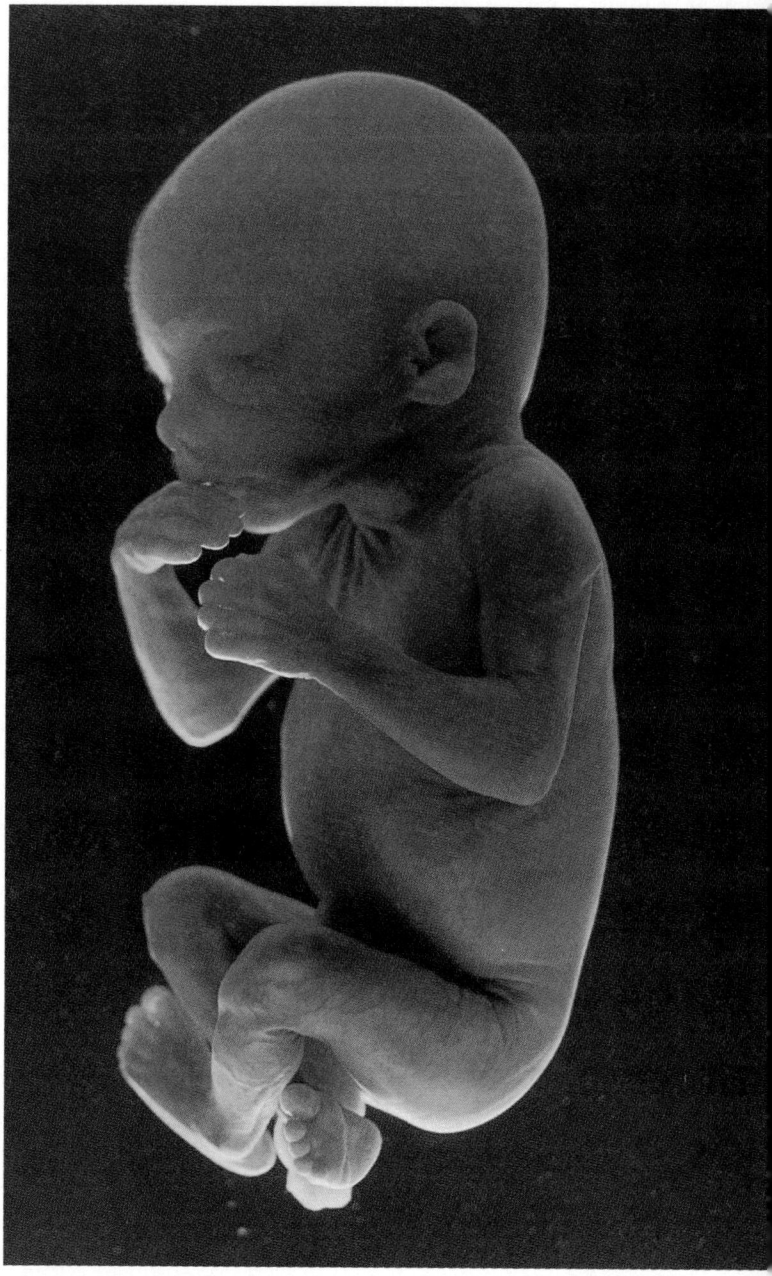

Figure 16.11

A 5- to 7-month-old fetus.

At the end of 9 months, the fetus is about 530 mm (21 in) long and weighs about 3,400 grams (7 1/2 lb) (fig. 16.12). Weight gain is due largely to an accumulation of fat beneath the skin. Full-term babies have a better chance of survival.

> From the fifth to the ninth month, the fetus continues to grow and to gain weight. Babies born after 6 or 7 months may survive, but full-term babies have a better chance of survival.

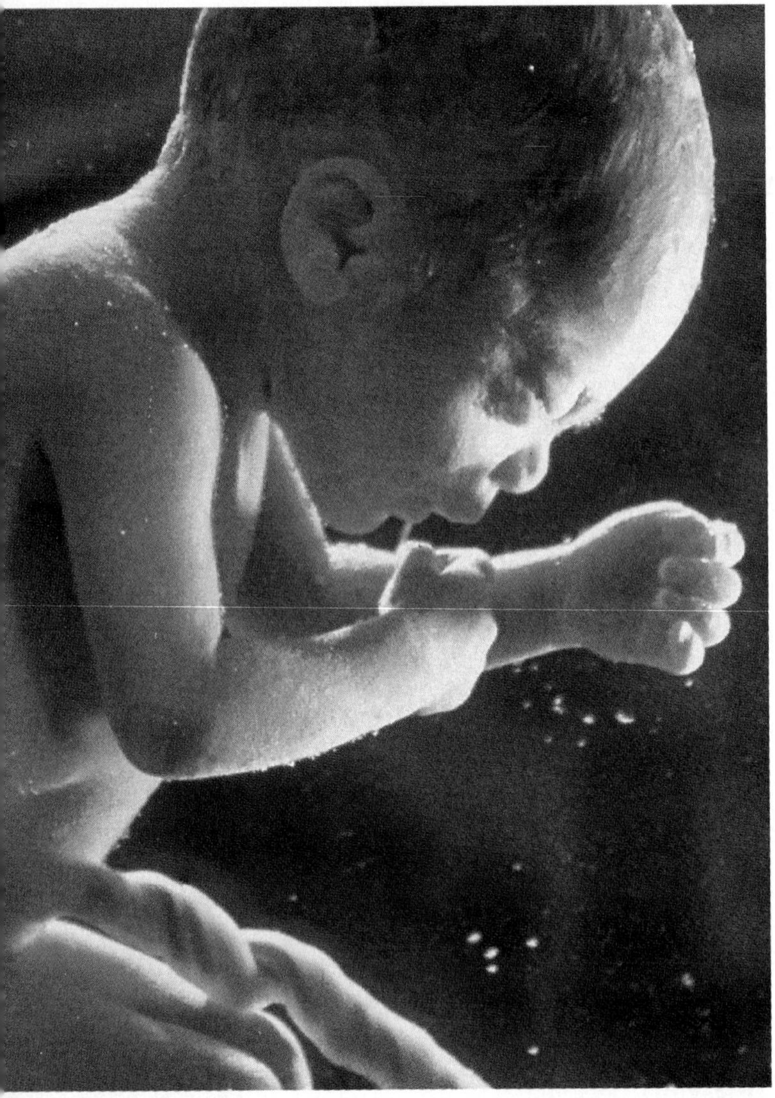

Figure 16.12

An 8- to 9-month-old fetus.

Effects of Pregnancy on the Mother

Table 16.1 outlines the major changes in the mother during pregnancy. When first pregnant, the mother may experience nausea and vomiting, loss of appetite, and fatigue. Other changes are swelling and tenderness of the breasts, increased urination, and irregular bowel movements. Some women, however, report increased energy levels and a general sense of well-being during this time.

The uterus enlarges greatly during pregnancy. In the nonpregnant female, the uterus is only about 50 mm (2 in) by 75 mm (3 in). Just before giving birth, the uterus almost fills the abdominal cavity and reaches the rib cage. Breasts increase as much as 25% in size. By the end of the sixth month, the nipples and areolae are darkly pigmented, and *colostrum* is produced.

Table 16.2

Approximate Weight Gain During Pregnancy

Increased blood volume	4.0 lb (1,814 g)
Increased fluid retention	2.7 lb (1,225 g)
Enlarged uterus	2.0 lb (907 g)
Enlarged breasts	.9 lb (408 g)
Maternal storage fat	3.5 lb (1,588 g)
Amniotic fluid	1.8 lb (816 g)
Placenta	1.4 lb (635 g)
Full-term infant	7.7 lb (3,490 g)
Total approximate gain	**24.0 lb (11 kg)**

Toward the end of pregnancy, the enlarged size of the baby and extra weight cause various difficulties. (Table 16.2 gives the approximate weight gain to be expected during pregnancy.) The mother may have trouble breathing and may have an increased need to urinate. The center of gravity is thrown forward, therefore standing and walking are difficult. Sleeping may be disturbed not only due to the kicking of the baby but also due to an inability to find a comfortable position.

Birth

The uterus characteristically contracts throughout pregnancy. At first, light, often indiscernible contractions lasting about 20–30 seconds occur every 15–20 minutes, but near the end of pregnancy, they become stronger and more frequent so that the woman may think that she is in labor. However, the onset of true labor is marked by uterine contractions that occur regularly every 15–20 minutes and last for 40 seconds or more. **Parturition,** which includes labor and expulsion of the fetus, usually is considered to have 3 stages.

The events that cause parturition still are not known entirely, but there is now evidence suggesting the involvement of prostaglandins. It may be, too, that the prostaglandins cause the release of oxytocin from the maternal posterior pituitary. Both prostaglandins and oxytocin cause the uterus to contract, and either hormone can be given to induce parturition.

Stages of Birth: One, Two, Three

During the *first stage* of parturition, the cervix dilates; during the *second stage,* the baby is born; and during the *third stage,* the afterbirth is expelled.

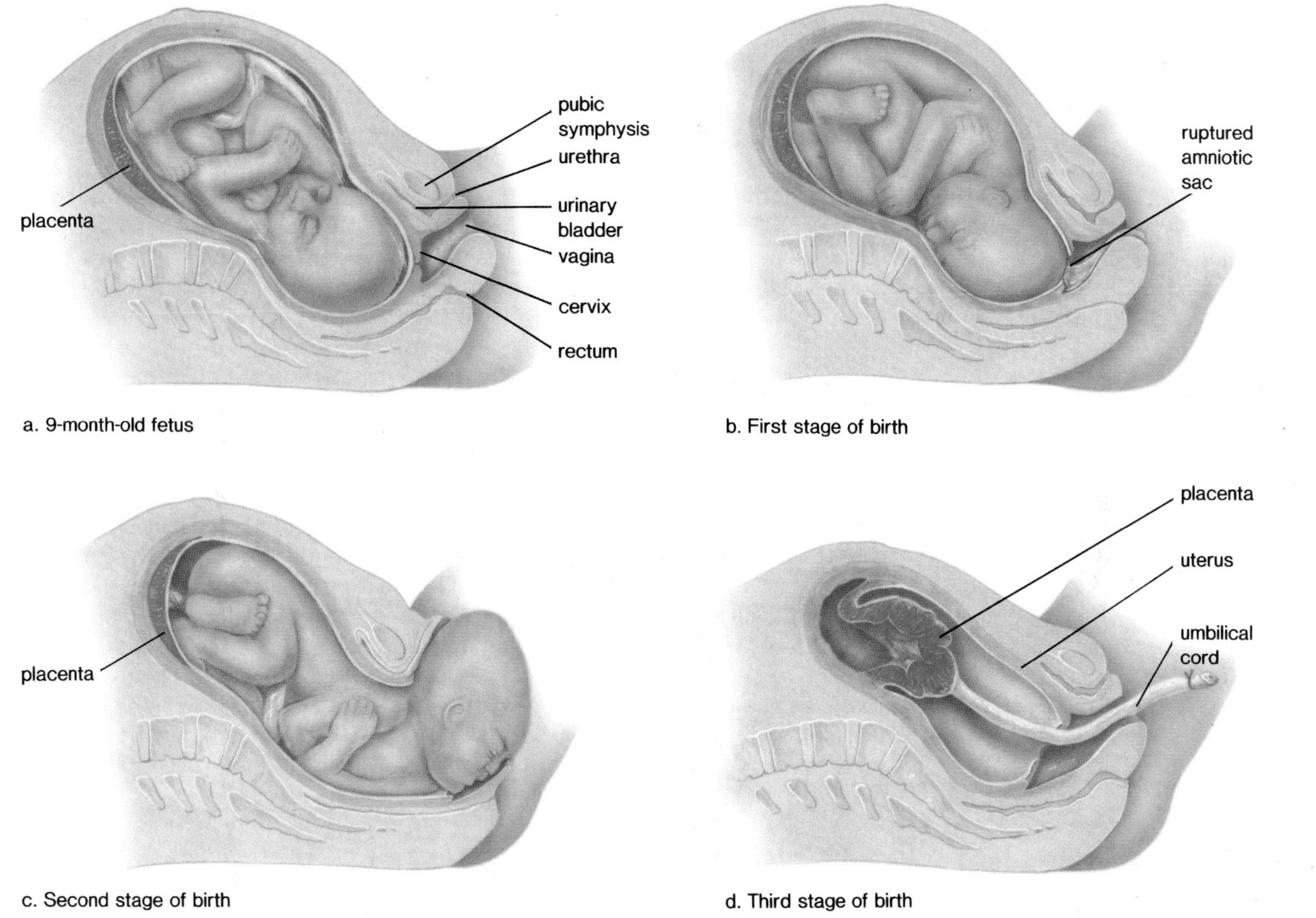

Figure 16.13

Three stages of parturition. **a.** Position of fetus just before birth begins. **b.** Dilation of cervix. **c.** Birth of baby. **d.** Expulsion of afterbirth.

Stage 1

Prior to or concomitant with the first stage of parturition, there can be a "bloody show" caused by the expulsion of a mucus plug from the cervical canal. This plug prevents bacteria and sperm from entering the uterus during pregnancy.

Uterine contractions during the first stage of labor occur in such a way that the cervical canal slowly disappears as the lower part of the uterus is pulled upward toward the baby's head (fig. 16.13*b*). This process is called *effacement*, or "taking up the cervix." With further contractions, the baby's head acts as a wedge to assist cervical dilation. The baby's head usually has a diameter of about 100 mm (4 in); therefore, the cervix has to dilate to this diameter to allow the head to pass through. If it has not occurred already, the amniotic membrane is apt to rupture during this stage, releasing the amniotic fluid, which leaks out the vagina. The first stage of labor ends once the cervix is dilated completely.

Stage 2

During the second stage of parturition, the uterine contractions occur every 1–2 minutes and last about one minute each. They are accompanied by a desire to push, or bear down. As the baby's head gradually descends into the vagina, the desire to push becomes greater. When the baby's head reaches the exterior, it turns so that the back of the head is uppermost (fig. 16.13*c*). Since the vaginal orifice may not expand enough to allow passage of the head without tearing, an **episiotomy** often is performed. This incision, which enlarges the opening, is sewn together later and heals more perfectly than a tear. As soon as the head is delivered, the baby's shoulders rotate so that the baby faces either to the right or the left. At this time, the physician may hold the head and guide it downward, while one shoulder and then the other emerges. The rest of the baby follows easily.

Once the baby is breathing normally, the umbilical cord is cut and tied, severing the child from the placenta. The stump of the cord shrivels and leaves a scar, which is the **umbilicus.**

Stage 3

The placenta, or *afterbirth,* is delivered during the third stage of labor (fig. 16.13*d*). About 15 minutes after delivery of the baby, uterine muscular contractions shrink the uterus and dislodge the placenta. The placenta then is expelled into the vagina. As soon as the placenta and its membranes are delivered, the third stage of labor is complete.

> During the first stage of birth, the cervix dilates; during the second stage, the child is born; and during the third stage, the afterbirth is expelled.

Female Breast and Lactation

A female breast contains 15 to 25 lobules, each with its own milk duct, which begins at the nipple and divides into numerous other ducts that end in blind sacs called *alveoli* (fig. 16.14).

During pregnancy, the breasts enlarge as the ducts and alveoli increase in number and size. The same hormones that affect the mother's breast can also affect the child's. Some newborns, including males, even secrete a small amount of milk for a few days.

Usually, there is no production of milk during pregnancy. The hormone *prolactin* is needed for lactation to begin, and the production of this hormone is suppressed because of the feedback control that the increased amount of estrogen and progesterone during pregnancy has on the pituitary. Once the baby is delivered, however, the pituitary begins secreting prolactin. It takes a couple of days for milk production to begin, and, in the meantime, the breasts produce colostrum.

The continued production of milk requires a suckling child. When a breast is suckled, the nerve endings in the areola are stimulated, and a nerve impulse travels along neural pathways from the nipples to the hypothalamus, which directs the pituitary gland to release the hormone *oxytocin.* When this hormone arrives at the breast, it causes contraction of the lobules so that milk flows into the ducts (called milk letdown), where it may be drawn out of the nipple by the suckling child. The more suckling, the more oxytocin released, and the more milk there is for the child (fig. 16.15). Some women choose to breast-feed and some do not. The Health Focus, "Deciding Between Feeding by Breast or by Bottle," discusses the benefits of both feeding methods.

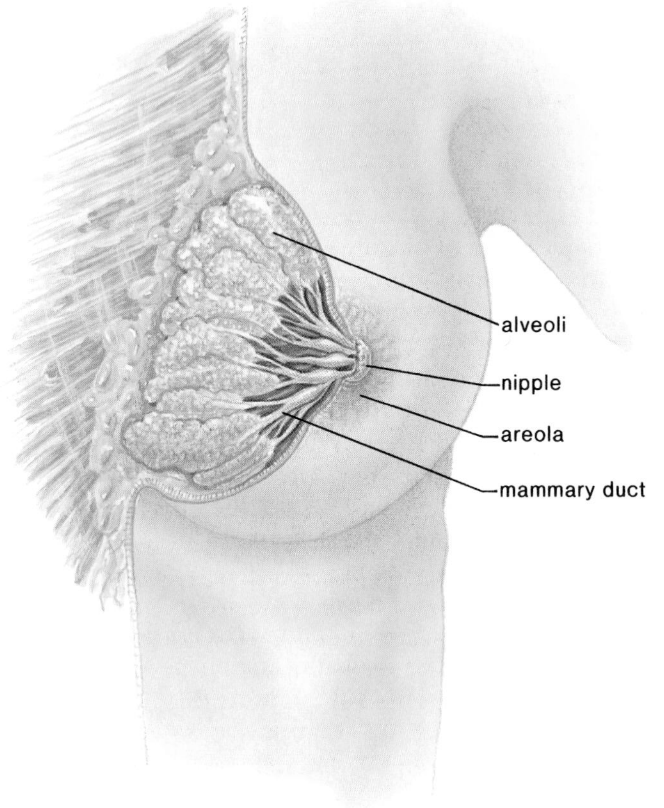
alveoli
nipple
areola
mammary duct

Figure 16.14

Female breast anatomy. The female breast contains lobules consisting of ducts and alveoli. The alveoli are lined by milk-producing cells in the lactating (milk-producing) breast.

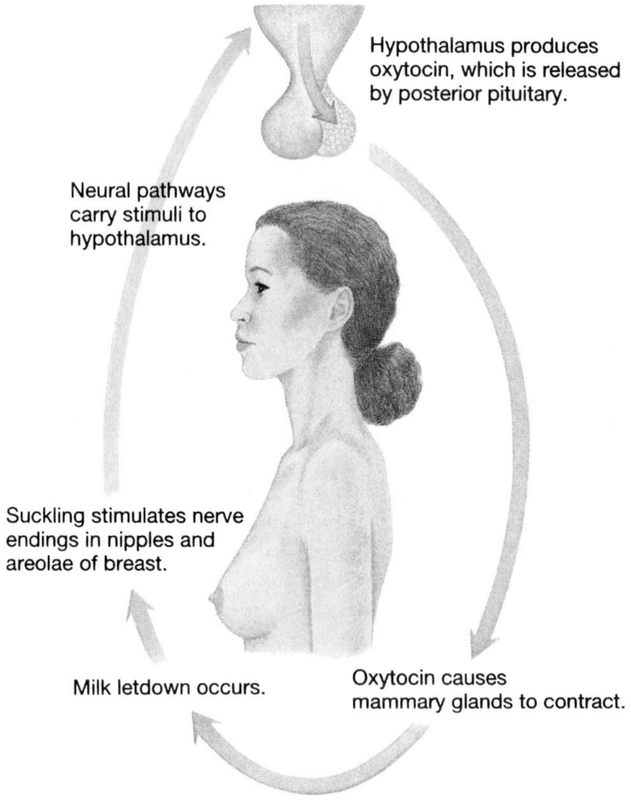
Hypothalamus produces oxytocin, which is released by posterior pituitary.
Neural pathways carry stimuli to hypothalamus.
Suckling stimulates nerve endings in nipples and areolae of breast.
Milk letdown occurs.
Oxytocin causes mammary glands to contract.

Figure 16.15

Suckling reflex. Suckling sets in motion the sequence of events that leads to milk letdown, the flow of milk into ducts of the breast.

Health Focus

Deciding Between Feeding by Breast or by Bottle

Breast-feeding has come back in vogue in a big way. Not 20 years ago, only one mother in 5 breast-fed her baby. Today, an estimated 2 out of every 3 newborns are being given breast milk. The shift has many advantages, not just for the infant but for the mother as well. One is that when the baby is ready to eat, the milk is ready for the drinking. There are no bottles to sterilize or formulas to measure. Breast milk is also less expensive than formula, even after you account for the cost of the extra food a mother needs to ensure that she will produce enough milk and that it will contain enough calories. In addition, hormones released during breast-feeding cause contractions in the uterus that assist it in shrinking closer to its prepregnancy size. And the calories a mother's body spends producing breast milk may help her to lose the 7 to 8 pounds—or more—of fat she puts on during pregnancy but does not "deliver" along with the baby, placenta, and amniotic and other fluids.

As for the baby, "human milk is unquestionably the best source of nutrition...during the first months of life," says the American Academy of Pediatrics. That's especially true if conditions in the home are unsanitary (breast milk, unlike formula, does not need to be kept "clean," because it goes directly from mother to child), or money is too scarce to ensure that formula will always be affordable, or the educational level of the parents is too low (or the emotional level of the household too "chaotic") for the family to read, fully understand, and consistently apply the rules of proper formula preparation and storage. But even full-term infants born to reasonably well-educated, financially secure parents who live in clean environments may benefit.

It isn't that today's formulas are not reliable. Healthy babies can grow well on breast or bottle. But breast milk is an amazingly sophisticated substance (it contains more zinc in the first few weeks than later on to meet the newborn's higher zinc needs). And although scientists have been able to imitate it safely enough in store-bought formula, "there may still be some subtle, not-yet-discovered ways in which breast milk is preferable to bottle milk," says Ronald Kleinman, M.D., head of the Committee on Nutrition at the American Academy of Pediatrics. "There remain differences between the 2," he adds, "whose significance isn't understood."

These differences may be behind such findings as one reported in the *British Medical Journal* that breast-fed infants appear better protected against wheezing during the first few months of life. The journal has also published a report suggesting that breast-fed newborns are less prone to develop stomach and intestinal illness during the first 13 weeks of life and, thus, suffer less than bottle-fed babies from vomiting and diarrhea.

Can the Bottle Ever Be a Better Choice?

Despite the many advantages of breast-feeding, it may not always be the appropriate way for a mother to nourish her child. Some women simply do not want to breast-feed because they are taking prescription drugs that may get into the milk and harm the baby, or cannot breast-feed because they will return to work soon after giving birth and would find "pumping" breast milk to be given to the infant while they are away from home too tiring. Others may not wish to breast-feed because they are uncomfortable with the sexuality of the process or with some other emotionally related aspect. All these reasons are considered valid. In other words, mothers who do not want to breast-feed should not be pressured or made to feel guilty about it. As family therapist and dietitian Ellyn Satter says in her book *Child of Mine: Feeding with Love and Good Sense* (Bull Publishing: Palo Alto, California), "You will have plenty of opportunities to feel guilty as a parent without feeling guilty about *that*, too." Besides, if a mother breast-feeds but hates doing it, her baby is going to sense that, and it will do greater harm than the breast milk will do good.

Indeed, more important than whether a mother breast-feeds, especially in countries like the United States, where the standard of living for the average family is such that breast-feeding is not a life-or-death matter as it is in some developing nations, is that she develops a relaxed, loving relationship with her child. "Even the sophisticated components of breast milk can't make up for that," Ms. Satter says. The closeness, warmth, and stimulation provided by an infant's caretakers are as important to his normal growth and development as the source of his food.

Source: Tufts University Diet and Nutrition Letter *(ISSN 0747–4105) is published monthly by* Tufts University Diet and Nutrition Letter, *53 Park Place, New York, NY 10007. This article extracted from a Special Report in the December 1990 issue. Reprinted by permission.*

Human Development After Birth

Development does not cease once birth has occurred but continues throughout the stages of life: infancy, childhood, adolescence, and adulthood.

Infancy lasts until about 2 years of age. It is characterized by tremendous growth and sensorimotor development. During *childhood,* the individual grows and the body proportions change. *Adolescence* begins with **puberty,** when the secondary sex characteristics appear and the sexual organs become functional. At this time, there is an acceleration of growth leading to changes in height, weight, fat distribution, and body proportions. Males commonly experience a growth spurt later than females; therefore, they grow for a longer period of time. Males are generally taller than females and have broader shoulders and longer legs relative to their trunk length.

Young adults are at their physical peak in muscle strength, reaction time, and sensory perception. The organ systems at this time are best able to respond to altered circumstances in a homeostatic manner. From now on, however, there is an almost imperceptible, gradual loss in certain of the body's abilities. **Aging** encompasses these progressive changes that contribute to an increased risk of infirmity, disease, and death (fig. 16.16).

Today, there is great interest in **gerontology,** the study of aging, because there are now more older individuals in our society than ever before and the number is expected to rise dramatically. In the next half-century, those over age 75 will rise from the present 13 million to 34–45 million, and those over age 80 will rise from 3 million to 6 million individuals. The human life span is judged to be a maximum of 110–115 years. The present goal of gerontology is not to necessarily increase the life span but to increase the health span, the number of years that an individual enjoys the full functions of all body parts and processes.

Figure 16.16

Aging is a slow process during which the body undergoes changes that eventually bring about death, even if no marked disease or disorder is present. Medical science is trying to extend the human life span and the health span, the length of time the body functions normally.

> Developmental changes keep occurring throughout infancy, childhood, adolescence, and adulthood. Aging is not noticeable during childhood.

Why We Age

There are many theories about what causes aging. Three of these are considered here.

Genetic in Origin

Several lines of evidence indicate that aging has a genetic basis. (1) The number of times a cell divides is species-specific. The maximum number of times human cells divide is around 50. Perhaps as we grow older, more and more cells are unable to divide any longer, and instead they undergo degenerative changes and die. (2) Some cell lines may become nonfunctional long before the maximum number of divisions has occurred. Whenever DNA replicates, mutations can occur, and this can lead to the production of nonfunctional proteins. Eventually, the number of inadequately functioning cells can build up, which contributes to the aging process. (3) The children of long-lived parents tend to live longer than those of short-lived parents. Recent work in lower animals suggests that when an animal produces fewer free radicals, it lives longer. *Free radicals* are molecules that are reduced, because they have taken electrons away from the macromolecules that make up a cell. In this way, production of free radicals leads to the destruction of the cell. There are genes that code for antioxidant enzymes that detoxify free radicals. This research suggests that animals with particular forms of these genes—and therefore more efficient antioxidant enzymes—live longer.

Whole-Body Process

A decline in the hormonal system can affect many different organs of the body. For example, type II diabetes is common in older individuals. The pancreas makes insulin, but the cells lack the receptors that enable them to respond. Menopause in women occurs for a similar reason. There is plenty of FSH in the bloodstream, but the ovaries do not respond. Perhaps aging results from the loss of hormonal activities and a decline in the functions they control.

The immune system, too, no longer performs as it once did, and this can affect the body as a whole. The thymus gland gradually decreases in size, and eventually most of it is replaced by fat and connective tissue. The incidence of cancer increases among the elderly, which may signify that the immune system is no longer functioning as it should. This idea is substantiated, too, by the increased incidence of autoimmune diseases in older individuals.

It is possible, though, that aging is not due to the failure of a particular system that can affect the body as a whole but to a specific type of tissue change that affects all organs and even the genes. It has been noticed for some time that proteins—such as collagen, which makes up the white fibers (p. 49) and is present in many support tissues—become increasingly cross-linked as people age. Undoubtedly, this cross-linking contributes to the stiffening and the loss of elasticity characteristic of aging tendons and ligaments. It may also account for the inability of such organs as the blood vessels, the heart, and the lungs to function as they once did. Some researchers have now found that glucose has the tendency to attach to any type of protein, which is the first step in a cross-linking process that ends with the formation of advanced glycosylation end products (AGEs).

AGEs not only explain why cataracts develop, they also may contribute to the development of atherosclerosis and to the inefficiency of the kidneys in diabetics and older individuals. Even DNA-associated proteins seem capable of forming glucose-derived cross-links, and perhaps this increases the rate of mutations as we age. These researchers are presently experimenting with the drug aminoguanidine, which can prevent the development of AGEs.

Extrinsic Factors

The current data about the effects of aging are often based on comparisons of the characteristics of the elderly to younger age groups, but perhaps today's elderly were not as aware when they were younger of the importance of, for example, diet and exercise to general health. It is possible, then, that much of what we attribute to aging is instead due to years of poor health habits.

Consider, for example, osteoporosis. This condition is associated with a progressive decline in bone density in both males and females so that fractures are more likely to occur after only minimal trauma. Osteoporosis is common in the elderly—by age 65, one-third of women will have vertebral fractures, and by age 81, one-third of women and one-sixth of men will have suffered a hip fracture. While there is no denying that a decline in bone mass occurs as a result of aging, certain extrinsic factors are also important. The occurrence of osteoporosis itself is associated with cigarette smoking, heavy alcohol intake, and perhaps inadequate calcium intake. Not only is it possible to eliminate these negative factors by personal choice, it is also possible to add a positive factor. A moderate exercise program has been found to slow down the progressive loss of bone mass.

Even more important, an exercise program helps eliminate cardiovascular disease, the leading cause of death today. Experts no longer believe that the cardiovascular system necessarily suffers a large decrease in functioning ability with age. Persons 65 years of age and older can have well-functioning hearts and open coronary arteries if their health habits are good and they continue to exercise regularly.

Rather than collecting data on the average changes observed between different age groups, it might be more useful to note the differences within any particular age group. If this type of comparison is done, extrinsic factors that contribute to a decline as well as those that promote the health of an organ can be identified.

How Aging Affects Body Systems

Keep in mind that data about how aging affects body systems should be accepted with reservations, however, we will still discuss this topic in general.

Skin

As aging occurs, skin becomes thinner and less elastic because the number of elastic fibers decreases and the collagen fibers undergo cross-linking, as discussed previously. Also, there is less adipose tissue in the subcutaneous layer; therefore, older people are more likely to feel cold. The loss of thickness accounts for skin sagging and wrinkling.

Homeostatic adjustment to heat is also limited because there are fewer sweat glands for sweating to occur. There are fewer hair follicles, so the hair on the scalp and the extremities thins out. The number of sebaceous glands is reduced, and the skin tends to crack. Older people also experience a decrease in the number of melanocytes, making hair gray and skin pale. In contrast, some of the remaining pigment cells are larger, and pigmented blotches appear in skin.

Processing and Transporting

Cardiovascular disorders are the leading cause of death among the elderly. The heart shrinks because there is a reduction in cardiac muscle cell size. This leads to loss of cardiac muscle strength and reduced cardiac output. Still, it is observed that the heart, in the absence of disease, is able to meet the demands of increased activity. It can increase its rate to double or triple the amount of blood pumped each minute even though the maximum possible output declines.

Because the middle coat of arteries contains elastic fibers, which most likely are subject to cross-linking, the arteries become more rigid with time, and their size is further reduced by plaque (p. 103). Therefore, blood pressure readings gradually rise. Such changes are common in individuals living in Western industrialized countries but not in agricultural societies. As mentioned earlier, diet has been suggested as a way to control degenerative changes in the cardiovascular system (p. 102).

There is reduced blood flow to the liver, and this organ does not metabolize drugs as efficiently as before. This means that as a person gets older, less medication is needed to maintain the same level in the bloodstream.

Circulatory problems often are accompanied by respiratory disorders and vice versa. Growing inelasticity of lung tissue means that ventilation is reduced. Because we rarely use the entire vital capacity, these effects are not noticed unless there is increased demand for oxygen.

There is also reduced blood supply to the kidneys. The kidneys become smaller and less efficient at filtering wastes. Salt and water balance are difficult to maintain, and the elderly dehydrate faster than young people. Difficulties involving urination include incontinence (lack of bladder control) and the inability to urinate. In men, the prostate gland may enlarge and reduce the diameter of the urethra, making urination so difficult that surgery is often needed.

The loss of teeth, which is frequently seen in elderly people, is more apt to be the result of long-term neglect than aging. The digestive tract loses tone, and secretion of saliva and gastric juice is reduced, but there is no indication of reduced absorption. Therefore, an adequate diet, rather than vitamin and mineral supplements, is recommended. There are common complaints of constipation, increased amount of gas, and heartburn, but gastritis, ulcers, and cancer can also occur.

Figure 16.17

The aim of gerontology is to allow the elderly to enjoy living. This requires studying the debilities that can occur with aging and then making recommendations as to how best to forestall or prevent their occurrence.

Integration and Coordination

It is often mentioned that while most tissues of the body regularly replace their cells, some at a faster rate than others, the brain and the muscles do not. No new nerve or skeletal muscle cells are formed in the adult. However, contrary to previous opinion, recent studies show that few neural cells of the cerebral cortex are lost during the normal aging process. This means that cognitive skills remain unchanged even though there is characteristically a loss in short-term memory. Although the elderly learn more slowly than the young, they can acquire and remember new material as well. It is noted that when more time is given for the subject to respond, age differences in learning decrease.

Neurons are extremely sensitive to oxygen deficiency, and if neuron death does occur, it may not be due to aging itself but to reduced blood flow in narrowed blood vessels. Specific disorders, such as depression, Parkinson disease, and Alzheimer disease (p. 198), are sometimes seen, but they are not common. Reaction time, however, does slow, and more stimulation is needed for hearing, taste, and smell receptors to function as before. After age 50, there is a gradual reduction in the ability to hear tones at higher frequencies, and this can make it difficult to identify individual voices and to understand conversation in a group. The lens of the eye does not accommodate as well and also may develop a cataract. Glaucoma is more likely to develop because of a reduction in the size of the anterior cavity of the eye.

Loss of skeletal muscle mass is not uncommon, but it can be controlled by a regular exercise program. There is a reduced capacity to do heavy labor, but routine physical work should be no problem. A decrease in the strength of the respiratory muscles and inflexibility of the rib cage contribute to the inability of the lungs to expand as before, and reduced muscularity of the urinary bladder contributes to difficulties with urination.

As noted before, aging is accompanied by a decline in bone density. Osteoporosis, characterized by a loss of calcium and mineral from bone, is not uncommon, but there is evidence that proper health habits can prevent its occurrence. Arthritis, which causes pain upon movement of the joint, is also seen.

Weight gain occurs because the basal metabolism decreases and inactivity increases. Muscle mass is replaced by stored fat and retained water.

The Reproductive System

Females undergo menopause and thereafter the level of female sex hormones in blood falls markedly. The uterus and the cervix are reduced in size, and there is a thinning of the walls of the oviducts and the vagina. The external genitals become less pronounced. In males, the level of androgens falls gradually over the age span of 50–90, but sperm production continues until death.

It is of interest that as a group, females live longer than males. Although their health habits may be better, it is also possible that the female sex hormone estrogen offers women some protection against circulatory disorders when they are younger. Males suffer a marked increase in heart disease in their forties, but an increase is not noted in females until after menopause. Then women lead men in the incidence of stroke. Men are still more likely than women to have a heart attack, however.

How To Age Well

We have listed many adverse effects due to aging, but it is important to emphasize that while such effects are seen, they are not a necessary occurrence (fig. 16.17). We must discover any extrinsic factors that precipitate these adverse effects and guard against them. Just as it is wise to make the proper preparations to remain financially independent when older, it is also wise to realize that biologically successful old age begins with the health habits developed when we are younger.

Summary

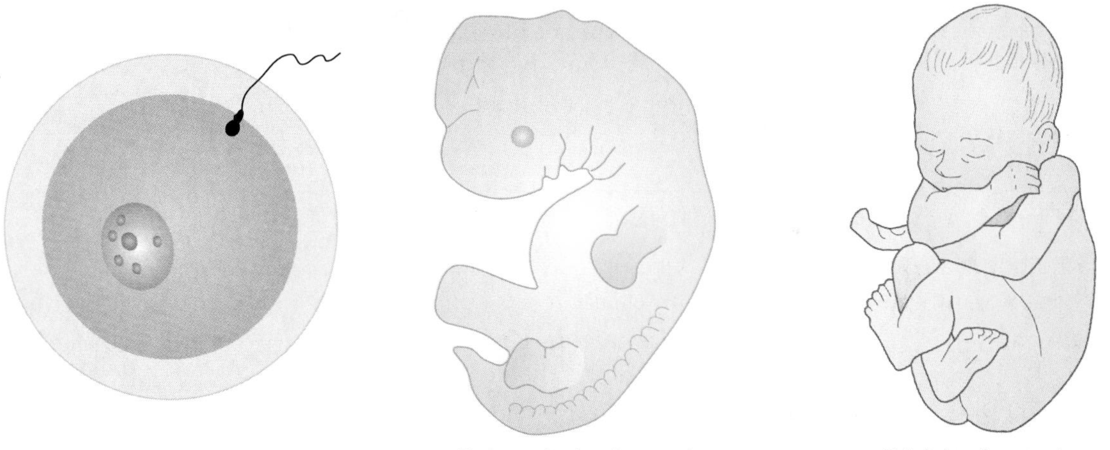

If sexual intercourse and ejaculation occur, sperm may penetrate cervical mucus, particularly at the time of ovulation. During their passage through the female reproductive tract, the sperm undergo capacitation so that acrosome enzymes that allow penetration of the egg are released. Only one sperm head actually enters the egg, and this sperm nucleus fuses with the egg nucleus. The zygote begins to develop into an embryo, which travels down the oviduct to embed itself in the uterine lining. Cells surrounding the embryo produce HCG, and the presence of this hormone indicates that the female is pregnant.

Human development consists of embryonic (first 2 months) and fetal (third to ninth month) development. During the embryonic period, the extraembryonic membranes appear and serve important functions: the embryo acquires organ systems. During the fetal period, there is a refinement of these systems.

Birth, or parturition, has 3 phases. During the first stage, the cervix dilates to allow passage of the baby's head and body. The amnion usually bursts sometime during this stage. During the second stage, the baby is born and the umbilical cord is cut. As the baby takes his or her first breath, anatomical changes convert fetal circulation to adult circulation. During the third stage, the placenta is delivered.

Milk is not produced during pregnancy because of hormonal suppression, but once the child is born milk production begins. Prolactin promotes the production of milk and oxytocin allows milk letdown.

Development after birth consists of infancy, childhood, adolescence, and adulthood. Young adults are at their prime, and then the aging process begins. Aging encompasses progressive changes from about age 20 on that contribute to an increased risk of infirmity, disease, and death. Perhaps aging is genetic in origin, perhaps it is due to a change that affects the whole body, or perhaps it is due to extrinsic factors.

Study and Thought Questions

1. Describe the process of fertilization and the events immediately following it. (pp. 317–318)
2. What is the basis of the pregnancy test? (p. 318)
3. Name the 4 extraembryonic membranes, and give a function for each. (p. 318)
4. Describe the structure and function of the placenta. (pp. 318–321)

   Critical Thinking Explain why neither the lungs nor the kidneys are operative in the fetus.
5. During which period of pregnancy does a woman have to be the most careful about the intake of medications and other drugs? (p. 319)
6. Describe the structure and function of the umbilical cord. (p. 319)

   Critical Thinking The umbilical cord is not attached to the stomach of the fetus. Where is it attached?
7. Specifically, what events normally occur during the first, second, third, and fourth weeks of development? What events normally happen during the second through the ninth months? (pp. 322–327)
8. In general, describe the physical changes in the mother during pregnancy. (p. 328)
9. What are the 3 stages of birth? Describe the events of each stage. (pp. 329–330)
10. Describe the suckling reflex. (p. 330)

    Critical Thinking Does figure 16.15 describe a negative or positive feedback system? Explain.
11. Discuss 3 theories of aging. What are the major changes in body systems that have been observed as adults age? (pp. 332–333)

    Ethical Issue Do you believe in euthanasia? If so, how should eligibility be regulated?

Writing Across the Curriculum

1. Would you expect a woman who produces a normal amount of estrogen but a limited amount of progesterone to menstruate and/or get pregnant? Explain.
2. How can an abdominal pregnancy occur and the fetus come to term and be born?
3. Knowing the adverse effects of various substances on the developing fetus, discuss the desirability of regulations or laws controlling their use by pregnant women.

Objective Questions

1. Fertilization occurs when the ___________ nucleus fuses with the ____________ nucleus.
2. The ____________ membranes include the chorion, the ____________, the yolk sac, and the allantois.
3. The zygote divides as it passes down a uterine tube. This process is called ______________.
4. Once the embryo arrives at the uterus, it begins to ______________ itself in the uterine lining.
5. When cells take on a specific structure and function, ______________ occurs.
6. During embryonic development, all major ______________ form.
7. Fetal development begins at the end of the ______________ month.
8. During development, the nutrient needs of the developing embryo (fetus) are served by the ______________.
9. In most deliveries, the ______________ appears before the rest of the body.
10. The hormone ______________ is required for milk letdown during the suckling reflex.

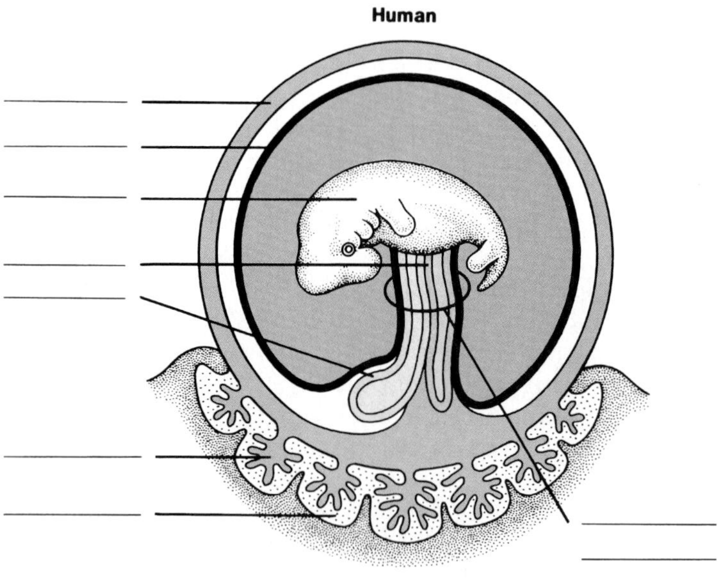

11. Label this diagram.
12. As we age, the proteins in the body undergo ______________, a process that causes body parts to become stiff and rigid.

Selected Key Terms

allantois (ah-lan-to-is) An extraembryonic membrane that serves as a source of blood vessels for the umbilical cord. 318

amniocentesis The removal of a small amount of amniotic fluid to examine the chromosomes and the enzymatic potential of fetal cells. 324

amnion (am´ne-on) One of the extraembryonic membranes; a fluid-filled sac around the embryo. 318

blastocyst An early stage of embryonic development that consists of a hollow ball of cells. 322

chorion (ko´re-on) An extraembryonic membrane that forms an outer covering around the embryo and contributes to the formation of the placenta. 318

conception Fertilization and implantation of an embryo resulting in pregnancy. 318

embryo The organism in its early stages of development; first week to 2 months. 322

embryonic development The period of development from the first week to 2 months, during which time all major organs form. 321

episiotomy (ĕ-piz˝e-ot´o-me) A surgical procedure performed during childbirth in which the opening of the vagina is enlarged to avoid tearing. 329

extraembryonic membranes Membranes that are not a part of the embryo but are necessary to the continued existence and health of the embryo. 318

fertilization The union of a sperm nucleus and an egg nucleus, which creates the zygote with the diploid number of chromosomes. 317

fetal development The period of development from the ninth week through birth. 321

gestation period (jes-ta´shun) The period of development measured from the start of the last menstrual cycle until birth; in humans, typically 280 days. 318

implantation The attachment and penetration of the embryo to the lining of the uterus (endometrium). 318

lanugo (lah-nu´go) Short, fine hair that is present during the later portion of fetal development. 327

morula (mor´u-lah) An early stage in development in which the embryo consists of a mass of cells, often spherical. 322

parturition Birth of a human and the expulsion of the extraembryonic membranes through the terminal portion of the female reproductive tract. 328

placenta A structure formed from the chorion and uterine tissue through which nutrient and waste exchange occur for the embryo and later the fetus. 318

prolactin (PRL) A hormone secreted by the anterior pituitary that stimulates the production of milk from the mammary glands. 330

umbilical cord Cord through which blood vessels pass, connecting the fetus to the placenta. 319

umbilicus The scar left after the umbilical cord stump shrivels and falls off. 330

vernix caseosa (ver´niks ka˝se-o´-sah) Cheeselike substance covering the skin of the fetus. 327

yolk sac An extraembryonic membrane that serves as the first site for blood cell formation. 318

zygote (zi´gōt) Diploid cell formed by the union of 2 gametes; the product of fertilization. 318

Further Readings for Part Four

Anderson, R. M., and May, R. M. May 1992. Understanding the AIDS pandemic. *Scientific American.*

Aral, S. L., and Holmes, K. K. February 1991. Sexually transmitted diseases in the AIDS era. *Scientific American.*

Eigen, M. July 1993. Viral quasispecies. *Scientific American.*

Fischetti, V. A. February 1992. Streptococcal M protein. *Scientific American.*

Gilbert, S. 1991. *Developmental biology.* 3d ed. Sunderland, Mass.: Sinauer Associates.

Hampton, J. K., Jr. 1991. *Biology of human aging.* Dubuque, Iowa: Wm. C. Brown Publishers.

Kalil, R. E. December 1989. Synapse formation in the developing brain. *Scientific American.*

Kart, C. S., et al. 1992. *Human aging and chronic disease.* Boston: Jones and Bartlett.

Kimura, D. September 1992. Sex differences in the brain. *Scientific American.*

Mills, J., and Masur, H. August 1990. AIDS-related infections. *Scientific American.*

Moore, K.L. 1988. *Essentials of human embryology.* Toronto: B. C. Decker, or St. Louis: Mosby Year Book.

Murray, A. W., and Kirschner, M. W. March 1991. What controls the life cycle. *Scientific American.*

Nathans, J. February 1989. The genes for color vision. *Scientific American.*

National Research Council, Institute of Medicine. 1990. *Developing new contraceptives: Obstacles and opportunities.* Washington, D.C.: National Academy Press.

Nilsson, L. 1977. *A child is born.* Rev. ed. New York: Delacorte Press.

Rietschel, E. T., and Brade, H. August 1992. Bacterial endotoxins. *Scientific American.*

Rusting, R. L. December 1992. Why do we age? *Scientific American.*

Scientific American. October 1988. Entire issue is devoted to AIDS.

Selkoe, D. J. September 1992. Aging brain, aging mind. *Scientific American.*

Shatz, C. J. September 1992. The developing brain. *Scientific American.*

Stine, G. J. 1992. *Biology of sexually transmitted diseases.* Dubuque, Iowa: Wm. C. Brown Publishers.

Ulmann, T., and Philibert, D. June 1990. RU 486. *Scientific American.*

Uvnas-Moberg, K. July 1989. The gastrointestinal tract in growth and reproduction. *Scientific American.*

Weissman, I. L., and Cooper, M. D. September 1993. How the immune system develops. *Scientific American.*

Winkler, W. G., and Bogel, K. June 1992. Control of rabies in wildlife. *Scientific American.*

Wistreich, G. A. 1992. *Sexually transmitted diseases: A current approach.* Dubuque, Iowa: Wm. C. Brown Publishers.

Wolpert, L. 1991. *The triumph of the embryo.* New York: Oxford University Press.

Part FIVE

Human Genetics

Human beings practice sexual reproduction, which requires gamete production. Gametes carry half the total number and various combinations of chromosomes and genes. It is sometimes possible to determine the chances of an offspring receiving a particular parental gene and, therefore, inheriting a genetic disorder.

Genes, now known to be constructed of DNA, control not only the metabolism of the cell but also, ultimately, the characteristics of the individual. The step-by-step procedure by which a DNA code controls protein synthesis has been discovered. Biotechnology is a new and burgeoning field that permits the extraction of DNA from one organism and its insertion in a different organism for a purpose useful to human beings.

Cancer is a cellular disease brought on by DNA mutations that transform a normal cell into a cancer cell. Cancer-causing genes are derived from normal genes, which keep cell division under control. Because environmental factors play a major role in causing or promoting cancer, the possibility exists of reducing its incidence. Knowledge about the detection and treatment of cancer is improving daily.

Chapter 17

Chromosomal Inheritance

A mother's love has been extolled by poets and religious leaders for centuries as selfless devotion. However, behavioral geneticists have a possible alternative explanation. First of all, a woman's chances of having children are not as great as a man's—she produces only one egg a month for a limited number of years, but a man produces millions of sperm each day even into old age. Children have some of the same chromosomes as their mother, and therefore a child represents a way for her to be immortal in a biological sense. Then, too, you have to consider that a woman is certain in a way that no man can be that a child is hers. Although this explanation of motherly love is based on a certain amount of selfishness, the result is the same. Most mothers will do all in their power to protect and teach their children so that they can become productive members of society.

Chapter Concepts

Chapter Outline

Genes carried on chromosomes determine what the cell is like and what the individual is like. An examination of the body cells of a multicellular organism shows that all the nuclei have the same number of chromosomes. This number is characteristic of the organism—corn plants have 20 chromosomes, houseflies have 12, and humans have 46.

Chromosomes

In a nondividing cell, there is indistinct and diffuse *chromatin* (see fig. 2.4), but in a dividing cell chromatin becomes the short and thick *chromosomes.*

Normal Karyotype: Pairs of Chromosomes

A cell may be photographed just prior to division so that a picture of the chromosomes is obtained. The chromosomes may be cut out of the picture and arranged by pairs of homologous chromosomes (fig. 17.1). **Homologous chromosomes** are recognized by the fact that each member of a pair is of the same size and has the same general appearance. The resulting display of pairs of chromosomes is called a **karyotype.** Although both males and females have 23 pairs of chromosomes, one of these pairs is of unequal length in males. The larger chromosome of this pair is called the X and the smaller is called the Y. Females have two X chromosomes in their karyotype. The X and Y chromosomes are called the **sex chromosomes** because they contain the genes that determine sex. The other chromosomes, known as **autosomes,** include all of the pairs of chromosomes except the X and Y chromosomes. Each pair of autosomes in the human karyotype is numbered.

> Each organism has a characteristic number of chromosomes; humans have 46. A human karyotype shows 22 pairs of autosomes and one pair of sex chromosomes. Males have an X and Y chromosome; females have two X chromosomes.

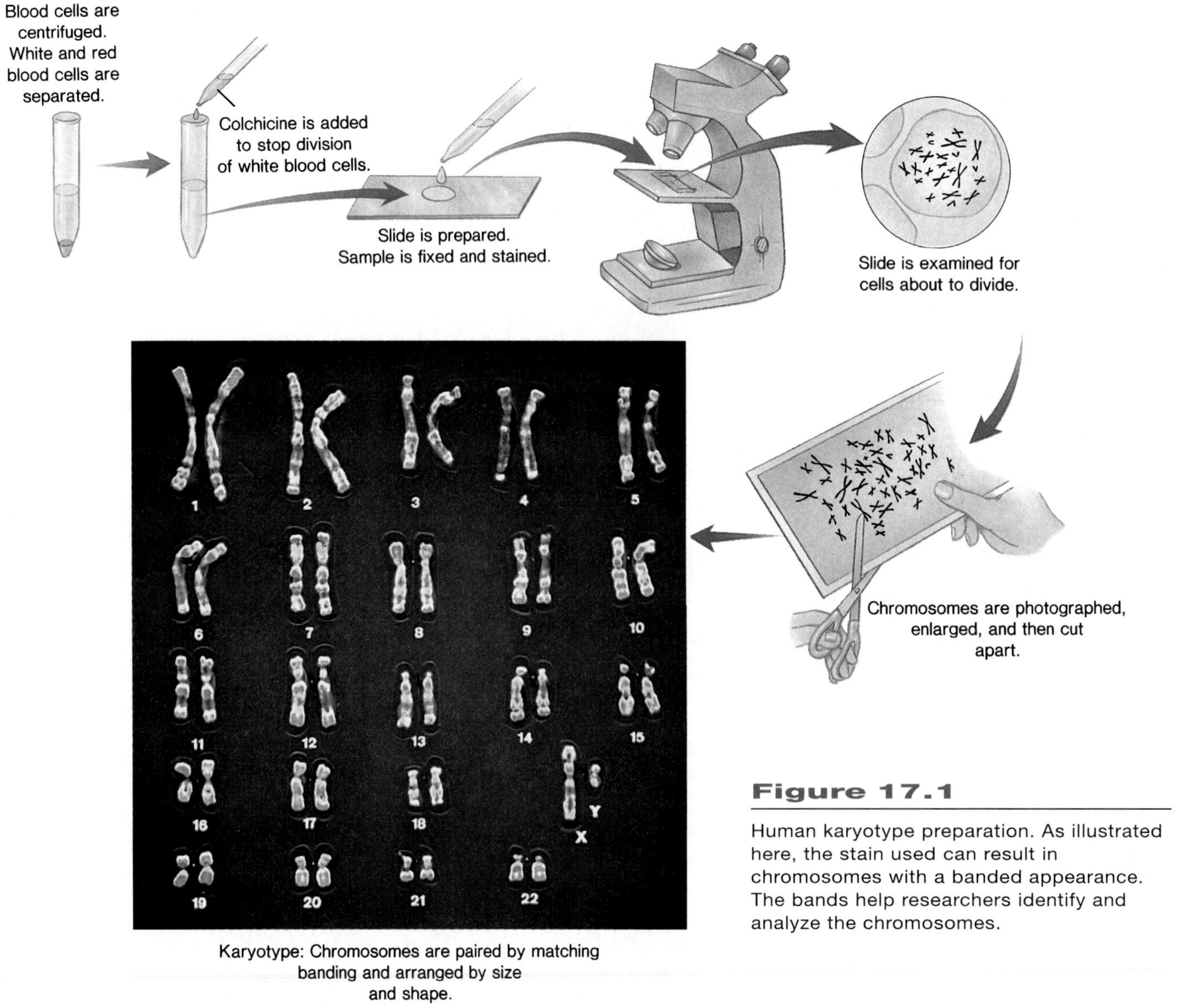
Blood cells are centrifuged. White and red blood cells are separated.
Colchicine is added to stop division of white blood cells.
Slide is prepared. Sample is fixed and stained.
Slide is examined for cells about to divide.
Chromosomes are photographed, enlarged, and then cut apart.
1
2
3
4
5
6
7
8
9
10
11
12
13
14
15
16
17
18
X
Y
19
20
21
22
Karyotype: Chromosomes are paired by matching banding and arranged by size and shape.

Figure 17.1

Human karyotype preparation. As illustrated here, the stain used can result in chromosomes with a banded appearance. The bands help researchers identify and analyze the chromosomes.

Down Syndrome: Extra Chromosome

The most common autosomal abnormality is seen in individuals with **Down syndrome** (fig. 17.2). This syndrome is easily recognized by these characteristics: short stature; an oriental-like fold of the eyelids; stubby fingers; a wide gap between the first and second toes; a large, fissured tongue; a round head; a palm crease, the so-called simian line; and, unfortunately, mental retardation, which can sometimes be severe.

Persons with Down syndrome usually have 3 copies of chromosome 21 in their karyotype because the egg had 2 copies instead of one. (In 23% of the cases studied, however, the sperm had the extra chromosome 21.) The chances of a woman having a Down syndrome child increase rapidly with age, starting at about age 40 (table 17.1). As a possible explanation, researchers suggest that older women may be more likely to carry a Down syndrome child to term because their bodies fail to recognize and abort abnormal embryos.

Although an older woman is more likely to have a Down syndrome child, most babies with Down syndrome are born to women younger than age 40 because this is the age group having the most babies. As discussed in the Health Focus in the previous chapter (p. 324), chorionic villi testing and amniocentesis followed by karyotyping can detect a Down syndrome child. However, young women are not encouraged to undergo such procedures because the risk of complications resulting from these tests is greater than the risk of having a Down syndrome child. Fortunately, there is now a test based on substances in the blood that can identify mothers who might be carrying a Down syndrome child, and only these individuals need to undergo further testing.

It is known that the genes that cause Down syndrome are located on the bottom third of chromosome 21 (fig. 17.2*b*), and extensive investigative work has been directed toward discovering the specific genes responsible for the characteristics of the syndrome. Thus far, investigators have discovered several genes that may account for various conditions seen in persons with Down syndrome. For example, they have located genes most likely responsible for the increased tendency toward leukemia, cataracts, accelerated rate of aging, and mental retardation. The gene for mental retardation, dubbed the *Gart* gene, causes an increased level of purines in blood, a finding associated with mental retardation. It is hoped that someday it will be possible to control the expression of the *Gart* gene even before birth so that at least this symptom of Down syndrome does not appear.

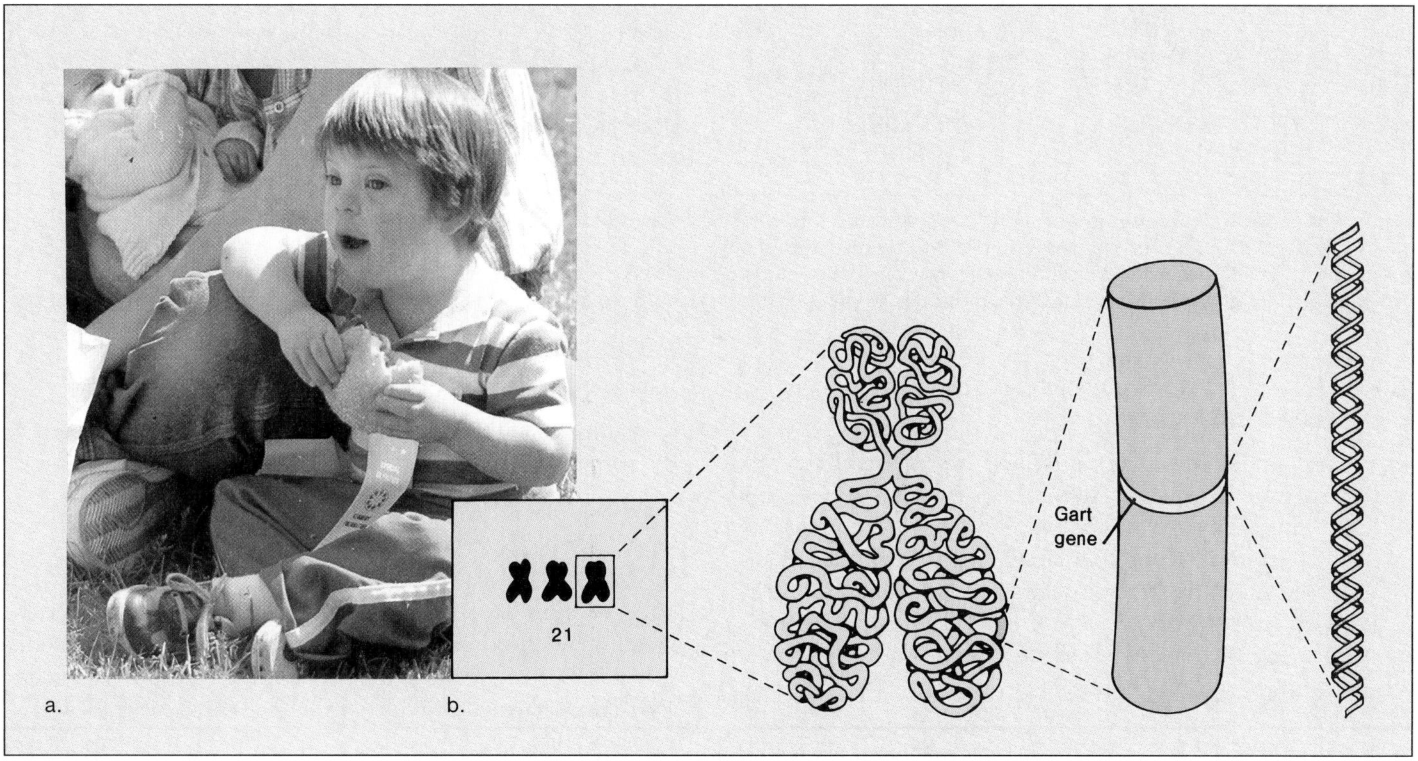

Figure 17.2

Down syndrome. **a.** Common characteristics of the syndrome include a wide, rounded face and a fold of the upper eyelids. Mental retardation, along with an enlarged tongue, makes it difficult for a person with Down syndrome to speak coherently. **b.** Karyotype of an individual with Down syndrome shows an extra chromosome 21. More sophisticated technologies allow investigators to pinpoint the location of specific genes associated with the syndrome. An extra copy of the *Gart* gene, which leads to a high level of purines, may account for the mental retardation seen in persons with Down syndrome.

Table 17.1

Incidence of Selected Chromosomal Abnormalities

Syndrome	Involved Chromosomes	Births
Down	Extra 21	
Mothers under 40		1/800
Mothers over 40		1/60
Fragile X	Broken X	1/1,000 (M) 1/2,500 (F)
Turner	X	1/2,500–10,000
Klinefelter	XXY	1/500–2,000
Metafemale	XXX	1/1,000–2,000
XYY	XYY	1/1,000

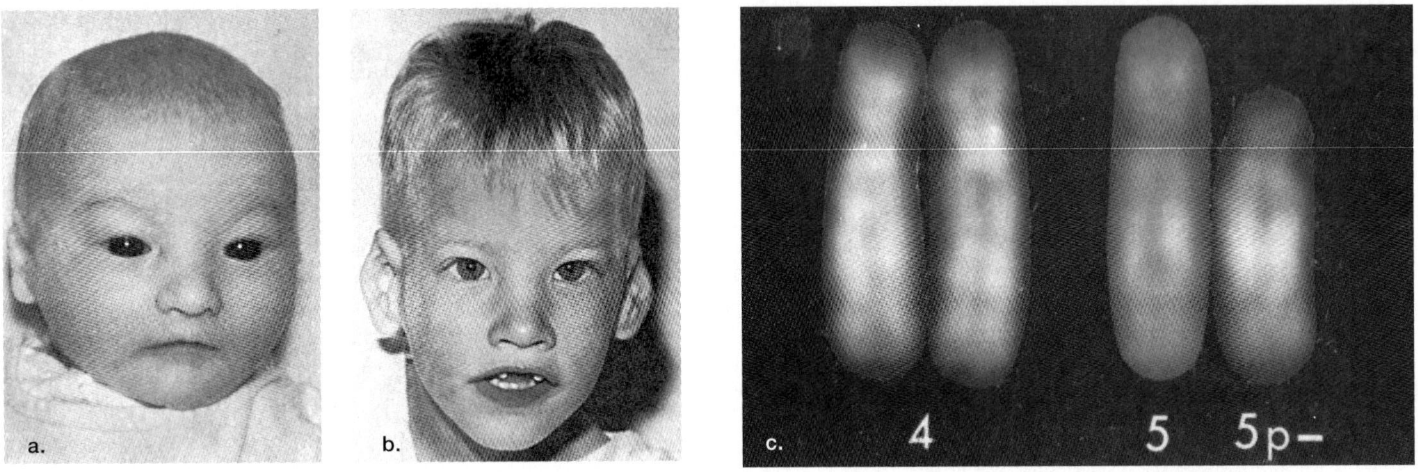

Figure 17.3

Cri du chat syndrome. **a.** An infant with this syndrome has a moon face, small head, and a cry that sounds like the meow of a cat. **b.** An older child has an oriental-like fold in the eyelid and misshapen ears, which are placed low on the head. Mental retardation is evident by this age. **c.** Chromosomes 4 and 5 are shown for comparison. Chromosome 4 is normal. Cri du chat is caused by a deletion of part of the short arm of chromosome 5, designated by 5p– (p = petite in French; the minus sign indicates the deletion).

Cri du Chat Syndrome: Portion Missing

A chromosomal deletion is responsible for **cri du chat** (cat's cry) **syndrome** (fig. 17.3). Affected individuals meow like a kitten when they cry, but more important perhaps is the fact that they tend to have a small head with malformations of the face and body and that mental defectiveness usually causes retarded development. A karyotype shows that a portion of one chromosome 5 is missing (deleted), while the other chromosome 5 is normal.

> Down syndrome is most often due to the inheritance of an extra chromosome 21, and cri du chat syndrome is due to the inheritance of a defective chromosome 5.

Sex Chromosome Problems

Karyotypes with abnormal sex chromosomes are due to the inheritance of an abnormal X or an abnormal number of sex chromosomes.

Fragile X Syndrome: More Common in Males

Males outnumber females by about 25% in institutions for the mentally retarded. In some of these males, the X chromosome is nearly broken, leaving the tip hanging by a flimsy thread. These males are said to have *fragile X syndrome* (fig. 17.4).

As children, fragile X syndrome individuals may be hyperactive or autistic; their speech is delayed in development and is often repetitive in nature. As adults, they have large testes and big, usually protruding ears. They are short in stature, but the jaw is prominent and the head is often large. Stubby hands, lax joints, and a heart defect may also occur.

Figure 17.4

Fragile X syndrome. Fragile X syndrome is due to the inheritance of a fragile X chromosome. An arrow points out the fragile site.

Fragile X chromosomes occur in both males and females (see table 17.1), but the syndrome is seen less often in females. When symptoms do appear in females, they tend to be less severe.

Too Many/Too Few Sex Chromosomes

From birth, an XO individual with **Turner syndrome** has only one sex chromosome, an X; the O signifies the absence of a second sex chromosome. Turner females are short, have a broad chest, and may have congenital heart defects. The ovaries never become functional and in many individuals they are simply white streaks. Turner females do not undergo puberty or menstruate, and there is a lack of breast development (fig. 17.5*a*). Although no overt mental retardation is reported, Turner females show reduced skills in interpreting spatial relationships.

A male with **Klinefelter syndrome** has 2 or more X chromosomes in addition to a Y chromosome. Affected individuals are sterile males; the testes are underdeveloped and there may be some breast development (fig. 17.5*b*). These phenotypic abnormalities are not apparent until puberty, although some evidence of subnormal intelligence may be apparent before this time.

A **metafemale** is an individual with more than two X chromosomes. It might be supposed that the XXX female is especially feminine, but this is not the case. Although in some cases there is a tendency toward learning disabilities, most metafemales have no apparent physical abnormalities except that they may have menstrual irregularities, including early onset of menopause.

XYY males also can result from nondisjunction during spermatogenesis. Affected males usually are taller than average, suffer from persistent acne, and tend to have barely normal intelligence. At one time, it was suggested that these men were likely to be criminally aggressive, but it has since been shown that the incidence of such behavior among them is no greater than among XY males.

Individuals sometimes are born with the sex chromosomes XO (Turner syndrome), XXY (Klinefelter syndrome), XXX (metafemale), and XYY. No matter how many X chromosomes there are, an individual with a Y chromosome is usually a male.

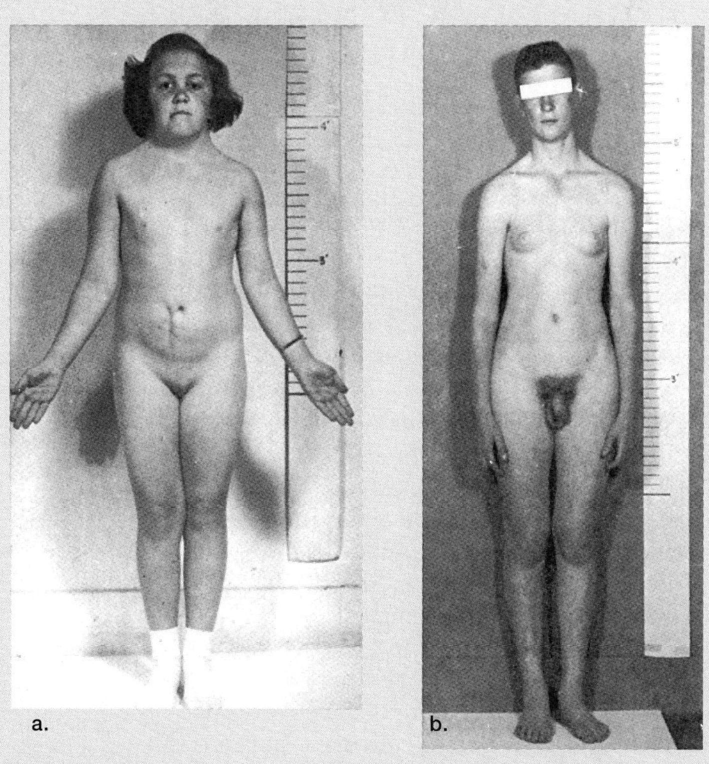

Figure 17.5

Abnormal sex chromosome inheritance. **a.** Female with Turner (XO) syndrome, which includes a bull neck, short stature, and immature sexual features. **b.** A male with Klinefelter (XXY) syndrome, which is marked by immature sex organs and some development of the breasts.

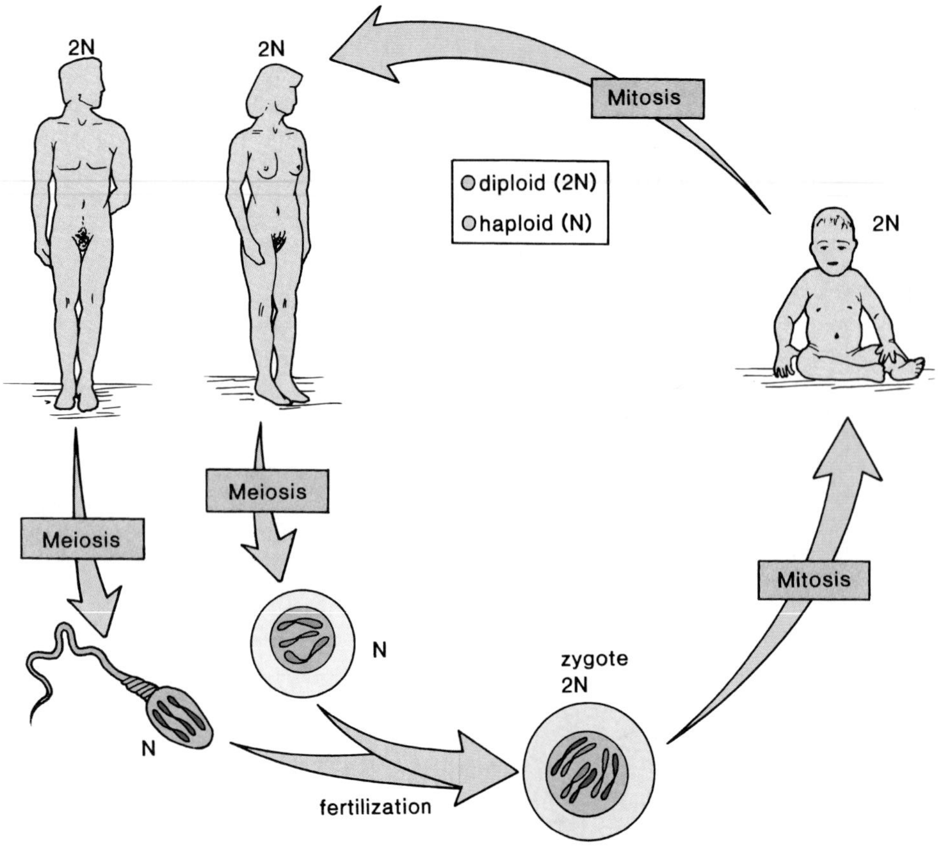

Figure 17.6

Life cycle of humans. Meiosis in males is a part of sperm production, and meiosis in females is a part of egg production. When a haploid sperm fertilizes a haploid egg, the zygote is diploid. The zygote undergoes mitosis as it develops into a newborn child. Mitosis continues after birth until the individual reaches maturity and then the life cycle begins again.

Human Life Cycle

The human life cycle involves growth and sexual reproduction (fig. 17.6). During growth, cells divide by a process called *mitosis,* which ensures that each and every cell has a complete number of chromosomes. Sexual reproduction requires the production of sex cells, which have half the number of chromosomes. A type of cell division called *meiosis* reduces the chromosome number by one-half.

Meiosis occurs in the sex organs. In males, it produces the cells that become sperm; in females, it produces the cells that become eggs. The sperm and the egg are the sex cells, or **gametes.** Gametes contain half the number of chromosomes in a karyotype—one chromosome from each of the pairs of chromosomes. This is called the **haploid (N)** number of chromosomes; the haploid number of chromosomes in humans is 23.

A new individual comes into existence when a sperm fertilizes an egg. The resulting zygote has the **diploid (2N)** number of chromosomes. Each parent contributes one chromosome to each of the pairs of chromosomes present. As the individual develops, *mitosis* occurs and each **somatic** (body) **cell** has the diploid number of chromosomes. In humans, the diploid number is 46 because there are 23 pairs of chromosomes.

Table 17.2 summarizes the major differences between mitosis and meiosis in multicellular animals.

> The life cycle of humans requires 2 types of cell division: mitosis and meiosis.

Table 17.2

Mitosis versus Meiosis

Location	Cell Division	Description	Result
Somatic (body) cells	Mitosis	2N (diploid) $\rightarrow$ 2N (diploid)	Growth and repair
Sex organs	Meiosis	2N (diploid) $\rightarrow$ N (haploid)	Gamete production

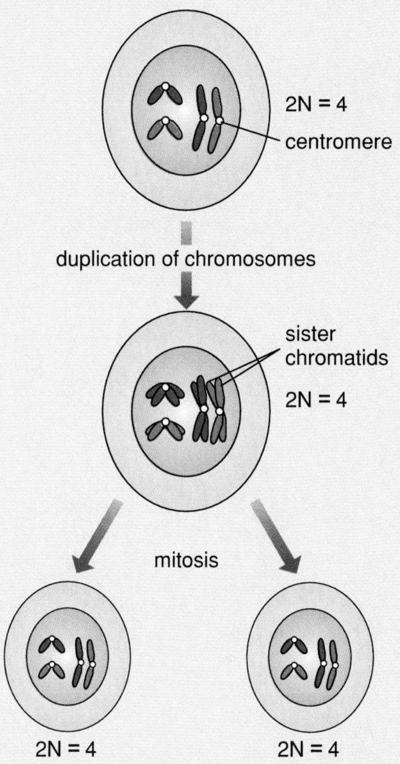

Figure 17.7

Overview of mitosis. Following duplication of chromosomes, each chromosome in the parent cell contains 2 sister chromatids. During mitotic division, the sister chromatids separate, and thereafter each is called a chromosome. Therefore, the daughter cells have the same number and kinds of chromosomes as the parent cell. Counting the number of centromeres tells the number of chromosomes in the nucleus. (The blue chromosomes were inherited from one parent and the red chromosomes were inherited from the other parent.)

Overview of Mitosis: 2N → 2N

Mitosis is cell division that produces *2 daughter cells, each with the same number and kinds of chromosomes as the parent cell, the cell that divides.*[1] Therefore, following mitotic cell division, the parent cell and the daughter cells are genetically identical.

Before mitosis begins, the parental cell is 2N. Each chromosome is composed of 2 identical parts, called **chromatids.** Therefore, these are sometimes called sister (twin) chromatids. The chromatids are held together in a region called the **centromere.** At the completion of mitosis, each chromosome is composed of a single chromatid:

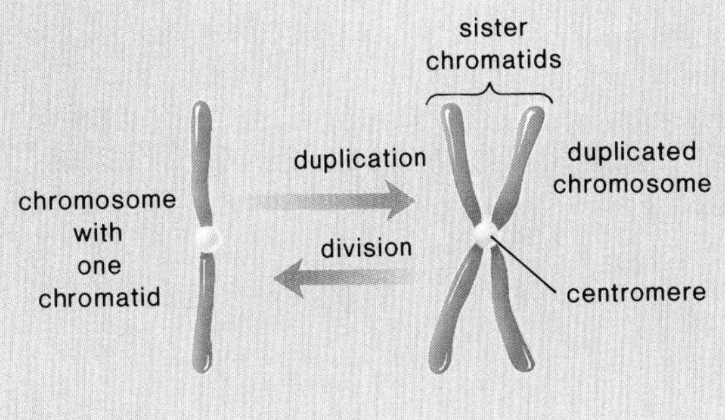

Figure 17.7 gives an overview of mitosis; for simplicity, only 4 chromosomes are depicted. (In determining the number of chromosomes, it is necessary to count only the number of independent centromeres.) During mitosis, the centromeres divide, the sister chromatids separate, and one of each kind goes into each daughter cell. Therefore, each daughter cell gets a complete set of chromosomes and is 2N. (Following separation, each chromatid is called a chromosome.) Since each daughter cell receives the same number and kinds of chromosomes as the parent cell, each is genetically identical to each other and to the parent cell.

Mitosis occurs in humans when tissues grow or when repair occurs. Following fertilization, the zygote begins to divide mitotically, and mitosis continues during development and the life span of the individual. In the adult, tissues differ as to their ability to divide; nervous and muscle tissue cells seem to lose the ability to divide. Epidermal cells, which line the respiratory tract and the digestive tract and form the outer layer of the skin, divide continuously. Stem cells in the red bone marrow divide to produce millions of blood cells every day. Whenever repair takes place, as when a broken bone is mended, mitosis has occurred.

Following mitosis, each of 2 daughter cells has the same number and kinds of chromosomes as the parental cell. Body (somatic) cells undergo mitosis, a process that is necessary to the growth and repair of tissues.

1. The terms *mitosis and meiosis* technically refer only to nuclear division, but for convenience, they are used here to refer to the division of the entire cell.

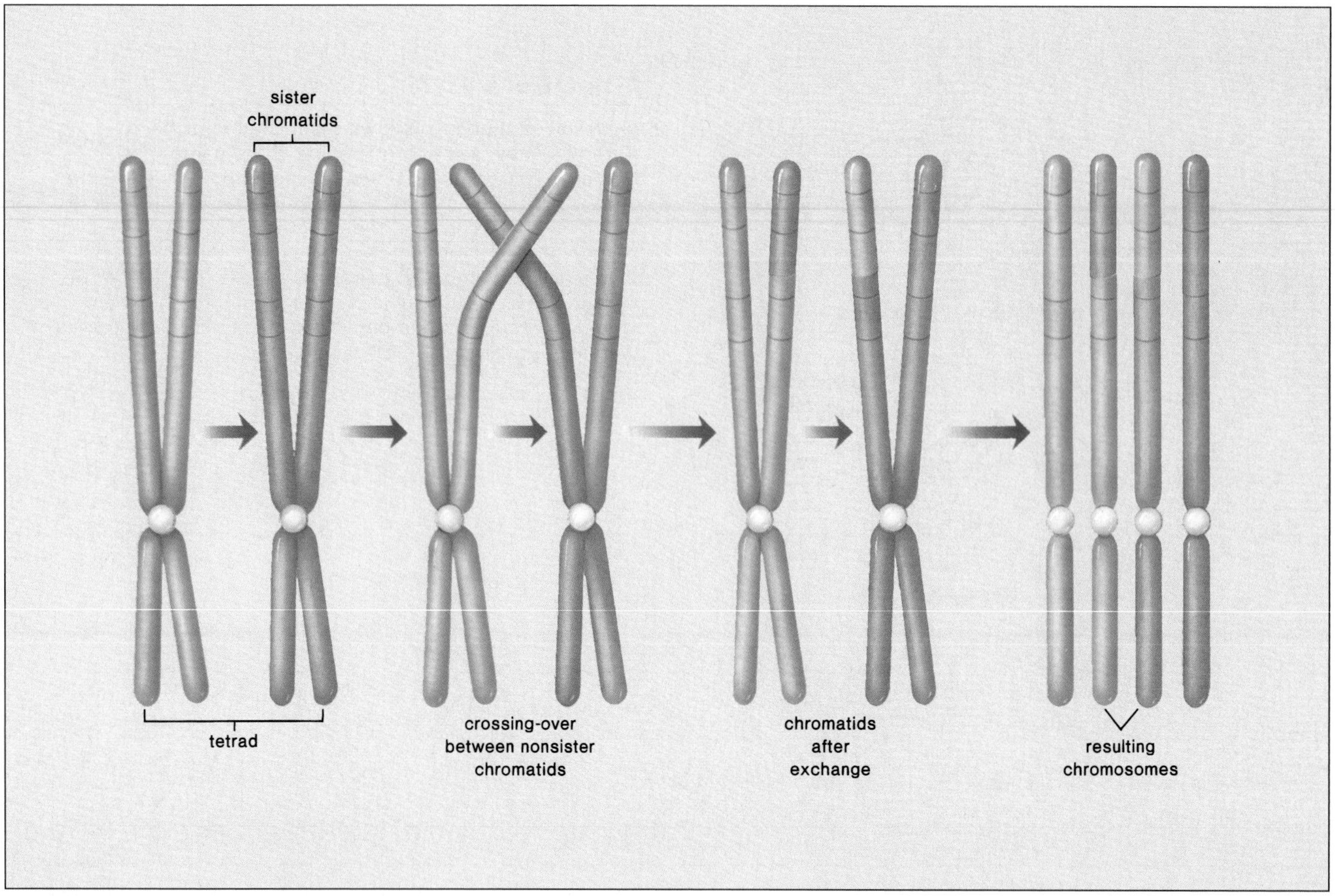

Figure 17.8

Crossing-over. When homologous chromosomes are in synapsis, the nonsister chromatids exchange genetic material. Following crossing-over, there is a different combination of genes on each chromatid.

Overview of Meiosis: 2N → N

Meiosis, which requires 2 cell divisions, results in *4 daughter cells, each having one of each kind of chromosome and therefore half the number of chromosomes as the parent cell.*[2] The parent cell has the 2N number of chromosomes, while the daughter cells have the N number of chromosomes. Therefore, meiosis is often called reduction division. Following meiotic cell division, the daughter cells are not genetically identical.

Meiosis results in 4 daughter cells because it consists of 2 divisions called **meiosis I** and **meiosis II.** Before meiosis I begins, each chromosome has duplicated and is composed of 2 chromatids. Therefore, each chromosome can be called a **dyad.** The parental cell is 2N. Recall that when a cell is 2N, the chromosomes occur in pairs. For example, the 46 chromosomes of humans occur in 23 pairs of chromosomes. These pairs are called homologous chromosomes.

During meiosis I, the homologous chromosomes of each pair come together and line up side by side due to a means of attraction still unknown. This so-called **synapsis** results in a **tetrad,** an association of 4 chromatids that stay in close proximity until the homologous chromosomes separate. During synapsis, nonsister chromatids exchange genetic material. The exchange of genetic material between chromatids is called **crossing-over.** Crossing-over recombines the genes of the parental cell (fig. 17.8).

2. See footnote 1.

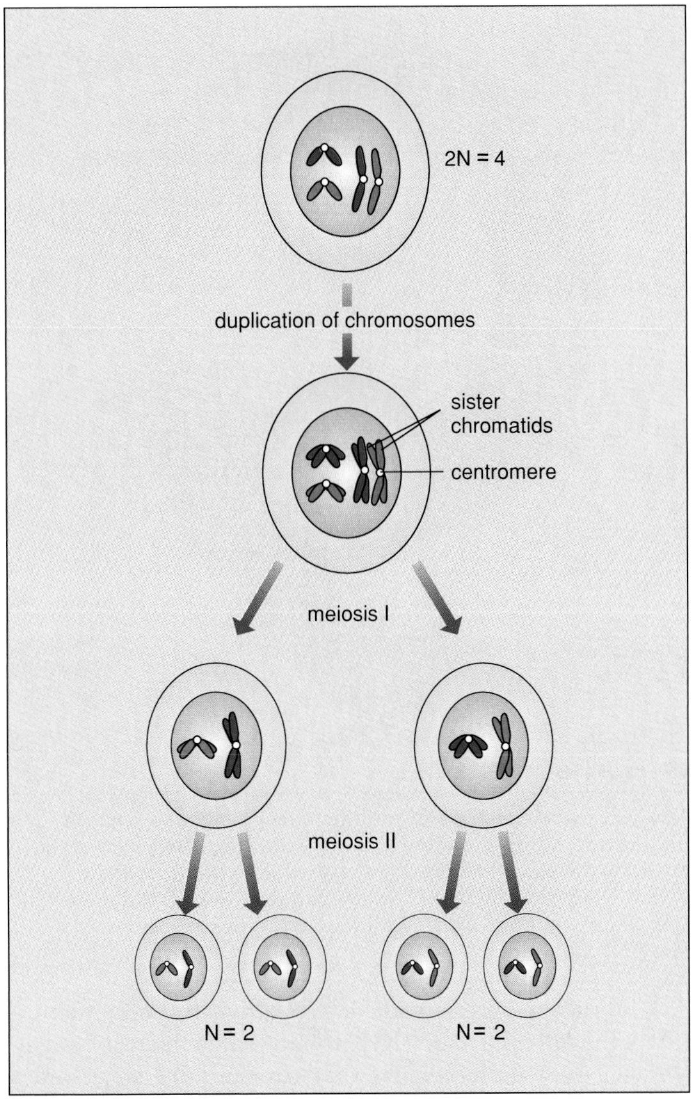

Figure 17.9

Overview of meiosis. Following duplication of chromosomes, the parent cell undergoes 2 divisions, meiosis I and meiosis II. During meiosis I, homologous chromosomes separate, and during meiosis II, chromatids separate. The final daughter cells are haploid. (The blue chromosomes were inherited from one parent and the red chromosomes were inherited from the other parent.)

Following synapsis during meiosis I, the homologous chromosomes of each pair separate. This separation means that one chromosome from each homologous pair reaches each daughter cell. There are no restrictions on the separation process, and either chromosome of a homologous pair can occur in a daughter cell with either chromosome of any other pair. Therefore, all possible combinations of chromosomes occur within the daughter cells.

Notice that following meiosis I, the daughter cells have half the number of chromosomes and the chromosomes are still duplicated (fig. 17.9). Again, counting the number of centromeres tells the number of chromosomes in each daughter cell.

> During meiosis I, homologous chromosomes separate and the daughter cells receive one of each pair. The daughter cells are not genetically identical. The chromosomes are still duplicated.

When meiosis II begins, the chromosomes are still duplicated. Therefore, no duplication of chromosomes is needed between meiosis I and meiosis II. The chromosomes are already dyads and are composed of 2 sister chromatids. During meiosis II, the sister chromatids separate in each of the cells from meiosis I. Each of the resulting 4 daughter cells has the haploid number of chromosomes (fig. 17.9).

> During meiosis II, the chromatids separate and the resulting 4 daughter cells are each haploid.

In humans, meiosis occurs in the testes and ovaries during the production of the gametes. The *gametes* are the sperm in males and eggs in females. Because of meiosis, the chromosome number stays constant in each generation of humans. When a haploid sperm fertilizes a haploid egg, the new individual has the diploid number of chromosomes. There are 3 ways in which the new individual is assured a different combination of genes than either parent:

1. Crossing-over recombines the genes on the sister chromatids of homologous pairs.
2. Following meiosis, each gamete has a different combination of chromosomes.
3. Upon fertilization, recombination of chromosomes occurs.

> Following meiosis, each of 4 daughter cells has the haploid number of chromosomes, whereas the parental cell was diploid. Meiosis takes place in the testes and ovaries during the production of the gametes. It ensures that the chromosome number stays constant in each generation and that the new individual has a different combination of genes than either parent.

Mitosis

Mitosis occurs as a part of the cell cycle.

Cell Cycle

The **cell cycle** consists of mitosis and interphase (fig. 17.10). **Interphase** is the interval of time between cell divisions. The length of time required for the entire cell cycle varies according to the organism and even the type of cell within the organism, but 18–24 hours is typical for animal cells. Mitosis lasts less than an hour to slightly more than 2 hours; for the rest of the cycle, the cell is in interphase.

During interphase, a human cell resembles that shown in figure 2.4 (p. 33). The nuclear envelope and the nucleoli are visible. The chromosomes, however, are not visible because the chromosome material is not compacted; only indistinct chromatin is seen.

It used to be said that interphase was a resting stage, but we now know that this is not the case. The organelles are metabolically active and are carrying on their normal functions. If the cell is going to divide, *DNA replication* occurs. During replication, DNA is copied and the chromosome becomes duplicated—it is now composed of 2 sister chromatids. *Sister chromatids* are genetically identical—they contain the same genes. Also, organelles, including the *centrioles,* duplicate. A nondividing cell has one pair of centrioles, but in a cell that is going to divide, this pair duplicates, and there are 2 pairs of centrioles outside the nucleus.

> The cell cycle includes mitosis and interphase. During interphase, DNA replication results in each chromosome having sister chromatids. The organelles, including centrioles, also duplicate during interphase.

There is usually a limit to the number of times an animal cell enters the cell cycle and divides before degenerative changes lead to the cell's death. Normal cells divide only about 50 times. Usually, they break out of the cell cycle and become specialized even before this number of divisions is reached. This can be contrasted to cancer cells, which can continue to divide indefinitely. Cancer cells have abnormal chromosomes and/or cell structure irregularities associated with their ability to keep repeating the cell cycle.

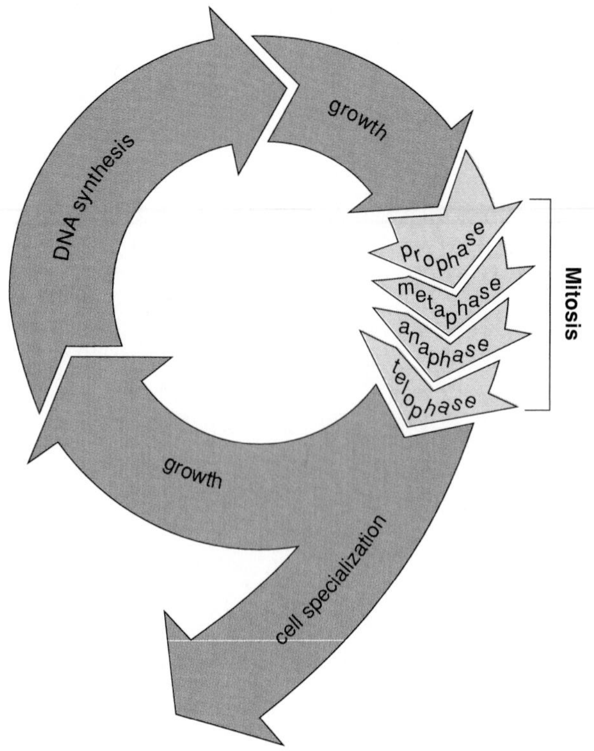

Figure 17.10

The cell cycle consists of mitosis and interphase. During interphase, there is growth before and after DNA synthesis. DNA synthesis is required for replication, the process by which DNA is duplicated. Some daughter cells "break out" of the cell cycle and become specialized cells performing a specific function.

Today, researchers are actively investigating the control of the cell cycle. They have found that in the early stages, the cell releases molecules that control later stages, and later stages release molecules that feed back to control early stages. A critical regulatory event occurs during the first stage—if the nutrient and control signals are positive, the cell always completes the rest of the cycle.

Stages of Mitosis

As an aid in describing the events of mitosis, the process is divided into 4 phases: prophase, metaphase, anaphase, and telophase (fig. 17.11). Although it is helpful to depict the stages of mitosis as if they could be separate, they are actually continuous and flow from one stage to another with no noticeable interruption.

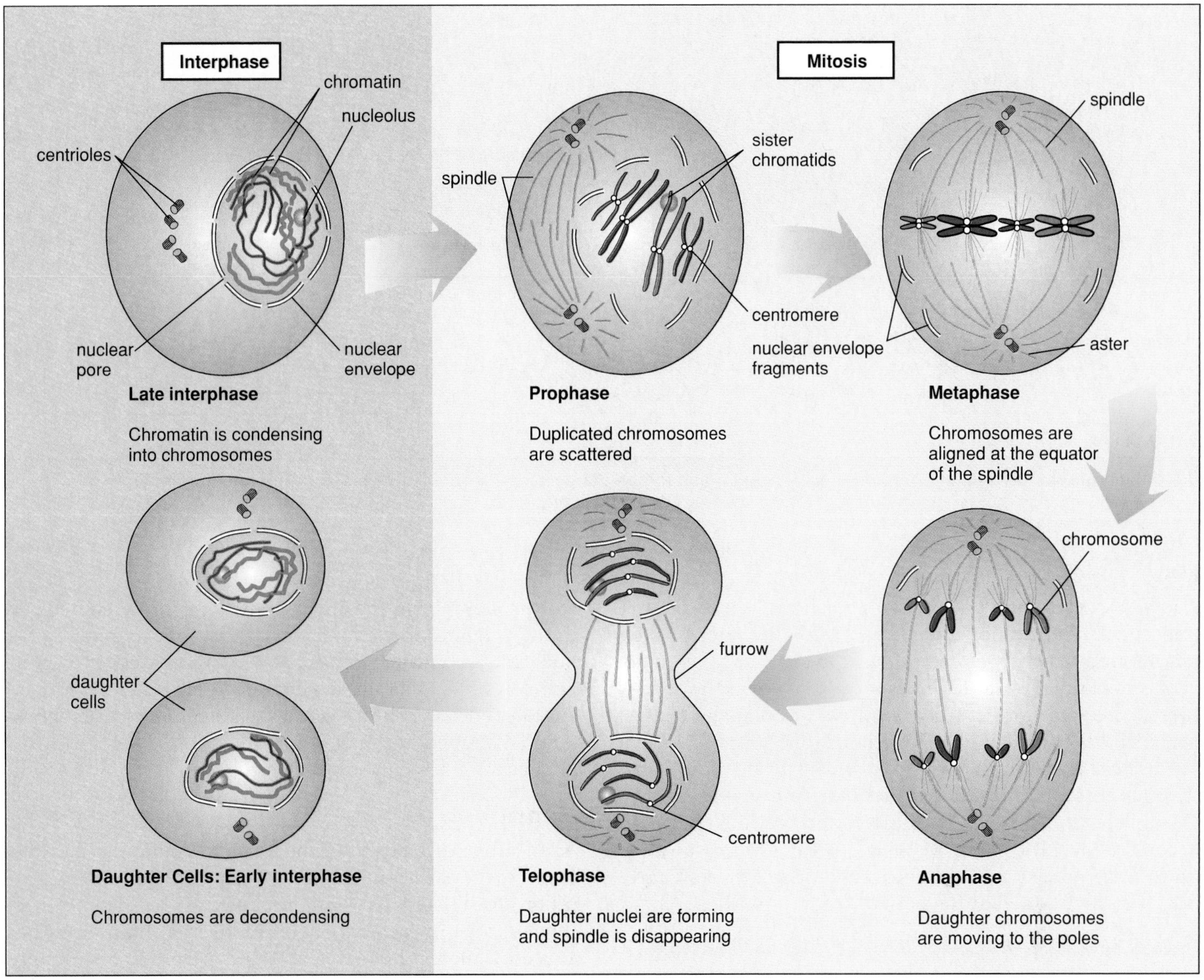

Figure 17.11

Interphase and mitosis. Mitosis has 4 stages: prophase, metaphase, anaphase, and telophase. Interphase, which occurs between cell divisions, is not part of mitosis. Notice that the centrioles duplicate during interphase so that there are 2 pairs at the start of mitosis—compare early interphase to late interphase. Centrioles are believed to help organize the spindle which is involved in chromosome movement. (The blue chromosomes were inherited from one parent and the red chromosomes were inherited from the other parent.)

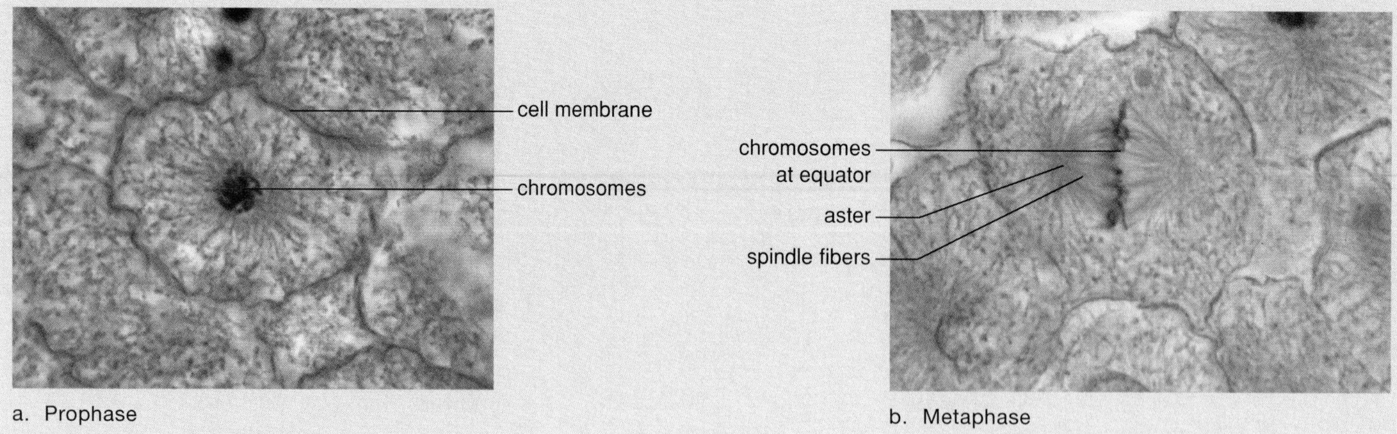

Figure 17.12

Photomicrographs of cells undergoing mitosis. **a.** Chromosomes are visible and randomly placed in the nucleus. **b.** During metaphase, the chromosomes are lined up along the equator of the spindle. **c.** During anaphase, the separation of sister

Prophase

It is apparent during **prophase** that cell division is about to occur. The 2 pairs of centrioles outside the nucleus begin moving away from each other toward opposite ends of the nucleus. **Spindle fibers** appear between the separating centriole pairs, the nuclear envelope begins to fragment, and the nucleolus begins to disappear.

The chromosomes are now visible. Each is composed of sister chromatids held together at a centromere. As the chromosomes continue to shorten and to thicken, they become attached by way of their centromeres to spindle fibers. At prophase chromosomes are randomly placed and have not yet aligned at the equator of the spindle.

Structure of the Spindle

At the end of prophase, a cell has a fully formed spindle. A **spindle** has poles, asters, and fibers. The **asters** are arrays of short microtubules that radiate from the poles, and the fibers are bundles of microtubules that stretch between the poles. Microtubule organizing centers (MTOC) are associated with the centrioles at the poles. It is well known that a MTOC organizes microtubules including, presumably, those of the spindle. It is possible that the centrioles assist in this function, but it could also be that their location at the poles of a spindle simply ensures that each daughter cell receives a pair of centrioles.

Metaphase

During **metaphase,** the nuclear envelope is fragmented and the spindle occupies the region formerly occupied by the nucleus. The chromosomes are now at the *equator* (center) of the spindle. Metaphase is characterized by a fully formed spindle, with the chromosomes, each having 2 sister chromatids, aligned at the equator (fig. 17.12). At the close of metaphase, the centromeres uniting the chromatids split.

Anaphase

At the start of **anaphase,** the sister chromatids separate. *Once separated, the chromatids are called chromosomes.* Separation of the sister chromatids ensures that each cell receives a copy of each type of chromosome and thereby has a full complement of genes. During anaphase, the daughter chromosomes move (*ana* means *up*) to the poles of the spindle. Anaphase is characterized by the diploid number of chromosomes at each pole.

Function of the Spindle

The spindle brings about chromosome movement. Two types of spindle fibers are involved in the movement of chromosomes during anaphase. One type extends from the poles to the equator of the spindle; there they overlap. As mitosis proceeds, these fibers increase in length and this helps push the chromosomes apart. The chromosomes themselves are

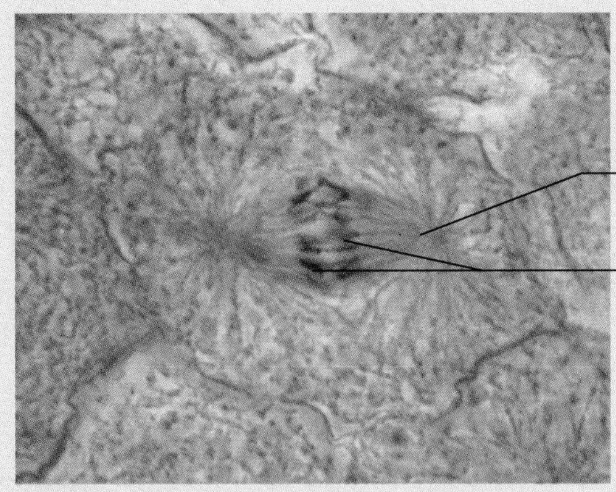

c. Anaphase

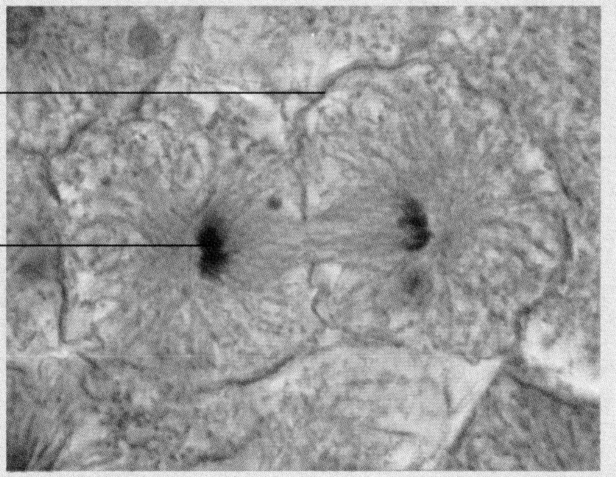

d. Telophase

chromatids results in chromosomes that are pulled by spindle fibers to opposite poles of the spindle. **d.** A full set of chromosomes is in each newly forming nucleus. Cell membrane forms between daughter cells.

attached to other spindle fibers that simply extend from their centromeres to the poles. These fibers get shorter and shorter as the chromosomes move toward the poles, and eventually disappear. These fibers pull the chromosomes apart.

Spindle fibers, as stated, are composed of microtubules. Microtubules can assemble and disassemble by the addition or subtraction of tubulin (protein) subunits. This is what enables spindle fibers to lengthen and shorten and what ultimately causes the movement of the chromosomes.

Telophase

Telophase begins when the chromosomes arrive at the poles. During telophase, the chromosomes become indistinct chromatin again. The spindle disappears as nucleoli appear, and nuclear envelopes form in each cell. Telophase is characterized by the presence of 2 daughter nuclei.

In animal cells, a slight indentation called a **cleavage furrow** passes around the circumference of the cell. Actin filaments form a contractile ring, and as the ring gets smaller and smaller the cleavage furrow pinches the cell in half. Now each cell is enclosed by its own cell membrane.

Following mitosis each daughter cell is 2N. When the sister chromatids separate during anaphase, each newly forming cell receives the same number and kinds of chromosomes as the parental cell.

Importance of Mitosis: For Growth

Mitosis assures that each body cell has the same number and kinds of chromosomes. It is important to the growth and repair of humans. When a baby develops in its mother's womb, mitosis occurs as a component of growth. As a wound heals, mitosis occurs to repair the damage.

Mitosis is necessary to *cloning,* the production of 2 or more individuals with the same chromosomes and genes. Cloning occurs during development when the first 2 embryonic cells separate and give rise to identical twins. Separation of embryonic cells can now be done in the laboratory, and the cells can be stored or used for in vitro fertilization. If researchers separate the cells of a 16-cell embryo, then 16 identical individuals are a possibility. Society must consider the morality of such a procedure, however.

The cloning of an adult individual for whom there are no stored embryonic cells is not yet a possibility. Although each cell of an adult has an entire set of chromosomes and genes, the cells are too specialized to begin the process of development.

Mitosis occurs in body (somatic) cells during the growth and repair of tissues. During cloning, mitosis produces 2 individuals with the same chromosomes and genes.

Meiosis

The same 4 stages in mitosis—prophase, metaphase, anaphase, and telophase—occur during both meiosis I and meiosis II.

The First Division

The stages of meiosis I as they appear in an animal cell are diagrammed in figure 17.13. During *prophase I,* the spindle appears while the nuclear envelope fragments and the nucleolus disappears. The homologous chromosomes, each having 2 sister chromatids, undergo synapsis, forming tetrads. Crossing-over occurs now, but for simplicity, this event has been omitted from figure 17.13. In *metaphase I,* tetrads line up at the equator of the spindle. During *anaphase I,* homologous chromosomes separate and dyads move to opposite poles of the spindle. During *telophase I,* nucleoli appear and nuclear envelopes form as the spindle disappears. In certain species, the cell membrane furrows to give 2 cells, and in others, the second division begins without benefit of complete furrowing. Regardless, each daughter cell contains only one chromosome from each homologous pair. The chromosomes are dyads, and each has 2 sister chromatids. No replication of DNA occurs during a period of time called *interkinesis.*

The Second Division

The stages of meiosis II for an animal cell are diagrammed in figure 17.14. At the beginning of *prophase II,* a spindle appears while the nuclear envelope fragments and the nucleolus disappears. Dyads (one dyad from each pair of homologous chromosomes) are present, and each attaches to the spindle independently. During *metaphase II,* the dyads are lined up at the equator. At the close of metaphase, the centromeres split. During *anaphase II,* the sister chromatids of each dyad separate and move toward the poles. Each pole receives the same number of chromosomes. In *telophase II,* the spindle disappears as nuclear envelopes form. The cell membrane furrows to give 2 complete cells, each of which has the haploid, or N, number of chromosomes. Since each cell from meiosis I undergoes meiosis II, there are 4 daughter cells altogether.

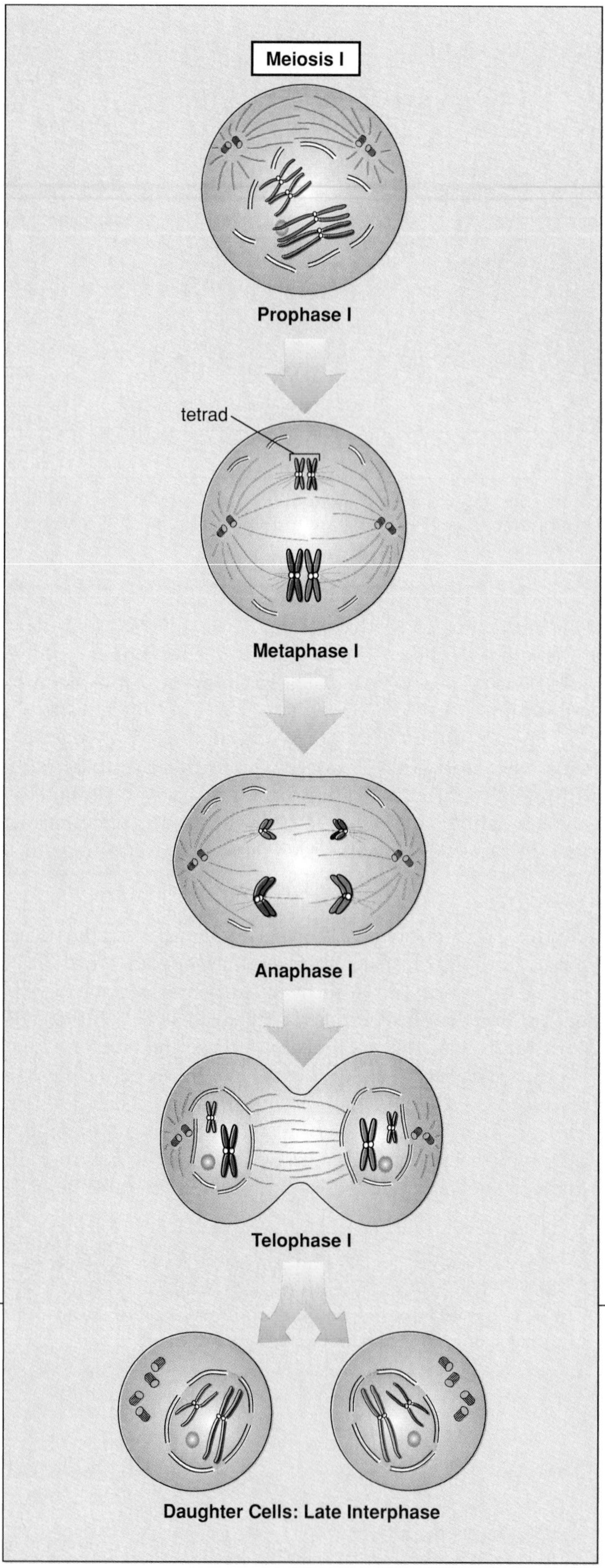

Figure 17.13

Meiosis I. During meiosis I, homologous chromosomes undergo synapsis and then separate so that each daughter cell has only one chromosome from each original homologous pair. For simplicity's sake, the results of crossing-over have not been depicted. Notice that each daughter cell is haploid and each chromosome has 2 chromatids. (The blue chromosomes were inherited from one parent, and the red chromosomes were inherited from the other parent.)

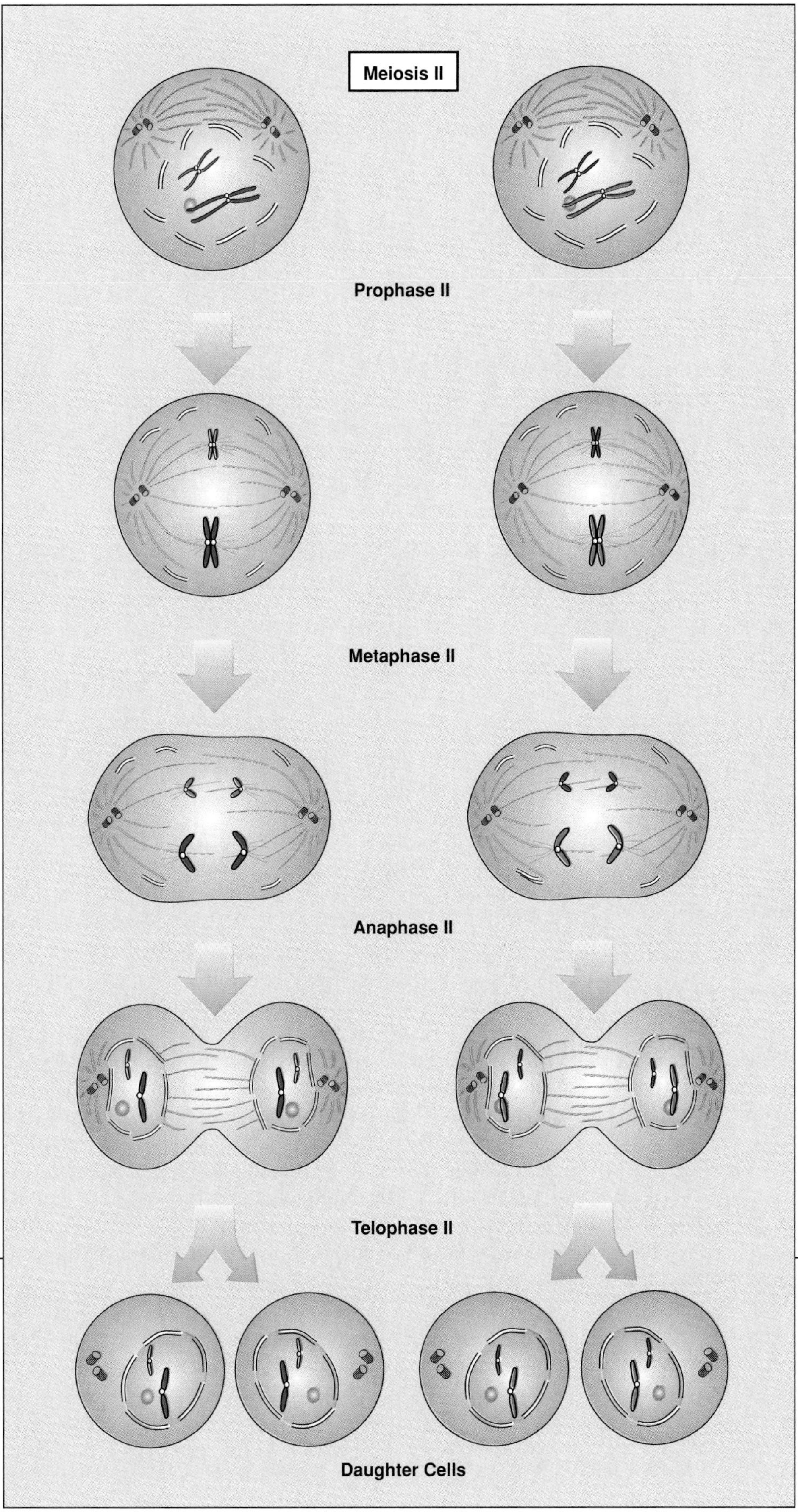

Meiosis involves 2 cell divisions. During meiosis I, tetrads form and crossing-over occurs. Homologous chromosomes separate and each daughter cell receives a pair of sister chromatids. During meiosis II, separation of chromatids in daughter cells from meiosis I results in 4 daughter cells, each with the haploid number of chromosomes.

Importance of Meiosis: For Reproduction

Meiosis is nature's way of keeping the chromosome number constant from generation to generation. It assures that the next generation will have a different genetic makeup than that of the previous generation. As a result of independent assortment of chromosomes and crossing-over, the gametes carry a new combination of genes. The egg carries half of the genes from the female parent and the sperm carries half of the genes from the male parent. When the sperm fertilizes the egg, the zygote has a different combination of genes than either parent. In this way, meiosis assures genetic variation generation after generation.

Figure 17.14

Meiosis II. During meiosis II, chromatids separate. Each daughter cell is haploid and each chromosome has one chromatid. (The blue chromosomes were inherited from one parent, and the red chromosomes were inherited from the other parent.)

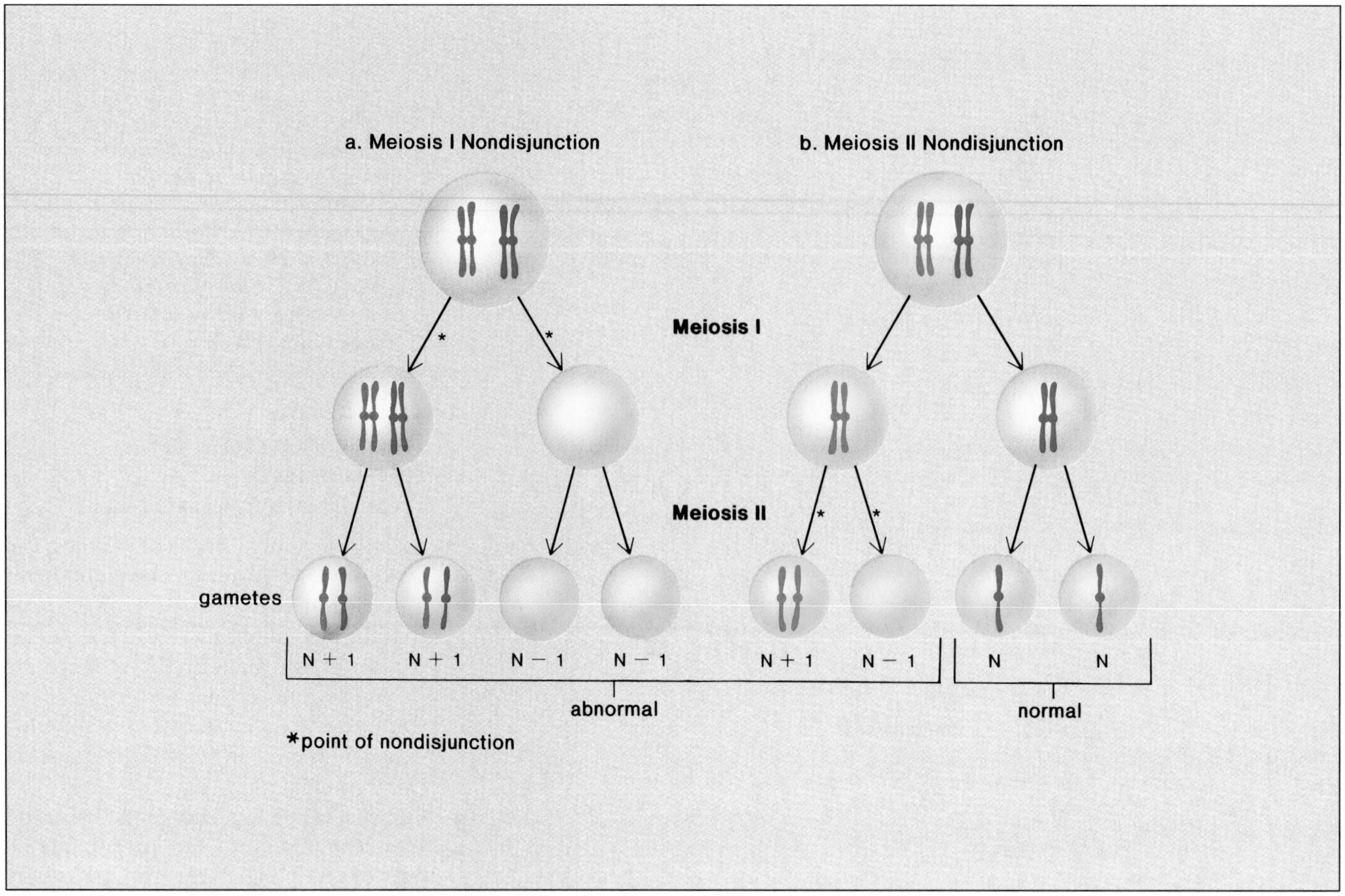

Figure 17.15

Nondisjunction of autosomes during oogenesis. **a.** Nondisjunction can occur during meiosis I if homologous chromosomes fail to separate and **b.** during meiosis II if the sister chromatids fail to separate completely. In either case, certain abnormal eggs carry an extra chromosome. Nondisjunction of chromosome 21 can lead to Down syndrome.

Nondisjunction: Misassortment

Abnormal chromosome constituencies (table 17.1) are due to nondisjunction. **Nondisjunction** is the failure of homologous chromosomes or sister chromatids to separate during the formation of gametes. Nondisjunction can occur during meiosis I if the homologous chromosomes fail to separate or during meiosis II if the sister chromatids fail to separate.

Figure 17.15*a* shows nondisjunction during meiosis I. The homologous chromosomes are duplicated in the parental cell. The daughter cells are abnormal because the daughter cell on the left received both members of the homologous pair and the daughter cell on the right received neither. Following meiosis II, all 4 daughter cells are abnormal. If the chromosomes are chromosome 21, the first 2 cells could result in Down syndrome after fertilization. If the chromosomes were X chromosomes, Klinefelter syndrome could result. If they were Y chromosomes, XYY syndrome would result following fertilization. The 2 cells on the right could result in Turner syndrome.

In figure 17.15*b,* nondisjunction occurs during meiosis II. Meiosis I occurred normally and each daughter cell has one of the members of the homologous chromosomes. During meiosis II, the chromatids fail to separate in the daughter cell on the left. The abnormal daughter cells can result in the same syndrome discussed in the preceding paragraph.

> Nondisjunction results in gametes with an abnormal number of chromosomes and, following fertilization, the syndromes listed in table 17.1.

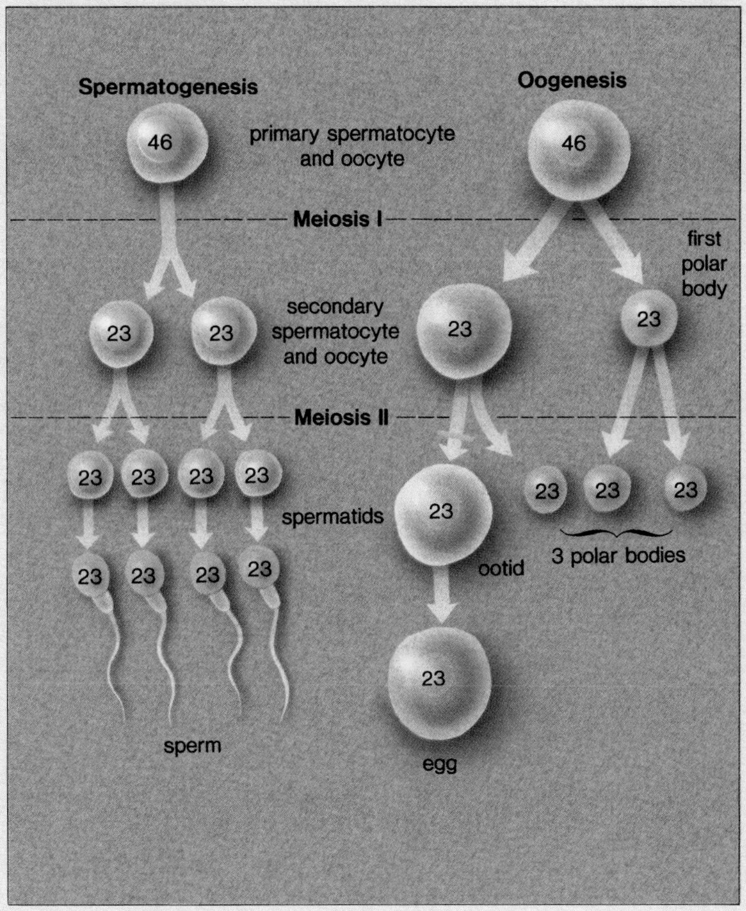

Figure 17.16

Spermatogenesis and oogenesis. Spermatogenesis produces 4 viable sperm, whereas oogenesis produces one egg and 3 polar bodies. Notice that oogenesis does not go to completion unless the secondary oocyte is fertilized. In humans, both sperm and egg have 23 chromosomes each; therefore, following fertilization, the zygote has 46 chromosomes.

Spermatogenesis and Oogenesis

Spermatogenesis and oogenesis occur in the sex organs, the testes in males and the ovaries in females. During **spermatogenesis,** sperm are produced, and during **oogenesis,** eggs are produced. The gametes appear differently in the 2 sexes (fig. 17.16), and meiosis is different, too. The process of meiosis in males always results in 4 cells that become sperm. Meiosis in females produces only one cell that goes on to become an egg. Meiosis I results in one large cell called a secondary oocyte and one polar body. After meiosis II there is one egg and 3 polar bodies. The **polar bodies** are a way to discard unnecessary chromosomes while retaining much of the cytoplasm in the egg. The cytoplasm serves as a source of nutrients for the developing embryo.

Spermatogenesis, once started, continues to completion and mature sperm result. In contrast, oogenesis does not necessarily go to completion. Only if a sperm fertilizes the secondary oocyte does it undergo meiosis II and become an egg. Regardless of this complication, however, both the sperm and the egg contribute the haploid number of chromosomes to the fertilized egg. In humans, each contributes 23 chromosomes. In addition, the egg contributes most of the cytoplasm. Figure 17.16 shows how the sperm and egg are adapted to their function. The sperm is a tiny flagellated cell while the egg is stationary and quite large.

Spermatogenesis, which occurs in the testes of males, produces sperm. Oogenesis, which occurs in the ovaries of females, produces eggs. Meiosis is a part of spermatogenesis and oogenesis; therefore, both sperm and egg are haploid.

SUMMARY

The cells of each organism have a characteristic number of chromosomes; humans have 46. A human karyotype ordinarily shows 22 pairs of autosomes and one pair of sex chromosomes. The sex pair is an X and a Y chromosome in males and two X chromosomes in females. Abnormalities do occur. The major inherited autosomal abnormality is Down syndrome, in which the individual inherits 3 copies of chromosome 21. Cri du chat occurs when part of chromosome 5 is deleted. Examples of abnormal sex chromosome inheritance are a fragile X chromosome and abnormal chromosome numbers: Turner syndrome (XO), Klinefelter syndrome (XXY), XYY males, and metafemales (XXX).

The life cycle of higher organisms requires 2 types of cell divisions, mitosis and meiosis. Mitosis assures that all cells in the body have the diploid number and same kinds of chromosomes. It is made up of 4 stages: prophase, metaphase, anaphase, and telophase. The cell cycle includes an additional stage termed *interphase.* During interphase, DNA replication causes each chromosome to have sister chromatids. When mitosis occurs, the chromatids separate, and each newly forming cell receives the same number and kinds of chromosomes as the original cell. The cytoplasm is partitioned by furrowing in human cells.

Meiosis involves 2 cell divisions. During meiosis I, the homologous chromosomes (following crossing-over between nonsister chromatids) separate, and during meiosis II the sister chromatids separate. The result is 4 cells with the haploid number of chromosomes in a single copy. Meiosis is a part of gamete formation in humans.

Spermatogenesis in males produces 4 viable sperm, while oogenesis in females produces one egg and 2 or 3 polar bodies. Each gamete is specialized for the job it does; the sperm is a tiny, flagellated cell that swims to the cytoplasm-laden egg.

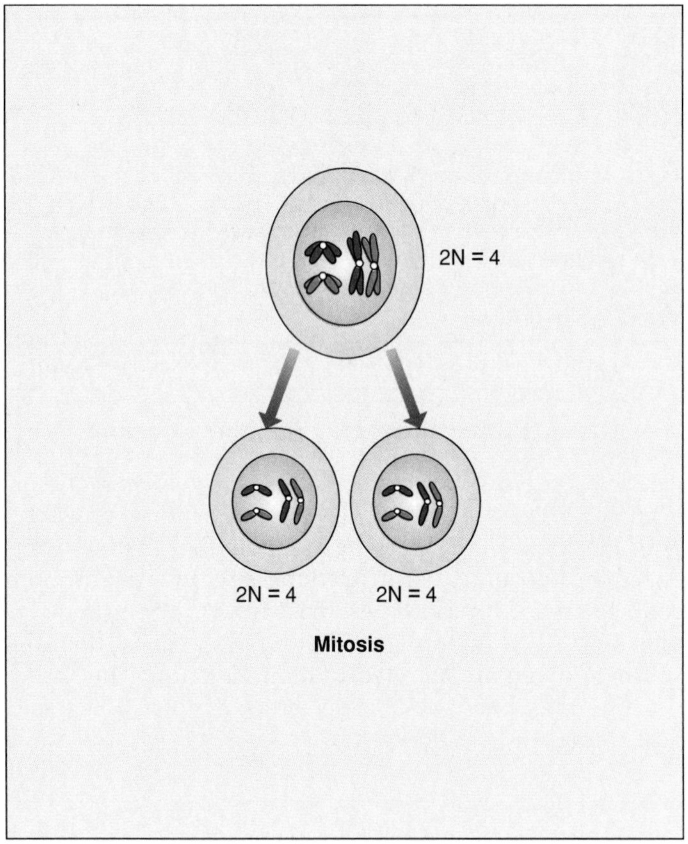

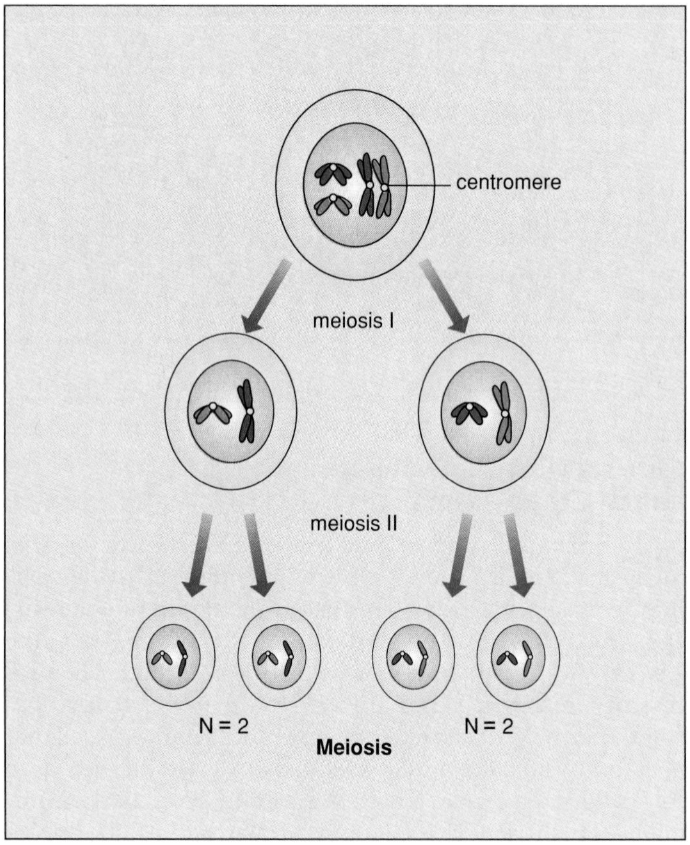

STUDY AND THOUGHT QUESTIONS

1. Describe the normal karyotype of a human being. What is the difference between a male and a female karyotype? (p. 342)
2. Describe Down syndrome, the most common autosome abnormality in humans. What is cri du chat syndrome? (p. 343)
3. Name 2 types of sex chromosome abnormalities in humans. List 4 that have to do with an abnormal number of sex chromosomes. (pp. 344–345)

   Critical Thinking In females, the "extra" X chromosome forms a condensed structure called a barr body. How many barr bodies would exist in an XXXY individual?
4. Draw a diagram describing the human life cycle. What is mitosis? What is meiosis? (p. 346)
5. What is the importance of mitosis and meiosis in the life cycle of humans? (p. 346)

   Ethical Issue Should couples be permitted to have identical offspring by means of cloning—even if one child is many years older than the other?
6. How do the terms diploid (2N) and haploid (N) pertain to meiosis? (pp. 346, 349)
7. Describe the stages of meiosis I, including the terms *tetrad* and *dyad* in your description. (pp. 348, 354)
8. Describe the stages of mitosis, including in your description the terms *centrioles, nucleolus, spindle,* and *furrowing.* (pp. 350–353)
9. Compare the stages of meiosis II to a mitotic division. (p. 354)

   Critical Thinking Meiosis I is the reduction division. Speculate why there is meiosis II.
10. What is nondisjunction, and how does it occur? (p. 356)

    Critical Thinking What error, during what stage, most likely accounts for nondisjunction in conjunction with meiosis I?
11. How does spermatogenesis in males compare to oogenesis in females? (p. 357)

    Ethical Issue Do you favor preselecting sperm based on its ability to produce male or female children?

WRITING ACROSS THE CURRICULUM

1. The spindle apparatus ensures the inheritance of the diploid number of chromosomes in each daughter cell. Why is there no need to ensure equal distribution of the organelles to the daughter cells?
2. Both egg and sperm contribute one member of each pair of chromosomes to the new individual, but the egg contributes most of the cytoplasm. Why does this seem appropriate, considering the manner in which humans procreate, and what implications does it have for inheritance?
3. Down syndrome children are subject to serious illness and mental retardation. The chance of having a Down syndrome child is greater in older women. The syndrome can be detected by chorionic villi sampling or amniocentesis in the early months of pregnancy. Do you believe an older woman should make use of these tests? Why or why not?

Objective Questions

1. The arrangement of an individual's chromosomes according to homologous pairs is called a ____________________.
2. The karyotype of males includes the sex chromosomes ______________, and the karyotype of females includes the sex chromosomes ____________________.
3. The diploid number of chromosomes is designated as the ______________ number, and the haploid number is designated as the ___________ number.
4. If the parent cell has 24 chromosomes, the daughter cells following mitosis will have ______________ chromosomes.
5. As the organelles called ______________ separate and move to the poles, the spindle fibers appear.
6. During meiosis I, the ______________ separate, and during meiosis II the ______________ separate.
7. Meiosis in males is a part of ______________, and meiosis in females is a part of ______________.
8. There is a ______________ chance of a newborn being a male or a female.

To answer questions 9–15, use this key:

a. Down syndrome
b. Turner syndrome
c. Klinefelter syndrome
d. cri du chat syndrome
e. metafemale

9. XXY _____
10. extra chromosome 21 _____
11. XXX _____
12. deletion in chromosome 5 _____
13. XO _____
14. due to an autosomal nondisjunction _____
15. due to a chromosomal mutation _____

Selected Key Terms

autosome Chromosome other than a sex chromosome. 342

centromere (sen´tro-mēr) A region of attachment of a chromosome to spindle fibers that is generally seen as a constricted area. 347

chromatid (kro´ma-tid) One of the 2 identical parts of a chromosome following replication of DNA. 347

crossing-over The exchange of corresponding segments of genetic material between nonsister chromatids of homologous chromosomes during synapsis of meiosis I. 348

diploid (2N) (dĭp´loid) The number of chromosomes in the body cells; twice the number of chromosomes found in gametes. 346

Down syndrome Human congenital disorder associated with an extra chromosome 21. 343

gamete (gam´ēt) One of 2 types of reproductive cells that join in fertilization to form a zygote; most often an egg or a sperm. 346

haploid (N) Half the diploid number; the number of chromosomes in the gametes. 346

homologous chromosome Similarly constructed; homologous chromosomes have the same shape and contain genes for the same traits. 342

karyotype (kar´e-o-tīp) The arrangement of all the chromosomes within a cell by pairs in a fixed order. 342

Klinefelter syndrome A condition caused by the inheritance of a chromosome abnormality in number; an XXY individual. 345

meiosis A type of cell division occurring during the production of gametes in animals by means of which the 4 daughter cells have the haploid number of chromosomes. 348

metafemale A female who has three X chromosomes. 345

mitosis Type of cell division in which daughter cells receive the exact chromosome and genetic makeup of the parent cell; occurs during growth and repair. 347

nondisjunction The failure of homologous chromosomes or sister chromatids to separate during the formation of gametes. 356

oogenesis (o″o-jen´ ĕ-sis) Production of an egg in females by the processes of meiosis and maturation. 357

sex chromosome Chromosome responsible for the development of characteristics associated with gender; an X or Y chromosome. 342

spermatogenesis (sper″ mah-to-jen´ĕ-sis) Production of sperm in males by the process of meiosis and maturation. 357

spindle fibers Microtubule bundles in eukaryotic cells that are involved in the movement of chromosomes during mitosis and meiosis. 352

synapsis (si-nap″sis) The attracting and pairing of homologous chromosomes during prophase I of meiosis. 348

tetrad A set of 4 chromatids resulting from the pairing of homologous chromosomes during prophase I of meiosis. 348

Turner syndrome A condition caused by the inheritance of an abnormality in chromosome number; an X chromosome lacks a homologous counterpart—XO. 345

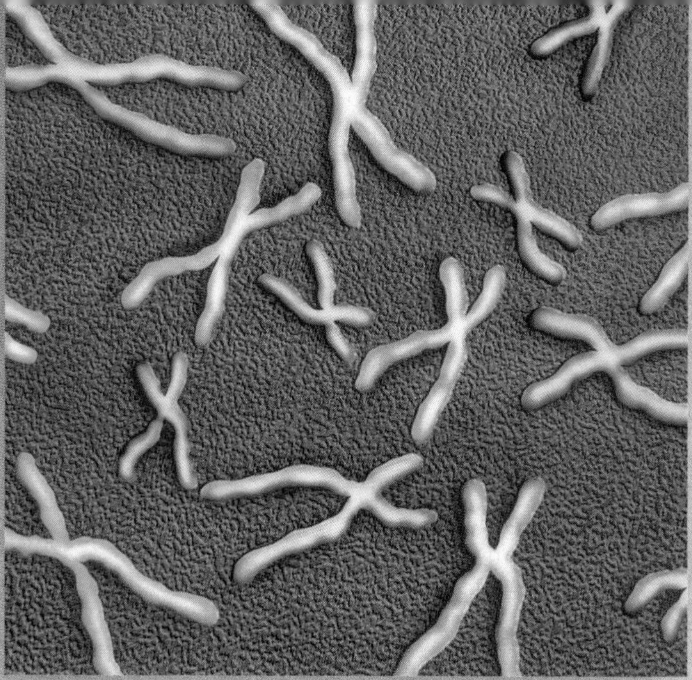

Chapter 18

Genes and Medical Genetics

The anatomical features of human beings are controlled by the genes. And, indeed, so are physiological and some behavioral features controlled by the genes. But what are the genes? This chapter shows that it is sometimes helpful to think of genes as being units of chromosomes, particularly when we want to calculate the chances of passing on a genetic disorder. But the next chapter shows that genes are actually DNA molecules that control protein synthesis. Only now are we beginning to decipher how this function can directly affect the structure of the cell and the organism.

Chapter Concepts

1. Inheritance of particular genes (dominant/recessive alleles) controls many of our traits. 363
2. Recessive genetic disorders require two alleles, one from each parent; dominant disorders require only one allele. 365–368
3. Some traits like sickle-cell disease are incompletely dominant. 370
4. Polygenic traits include skin color, behavior, and various syndromes. 371
5. Blood type is controlled by multiple alleles. 372
6. Sex-linked traits are usually carried on the X chromosome. Males, with only one X, are more likely to inherit X-linked disorders. 373
7. Some traits are sex-influenced rather than sex-linked. 375

Chapter Outline

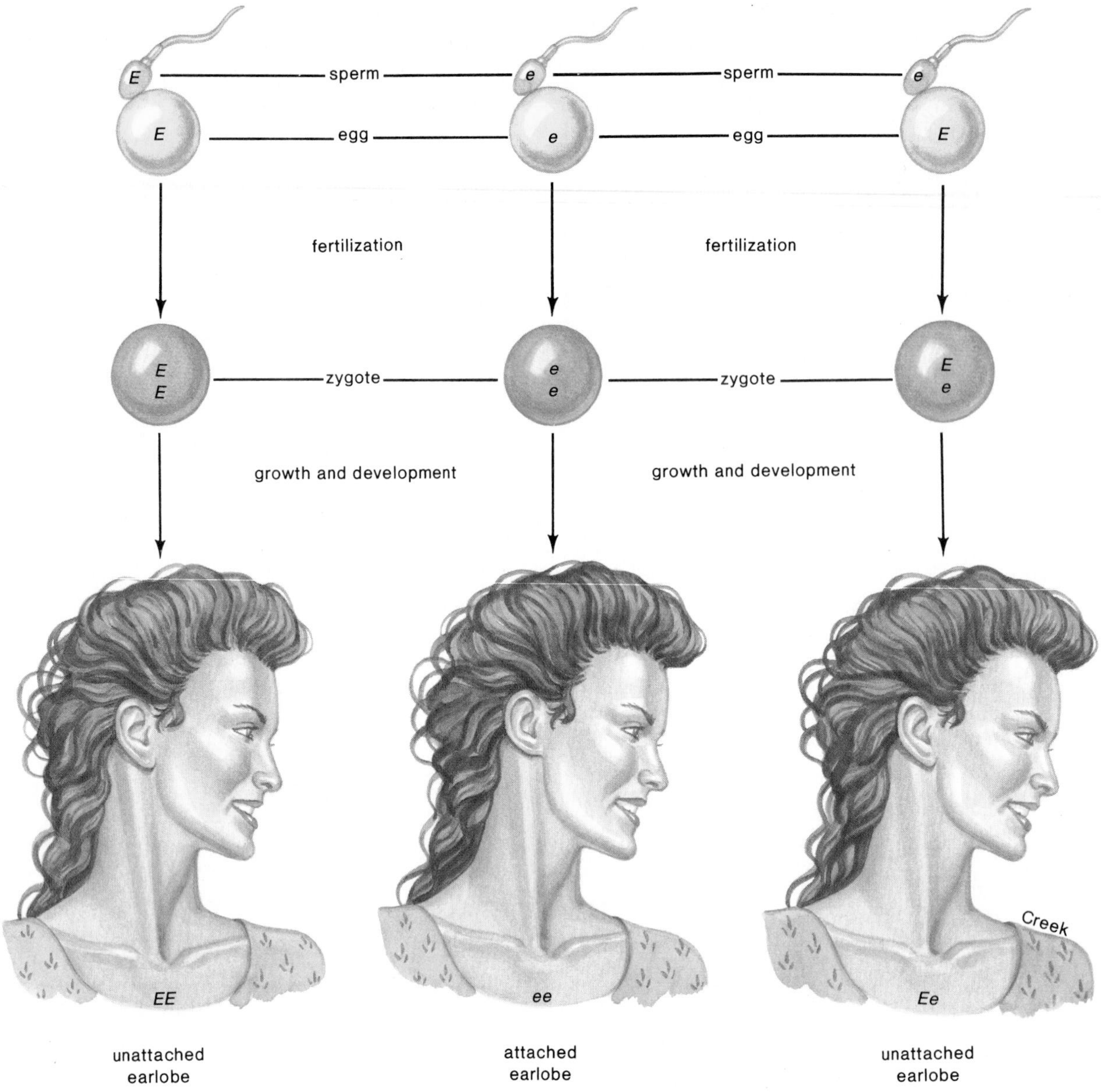

Figure 18.1

Genetic inheritance. Individuals inherit a minimum of 2 alleles for every characteristic of their anatomy and physiology. The inheritance of a single dominant allele (*E*) causes an individual to have unattached earlobes; 2 recessive alleles (*ee*) cause an individual to have attached earlobes. Notice that each individual receives one allele from the father (by way of a sperm) and one allele from the mother (by way of an egg).

Alternate forms of a gene having the same position (locus) on a pair of chromosomes and affecting the same **trait** are called **alleles.** It is customary to designate an allele by a letter, which represents the specific characteristic it controls; a **dominant allele** is assigned an uppercase (capital) letter, while a **recessive allele** is given the same letter, lowercased. In humans, for example, unattached (free) earlobes are dominant over attached earlobes, so a suitable key would be *E* for unattached earlobes and *e* for attached earlobes.

Table 18.1

Genotype and Phenotype

Genotype (in letters)	Genotype (in words)	Phenotype
EE	Homozygous (pure) dominant	Unattached earlobes
ee	Homozygous (pure) recessive	Attached earlobes
Ee	Heterozygous (hybrid)	Unattached earlobes

Since autosomal alleles occur in pairs, the individual normally has 2 alleles for a characteristic. Just as one of each pair of chromosomes is inherited from each parent, so too is one of each pair of alleles inherited from each parent.

Figure 18.1 shows 3 possible fertilizations and the resulting genetic makeup of the zygote and, therefore, the individual. In the first instance, the chromosome of both the sperm and egg carries an *E*. Consequently, the zygote and subsequent individual have the alleles *EE*, which may be called a **homozygous** (pure) dominant genotype (table 18.1). The word **genotype** refers to the genes of the individual. A person with genotype *EE* obviously has unattached earlobes. The physical appearance of the individual, in this case unattached earlobes, is called the **phenotype.**

In the second fertilization, the zygote received 2 recessive alleles (*ee*), and the genotype is called homozygous (pure) recessive. An individual with this genotype has attached earlobes. In the third fertilization, the resulting individual has the alleles *Ee*, which is called a **heterozygous** genotype. A heterozygote shows the dominant characteristic; therefore, the phenotype of this individual is unattached earlobes.

These examples show that a dominant allele contributed from only one parent can bring about a particular phenotype. A recessive allele must be received from both parents to bring about the recessive phenotype.

The genotype, whether homozygous recessive (*ee*), homozygous dominant (*EE*), or heterozygous (*Ee*), tells what genes a person has. The phenotype, for example, attached or free earlobes, tells what the person looks like.

Dominant/Recessive Traits

Many times parents would like to know the chances of an individual child having a certain genotype and, therefore, a certain phenotype. If one of the parents is homozygous dominant (*EE*), it is obvious that the chances of having a child with unattached earlobes is 100%, because this parent has only a dominant allele (*E*) to pass on to the offspring. On the other hand, if both parents are homozygous recessive (*ee*), there is a 100% chance that each child will have attached earlobes. However, there are instances in which the expected phenotype is not so easily ascertained. If both parents are heterozygous, then what are the chances that a child will have unattached or attached earlobes? To solve a problem of this type, it is customary *first* to indicate the genotype of the parents and their possible gametes.

Genotypes	*Ee*	*Ee*
Gametes:	*E* and *e*	*E* and *e*

Second, a **Punnett square** is used to determine the phenotype ratio among the offspring when all possible sperm are given an equal chance to fertilize all possible eggs (fig. 18.2). The possible sperm are lined up along one side of the square, and the possible eggs are lined up along the other side of the square (or vice versa). The ratio among the offspring in this case is 3:1 (3 children with unattached earlobes to one with attached earlobes). This means that there is a 3/4 chance (75%) for each child to have unattached earlobes and a 1/4 chance (25%) for each child to have attached earlobes.

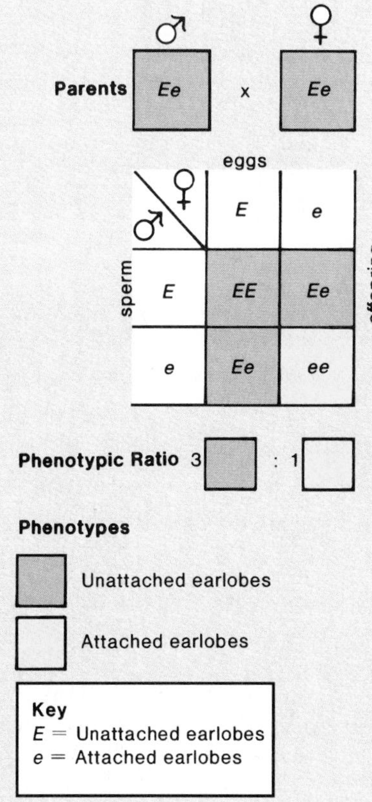

Figure 18.2

When the parents are heterozygous, each child has a 75% chance of having the dominant phenotype and a 25% chance of having the recessive phenotype.

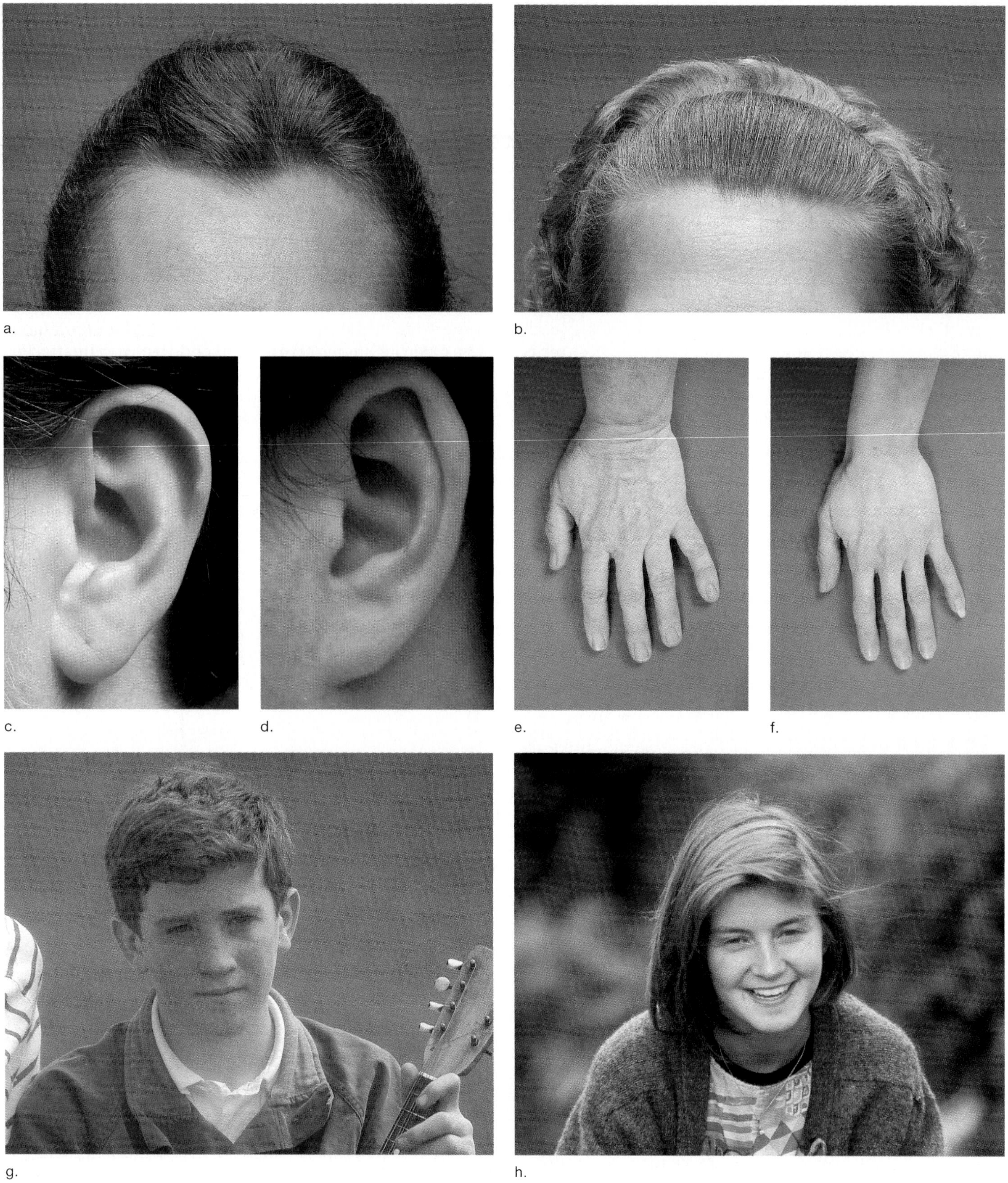

Figure 18.3

Common inherited characteristics in human beings. Widow's peak **a.** is dominant over **b.** continuous hairline. Unattached earlobe **c.** is dominant over **d.** attached earlobe. Short fingers **e.** are dominant over long fingers **f.** Freckles **g.** are dominant over **h.** lack of freckles.

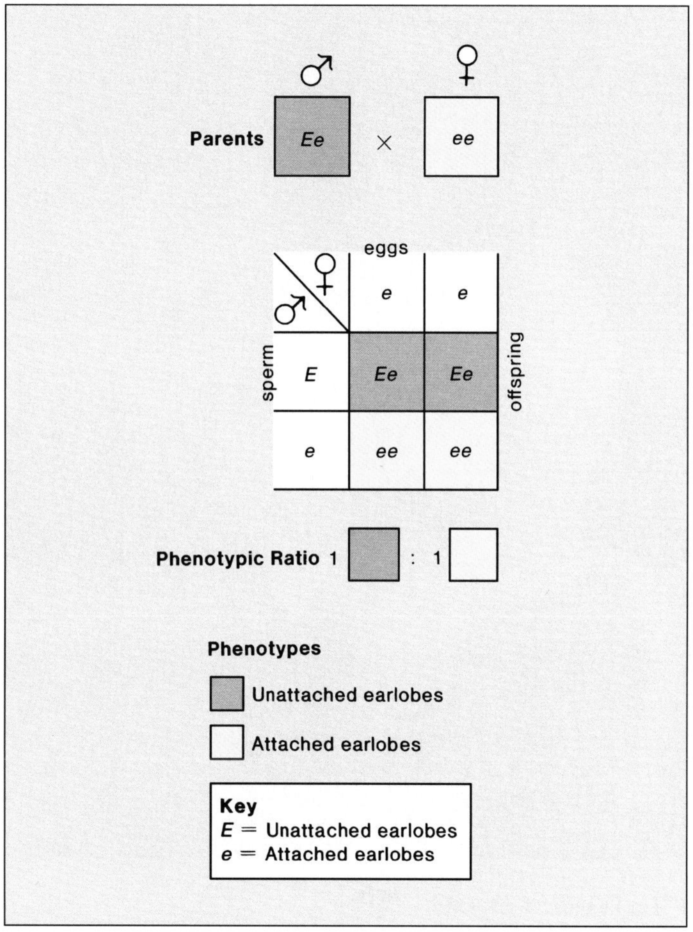

Figure 18.4

When one parent is heterozygous and the other recessive, each child has a 50% chance of having the dominant phenotype and a 50% chance of having the recessive phenotype.

Another cross of particular interest is that between a heterozygous individual (*Ee*) and a pure recessive (*ee*). In this case, the Punnett square shows that the ratio among the offspring is 1:1, and the chance of the dominant or recessive phenotype is 1/2, or 50% (fig. 18.4).

Comparing the two crosses (fig. 18.2 and fig. 18.4), each child has a 75% chance of having the dominant phenotype if the 2 parents are heterozygous and a 50% chance if one parent is heterozygous and the other is recessive. Each child has a 25% chance of having the recessive phenotype if the parents are heterozygous and a 50% chance if one parent is heterozygous and the other is pure recessive.

If the genotype of the parents is known, it is possible to calculate the chances of a child having a particular characteristic. If both parents are heterozygous, each child has a 25% chance of the recessive phenotype. If one parent is heterozygous and the other is homozygous recessive, each child has a 50% chance of the recessive phenotype.

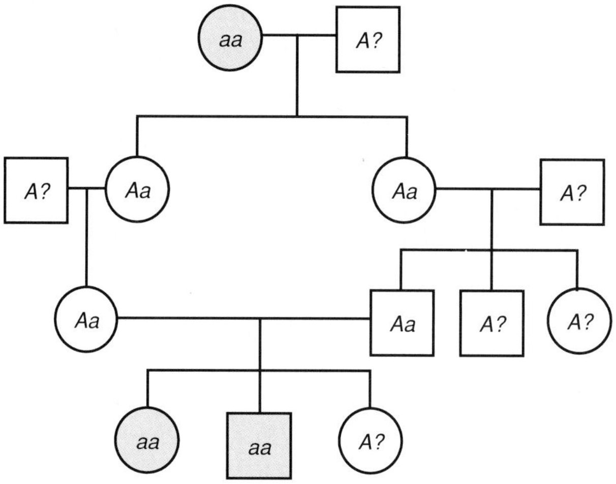

Autosomal Recessive Genetic Disorders

- Most affected children have normal parents.
- Heterozygotes have a normal phenotype.
- Two affected parents will always have affected children.
- Affected individuals who have noncarrier spouses will have normal children.
- Close relatives who reproduce are more likely to have affected children.
- Both males and females are affected with equal frequency.

Key
aa = Affected
Aa = Carrier (appears normal)
AA = Normal

Figure 18.5

Sample pedigree chart for an autosomal recessive genetic disorder. Only those affected are shaded. Which persons in the chart are carriers?

Recessive Disorders: Takes Two Alleles

When studying human genetic disorders, biologists often construct pedigree charts to show the pattern of inheritance for a characteristic within a group of people. In a pedigree chart, males are designated by squares and females are designated by circles (fig. 18.5). Shaded circles and squares indicate affected individuals. A line between a square and a circle represents a couple who have mated. A vertical line going downward leads to a single child. (If there is more than one child, they are placed off a horizontal line.)

Figure 18.5 shows a typical pedigree chart for a recessive genetic disorder. It is obvious that a genetic disorder is recessive when an affected child is born to parents who appear to be normal. Other ways of recognizing a recessive disorder are also given in the figure.

Parents with a heterozygous genotype, that is, those who carry a hidden faulty gene, are called **carriers.** It is important to realize that "chance has no memory," therefore, each child of these parents has a 25% chance of having the disorder. In other words, it is possible that if a heterozygous couple had 4 children, each child might have the condition.

Recessively inherited genetic disorders are sometimes more prevalent among members of a particular ethnic group. Members of the same ethnic group are more apt to be carriers for the same recessive disorder. Members of the same family are even more likely to be carriers for the same recessive disorder. Parents who are concerned about passing on a genetic disorder sometimes go for counseling, as discussed in this chapter's Health Focus.

Cystic Fibrosis: Thick Mucus

Cystic fibrosis is the most common lethal genetic disease among Caucasians in the United States. About one in 5 Caucasians is a carrier, and about one in 2,500 children born to this group has the disorder. In these children, the mucus in the lungs and the digestive tract is particularly thick and viscous. In the lungs, the mucus interferes with gas exchange (fig. 18.6).

Thick mucus also impedes the secretion of pancreatic juices, and food cannot be properly digested. This results in large, frequent, and foul-smelling stools.

In the past few years, much progress has been made in our understanding of cystic fibrosis, and new treatments have raised the average life expectancy to 17–20 years of age. Research has demonstrated that chloride ions (Cl^-) fail to pass through cell membrane channel proteins in these patients. Ordinarily, after chloride ions have passed through the membrane, water follows. It is believed that lack of water in the lungs is what causes the mucus to be abnormally thick. The cystic fibrosis gene, which is located on chromosome 7, has been isolated and inserted into the lungs of living animals. The hope is that one day it will be possible to use an inhaler to carry copies of the normal gene into the lungs of cystic fibrosis patients.

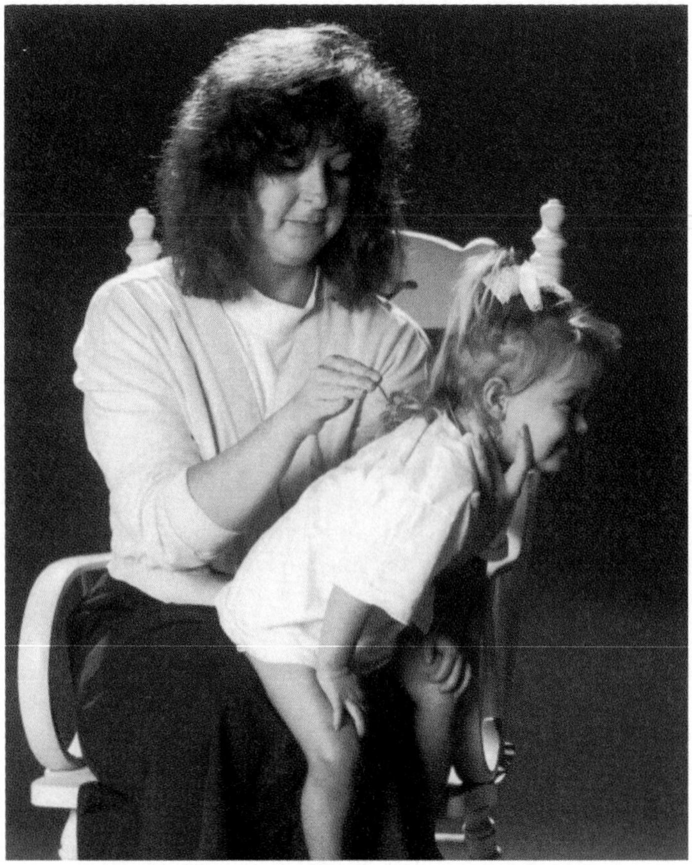

Figure 18.6

The mucus in the lungs of a child with cystic fibrosis should be periodically loosened by clapping the back. A new treatment destroys the cells that tend to build up in the lungs. The white blood cells are destroyed by one drug and another drug does away with the DNA the cells leave behind. Judicious use of antibiotics controls pulmonary infection, and aggressive use of other drugs thins mucous secretions.

Tay-Sachs Disease: No Hex A

Tay-Sachs disease is a well-known genetic disease of high incidence among Jewish people in the United States, most of whom are of central and eastern European descent. At first, it is not apparent that a baby has Tay-Sachs disease. However, development begins to slow down between 4 months and 8 months of age—neurological impairment and psychomotor difficulties then become apparent. The child gradually becomes blind and helpless, develops uncontrollable seizures, and eventually becomes paralyzed. There is no treatment or cure for Tay-Sachs disease, and most affected individuals die by the age of 3 or 4.

So-called late-onset Tay-Sachs disease occurs in adults. The symptoms are progressive mental and motor deterioration, depression, schizophrenia, and premature death. The gene for late-onset Tay-Sachs disease has been sequenced, and this form of the disorder apparently is due to one changed pair of bases in the DNA of chromosome 1.

Tay-Sachs disease results from a lack of the enzyme hexosaminidase A (Hex A) and the subsequent storage of its substrate, a glycosphingolipid, in lysosomes. Although more and more lysosomes build up in many body cells, the primary sites of storage are the cells of the brain, which accounts for the onset and the progressive deterioration of psychomotor functions.

There is a test to detect carriers of Tay-Sachs. The test uses a sample of serum, white blood cells, or tears to determine if Hex A activity is present. Affected individuals have no detectable Hex A activity. Carriers have about half the level of Hex A activity found in normal individuals. Prenatal diagnosis of the disease also is possible following either amniocentesis or chorionic villi sampling.

Health Focus

Genetic Counseling

Now that potential parents are becoming aware that many illnesses are caused by faulty genes, more couples are seeking genetic counseling. The counselor studies the background of the couple and tries to determine if any immediate ancestor may have had a genetic disorder. A pedigree chart may be constructed. Then the counselor studies the couple themselves. As much as possible, laboratory tests are performed on all persons involved.

Tests are now available for a large number of potential genetic diseases. For example, chromosome tests are available for cystic fibrosis, neurofibromatosis, and Huntington disease. Blood tests can identify carriers of thalassemia and sickle-cell disease. By measuring enzyme levels in blood, tears, or skin cells, carriers of enzyme defects can also be identified for certain inborn metabolic errors, such as Tay-Sachs disease. From this information, the counselor can sometimes predict the chances of a child having the disorder.

Whenever the woman is pregnant, chorionic villi sampling can be done early and amniocentesis can be done later in the pregnancy. These procedures, which were illustrated in the previous chapter (p. 324), allow the testing of embryonic and fetal cells, respectively, to determine if the child has a genetic disorder. Sometimes treatment can be started even before birth (table 18A).

Table 18A

Treatment for Some Human Genetic Disorders

Name	Description	Chromosome	Incidence Among Newborns	Status
Autosomal Recessive Disorders				
Cystic fibrosis	Mucus in the lungs and digestive tract is thick and viscous, making breathing and digestion difficult	7	One in 2,500 Caucasians	Allele located; chromosome test now available;* treatment and cure a possibility soon
Tay-Sachs disease	Neurological impairment and psychomotor difficulties develop early, followed by blindness and uncontrollable seizures before death occurs	15	One in 3,600 eastern European Jews	Biochemical test now available*
Phenylketonuria (PKU)	Inability to metabolize phenylalanine, and if a special diet is not begun mental retardation develops	12	One in 5,000 Caucasians	Biochemical test now available; treatment available
Autosomal Dominant Disorders				
Neurofibromatosis (NF)	Benign tumors occur under the skin or deeper	17	One in 3,000	Allele located; chromosome test now available*
Huntington disease	Minor disturbances in balance and coordination develop in middle age and progress towards severe neurological disturbances before death occurs	4	One in 20,000	Allele located; chromosome test now available*
Incomplete Dominance				
Sickle-cell disease	Poor circulation, anemia, internal hemorrhaging, due to sickle-shaped blood cells	11	One in 500 African Americans	Chromosome test now available*
X-Linked Recessive				
Hemophilia A	Propensity for bleeding, often internally, due to the lack of a blood clotting factor	X	One in 15,000 live male births	Treatment available
Duchenne muscular dystrophy	Muscle weakness develops early and progressively intensifies until death occurs	X	One in 5,000 live male births	Allele located; biochemical tests of muscle tissue available; treatment available soon

**Prenatal testing is done.*

Phenylketonuria (PKU): Buildup Damages Brain

Phenylketonuria (PKU) occurs in one in 5,000 births and so is not as frequent as the disorders previously discussed. When it does occur, the parents are very often close relatives. Affected individuals lack an enzyme that is needed for the normal metabolism of the amino acid phenylalanine, and an abnormal breakdown product, a phenylketone, accumulates in the urine. Newborns are routinely tested, and if they lack the necessary enzyme, they are placed on a diet low in phenylalanine. This diet must be continued until the brain is fully developed or else severe mental retardation develops.

To avoid the high risk of having a microcephalic child—one with an abnormally small head and severe mental retardation—a PKU woman should resume her limited diet several months before getting pregnant. The diet must not include the artificial sweetener NutraSweet™ because it contains phenylalanine.

Dominant Disorders: Takes One Allele

Figure 18.7 shows a pedigree chart for a dominant genetic disorder. A child who has the condition must have a parent with the condition unless, of course, a **mutation** (genetic change) has occurred in the inherited chromosome. Figure 18.7 gives other ways of recognizing a dominant disorder.

Usually an offspring with a dominant genetic disorder has one heterozygous parent and one homozygous recessive parent. Therefore, the child had a 50% chance of getting the faulty gene or escaping it completely. Again, it must be remembered that chance has no memory, and since each child has the same genetic chance it would be possible for several children in the same family to inherit a dominant genetic disease.

Neurofibromatosis: Common

Neurofibromatosis (NF), sometimes called von Recklinghausen[1] disease, is one of the most common genetic disorders. It affects roughly one in 3,000 people, including an estimated 100,000 in the United States. It is seen equally in every racial and ethnic group throughout the world.

At birth or later, the affected individual may have 6 or more large tan spots on the skin. Such spots may increase in size and number and get darker. Small benign tumors (lumps) called neurofibromas may occur under the skin or in the muscles. Neurofibromas are made up of nerve cells and other cell types.

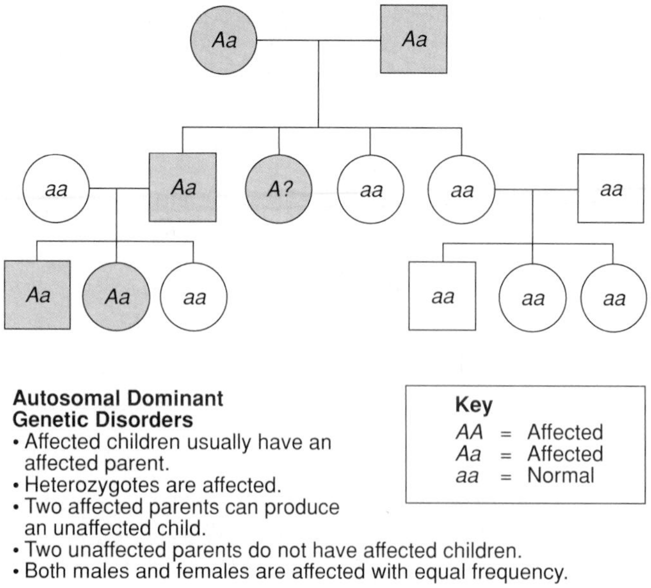
Aa
Aa
aa
Aa
A?
aa
aa
aa
Aa
Aa
aa
aa
aa
aa
Autosomal Dominant Genetic Disorders
• Affected children usually have an affected parent.
• Heterozygotes are affected.
• Two affected parents can produce an unaffected child.
• Two unaffected parents do not have affected children.
• Both males and females are affected with equal frequency.
Key
AA = Affected
Aa = Affected
aa = Normal

Figure 18.7

Sample pedigree chart for an autosomal dominant genetic disorder. Are there any carriers?

This genetic disorder shows *variable expressivity.* In most cases, symptoms are mild and patients live a normal life. In some cases, however, the effects are severe. Skeletal deformities, including a large head, are seen, and eye and ear tumors can lead to blindness and hearing loss. Many children with NF have learning disabilities and are hyperactive.

The NF gene is known to be on chromosome 17, and researchers have developed a test to diagnose the disorder. When the gene is functioning properly, it suppresses abnormal cell division. Cells that divide and produce tumors have a nonfunctioning gene.

Huntington Disease: Begins in Middle Age

One in 20,000 persons in the United States has **Huntington disease (HD),** a neurological disorder that affects specific regions of the brain. Most individuals who inherit the allele appear normal until middle age. Then, minor disturbances in balance and coordination lead to progressively worse neurological disturbances (fig. 18.8). The individual becomes insane before death occurs.

1. Although neurofibromatosis is commonly associated with Joseph Merrick, the severely deformed nineteenth-century Londoner depicted in *The Elephant Man*, researchers today believe Merrick actually suffered from a much rarer disorder called Proteus syndrome.

Figure 18.8

Huntington disease. Persons with this condition gradually lose psychomotor control of the body. At first there are only minor disturbances, but the symptoms become worse over time.

Much has been learned about Huntington disease. The gene for the disease is located on chromosome 4, and there is a test, of the type described on page 398, to determine if the dominant gene has been inherited. Because treatment is not available, however, few may want to have this information.

Research is being conducted, though, to determine the underlying cause of the disorder. It is known that the brain of a person with Huntington disease produces more than the usual amount of quinolinic acid, an excitotoxin that can overstimulate certain nerve cells. It is believed to lead to the death of these cells and to the subsequent symptoms of Huntington disease. Researchers are looking for chemicals that block quinolinic acid's action or inhibit quinolinic acid synthesis.

> Some of the best-known genetic disorders in humans are either autosomal recessive or autosomal dominant disorders. A pedigree charts shows the pattern of inheritance in a particular family.

Incompletely Dominant Traits

It is possible for 2 genes to be partially expressed, in which case both equally affect the phenotype. For example, a union between a straight-haired person and a curly-haired person produces children with wavy hair:

SS = straight hair

ss = curly hair

Ss = wavy hair

(In this instance, capital and lowercase letters are used for the key. If this type of key is chosen, remember that *Ss* will have an intermediate genotype.)

Figure 18.9 shows that if 2 wavy-haired individuals mate, any of the 3 phenotypes is possible. The chances for straight hair are 25%, the chances for wavy hair are 50%, and the chances for curly hair are 25%.

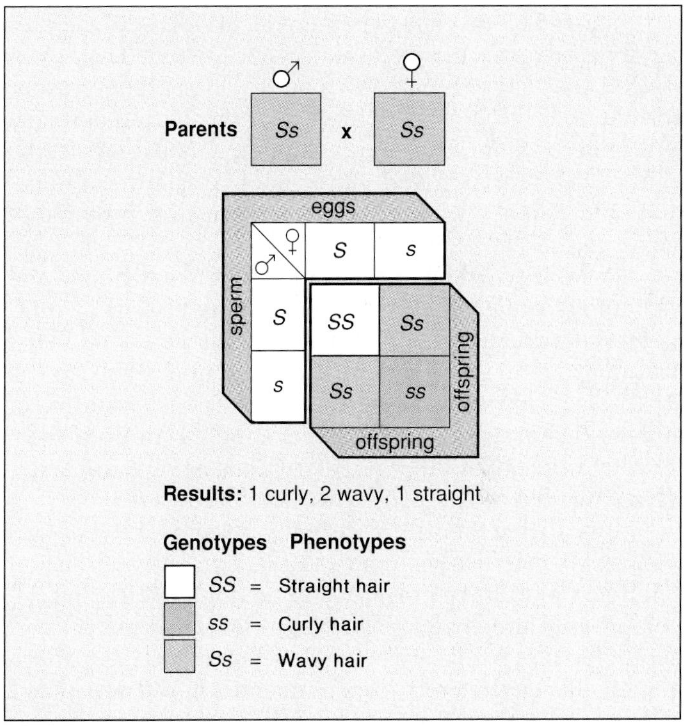

Figure 18.9

Incomplete dominance. Among Caucasians, neither straight nor curly hair is dominant. When 2 wavy-haired individuals reproduce, each offspring has a 25% chance of having either straight or curly hair and a 50% chance of having wavy hair, the intermediate phenotype.

Sickle-cell disease is an example of a human disorder that is controlled by incompletely dominant alleles. Individuals with the genotype $Hb^{A}Hb^{A}$ are normal, those with the $Hb^{S}Hb^{S}$ genotype have sickle-cell disease, and those with the $Hb^{A}Hb^{S}$ genotype have **sickle-cell trait,** a condition in which the cells are sometimes sickle shaped. Two individuals with sickle-cell trait can produce children with all 3 phenotypes, as indicated in figure 18.10.

Among regions of malaria-infested Africa, infants with sickle-cell disease die, but infants with sickle-cell trait have a better chance of survival than the normal homozygote. Their sickle cells give protection against the malaria-causing parasite, which uses red blood cells during its life cycle. The parasite dies when potassium leaks out of the red blood cells as the cells become sickle shaped. The protection afforded by sickle-cell trait keeps the allele prevalent in populations exposed to malaria. As many as 60% of blacks in malaria-infected regions of Africa have the allele. In the United States, about 10% of the black population carries the allele.

The red blood cells in persons with sickle-cell disease cannot easily pass through small blood vessels. The sickle-shaped cells either break down or they clog blood vessels, and the individual suffers from poor circulation, anemia, and sometimes internal hemorrhaging. Jaundice, episodic pain in the abdomen and joints, poor resistance to infection, and damage to internal organs are all symptoms of sickle-cell disease.

Persons with sickle-cell trait do not usually have any difficulties unless they experience dehydration or mild oxygen deprivation. Although a recent study found that army recruits with sickle-cell trait are more likely to die when subjected to extreme exercise, previous studies of athletes do not substantiate these findings. At present, most investigators believe that no restrictions on physical activity are needed for persons with sickle-cell trait.

Innovative therapies are being attempted in persons with sickle-cell disease. For example, persons with sickle-cell disease produce normal fetal hemoglobin during development, and drugs that turn on the genes for fetal hemoglobin in adults are being developed. Mice have been genetically engineered to produce sickle-cell red blood cells in order to test new antisickling drugs and various genetic therapies.

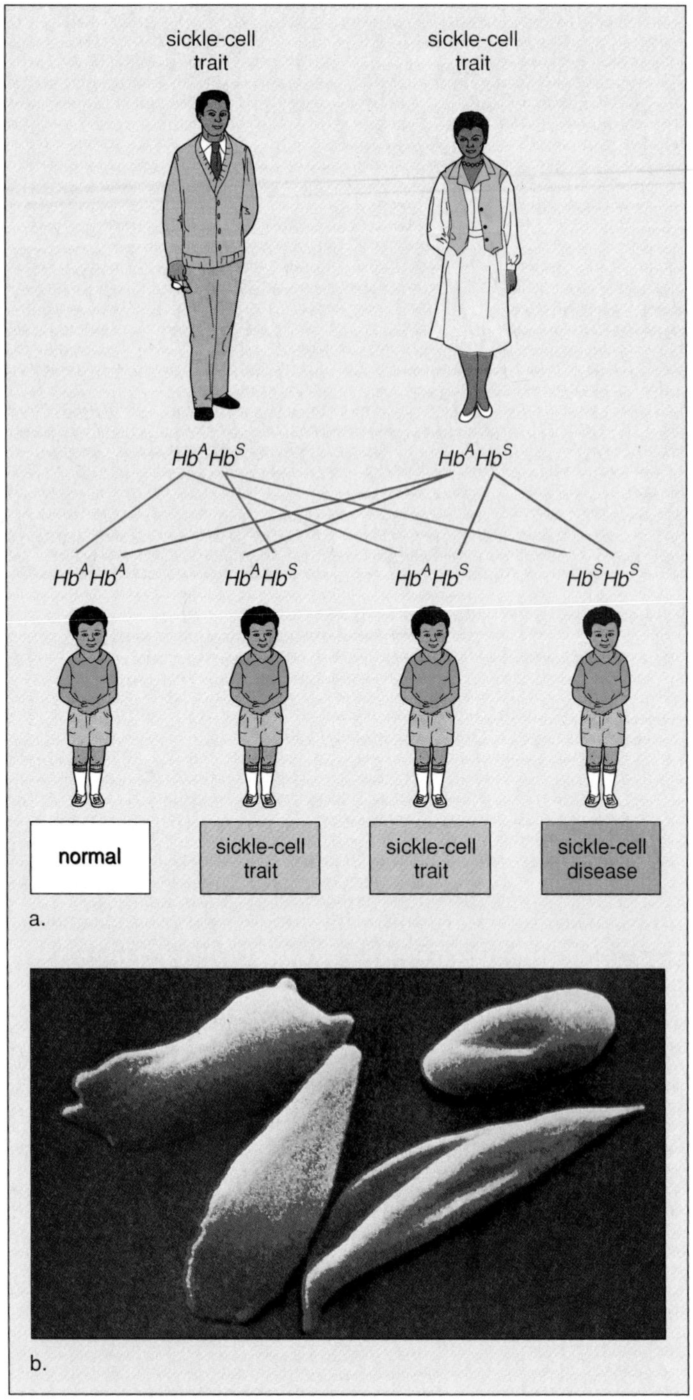

Figure 18.10

Inheritance of sickle-cell disease. **a.** In this example, both parents have the sickle-cell trait. Therefore, each child has a 25% chance of having sickle-cell disease or of being perfectly normal and a 50% chance of having the sickle-cell trait. **b.** Sickled cells. Individuals with sickle-cell disease have sickled red blood cells that tend to clump, as illustrated here.

Figure 18.11

Inheritance of skin color. This white husband (*aabb*) and his intermediate wife (*AaBb*) had fraternal twins, one of whom is white and one of whom is intermediate.

Figure 18.12

Jack Yufe (*left*) and Oskar Stohr (*right*) are identical twins with remarkably similar behavior even though they were reared separately.

Polygenic Traits

Two or more pairs of alleles may affect the same trait, sometimes in an additive fashion. *Polygenic inheritance* can cause the distribution of human characteristics according to a bell-shaped curve, with most individuals exhibiting the average phenotype. The more genes that control the trait, the more continuous the distribution will be.

Skin Color: Several Allele Pairs

Just how many pairs of alleles control skin color is not known, but a range in colors can be explained on the basis of 2 pairs. When a black person reproduces with a white person, the children have medium-brown skin, and when 2 people with medium-brown skin reproduce with one another, the children range in skin color from black to white. This can be explained by assuming that skin color is controlled by 2 pairs of alleles and that each capital letter contributes to the color of the skin:

Phenotype	Genotypes
Black	*AABB*
Dark	*AABb* or *AaBB*
Medium-brown	*AaBb* or *AAbb* or *aaBB*
Light	*Aabb* or *aaBb*
White	*aabb*

Notice again that there is a range in phenotypes and that there are several possible phenotypes in between the 2 extremes (fig. 18.11). Therefore, the distribution of these phenotypes is expected to follow a bell-shaped curve—few people have the extreme phenotypes and most people have the phenotype that lies in the middle between the extremes.

Behavior: Nature or Nurture?

Is behavior primarily inherited, or is it shaped by environmental influences? This nature (inherited) versus nurture (environment) question has been asked for a long time, and twin studies have been employed to attempt to find the answer. Identical twins are derived from a single fertilized egg, and therefore they have inherited exactly the same chromosomes and genes. Fraternal twins are derived from 2 different fertilized eggs and, therefore, have no more genes in common than do any other brother and sister.

Twin studies help determine to what extent behavior is inherited. It has been found that fraternal twins even when raised in the same environment are not remarkably similar in behavior, whereas identical twins raised separately are sometimes remarkably similar. For example, Oskar Stohr was raised as a Catholic by his grandmother in Nazi Germany; Jack Yufe was raised by his Jewish father in the Caribbean (fig. 18.12). Yet these 2 men "like sweet liqueurs, . . . store rubber bands on their wrists, read magazines from back to front, dip buttered toast in their coffee, and have similar personalities."[2]

Responses to a questionnaire designed to provide additional information about behavioral traits showed identical twins reared separately tend to have more similar personalities than fraternal twins reared together. Altogether the data seemed to show that about 50% of the *differences* in human personality traits was due to polygenic inheritance and 50% was due to environmental influence.

2. C. Holden, "Identical Twins Reared Apart," *Science* 207 (1980):1323–28.

Polygenic Disorders

A number of serious genetic disorders, such as *cleft lip* (or palate), *clubfoot, congenital dislocation of the hip,* and certain spinal conditions, are traditionally believed to be controlled by a combination of genes on autosomal chromosomes. This belief is being challenged by researchers who studied the inheritance of cleft palate in a large family in Iceland. These researchers found a gene on the X chromosome that alone can cause cleft palate.

Neural tube defect is a spinal condition that is inherited as a polygenic disorder. A blood test is now available to couples concerned about the birth of a child with a neural tube defect. If the mother has a high serum level of *α-feto protein*, further testing is advised. An analysis of the amniotic fluid following amniocentesis (p. 324) can reveal if there has been a leakage of neural tube substance into the fluid. If such a leakage has taken place, it is usually possible to diagnose the condition of the fetus.

Multiple Alleles

Some traits are controlled by **multiple alleles,** that is, an odd number of alleles. Each person inherits only 2 of the total possible number of alleles.

ABO Blood Types: Four Types

Three alleles for the same gene control the inheritance of ABO blood types. These alleles determine the presence or absence of antigens on the red blood cells.

A = A antigen on red blood cells
B = B antigen on red blood cells
O = no antigens on red blood cells

Each person has only 2 of the 3 possible alleles, and both *A* and *B* are dominant over *O*. Therefore, there are 2 possible genotypes for type A blood and 2 possible genotypes for type B blood. On the other hand, alleles *A* and *B* are fully expressed in the presence of the other. Therefore, if a person inherits one of each of these alleles, that person will have type AB blood. Type O blood can only result from the inheritance of 2 *O* alleles:

Phenotype	Possible Genotype
A	*AA, AO*
B	*BB, BO*
AB	*AB*
O	*OO*

An examination of possible matings between different blood types sometimes produces surprising results; for example,

Genotypes of parents: *AO* × *BO*

Possible genotypes of children: *AB, OO, AO, BO*
Therefore, from this particular mating, every possible phenotype (types AB, O, A, B blood) is possible.

Blood typing can sometimes aid in paternity suits. However, a blood test of a supposed father only can suggest that he *might* be the father, not that he definitely *is* the father. For example, it is possible, but not definite, that a man with type A blood (having genotype *AO*) is the father of a child with type O blood. On the other hand, a blood test sometimes can definitely prove that a man is not the father. For example, a man with type AB blood cannot possibly be the father of a child with type O blood. Therefore, blood tests can be used legally only to exclude a man from possible paternity.

Rh Factor

The Rh factor is inherited separately from A, B, AB, or O blood types. In each instance, it is possible to be Rh positive (Rh^+) or Rh negative (Rh^-). When you are Rh positive, there is a particular antigen on the red blood cells, and when you are Rh negative, it is absent. It can be assumed that the inheritance of this antigen is controlled by a single allelic pair in which simple dominance prevails: the Rh-positive allele is dominant over the Rh-negative allele. Complications arise when an Rh-negative woman reproduces with an Rh-positive man and the child in the womb is Rh positive. Under certain circumstances, the woman may begin to produce antibodies that will attack the red blood cells of this baby or of a future Rh-positive baby. As discussed on page 122, the Rh problem can be eliminated by giving an Rh-negative woman an Rh immunoglobulin injection either midway through a pregnancy or no later than 72 hours after giving birth to any Rh-positive child.

If an Rh-negative woman is given Rh immunoglobulin within 72 hours of exposure to Rh-positive erythrocytes, she will not produce antibodies against them, and the next Rh-positive baby will be protected. The injected immunoglobulin remains in the mother's bloodstream long enough to prevent her immune system from producing its own anti-Rh antibodies, but not long enough to affect subsequent offspring. The mother must be given Rh immunoglobulin injections after the birth of each Rh-positive baby and with any abortion.

Sex-Linked Traits

The sex chromosomes contain genes just as the autosomal chromosomes do. Some of these genes determine whether the individual is a male or a female. Investigators have found that the Y chromosome has an *SRY* gene (sex-determining region Y gene). When this gene is lacking from the Y chromosome, the individual is a female even though the chromosomal inheritance is XY.

Traits controlled by alleles on the sex chromosomes are said to be **sex-linked;** an allele that is only on the X chromosome is **X-linked** and an allele that is only on the Y chromosome is Y-linked. Most sex-linked alleles are on the X chromosome, and the Y chromosome is blank for these. Very few alleles have been found on the Y chromosome, as you might predict, since it is much smaller than the X chromosome.

The X chromosomes carry many genes unrelated to the sex of the individual, and we will look at a few of these in depth. It would be logical to suppose that a sex-linked trait is passed from father to son or from mother to daughter, but this is not the case. A male always receives a sex-linked condition from his mother, from whom he inherited an X chromosome. The Y chromosome from the father does not carry an allele for the trait. Usually the trait is recessive; therefore a female must receive 2 alleles, one from each parent, before she has the condition.

X-Linked Alleles: Only on the X

When examining X-linked traits, the allele on the X chromosome appears as a letter attached to the X chromosome. For example, the key for color blindness is as follows:

X^B = normal vision

X^b = color blindness

The possible genotypes in both males and females are as follows:

X^BX^B = female who has normal color vision

X^BX^b = carrier female who has normal color vision

X^bX^b = female who is color blind

X^BY = male who has normal vision

X^bY = male who is color blind

Note that the second genotype is a *carrier* female because although a female with this genotype appears normal, she is capable of passing on an allele for color blindness. Color-blind females are rare because they must receive the allele from both parents; color-blind males are more common since they need only one recessive allele to be color blind. The allele for color blindness has to be inherited from their mother because it is on the X chromosome; males only inherit the Y chromosome from their fathers.

Figure 18.13 calculates the chances of a color-blind son when the mother is a carrier and the father has normal vision. Figure 18.14 gives a pedigree chart for an X-linked recessive allele. It also lists ways to recognize this pattern of inheritance. Three well-known X-linked recessive disorders with this pattern are color blindness, hemophilia, and muscular dystrophy.

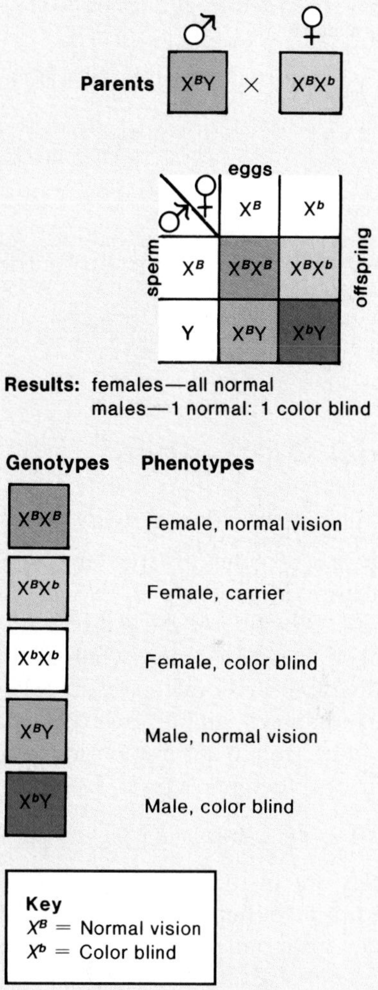

Figure 18.13

Cross involving an X-linked allele. The male parent is normal, but the female parent is a carrier; an allele for color blindness is located on one of her X chromosomes. Therefore, each son stands a 50% chance of being color blind.

X-linked Recessive Genetic Disorders

- More males than females are affected.
- An affected son can have parents who have the normal phenotype.
- For a female to have the characteristic, her father must also have it. Her mother must have it or be a carrier.
- The characteristic often skips a generation from the grandfather to the grandson.
- If a woman has the characteristic, all of her sons will have it.

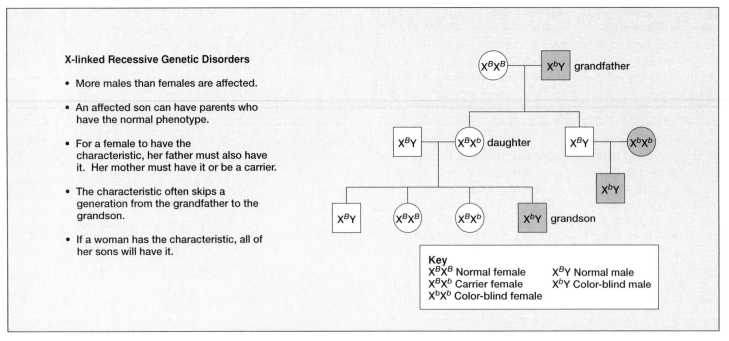

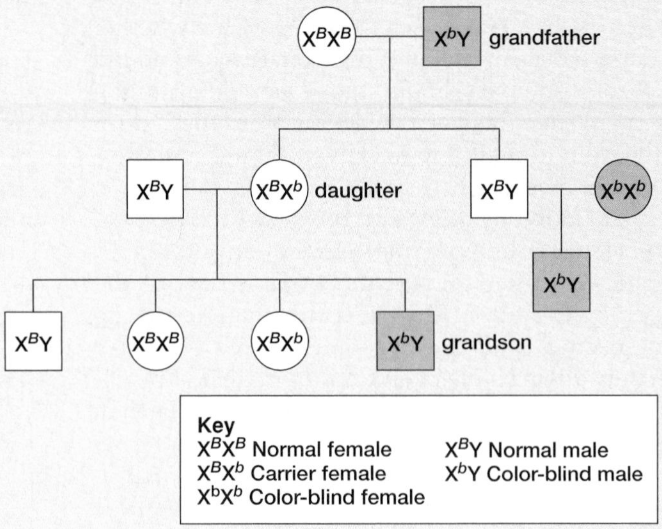

Figure 18.14

Sample pedigree chart for an X-linked recessive genetic disorder. Which females in the chart are carriers?

X-Linked Disorders

Color Blindness: Three Genes

In humans, there are 3 genes involved in distinguishing color because there are 3 different types of cones, the receptors for color vision (p. 235). Two of these are X-linked genes; one affects the green-sensitive cones, whereas the other affects the red-sensitive cones. About 5% of Caucasian men are color blind due to a mutation involving green perception, and about 2% are color blind due to a mutation involving red perception. In either case the genotype is designated as X^bY.

Hemophilia: Bleeder's Disease

Approximately one in 15,000 males are hemophiliacs. The most common type of hemophilia is hemophilia A, due to the absence or minimal presence of a particular clotting factor called factor VIII. *Hemophilia* is called the bleeder's disease because the affected person's blood does not clot. Although hemophiliacs bleed externally after an injury, they also suffer from internal bleeding, particularly around joints. Hemorrhages can be checked with transfusions of fresh blood (or plasma) or concentrates of the clotting protein. Unfortunately, some hemophiliacs have contracted AIDS after using concentrated blood from untested donors, but this cannot occur if a purified form of the concentrate from donors who have been tested is used.

At the turn of the century, hemophilia was prevalent among the royal families of Europe, and all of the affected males could trace their ancestry to Queen Victoria of England (fig. 18.15). Because none of Queen Victoria's forebearers or relatives were affected, it seems that the gene she carried arose by mutation either in herself or in one of her parents. Her carrier daughters, Alice and Beatrice, introduced the gene into the ruling houses of Russia and Spain, respectively. Alexis, the last heir to the Russian throne before the Russian Revolution, was a hemophiliac. There are no hemophiliacs in the present British royal family because Victoria's eldest son, King Edward VII, did not receive the gene and therefore could not pass it on to any of his descendants.

Muscular Dystrophy: Muscles Waste Away

Muscular dystrophy, as the name implies, is characterized by a wasting away of the muscles. The most common form, *Duchenne muscular dystrophy*, is X-linked and occurs in about one out of every 5,000 male births. Symptoms, such as waddling gait, toe walking, frequent falls, and difficulty in rising, may appear as soon as the child starts to walk. Muscle weakness intensifies until the individual is confined to a wheelchair. Death usually occurs by age 20; therefore, affected males are rarely fathers. The recessive allele remains in the population by passage from carrier mother to carrier daughter.

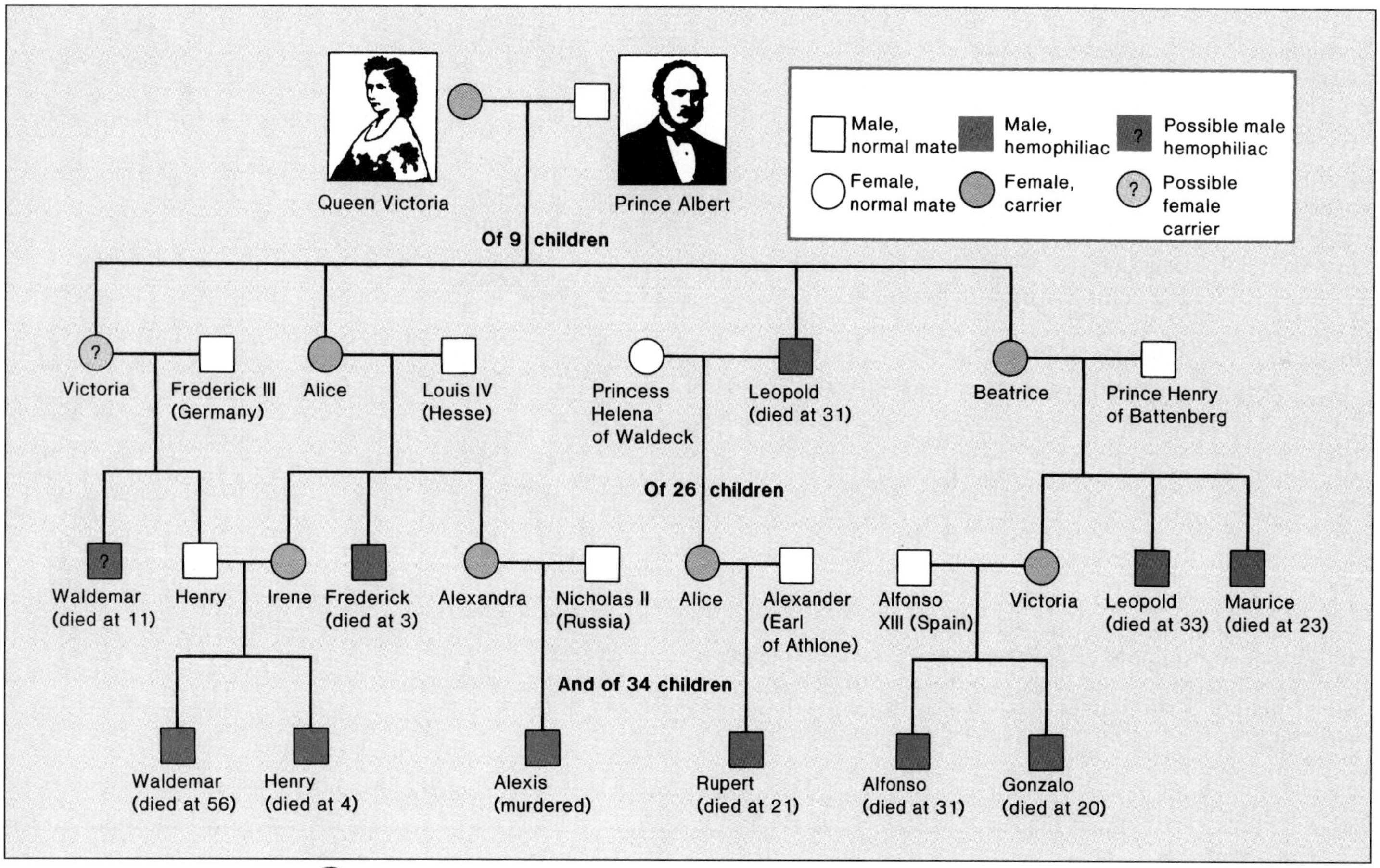

Figure 18.15

A simplified pedigree showing the X-linked inheritance of hemophilia in European royal families. Because Queen Victoria was a carrier, each of her sons had a 50% chance of having the disease and each of her daughters had a 50% chance of being a carrier. This pedigree shows only the affected individuals. Many others are unaffected, such as the members of the present British royal family.

Recently, the gene for muscular dystrophy was isolated, and it was discovered that the absence of a protein, now called dystrophin, is the cause of the disorder. Much investigative work determined that dystrophin is involved in the release of calcium from the calcium-storage sacs (p. 221) in muscle fibers. The lack of dystrophin causes calcium to leak into the cell, which promotes the action of an enzyme that dissolves muscle fibers. When the body attempts to repair the tissue, fibrous tissue forms, and this cuts off the blood supply so that more and more cells die.

A test is now available to detect carriers for Duchenne muscular dystrophy. Also, various treatments are being attempted. Immature muscle cells can be injected in muscles, and for every 100,000 cells injected, dystrophin production occurs in 30%–40% of muscle fibers. The allele for dystrophin has been inserted into the thigh muscle cells of mice, and about 1% of these cells produced dystrophin.

> There are sex-linked alleles on the X chromosome that have nothing to do with sexual characteristics. Males have only one copy of these alleles, and if they inherit a recessive allele, it is expressed.

Sex-Influenced Traits

Not all traits we associate with the sex of the individual are due to sex-linked alleles. Some are simply sex-influenced traits. Sex-influenced traits are characteristics that often appear in one sex but only rarely appear in the other. It is believed that these traits are governed by genes that are turned on or off by hormones. For example, the secondary sexual characteristics, such as the beard of a male and the developed breasts of a female, probably are controlled by the balance of hormones.

Baldness is believed to be caused by the male sex hormone testosterone because males who take the hormone to increase masculinity begin to lose their hair (fig. 18.16). A more detailed explanation has been suggested by some investigators. It has been reasoned that due to the effect of hormones, males require only one gene for the trait to appear, whereas females require 2 genes. In other words, the gene acts as a dominant in males but as a recessive in females. This means that males born to a bald father and a mother with hair *at best* have a 50% chance of going bald. Females born to a bald father and a mother with hair *at worst* have a 25% chance of going bald.

Another sex-influenced trait of interest is the length of the index finger. In women, the index finger is at least equal to if not longer than the fourth finger. In males, the index finger is shorter than the fourth finger.

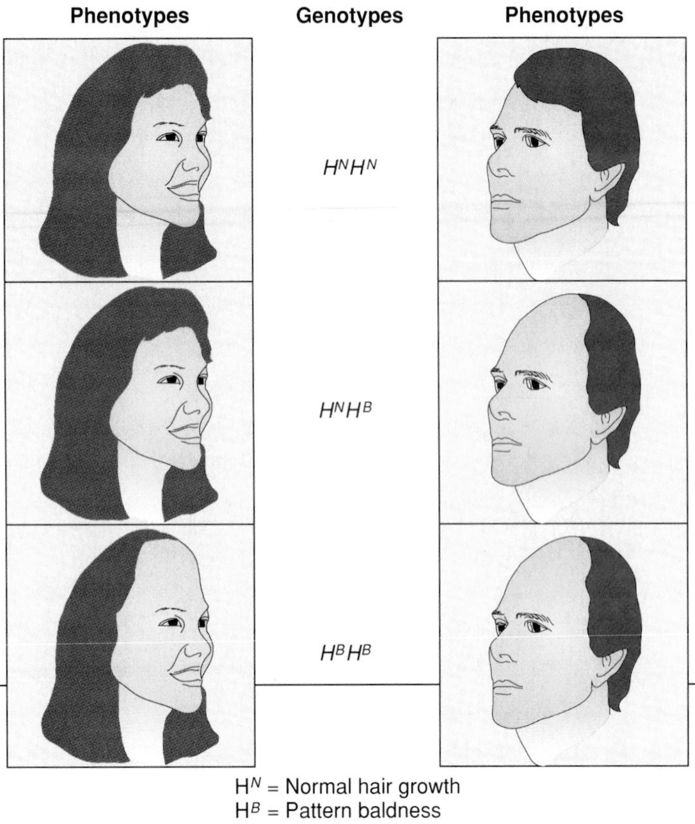

Figure 18.16

Baldness, a sex-influenced characteristic. Due to hormonal influences, the presence of only one gene for baldness causes the condition in the male, whereas the condition does not occur in the female unless she possesses both genes for baldness.

Summary

A gene normally has 2 alleles, which are designated by a letter. The genotype of an individual is either homozygous dominant (2 capital letters); homozygous recessive (2 lowercase letters); or heterozygous (both a capital and lowercase letter). The phenotype of an individual is determined by the alleles inherited, one from each parent. An individual with the heterozygous genotype has the dominant phenotype. Figure 18.3 illustrates several traits controlled by a simple dominance.

A Punnett square can help to determine the phenotype ratio among offspring because all possible sperm types of a particular father are given an equal chance to fertilize all possible egg types of a particular mother. When a heterozygous individual reproduces with another heterozygote, there is a 75% chance the offspring will have the dominant phenotype and a 25% chance the child will have the recessive phenotype. When a heterozygous individual reproduces with a pure recessive, the offspring has a 50% chance of either phenotype.

Determination of inheritance is sometimes complicated by the fact that characteristics are controlled by incompletely dominant alleles, more than one pair of alleles (polygenic inheritance), or by multiple alleles. Degree of hair curl illustrates incomplete dominance. Skin color illustrates polygenic inheritance, and ABO blood types illustrate inheritance by multiple alleles.

Sex-linked genes are located on the sex chromosomes, and most alleles are X-linked because they are on the X chromosome. Since the Y chromosome does not contain these alleles, only the allele received from the mother determines the phenotype in males. This means that males are more apt to have a phenotype controlled by an X-linked recessive allele. It also indicates that if a female does have the phenotype, then her father must also have it and her mother must be a carrier. Usually X-linked traits skip a generation and go from maternal grandfather to grandson by way of a carrier daughter.

Some conditions are sex influenced and are apt to appear more often in one sex than another. Sex-influenced traits are controlled by genes that are believed to be turned on and off by hormones.

It is now known that many disorders of humans are genetic in origin. Pedigree charts have helped work out the pattern of inheritance, and it is possible to calculate the chances of a child having a particular disorder, such as those listed in table 18A.

Study and Thought Questions

1. What is the difference between the genotype and the phenotype of an individual? For which phenotype are there 2 possible genotypes? (p. 363)
2. What is the chance of a child with the dominant phenotype for these crosses?? (p. 363)

   AA X *AA*
   Aa X *AA*
   Aa X *Aa*
   aa X *aa*

3. From which of the crosses in question 2 can there be an offspring with the recessive phenotype? Explain.? (p. 363)

   Critical Thinking In fruit flies, long wings are dominant to short wings. A student is observing the results of a cross and counts 300 long-winged flies and 98 short-winged flies. a. What was the genotype of the parent flies? How do you know? b. Of the long-winged flies, how many do you predict are homozygous dominant and how many are heterozygous? Why do you say so?

4. List ways it is possible to tell an autosomal recessive genetic disorder and an autosomal dominant genetic disorder by examining a pedigree chart.? (pp. 365–368)
5. Give examples of genetic disorders caused by inheritance of 2 recessive alleles and disorders caused by inheritance of a single dominant allele. (pp. 365, 368)

   Ethical Issue Should genetic disorders be prevented by restricting the reproduction of those who have a dominant disorder or are carriers of a recessive disorder?

6. Give examples of these patterns of inheritance: incomplete dominance, polygenic inheritance, and multiple alleles. (pp. 369–372)

   Ethical Issue Should inheritance be used as a possible excuse for a person's behavior, such as alcoholism or violent behavior?

7. Give examples of genetic disorders caused by the patterns of inheritance in question 6. (pp. 369–372)
8. If a trait is on an autosome, how would you designate a homozygous dominant female? If a trait is on an X chromosome, how would you designate a homozygous dominant female? (p. 373)
9. List ways it is possible to tell an X-linked recessive disorder. Why do males exhibit such disorders more often than females? (p. 374)

   Critical Thinking Persons with sickle-cell disease tend to die at a young age and leave no offspring. Why doesn't sickle-cell disease cease to be a threat in future generations?

10. Give examples of genetic disorders caused by X-linked recessive alleles. (pp. 374–375)

Writing Across the Curriculum

1. List the ways in which chromosomes and alleles behave similarly.
2. There are 4 ABO blood types determined by the *A*, *B*, or *O* genes. The *O* gene is recessive to the *A* and *B* genes and can only be expressed when homozygous (*OO*). Explain why the O blood type does not rapidly cease to exist when breeding occurs between persons who are blood type O and persons who are either blood type A, B, or AB.
3. There is now a test to determine those who have the gene for Huntington disease. Why might a young person want or not want to be tested?

Objective Questions

1. Whereas an individual has 2 genes for every trait, the gametes have ________________ gene for every trait.
2. The recessive allele for the dominant gene *W* is ________________.
3. Mary has a widow's peak and John has a continuous hairline. This would be a description of their ________________.
4. *W* = widow's peak and *w* = continuous hairline; therefore, only the phenotype ________________ could be heterozygous.
5. Two heterozygotes, each having a widow's peak, already have a child with a continuous hairline. The next child has what chance of having a continuous hairline? ________________
6. Most sex-linked genes are attached to the ________________ chromosome.
7. If a male is color blind, he inherited the allele for color blindness from his ________________.
8. What is the genotype of a female who has a color-blind father but a homozygous normal mother? ________________
9. In a pedigree chart, it is observed that although the children have a characteristic, neither parent has it. The characteristic must be inherited as a ________________ gene.
10. Determine if the characteristic possessed by the shaded males and females in the following pedigree chart is an autosomal dominant, autosomal recessive, or X-linked condition:

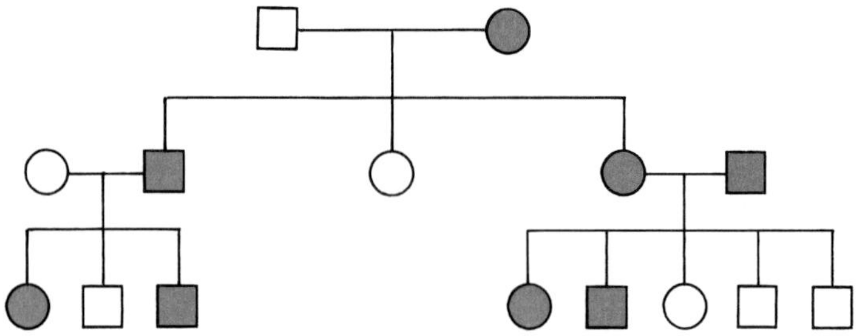

11. Identify each of these genetic disorders.
    a. Mucus in lungs and digestive tract is thick and viscous.
    b. Neurological impairment and psychomotor difficulties develop early.
    c. Benign tumors occur under the skin or deeper.
    d. Minor disturbances in balance and coordination develop in middle age and progress toward severe mental disturbances.
    e. Propensity for bleeding due to lack of blood clotting factor.
    f. Muscle weakness develops early and intensifies until death occurs.

Genetics Problems

The trait is dominant for questions 1–5:

1. A woman heterozygous for polydactyly, a condition that produces 6 fingers and toes, reproduces with a man without this condition. What are the chances that their children will have 6 fingers and toes?
2. A young man's father has just been diagnosed as having Huntington disease. What are the chances that the son will inherit this condition?
3. Black hair is dominant over blond hair. A woman with black hair whose father had blond hair reproduces with a blond-haired man. What are the chances of this couple having a blond-haired child?
4. Your maternal grandmother Smith had Huntington disease. Aunt Jane, your mother's sister, also had the disease. Your mother dies at age 75 with no signs of Huntington disease. What are your chances of getting the disease assuming that your father is perfectly normal?
5. Could a person who can curl her tongue have parents who cannot curl their tongues? Explain your answer.

The trait is recessive for questions 6–8:

6. Parents who do not have Tay-Sachs disease produce a child who has Tay-Sachs disease. What is the genotype of each parent? What are the chances each child will have Tay-Sachs disease?
7. One parent has lactose intolerance, the inability to digest lactose, the sugar found in milk, and the other parent is heterozygous. What are the chances that their child will have lactose intolerance?
8. A child has cystic fibrosis. His parents are normal. What is the genotype of all persons mentioned?

The trait is incompletely dominant for questions 9–12:

9. What are the chances that a person homozygous for straight hair who reproduces with a person homozygous for curly hair will have children with wavy hair?
10. One parent has sickle-cell disease and the other is perfectly normal. What are the phenotypes of their children?
11. A child has sickle-cell disease but her parents do not. What is the genotype of each parent?
12. Both parents have the sickle-cell trait. What are their chances of having a perfectly normal child?

The trait is either controlled by multiple alleles or by more than one pair of alleles for problems 13–16:

13. The genotype of a woman with type B blood is *BO*. The genotype of her husband is *AO*. What could be the genotypes and phenotypes of the children?
14. A man has type AB blood. What is his genotype? Could this man be the father of a child with type B blood? If not, why not? If so, what blood types could the child's mother have?
15. Baby Susan has type B blood. Her mother has type O blood. What type blood could her father have?
16. Fill in the following pedigree chart to give the probable genotypes of the twins:

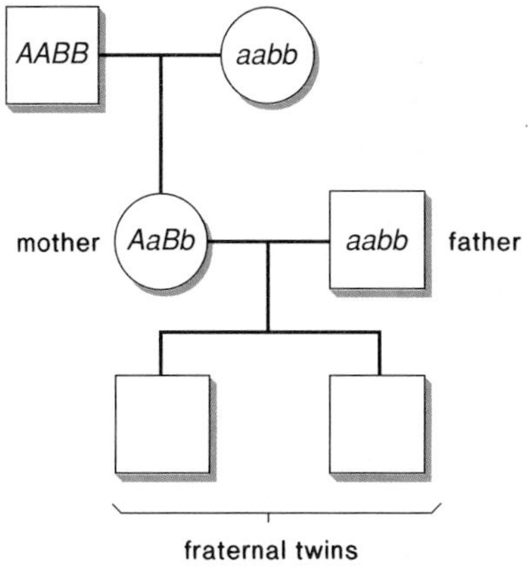

SELECTED KEY TERMS

allele (ah-lēl´) An alternative form of a gene located at a particular chromosome site (locus). 363

dominant allele The hereditary factor that expresses itself in the phenotype when the genotype is heterozygous. 363

genotype (ge´nə-tīp) The genes of any individual for (a) particular trait(s). 363

heterozygous Having 2 different alleles (as *Aa*) for a given trait. 363

homozygous Having identical alleles for a given trait; homozygous dominant (*AA*) or homozygous recessive (*aa*). 363

phenotype (fe´no-tīp) The outward appearance of an organism caused by the genotype and environmental influences. 363

Punnett square A gridlike device used to calculate the expected results of simple genetic crosses. 363

recessive allele A hereditary factor that expresses itself in the phenotype only when the genotype is homozygous. 362

sex-linked Allele located on the sex chromosomes. 373

trait Specific term for a distinguishing phenotypic feature studied in heredity. 362

X-linked Allele is located on the X chromosome. 373

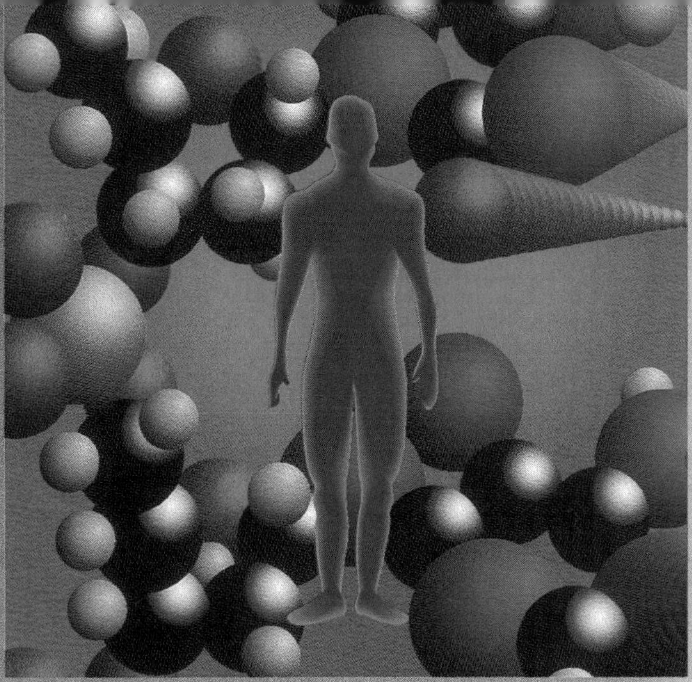

Chapter 19

DNA and Biotechnology

Molecular genetics and biotechnology provide evidence of organic evolution. The same 4 types of nucleotides found in human DNA make up the DNA of all organisms. Further, you can put a human gene into a bacterium, and the gene will perform its normal cellular function. It is not surprising, then, that you can also transfer genes between plants and animals. For certain, the very first cell or cells must have evolved into various forms of life, and the evolutionary process must be dependent upon slight changes in the DNA. Only in recent times have we come to realize that to truly understand the evolution of plants and animals, we have to look inside the workings of the cell.

Chapter Concepts

1. DNA is the genetic material, and therefore its structure and function constitute the molecular basis of inheritance. 382
2. DNA is able to replicate and control protein synthesis. 384
3. Protein synthesis is a process that requires the participation of RNA in varied forms. 385
4. Recombinant DNA technology is the basis for biotechnology, an industry that produces many products. 391
5. Modern-day biotechnology is expected to revolutionize medical care and agriculture. It also permits gene therapy and the mapping of the human chromosomes. 394–398

Chapter Outline

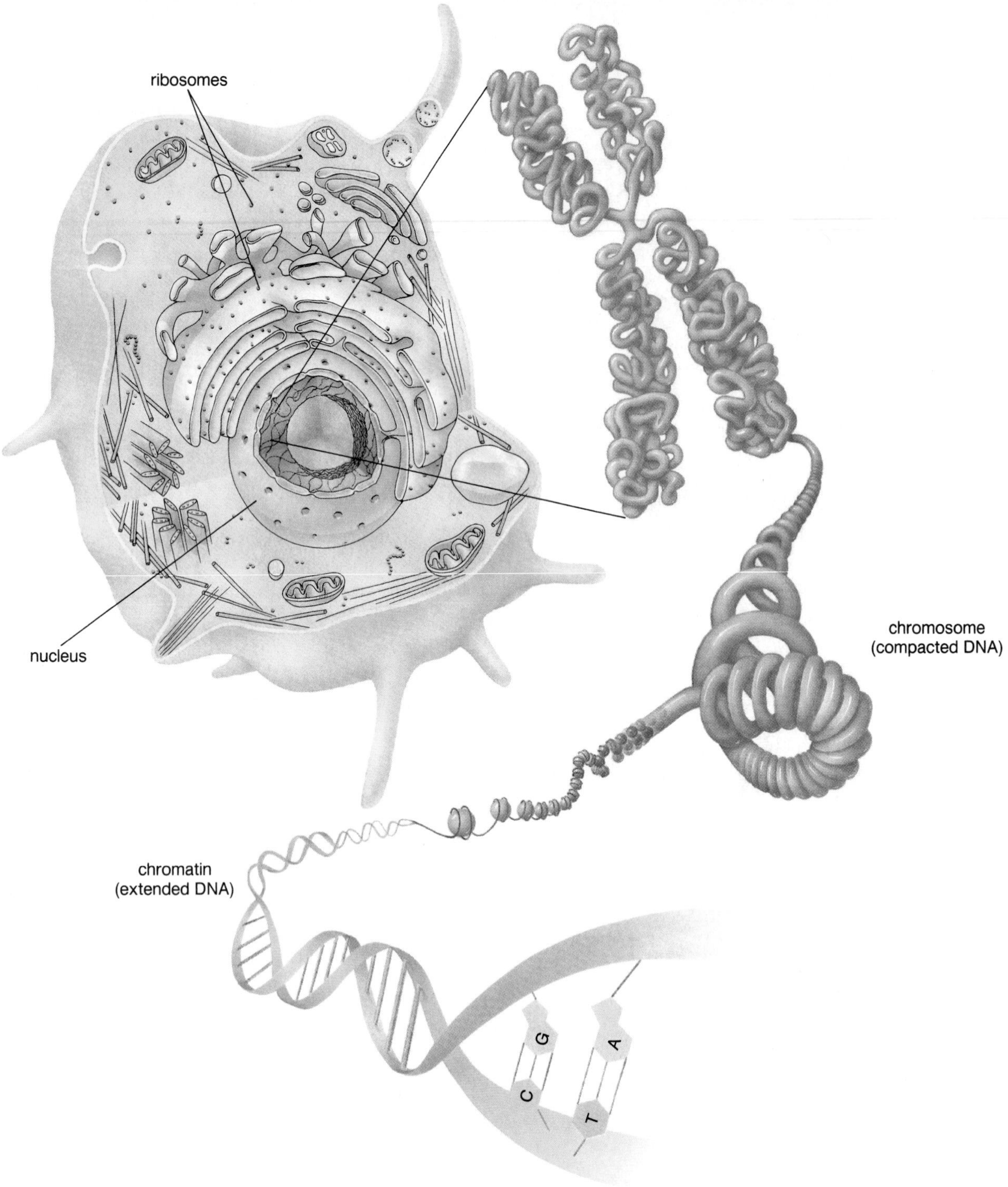

Figure 19.1

DNA location and structure. DNA is highly compacted in chromosomes, but it is extended as chromatin during interphase. It is during interphase that DNA can be extracted from a cell and its structure and function studied.

Our approach to genetics thus far has been to consider the genes as particulate units of a chromosome. In contrast, molecular genetics considers the chemical nature of the gene and the biochemical function of genes in the cell.

DNA and RNA Structure and Function

Genes are made up of a molecule called **DNA (deoxyribonucleic acid).** This means that DNA is the genetic material, and that chromosomes contain DNA. DNA is found principally in the nucleus of a cell (fig. 19.1), but **RNA (ribonucleic acid)** is found both in the nucleus and the cytoplasm. A type of RNA called ribosomal RNA (rRNA) is found within the **ribosomes,** which are small structures whose subcomponents are assembled at the nucleolus within the nucleus. Ribosomes occur on rough endoplasmic reticulum within the cytoplasm.

DNA and RNA Structure

DNA is a type of nucleic acid, and like all nucleic acids, it is formed by the sequential joining of molecules called nucleotides. Nucleotides, in turn, are composed of 3 smaller molecules—a *phosphate,* a *sugar,* and a *base.* The sugar in DNA is deoxyribose, which accounts for its name, *deoxyribonucleic acid.* There are 4 different nucleotides in DNA, each having a different base: **adenine (A), thymine (T), cytosine (C),** and **guanine (G)** (fig. 19.2).

When nucleotides join, the sugar and the phosphate molecules become the backbone of a strand and the bases project to the side. In DNA, there are 2 strands of nucleotides; consequently, DNA is double-stranded. Weak hydrogen bonds between the bases hold the strands together. Each base is bonded to another particular base, called **complementary base pairing**—adenine (A) is always paired with thymine (T), and cytosine (C) is always paired with guanine (G), and vice versa. The dotted lines in figure 19.3*a* represent the hydrogen bonds between the bases. The structure of DNA is said to resemble a ladder; the sugar-phosphate backbone makes up the sides of a ladder and the paired bases are the steps. The ladder structure of DNA twists to form a spiral staircase called a *double helix* (fig. 19.3*b*).

DNA is a double helix with a phosphate-sugar backbone on the outside and paired bases on the inside. Complementary base pairing occurs: adenine (A) pairs with thymine (T) and guanine (G) pairs with cytosine (C).

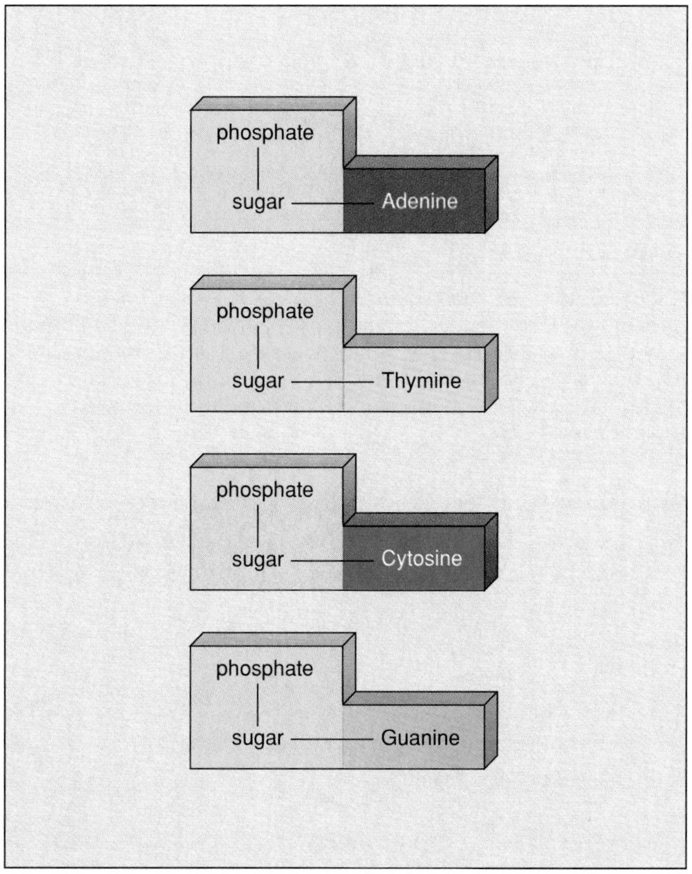

Figure 19.2

DNA structure. DNA contains 4 different kinds of nucleotides, molecules that in turn contain a phosphate, a sugar, and a base. The base of a DNA nucleotide can be either adenine (A), thymine (T), cytosine (C), or guanine (G).

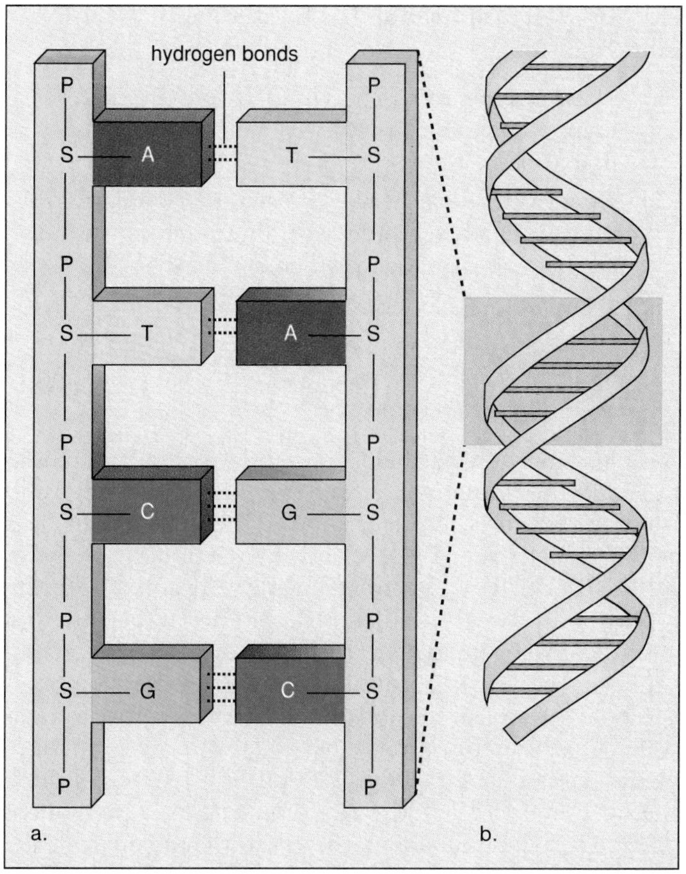

Figure 19.3

DNA is double-stranded. **a.** When the DNA nucleotides join, they form 2 strands so that the structure of DNA resembles a ladder. The phosphate (P) and sugar (S) molecules make up the sides of the ladder, and the bases make up the rungs of the ladder. Each base is weakly bonded (*dotted lines*) to its complementary base (T is bonded to A and vice versa; G is bonded to C and vice versa). **b.** The DNA ladder twists to give a double helix. Each chromatid of a duplicated chromosome contains one double helix.

RNA is a nucleic acid made up of nucleotides containing the sugar ribose. This sugar accounts for its name, *ribonucleic acid.* RNA also contains 4 different nucleotides, each with a different base: adenine (A), uracil (U), cytosine (C), and guanine (G) (fig. 19.4). In RNA, the base uracil replaces the base thymine.

RNA, unlike DNA, is always single-stranded in humans. Similarities and differences between these 2 nucleic acid molecules are given in table 19.1.

Functions of DNA: Master Controller

The 2 primary functions of DNA are replication and control of protein synthesis.

Replication: Unzipping and Molding

Between cell divisions, chromosomes must duplicate before cell division can occur again. Actually, when chromosomes duplicate, DNA is replicating, that is, it is making a copy of itself. **Replication** has been found to require the following steps:

1. The 2 strands that make up DNA unwind and "unzip" (i.e., the hydrogen bonds between the paired bases break).
2. New nucleotides, always present in the nucleus,[1] move into place beside each old (parental) strand by the process of complementary base pairing.
3. These new nucleotides become joined.
4. When the process is finished, 2 complete DNA molecules are present, identical to each other and to the original molecule.

Each new double helix is composed of an old (parental) and a new (daughter) strand (fig. 19.5). The new strand is complementary to the old strand. Because of this, it is said that each strand of DNA serves as a **template,** or mold, for the production of a complementary strand. The entire replication process is called *semiconservative* because old strands are always paired with new strands. Because of semiconservative replication, the new double helix molecules are identical to each other and to the original molecule. Rarely a replication error occurs—the new sequence of the bases is not exactly like that of a parental strand. Replication errors are a source of *mutations,* a permanent change in a gene causing a change in the phenotype.

> DNA replication results in 2 double helixes. DNA unwinds and unzips, and new (daughter) strands, each complementary to an old (parental) strand, form. This is called semiconservative replication.

1. The food we eat is digested into molecules such as nucleotides and amino acids, which are carried in the bloodstream to the cells.

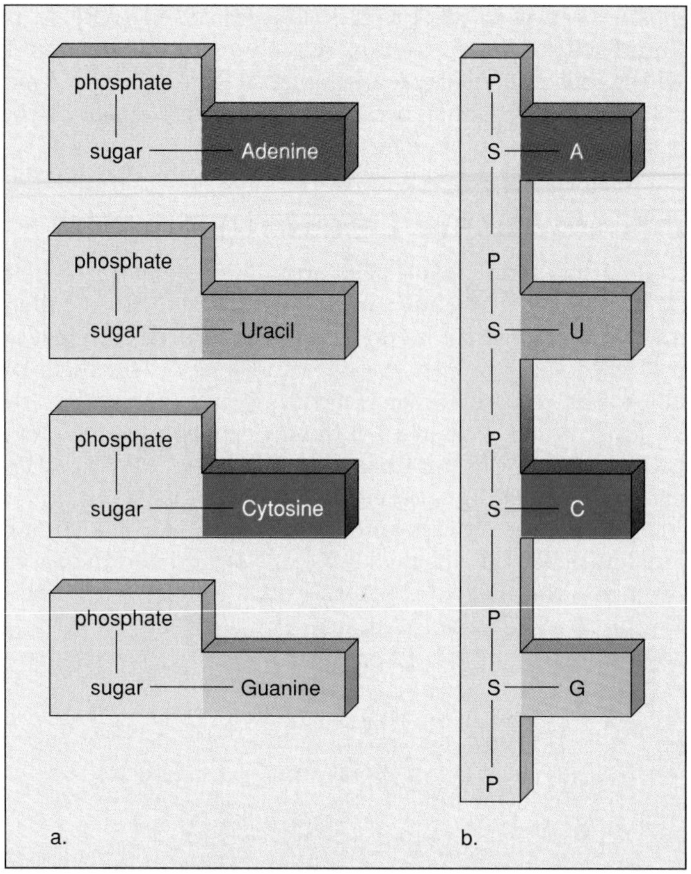

Figure 19.4

RNA structure. **a.** The 4 nucleotides in RNA each have a phosphate (P) molecule; the sugar (S) is ribose; and the base may be either adenine (A), uracil (U), cytosine (C), or guanine (G). **b.** RNA is single-stranded. The sugar and phosphate molecules join to form a single backbone, and the bases project to the side.

Table 19.1

DNA-RNA Similarities and Differences

DNA-RNA Similarities	
Both are nucleic acids	
Both are composed of nucleotides	
Both have a sugar-phosphate backbone	
Both have 4 different type bases	
DNA-RNA Differences	
DNA	*RNA*
Found in nucleus	Found in nucleus and cytoplasm
The genetic material	Helper to DNA
Sugar is deoxyribose	Sugar is ribose
Bases are A, T, C, G	Bases are A, U, C, G
Double-stranded	Single-stranded
Is transcribed (to give mRNA)	Is translated (to give proteins)

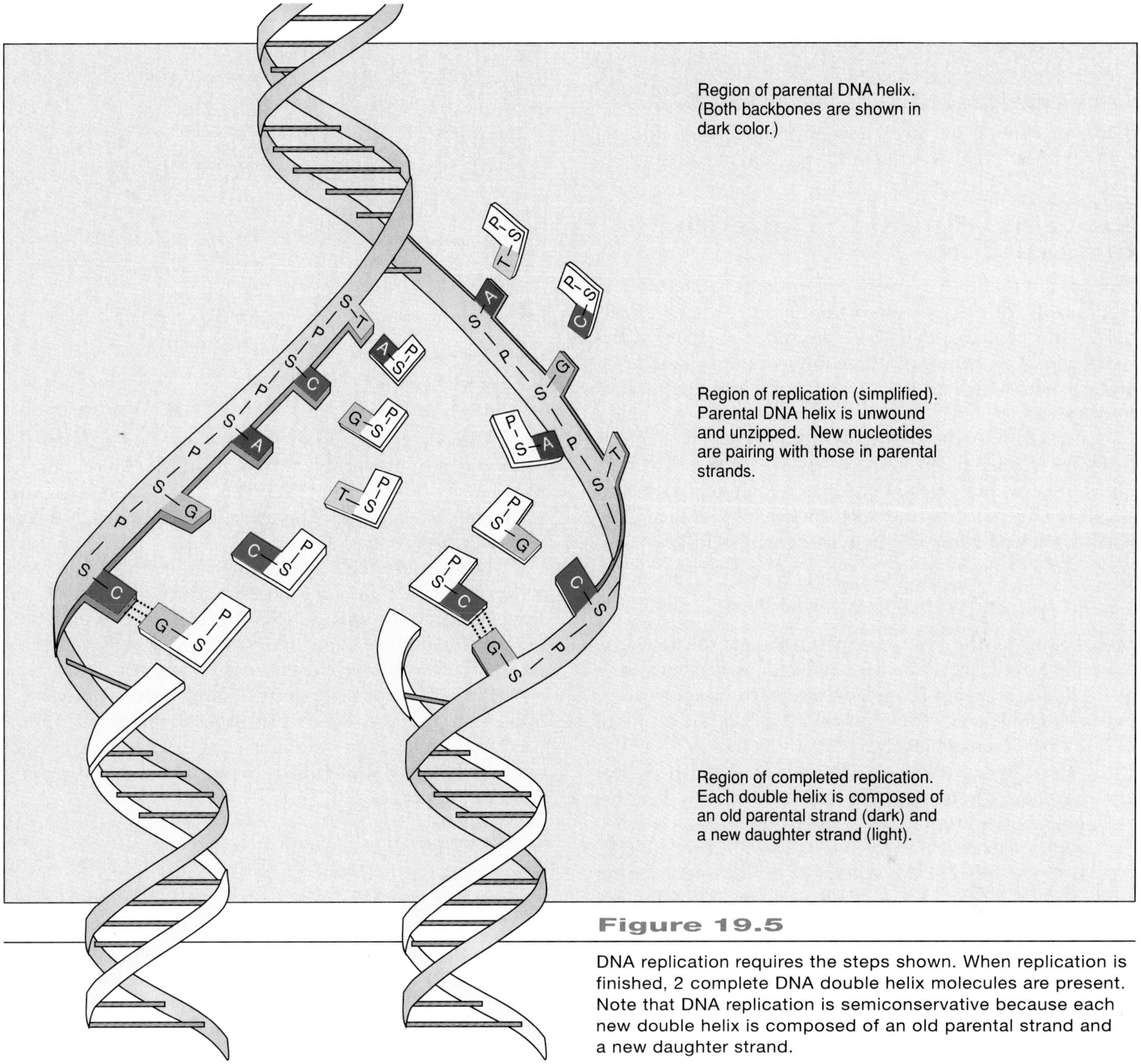
Region of parental DNA helix. (Both backbones are shown in dark color.)
Region of replication (simplified). Parental DNA helix is unwound and unzipped. New nucleotides are pairing with those in parental strands.
Region of completed replication. Each double helix is composed of an old parental strand (dark) and a new daughter strand (light).

Figure 19.5

DNA replication requires the steps shown. When replication is finished, 2 complete DNA double helix molecules are present. Note that DNA replication is semiconservative because each new double helix is composed of an old parental strand and a new daughter strand.

Protein Synthesis: Occurs in Cells

The protein synthesis function of DNA, a rather complicated process, is discussed shortly in some detail. Because DNA controls protein synthesis, it controls the structure and function of the cell. To bring about protein synthesis, DNA requires the help of RNA.

Functions of RNA: Varied Helper

RNA is a helper to DNA, allowing protein synthesis to occur in the manner DNA directs. RNA is more intimately involved in protein synthesis because protein synthesis takes place at the ribosomes in the cytoplasm.

There are 3 types of RNA, each with a specific function in protein synthesis:

Ribosomal RNA (rRNA) is found within the ribosomes where protein synthesis occurs. Ribosomes, you'll recall, are often attached to the endoplasmic reticulum, within the cytoplasm of the cell.

Messenger RNA (mRNA) carries genetic information from chromosomes to the ribosomes and directs the synthesis of proteins.

Transfer RNA (tRNA) transfers amino acids to the ribosomes, where they are joined.

Proteins and Protein Synthesis

Before describing the steps needed for protein synthesis, we will review the structure and function of proteins.

Proteins: Function Depends on Structure

Proteins are found in all cells. The protein hemoglobin is responsible for the red color of red blood cells. The antibodies are proteins, and there are other proteins in the plasma as well. Muscle cells contain the proteins actin and myosin, which give them substance and their ability to contract.

Certain proteins are *enzymes,* which speed chemical reactions in cells (p. 39). Enzymes allow reactions to occur quickly, even though the body has a relatively low temperature. The reactions in cells form chemical or metabolic pathways. A pathway can be represented as follows:

$$A \xrightarrow{E_A} B \xrightarrow{E_B} C \xrightarrow{E_C} D \xrightarrow{E_D} E$$

In this pathway, the letters are molecules and the notations over the arrows are enzymes: molecule *A* becomes molecule *B,* and enzyme E_A speeds the reaction; molecule *B* becomes molecule *C,* and enzyme E_B speeds the reaction, and so forth. Notice that each reaction in the pathway has its own enzyme: enzyme E_A can only convert *A* to *B,* enzyme E_B can only convert *B* to *C,* and so forth. For this reason, enzymes are said to be *specific.*

Proteins are very large molecules composed of individual units called *amino acids.* Twenty different amino acids are commonly found in proteins. Proteins differ because the number and order of their amino acids differ (fig. 19.6). The sequence of amino acids results in a protein with a particular shape, and shape helps determine the function of a protein.

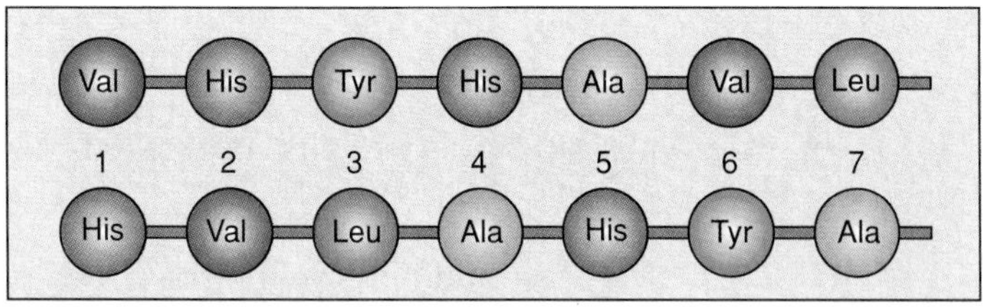

Figure 19.6

Amino acid sequences from 2 different proteins are represented. Proteins differ in the number and sequence of their amino acids.

Table 19.2

Some DNA Codes and RNA Codons

DNA Code	mRNA Codon	tRNA Anticodon	Amino Acid
TTT	AAA	UUU	Lysine
TGG	ACC	UGG	Threonine
CCG	GGC	CCG	Glycine
CAT	GUA	CAU	Valine

Protein Synthesis: DNA Directs, RNA Helps

DNA directs protein synthesis because it determines the order of the amino acids in proteins. DNA can do this because every 3 bases in DNA codes for, or represents, one amino acid (table 19.2). Therefore, it is said the DNA contains a triplet code.

The genetic code is essentially universal. The same codons stand for the same amino acids in most organisms from bacteria to humans. This illustrates the remarkable biochemical unity of living things and suggests that all living thing have a common evolutionary ancestor.

Protein synthesis requires 2 steps, called transcription and translation. It is helpful to recall that **transcription** is often used to signify making a copy of certain information, while **translation** means to put this information into another language.

Transcription

Referring again to the structure of the cell, recall that DNA is found in the nucleus (see fig. 19.1). Protein synthesis occurs at the ribosomes, which are located in the cytoplasm or on the endoplasmic reticulum. Proteins produced by ribosomes free in the cytoplasm stay in the cell. Those produced by ribosomes attached to the ER are for transport outside the cell. Therefore, it is obvious that there must be some way of getting the DNA message, or code, to the ribosomes. *Messenger RNA* (mRNA) takes the message from the DNA in the nucleus to the ribosomes in the cytoplasm.

Just as DNA can serve as a template for the production of itself, it can also serve as a template for the production of mRNA. During *transcription,* RNA nucleotides complementary to a portion of one DNA strand join: G pairs with C and vice versa; U pairs with A and A with T in DNA. The mRNA that results, therefore, has a sequence of bases complementary to those of a gene (fig. 19.7).

While DNA contains a **triplet code** in which every 3 bases stand for one amino acid, mRNA contains **codons,** each of which is made up of 3 bases that also stand for an amino acid (table 19.2). After formation, mRNA moves to the cytoplasm, where ribosomes attach to it.

Following transcription in the nucleus, mRNA has a sequence of bases complementary to one of the DNA strands. mRNA moves into the cytoplasm and becomes associated with the ribosomes.

Translation

During *translation,* the order of codons in mRNA brings about a particular order of amino acids in a protein. For this to happen, free (or unattached) amino acids in the cytoplasm must move to a ribosome.[2] *Transfer RNA* (tRNA) molecules combine with and bring amino acids to the ribosomes (fig. 19.8). There is at least one kind of tRNA for each of the 20 amino acid molecules. An amino acid attaches to one end of a tRNA, and at the other end a tRNA has 3 particular nucleotides whose bases make up an **anticodon.** Each anticodon is complementary to a particular codon (table 19.2).

As a ribosome moves along a mRNA, the codons become prominent. When the anticodons of tRNAs pair with the codons, a particular sequence of amino acids join as protein synthesis occurs. Therefore, a particular sequence of bases (codons) has been translated into a particular sequence of amino acids within a protein. This is the indirect process by which the DNA code directs the synthesis of a protein. The central dogma of molecular genetics is as follows:

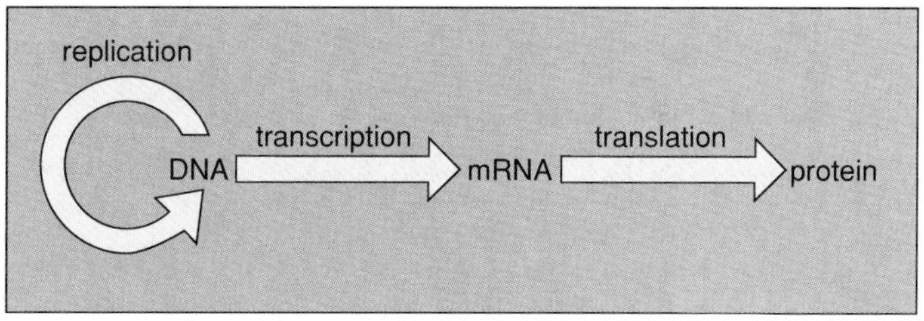

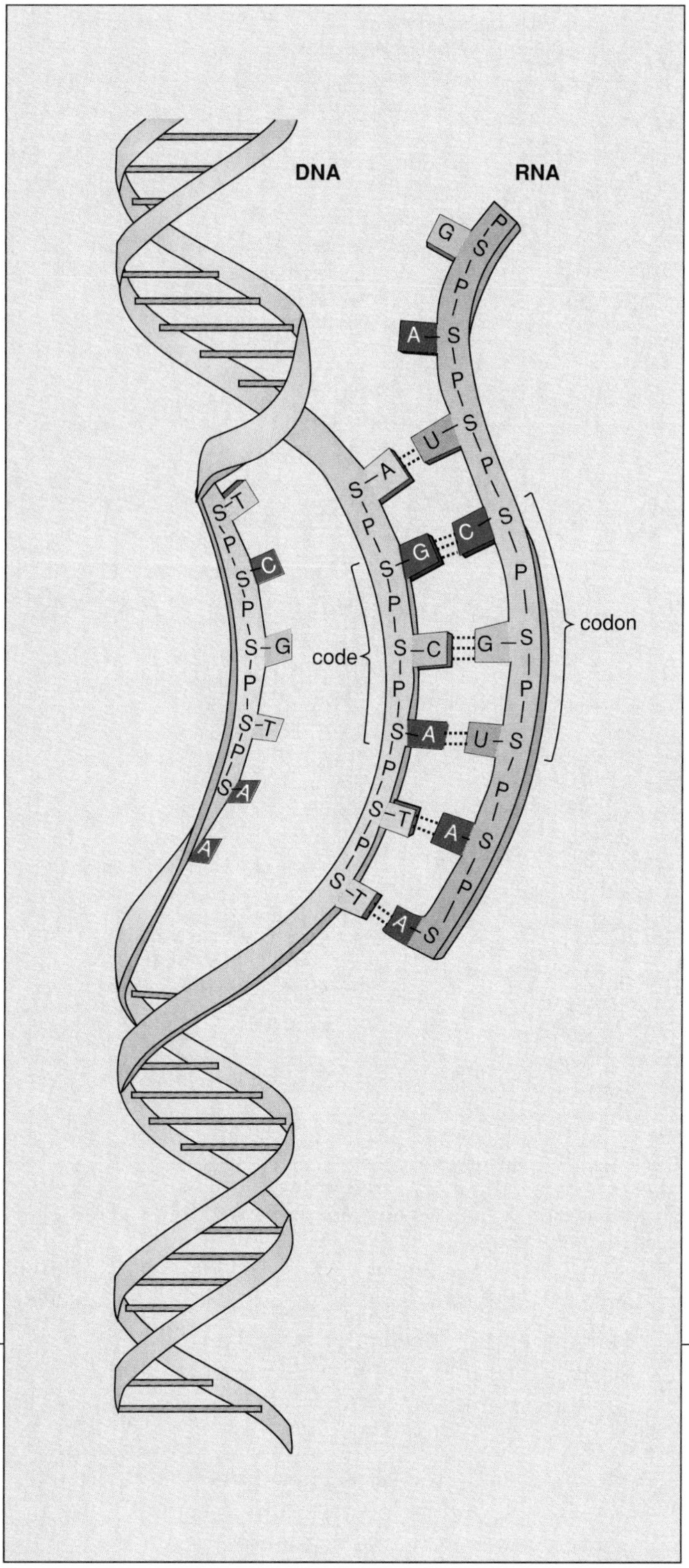

Figure 19.7

Transcription. During transcription, mRNA is formed when nucleotides complementary to the sequence of bases in a portion of DNA (i.e., a gene) join. Note that DNA contains a code (sequences of 3 bases), and that mRNA contains codons (sequences of 3 complementary bases). *Top*—mRNA transcript is ready to move into the cytoplasm. *Middle*—transcription has occurred and mRNA nucleotides have joined together. *Bottom*—rest of DNA molecule.

2. See footnote 1.

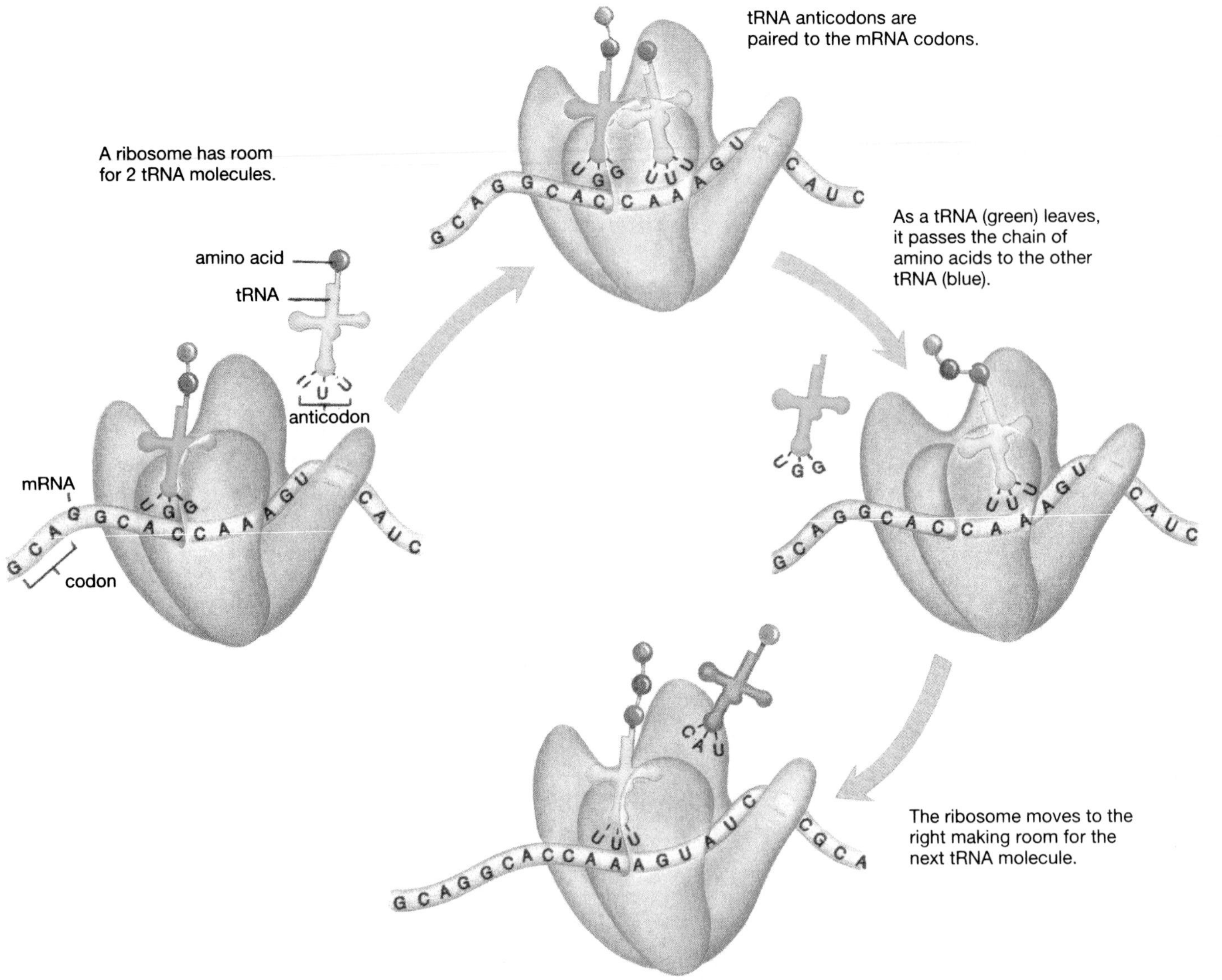
tRNA anticodons are paired to the mRNA codons.
A ribosome has room for 2 tRNA molecules.
As a tRNA (green) leaves, it passes the chain of amino acids to the other tRNA (blue).
amino acid
tRNA
anticodon
mRNA
codon
The ribosome moves to the right making room for the next tRNA molecule.

Figure 19.8

Translation. During translation, tRNA molecules bring amino acids to the ribosomes in the order dictated by mRNA codons and originally the DNA code. If the codon is AAA, the anticodon is UUU, and the amino acid is lysine (see table 19.2) represented by a red ball. Note that as tRNA departs, the amino acid chain is shifted to newly arrived tRNA.

Summary of Protein Synthesis

It is now possible to list the steps involved in protein synthesis (table 19.3 and figure 19.9).

1. DNA, which remains in the nucleus, contains a series of bases that serve as a triplet code (a sequence of 3 bases).
2. One strand of DNA serves as a template for the formation of messenger RNA (mRNA), which contains triplet codons (sequences of 3 complementary bases).
3. Messenger RNA goes into the cytoplasm and becomes associated with the ribosomes, which are composed of ribosomal RNA (rRNA).
4. Transfer RNA (tRNA) molecules, each of which is bonded to a particular amino acid, have anticodons that pair complementarily to the codons in mRNA.
5. Transfer RNA molecules, along with their amino acids, come to the ribosomes in the order dictated by mRNA, and in this way amino acids become ordered in a particular sequence.
6. As the amino acid chain lengthens, a specific protein begins to form.
7. The transfer RNA molecules repeat the process of transporting amino acids to the ribosomes until the protein molecule is completed.

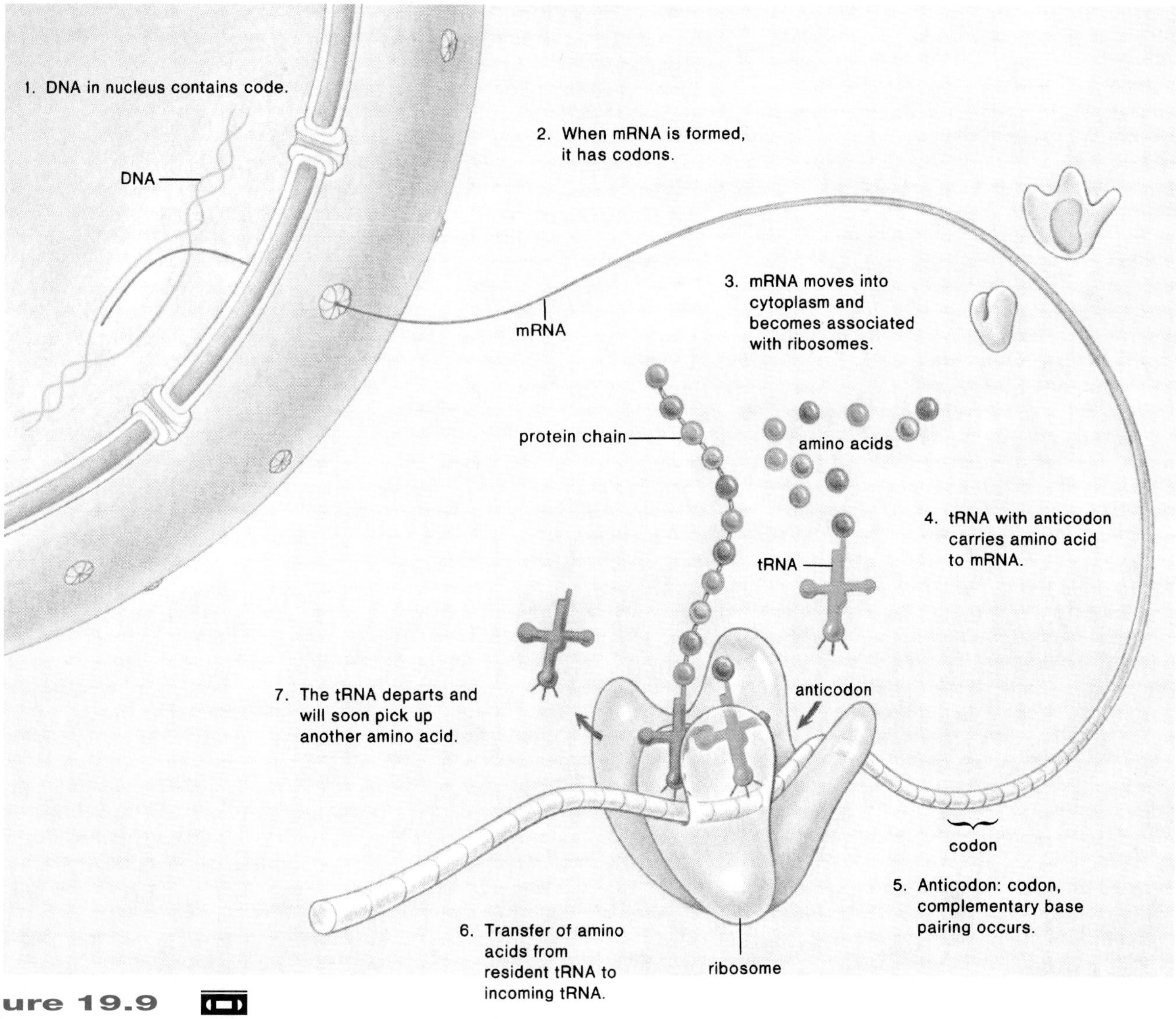
1. DNA in nucleus contains code.
DNA
2. When mRNA is formed, it has codons.
mRNA
3. mRNA moves into cytoplasm and becomes associated with ribosomes.
protein chain
amino acids
4. tRNA with anticodon carries amino acid to mRNA.
tRNA
anticodon
7. The tRNA departs and will soon pick up another amino acid.
codon
5. Anticodon: codon, complementary base pairing occurs.
6. Transfer of amino acids from resident tRNA to incoming tRNA.
ribosome

Figure 19.9

Summary of protein synthesis. Transcription occurs in the nucleus, and translation occurs in the cytoplasm (light blue background). During translation, as a ribosome moves along the mRNA, the tRNAs pair with the codons so that the amino acids become sequenced in a particular order.

mRNA, transcribed in the nucleus, contains a sequence of triplet codons complementary to the triplet DNA code. Translation occurs at a ribosome in the cytoplasm, where tRNA anticodons bind to the mRNA codons. The sequence of codons determine the sequence of tRNA binding and therefore the sequence of amino acids in a protein.

Gene and Gene Mutations: Redefined

In the previous chapter a gene was thought of as a particle on a chromosome. To molecular geneticists, however, a gene is a sequence of DNA nucleotide bases that codes for a protein. Early geneticists understood that genes undergo mutations, but they didn't know what causes mutations. It is apparent today that a gene mutation is a change in the nucleotide sequence of a gene.

Table 19.3

Steps in Protein Synthesis

Molecule	Special Significance	Definition
DNA	Code	Sequence of bases in 3s
mRNA	Codon	Complementary sequence of bases in 3s
tRNA	Anticodon	Sequence of 3 bases complementary to codon
Amino acids	Building blocks	Transported to ribosomes by tRNAs
Protein	Enzymes and structural proteins	Amino acids joined in a predetermined order

Table 19.4

Representative Biotechnology Products

Hormones and Similar Types of Proteins		Vaccines
Treatment in Humans	*For*	*Use in Humans*
Insulin	Diabetes	AIDS*
Growth hormone	Pituitary dwarfism	Herpes (oral and genital)*
tPA (tissue plasminogen activator)	Heart attack	Hepatitis A, B, and C
Interferons	Cancer	Lyme disease
Erythropoietin	Anemia	Whooping cough
Interleukin-2	Cancer	Chlamydia*
Clotting factor VIII	Hemophilia	
Human lung surfactant	Respiratory distress syndrome	
Atrial natriuretic factor	High blood pressure	
Tumor necrosis factor	Cancer	
Ceredase	Gaucher disease	

*Experimental

Figure 19.10

Biotechnology, an industrial endeavor. **a.** Laboratory procedures are adapted to mass-produce the product. **b.** Bacteria are grown in huge tanks. These tanks are called fermenters because they were first used for yeast fermentation in the production of wine. **c.** The product is purified and **d.** packaged.

Biotechnology

Biotechnology, the use of a natural biological system to produce a product or to achieve an end desired by human beings, is not new. Plants and animals have been bred to yield a particular phenotype since the dawn of civilization. In addition, the biochemical capabilities of microorganisms have been exploited for a very long time. For example, the baking of bread and the production of wine are dependent on yeast cells to carry out fermentation reactions.

Today, however, biotechnology is first and foremost an industrial process in which we are able to **genetically engineer** (bioengineer) bacteria to produce human proteins or other protein products of interest (fig. 19.10). Biotechnology is also being used as a way to alter the genotype, and subsequently the phenotype, of animals and plants. Genetically engineered organisms contain a foreign gene. Then they are capable of producing a new and different protein. Gene therapy in humans is the genetic engineering of humans to correct an inborn error of metabolism. Once a working gene replaces a faulty gene, a missing protein can be produced.

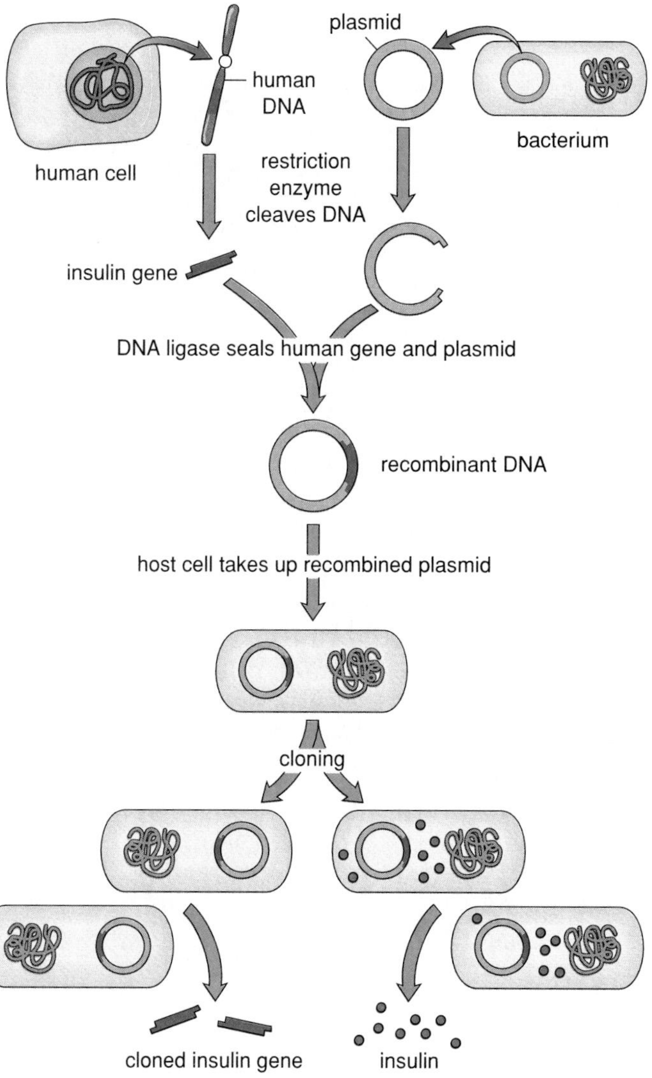

Figure 19.11

Cloning of the human insulin gene. Restriction enzyme is used to remove the insulin gene from human DNA and to open a plasmid. Second, the human gene and the plasmid are sealed together by DNA ligase. Gene cloning is achieved when a host cell takes up the recombined plasmid and the host and plasmid reproduce. Multiple copies of the gene are now available to an investigator. If the insulin gene functions normally as expected, the product (insulin) may also be retrieved.

Recombinant DNA: Uses Two Sources

Recombinant DNA, which contains DNA from 2 different sources, is used in particular to carry a foreign gene into bacteria. For example, recombinant DNA might contain bacterial DNA and human DNA.

A bacterial cell does not have a nucleus like a human cell, but it does have a single chromosome and often a **plasmid,** which is a small accessory ring of DNA outside the chromosome. The introduction of a human gene into a plasmid to produce recombinant DNA requires 2 different types of enzymes (fig. 19.11). A **restriction enzyme** is used to cut open plasmid DNA and to slice out the human gene from human DNA. **DNA ligase enzyme** is used to seal the cut ends so that the human gene can be inserted into the plasmid. Bacteria will take up recombined plasmids. The plasmid replicates, and then eventually there are many copies of not only the plasmid but also the human gene, such as the gene that codes for insulin. The cloned gene or its product can now be retrieved.

Other Biotechnology Techniques

Aside from the use of recombinant DNA, other techniques are common in biotechnology, including the following:

1. Gene sequencing, or determining the order of nucleotides in a gene. There are automated DNA sequences that make use of computers to sequence genes at a fairly rapid rate.
2. Manufacture of a gene by using a DNA synthesizer (fig. 19.12*b*). The nucleotides are joined in the correct sequence, or if desired, a mutated gene is prepared by altering the sequence.
3. Insertion of a gene directly into a host cell without using a plasmid. Animal cells, in particular, do not take up plasmids, but DNA can be microinjected into them.

Many Biotechnology Products

Table 19.4 shows some of the biotechnology products either available now or still experimental. Monoclonal antibodies produced by bacteria are in the experimental stage. In the meantime, monoclonal antibodies, while still considered a biotechnology product, are produced by the procedure described on page 141.

Mass Producing Hormones and Similar Proteins

One impressive advantage of biotechnology is that it facilitates the mass-production of proteins that are very difficult to otherwise obtain. For example, growth hormone (GH) was previously extracted from the pituitary glands of cadavers, and it took 50 glands to obtain enough of the hormone for one dose. Now growth hormone produced by biotechnology is used to treat growth abnormalities. Insulin was previously extracted from the pancreas glands of slaughtered cattle and pigs; it was expensive to procure and sometimes caused allergic reactions in recipients. Now insulin produced by biotechnology is used to treat diabetes. Not long ago, few of us knew of tPA (tissue plasminogen activator), a blood protein that activates an enzyme to dissolve blood clots. Now tPA is a biotechnology product that is used to treat heart attacks.

A study of table 19.4 lists other troublesome and serious afflictions in humans that are now treatable by biotechnology products. Clotting factor VIII treats hemophilia; human

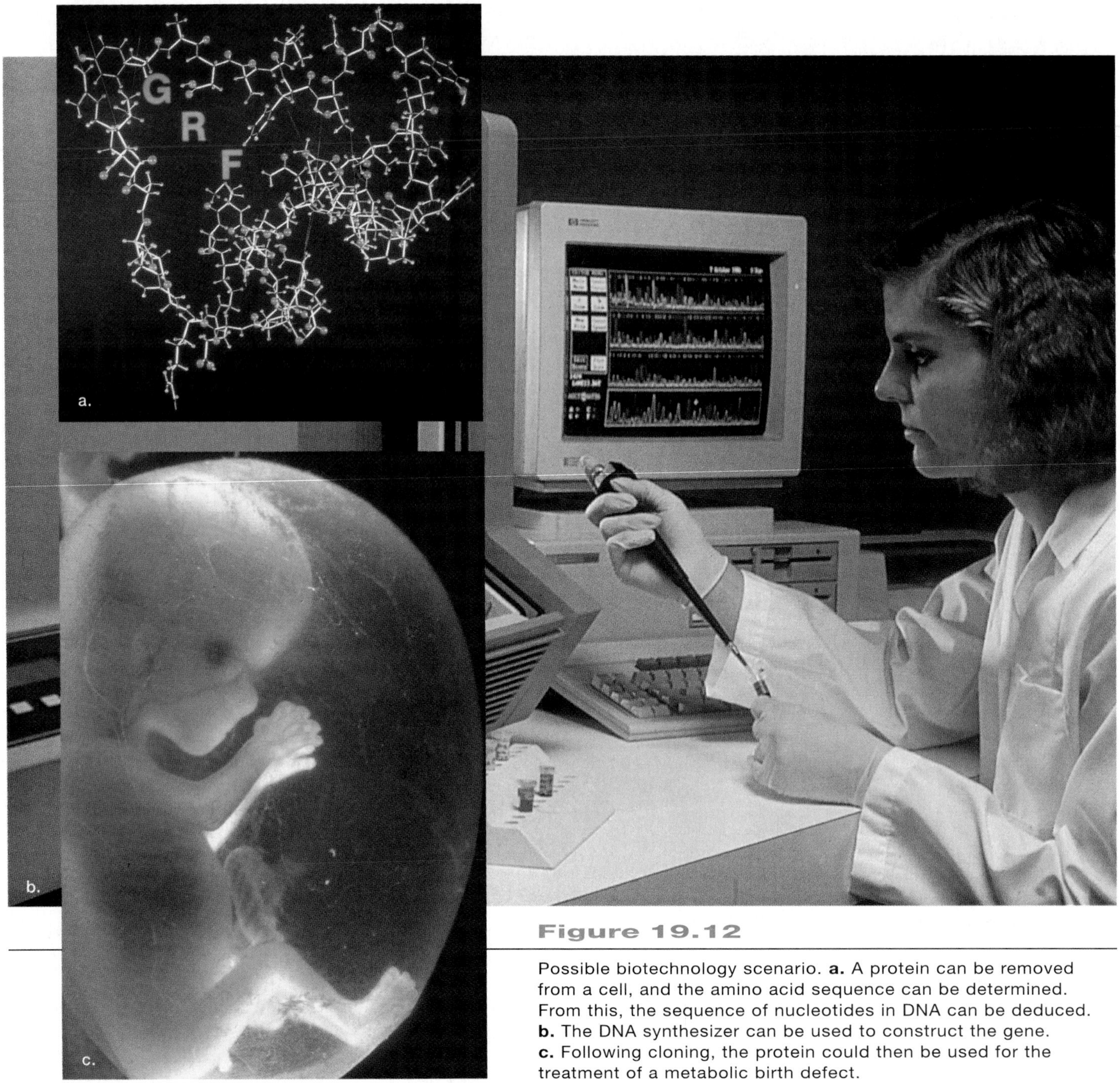

Figure 19.12

Possible biotechnology scenario. **a.** A protein can be removed from a cell, and the amino acid sequence can be determined. From this, the sequence of nucleotides in DNA can be deduced. **b.** The DNA synthesizer can be used to construct the gene. **c.** Following cloning, the protein could then be used for the treatment of a metabolic birth defect.

lung surfactant treats respiratory distress syndrome in premature infants; and atrial natriuretic factor helps control hypertension. This list will grow because bacteria (or other cells) can be engineered to produce virtually any protein.

Several hormones are products of biotechnology used in animals. It is no longer necessary to feed steroids to farm animals; they can be given growth hormone, which produces a leaner meat that is more healthful for humans. Cows given bovine growth hormone (bGH) produce 25% more milk than usual, which should make it possible for dairy farmers to maintain fewer cows and to cut down on overhead expenses. Even so there are some who do not wish to drink milk from cows given a biotechnology product.

Making Safer Vaccines

Vaccines are used to make people immune to an infection so they do not become ill when exposed to an infectious organism (p. 138). In the past, vaccines were made from treated bacteria or viruses, and on occasion they caused the illness they were supposed to prevent.

Vaccines produced through biotechnology do not cause illness. Bacteria and viruses have surface proteins, and a gene for just one of these can be used to genetically engineer bacteria. The copies of the surface protein that result can be used as a vaccine. Since only a portion of the microbe is present in the vaccine, no illness can result. A vaccine for hepatitis B is now available, and vaccines for chlamydia, malaria, and AIDS are in experimental stages.

Vaccines are available through biotechnology for the inoculation of farm animals, too. There are vaccines for such illnesses as hoof-and-mouth disease and scours. These animal ailments were once a severe drain on the time, energy, and resources of farmers.

> Biotechnology products include hormones and similar types of proteins and vaccines. These products are of enormous importance to the fields of medicine and animal husbandry.

Polymerase Chain Reaction Multiplies DNA

The *polymerase chain reaction* (*PCR*) can create millions of copies of a single gene or any specific piece of DNA in a test tube. PCR is very specific—the *targeted DNA sequence* can be less than one part in a million of the total DNA sample! This means that a single gene among all the human genes can be amplified (copied) using PCR.

PCR takes its name from DNA polymerase, the enzyme that carries out DNA replication in a cell. It is considered a chain reaction because DNA polymerase will carry out replication over and over again, until there are millions of copies of the targeted DNA. PCR does not replace gene cloning; cloning provides many more copies of a gene, and it still is used whenever a large quantity of a gene or a protein product is needed.

Before carrying out PCR, *primers*—sequences of about 20 bases that are complementary to the bases on either side of the "target DNA"—must be available. The primers are needed because DNA polymerase does not start the replication process—it only continues or extends the process. After the primers bind by complementary base pairing to the DNA strand, DNA polymerase copies the target DNA (fig. 19.13*a* and *b*).

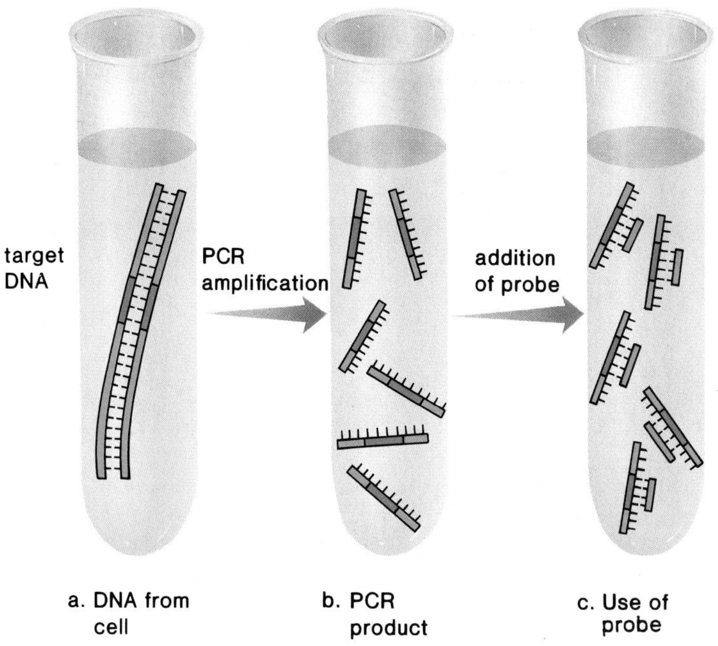

Figure 19.13

PCR amplification and analysis. **a.** DNA is removed from a cell and placed in a test tube along with appropriate primers, DNA polymerase, and a supply of nucleotides. **b.** Following PCR amplification, many copies of target DNA (red) are present. **c.** Binding of a labeled DNA probe (blue) allows the scientist to determine that a particular DNA segment was indeed present in the original sample.

PCR has been in use for several years, but the introduction of automated PCR machines is a recent advance. Now almost any laboratory can carry out the procedure. Automation became possible after a temperature insensitive (thermostable) DNA polymerase was extracted from a bacterium. The availability of this enzyme means that there is no need to add more DNA polymerase each time a high temperature is used to separate double-stranded DNA so that replication can reoccur.

Analyzing DNA Strands

The DNA resulting from PCR amplification can be analyzed using the following procedures involving DNA probes. A **DNA probe** is a single strand of radioactive DNA nucleotides that can be made by a DNA synthesizer. Because a DNA probe is single-stranded, it seeks out and binds to a complementary DNA strand (fig. 19.13*c*). A DNA probe that has a sequence of nucleotides complementary to the gene of a microbe can be used to diagnose an infection caused by the microbe. Specific DNA probes can also be used to diagnose tuberculosis or an HIV infection. In addition, a DNA probe can tell us whether a gene coding for a hereditary defect or causing a cell to be cancerous is present.

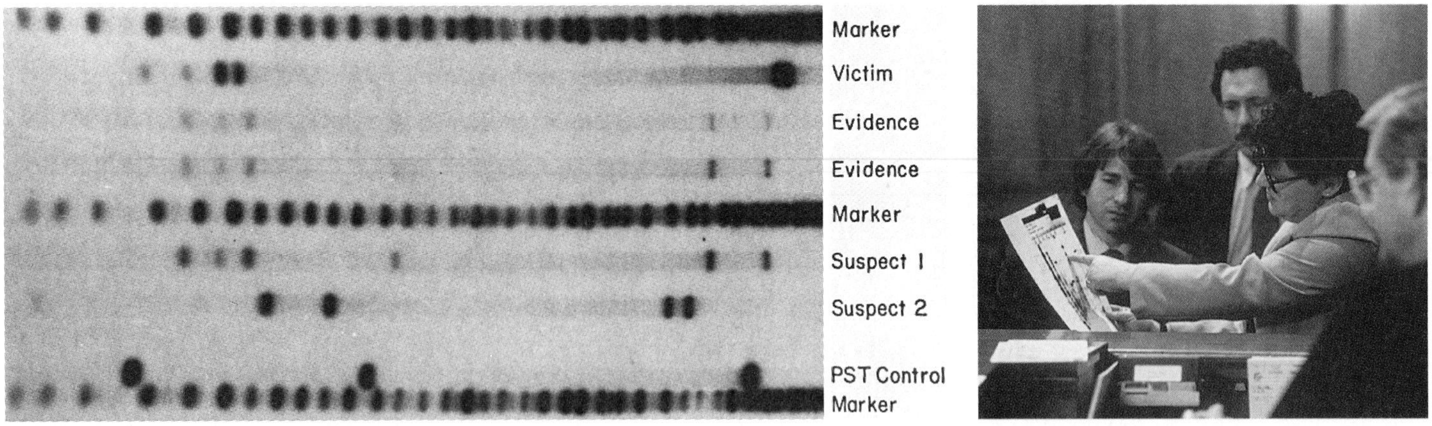

Figure 19.14

Results from DNA fingerprinting. Compare the pattern of fragments of the victim's DNA; the evidence (DNA fragments from sperm found in or on the victim); the DNA fragments from suspect 1 and suspect 2. (The other rows are controls.) From these results, which suspect is the rapist?

DNA probes are used during **DNA fingerprinting.** The DNA is first treated with restriction enzymes, which cut it into fragments. Each individual's DNA results in a collection of different-sized fragments. During a process called gel electrophoresis, the fragments are separated according to their lengths, and the result is a pattern of bands that is different for each individual. The use of radioactive probes allows the pattern to be recorded on X-ray film.

DNA fingerprinting can be done on any piece of DNA. It was used to successfully identify a teenage murder victim from 8-year-old remains. Skeletal DNA was compared to that obtained from blood samples donated by the victim's parents. A DNA fingerprint resembles that of the parents because it is inherited. DNA from a single sperm is enough to identify a suspected rapist when PCR amplification precedes DNA fingerprinting (fig. 19.14).

Genetically Engineered Organisms

Free-living organisms in the environment that have had a foreign gene inserted into them are called **transgenic organisms.**

Transgenic Bacteria

Genetically engineered bacteria can be used to promote the health of plants. For example, bacteria that normally cause ice to crystallize on plants have been changed from frost-plus to frost-minus bacteria. Field tests showed that these genetically engineered bacteria now protect the vegetative parts of plants from frost damage. Also, a bacterium that normally colonizes the roots of corn plants has now been endowed with genes (from another bacterium) that code for an insect toxin.

Many other recombinant DNA applications in agriculture are thought to be possible. For example, bacteria in the genus *Rhizobium* form nodules on the roots of leguminous plants such as bean plants. Here, the bacteria capture atmospheric nitrogen in a form that can be used by the plant. Perhaps the necessary genes could be transferred to other types of bacteria capable of infecting nonleguminous plants. This might eventually reduce the amount of fertilizer needed on agricultural fields.

There are naturally occurring bacteria that can degrade almost any type of chemical or material. This ability can be enhanced by genetic engineering and the bacteria can be used for *remediation,* the cleanup of the environment by biological means. For example, naturally occurring bacteria that eat oil can be genetically engineered to do an even better job of cleaning up beaches after oil spills (fig. 19.15).

As discussed in the Ecology Focus for this chapter, there are those who are very concerned about the deliberate release of genetically engineered microbes (GEMs) into the environment. Ecologists point out that these bacteria might displace those that normally reside in an ecosystem, and the effects could be deleterious. Others rely on past experience with GEMs, primarily in the laboratory, to suggest that these fears are unfounded. Tools are now available to detect, measure, and even stop cell activity in the natural environment. It is hoped that these tools will eventually pave the way for GEMs to play a significant role in agriculture and in environmental protection.

Transgenic Plants

It is hoped that one day genetically engineered plants will include the following:

Crops that have an increased ability to resist insect attacks (fig. 19.16) or to grow under unfavorable environmental conditions, such as in the presence of herbicides. Heat-, cold-,

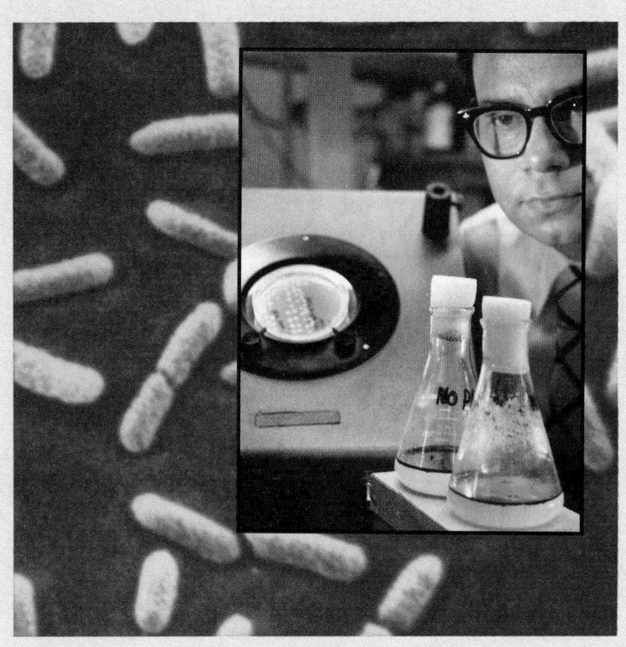

Figure 19.15

Oil-eating bacteria, genetically engineered and patented by investigator Dr. Chakrabarty. In the inset, the flask toward the front contains oil and no bacteria; the flask toward the rear contains the bacteria and is almost clear of oil. Now that genetically engineered organisms (e.g., bacteria and plants) can be patented, there is even more impetus to create them.

drought-, and salt-tolerant and herbicide-resistant crops are expected in the near future.

Crops that are more nutritious because the seed contains all the amino acids required by humans. Protein-enhanced beans, corn, soybeans, and wheat are now being developed. The food processing industry is also interested in plants that can be stored and transported without fear of damage. Already, bruise-resistant tomatoes in which fruit ripening is delayed have been produced.

Crops that require less fertilizer because they are able to make use of nitrogen from the atmosphere. (Plants ordinarily take nitrogen from the soil only.) Certain bacteria of the genus *Rhizobium* can make use of atmospheric nitrogen. If the required genes were cloned, perhaps they could be transferred directly to plants, bypassing the need for the transgenic bacteria mentioned earlier.

Plants that produce chemicals and drugs of interest to humans. Potatoes have been genetically engineered to produce human albumin, and one day it is believed that plants will produce human proteins, such as hormones, in their seeds. Tobacco plants have been infected with genetically engineered viruses that cause them to produce all manner of human proteins, including hemoglobin and α-amylase.

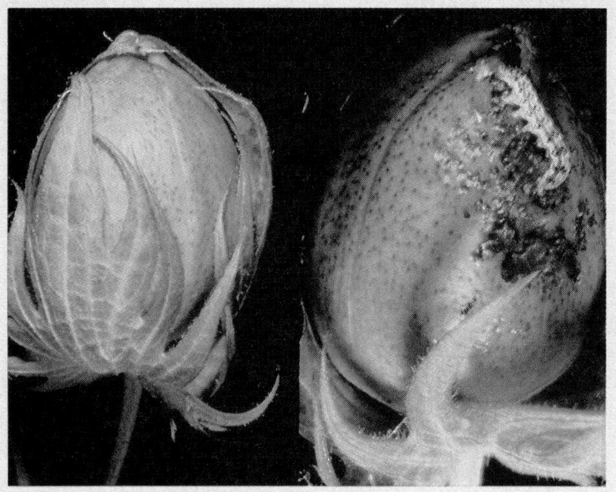

Figure 19.16

Genetically engineered plants. The cotton boll on the right is from a plant that has not been genetically engineered to resist cotton bollworm larvae and is heavily infested. The boll on the left is from a plant that was genetically engineered to resist cotton bollworm larvae and will go on to give a normal yield of cotton.

Transgenic Animals: Bigger Fishes and tPA Sheep

Animals, too, are being genetically engineered. Previously, the most common method was to microinject foreign genes into eggs before they were fertilized. During vortex mixing, a time-saving procedure, eggs are placed in an agitator with DNA and silicon-carbide needles. The needles make tiny holes through which the DNA can enter the eggs. Using this technique, many types of animal eggs have been injected with bovine growth hormone (bGH). The procedure has been used to produce larger fishes, cows, pigs, rabbits, and sheep. Genetically engineered fishes are now being kept in ponds that offer no escape to the wild because there is much concern that they will upset or destroy natural ecosystems.

Transgenic farm animals are being developed to produce biotechnology products. For example, the milk of a transgenic cow now contains lactoferin, a protein that is involved in iron transport and has antibacterial activity. Similarly, the milk of a transgenic sheep contains tPA, and there are now sheep that produce human factor VIII, needed by hemophiliacs in their blood.

> Transgenic bacteria, plants, and animals are now a reality. These organisms have been genetically engineered to serve all sorts of purposes.

Ecology Focus

Biotechnology: Friend or Foe?

This notice recently appeared in a popular science magazine:

> In a feat that could boost wheat production worldwide, plant biologists have for the first time permanently transferred a foreign gene into wheat. The genes make wheat resistant to the herbicide phosphinothricin which normally kills any plant it touches, weed or crop. Plant breeders say the herbicide-resistant wheat should enable farmers to spray their fields with the powerful herbicide to eradicate weeds without harming their harvest.*

In a hungry world you would expect such news to be greeted with enthusiasm—this new strain of wheat offers the possibility of a more bountiful harvest and the feeding of many more people. There are some, however, who see dangers lurking in the use of biotechnology to develop new and different strains of plants and animals. And these doomsayers are not just anybody, they are ecologists.

Suppose, the argument goes, this new form of wheat is better able to compete in the wild. Certainly we know of plants that have become pests when transported to a new environment: prickly pear cactus took over many acres of Australia; an ornamental tree, the melaleuca, has invaded and is drying up many of the swamps in Florida, for example. Such plants spread because they are able to overrun the native plants of an area. Perhaps genetically engineered plants will also spread everywhere and be out of control. Or worse, suppose herbicide-resistant wheat were to hybridize with a weed, making the weed also resistant and able to take over other agricultural fields. As more and different herbicides are used to kill off the weed, the environment would be degraded. And similar concerns pertain to any transgenic organism, whether a bacterium, plant, or animal (fig. 19A).

Figure 19A

Would it be irresponsible to spray fruit trees with ice-minus bacteria, releasing vast quantities that could drift into natural areas and perhaps upset the ecological balance there? Ice-minus, unlike wild-type, bacteria lack a protein that encourages ice formation.

In the past, humans have been quick to believe that a new advance was the answer to a particular problem. The new pesticide DDT was going to kill off mosquitoes, making malaria a disease of the past. Instead, mosquitoes become resistant and DDT accumulates in the tissues of humans, possibly contributing to all manner of health problems from reduced immunity to reproductive infertility. When antibiotics were first introduced it was hoped a disease like tuberculosis would be licked forever. Resistant strains of tuberculosis have now evolved to threaten us all.

Clearly, we should proceed with caution, and thus far the scientific community has done so. At first, recombinant DNA technology was under strict controls until scientists were satisfied that transgenic bacteria would not be able to take residence in the wild or in humans. Limited release has shown that transgenic bacteria do not spread beyond their boundaries and genetically engineered plants do not show any evidence of being able to overrun their neighbors. However, there are those who still maintain that the absence of a negative effect today does not mean that there could not be one in the future.

*Research Notes, "Biology," *Science News* 141(1992): 379.

The same ecological concerns regarding genetically engineered bacteria also pertain to genetically engineered plants and animals. The Ecology Focus for this chapter uses herbicide-resistant wheat as a point of reference, but the very same arguments can be waged against any genetically engineered organism.

Gene Therapy in Humans

Gene therapy replaces defective genes with healthy genes. Gene therapy also includes the use of added genes to treat such human ills as diabetes and AIDS.

During *ex vivo* (outside living organism) *therapy,* cells are removed from a patient, treated, and returned to the patient. A retrovirus, which has RNA genes instead of DNA genes, is used as a vector to carry normal genes into the cells of the patient, where they become incorporated into the genome. When a retrovirus is used for gene therapy, it has been equipped with recombinant RNA—viral RNA plus an RNA copy of the normal gene. After recombinant RNA enters a human cell, such as a bone marrow stem cell, reverse transcription occurs, and then recombinant DNA carrying the normal gene enters a human chromosome (fig. 19.17).

Ex vivo therapy clinical trials are underway. In one such trial, the gene needed by 2 girls with severe combined immune deficiency syndrome (SCID) was introduced into their lymphocytes. These girls lack an enzyme that is involved in the maturation of T and B cells; and as a precaution, they also received the enzyme (isolated from cows)

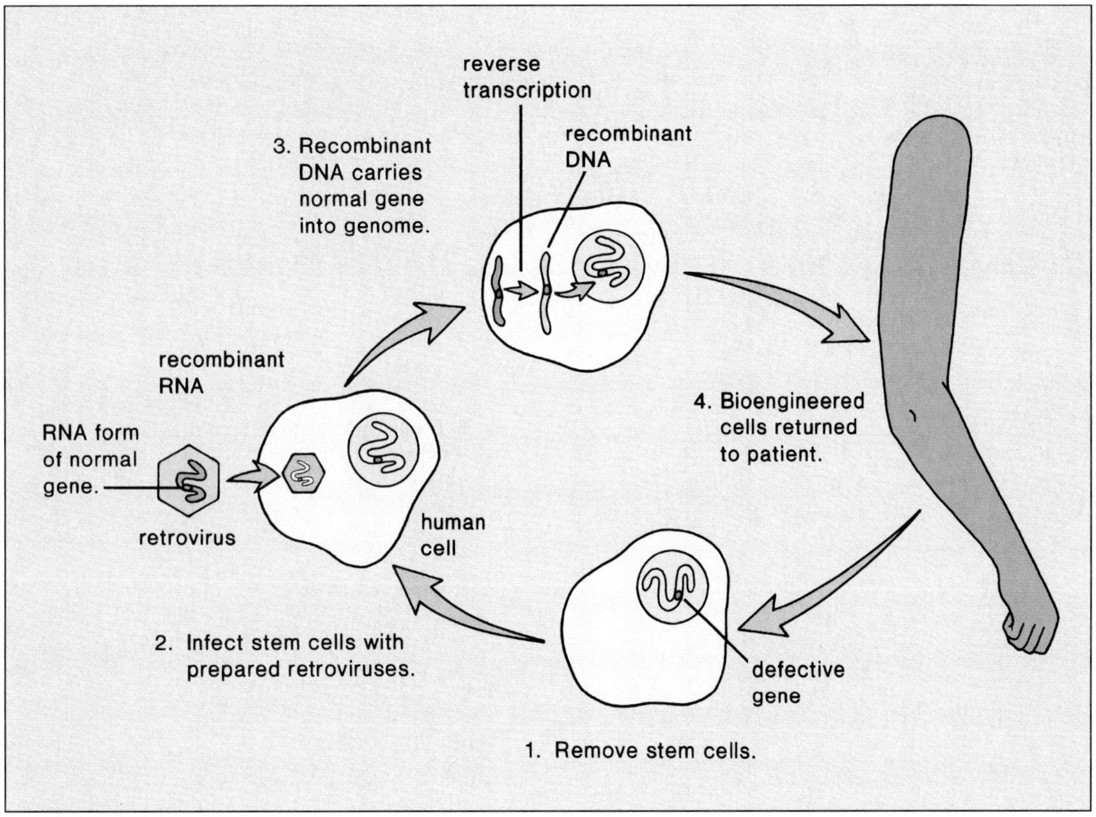
reverse
transcription
3. Recombinant
DNA carries
normal gene
into genome.
recombinant
DNA
recombinant
RNA
RNA form
of normal
gene.
retrovirus
human
cell
4. Bioengineered
cells returned
to patient.
2. Infect stem cells with
prepared retroviruses.
defective
gene
1. Remove stem cells.

Figure 19.17

Ex vivo gene therapy in humans. Stem cells are withdrawn from the body, a normal gene is inserted into them, and they are then returned to the body.

as a medication. Recently, one of these girls received genetically engineered bone marrow stem cells; this is preferred because stem cells are long-lived and their use may result in a permanent cure. In another proposed clinical trial, investigators will treat the patients' liver cells so that these cells will remove cholesterol from blood. The participants in this trial are at risk for a heart attack early in life because they have too much cholesterol in their blood.

Other gene therapy procedures use viruses, laboratory-grown cells, or even synthetic carriers to introduce genes directly into the patient. If *in vivo* (inside living organism) *therapy* is used, no cells are removed from the patient. For example, an adenovirus that contains a gene to treat cystic fibrosis patients has been placed in an aerosol spray. When laboratory animals inhale the spray, the virus enters the cells lining the lungs.

Perhaps it will be possible to use in vivo therapy to cure hemophilia, diabetes, Parkinson disease, or AIDS. To cure hemophilia, patients would get regular doses of cells containing normal clotting-factor genes. Or cells could be placed in organoids, artificial organs that can be implanted in the abdominal cavity. To cure Parkinson disease, dopamine-producing cells could be grafted directly into the brain. These procedures will use laboratory-grown cells that have been stripped of antigens (to decrease the possibility of an immune system attack). To cure AIDS, decoys that display the CD4 receptor, which is attractive to the gp120 protein in the HIV envelope, could be injected into the body. The HIV virus would combine with these cells instead of with helper T lymphocytes.

Some investigators are pursuing gene therapy by direct injection of a protein-DNA complex, which is expected to enter liver cells for the purpose of treating diseases associated with the liver, including cancer.

Mapping Human Chromosomes

If investigators knew the order and precise location of the genes on the human chromosomes, it would facilitate laboratory research and medical diagnosis and treatment. Several methods have been used to attempt mapping the human chromosomes, and we will discuss one of these.

The DNA that a person inherits from a parent has a unique sequence of base pairs. A *genetic marker* is a place on the chromosome where the sequence of base pairs differs from one person to another. These differences are usually seen in "filler DNA," sequences of bases between the genes. Genetic markers are discovered by using restriction enzymes, which cleave DNA into fragments. Because each person has their own restriction enzyme cleavage sites, each person produces a particular pattern of different-length fragments, as discussed formerly (p. 394).

Genes can sometimes be assigned a location on a chromosome according to their relative relationship to genetic markers. Therefore, they have helped scientists develop maps of the human chromosomes. Genetic markers can also be used to test for a genetic disorder when a particular marker is always inherited with a particular gene. Individuals are tested for the presence of the marker instead of the specific faulty allele. For a marker to be dependable, it should be inherited with the faulty allele at least 98% of the time. The available tests for sickle-cell disease, Huntington disease, and Duchenne muscular dystrophy all are based on the presence of a marker (fig. 19.18).

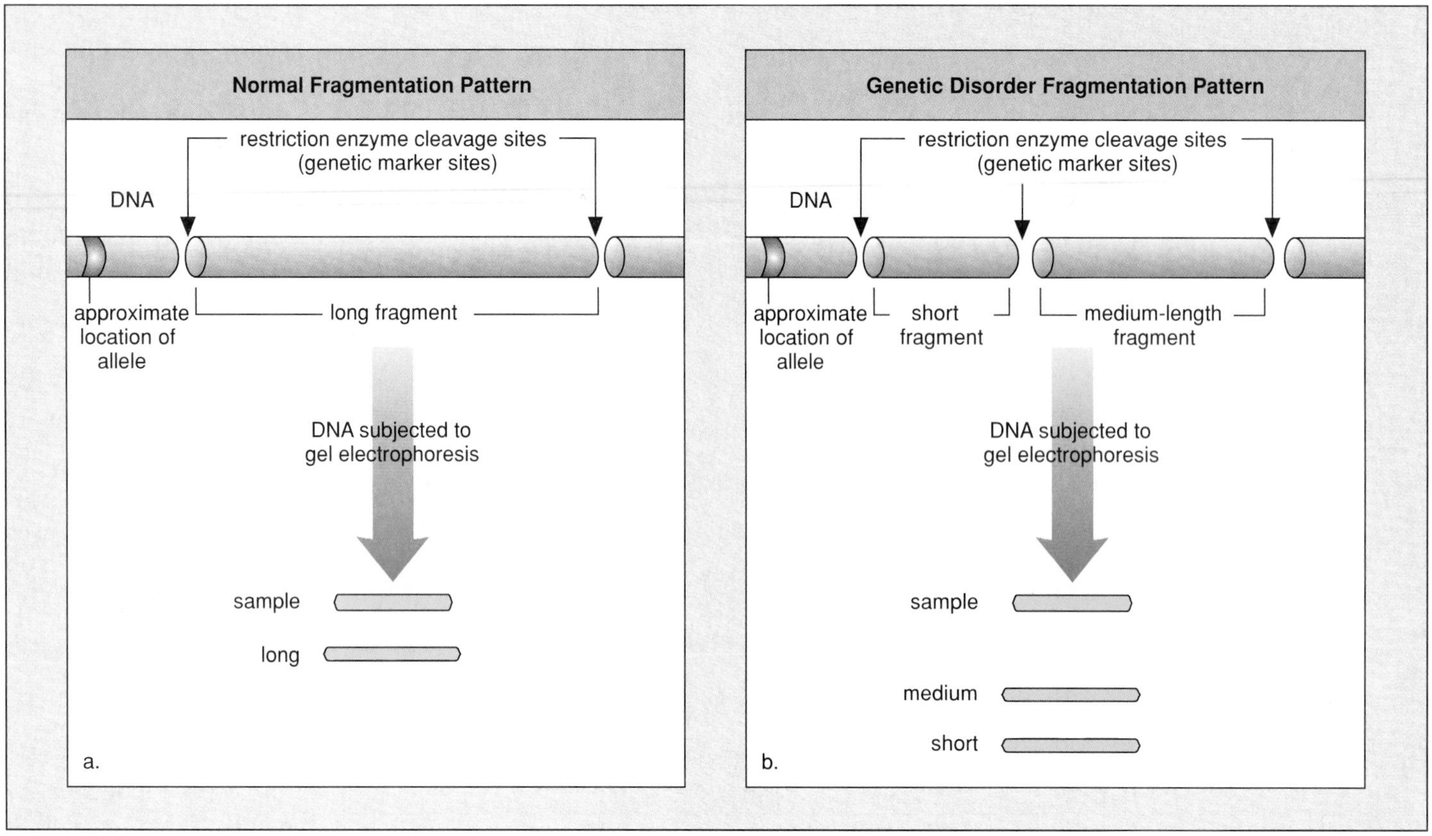

Figure 19.18

Use of genetic markers to test for a genetic disorder. **a.** DNA from the normal individual has certain restriction enzyme cleavage sites near the gene in question. The resulting fragmentation pattern is discovered by gel electrophoresis, a laboratory process that separates the fragments according to length. **b.** DNA from another individual has an additional cleavage site, and this addition indicates he or she has the genetic disorder. Again, gel electrophoresis reveals the fragmentation pattern.

Human Genome Project: Locating Genes and Bases

The goal of the *Human Genome Project* is to identify the location of the approximately 100,000 human genes on all the chromosomes. To create this *genetic map,* genetic markers are used to index the chromosomes. Known and newly discovered genes are assigned locations between the markers (fig. 19.19).

Eventually, those involved in the project also want to determine the sequence of the 3 billion bases in the human genome. To create this *physical map,* researchers use laboratory procedures to determine the sequence of the DNA bases.

It has recently been discovered that genetic markers contain unique stretches of DNA called sequence-tagged sites, or STSs. It is hoped that these can be used to create links between the genetic map and the physical map.

The human genome project will require millions of dollars and several years to complete. Just how successful and useful it will be cannot be determined yet.

Genetic markers now make it possible to map the human chromosomes at a faster rate than was formerly possible.

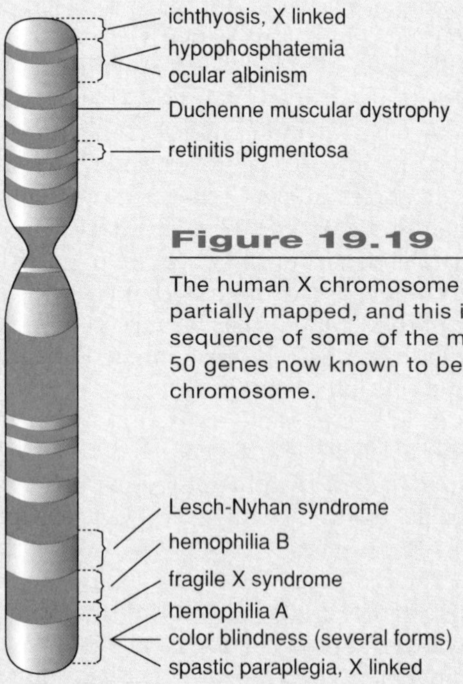

Figure 19.19

The human X chromosome has been partially mapped, and this is the sequence of some of the more than 50 genes now known to be on this chromosome.

Summary

DNA is a double helix composed of 2 nucleic acid strands that are held together by weak hydrogen bonds between the bases: A is bonded to T and C is bonded to G. RNA is a single-stranded nucleic acid that occurs in 3 versions: mRNA, rRNA, and tRNA.

DNA is the genetic material; it replicates and directs protein synthesis. During replication, the DNA strands unzip and then a new complementary strand forms opposite to each old strand. In the end there are 2 identical DNA molecules. Protein synthesis requires transcription and translation. During transcription, the DNA code (triplet of 3 bases) is passed to a mRNA that is complementary to one of the DNA strands and therefore contains codons. During translation, a protein with a particular sequence of amino acids forms: tRNA molecules each carrying an amino acid have anticodons that pair with the mRNA codons.

Biotechnology uses genes to produce protein products and to genetically engineer organisms. Recombinant DNA contains DNA from 2 different sources. A human gene can be inserted into a plasmid, which is taken up by bacteria. When the plasmid replicates, the gene is cloned and its protein is produced. This is the basis for the production of biotechnology products, such as hormones and vaccines.

The polymerase chain reaction (PCR) uses the enzyme DNA polymerase to carry out replication of a target piece of DNA over and over again (amplification), until there are a million or so copies. A suitable radioactive probe can then be used to determine if the DNA of an infectious organism or any particular sequence of DNA is present.

Alternately, the copies of DNA can also be subjected to DNA fingerprinting, which can be used to identify an individual: (1) the DNA is cleaved by restriction enzymes; (2) each person's DNA has its own sequence of cleavage sites and therefore a different fragment-length pattern; (3) gel electrophoresis and the use of radioactive probes reveals the pattern.

Transgenic organisms also have been made. More nutritious crops resistant to pests and herbicides will soon be commercially available. Transgenic animals have been supplied with various genes, in particular the one for bovine growth hormone (bGH). Animals are also being used to produce protein products of interest.

Human gene therapy is undergoing clinical trials. Ex vivo therapy involves withdrawing cells from the patient, inserting a functioning gene, usually via a retrovirus, and then returning the treated cells to the patient. Many investigators are trying to develop in vivo therapy, in which viruses, laboratory-grown cells, or synthetic chemicals will be used to carry healthy genes into the patient.

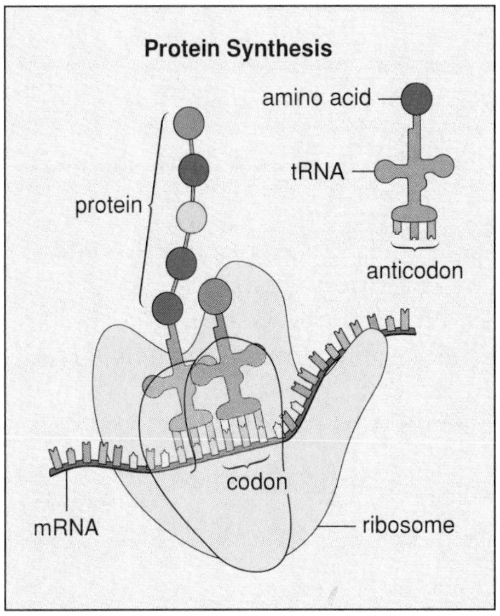

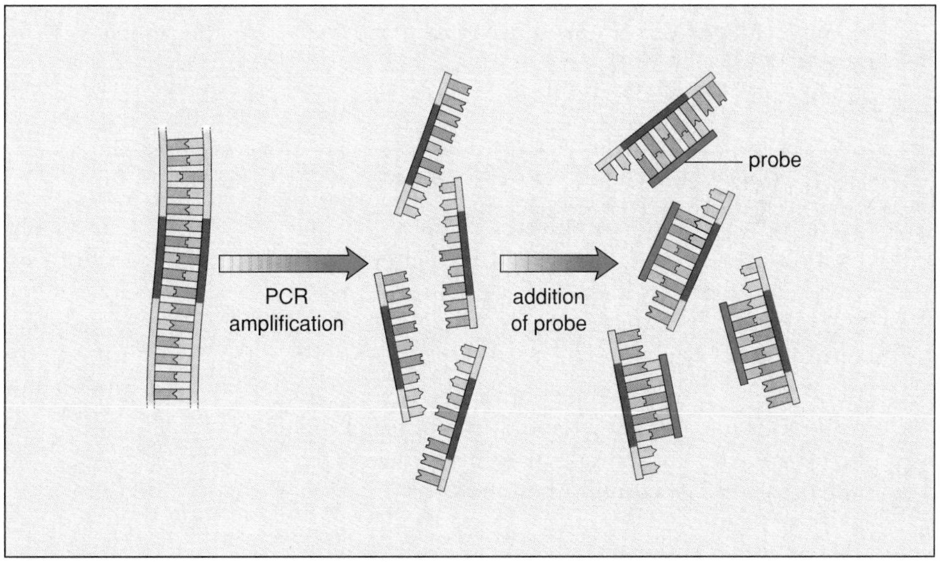

Study and Thought Questions

1. Compare and contrast the structure of DNA and RNA. (pp. 383–384)

   Critical Thinking Enzymatic RNA has been discovered in the nucleus, and this has led some researchers to believe that of the 3 molecules—DNA, RNA, and protein—the first cell required only RNA. Why should that be?
2. Explain how DNA replicates. Why is this replication called semiconservative? (p. 384)
3. List the steps involved in protein synthesis, mentioning the process of transcription and translation and the roles of DNA, mRNA, rRNA, and tRNA. (pp. 386–389)
4. If the code is TTT;CAT;TGG;CCG, what are the codons and what is the sequence of amino acids? Show how a deletion or duplication could affect the code. (p. 386)

   Critical Thinking The genetic code is degenerate, which means that certain amino acids are coded for by more than one nucleotide triplet. How might this protect the amino acid sequence of a protein?
5. You are a scientist who has decided to "clone a gene." Tell precisely how you would proceed. (p. 391)
6. Name 2 categories of biotechnological products that are now available, and discuss their advantages. (pp. 391–393)

   Ethical Issue Some parents want their children to take GH (growth hormone) because they believe larger persons compete better in sports and business. Some adults want to take GH because they believe it will keep them younger longer. Should there be any restrictions to the use of a drug like hGH?
7. Naturally occurring bacteria have been genetically engineered to perform what services? What are the ecological concerns regarding their release into the environment? (p. 394)
8. What types of genetically engineered plants are expected in the near future? What types of animals have been genetically engineered and for what purpose? (pp. 394–395)
9. How is gene therapy in humans currently being done? (p. 397)

   Ethical Issue Researchers may soon be able to genetically engineer human zygotes before they are implanted in the uterus. Does the prospect of children who are free of significant genetic diseases justify prior experimentation with human embryos?
10. What is the Human Genome Project, and in what ways might the project be useful? (p. 398)

Writing Across the Curriculum

1. Drawing from your knowledge of the reproductive process, support the suggestion that we are nothing but transport vehicles or hosts for DNA. What arguments can you think of to counter this suggestion?
2. Both mRNA and tRNA are transcribed from DNA in the nucleus, but they have different functions in the cytoplasm. How might they be "guided" to take up their respective functions?
3. Commercial laboratories are attempting to patent sections of DNA. Their permission, plus payment of a fee, would then be required if these sections were later found to contain information to cure human ills. How do you think access to human genetic material should be handled?

Objective Questions

1. In DNA, the base G is always paired with the base ________________, and the base A is always paired with the base________________.
2. Replication of DNA is semiconservative, meaning that each new double helix is composed of an ______________ strand and a ______________ strand.
3. The DNA code is a ______________ code, meaning that every 3 bases stands for an ________________.
4. The 3 types of RNA that are necessary to protein synthesis are ______________, ______________, and ________________.
5. Which of the types of RNA from question 4 carries amino acids to the ribosomes? ________________
6. The sequence of mRNA codons dictates the sequence of amino acids in a protein. This step in protein synthesis is called ________________.
7. The 2 types of enzymes needed to make recombinant DNA are ________________ and ________________.
8. Many types of transgenic animals have now received a(n) ______________ gene, which makes them grow larger.
9. The current ex vivo gene therapy clinical trials use a(n) ______________ as a vector to insert healthy genes in the patient's cells.
10. This is a segment of a DNA molecule. (Remember that only the transcribed strand serves as the template.) What are (a) the RNA codons, (b) the tRNA anticodons, and (c) the sequence of amino acids in a protein?

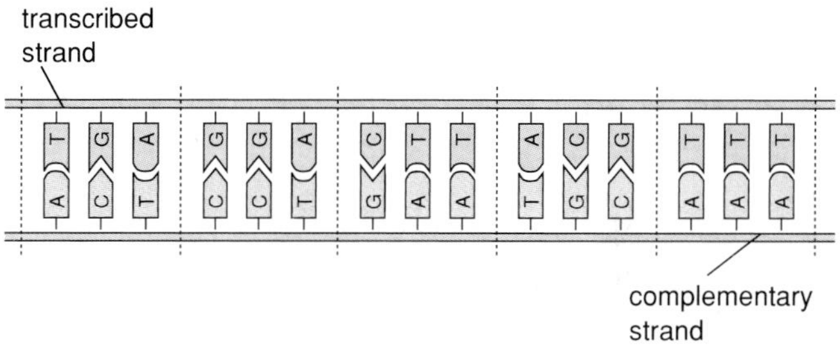

Selected Key Terms

anticodon A "triplet" of 3 nucleotides in transfer RNA that pairs with a complementary triplet (codon) in messenger RNA. 387

biotechnology Use of a natural biological system to produce a commercial product or achieve an end desired by humans. 390

codon A "triplet" of 3 nucleotides in messenger RNA that directs the placement of a particular amino acid into a protein. 387

complementary base pairing Pairing of bases between nucleic acid strands; adenine (A) pairs with either thymine (T) if DNA, or uracil (U) if RNA, and cytosine (C) pairs with guanine (G). 383

DNA fingerprinting Using fragment lengths resulting from restriction enzyme cleavage to identify particular individuals. 394

DNA ligase enzyme (li´gās) An enzyme that links DNA from 2 sources; used in genetic engineering to put a gene into plasmid DNA. 391

DNA probe Single strand of radioactive DNA that can be used to locate a particular stretch of DNA. 393

gene therapy The use of transplanted genes to overcome an inborn error of metabolism or to otherwise treat human ills. 396

genetically engineered To alter an organism's DNA or insert a foreign gene by a technological process; bioengineering. 390

messenger RNA (mRNA) Ribonucleic acid complementary to DNA; has codons that direct protein synthesis at the ribosome. 385

plasmid A circular DNA segment that is present in bacterial cells but is not part of the bacterial chromosome. 391

recombinant DNA DNA with genes from 2 different sources, often produced in the laboratory by introducing a foreign gene into a bacterial plasmid. 391

replication The duplication of DNA; occurs during interphase of the cell cycle. 384

restriction enzyme Enzyme that stops viral reproduction by cutting viral DNA; used in genetic engineering to cut DNA at specific points. 391

ribosomal RNA (rRNA) RNA occurring in ribosomes, structures involved in protein synthesis. 385

template A pattern that serves as a mold for the production of an oppositely shaped structure; one strand of DNA is a template for a complementary strand. 384

transcription The process resulting in the production of a strand of mRNA that is complementary to a segment of DNA. 386

transfer RNA (tRNA) Molecule of RNA that carries an amino acid to a ribosome engaged in the process of protein synthesis. 385

translation The process by which the sequence of codons in mRNA dictates the sequence of amino acids in a protein. 386

Chapter 20

Cancer

Cancer is not a new disease: fossil bones of dinosaurs and early humans sometimes show evidence of the disease. Its prominence among modern humans is due, in general, to the fact that many humans now live long enough to get cancer. Knowledge about the development of tumors like the one shown in this photo of a lung is progressing, and new types of treatments are expected soon. It is only in the past 50 years that generalized hypotheses about cancer have led to specific findings that shape the treatment people receive. This chapter explores what we know about cancer.

Chapter Concepts

1. Cancer cells have characteristics that can be associated with their ability to grow uncontrollably. 404
2. A number of agents are known to bring about the mutations that cause cancer. The immune system is at fault when cancer develops. 406–408
3. Mutations of proto-oncogenes and tumor suppressor genes are now known to cause cancer. 409–411
4. There are few routine screening tests for specific cancers, and cancer often goes undetected until it is advanced. New ways to diagnose cancer are under investigation. 414
5. Surgery, radiation, and chemotherapy are presently the standard ways of treating cancers. New treatment methods are under investigation. 415–418
6. It is possible to avoid certain agents that contribute to the development of cancer and to take protective steps to reduce the risk of cancer. 418–419

Chapter Outline

Nature and Causes of Cancer

Cancer cells have characteristics that set them apart from normal cells, and these characteristics are the result of various causes that we will explore in this chapter.

Characteristics of Cancer Cells

Cancer cells exhibit characteristics that distinguish them from normal cells (table 20.1). In general, cancer cells undergo uncontrolled growth and are not regulated by the signals and constraints that usually control the growth of cells.

Cancer Cells Lack Differentiation

Normal cells are specialized; they have a specific form and function that suits them to the role they play in the body. Cancer cells are nonspecialized and do not contribute to the functioning of a body part. A cancer cell does not look like a differentiated epithelial, muscular, nervous, or connective tissue cell and instead has a shape and form that is distinctly abnormal (fig. 20.1). Normal cells can undergo the cell cycle for about 50 times and then they die. Cancer cells can undergo the cell cycle repeatedly, and in this way, they are immortal. In culture, they die only because they run out of nutrients or are killed by their own toxic waste products.

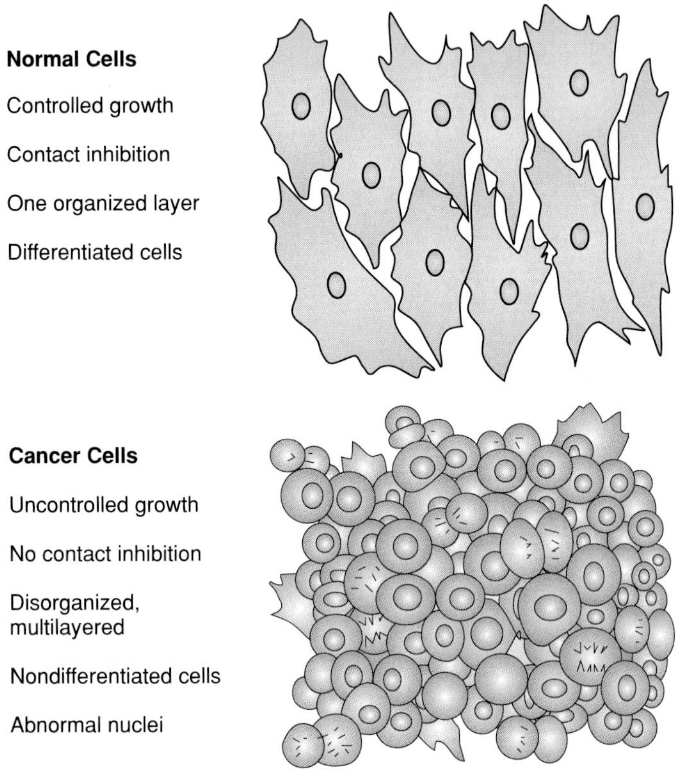

Figure 20.1

Cancer cells differ from normal cells in the ways noted.

Table 20.1

Characteristics of Normal Cells versus Cancer Cells

Characteristic	Normal Cells	Cancer Cells
Differentiation	Yes	No
Nuclei	Normal	Abnormal
Growth	Controlled	Uncontrolled
Contact inhibition	Yes	No
Growth factors	Required	Not required
Angiogenesis	No	Yes
Metastasis	No	Yes

Cancer Cells Have Abnormal Nuclei

The nuclei of cancer cells are enlarged and there may be an abnormal number of chromosomes. The chromosomes have mutated; some parts may be duplicated and some may be deleted, for example. In addition, *gene amplification* (extra copies of specific genes) is seen much more frequently than in normal cells.

Cancer Cells Form Tumors

Normal cells anchor themselves to a substratum and/or adhere to their neighbors. They exhibit *contact inhibition*—when they come in contact with a neighbor they stop dividing. In culture, normal cells form a single layer that covers the bottom of a petri dish. Cancer cells have lost all restraint; they pile up on top of each other and grow in multiple layers. They have a reduced need for growth factors, such as epidermal growth factor. Growth factors are hormones needed by normal cells to grow. For example, epidermal growth factor stimulates the growth of many types of cells.

In the body, a cancer cell divides to form an abnormal growth, or **tumor,** that invades and destroys neighboring tissue (fig. 20.2). This new growth, termed *neoplasia,* is made up of cells that are disorganized, a condition termed *anaplasia.* A *benign tumor* is a disorganized, usually encapsulated, mass that does not invade adjacent tissue.

Cancer Cells Undergo Angiogenesis and Metastasis

Angiogenesis, the formation of new blood vessels, is required to bring nutrients and oxygen to a cancerous tumor whose growth is not contained within a capsule. Cancer cells release a growth factor that causes neighboring blood vessels to branch into the cancerous tissue. Some modes of cancer treatment are aimed at preventing angiogenesis from occurring.

Cancer *in situ* is cancer found in one place, before invasion of normal tissue. **Cancer** is a malignant tumor that undergoes **metastasis,** the establishment of new tumors distant from the primary tumor. To accomplish metastasis, cancer cells must first make their way across a basement membrane and into a blood vessel or lymphatic vessel. It has been discovered that cancer cells have receptors that

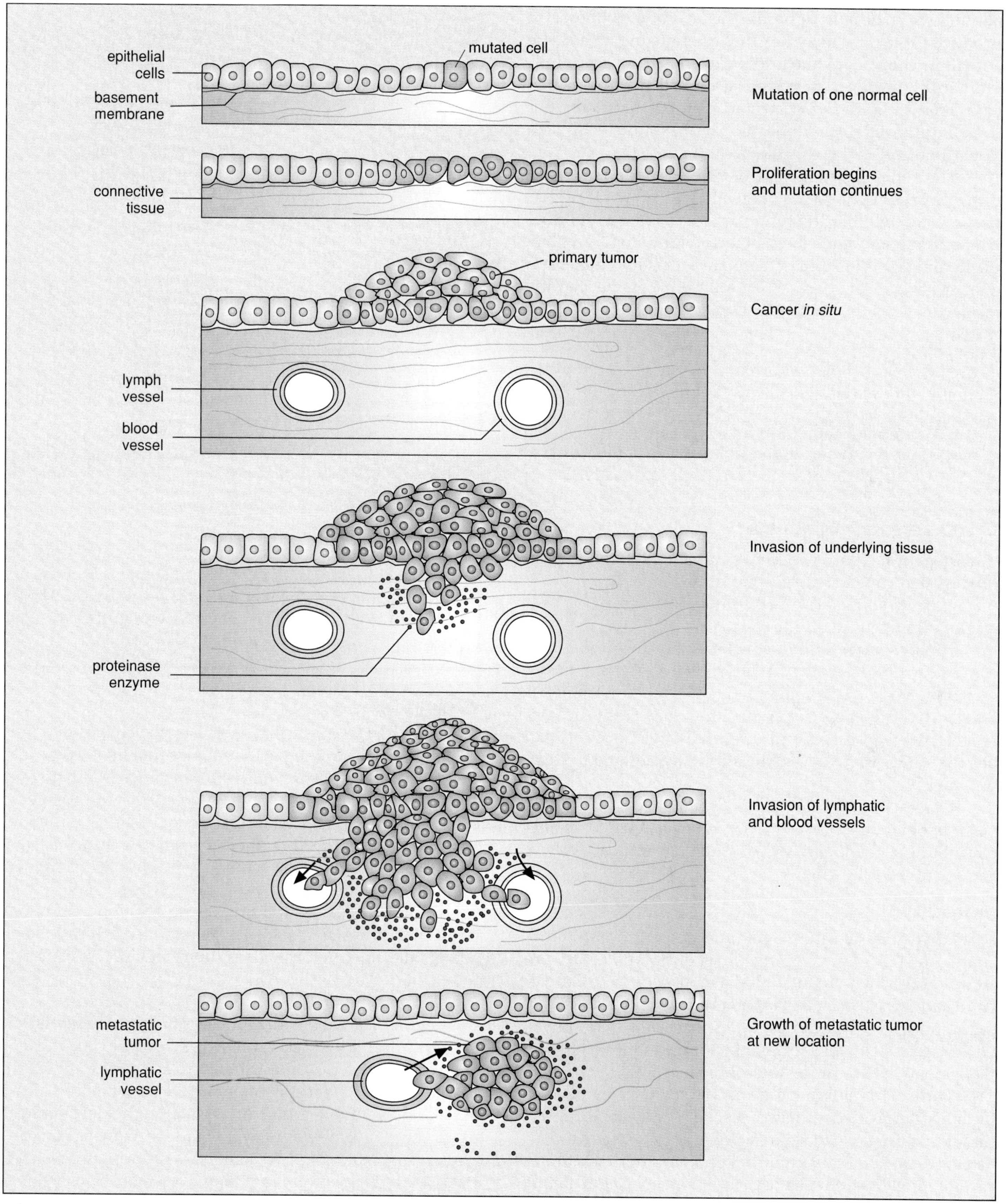

Figure 20.2

Carcinogenesis. The development of cancer requires a series of mutations leading to first a primary tumor and then metastatic tumors.

allow them to adhere to basement membranes; they also produce proteinase enzymes that degrade the membrane and allow them to invade underlying tissues. Cancer cells tend to be motile. They have a disorganized internal cytoskeleton and lack intact actin filament bundles. After traveling through the blood or lymph, cancer cells then start tumors elsewhere in the body (table 20.1).

The patient's prognosis (foretelling the probable outcome) is dependent on the degree to which the cancer has progressed: whether the tumor has invaded surrounding tissues; if so, whether there is any lymph node involvement, and finally whether there are metastatic tumors in distant parts of the body. With each progressive step the prognosis becomes less favorable.

> Cancer cells are nondifferentiated, have abnormal nuclei, do not require growth factors, and divide repeatedly. Because they are not constrained by their neighbors, they form a tumor. Finally, they metastasize and start tumors elsewhere in the body.

Causes of Cancer

Carcinogenesis (the development of cancer) is a multistage process that can be divided into these 3 phases:

Initiation: A single cell undergoes a mutation that causes it to begin to divide repeatedly (fig. 20.2).

Promotion: A tumor develops and the tumor cells continue to divide. As they divide they undergo mutations.

Progression: One cell undergoes a mutation that gives it a selective advantage over the other cells. This process is repeated several times, and eventually there is a cell that has the ability to invade surrounding tissues and to carry out metastasis.

Carcinogenesis can require many years; therefore, older people are more apt to have cancer. Various factors contribute to the likelihood of carcinogenesis, and these are discussed following.

Heredity

Particular types of cancer seem to run in families. For instance, the risk of developing breast, lung, and colon cancers increases two- to threefold when first-degree relatives have had these cancers. Investigators are in the process of pinpointing the location of a gene called BRCA1 (Breast Cancer gene #1), which occurs in a large family whose female members are prone to breast cancer.

Particularly childhood cancers seem to be controlled by the inheritance of a dominant gene. Retinoblastoma is an eye tumor that usually develops by age 3; Wilm's tumor is characterized by numerous tumors in both kidneys. In adults, several family syndromes (e.g., Li-Fraumeni cancer family syndrome, Lynch cancer family syndrome, and Warthin cancer family syndrome) are known. Those who inherit a dominant allele develop tumors in various parts of the body.

Table 20.2

Carcinogens in Cigarette Smoke

Aminostilbene	N-Dibutylnitrosamine
Arsenic	2, 3-Dimethylchrysene
Benz (a) anthracene	Indenol (1, 2, 3-cd) pyrene
Benz (a) pyrene	5-Methylchrysene
Benzene	Methylfluoranthene
Benzo (b) fluoranthene	B-Napthylamine
Benzo (c) phenanthrene	Nickel compounds
Cadmium	N-Nitrosodiethylamine
Chrysene	N-Nitrosodimethylamine
Dibenz (a,c) anthracene	N-Nitrosomethylethylamine
Dibenz (a,e) fluoranthene	N-Nitrosonanabasine
Dibenz (a,h) acridine	Nitrosonornicotine
Dibenz (a,j) acridine	N-Nitrosopiperidine
Dibenz (c,g) carbazone	N-Nitrosopyrrolidine
	Polonium-210

From Steven B. Oppenheimer "Advances in Cancer Biology" in American Biology Teacher, 49(1):13, January 1987. Copyright © 1987 National Association of Biology Teachers, Reston, VA. Reprinted by permission.

> The development of cancer is a multistage process involving initiation, promotion, and progression. The inheritance of and/or the occurrence of mutated genes leads to the characteristics of cancer.

Carcinogens

A **mutagen** is an agent that increases the chance of a mutation, while a **carcinogen** is an environmental agent that can contribute to the development of cancer. Carcinogens are often mutagenic. Among the best-known mutagenic carcinogens are (1) certain organic chemicals, (2) radiation, and (3) viruses. Some organic chemicals are also *promoters* of cancer. Promoters are agents that increase the likelihood of cell division during carcinogenesis. The hormone estrogen is known to be a promoter in the development of endometrial cancer, for example. Dietary fat is believed to be a promoter for colon cancer and possibly breast cancer.

Organic Chemicals

Cigarette smoke contains a number of organic chemicals that are carcinogens, and it is estimated that nearly one-third of all cancer deaths can be attributed to smoking (table 20.2). Lung cancer is the most frequent lethal cancer in the United States, and smoking is also implicated in the development of cancers of the mouth, larynx, bladder, kidney, and pancreas. When smoking is combined with drinking alcohol, the risk of these cancers increases even more (fig. 20.3*a*).

Animal testing of certain food additives such as red dye #2 and the synthetic sweetener saccharin has shown that these substances are cancer-causing in very high doses.

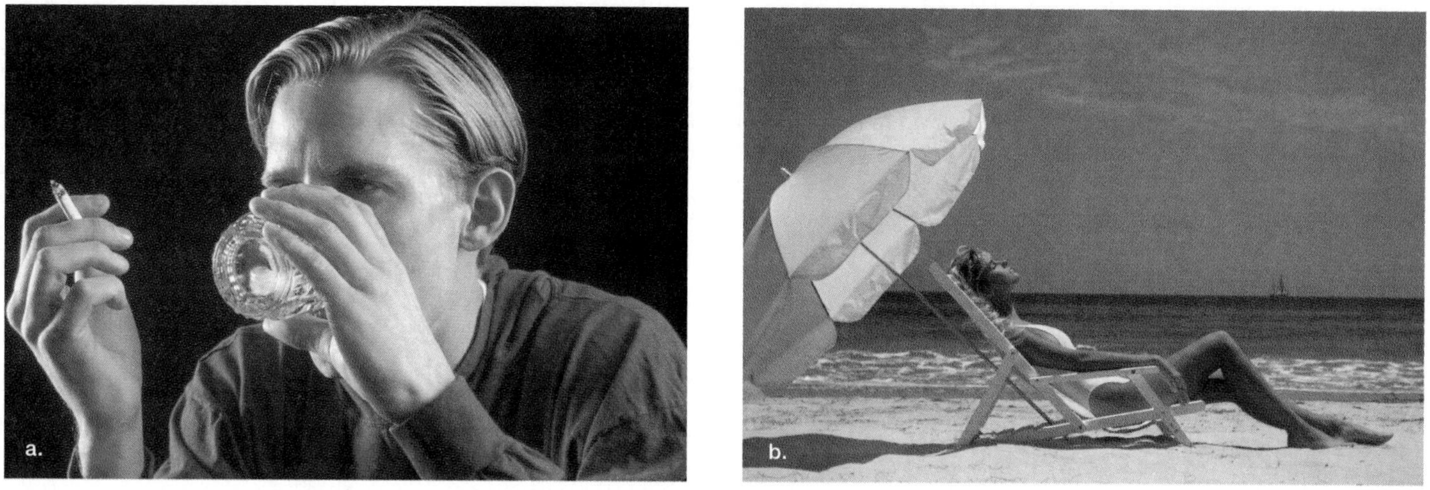

Figure 20.3

Two well-known causes of cancer. **a.** Smoking cigarettes, especially when combined with drinking of alcohol, leads to cancer of the lungs, mouth, larynx, kidney, bladder, pancreas, and many other cancers. **b.** Sunbathing leads to skin cancer. Melanoma is a particularly malignant skin cancer.

Industrial chemicals such as benzene and carbon tetrachloride and industrial materials such as vinyl chloride and asbestos fibers are also associated with the development of cancer. Pesticides and herbicides are dangerous not only to pests and plants but also to our own health because they contain organic chemicals that can cause mutations. A 16-member panel of experts assembled by the National Academy of Sciences' Institute of Medicine has found conclusive evidence that exposure to dioxin, a contaminant of the herbicide Agent Orange used during the Vietnam War, can be conclusively linked to cancers of lymphoid tissues and those of muscles and certain connective tissues. Similarly, another study has found that these and other cancers were found more often in those exposed to dioxin accidently released after an explosion in a chemical plant over Seveso, Italy, on July 10, 1976, than in the rest of the population.

Radiation

Radiation is also carcinogenic. Human beings are exposed to natural radiation and artificial radiation. Ultraviolet radiation in sunlight and tanning lamps is responsible for the dramatic increases seen in skin cancer the past several years (fig. 20.3*b*). There are at least 6 cases of skin cancer for every one case of lung cancer, but then the nonmelanoma skin cancers are usually curable. Melanoma skin cancer, on the other hand, tends to metastasize and is responsible for 1%–2% of total cancer deaths in the United States.

The public is now aware that naturally occurring radioactive radon gas can cause cancer, particularly if concentrated inside homes. In the very rare house with an extremely high level of exposure, the risk of developing lung cancer is thought to be the equivalent of smoking a pack of cigarettes a day. No wonder that the combination of radon gas and smoking cigarettes can be particularly dangerous. Radon gas is thought to cause about 2% of total cancer deaths.

Most of us are familiar with the damaging effects of the nuclear bomb explosions or accidental emissions from nuclear power plants. For example, there was an increased rate of cancer among the survivors of World War II atomic bombings of Hiroshima and Nagasaki, and certainly more cancer deaths are expected in the vicinity of the Chernobyl Power Station (in the former U.S.S.R.), which suffered a terrible accident in 1986. Usually, however, diagnostic X rays account for most of our exposure to artificial sources of radiation. The benefits of these procedures can far outweigh the possible risk, but it is still wise to avoid any X-ray procedures that are not medically warranted.

Viruses

Three DNA viruses have been linked to human cancers: hepatitis B virus to liver cancer, human papillomavirus to cancer of the cervix, and Epstein-Barr virus to Burkitt's lymphoma and nasopharyngeal cancer. In China, almost all persons have been infected with the hepatitis B virus, and this correlates with the high incidence of liver cancer in that country. For a long time, circumstances suggested that cervical cancer was a sexually transmitted disease, and now human *papillomaviruses* are routinely isolated from cervical cancers. Burkitt's lymphoma (a cancer of the lymphoid tissues) occurs frequently in Africa, where virtually all children are infected with the *Epstein-Barr virus.* In China, the Epstein-Barr virus is isolated in nearly all nasopharyngeal cancer specimens. It's

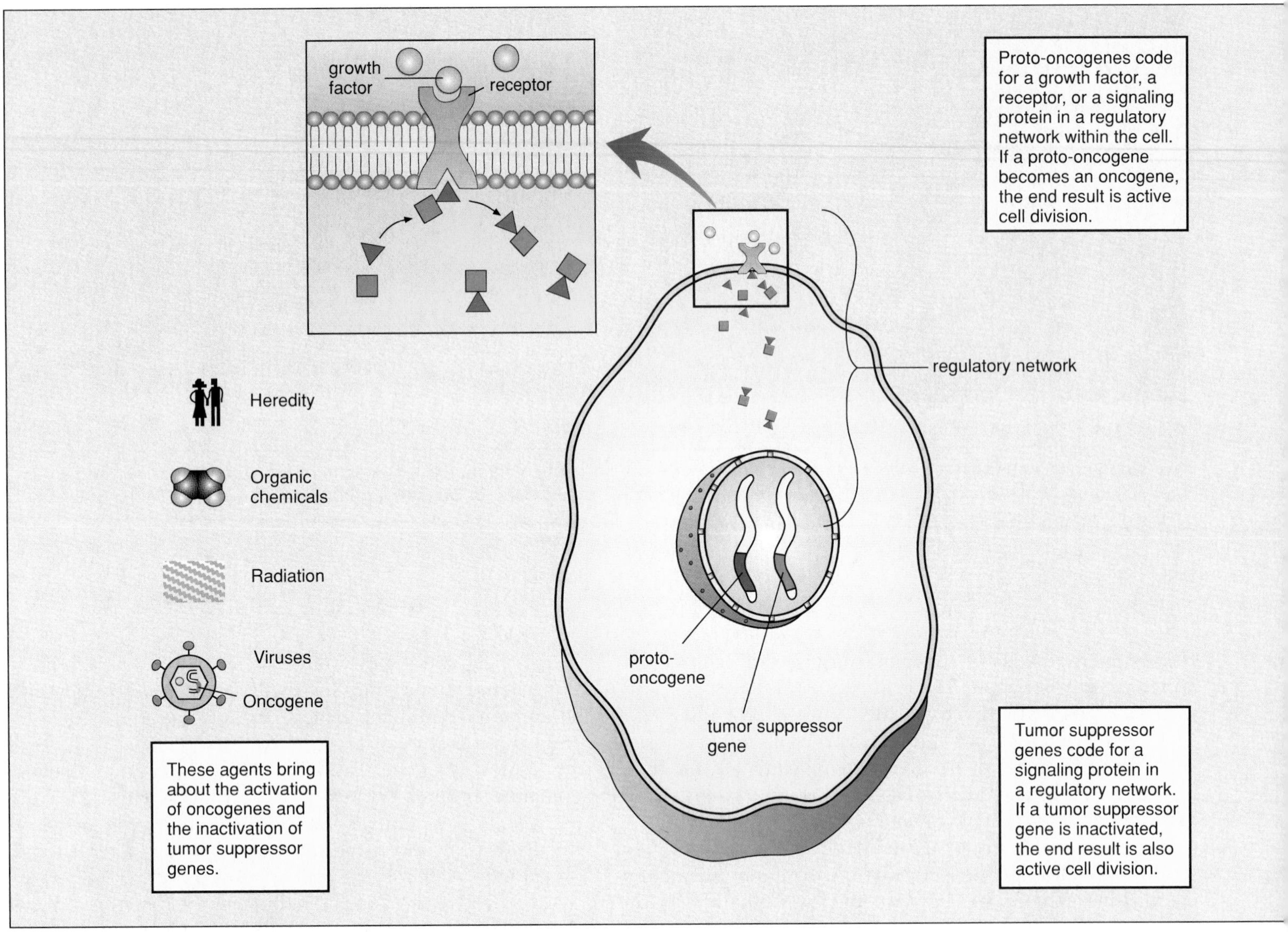

Figure 20.4

Causes of cancer. Hereditary and environmental factors cause mutations of proto-oncogenes and tumor suppressor genes. A regulatory network, which includes cell membrane receptors for growth factors, intracellular reactions, and the genes, is

believed that environmental factors must be involved in determining the final effect of an Epstein-Barr viral infection because the virus is associated with different diseases in different countries (in the United States, this virus causes mononucleosis).

Retroviruses, in particular, are known to cause cancers in animals. In humans, the retrovirus HTLV (human T-cell lymphotropic virus, type 1) has been shown to cause adult T-cell leukemia. This disease occurs frequently in parts of Japan, the Caribbean, and Africa, particularly in those regions where people are known to be infected with the virus.

Immunodeficiencies

Cancer is apt to develop in individuals who exhibit an immunodeficiency. For example, people with AIDS develop a cancer of the blood vessels called Kaposi's sarcoma. Transplant patients who are on immunosuppressive drugs are more apt to develop lymphomas and Kaposi's sarcoma. Persons who inherit an immunodeficiency are also more apt to develop these cancers.

It appears, then, that an active immune system can help protect us from cancer. Mutated cells may display antigens that normally subject them to attack by T lymphocytes and possibly also antibodies. Perhaps, cancer is seen more often among the elderly because the immune system weakens as we age.

> Specific mutations cause cancer. A cancer-causing mutation can be inherited or can arise due to exposure to certain organic chemicals, radiation, and a few types of viruses. Still, cancer is most apt to occur in the immunodeficient individual.

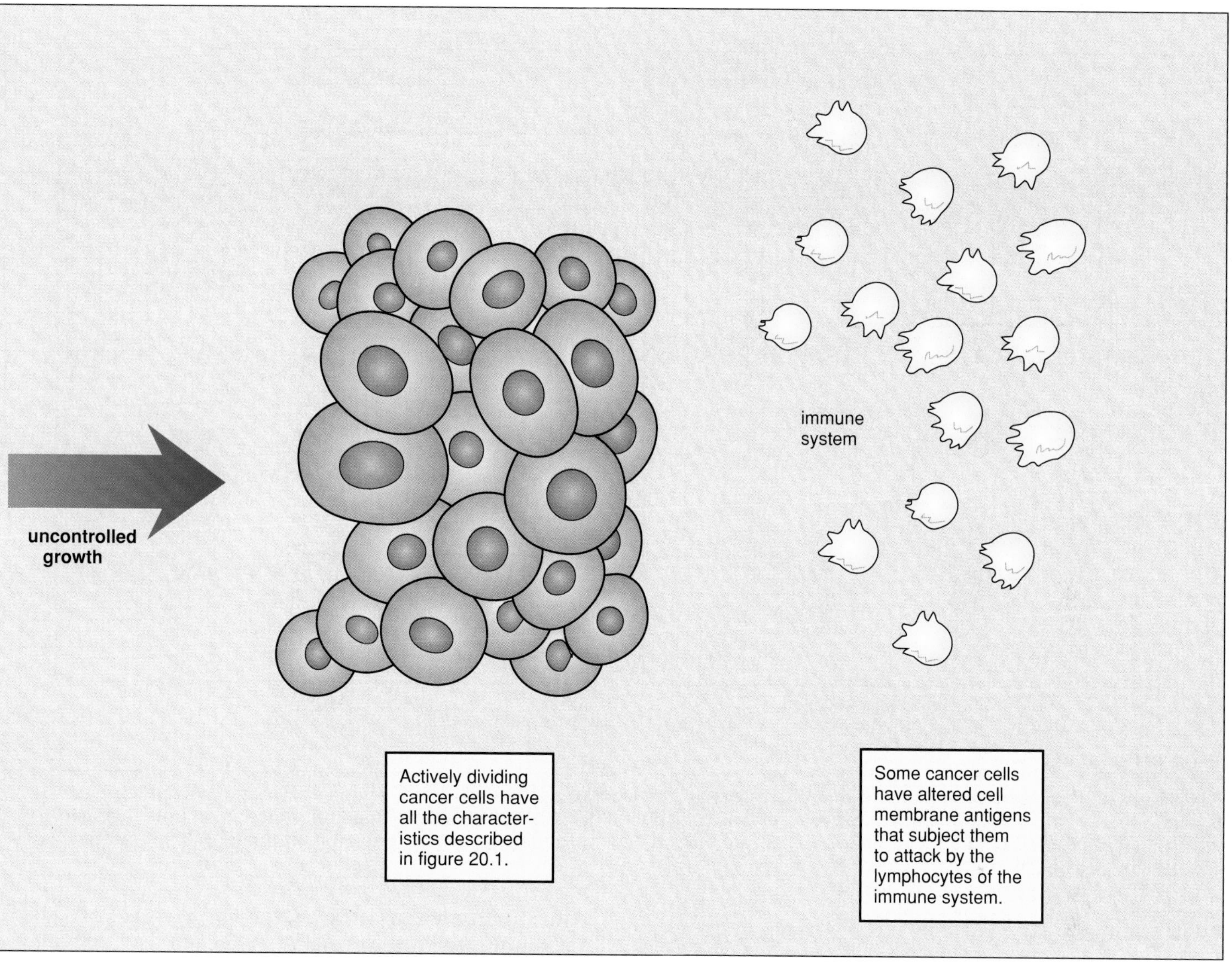
uncontrolled growth
immune system
Actively dividing cancer cells have all the characteristics described in figure 20.1.
Some cancer cells have altered cell membrane antigens that subject them to attack by the lymphocytes of the immune system.

unbalanced. This leads to uncontrolled growth and a tumor. Cancer still does not develop unless the immune system fails to respond and kill these abnormal cells.

Figure 20.4 illustrates the causes of cancer. It also indicates that activation of oncogenes and inactivation of tumor suppressor genes lead to cancer. An **oncogene** is a cancer-causing gene. A **tumor suppressor gene** is so called because it prevents cancer from occurring.

Oncogenes and Tumor Suppressor Genes

Normally, cell growth and differentiation are regulated to meet the needs of the body, and cell division is turned on only when required. A *regulatory network* that controls cell division has been discovered. This network involves extracellular *growth factors,* cell membrane *growth factor receptors,* various proteins within the cytoplasm, and various genes within the nucleus (fig. 20.4). In many cells such as muscle and nerve cells, the regulatory network is always shut down and the cell does not divide. In other cells, such as skin epidermal cells, the pathway is closely controlled so that cell division is orderly. In cancer cells, some of the genes that code for proteins in the network have undergone mutations and growth is unregulated. A tumor results.

A cell contains many **proto-oncogenes,** so called because a mutation can cause them to become an oncogene. An oncogene can produce an abnormal protein product or else abnormally high levels of a normal product. In either case, uncontrolled growth threatens. When a tumor suppressor gene undergoes a mutation, a reduced amount or a defective product is present; uncontrolled growth cannot be stopped, and the mutated cell begins to proliferate wildly.

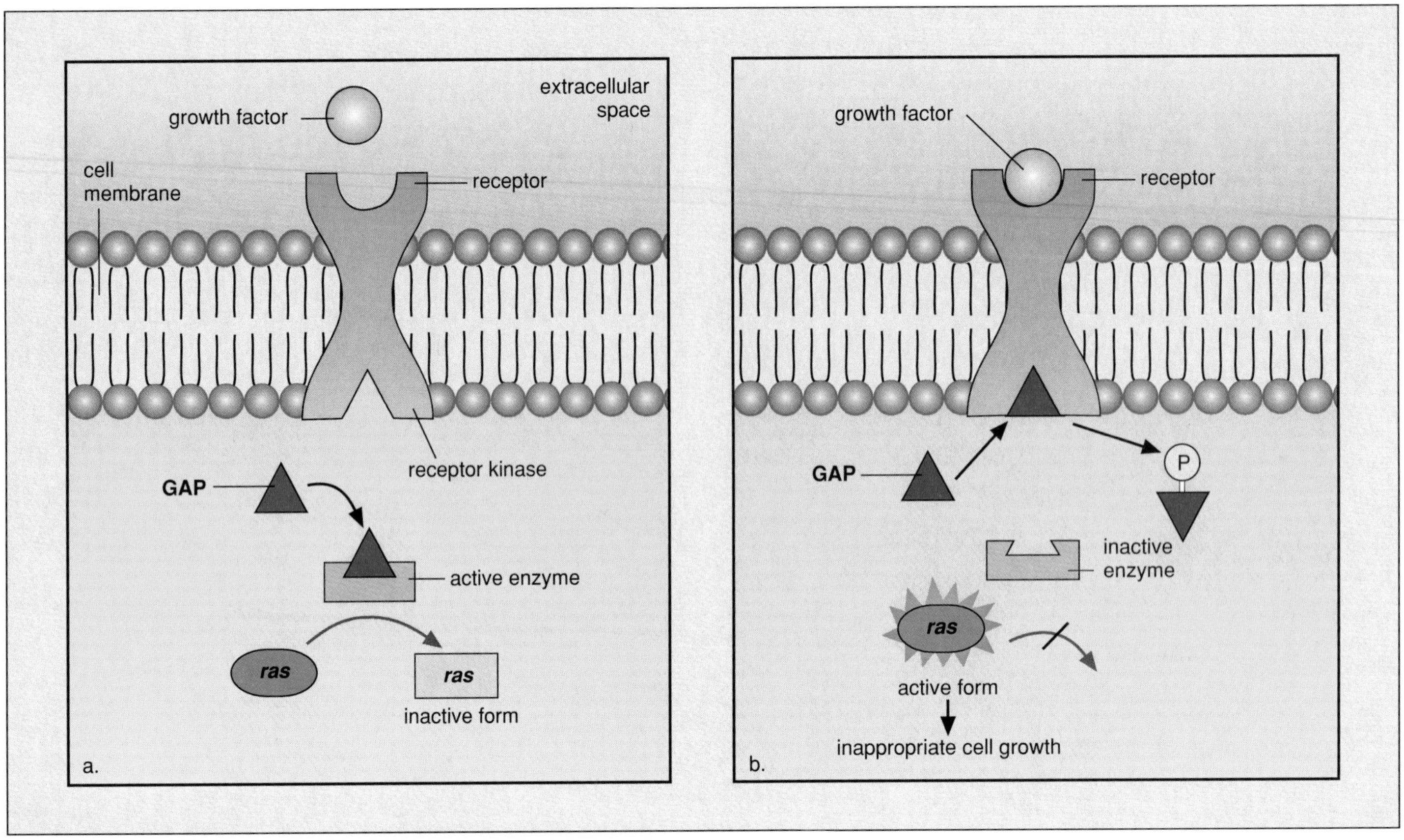

Figure 20.5

Regulatory network involving the *ras* protein. **a.** Normally, the signaling protein GAP causes an enzyme to inactivate *ras*. **b.** When a growth factor is present, a receptor kinase phosphorylates GAP; the enzyme does not function and *ras* is active. If *ras* is always turned on, as might happen when a mutation occurs to unbalance this regulatory pathway, inappropriate cell growth can occur.

Oncogenes

Researchers have identified perhaps 100 oncogenes that can cause increased growth and lead to a tumor. The oncogenes most frequently involved in human cancers belong to the *ras* gene family. An alteration of only a single nucleotide pair is sufficient to convert a normally functioning *ras* proto-oncogene to an oncogene. The *ras*K oncogene is found in about 25% of lung cancers, 50% of colon cancers, and 90% of pancreatic cancers. The *ras*N oncogene is associated with leukemias (cancer of blood-forming cells) and lymphomas (cancers of lymphoid tissue), and both *ras* oncogenes are frequently found in thyroid cancers.

Figure 20.5 illustrates that the *ras* protein (coded for by the *ras* gene) can exist in an active or inactive form. An active *ras* protein is associated with cell growth, although the precise role of the *ras* protein in the regulatory network has not yet been determined. The activity of the *ras* protein is controlled in part by a protein called GAP (GTPase activating protein). Normally, GAP causes an enzyme to inactivate *ras*[1] and cell growth does not occur. However, when a growth factor attaches to a receptor that phosphorylates GAP, GAP becomes inactive and *ras* becomes active. It turns out that phosphorylation (the addition of a phosphate group) is a common way by which *signaling proteins* (so called because they regulate enzymatic reactions) in the regulatory network are turned on or off. Notice in figure 20.5 that a portion of the receptor, called receptor kinase, is an enzyme that phosphorylates GAP. A *kinase* is any enzyme that transfers a phosphate group from one molecule (often ATP) to another.

Many investigators are working to discover other ways in which the *ras* protein is regulated and to work out the precise functions of the *ras* protein. The *ras* oncogene could perhaps cause the *ras* protein to be unresponsive to ordinary controls.

1. Active *ras* is attached to guanine triphosphate; inactive *ras* is attached to guanine diphosphate.

Tumor Suppressor Genes

Ordinarily, a tumor suppressor gene codes for proteins whose presence prevents uncontrolled growth. In the pathway just described (fig. 20.5), GAP is such a protein. Investigation is going forward to determine if GAP is coded for by a tumor suppressor gene. A mutation in this tumor suppressor gene would cause GAP to be ineffective, and *ras* would be active even when no growth factor was present.

Researchers have identified about a half-dozen or so tumor suppressor genes that, when they malfunction, can produce increased growth. The *RB* tumor suppressor gene was discovered by studying the inherited condition retinoblastoma. When a child receives only one normal gene and that gene mutates, eye tumors develop in the retina by age 3. *RB* has now been found to malfunction in other cancers such as breast, prostate, and bladder cancers. Loss of *RB* through chromosome deletion is particularly frequent in a type of lung cancer called small cell lung carcinoma.

We now know that signaling proteins often regulate the transcription of genes whose products are needed for cell division, and this appears to be true of the *RB* protein. Figure 20.6 suggests that when the growth factor[2] TGFβ attaches to a receptor, the *RB* protein is activated. An active *RB* protein turns off the expression of the proto-oncogene *c-myc*. When the *RB* protein is not present, the protein product of the *c-myc* gene is thought to turn on the expression of other genes whose products lead to cell division and cancer.

Tumor suppressor genes and oncogenes are a part of the regulatory network we have been discussing. Reception of a growth factor sets in motion a whole series of responses in which one metabolic reaction leads to another, until finally a gene is turned on or off. Growth factor receptors, once important during prenatal development or even childhood, are not ordinarily present on the cell membrane of an adult. But when certain tumor suppressor genes do not work or proto-oncogenes become oncogenes, growth factor receptors appear on the cell membrane and the regulatory network is again activated even when no growth factor has been received and the result is abnormal cellular growth. In other words, a change occurring anywhere along the pathway of the regulatory network can possibly contribute to the occurrence of abnormal cellular growth.

From this discussion you can see why carcinogenesis is a multistage process that involves the mutation of various proto-oncogenes and tumor suppressor genes. The sequence of mutations believed to be required for colorectal cancer to develop is depicted in figure 20.7.

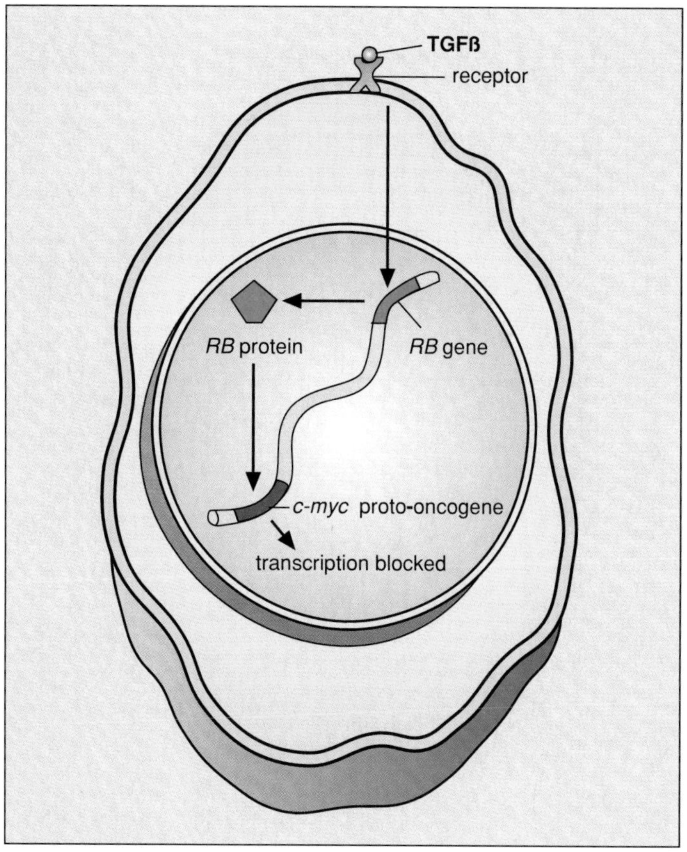

Figure 20.6

Regulatory network involving the *RB* protein. When TGFβ attaches to a receptor, the *RB* gene is turned on and the *RB* protein prevents the proto-oncogene *c-myc* from being transcribed. When a mutation occurs and the *RB* protein is not present, *c-myc* becomes an oncogene, leading to inappropriate cell growth.

> Each cell contains a regulatory network involving proto-oncogenes and tumor suppressor genes, which code for a growth factor receptor, and signaling proteins active in intracellular reactions. When proto-oncogenes mutate becoming oncogenes and tumor suppressor genes mutate, the regulatory network no longer functions as it should and uncontrolled growth results.

2. TGFβ brings about differentiation of cells rather than cell division.

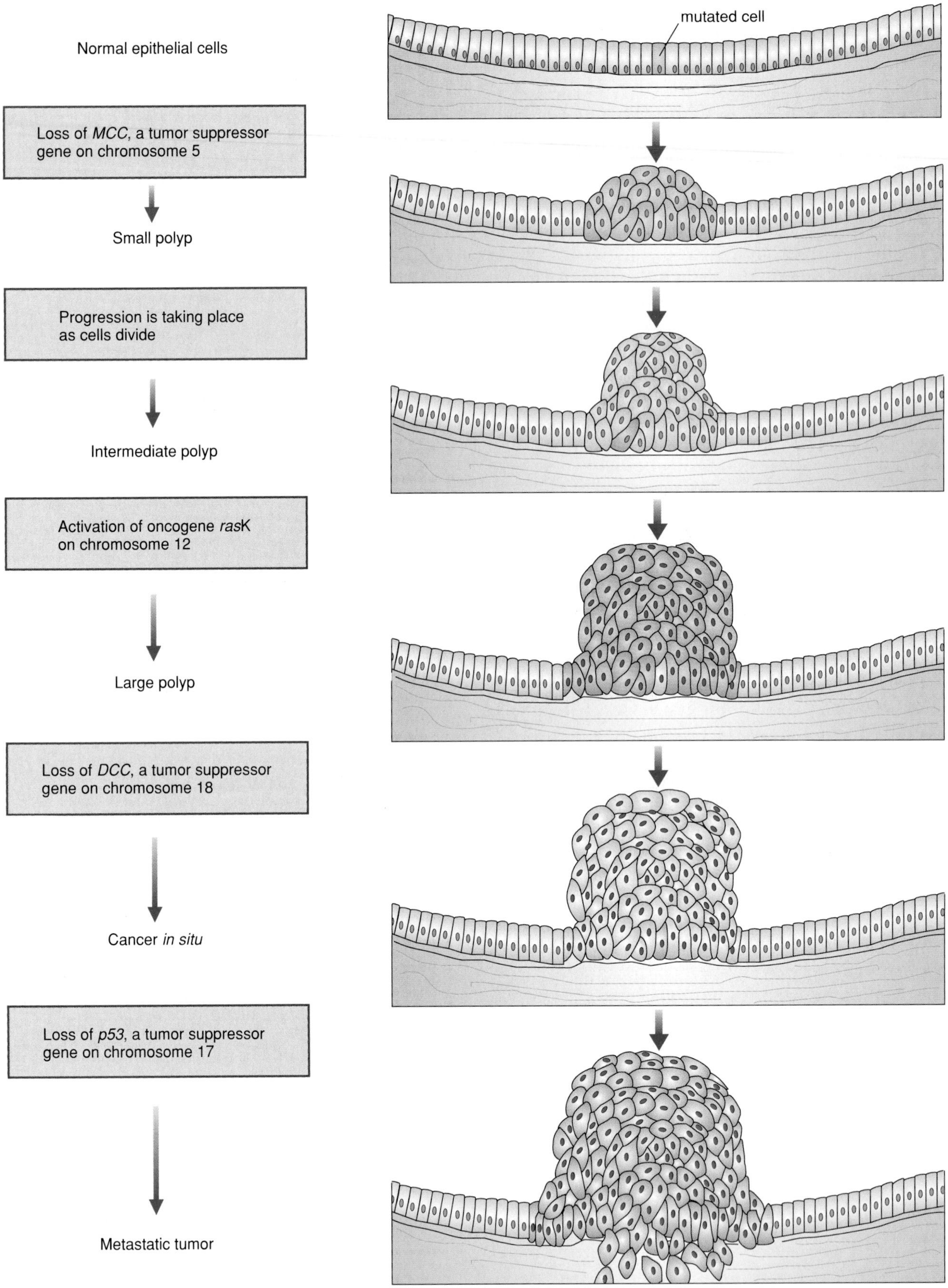

Figure 20.7

Development of colorectal cancer. A series of mutations causes ever-increasing abnormal growth and finally metastasis.

Table 20.3

Most Frequent Cancers in the United States

Cancer Site	Cases per Year		Deaths per Year	
Breast	183,000	(16%)	46,300	(9%)
Lung	170,000	(15%)	149,000	(28%)
Prostate	165,000	(14%)	35,000	(7%)
Colon/rectum	152,000	(13%)	57,000	(11%)
Bladder	52,300	(5%)	9,900	(2%)
Lymphoma	50,900	(4%)	22,000	(4%)
Uterus	44,500	(4%)	10,100	(2%)
Skin (melanoma)	32,000	(3%)	6,800	(1%)
Oral cancer	29,800	(3%)	7,700	(2%)
Leukemia	29,300	(3%)	18,600	(4%)
Pancreas	27,700	(2%)	25,000	(5%)
Ovary	22,000	(2%)	13,300	(3%)
	958,500	(84%)	400,700	(76%)
All sites:	1,170,000	(100%)	526,000	(100%)

Data are for the year 1993. Nonmelanoma skin cancers (approximately 700,000 cases per year) and carcinomas of the uterine cervix diagnosed in situ *(approximatey 55,000 cases per year) are not included in incidence figures.*

Source: Data from American Cancer Society, Inc., Cancer Facts and Figures, *1993.*

Classification, Diagnosis, and Treatment

It is estimated that about one out of every 3 people in the United States will develop cancer and one out of every 4 will die from it. Table 20.3 lists the most frequent cancers in the United States along with the number of cases per year and the number of deaths per year. Appendix D gives the warning signs, risk factors, methods for detection, and the suggested treatment for each of these types of cancer.

Classification of Cancers

Cancers are classified according to the type of tissue from which they arise. **Carcinomas** are cancers of the epithelial tissues, and adenocarcinomas are cancers of glandular epithelial cells. Carcinomas include cancer of the skin, breast, liver, pancreas, intestines, lung, prostate, and thyroid. **Sarcomas** are cancers that arise in muscles and connective tissue such as bone and fibrous connective tissue. **Leukemias** are cancers of the blood, and **lymphomas** are tumors of lymphoid tissue.

The size of a tumor and the degree of metastasis is of critical importance when deciding on a plan of treatment. Currently, different ways to describe staging—the extent of the disease—are frequently used for different cancers. However, the TNM system is applicable to all cancers. In this system, the extent of disease is described in terms of 3 parameters: T, the condition of the primary tumor, that is, the extent to which the tumor has invaded proximal tissues; N, the extent of lymph node involvement; and M, the extent of distant metastases.

> Cancers are classified as carcinomas (epithelial tissue cancers), sarcomas (muscles and connective tissue), leukemias (blood), and lymphomas (lymphoid tissue). Appendix D gives pertinent information about cancers at specific sites.

Diagnosis of Cancer

It is sometimes difficult to diagnose cancer before metastasis has taken place and treatment is more likely to be successful. The American Cancer Society, Inc. publicizes 7 warning signals that spell out the word *CAUTION* and that everyone should be aware of:

C hange in bowel or bladder habits;
A sore that does not heal;
U nusual bleeding or discharge;
T hickening or lump in breast or elsewhere;
I ndigestion or difficulty in swallowing;
O bvious change in wart or mole;
N agging cough or hoarseness.

Unfortunately, some of these symptoms are not obvious until cancer has progressed to one of its later stages.

Routine Screening Tests

The aim of medicine is to develop tests for cancer that are relatively easy to do, cost little, and are fairly accurate. So far, only the *Pap smear* for cervical cancer fulfills these 3 requirements. A physician merely takes a sample of cells from the cervix, which are examined microscopically for signs of abnormality. Regular Pap smears are credited with preventing over 90% of deaths from cervical cancer.

Breast cancer is not as easily detected, but 3 procedures are recommended. Every woman should do a monthly breast self-examination (see the Health Focus for this chapter). During an annual physical examination, recommended especially for women above age 40, a physician does this same procedure. While helpful, an examination may not detect lumps before metastasis has already taken place. The third recommended procedure, *mammography*, which is an X-ray study of the breast, is expected to do this. However, mammograms do not show all cancers, and cancer may develop during the interval between mammograms. The hope is that a mammogram will reveal a lump that is too small to be felt and at a time when the cancer is still highly curable.

Screening for colon cancer is also dependent upon 3 types of testing. A digital rectal examination performed by a physician is actually of limited value because only a limited portion of the rectum can be reached by finger. With flexible *sigmoidoscopy*, the second procedure, a much larger portion of the colon can be examined by using a thin, pliable lighted tube. Finally, a stool blood test (fecal occult blood test) consists of examining a stool sample to detect any hidden blood. The sample is smeared on a slide, and a chemical is added that changes color in the presence of hemoglobin. This procedure is based on the supposition that a cancerous polyp bleeds, but actually some polyps do not bleed and some bleeding is not due to a polyp. Therefore, the percentage of false negatives and false positives is high. All positive tests are followed up by a *colonoscopy*, an examination of the entire colon or by X ray after a barium enema.

Other tests in routine use are blood tests to detect leukemia and urinalysis for the detection of bladder cancer. Newer tests under consideration are tumor marker tests and tests for oncogenes.

Tumor Marker Tests

Blood tests for tumor antigens/antibodies are called *tumor marker tests*. They are possible because tumors release substances that provoke an antibody response in the body. For example, if an individual has already had colon cancer it is possible to use the presence of an antigen, called CEA (for carcinoembryonic antigen), to detect any relapses. When the CEA level rises, additional tumor growth has occurred.

There are also tumor marker tests that can be used as an adjunct procedure to detect cancer in the first place. They are not reliable enough to count on solely, but in conjunction with physical examination and ultrasound (see following) they are considered useful. There is a prostate-specific antigen (PSA) test for prostate cancer, a CA 125 test for ovarian cancer, and an alpha-fetoprotein (AFP) test for liver tumors, for example.

Oncogene Tests

The Pap smear has proven to be a reliable and helpful test for cervical cancer, but more reliable tests are needed to detect other cancers at an early stage. Researchers are now working on the possibility of using the presence of the *ras* gene to detect colon cancer. They have found that when a person has a tumor that contains a *ras* oncogene, this same oncogene can be detected in stool samples. However, to catch all sorts of colon cancers you would have to screen stool samples for all possible oncogenes. Right now that is not feasible.

Genetic testing for inherited breast cancer may also be on the horizon. Researchers have been able to develop a genetic marker test for the oncogene BRCA1 (Breast Cancer gene #1) mentioned earlier. Those who test positive for inheritance of the gene can choose either to have prophylactic surgery or to be frequently examined for signs of breast cancer. Perhaps genetic analysis for breast cancer in the general populace may one day be routine.

Confirming the Diagnosis

There are ways to confirm a diagnosis of cancer without major surgery. Needle biopsies allow removal of a few cells for examination, and sophisticated imaging techniques including laparoscopy (p. 284) permit a viewing of body parts. Computerized axial tomography (or *CAT scan*) uses computer analysis of scanning X-ray images to create cross-sectional pictures that portray a tumor's size and location. Magnetic resonance imaging (*MRI*) is another type of imaging technique that depends on computer analysis. MRI is particularly useful for analysis of tumors in tissues surrounded by bone, such as tumors of the brain or spinal cord (fig. 20.8). A radioactive scan obtained after a radioactive isotope is administered can reveal any abnormal isotope accumulation due to a tumor. During *ultrasound*, echoes of high-frequency sound waves directed at a part

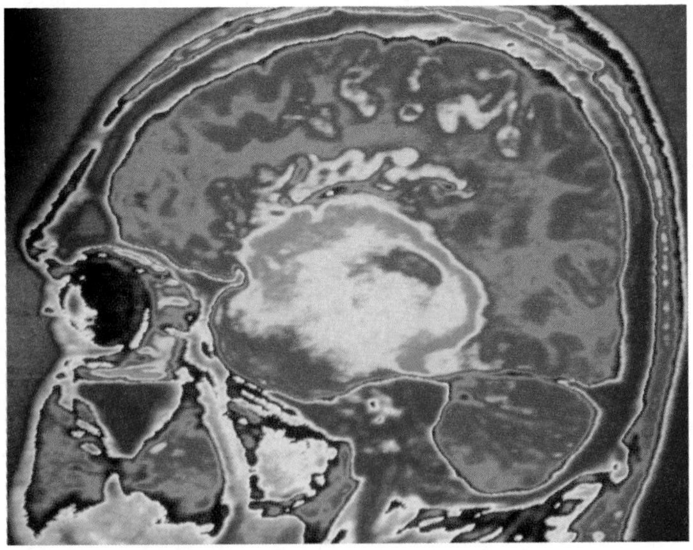

Figure 20.8

MRI image of brain; the yellow region is a tumor.

Health Focus

Shower Check for Cancer

The American Cancer Society urges women to do a breast self-exam and men to do a testicle self-exam every month. Breast cancer and testicular cancer are far more curable if found early, and we must all take on the responsibility of checking for it.

Breast Self-Exam for Women

1. Check your breasts for any lumps, knots, or changes about one week after your period.
2. Place your right hand behind your head. Press firmly with the pads of your fingers (fig. 20A). Move your *left* hand over your *right* breast in a circle. Also check the armpit.
3. Now place your left hand behind your head and check your *left* breast with your *right* hand in the same manner as before. Also check the armpit.
4. Check your breasts while standing in front of a mirror right after you do your shower check. First, put your hands on your hips and then raise your arms above your head (fig. 20B). Look for any changes in the way your breasts look; dimpling of the skin, changes in the nipple, or redness or swelling.
5. If you find any changes during your shower or mirror check, see your doctor right away.

You should know that the best check for breast cancer is a mammogram. When your doctor checks your breasts, ask about this. See appendix D, which gives the warning signals for breast cancer.

Testicle Self-Exam for Men

1. Check your testicles once a month.
2. Roll each testicle between your thumb and finger as shown in figure 20C. Feel for hard lumps or bumps.
3. If you notice a change or have aches or lumps, tell your doctor right away so he or she can recommend what should be done.

Cancer of the testicles can be cured if you find it early. You should also know that prostate cancer is the most common cancer in men. Men over age 50 should have an annual health checkup that includes a prostate examination. See appendix D, which gives the warning signs for prostate cancer.

This information was provided by the American Cancer Society.

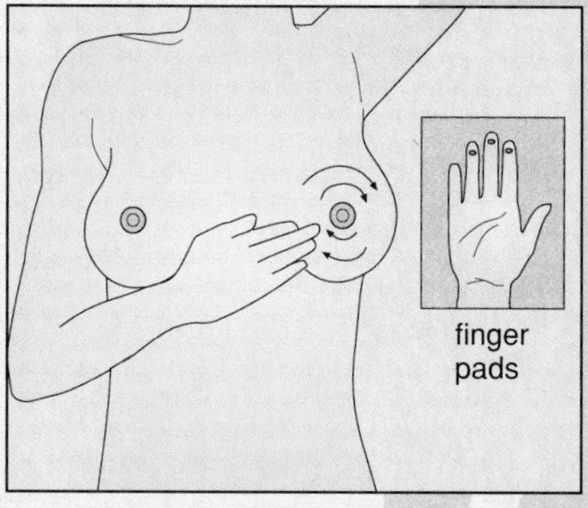

Figure 20A

Shower check for breast cancer.

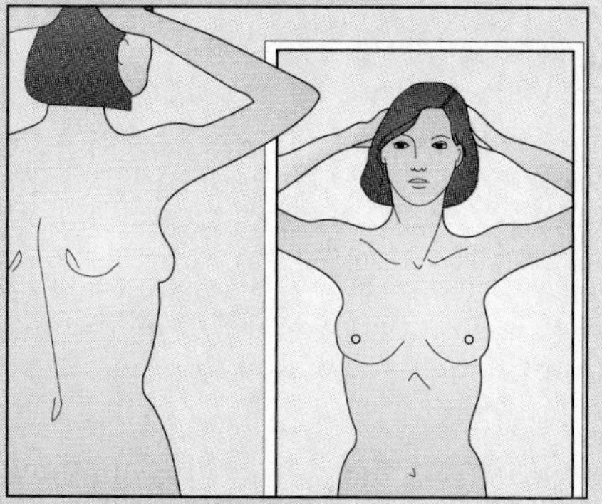

Figure 20B

Mirror check for breast cancer.

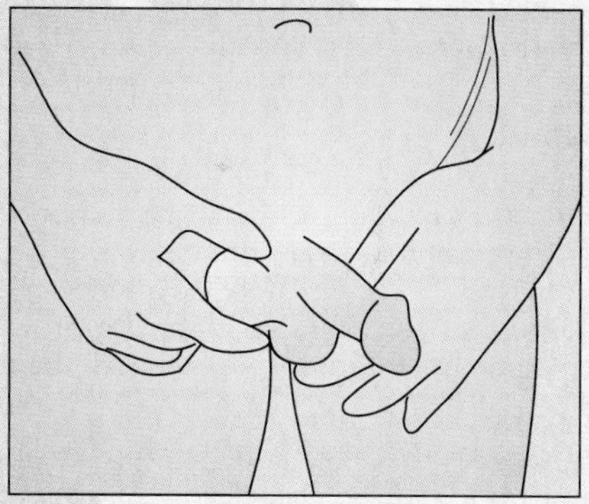

Figure 20C

Shower check for testicular cancer.

of the body are used to reveal the size, shape, and location of tissue masses. Ultrasound can confirm tumors of the stomach, prostate, pancreas, kidney, uterus, and ovary.

> There are standard procedures to detect specific cancers, for example, the Pap smear for cervical cancer, mammograms for breast cancer, and the stool blood test for colon cancer. Tumor marker tests and oncogene tests are new ways to test for cancer. Biopsy and imaging are used to confirm the diagnosis of cancer.

Treatment of Cancer

Surgery, radiation, and chemotherapy are the standard methods of treatment (fig. 20.9). Surgery alone is sufficient for cancer *in situ.* But because there is always the danger that undetected metastasis has occurred or will occur, surgery is often preceded by or followed by radiation therapy. Radiation is mutagenic and affects dividing cells more than those cells that are not dividing. The cancer cells are dividing wildly; therefore, the hope is that they will mutate to the point of self-destruction. X rays can be used, but the use of radioactive protons is preferred

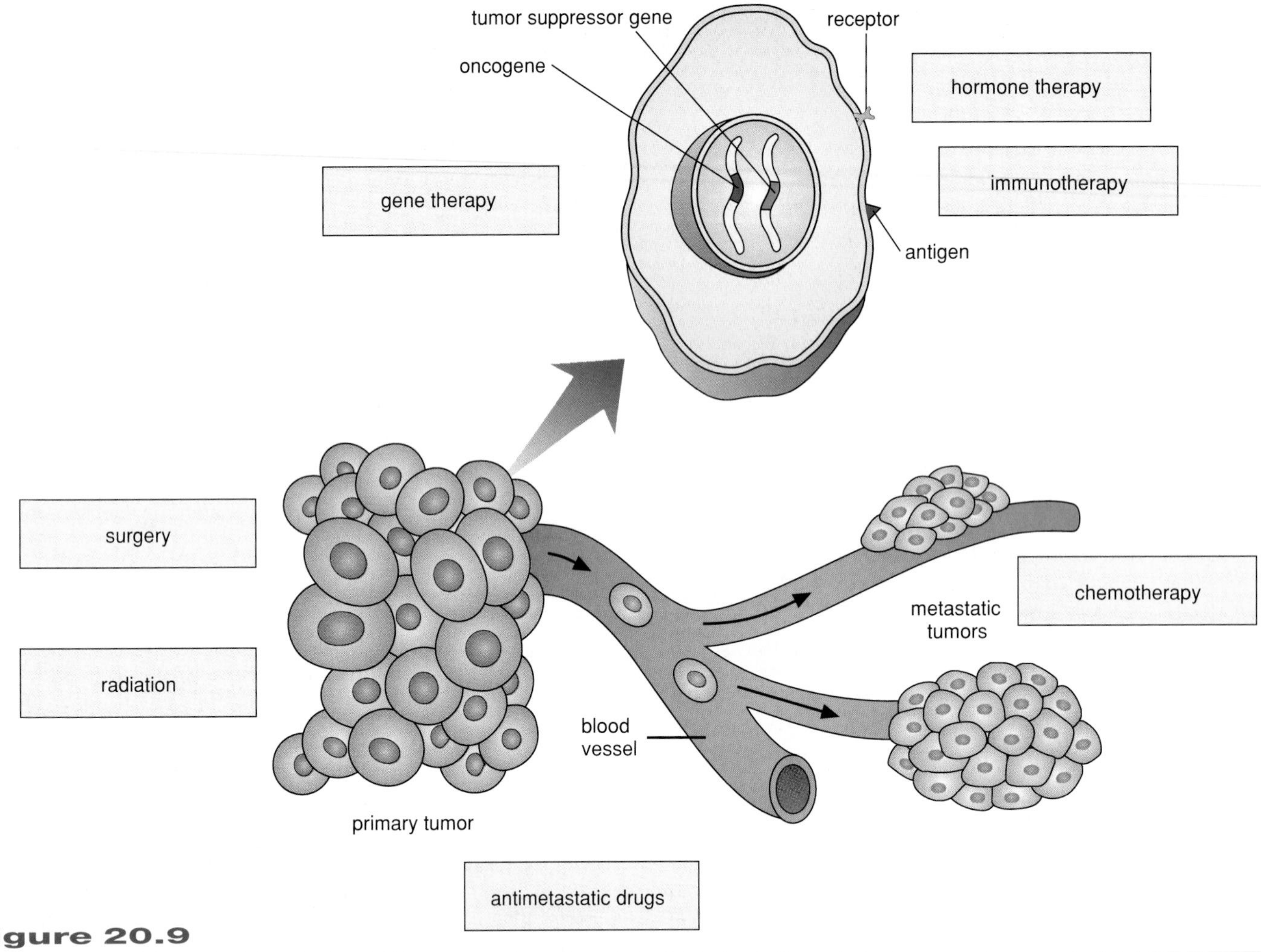

Figure 20.9

Treatment of cancer. Surgery and radiation are used to rid the body of a localized primary tumor. Chemotherapy is more effective and used when the cancer has metastasized. Hormone therapy is used to inactivate the cell membrane receptors that are active when cancer develops. Immunotherapy zeros in on cell membrane receptors and antigens as a way to identify cancer cells. Gene therapy seeks to prevent the action of an oncogene or make up for the inactivity of a tumor suppressor gene. Antimetastatic drugs may soon be used to prevent metastasis.

over X ray whenever a hospital provides this form of treatment. Proton beams can be aimed at the tumor like a rifle hitting the bull's eye of a target.

Chemotherapy is a way to catch cancer cells that have spread throughout the body. Most chemotherapeutic drugs kill cells by damaging their DNA or interfering with DNA synthesis. The hope is that all cancer cells will be killed while leaving untouched enough normal cells to allow the body to keep functioning. Whenever possible, chemotherapy is specifically designed for the particular cancer. For example, in Allen's lymphoma/leukemia, it is known that a small portion of a chromosome 9 is missing, and, therefore, DNA metabolism differs in the cancerous cells compared to normal cells. Specific chemotherapy for this cancer provides the patient with a drug designed to exploit this metabolic difference and destroy the cancerous cells.

Chemotherapy has produced a few dramatic successes. Almost 75% of children with childhood leukemia are completely cured. Hodgkin's disease, a lymphoma, once killed 2 out of 3 patients. Now, a combination therapy of 4 different drugs can wipe out the disease in a matter of months in 3 out of 4 patients, even when the cancer is not diagnosed immediately. In other cancers—most notably breast and colon cancer—chemotherapy can reduce the chance of recurrence after surgery has removed all detectable traces of the disease.

The new drug taxol, extracted from bark of the Pacific yew tree, seems to be particularly effective against advanced ovarian cancers as well as breast, head, and neck tumors. Taxol interferes with microtubules needed for cell division. Chemists are trying to devise a molecule, called taxotere, that is more potent and less toxic by modifying a chemical produced by the European yew tree.

Chemotherapy sometimes fails because cancer cells become resistant to one or several chemotherapeutic drugs. One mechanism for resistance is amplification of a gene. For example, the chemotherapeutic drug methotrexate inhibits the enzyme dihydrofolate reductase; resistance occurs when there are multiple copies of the gene for this enzyme. So much dihydrofolate reductase is now produced that methotrexate has no effect. When cancer cells become resistant to combinations of drugs, it is called *multidrug resistance.* This occurs because all the drugs are capable of interacting with a cell membrane carrier that pumps them out of the cell. Researchers are testing drugs known to poison the pump in an effort to restore effectivity of the drugs. In the meantime, combinations of drugs with nonoverlapping patterns of toxicity are still helpful because cancer cells can't become resistant to all different types at once. Also, when smaller doses of each type drug are used, more normal cells survive.

Bone Marrow Transplants

The red bone marrow contains large populations of dividing cells; therefore, red bone marrow is particularly prone to destruction by chemotherapeutic drugs. In bone marrow autotransplantation, a patient's own bone marrow is harvested, treated to remove cancer cells, and stored before chemotherapy begins. Quite high doses of radiation or chemotherapeutic drugs are now given within a relatively short period of time. This prevents multidrug resistance from occurring, and the treatment is more likely to catch each and every cancer cell. Then, the stored bone marrow is returned to the patient.

Antihormone Therapy

As mentioned previously, hormones are promoters for certain types of cancers. Breast and uterine cancer cells have receptors for estrogen, which can be blocked by the administration of a drug called tamoxifen. Hormone therapy is also standard treatment for metastatic prostate cancer because this type of tumor cell is stimulated by testosterone. GnRH analogs are used to prevent the production of testosterone in the first place, or antiandrogens are used to block the reception of testosterone by tumor cells.

In promyelocytic leukemia, cells have too many retinoic acid (a cousin of vitamin A) receptors. Strangely enough the treatment is the administration of retinoic acid because this leads to differentiation and the concomitant cessation of cell growth.

Immunotherapy

Immunotherapy is the use of the body's immune system to promote the health of the body. Figure 20.9 shows an antigen on the cancer cell enlargement to suggest that immunotherapy could possibly be successful in the treatment of cancer.

Various approaches have been used but none have been highly successful. The drug levamisole stimulates the immune system, and there is a 30% reduction in tumor recurrence and mortality due to colon cancer. The administration of *interferon,* a lymphokine (p. 140), has been found to be effective only against a type of leukemia known as hairy-cell leukemia.

Lymphocytes removed from tumor cells (called tumor-infiltrating lymphocytes) have been genetically engineered to express a gene for tumor necrosis factor, also a lymphokine, before being returned to the patient. Such treatments have been about 20% effective in patients with renal cancers and melanoma.

Monoclonal antibodies produced by the method described on page 141 can be designed to zero in on cell membrane receptors of cancer cells, but alone they are not effective at killing cancer cells. One idea is to combine them with a radiation source called yttrium 90, which travels only a very short distance. Another is to produce monoclonal antibodies linked with a chemotherapeutic drug such as doxorubicin.

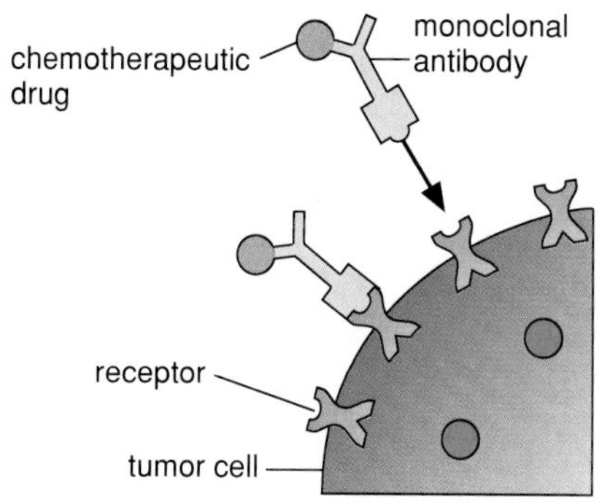

Antimetastatic Drugs

It is often pointed out that the process of metastasis makes the treatment of cancer very difficult. Usually primary tumors can be successfully treated by surgery and radiation, but more accurate methods are needed for metastatic tumors. It has come as a surprise to learn that cancer cells themselves produce inhibitors of proteinase enzymes that allow invasion of nearby tissues. Proteinase action occurs only if the number of enzyme molecules is greater than the number of TIMP (tissue inhibitor of metalloproteinase) molecules. This means that TIMPs or drugs that act like them may offer an approach to prevent metastasis.

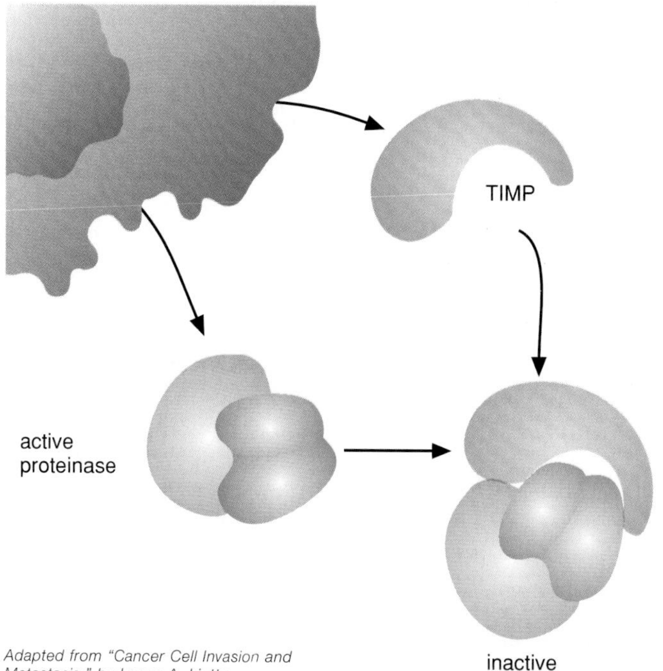

Adapted from "Cancer Cell Invasion and Metastasis," by Lance A. Liotta,

In other studies, investigators found that cells having a high level of a protein called nm23 (nonmetastatic 23) do not metastasize. Someday this protein might be used as a drug. In the meantime, human trials of the drug CAI (carboxyamide aminoimidazoles) are underway because this drug prevents metastasis in some unknown way.

Gene Therapy

With greater understanding of genes and carcinogenesis, it is not unreasonable to anticipate that gene therapy may eventually be applied in curing human cancer. For example, researchers think one therapy might be to inject tumors with genetically engineered viruses carrying an antisense nucleotide to combine with and shut down the action of *ras*K and a second gene to replace a defective *p53* gene in tumors. These genes will become effective once the viruses have entered cancer cells.

> Surgery, followed by radiation and/or chemotherapy, is the standard method of treating cancer. Bone marrow transplants and hormone therapy are also used. New approaches being investigated are immunotherapy, antimetastatic drugs, and gene therapy.

Prevention of Cancer

There are many who believe that more emphasis should be placed on prevention of cancer. Everyone needs to be aware that despite great effort by the medical and scientific community, the survival rate for all sites of cancer is only 53% (Caucasian) and 38% (African American).[3] Prevention works and saves time and money. Forty thousand fewer individuals will die this year from stomach cancer compared to the 1930s. The drop can be attributed to the reduced need to preserve food by pickling, smoking, or salt-curing because of the universal availability of home refrigerators. On the other hand, there has been a steady rise in lung cancer deaths, which may be associated with an increase in smoking among women since World War II. These behaviors help prevent cancer:

DON'T SMOKE. Cigarette smoking accounts for about 30% of all cancer deaths. Smoking is responsible for 90% of lung cancer cases among men and 79% among women—about 87% altogether. Those who smoke 2 or more packs of cigarettes a day have lung cancer mortality rates 15 to 25 times greater than nonsmokers. Smokeless tobacco (chewing tobacco or snuff) increases the risk of cancer of the mouth, larynx, throat, and esophagus. (See the Health Focus for chapter 8, "The Most Often Asked Questions About Smoking, Tobacco, and Health, and . . . The Answers.")

DON'T SUNBATHE. Almost all cases of basal and squamous cell skin cancers are considered to be sun-related. Further sun exposure is a major factor in the development of melanoma, and the incidence of this cancer increases for those living near the equator. (See the Health Focus for chapter 3, "Start Saving Your Skin.")

AVOID ALCOHOL. Cancers of the mouth, throat, esophagus, larynx, and liver occur more frequently among heavy drinkers, especially when accompanied by cigarette smoking or chewing tobacco. (See the Health Focus for chapter 10, "The Bad Effects of Alcohol.")

AVOID RADIATION. Excessive exposure to ionizing radiation can increase cancer risk. Even though most medical and dental X rays are adjusted to deliver the lowest dose possible, unnecessary X rays should be avoided. Excessive radon exposure in homes increases the risk of lung cancer, especially in cigarette smokers. It is best to test your home and take the proper remedial actions.

BE TESTED FOR CANCER. Do the shower check for breast cancer or testicular cancer. Have other exams done regularly by a physician.

BE AWARE OF OCCUPATIONAL HAZARDS. Exposure to several different industrial agents (nickel, chromate, asbestos, vinyl chloride, etc.) and/or radiation increases the risk of various cancers. Risk from asbestos is greatly increased when combined with cigarette smoking.

BE AWARE OF HORMONE THERAPY. Estrogen therapy to control menopausal symptoms increases the risk of endometrial cancer. However, including progesterone in estrogen replacement therapy helps to minimize this risk. (See the Health Focus for chapter 13, "Dangers of Anabolic Steroids.")

DO FOLLOW THE DIETARY GUIDELINES. Statistical studies have suggested that persons who follow certain dietary guidelines are less likely to have cancer.

Dietary Guidelines

The following dietary guidelines greatly reduce your risk of developing cancer:

1. **Avoid obesity.** Women who are obese have a 55% greater risk, and men have a 33% greater risk of cancer than those of normal weight. Risk for colon, breast, and uterine cancers increases in obese people.
2. **Lower total fat intake.** A high-fat intake has been linked to development of colon, prostate, and possibly breast cancer.

3. When cancer patients live 5 years beyond the time of diagnosis and treatment, it is generally considered that they are cured.

3. **Eat plenty of high-fiber foods.** These include whole-grain cereals, fruits, and vegetables. Studies have indicated that a high-fiber diet is protective against colon cancer, a frequent cause of cancer deaths (see table 20.3). It is worth noting that foods high in fiber tend to be low in fat!
4. **Increase consumption of foods that are rich in vitamins A and C.** Beta-carotene, a precursor of vitamin A, is found in dark green leafy vegetables, carrots, and various fruits. Vitamin C is present in citrus fruits. These vitamins are called antioxidants because in cells they prevent the formation of free radicals (organic ions that have a nonpaired electron) that can possibly damage DNA. Vitamin C also prevents the conversion of nitrates and nitrites into carcinogenic nitrosamines in the digestive tract. Processed foods may have had nitrates and nitrites added as preservatives.

Figure 20.10

Some data suggest that diet can influence the development of cancer. Fresh fruits, especially those high in vitamin A and C, and vegetables, especially those in the cabbage family, are believed to reduce the risk of cancer.

5. **Cut down on consumption of salt-cured, smoked, or nitrite-cured foods.** Salt-cured or pickled foods may increase the risk of stomach and esophageal cancer. Smoked foods like ham and sausage contain chemical carcinogens similar to those in tobacco smoke. Nitrites are sometimes added to processed meats (e.g., hot dogs and cold cuts) and other foods to protect them from spoilage; as mentioned previously, nitrites are converted to nitrosamines in the digestive tract.
6. **Include vegetables of the cabbage family in the diet.** Besides cabbage, broccoli, brussels spouts, kohlrabi, and cauliflower are in the cabbage family (fig. 20.10). These vegetables may reduce the risk of gastrointestinal and respiratory tract cancers.
7. **Be moderate in the consumption of alcohol.** People who drink and smoke are at an unusually high risk for cancer of the mouth, larynx, and esophagus.

> Knowing that the incidence of cancer is high, and treatment is not always successful, increased effort should be placed on preventing cancer in the first place. Prevention includes these behaviors: don't smoke, don't sunbathe, avoid alcohol, avoid radiation, and be aware of occupational hazards and hormone therapy and do follow the dietary guidelines provided.

Summary

Cancer cells have characteristics that distinguish them from normal cells. They are nondifferentiated, have abnormal nuclei, do not require growth factors, and divide repeatedly. Because they are not constrained by their neighbors they form a tumor. Finally, they metastasize and start new tumors elsewhere in the body. Cancer-causing mutations accumulate during the development of cancer, which is a multistage process involving initiation, promotion, and progression. A cancer-causing mutation can be inherited or arise due to exposure to certain organic chemicals, radiation, and a few types of viruses. Still, cancer is most apt to occur in the immunodeficient individual.

Each cell contains a regulatory network involving proto-oncogenes and tumor suppressor genes, which code for a growth factor receptor, and signaling proteins active in intracellular reactions that lead to cell division and growth. When proto-oncogenes mutate, becoming oncogenes, and tumor suppressor genes mutate the regulatory network no longer functions as it should and uncontrolled growth results.

Cancers are classified as carcinomas (epithelial tissue cancers), sarcomas (muscles and connective tissue), leukemias (blood), and lymphomas (lymphoid tissue). Appendix D gives pertinent information about cancers at specific sites. There are standard procedures to detect specific cancers, for example, the Pap smear for cervical cancer, mammograms for breast cancer, and the stool blood test for colon cancer, to name a few. But new ways such as tumor marker tests and oncogene tests are being developed. Biopsy and imaging are used to confirm the diagnosis of cancer.

Table 20.4

Specific Therapy

Cancer Cell Anatomy and Physiology	Associated Therapy	State of Therapy
Active oncogenes	Gene therapy to inactivate	Human trials soon
Inactive tumor suppressor genes	Gene therapy to supply an active gene	Human trials soon
Growth factor receptors at cell membrane	Monoclonal antibodies to seek out cells Antihormone drugs	Human trials soon Standard therapy
Antigens on cell membrane	Potentiation of the immune system	Human trials now
Active signaling proteins	Interrupt and stop metabolic reactions	Test tube studies
Frequent cell division	Radiation and chemotherapy	Standard therapy

Surgery followed by radiation and/or chemotherapy is the standard method of treating cancer. Chemotherapy followed by bone marrow transplants and hormone therapy is also used. These standard methods work because cancer cells are rapidly dividing. New therapies under investigation are gene therapy, immunotherapy, and antimetastatic drugs. Each type of therapy is based on the unique anatomy and physiology of the cancer cell, as described in table 20.4.

Knowing that the incidence of cancer is high and treatment is not always successful, all effort should be placed on preventing cancer in the first place. Prevention includes these behaviors: don't smoke, don't sunbathe, avoid alcohol, avoid radiation, and be aware of occupational hazards and the danger of taking estrogen without also taking progesterone.

Certain dietary guidelines are believed to be helpful in avoiding cancer. A high-fat diet and salt-cured, smoked, and nitrite-cured foods are associated with the development of specific cancers. High-fiber foods are thought to help reduce the risk of colon cancer, and a diet containing plenty of vegetables and fruits is protective against the development of cancers in general.

Study and Thought Questions

1. List and discuss 4 characteristics of cancer cells that distinguish them from normal cells. (p. 404)
2. Name 3 stages of carcinogenesis and describe each phase. (p. 406)
3. What role does heredity, carcinogens, and immunodeficiency play in the development of cancer? Name 3 types of carcinogens and give examples of each type. (pp. 406–408)

   Ethical Issue Soon, it may be possible to determine which persons are genetically susceptible to the development of cancer. Should these individuals be required to protect themselves against the development of cancer so that society doesn't have to pay the cost of their treatment?
4. What are oncogenes and tumor suppressor genes? What role do they play in a regulatory network that controls cell division and involves cell membrane receptors and signaling proteins? (pp. 409–411)

   Critical Thinking Develop an example in which a hypothetical "C protein" is involved in controlling the growth of cells, and then tell what happens when the C gene becomes an oncogene.
5. Draw and describe a diagram that explains how the activity of the *ras* protein is controlled by a regulatory network involving a cell membrane receptor and the GAP protein. (p. 410)
6. Draw and describe a diagram that explains why the *RB* gene is called a tumor suppressor gene. (p. 411)
7. Tell in general how cancers are classified, and give a system of clinical staging that can apply to all cancers. (p. 413)
8. What are the standard ways to detect cervical cancer, breast cancer, and colon cancer? (p. 414) What is taxol? (p. 416)
9. Describe and give examples of tumor marker tests and oncogene tests. (p. 414)
10. What are the standard methods of treatment for cancer? Explain the rationale for bone marrow transplants. (pp. 416–417)

    Ethical Issue Cancer therapy sometimes prolongs the life span but not the health span of an individual. Should all persons be treated regardless of their prognosis?
11. List and describe 3 different ways that immunotherapy is being used to treat cancer. How will antimetastatic drugs and gene therapy be used to treat cancer? (pp. 417–418)

    Critical Thinking Categorize cancer therapies into those that rely on normal body functions and those that rely on external measures to treat cancer.
12. List 4 carcinogens to be avoided to prevent the development of cancer. Give the dietary guidelines that can possibly prevent the development of cancer. (pp. 418–419)

Writing Across the Curriculum

1. A number of retroviruses carry oncogenes in animals. Oncogenes are not viral in origin; they are human in origin. Explain.
2. Unsaturated fatty acids have a tendency to form free radicals (organic ions that have a nonpaired electron), which may damage DNA. With this knowledge in mind, hypothesize why a low-fat diet reduces the risk of breast, colon, and prostate cancers.
3. Many foods contain known carcinogens and other substances suspected of being carcinogens. What do you think can or should be done to reduce these risks of cancer?

Objective Questions

1. Cancer cells ______________; they travel to distant body parts and start new tumors.
2. The mutation of ____________ and ____________ genes leads to uncontrolled growth and cancer.
3. The ____________ virus is a carcinogen for cancer of the cervix.
4. ________________ contains many organic chemical carcinogens and is associated with one-third of all cancers.
5. The *RB* gene is a ____________; it normally keeps *c-myc* from being an oncogene.
6. Cancer cells are sensitive to radiation therapy and chemotherapy because they are constantly ______________.
7. Autotransplants of bone marrow permit a much higher dosage of ___________ than otherwise.
8. Immunotherapy includes genetic engineering of tumor-infiltrating cells to carry tumor necrosis factor, a ______________.
9. To prevent cancer, you should avoid __________, environmental agents associated with the development of cancer.
10. Vitamins __________ and __________ are associated with the avoidance of cancer.
11. Identify these portions of a cancer cell. Tell what type of therapy is appropriate for each abnormality in anatomy and physiology.

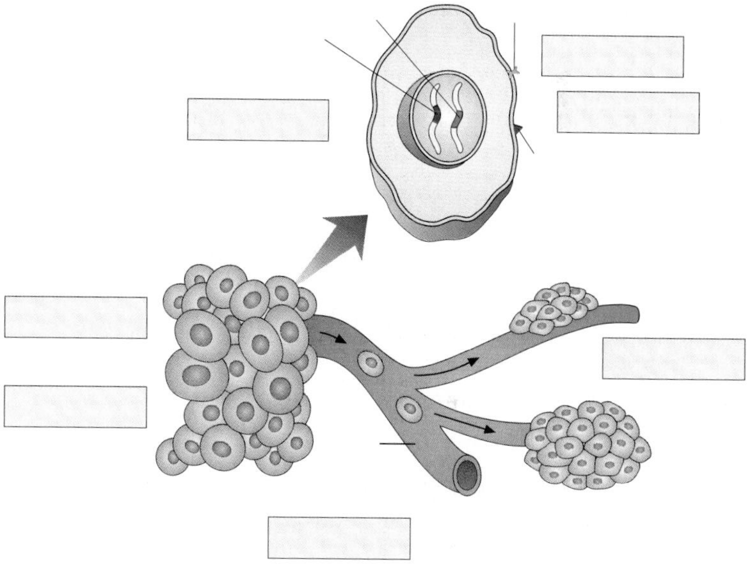

Selected Key Terms

cancer A malignant tumor that metastasizes. 404

carcinogen An environmental agent that contributes to the development of cancer. 406

carcinogenesis (kar″- sī-no-jen′-ĕ-sis) Development of cancer. 406

chemotherapy The use of a drug to selectively kill off cancer cells as opposed to normal cells. 416

immunotherapy The use of any immune system component such as antibodies, cytotoxic T cells, or lymphokines to promote the health of the body, such as curing cancer. 417

leukemia Cancer of the blood-forming tissues leading to the overproduction of abnormal white blood cells. 413

lymphoma Cancer of the lymphoid organs such as lymph nodes, spleen, and thymus gland. 413

metastasis (mĕ-tas′-tah-sis) The spread of cancer from the place of origin throughout the body caused by the ability of cancer cells to migrate and invade tissues. 404

mutagen An environmental agent that induces mutations. 406

oncogene (ong′-ko-jen) A gene that contributes to the transformation of a normal cell into a cancer cell. 409

proto-oncogene A normal gene that becomes an oncogene through mutation. 409

sarcoma A cancer that arises in connective tissue, such as muscle, bone, and fibrous connective tissue. 413

tumor Growth containing cells derived from a single mutated cell that has repeatedly undergone cell division; benign tumors remain at the site of origin and malignant tumors metastasize. 404

tumor suppressor gene Genes that, when expressed, prevent abnormal cell division and cancer. 409

Further Readings for Part Five

Barton, J. H. March 1991. Patenting life. *Scientific American.*

Beardsley, T. August 1991. Smart genes. *Scientific American.*

Boon, T. March 1993. Teaching the immune system to fight cancer. *Scientific American.*

Bugg, C. E., et al. December 1993. Drugs by design. *Scientific American.*

Cooper, G. 1992. *Elements of human cancer.* Boston: Jones and Bartlett Publishers.

Cooper, G. M. 1993. *The Cancer Book.* Boston: Jones and Bartlett Publishers.

Cummings, M. R. 1991. *Human heredity.* 2d ed. St. Paul, Minn.: West Publishing.

Drlica, K. 1991. *Understanding DNA and gene cloning: A guide for the curious.* 2d ed. New York: John Wiley and Sons.

Edlin, G. 1990. *Human genetics: A modern synthesis.* Boston: Jones and Bartlett Publishers.

Erikson, D. April 1992. Hacking the genome. *Scientific American.*

Feldman, M., and Eisenback, L. November 1988. What makes a tumor cell metastatic? *Scientific American.*

Gardner, E. J., et al. 1991. *Principles of genetics.* 8th ed. New York: John Wiley and Sons.

Gasser, C. S., and Fraley, R. T. June 1992. Transgenic crops. *Scientific American.*

Grunstein, M. October 1992. Histones as regulators of genes. *Scientific American.*

Holliday, R. June 1989. A different kind of inheritance. *Scientific American.*

Horgan, J. June 1993. Eugenics revisited. *Scientific American.*

Joyce, G. F. December 1992. Directed molecular evolution. *Scientific American.*

Kartner, N., and Ling, V. March 1989. Multidrug resistance in cancer. *Scientific American.*

Kieffer, G. H. 1987. *Biotechnology, genetic engineering, and society.* Reston, Va.: National Association of Biology Teachers.

Liotta, L. A. February 1992. Cancer cell invasion and metastasis. *Scientific American.*

McNight, S. L. April 1991. Molecular zippers in gene regulation. *Scientific American.*

Mange, A. P., and Mange, E. J. 1990. *Genetics: Human aspects.* 2d ed. Sunderland, Mass.: Sinauer Associates.

Moyzis, R. K. August 1991. The human telomere. *Scientific American.*

Mullis, K. B. April 1990. The unusual origin of the polymerase chain reaction. *Scientific American.*

Nathans, J. February 1989. The genes for color vision. *Scientific American.*

Neufeld, P. J., and Colman, N. May 1990. When science takes the witness stand. *Scientific American.*

Ptashne, M. January 1989. How gene activators work. *Scientific American.*

Radman, M., and Wagner, R. August 1988. The high fidelity of DNA duplication. *Scientific American.*

Rennie, J. March 1993. DNA's new twists. *Scientific American.*

Rhodes, D., and Klug, A. February 1993. Zinc fingers. *Scientific American.*

Rosenberg, S. A. May 1990. Adoptive immunotherapy for cancer. *Scientific American.*

Ross, J. April 1989. The turnover of messenger RNA. *Scientific American.*

Ross, P. E. May 1992. Eloquent remains. *Scientific American.*

Rusting, R. L. December 1992. Why do we age? *Scientific American.*

Sapienza, C. October 1990. Parental imprinting of genes. *Scientific American.*

Tamarin, R. 1991. *Principles of genetics.* 3d ed. Dubuque, Iowa: Wm. C. Brown Publishers.

Verma, I. M. November 1990. Gene therapy. *Scientific American.*

Weinberg, R. A. September 1988. Finding the anti-oncogene. *Scientific American.*

Weintraub, H. M. January 1990. Antisense RNA and DNA. *Scientific American.*

White, R., and Lalouel, J. February 1988. Chromosome mapping with DNA markers. *Scientific American.*

Part SIX

Human Evolution and Ecology

Evidence for the theory of evolution is drawn from many areas of biology. Charles Darwin was the first to present extensive evidence, and he suggested that those organisms best suited to the environment are the ones that survive and reproduce most successfully. Evolution causes life to have a history, and it is possible to trace the ancestry of humans even from the first cell or cells.

The diverse forms of life that are now present on earth live within ecosystems, where energy flows and chemicals cycle. Because population sizes remain constant, natural ecosystems tend to require the same amount of energy and chemicals each year. Humans have created their own ecosystems, consisting of the country and city, which differ from natural ecosystems in that the populations constantly increase in size and ever-greater amounts of energy and raw materials are needed each year. Since 1850, the human population has expanded so rapidly that some doubt there will be sufficient energy and food to permit the same degree of growth in the future. The human ecosystem depends on the natural ecosystems not only because they absorb pollutants but also because natural ecosystems are inherently stable. Every possible step should be taken to protect natural ecosystems and help ensure the continuance of the human ecosystem.

Chapter 21

Evolution

Which living things are able to shape the course of evolution? Human beings, of course. When human beings alter the environment, they influence which organisms will survive and which will die out in a particular locale. More directly, humans carry out breeding programs to select which plants or animals will reproduce more than others. The end result can be specific, discrete types of organisms; all of the different forms of domesticated dogs belong to the same species—*Canis familiaris*—and selection has brought about the many varieties. It's obvious to us that genes control the characteristics we wish to see selected. One of the great accomplishments of twentieth-century science has been to give the concept of evolution a genetic basis.

Chapter Concepts

1. The fossil record and comparative anatomy, embryology, biochemistry, and biogeography all provide evidence for evolution. 426–428
2. Charles Darwin formulated a mechanism for an evolutionary process that results in adaptation to the environment. 429
3. Organic evolution accounts for the diversity of life from the first cell(s) to all other living things, including humans. 430
4. Humans are primates, and many of their physical traits are the result of their ancestors' adaptations to living in trees. 432
5. It is possible to trace the evolution of humans using a series of fossils found in Africa, Europe, and Asia. 436
6. All human races are classified as *Homo sapiens*. 437

Chapter Outline

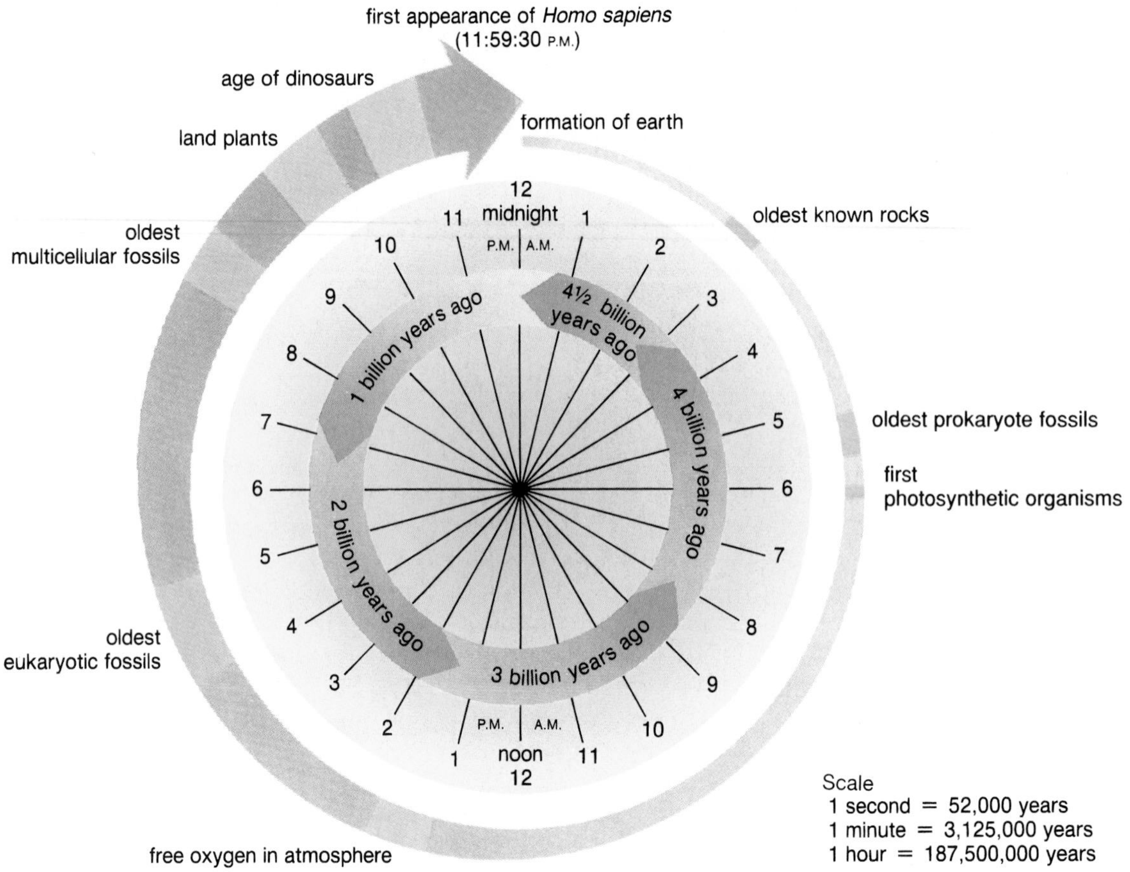

Figure 21.1

History of earth measured on a 24-hour time scale compared to actual years. A very large portion of life's history is devoted to the evolution of single-celled organisms. The first multicellular organisms do not appear until just after 8 P.M., and humans are not on the scene until less than a minute before midnight.

Evidence for Evolution

The fossil record, comparative anatomy, comparative embryology, comparative biochemistry, and biogeography all support the theory of evolution.

Fossils Show the History of Life

Our knowledge of the history of life is based primarily on the fossil record. **Fossils** are the remains or evidence of some organism that lived long ago. The oldest fossils found are of prokaryotic cells dated some 3.5 billion years ago. Thereafter the fossils get more and more complex. For example, prokaryotic cells were followed by eukaryotic cells, and among animals, invertebrates were followed by vertebrates. Humans began evolving about 5 million years ago, but modern humans (*Homo sapiens*) do not appear in the fossil record until about 100,000 years ago. This is clear evidence that there has been a history of life based on evolutionary events (fig. 21.1).

Related Species Share Anatomy

A comparative study of the anatomy of groups of organisms has shown that each has a *unity of plan.* For example, all vertebrate animals have essentially the same type of skeleton. Unity of plans allows organisms to be classified into various groups. Organisms most similar to one another are placed in the same **species,** similar species are placed in a **genus,** similar genera in a family; thus, we proceed from **family** to **order** to **class** to **phylum** to **kingdom.** The classification of any particular organism indicates to which kingdom, phylum, class, order, family, genus, and species the organism belongs. According to the **binomial system** of naming organisms, each organism is given a 2-part name, which consists of the genus and species to which it belongs (table 21.1). Thus, for example, a human is *Homo sapiens*[1] and the domesticated cat is *Felis domestica.* **Taxonomy** is the branch of biology that is concerned with classification, and biologists who specialize in classifying organisms are called taxonomists.

1. Varieties of the same species are also given a subspecies designation. Neanderthals are *Homo sapiens neanderthalensis* and modern humans are *Homo sapiens sapiens.*

Table 21.1

Classification of Modern Humans

Kingdom	Animalia	Includes all animals
Phylum	Chordata	Includes vertebrates: fishes, amphibians, reptiles, birds, and mammals
Class	Mammalia	Mammals: cats, dogs, horses, mice, sheep, cows, etc.
Order	Primates	Monkeys, apes, humans
Family	Hominidea	Hominids: forms closely related to humans, extinct humans, and modern humans
Genus	*Homo*	Extinct humans and modern humans
Species	*sapiens*	Modern humans

A *unity of plan* is explainable by descent from a **common ancestor,** an ancestor of 2 or more branches of evolution. Species that share a recent common ancestor will share a large number of the same genes and therefore will be quite similar to each other and to this ancestor. Species that share a more distant common ancestor will have fewer genes in common and will be less similar to each other and to this ancestor because differences arise as organisms continue on their own evolutionary pathways. Even after related organisms have become adapted to different ways of life, they may continue to show similarities of structure. For example, vertebrate forelimbs are used for flight (birds and bats), orientation during swimming (whales and seals), running (horses), climbing (arboreal lizards), or swinging from tree branches (monkeys). Yet, all vertebrate forelimbs contain the same sets of bones organized in similar ways despite their dissimilar functions (fig. 21.2).

Vestigial structures are anatomical features that are fully developed and functional in one group of organisms but are reduced and functionless in similar groups. Most birds, for example, have well-developed wings used for flight. Some bird species, however, have wings that are greatly reduced, prohibiting flight. Similarly, snakes have no use for hindlimbs, yet some have remnants of a pelvic girdle and legs. Humans have a tailbone (coccyx) but no tail.

A common evolutionary history explains the presence of vestigial structures. Vestigial structures occur because organisms inherit their anatomy from their ancestors; they are traces of an organism's evolutionary history.

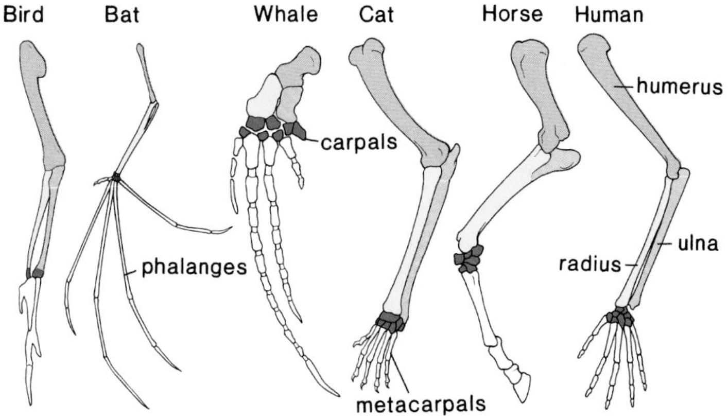

Figure 21.2

Bones in vertebrate forelimbs. The same bones are present (they are color-coded) but designed for different functions. The unity of plan is evidence of a common ancestor.

Related Species Share Development

Some groups of organisms share the same type of embryonic stages. As would be expected if all vertebrates are related, their embryonic stages are similar. During development, a human embryo at one point has gill arches—even though it will never breathe by means of gills, as do fishes—and a rudimentary tail—even though it will never have a long tail, as do some 4-legged vertebrates. These embryological observations suggest we are related to other vertebrates.

Related Organisms Have Similar DNA

Almost all living organisms use the same basic biochemical molecules, including DNA, ATP, and many identical or nearly identical enzymes. For example, the metabolic enzymes involved in aerobic cellular respiration and in the synthesis of cellular macromolecules are the same in all organisms. It would seem, then, that these enzymes appeared very early in the evolution of life and their genes have been passed on ever since.

Analyses of amino acid sequences in certain proteins such as hemoglobin and cytochrome *c* and analyses of DNA nucleotide differences are used to determine how closely related animals are. It is assumed that the number of differences will reflect how long ago the 2 species shared a common ancestor.

Figure 21.3

Parallel evolution. The grasslands in Africa and America have similarly adapted but different animals. Both the pronghorn antelope and zebra have sharp incisors for clipping and large molars for chewing grasses. The toes are hoofed and legs are elongated for fast running. Both the jackal and coyote are predators with well-developed jaws, shearing canine teeth, and crushing molars. The legs are long and slender for fast running.

Geography and Evolutionary Relationships

It is observed that environmentally similar but geographically separate environments have different plants and animals that are similarly adapted. For example, figure 21.3 gives examples of North American animals and African animals that are adapted to living in a grassland environment. First, you will notice that although each type of animal could live in the other's biogeographic region, they do not. Why? They do not because geographic separation made it impossible for a common ancestor to produce descendants for both regions. On the other hand, notice that the same type of adaptations are seen in both groups of animals. This phenomenon, called **convergent evolution,** supports the belief that the evolutionary process causes organisms to be adapted to their environments.

Evolution explains the history and diversity of life. Evidence for evolution can be taken from the fossil record, comparative anatomy, embryology, and biochemistry. Biogeography also supports the occurrence of evolution.

The Evolutionary Process

Based on the type of evidence we have just outlined, Charles Darwin formulated a theory of **natural selection** to explain the evolutionary process around 1860. The following are critical to understanding natural selection.

Variations Come First

Individual members of a species vary in physical characteristics. Such variations can be passed on from generation to generation. Darwin was never able to determine the cause of variations or how they are passed on. Today, we realize that genes determine the appearance of an organism and that mutations can cause new variations to arise.

Struggle for Existence

In Darwin's time, a socioeconomist, Thomas Malthus, stressed the reproductive potential of human beings. He proposed that death and famine were inevitable because the human population tended to increase faster than the supply of food. Darwin applied this concept to all organisms and saw that members of any plant or animal population must compete with one another for available resources. Competition must of necessity occur because reproduction produces more members of a population than can be sustained. Darwin calculated the reproductive potential of elephants. Assuming a life span of about 100 years and a breeding span of from 30–90 years, a single female will probably bear no fewer than 6 young. If all these young survived and continued to reproduce at the same rate, after only 750 years the descendants of a single pair of elephants would number about 19 million! Such reproductive potential necessitates a *struggle for existence,* and only certain organisms survive and reproduce.

Survival of the Fittest

Darwin noted that in *artificial selection* humans choose which plants or animals will reproduce. This selection process brings out certain traits. For instance, there are many varieties of dogs, each of which is derived from the wild wolf. In a similar way, several varieties of vegetables can be traced to a single type.

In nature, it is the environment that selects those members of the population that will reproduce to a greater degree than other members. In contrast to artificial selection, *natural selection* occurs only because certain members of a population happen to have a variation that makes them more suited to the environment. For example, any variation that increases the speed of a hoofed animal will help it escape predators and live longer; a variation that reduces water loss will help a desert plant survive; and one that increases the sense of smell will help a coyote find its prey. Therefore, we would expect organisms with these traits to eventually reproduce to a greater extent.

Whereas Darwin emphasized that only certain organisms survived to reproduce, modern evolutionists emphasize that competition results in *differential reproduction.* If certain organisms can acquire a greater share of available resources and if they have the ability to reproduce, then their chances of reproduction are greater than those around them.

Adaptation to the Environment

An **adaptation** is a trait that helps an organism be more suited to its environment. We can especially recognize an adaptation when unrelated organisms living in a particular environment display similar characteristics. For example, manatees, penguins, and sea turtles all have flippers, which help them move through the water. Similarly, fishes, including sea skates, have flattened bodies for the same function (fig. 21.4).

Natural selection results in the adaptation of populations to their specific environments. Because of differential reproduction generation after generation, adaptive traits are more and more common in each succeeding generation.

The following listing summarizes the theory of evolution as developed by Darwin.

1. There are inheritable variations among the members of a population.
2. Many more individuals are produced each generation than can survive and reproduce.
3. Individuals with adaptive characteristics are more likely to be selected to reproduce by the environment.
4. Gradually, over long periods of time, a population can become well adapted to a particular environment.
5. The end result of organic evolution is many different species, each adapted to specific environments.

Figure 21.4

Animals living in the same type of environment have some of the same adaptations. Manatees, penguins, and sea turtles all have flippers. Fishes, including sea skates, have flattened bodies.

Organic Evolution

Figure 21.1 outlines the evolutionary history of life on earth. Organic evolution began with the evolution of the first cell or cells. Scientists have pieced together a reasonable hypothesis concerning the evolution of the first cell(s), which they call the origin of life.

Life Evolves and Diversifies

The sun and the planets probably formed from aggregates of dust particles and debris about 4.6 billion years ago. Intense heat produced by gravitational energy and radioactivity of some atoms caused the earth to become stratified into a core, a mantle, and a crust. Heavier atoms of iron and nickel became the molten liquid core, and dense silicate minerals became the semiliquid mantle. The unstable mantle caused the thin crust to move continually, so there were no stable land masses during the time life was evolving.

The *primitive atmosphere* was not the same as today's atmosphere (fig. 21.5). It is now thought that the primitive atmosphere was produced after the earth formed by outgassing from the interior, particularly by volcanic action. The atmosphere would have consisted mostly of water vapor (H_2O), nitrogen gas (N_2), and carbon dioxide (CO_2), with only small amounts of hydrogen (H_2) and carbon monoxide (CO). The primitive atmosphere with little, if any, free oxygen was a reducing atmosphere as opposed to the oxidizing atmosphere of today. This was fortuitous because oxygen (O_2) attaches to organic molecules, preventing them from joining to form larger molecules.

Water, present at first as vapor in the atmosphere, formed dense, thick clouds, but cooling eventually caused the vapor to condense to liquid, and rain began to fall. This rain was in such quantity that it produced the oceans of the world. The gases, dissolved in the rain, were carried down into the newly forming oceans.

The dissolved gases, although relatively inert, are believed to have reacted together to form simple organic molecules when they were exposed to the strong outside *energy sources* present on the primitive earth (fig. 21.5*b*). These energy sources included heat from volcanoes and meteorites, radioactivity from the earth's crust, powerful electric discharges in lightning, and solar radiation, especially ultraviolet radiation. Now, the primitive oceans became a thick organic soup and the simple organic molecules condensed to produce the macromolecules characteristic of living things: proteins, nucleic acids, and lipids. As soon as a lipid membrane enclosed some of these macromolecules, a *protocell* would have formed (fig. 21.5*c*). Eventually, the self-reproducing system of today (DNA → RNA → protein) developed and the first cell or cells came into being.

A chemical evolution produced the first cell(s):

Gases ↓	(e.g., water (H_2O), nitrogen (N_2), carbon dioxide (CO_2), hydrogen (H_2))
Small organic molecules ↓	(amino acids, nucleotides, fatty acids)
Macromolecules ↓	(proteins, nucleic acids, lipids)
Protocell ↓	
Cell	

The first cell(s) must have been a heterotroph living off the preformed organic molecules in the primitive ocean, and it must have carried on anaerobic fermentation since there was no oxygen (O_2) in the atmosphere. It was a *heterotrophic fermenter.*

Once the preformed organic molecules were depleted, organic evolution would have favored any cell capable of making its own food. Eventually, photosynthesizers arose and gave off oxygen as a by-product (fig. 21.5*d*).

The presence of free oxygen (O_2) changed the character of the atmosphere; it became an oxidizing atmosphere instead of a reducing atmosphere. Abiotic synthesis of organic molecules was no longer possible because any organic molecules that happened to form would have been broken down by oxidation. As oxygen levels increased, cells capable of aerobic cellular respiration evolved. This is the type of respiration used by the vast majority of organisms today.

The buildup of oxygen in the atmosphere caused the development of an ozone (O_3) layer. This layer filters out ultraviolet rays, shielding the earth from dangerous radiation. Prior to this time, organisms probably lived only deep in the oceans, where they were not exposed to the intense radiation striking the earth's surface. Now life could safely spread to shallower waters and eventually move onto the land.

Plants were probably the first living organisms on land. This is reasonable because animals are dependent upon plants as a food source. The first vertebrates (animals with backbones) to dominate the land were the amphibians, whose name implies that their life cycle has 2 phases—one spent in the water and one spent on land. Frogs are amphibians, and during their tadpole stage, they metamorphose into an adult frog. Next came the reptiles, which have no need to return to water for reproductive purposes because the female lays a shelled egg. Within the egg are the extraembryonic membranes (see fig. 16.3) that service all the needs of the developing organism. Both birds and mammals evolved from the reptiles. We are mammals, animals characterized by the presence of hair and mam-

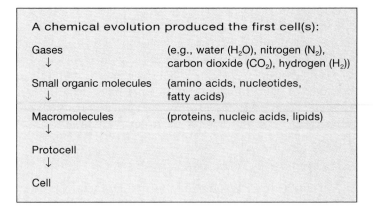

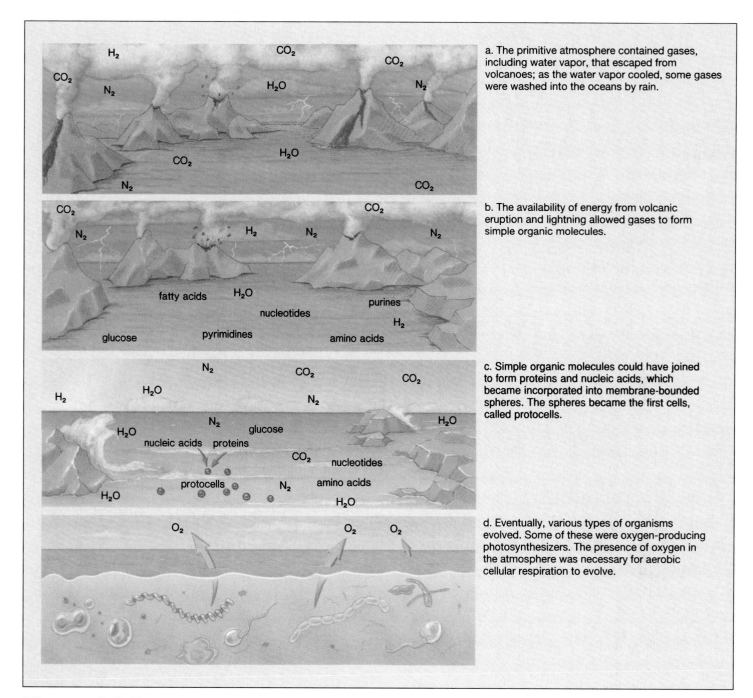

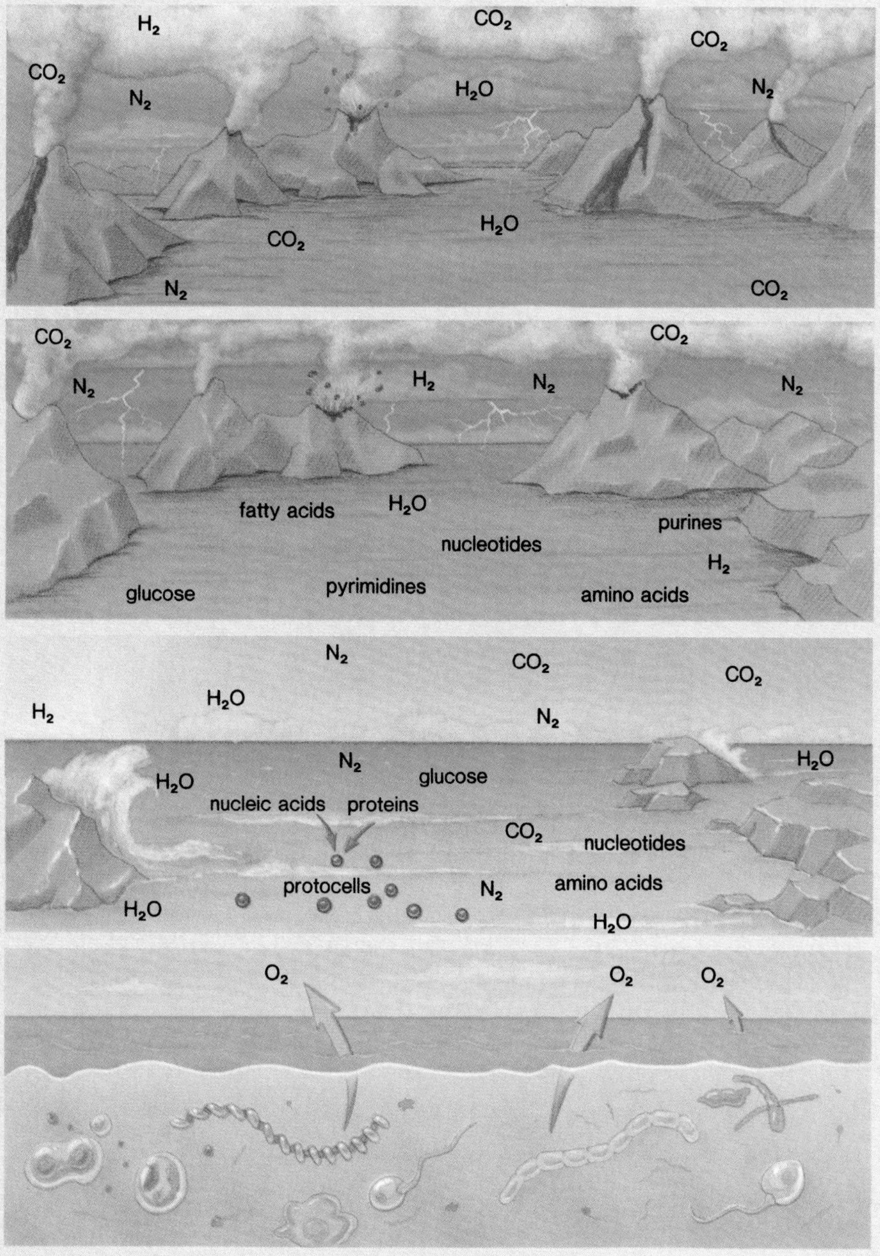

a. The primitive atmosphere contained gases, including water vapor, that escaped from volcanoes; as the water vapor cooled, some gases were washed into the oceans by rain.

b. The availability of energy from volcanic eruption and lightning allowed gases to form simple organic molecules.

c. Simple organic molecules could have joined to form proteins and nucleic acids, which became incorporated into membrane-bounded spheres. The spheres became the first cells, called protocells.

d. Eventually, various types of organisms evolved. Some of these were oxygen-producing photosynthesizers. The presence of oxygen in the atmosphere was necessary for aerobic cellular respiration to evolve.

Figure 21.5

A model for the origin of life.

mary glands. The brain is most highly evolved among mammals and reaches its largest size and complexity in the primates, the order to which humans belong (fig. 21.6).

The evolution of humans began about 5 million years ago; therefore, we have been evolving only 0.1% of the total history of the earth.

> Organic evolution involves the evolution of all life forms, including vertebrates. The first land vertebrates were the amphibians, which gave rise to the reptiles, from which the birds and mammals arose. Humans are primates, a type of mammal.

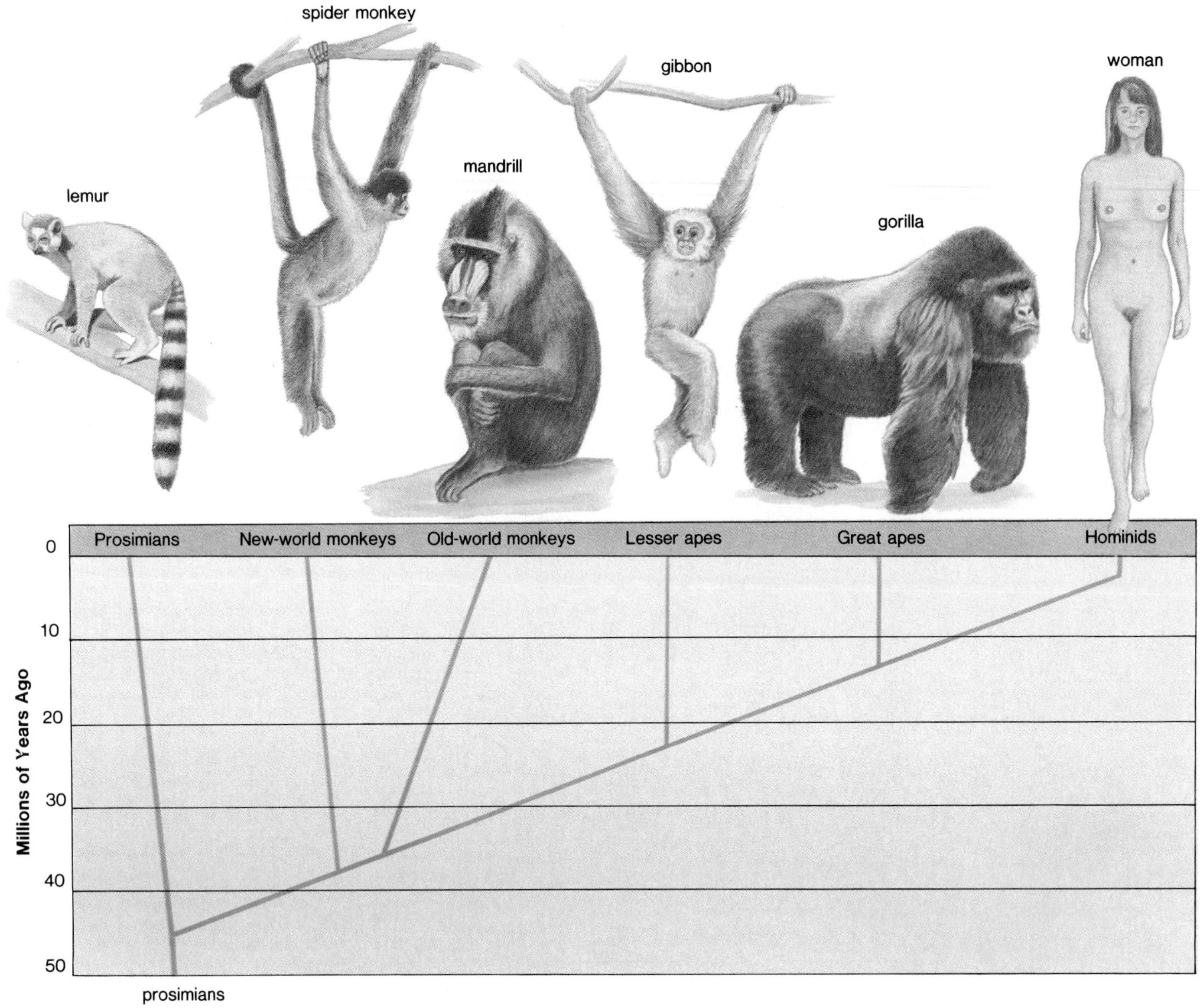

Figure 21.6

Primate evolution. There are at least 4 lines of evolution among the primates: prosimians, including lemurs and tarsiers; monkeys, including new-world and old-world monkeys; apes, including gibbons and great apes (orangutans, gorillas, and chimpanzees); and humans.

Modern Humans Evolve

Primates are placental mammals that in addition to having hair and mammary glands are adapted to living in trees. The limbs are mobile, as are the hands because the thumb (and in nonhuman primates, the big toe as well) is opposable; that is, the thumb can touch each of the other fingers. Therefore, a primate can easily reach out and bring food, such as fruits, to the mouth. A tree limb can be grasped and released freely because claws are replaced by nails. In humans, these features have led to the ability to use tools.

A snout is common in animals in which a sense of smell is of primary importance. In primates, the sense of sight is more important, and the snout is shortened considerably, allowing the eyes to move to the front of the head. This results in stereoscopic vision (or depth perception), permitting primates to make accurate judgments about the distance and position of adjoining tree limbs. All primates have well-developed cerebrums; apes and humans have highly convoluted cerebral hemispheres. Olfactory centers have declined in importance, while the visual portion of the brain is proportionately large, as are those centers responsible for hearing and touch.

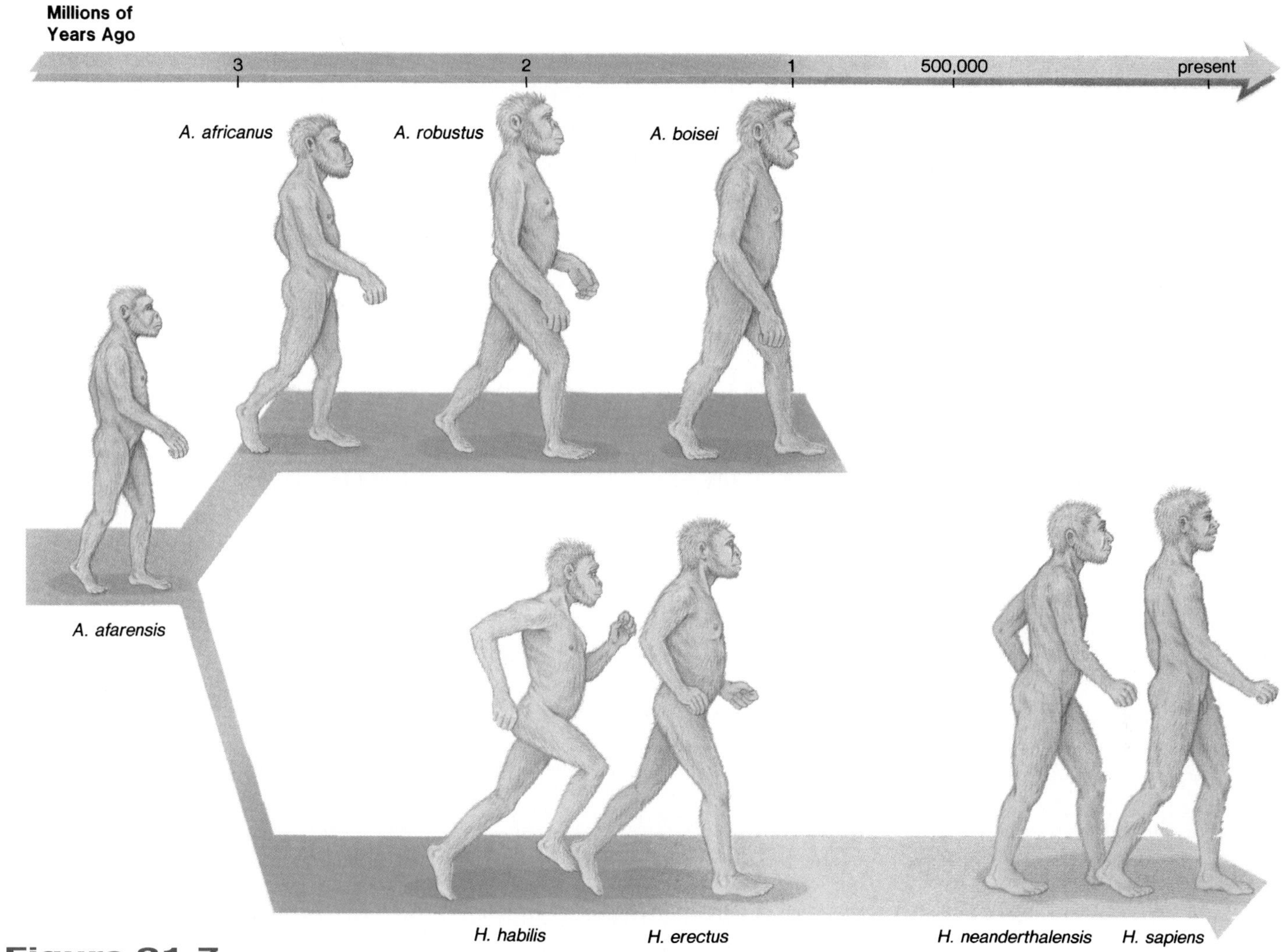

Figure 21.7

Hominid evolutionary tree. The tree is branched; members of the genus *Australopithecus* and *Homo* existed at the same time. All forms became extinct except *Homo sapiens sapiens.*

One birth at a time is the norm in primates; it would have been difficult to care for several offspring as large as primates while moving from limb to limb. The cerebrum is especially large, and there is an emphasis on learned behavior. The juvenile period of dependency is extended, and most primates tend to form social units that include individuals of all ages and sexes. A single male has exclusive rights to or several males share a group of females. Monogamous pairs are often seen in human societies.

These characteristics especially distinguish primates from other mammals:

Opposable thumb (and in some cases big toe)	Expanded cerebrum
Fingernails (not claws)	Emphasis on learned behavior
Single birth	Extended period of parental care

Molecular data tell us that humans are genetically very similar to certain apes—the chimpanzees and gorillas. Nucleotide sequences in genes, amino acid sequences in proteins, and immunological properties of various molecules all indicate our close relationship to these apes. Humans are distinguishable from apes by several features, including walking erect, as discussed in the Ecology Focus for this chapter. It's possible that understanding ourselves in an evolutionary sense will better enable us to accept our ecological responsibilities. Our dependency on nature is the same as those of apes, but this need is disguised by human culture. Our distance from the source of our basic needs often results in an absence of ecological awareness.

Ecology Focus

Human Characteristics and Ecological Awareness

Most agree that the anatomical differences between us and the apes stem from our habit of walking upright all the time. Apes tend to knuckle-walk, with hands bent, while on the ground. In keeping with walking on all fours, the ape pelvis is long and narrow and the hands and feet tend to resemble each other. In humans, the shorter pelvis is designed to bear the weight of the trunk, and it also better serves as a place of attachment for the muscles of the longer and more muscular legs. The human foot is markedly different from the hand; it supports the body on a broad heel and a thick-skinned ball cushioned by fat. Still, the foot has a springy arch needed for our striding gait.

The upright stance of humans frees the hands so that they are ready to do manipulative work. Although both apes and humans have a thumb that is "opposable" in the sense that it can reach over and touch the fingers, the human thumb is longer and rotates. It is this remarkable thumb that allows us to be such great tool users. Tool use and the human brain are believed to have evolved together, each one promoting an increase in the other:

tool use $\rightleftharpoons$ intelligence

Apes use tools but they don't make highly specialized tools for specific purposes. They also don't carry tools around with them for when they might be needed.

The human brain really sets us apart from the apes. If an ape were the size of a human, its brain would still be only one-third as large. Not surprising is the fact that the lobes of the brain needed for thinking and language are proportionately larger in humans. Apes make faces and sounds that serve to communicate with others, but only humans are capable of communicating many different types of ideas. It's possible that human intelligence and language evolved as a means to understand and function within a complex social system.

Because the large brain of humans develops after birth, infants are born in an immature state and are dependent upon their mothers for quite some time. Perhaps this is why women stayed home while men went out to hunt for food, which they brought home to share with the other members of their group. Again, apes form social groups but an organized system of food sharing is not characteristic of the group.

Learning how to think and use tools within a social group is the mainstay of human culture. Today, we humans are enveloped by our culture; most of us live in cities, existing on food and using materials grown or made far away. We no longer have a sense of our evolutionary past nor of our place in nature. We don't realize that even though we can control our environment, to a large extent we are still dependent upon natural ecological systems. Consider, for example, our dependence on the rains to bring us fresh water, on plants to supply us with food and to purify the air, and on the past remains of plants and animals to give us a supply of energy. Only when we fully appreciate our place in nature will it be possible for us to wisely guide our future destiny. If we greatly interfere with the cycles of nature, as when we overdam rivers, cut down forests, or burn fossil fuels to excess, it can only be to our detriment.

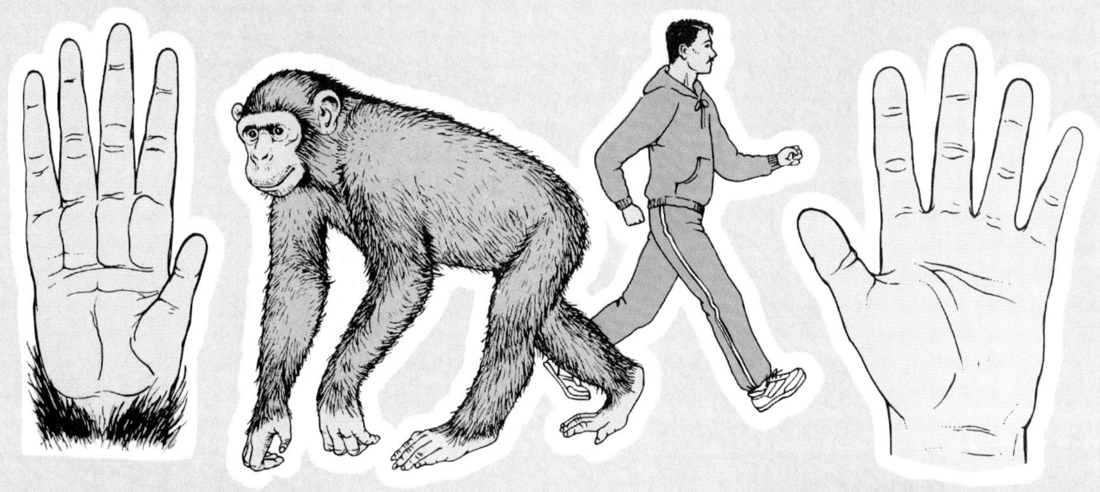

a.

Figure 21A

Was it the human hand that led to evolution of culture?

a. Apes use their hands primarily to climb trees.

b. They live in social groups where offspring are cared for.

c. Humans use their hands primarily to manipulate tools.

d. They live in a society where children learn the culture of that society.

c.

b.

d.

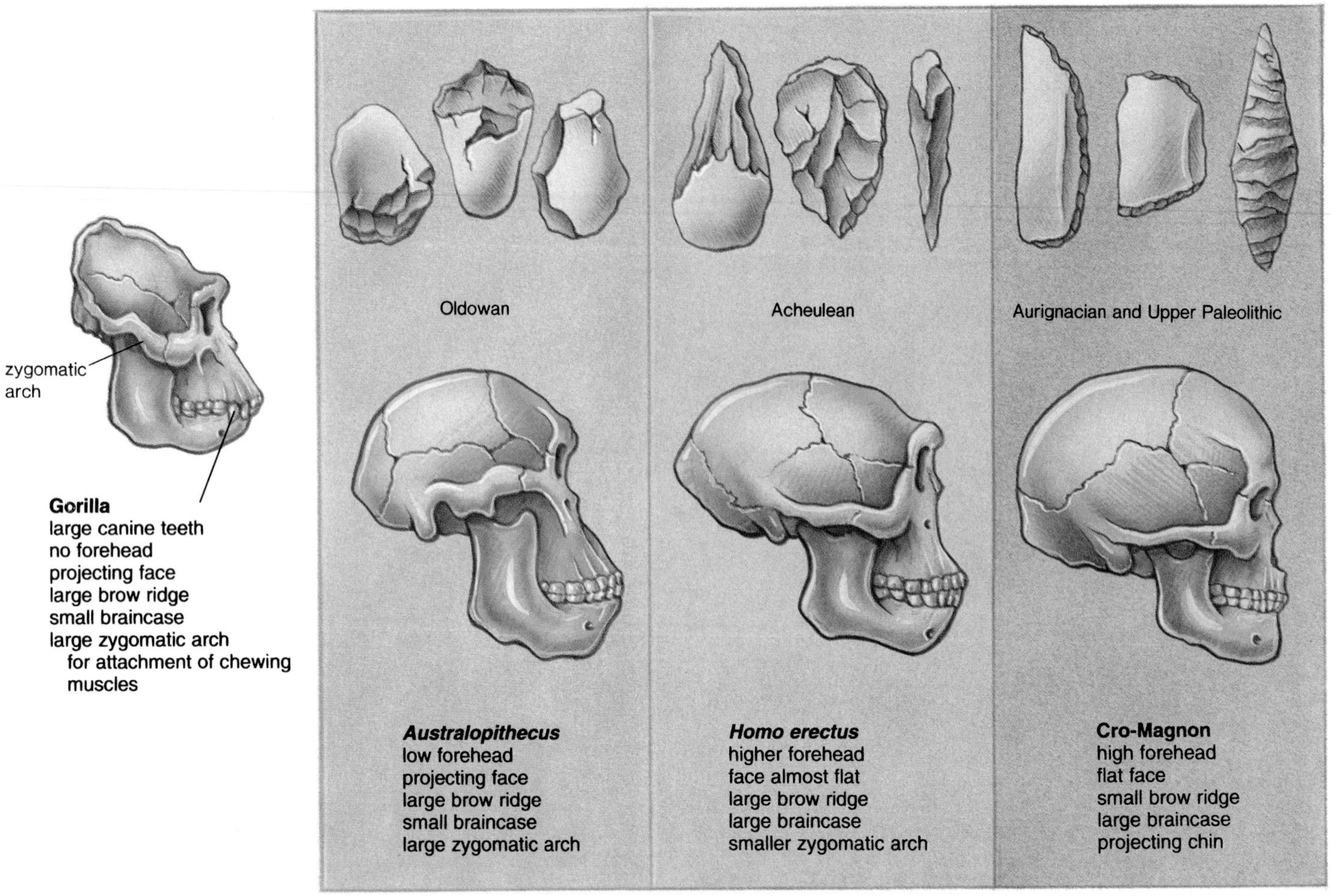

Figure 21.8

Comparative hominid skull anatomy and tools. Oldowan tools are crude; Acheulan tools are better made and more varied; Aurignacian tools are well designed for their specific purposes.

Hominids: What Separates Humans

The classification of humans (see table 21.1) indicates that modern humans are in the family Hominidae (**hominids**), along with other closely related forms, as depicted in figure 21.7.

Australopithecines (*Australopithecus*) are hominids. Their fossils, which date from about 4 million years ago, have been found in East Africa and South Africa. At this time, the weather was turning cooler and drier and the tropical forests were shrinking. The animals found with australopithecine remains are grassland, not forest, animals. The oldest australopithecine, called *A. afarensis,* was bipedal and walked erect. This can be deduced by a study of the skeletal bones; short but broad pelvic bones are evident, for example. The skull retained many apelike features, but the canine teeth are smaller than those of apes (fig. 21.8). The brain size is about that of an ape (400 cc), however, and there is no evidence of tools. It seems, then, that bipedalism was the first distinctly human trait to evolve.

On the basis of differences in the teeth, the jaws, and the skull, several other species of australopithecines have been named. In general, the body size of *A. africanus* was smaller than *A. robustus* and *A. boisei.* The larger forms had small front teeth and very large back teeth, indicating that they were most likely herbivores. Brain size in these later fossils increases to about 600 cc, and tools have been found with their bones.

Fossils now classified as ***Homo habilis*** date between 1.6 million and 2 million years ago and, like *Australopithecus,* are found in East Africa and South Africa. Although the face still projects, this species has a larger brain

(700 cc) and smaller teeth than those of many australopithecines. Some believe that these people were the first to fashion crude stone tools; in fact, the species' name means *handy man*. Other experts suggest that the fossils designated as *Homo habilis* are actually australopithecine fossils. They also say that tool production began with the australopithecines.

African fossils of ***Homo erectus,*** which have been dated at about 1.5 million years ago, indicate that at least some members of this species were about the size of modern humans and had our brain size (1,300 cc). The *Homo erectus* skull has a mixture of more advanced and primitive features (fig. 21.8). Skeletal features suggest greater endurance for walking and running than we have. The pelvis was narrower and the femur had a longer neck (portion between ball and shaft). A wider pelvis is not as adaptive for walking, but it accommodates the birth of offspring with larger brains and heads.

Homo erectus left Africa and ranged widely in Euroasia. Just when this occurred is being debated, but fossils have been found in Europe, China (called Peking man), and Java (called Java man). New techniques date Java man at the same age as *Homo erectus* in Africa. These hominids made more advanced tools, including hand axes, and it is possible they were big-game hunters. They had knowledge of fire and could have cooked their meat before eating it.

Fossils of the species *Homo sapiens neanderthalensis* are dated from about 100,000 years ago to about 35,000 years ago. The term **Neanderthal** is derived from the Neader Valley in Germany, where the species was first found. The Neanderthal skull resembles that of *Homo erectus,* except that Neanderthal's brain size (1,330 cc) is slightly larger than that of modern humans. The body is shorter and more massive than that of modern humans; the larger brain may correlate with the need for control over larger muscles. It is now believed that the Neanderthals stood fully erect, but they possibly moved more slowly than we do. The Neanderthals died out about 50,000–40,000 years ago for reasons unknown. Perhaps they interbred with or were replaced by *Homo sapiens sapiens.*

Where and when *Homo sapiens sapiens* evolved is also being researched. It's possible modern humans evolved only in Africa about 100,000 years ago, but new findings in China suggest that modern humans may have evolved there also. Fossils of modern humans are found in Africa, and also in Europe and Asia. Sometimes these fossils are collectively referred to as **Cro-Magnon,** after a cave site in southern France where remains were found. The facial features and brain size of *Cro-Magnon* distinctly resemble modern humans (fig. 21.8). Both Neanderthals and Cro-Magnon peoples were excellent hunters. Big-game hunting requires cooperation, which may have fostered societal living and the beginning of language. When the Neanderthals buried their dead, they seemed to have prepared them for a future life by supplying flint, tools, and food. Evidence of culture, however, is particularly evident with Cro-Magnon. For example, Cro-Magnon people painted beautiful drawings of animals on cave walls in Spain and France and sculpted many small figurines.

Studies based on mitochondrial DNA, which are now in dispute, suggest that modern humans originated in Africa and then spread out into Europe, giving rise to the various human races. Others believe that the human races originated in several geographical regions, but they became one species because of gene flow.

> We have seen that these features in particular distinguish humans from the apes:
>
> Bipedalism
> Flat (nonprojecting) face
> Brain size (400 cc for apes compared to 1,330 cc for humans)
> Toolmaking
> Language
> More complex culture

One Species, Several Races

All human races of today are also classified as *Homo sapiens sapiens* (fig. 21.9). This is consistent with the biological definition of species because it is possible for all types of humans to interbreed and bear fertile offspring. The close relationship between the races is supported by biochemical data showing that differences in amino acid sequence between 2 individuals of the same race are as great as those between 2 individuals of different races.

It is generally accepted that racial differences developed as adaptations to climate. Although it might seem as if dark skin is a protection against the hot rays of the sun, actually it is a protection against ultraviolet ray absorption. Dark-skinned persons living in southern regions and white-skinned persons living in northern regions absorb the same amount of radiation. (Some absorption is required for vitamin D production.)

Differences in body shape represent adaptations to temperature. A squat body with shortened limbs and nose retains more heat than an elongated body with longer limbs and nose. Also, the "almond" eyes, flattened nose and forehead, and broad cheeks of the Asian are believed to be adaptations to the extremely cold weather of the Ice Age.

Although it has always seemed to some that physical differences might warrant assigning human races to different species, this contention is not borne out by the biochemical data mentioned previously and by the fact that members of all races are able to reproduce with one another.

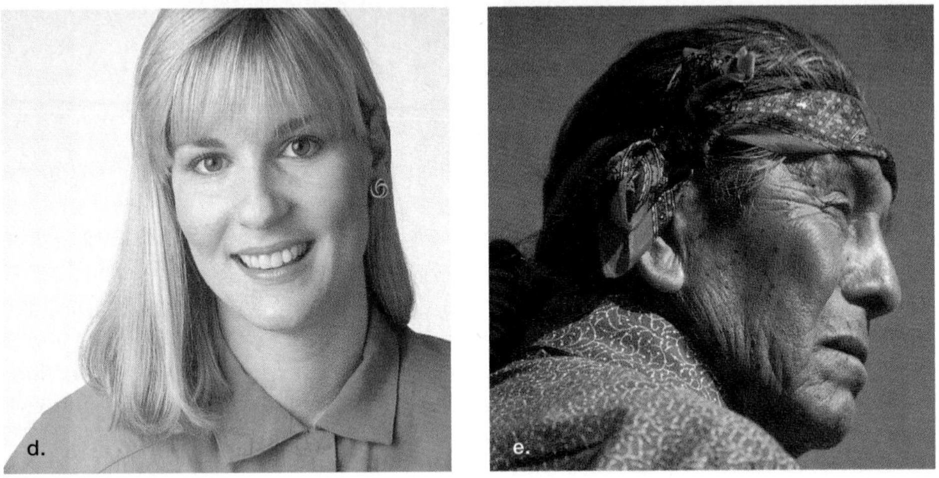

Figure 21.9

All human beings belong to one species, but there are several races. **a.** Negroid. **b.** Mongoloid. **c.** Australian. **d.** Caucasian. **e.** American Indian.

SUMMARY

The theory of evolution explains the history and diversity of life. Evidence for evolution can be taken from the fossil record, comparative anatomy, embryology, and biochemistry. Also, biogeography supports the occurrence of evolution.

Darwin not only presented evidence in support of organic evolution, he showed that evolution was guided by natural selection. Due to reproductive potential, there is a struggle for existence between members of the same species. Those members that possess variations more suited to the environment will most likely acquire more resources and so have more offspring than other members. Because of this natural selection process, there is a gradual change in species composition, which leads to adaptation to the environment.

Life came into being with the first cell(s), and thereafter diversification accounts for the evolution of all other life forms. The classification of humans reveals their evolutionary history:

Kingdom	Animalia	Includes all animals
Phylum	Chordata	Includes vertebrates: fishes, amphibians, reptiles, birds, and mammals
Class	Mammalia	Mammals: cats, dogs, horses, mice, sheep, cows, etc.
Order	Primates	Monkeys, apes, humans
Family	Hominidea	Hominids: forms closely related to humans, extinct humans, and modern humans
Genus	*Homo*	Extinct humans and modern humans
Species	*sapiens*	Modern humans

Humans are distinguishable from apes by their dendition, bipedalism, brain size, toolmaking, language, and culture. Only australopithecines and humans are hominids, a family designation. *Australopithecus afarensis* is a common ancestor to 3 other species of this genus and *Homo habilis,* the first fossil to be placed in the genus *Homo. Homo habilis* peoples may have retained apelike features and may not have been highly intelligent, but they did make tools.

Homo erectus had a large brain and walked as we do—perhaps better. These people used fire and probably were the first true hunters. *Homo sapiens neanderthalensis* were more massive than modern humans. The Neanderthals are believed to have interbred with or to have been replaced by Cro-Magnon, the first *Homo sapiens sapiens.* Cro-Magnon was an expert hunter. Hunting promoted language, socialization, and the development of culture.

Modern humans spread to all parts of the earth and became adapted to different climates. Even so, all human races belong to the same species, as witnessed by the fact they can all reproduce with one another.

Study and Thought Questions

1. Show that the fossil record, comparative anatomy, comparative embryology, vestigial structures, comparative biochemistry, and biogeography all give evidence of evolution. (pp. 426–428)

   Critical Thinking How does comparative anatomy support both common descent (organisms are related) and adaptation to the environment?
2. What are the 5 aspects of Darwin's theory of evolution? (p. 429)

   Ethical Issue Should all biology students be required to study Darwin's theory of evolution? Why or why not?
3. Describe the events that led to the origin of the first cell(s). (pp. 430–431)
4. Name several primate characteristics still retained by humans. (pp. 432–433)

   Critical Thinking Both apes and humans have opposable thumbs. What features of human anatomy have led to our greater use of the opposable thumb?
5. Draw a hominid evolutionary tree. (p. 433)

   Ethical Issue Should biology textbooks be required to mention the biblical account of human origins? Why or why not?
6. How do the australopithecines, *Homo erectus* and Cro-Magnon differ anatomically from one another and from apes? (pp. 436–437)
7. Which humans were tool users? walked erect? used fire? drew pictures? (pp. 436–437)
8. Name several races of humans. (p. 438)

   Critical Thinking What evidence do we have that all races of humans belong to the same species?

Writing Across the Curriculum

1. The evidence for evolution is not convincing to everyone. What types of problems are involved in producing convincing evidence?
2. Does the evolutionary process of evolution discussed on pages 428–429 apply to human beings today?
3. How does the size and life expectancy of humans today compare with humans of 500 years ago? Are these changes the result of evolution? Give reasons for your answer.

Objective Questions

Match the phrases in questions 1–4 with those in this key:

- **a.** biogeography
- **b.** fossil record
- **c.** biochemistry
- **d.** anatomy

1. Species change over time.
2. Forms of life are variously distributed.
3. A group of related species has a unity of plan.
4. The same types of molecules are found in all living things.
5. Evolutionary success is judged by ____________ success, or the number of offspring.
6. The end result of natural selection is ____________ to the environment.
7. A ____________ evolution is believed to have produced the first cell(s).
8. The australopithecines could probably walk ____________, but they had a ____________ brain.
9. The 2 varieties of *Homo sapiens* from the fossil record are ____________ and ____________.
10. Complete this table to describe the classification of humans:

Category	Name	Examples
Kingdom	________	________
Phylum	________	________
Class	________	________
Order	________	________
Family	________	________
Genus	________	________
Species	________	________

Selected Key Terms

adaptation The fitness of an organism for its environment, including the process by which it becomes fit and is able to survive and reproduce. 429

australopithecine (aw″ strah-lo-pith´ə-sin) The first generally recognized hominid. 436

common ancestor An ancestor to 2 or more branches of evolution. 427

Cro-Magnon The common name for the first fossils to be accepted as representative of modern humans. 437

fossil Any remains of an organism that have been preserved in the earth's crust. 426

hominid (hom ĭ-nid) Common term for member of a family of upright, bipedal primates that includes australopithecines and humans. 436

Homo erectus The earliest nondisputed species of humans, named for their erect posture, which allowed them to walk as we do. 437

Homo habilis An extinct species that may include the earliest humans, having a small brain but made quality tools. 436

natural selection The process by which populations become adapted to their environment. 428

Neanderthal The common name for an extinct subspecies of humans whose remains are found in Europe and Asia. 437

primate Animal that belongs to the order Primates; the order of mammals that includes monkeys, apes, and humans. 432

taxonomy The science of naming and classifying organisms. 426

vestigial structure The remains of a structure that was functional in some ancestor but is no longer functional in the organism in question. 427

Chapter 22

Ecosystems

Ecologists like to use the catchy phrase "everything is connected to everything else." By this, they mean that you cannot affect one part of the environment without affecting another part. For example, would you suppose that trees have anything to do with the daily environmental temperature? Well, they do and in perhaps an unexpected way. The accumulation of carbon dioxide (and other gases) in the air acts like the glass of a greenhouse and allows solar heat to be trapped near the surface of the earth. Environmentalists predict a gradual rise in the earth's daily temperature, especially if we continue to destroy the world's tropical rain forests. They act like a giant sponge that absorbs carbon dioxide, slowing down the greenhouse effect caused by all the carbon dioxide we pour into the environment from the burning of fossil fuels.

Chapter Concepts

Chapter Outline

a. b. c.

Figure 22.1

Three major communities in the United States, each containing its own mix of plants and animals. Temperature and amount of rainfall largely determine which type of community is found. **a.** A deciduous forest is typical of eastern United States. **b.** A prairie community is found in the Midwest. **c.** A semidesert is located in the Southwest.

Table 22.1

Ecological Terms

Term	Definition
Ecology	Study of the interactions of organisms with each other and with the physical environment
Population	All the members of the same species that inhabit a particular area
Community	All the populations that are found in a particular area
Ecosystem	A community and its physical environment; has nonliving (abiotic) and living (biotic) components
Biosphere	The portion of the surface of the earth (air, water, and land) where living things exist

Ecosystems

When the earth was first formed, the outer crust was covered by ocean and barren land. Over time, plants colonized the land so that eventually it supported many complex communities of living things. A community is made up of all the populations in a particular area (table 22.1). When we study a community, we are considering only the populations of organisms that make up that community, but when we study an **ecosystem,** we are concerned with the community and its physical environment. **Ecology** is the study of the interactions of organisms with each other and with the physical environment.

Succession: Change Over Time

Succession is a sequence of communities replacing one another in an orderly and predictable sequence, until finally there is a *climax community,* a mix of plants and animals typical of that area that remains stable year after year. For example, in the United States not too long ago, a deciduous forest was typical of the East, a prairie was common to the Midwest, and a semidesert covered the Southwest (fig. 22.1).

Primary succession is a series of sequential events by which bare rock becomes capable of sustaining many organisms. *Secondary succession* also is observed when a climax community that has been disturbed once again takes on its former state. Figure 22.2 shows a possible secondary succession for abandoned farmland in the eastern United States.

> The process of succession leads to a climax community, which contains the plants and animals characteristic of the area.

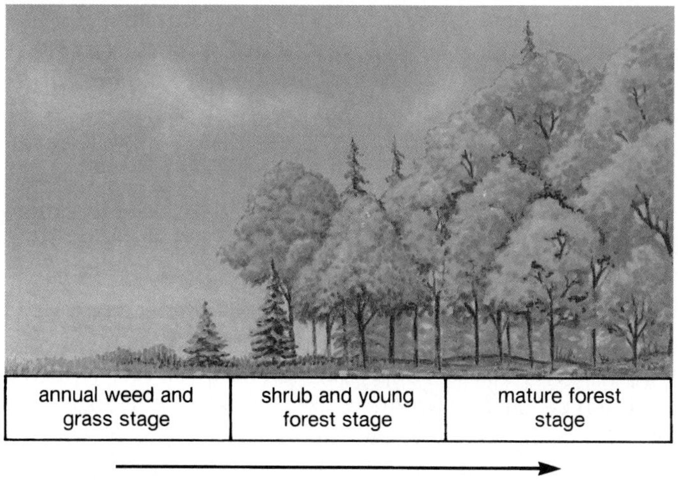

Figure 22.2

Secondary succession. This drawing shows a possible sequence of events by which abandoned farmland becomes a climax community again. During secondary succession in the Northeast, grass and weeds are followed by shrubs and trees and finally by a mature climax forest.

Community Composition: Each has a Role

Each population in an ecosystem has a habitat and a niche. The **habitat** of an organism is its place of residence (i.e., where it can be found), such as under a log or at the bottom of the pond. The **niche** of an organism is its profession or total role in the community. A description of an organism's niche includes its interactions with the physical environment and with the other organisms in the community (table 22.2). One important aspect of niche is how the organism acquires its food.

Producers are autotrophic organisms with the ability to carry on photosynthesis and to make food for themselves (and indirectly for the other populations as well). In terrestrial ecosystems, the producers are predominantly green plants, while in freshwater and saltwater ecosystems, the dominant producers are various species of algae.

Table 22.2

Aspects of Niche

Plants	Animals
Season of year for growth and reproduction	Time of day for feeding and season of year for reproduction
Sunlight, water, and soil requirements	Habitat and food requirements
Relationships with other organisms	Relationships with other organisms
Effect on abiotic environment	Effect on abiotic environment

Consumers are heterotrophic organisms that use preformed food. It is possible to distinguish 4 types of consumers, depending on their food source. **Herbivores** feed directly on green plants; they are termed *primary consumers.* **Carnivores** feed only on other animals and are, therefore, secondary or tertiary consumers. **Omnivores** feed on both plants and animals. Therefore, a caterpillar feeding on a leaf is a herbivore; a green heron feeding on a fish is a carnivore; a human eating both leafy green vegetables and beef is an omnivore. **Decomposers**, such as bacteria and fungi, are organisms of decay that break down detritus, nonliving organic matter. When detritus is broken down, inorganic molecules are released and these can be used by producers. In this way, the same chemical elements can be used over and over again in an ecosystem.

When we diagram the components of an ecosystem, as in figure 22.3, it is possible to illustrate that every ecosystem is characterized by 2 fundamental phenomena: energy flow and chemical cycling. Energy flow begins when producers absorb solar energy, and chemical cycling begins when producers take in inorganic nutrients from the physical environment. Thereafter, producers make food for themselves and indirectly for the other populations of the ecosystem. Energy flow occurs because all the energy content of organic food eventually is lost to the environment as heat. Therefore, most ecosystems cannot exist without a continual supply of solar energy. However, the original inorganic elements are cycled back to the producers, and no new input is required.

Within an ecosystem, energy flows and chemicals cycle.

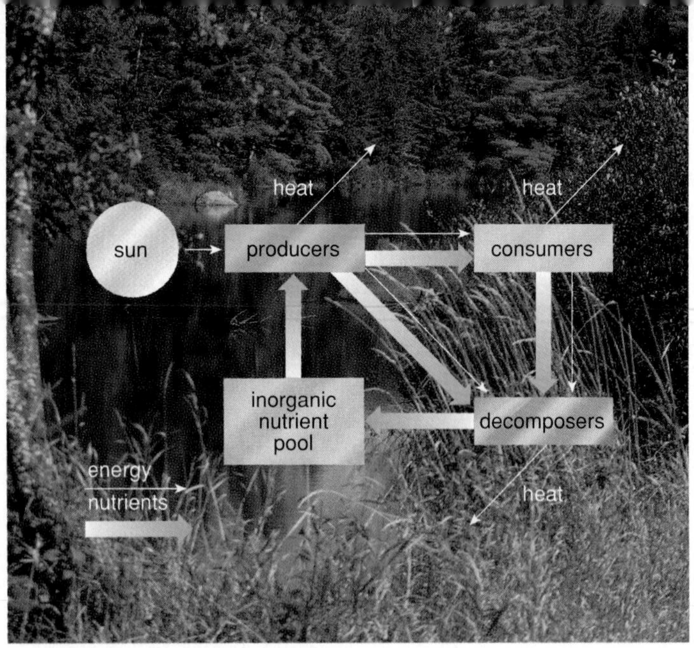

Figure 22.3

Ecosystem composition. The diagram illustrates energy flow and chemical cycling through an ecosystem. Energy does not cycle because all the energy that is derived from the sun eventually dissipates as heat.

Energy Flow: Till There is None

Energy flows through an ecosystem because when one form of energy is transformed into another form, there is always a loss of some usable energy as heat. For example, the conversion of energy in one molecule of glucose to 36 molecules of ATP represents only 50% of the available energy in a glucose molecule. The rest is lost as heat. This means that as one population feeds on another and as decomposers work on detritus, all of the captured solar energy that was converted to chemical-bond energy by algae and plants is returned to the atmosphere as heat. Therefore, energy flows through an ecosystem and does not cycle.

Food Chains: Join and Overlap

A **food chain** indicates who eats whom in an ecosystem. Figure 22.4 depicts examples of a terrestrial food chain and an aquatic food chain. It is important to realize that each represents just one path of energy flow through an ecosystem. Natural ecosystems have numerous food chains, each linked to others to form a complex **food web.** For example, figure 22.5 shows a deciduous forest ecosystem in which plants are eaten by a variety of insects, and in turn, these are eaten by several different birds, while any one of the latter may be eaten by a larger bird, such as a hawk. Therefore, energy flow is better described in terms of **trophic** (feeding) **levels,** each one further removed from the producer population, the first (photosynthetic) trophic level. All animals acting as primary consumers are part of a second trophic level, and all animals acting as secondary consumers are part of the third level, and so on.

The populations in an ecosystem form food chains, in which the producers make food for the other populations, which are consumers. While it is convenient to study food chains, the populations in an ecosystem actually form a food web, in which food chains join with and overlap one another.

One of the food chains depicted in figure 22.4 is part of the forest food web shown in figure 22.5, and the other food chain is part of the aquatic food web from the freshwater pond ecosystem shown in figure 22.6. Both of these are called *grazing food chains* because the primary consumer feeds on a photosynthesizer. In some ecosystems (forests, rivers, and marshes), the primary consumer feeds mostly on detritus. The *detritus food chain* accounts for more energy flow than the grazing food chain whenever most organisms die before they are eaten. In the forest, an example of a detritus food chain is

detritus → soil bacteria → earthworms

A detritus food chain is often connected to a grazing food chain, as when earthworms are eaten by a robin. Eventually, however, as dead organisms decompose, all the solar energy that was taken up by the producer populations dissipates as heat. Therefore, energy does not cycle.

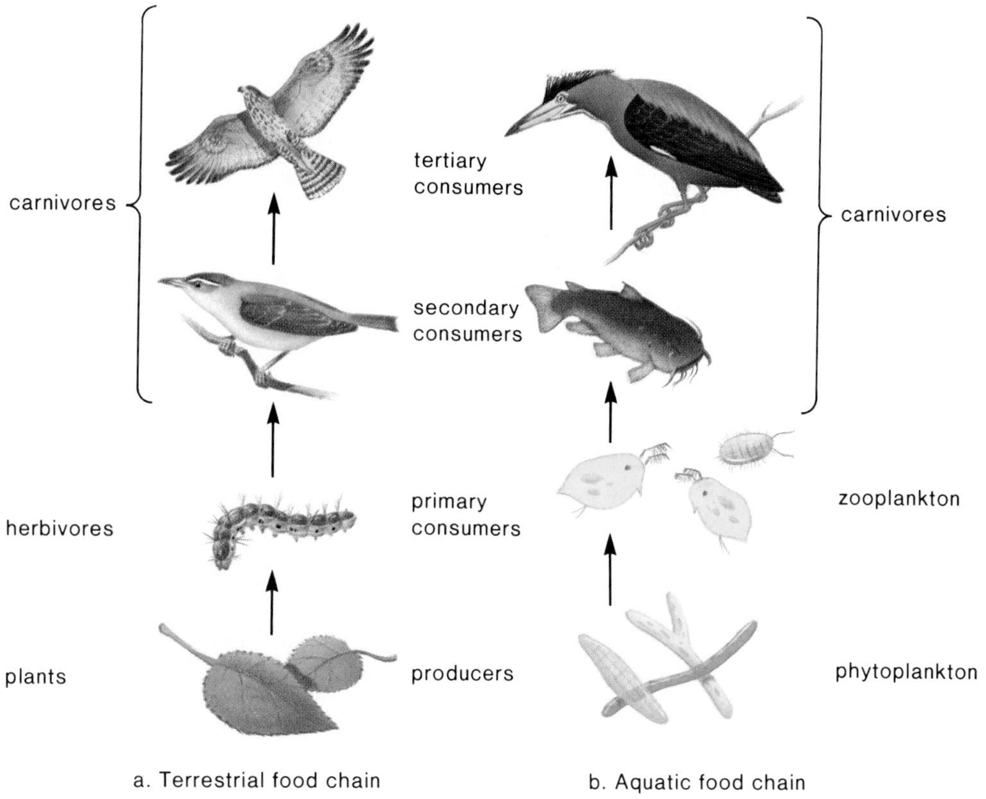

Figure 22.4

Examples of food chains. **a.** Terrestrial. **b.** Aquatic.

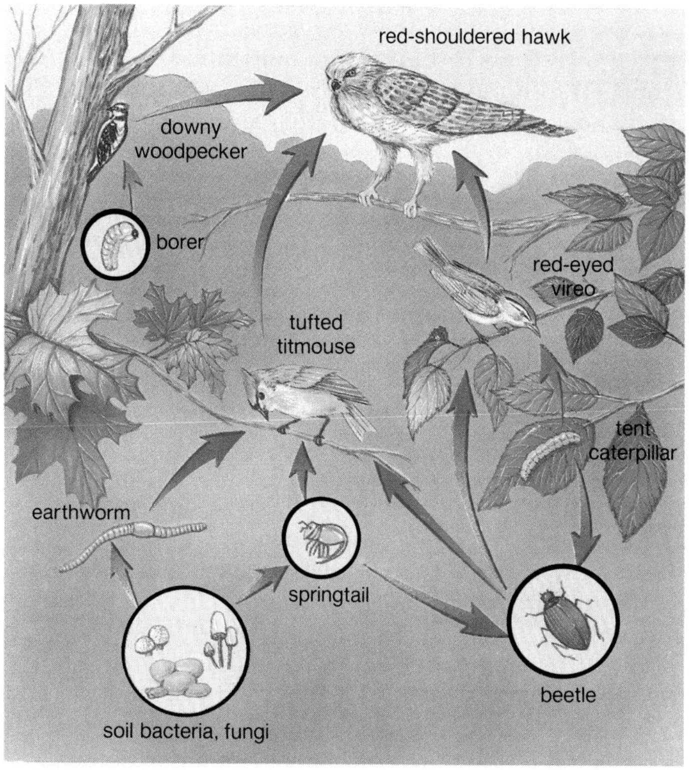

Figure 22.5

A deciduous forest ecosystem. The arrows indicate the flow of energy in a food web.

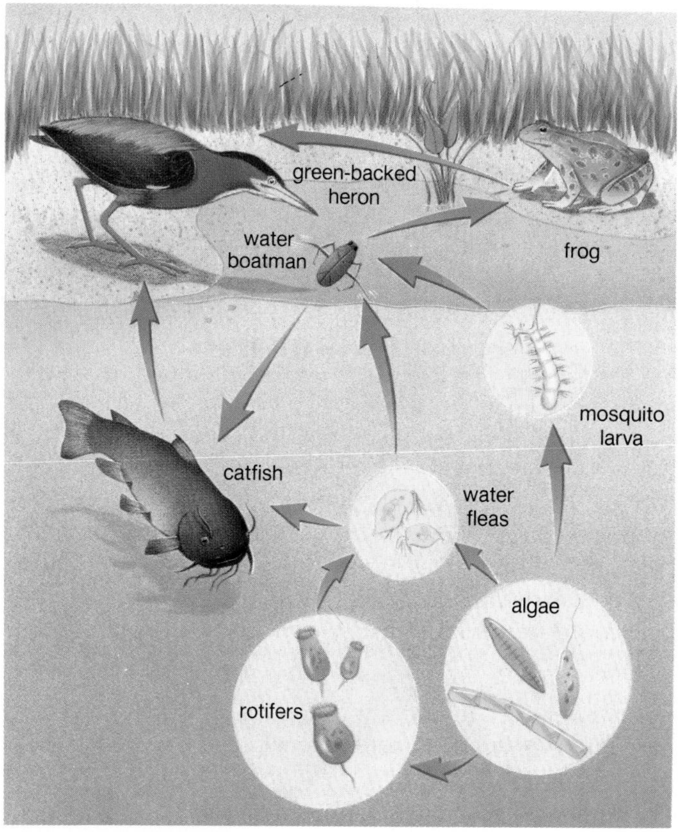

Figure 22.6

A freshwater pond ecosystem. The arrows indicate the flow of energy in a food web.

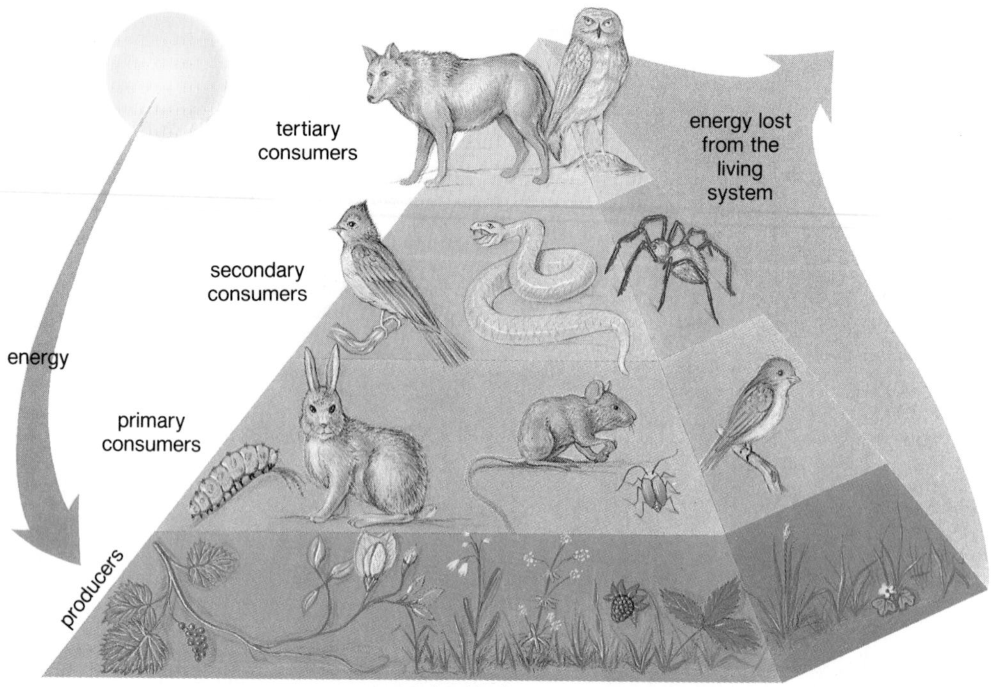

Figure 22.7

Pyramid of energy. At each step in the pyramid, an appreciable portion of energy originally trapped by the producer is dissipated as heat. Accordingly, organisms in each trophic level pass on less energy than they received.

Ecological Pyramids: Several Kinds

The trophic structure of an ecosystem can be summarized in the form of an **ecological pyramid.** The base of the pyramid represents the producer trophic level, and the apex is the highest-level consumer, called the top predator. The other consumer trophic levels are in between the producer and the top-predator levels.

There are 3 kinds of pyramids. One is a *pyramid of numbers,* based on the number of organisms at each trophic level. A second is the *pyramid of biomass. Biomass* is the weight of living material at some particular time. To calculate the biomass for each trophic level, an average weight for the organisms at each level is determined and then the number of organisms at each level is estimated. Multiplying the average weight by the estimated number gives the approximate biomass for each trophic level. A third pyramid, the *pyramid of energy* (fig. 22.7), shows that there is a decreasing amount of energy available at each successive trophic level. Less energy is available at each step for the following reasons:

1. Only a certain amount of food is captured and eaten by the next trophic level.
2. Some of the food that is eaten cannot be digested and exits the digestive tract as waste.
3. Only a portion of the food that is digested becomes part of the organism's body. The rest is used as a source of energy.

In regard to the last point, we have to realize that a significant portion of food molecules is used as an energy source for ATP buildup in mitochondria. This ATP is needed to build the proteins, carbohydrates, and lipids that compose the body. ATP is also needed for such activities as muscle contraction, nerve conduction, and active transport.

The energy considerations associated with ecological pyramids have implications for the human population. It is generally stated that only about 10% of the energy available at a particular trophic level is incorporated into the tissues of animals at the next level. This being the case, it can be estimated that 100 kilograms of grain could, if consumed directly, result in 10 human kilograms; however, if fed to cattle, the 100 kilograms of grain would result in only 1 human kilogram. Therefore, a larger human population can be sustained by eating grain than by eating grain-fed animals. However, humans generally need some meat in their diet, because this is the most common source of the essential amino acids, as discussed in chapter 4.

> In a food web, each successive trophic level has less total energy content. This is because some energy is lost at each level and so is not transferred from one trophic level to the next.

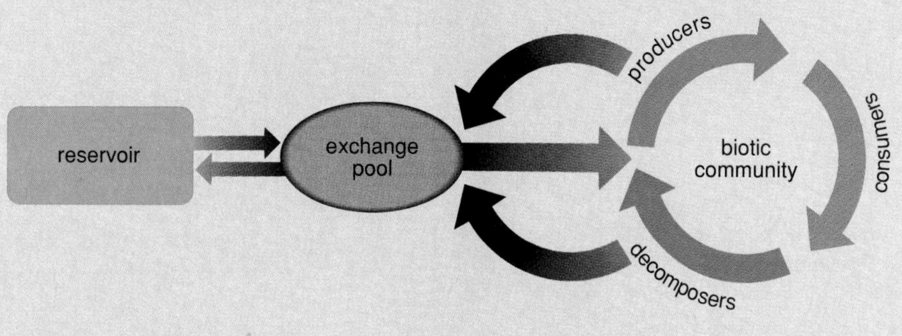

Figure 22.8

Components of a chemical cycle. The reservoir stores the chemical, and the exchange pool makes it available to producers. The chemical then cycles through food chains, which are the biotic community. Decomposition returns the chemical to the exchange pool once again if it has not already returned by another process. (See figs. 22.9 and 22.10.)

Chemical Cycling

In contrast to energy, inorganic nutrients (chemicals) do cycle through large natural ecosystems. Because there is minimal input from the outside, the various chemical elements essential for life are used over and over. For each element, the cycling process (fig. 22.8) involves (1) a reservoir—that portion of the earth that acts as a storehouse for the element; (2) an exchange pool—that portion of the environment from which the producers take their nutrients; and (3) the biotic community, through which elements move along food chains to and from the exchange pool.

Carbon Cycle: Photosynthesizing and Respiring

The relationship between photosynthesis and aerobic cellular respiration should be kept in mind when discussing the carbon cycle. Recall that for simplicity's sake, this equation in the forward direction represents aerobic cellular respiration, and in the other direction, it is used to represent photosynthesis.

$$C_6H_{12}O_6 + 6O_2 \underset{\text{photosynthesis}}{\overset{\text{aerobic cellular respiration}}{\rightleftharpoons}} 6CO_2 + 6H_2O$$

The equation tells us that aerobic cellular respiration releases carbon dioxide (CO_2), the molecule needed for photosynthesis. However, photosynthesis releases oxygen (O_2), the molecule needed for aerobic cellular respiration. Animals are dependent on green organisms, not only to produce organic food and energy but also to supply the biosphere with oxygen. However, since producers both photosynthesize and respire, they can function independently of the animal world.

In the carbon cycle, organisms in both terrestrial and aquatic ecosystems (fig. 22.9) exchange carbon dioxide with the atmosphere. On land, plants take up carbon dioxide from the air, and through photosynthesis, they incorporate carbon into food that is used by themselves and heterotrophs alike. When any organism respires, a portion of this carbon is returned to the atmosphere as carbon dioxide.

In aquatic ecosystems, the exchange of carbon dioxide with the atmosphere is indirect. Carbon dioxide from the air combines with water to give carbonic acid, which breaks down to bicarbonate ions (HCO_3^-). Bicarbonate ions are a source of carbon for algae, which produce food for themselves and for heterotrophs. Similarly, when aquatic organisms respire, the carbon dioxide they give off becomes bicarbonate. The amount of bicarbonate in the water is in equilibrium with the amount of carbon dioxide in the air.

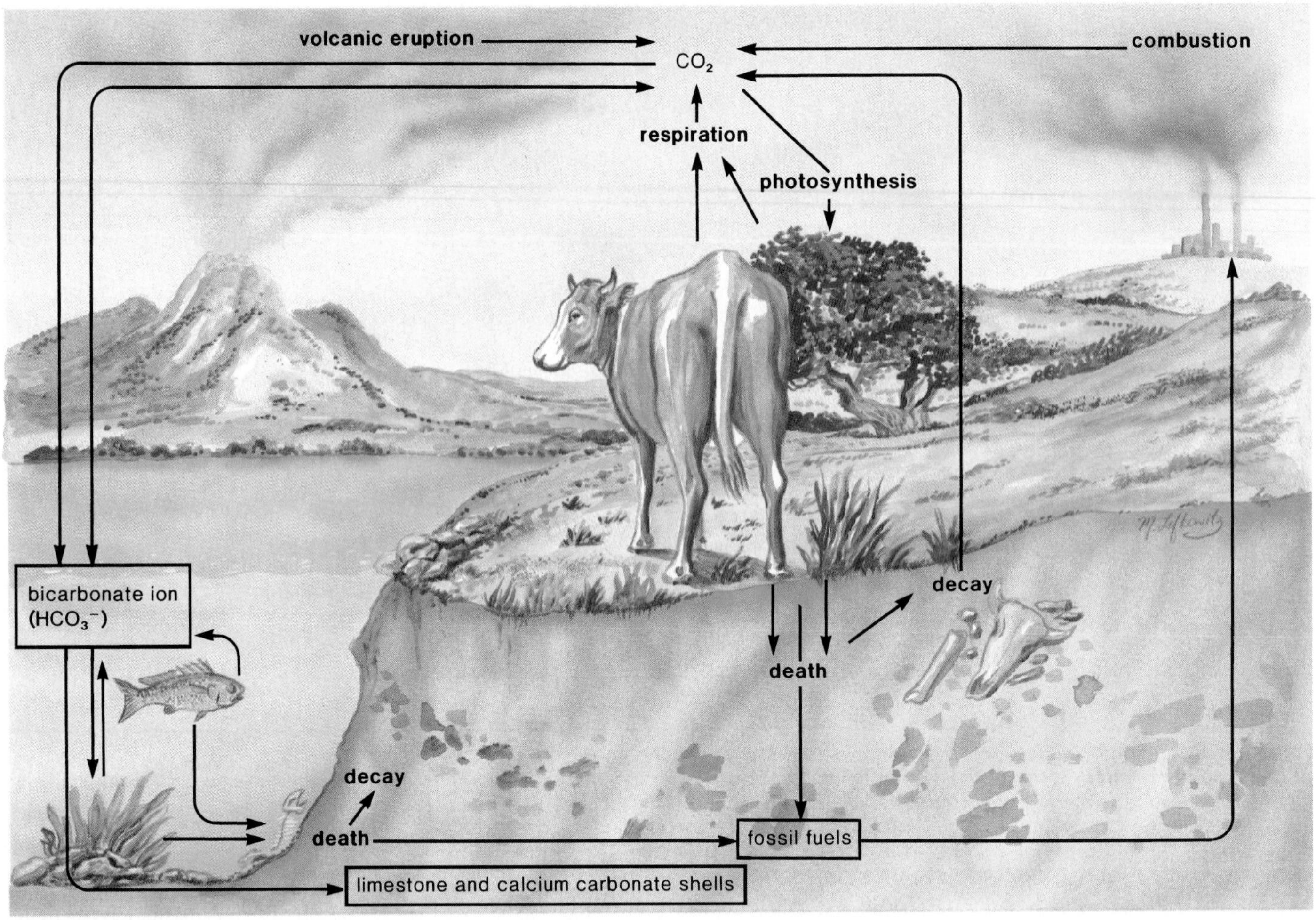

Figure 22.9

Carbon cycle. Photosynthesizers take up carbon dioxide (CO_2) from the air or bicarbonate ion (HCO_3^-) from the water. They and all other organisms return carbon dioxide to the environment. The carbon dioxide level is also increased when volcanoes erupt and fossil fuels are burned. Presently, the oceans are a primary reservoir for carbon in the form of limestone and calcium carbonate shells.

Carbon Reservoirs: The World's Trees, for Example

Living and dead organisms contain organic carbon and serve as one of the reservoirs for the carbon cycle. The world's biota (all living things), particularly trees, contain 800 billion tons of organic carbon, and an additional 1,000–3,000 billion tons are estimated to be held in the remains of plants and animals in the soil. Before decomposition could occur, some of these remains were subjected to physical processes that transformed them into coal, oil, and natural gas. We call these materials the fossil fuels. Most of the fossil fuels were formed during the Carboniferous period, 280–350 million years ago, when an exceptionally large amount of organic matter was buried before decomposing. Another reservoir is the calcium carbonate that accumulates in limestone and in calcium carbonate shells. The oceans abound with organisms, some microscopic, that have calcium carbonate shells. After these organisms die, their shells accumulate in ocean bottom sediments. Limestone is formed from these sediments by geological transformation.

Figure 22.10

Burning of trees in a tropical rain forest. When trees are burned, carbon dioxide (CO_2) is released from one of the reservoirs in the carbon cycle.

How People Influence the Carbon Cycle

The activities of human beings have increased the amount of carbon dioxide (CO_2) and other gases in the atmosphere. Data from monitoring stations recorded an increase of 20 ppm (parts per million) in carbon dioxide in only 22 years. (This is equivalent to 42 billion tons of carbon.) This buildup is attributed primarily to the burning of fossil fuels and the destruction of the world's tropical rain forests (fig. 22.10). When forests are destroyed, a reservoir that takes up excess carbon dioxide is reduced. At this time, the oceans are believed to be taking up most of the excess carbon dioxide; the burning of fossil fuels in the last 22 years has probably released 78 billion tons of carbon, yet the atmosphere registers an increase of "only" 42 billion tons.

As discussed on page 470, there is much concern that an increased amount of carbon dioxide (and other gases) in the atmosphere is causing global warming. These gases allow the sun's rays to pass through to the earth, but they also absorb and reradiate heat to the earth, a phenomenon called the *greenhouse effect.*

> In the carbon cycle, carbon dioxide is removed from the atmosphere by photosynthesis but is returned by aerobic cellular respiration and burning of fossil fuels. Living things and dead matter are carbon reservoirs. The oceans, because they abound with calcium carbonate shells and limestone, are also major carbon reservoirs.

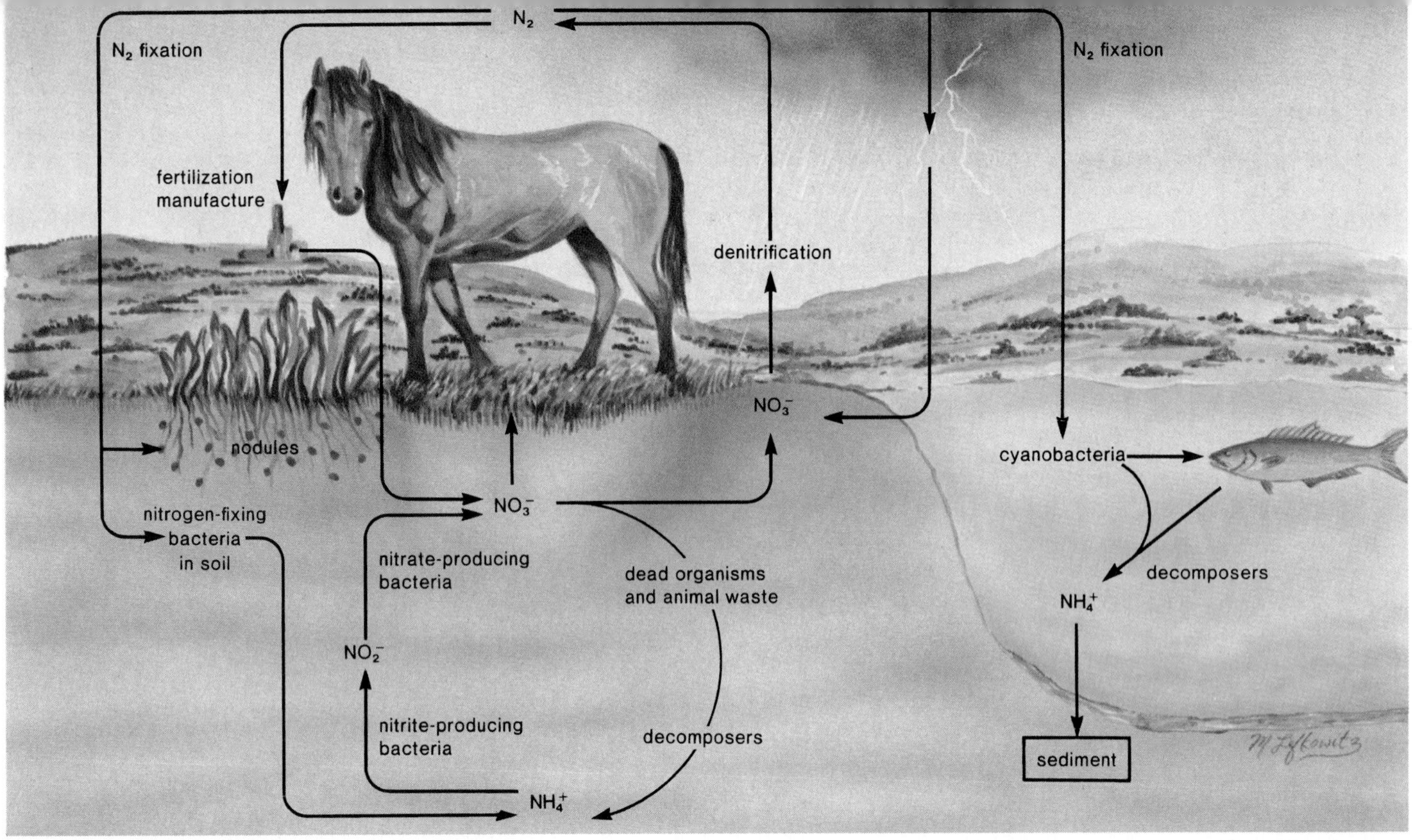

Figure 22.11

Nitrogen cycle. Several types of bacteria are at work: nitrogen-fixing bacteria reduce nitrogen gas (N_2); nitrifying bacteria, which include both nitrite-producing and nitrate-producing bacteria, convert ammonium (NH_4^+) to nitrate; and the denitrifying bacteria convert nitrate back to nitrogen gas. Humans contribute to the cycle by using nitrogen gas to produce nitrate for fertilizers.

Nitrogen Cycle: Important Bacteria

Nitrogen is an abundant element in the atmosphere. Nitrogen (N_2) makes up about 78% of the atmosphere by volume, yet nitrogen deficiency commonly limits plant growth. Plants cannot incorporate nitrogen into organic compounds and therefore depend on various types of bacteria to make nitrogen available to them (fig. 22.11).

Nitrogen Fixation: Bacteria Reduce Gas

Nitrogen fixation occurs when nitrogen (N_2) is reduced and added to organic compounds. Some cyanobacteria in aquatic ecosystems and some free-living bacteria in soil are able to reduce nitrogen gas to ammonium (NH_4^+). Other *nitrogen-fixing bacteria* infect and live in nodules on the roots of legumes (fig. 22.11). They make reduced nitrogen and organic compounds available to the host plant.

Nitrogen fixation also occurs after plants take up nitrates (NO_3^-) from the soil and produce amino acids and nucleic acids.

Nitrification Almost Balances Denitrification

Nitrification is the production of nitrates. Nitrogen gas (N_2) is converted to nitrate (NO_3^-) in the atmosphere when cosmic radiation, meteor trails, and lightning provide the high energy needed for nitrogen to react with oxygen. Also, humans make a most significant contribution to the nitrogen cycle when they convert nitrogen gas to nitrate for use in fertilizers.

Ammonium (NH_4^+) in the soil is converted to nitrate by certain soil bacteria in a 2-step process. First, nitrite-producing bacteria convert ammonium to nitrite (NO_2^-), and then nitrate-producing bacteria convert nitrite to nitrate. These 2 groups of bacteria are called the *nitrifying bacteria.* Notice the subcycle in the nitrogen cycle that involves only ammonium, nitrites, and nitrates. This subcycle does not depend on the presence of nitrogen gas at all (fig. 22.11).

Denitrification is the conversion of nitrate to nitrogen gas. There are *denitrifying bacteria* in both aquatic and terrestrial ecosystems. Denitrification counterbalances nitrogen fixation but not completely. More nitrogen fixation occurs, especially due to fertilizer production.

> In the nitrogen cycle: nitrogen-fixing bacteria (in nodules and in the soil) reduce nitrogen gas, and thereafter nitrogen can be incorporated into organic compounds; nitrifying bacteria convert ammonium to nitrate; denitrifying bacteria convert nitrate back to nitrogen gas.

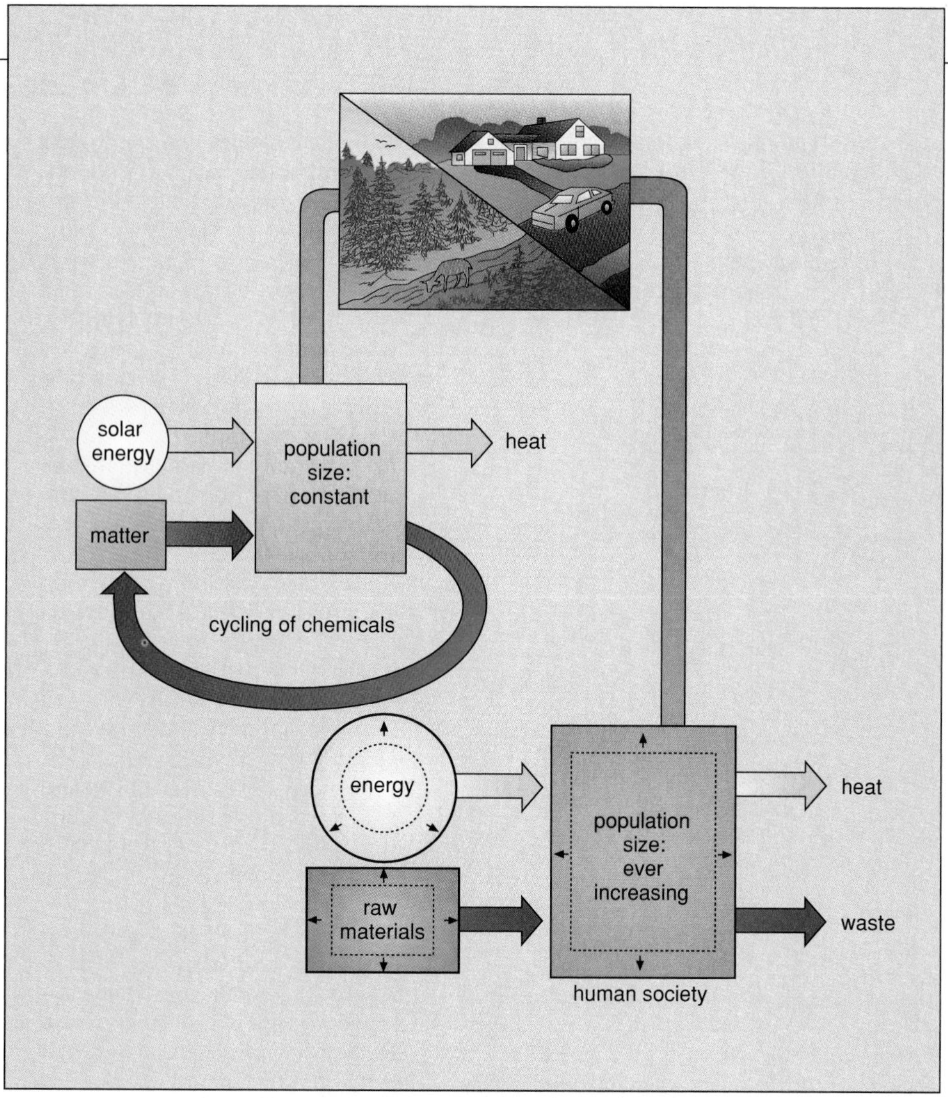

Figure 22.12

Human ecosystem versus a natural ecosystem. In natural ecosystems, population sizes remain about the same year after year, and materials cycle and energy is used efficiently. In the human ecosystem, the population size consistently increases, resulting in much pollution because of inadequate cycling of materials and inefficient use of supplemental energy.

Human Ecosystem

Mature natural ecosystems tend to be stable. The sizes of the many and varied populations are held in check by the interactions between species, such as predation and parasitism; the energy that enters and the amount of matter that cycles is appropriate to support these populations. **Pollution,** defined as any undesirable change in the environment that can be harmful to humans and other life, does not normally occur. Human beings have replaced natural ecosystems with one of their own making, as depicted in figure 22.12. This ecosystem essentially has 2 parts: the *country,* where agriculture and animal husbandry are found, and the *city,* where most people live and where industry is located. This representation of the human ecosystem, although simplified, allows us to see that the system requires 2 major inputs: *fossil fuel energy* and *raw materials* (e.g., metals, wood, synthetic materials). The use of these necessarily results in *pollution* and *waste.*

Just as the city is not self-sufficient and depends on the country to supply it with food, so the whole human ecosystem is dependent on the natural ecosystems to provide resources and to absorb wastes. Fuel combustion by-products, sewage, fertilizers, pesticides, and solid wastes are all added to natural ecosystems in the hope that these systems will cleanse the biosphere of these pollutants. But we have replaced natural ecosystems with our human ecosystem and have exploited natural ecosystems for resources, adding even more pollutants, to the extent that the remaining natural ecosystems are overloaded.

How to Save Natural Ecosystems

The contrast between the instability of the human ecosystem and the stability of the natural ecosystem suggests it is in the best interests of the human species to preserve what is left of natural ecosystems. This is often called "working with nature" rather than against nature. It is certainly possible, and it may prove essential to our survival, to stop working against nature and start working in harmony with our natural environment. This principle is explored in the Ecology Focus for this chapter.

Ecology Focus

Preservation of the Everglades

Originally, the Everglades encompassed the whole of southern Florida from Lake Okeechobee down to Florida Bay (fig. 22A). Now largely in the Everglades National Park alone do we find the vast saw grass prairie, interrupted occasionally by a cypress dome or hardwood tree island (fig. 22B). Within these islands, both temperate and tropical evergreen trees grow amongst the dense and tangled vegetation. Mangrove, or salt tolerant trees, are found along sloughs (creeks) and at the shoreline. Only the roots of the red mangrove can tolerate the sea constantly (fig. 22C). The prop roots of this tree protect over 40 different types of juvenile fishes as they grow to maturity. During the wet season, from May to November, animals are dispersed throughout the region, but in the dry season, from December to April, they congregate wherever pools of water are found. Alligators are famous for making "gator holes" where water collects (fig. 22D), and fish, shrimps, crabs, birds, and a host of living things survive until the rains come again. Almost everyone is captivated by the birds that find the ready supply of fish they need for daily existence at these holes. The large and beautiful herons, egrets, roseate spoonbill, and anhinga are awesome. These birds once numbered in the millions; now they number only in the thousands. Why is this?

At the turn of the century, settlers began to drain the land just south of Lake Okeechobee to grow crops on the soil enriched by partially decomposed saw grass. The large dike that now rings the lake prevents the water from taking its usual course: over the banks of Lake Okeechobee and slowly southward. In times of flooding, water can be shunted through the St. Lucie Canal to the Atlantic Ocean or through the canalized Caloosahatchee River to the Gulf of Mexico. In times of drought, water is contained not only in the lake but also in 3 so-called conservation areas established to the south of the lake. Water must be conserved for the irrigation of the farmland and to recharge the Biscayne aquifer (underground river), which supplies drinking water for the cities on the east coast of Florida. Containing and moving the water from place to place has required the construction of over 2,250 kilometers of canals, 125 water control stations, and 18 large pumping stations. Now the Everglades National Park receives water only when it is discharged artificially from a conservation area. This disruption of the natural flow of water has affected the reproduction pattern of the birds, which is attuned to the natural wet-dry season turnover.

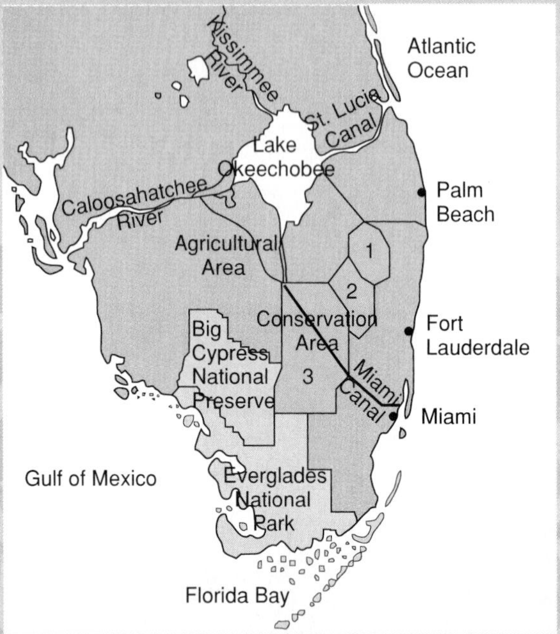

Figure 22A

The Everglades once extended from Lake Okeechobee south to Florida Bay. Now it only encompasses 3 conservation areas and Everglades National Park.

It took considerable human effort and a huge financial investment to control nature and to establish the Everglades Agricultural Area. Has this attempt to bend nature to human will been worthwhile? The area does, in fact, produce more sugar than Hawaii and a large proportion of the vegetables consumed in the United States each winter. But this has not been without a price. The rich soil, built up over thousands of years, is disappearing and most likely will be unable to sustain conventional agriculture after the year 2000. It has been suggested that *at that time* we might use the Everglades Agricultural Area for the growth of aquatic plants. Perhaps it should have been decided in the beginning to *work with nature* by growing aquatic plants instead of conventional plants. Then all the canals and pumping stations would have been unnecessary, the water would still flow from Lake Okeechobee to the glades as it had for eons, and the birds today would still number in the millions.

Figure 22B

The heart of the Everglades is a vast saw grass prairie interrupted occasionally by hardwood tree islands that contain a variety of tropical and temperate trees.

Figure 22C

Prop roots of the red mangrove provide protective cover for fishes and other sea life during maturation.

Figure 22D

An alligator beside his "gator hole," which also supplies food for the great egret in the background.

Another way to preserve natural ecosystems is to stop the steady increase in size of the human ecosystem and change it so that it is more cyclical. This approach would require us to achieve zero population growth and to conserve raw materials and energy.

Conservation can be achieved in 3 ways: (1) wise use of only what is actually needed; (2) recycling of nonfuel minerals, such as iron, copper, lead, and aluminum; and (3) use of renewable energy resources and development of more efficient ways to utilize all forms of energy. As a practical example, consider the plant built in Lamar, Colorado, that produces methane from feedlot animals' wastes. The methane is burned in the city's electrical power plant, and the heat given off, is used to incubate the anaerobic digestion process that produces the methane. In addition, a protein feed supplement is produced from the residue of the digestion process. This system represents a cyclical use of material and an efficient use of energy, similar to that found in nature. Many other such processes for achieving this end have been and will be devised. However, as long as the human ecosystem on the whole remains inefficient and noncyclical, it will continue to cause pollution.

Summary

The process of succession from either bare rock or disturbed land results in a climax community. Each population in an ecosystem has a habitat and a niche. Some populations are producers and some are consumers. Energy flow and chemical cycling are important aspects of ecosystems.

Food chains are paths of energy flow through an ecosystem. Both grazing food chains and detritus food chains exist. Eventually, the very same chemicals are made available to producer populations, but the energy dissipates as heat. The various food chains form an intricate food web, in which there are various trophic (feeding) levels. A pyramid of energy illustrates that each succeeding trophic level has less energy than the preceding level.

Each chemical cycle involves a reservoir, where the element is stored; an exchange pool, from which the populations take and return nutrients; and the populations themselves. In the carbon cycle, the reservoir is organic matter, calcium carbonate shells, and limestone. The exchange pool is the atmosphere: photosynthesis removes carbon dioxide (CO_2), and respiration and combustion add carbon dioxide.

In the nitrogen cycle, the reservoir is the atmosphere, but nitrogen gas (N_2) must be converted to nitrate (NO_3^-) for use by producers. Nitrogen-fixing bacteria, particularly in root nodules, make organic nitrogen available to plants. Other bacteria active in the nitrogen cycle are the nitrifying bacteria, which convert ammonium to nitrate, and denitrifying bacteria, which convert nitrate to nitrogen.

In general, humans affect the carbon and nitrogen cycles by withdrawing substances from the reservoirs. For example, when fossil fuels are burned, organic carbon is removed from a reservoir. Nitrogen (N_2) is also converted to ammonium (NH_4^+) and thereby nitrogen is removed from the air.

In mature natural ecosystems, the populations usually remain the same size and need the same amount of energy each year. Additional material inputs are minimal because matter cycles. In the human ecosystem, the population size constantly increases, more energy is needed each year, and additional material inputs are necessary. Therefore, there is much pollution. It would be beneficial for us and for future generations to find ways to use excess heat and to recycle materials.

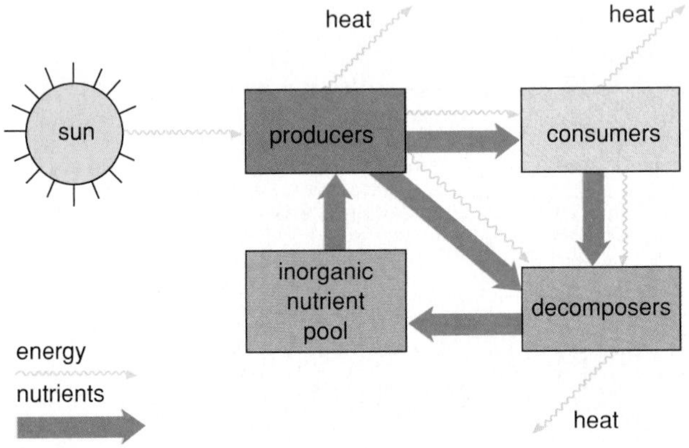

Study and Thought Questions

1. What is succession, and how does it result in a climax community? (p. 443)
2. Define habitat and niche. (p. 443)
3. What is the difference between a food chain and a food web? Define a trophic level. (p. 444)

Critical Thinking Where should humans be added to the food chains that can be discerned in figures 22.5 and 22.6?

4. Give an example of a grazing food chain and a detritus food chain for a terrestrial and an aquatic ecosystem. (p. 444)

5. Name 4 different types of consumers found in natural ecosystems. (p. 445)
6. Draw a pyramid of energy, and explain how such a pyramid can be used to verify that energy does not cycle. (p. 446)

   Ethical Issue Agricultural plants are at the base of the human energy pyramid. Should farming practices be regulated so that they are less polluting? Why or why not?
7. What are the reservoir and the exchange pool of a chemical cycle? (p. 447)
8. Describe the carbon cycle. How do humans contribute to this cycle? (pp. 447–449)
9. Describe the nitrogen cycle. How do humans contribute to this cycle? (p. 450)

   Critical Thinking Nitrate runoff pollutes waters. With reference to the nitrogen cycle, give two ways that humans increase the nitrate content of soil.
10. Contrast the characteristics of mature natural ecosystems with those of the human ecosystem. (p. 451)

    Ethical Issue What actions, if any, should be taken by industrial countries to prevent destruction of tropical rain forests?

Writing Across the Curriculum

1. Sometimes we are under the impression that all bacteria are harmful. What evidence can you give from this chapter to counter this belief?
2. Explain how humans and animals are an essential part of the carbon cycle.
3. What suggestions can you make to save tropical rain forests when they are primarily cleared by farmers who must provide food for their families?

Objective Questions

1. Chemicals cycle through the populations of an ecosystem, but energy is said to ______________ because all of it eventually dissipates as heat.
2. When organisms die and decay, chemical elements are made available to ____________ populations once again.
3. Organisms that feed on plants are called ______________.
4. A pyramid of energy illustrates that there is a loss of energy from one ___________ level to the next.
5. There is a loss of energy because one form of energy can never be completely ___________ into another form.
6. Forests are a(n) ____________ for carbon in the carbon cycle.
7. In the carbon cycle, when living organisms ____________, carbon dioxide (CO_2) is returned to the exchange pool.
8. Humans make a significant contribution to the nitrogen cycle when they convert nitrogen (N_2) to ____________ for use in fertilizers.
9. During the process of denitrification, nitrate is converted to ______________.
10. Natural ecosystems utilize the same amount of energy per year, but the human ecosystem utilizes a(n) ______________.
11. In reference to figure 22.6, which organisms would you place at a–e in this diagram?

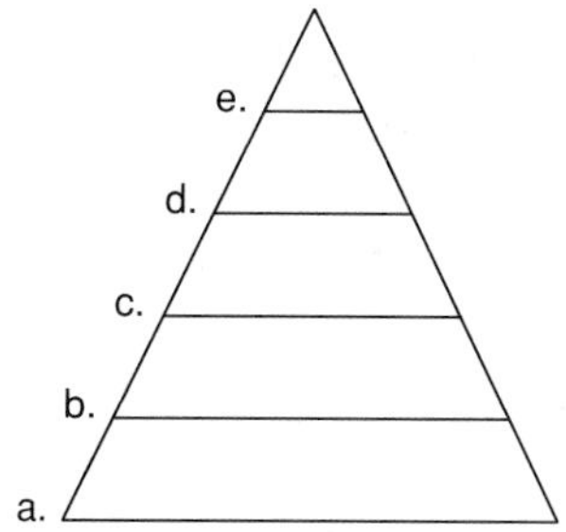

12. In this simplified diagram of the carbon cycle, label each lettered arrow as combustion, photosynthesis, or respiration.

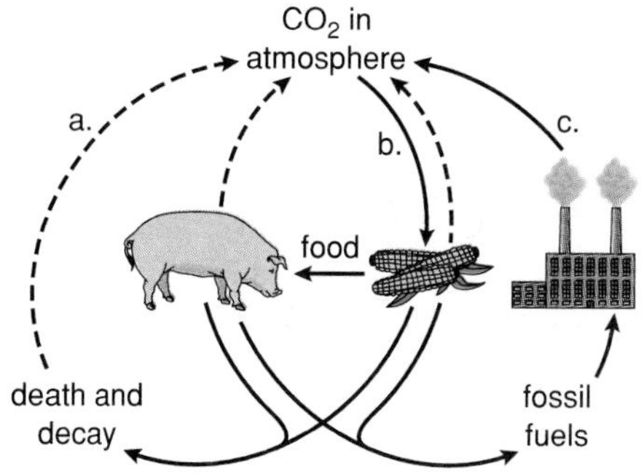

SELECTED KEY TERMS

consumer A member of a population that feeds on members of other populations in an ecosystem. 444

decomposer Organism of decay (fungus and bacterium) in an ecosystem. 444

detritus (di-tri´tus) Nonliving organic matter. 444

ecological pyramid Pictorial graph representing the biomass, organism number, or energy content of each trophic level in a food web, from the producer to the final consumer populations. 446

ecology The study of the interactions of organisms with each other and with the physical environment. 443

ecosystem A setting in which populations interact with each other and with the physical environment. 443

food chain A succession of organisms in an ecosystem that are linked by an energy flow and the order of who eats whom. 444

food web The complete set of food links between populations in a community. 444

habitat The natural abode of an animal or plant species. 443

niche (nich) Total description of an organism's functional role in an ecosystem, from activities to reproduction. 443

nitrogen fixation A process whereby nitrogen is reduced and incorporated into organic compounds. 450

pollution Detrimental alteration of the normal constituents of air, land, and water due to human activities. 451

producer Organism that produces food and is capable of synthesizing organic compounds from inorganic constituents of the environment; usually the green plants and algae in an ecosystem. 443

succession A series of ecological stages by which the community in a particular area gradually changes until there is a climax community that can maintain itself. 443

trophic level (tro´fik lev´el) A categorization of species in a food web according to their feeding relationships from the first-level autotrophs through succeeding levels of herbivores and carnivores. 444

Chapter 23

Population Concerns

Imagine a watering hole that can accommodate 100 zebras. If, at first, there are only 2 zebras and each pair of zebras only produces 4 zebras, how many doublings could there be without overtaxing the watering hole? You're correct if you say 4 and incorrect if you say 5. The problem is that you can't just consider the newly arrived zebras—you have to add in the number of zebras already there.

2 – 4 – 8 – 16 – 32 – 64 – 128

Also, notice that when it is time to stop, there are only 62 zebras (30 + 32). That's one of the unusual things about population growth—at one point in time it seems as if there is plenty of room and then all of a sudden there's not enough room.

Chapter Concepts

1 A population undergoing exponential growth has an ever-greater increase in numbers and a shorter doubling time, and it may outstrip the carrying capacity of the environment. 458–460

2 The world is divided into the more-developed countries and the less-developed countries; mainly the less-developed countries are presently undergoing exponential population growth. 461–462

3 Human activities cause land, water, and air pollution and threaten the integrity of the biosphere. 463–472

4 A sustainable world is possible if economic growth is accompanied by environmental preservation. 472

Chapter Outline

Figure 23.1

More-developed countries versus less-developed countries. **a.** In the more-developed countries, most people enjoy a high standard of living. **b.** In the less-developed countries, the majority of people are poor and have few amenities.

The countries of the world today are divided into 2 groups (fig. 23.1). The more-developed countries (MDCs), typified by countries in North America and Europe, are those in which population growth is under control and the people enjoy a good standard of living. The less-developed countries (LDCs), typified by countries in Latin America, Africa, and Asia, are those in which population growth is out of control and the majority of people live in poverty. (Sometimes the term *third-world countries* is used to mean the less-developed countries. This term was introduced by those who thought of the United States and Europe as the first world and the former USSR as the second world.)

Before we explore the reasons that the world is now divided into MDCs and LDCs, it is necessary to study exponential population growth in general.

Exponential Population Growth

The human population growth curve is a J-shaped growth curve (fig. 23.2). In the beginning, growth of the human population was relatively slow, but as more reproducing individuals were added, growth increased, until the curve began to slope steeply upward. It is apparent from the position of 1995 on the growth curve in figure 23.2 that growth is quite rapid now. The world population increases at least the equivalent of a medium-sized city every day (200,000) and the equivalent of the combined populations of the United Kingdom, Norway, Ireland, Iceland, Finland, and Denmark every year. These startling figures are a reflection of the fact that a very large world population is undergoing exponential growth.

Mathematically speaking, **exponential growth,** or geometric increase, occurs in the same manner as compound interest; that is, the percentage increase is added to the principal before the next increase is calculated. Referring specifically to populations, consider the hypothetical population sizes in table 23.1. This table illustrates the circumstances of world population growth at the moment: the percentage increase has decreased, yet the size of the population grows by a greater amount each year. The increase in size is dramatically large because the world population is very large.

In our hypothetical examples, an initial increase of 2% added to the original population size followed by a 1.99% increase results in the third-generation size listed in the last column. Notice that

1. in each instance, the second generation has a larger increase than the first generation even though the growth rate decreased from 2% to 1.99%;
2. the larger the population, the larger the increase for each generation.

The percentage increase is termed the **growth rate,** which is calculated per year.

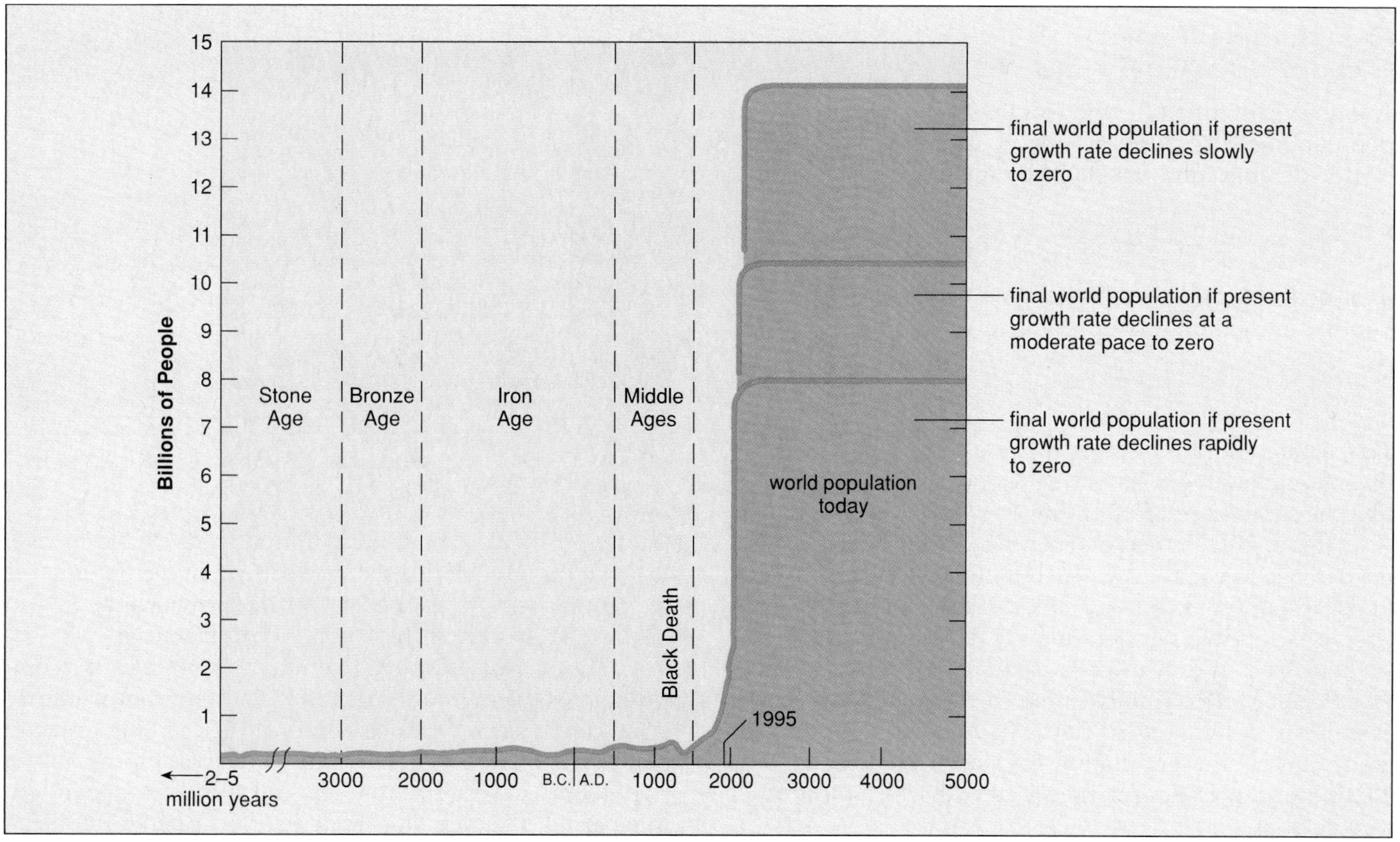

Figure 23.2

Growth curve for human population. The human population is now undergoing rapid exponential growth. Since the growth rate is declining, it is predicted that the population size will level off at 8, 10.5, or 14.2 billion, depending upon the speed with which the growth rate declines.

Table 23.1

Exponential Growth of Hypothetical Populations

Population Size	Increase (%)	Actual Increase in Numbers	Population Size	Increase (%)	Actual Increase in Numbers	Population Size
500,000,000	2.00	10,000,000	510,000,000	1.99	10,149,000	520,149,000
3,000,000,000	2.00	60,000,000	3,060,000,000	1.99	60,894,000	3,120,894,000
5,000,000,000	2.00	100,000,000	5,100,000,000	1.99	101,490,000	5,201,490,000

Growth Rate: Comparing Births and Deaths

The growth rate of a population is determined by considering the difference between the number of persons born (birthrate, or natality) and the number of persons who die per year (death rate, or mortality). It is customary to record these rates per 1,000 persons. For example, Canada at the present time has a birthrate of 15 per 1,000 per year, but it has a death rate of 7 per 1,000 per year. This means that Canada's population growth, or simply its growth rate, is:

$$\frac{15-7}{1{,}000} = \frac{8}{1{,}000} = \frac{0.8}{100} = 0.8\%$$

Notice that while birthrate and death rate are expressed in terms of 1,000 persons, the growth rate is expressed per 100 persons, or as a percentage.

After 1750, the world population growth rate steadily increased, until it peaked at 2% in 1965. It has fallen slightly since then to 1.7%. Yet, there is an ever-greater increase in the world population each year because of exponential growth. The explosive potential of the present world population can be appreciated by considering the doubling time.

Doubling Time: Could Be Only 39 Years

The **doubling time** (d)—the length of time it takes for the population size to double—can be calculated by dividing 70 (the demographic constant) by the growth rate (gr):

$$d = \frac{70}{gr}$$

If the present world growth rate of 1.8% continues, the world population will double in 39 years.

$$d = \frac{70}{1.8} \quad 39 \text{ years}$$

This means that in 39 years, the world will need double the amount of food, jobs, water, energy, and so on to maintain the same standard of living.

It is of grave concern to many that the amount of time needed to add each additional billion persons to the world population has taken less and less time (table 23.2). However, if the growth rate continues to decline, this trend will reverse itself, and eventually there will be *zero population growth* when births equal deaths. Then population size will remain steady. Therefore, figure 23.2 shows 3 possible logistic curves: the population may level off at 8, 10.5, or 14.2 billion, depending on the speed with which the growth rate declines.

Table 23.2

World Population Increase

Billions of People	Time Needed*	Year of Increase
First	2–5 million	1800
Second	130	1930
Third	30	1960
Fourth	15	1975
Fifth	12	1987
Sixth (projected)	11	1998

*Measured in years.

Source: Data from Elaine M. Murphy, World Population: Toward the Next Century. *Washington, D.C.: Population Reference Bureau, November 1981, page 3.*

Carrying Capacity: Population Levels Off

The growth curve for many nonhuman populations is S-shaped—the population tends to level off at a certain size. For example, figure 23.3 is based on actual data for the growth of a fruit fly population reared in a culture bottle. Because the fruit flies were adjusting to their new environment, growth was slow in the beginning. Then, because food and space were plentiful, the flies began to multiply rapidly. Notice that the curve began to rise dramatically, just as the human population curve does now. At this time, a population is demonstrating its biotic potential. **Biotic potential** is the maximum growth rate under ideal conditions. Biotic potential usually is not demonstrated for long because of an opposing force called environmental resistance. **Environmental resistance** includes all the factors that cause early death of organisms and therefore prevent the population from producing as many offspring as it might otherwise do. As far as the fruit flies are concerned, we can speculate that environmental resistance included the limiting factors of food and space. The waste given off by the fruit flies also may have limited the population size. When environmental resistance sets in, biotic potential is overcome, and the slope of the growth curve begins to decline. This is the inflection point of the curve.

The eventual size of any population represents a compromise between biotic potential and environmental resistance. This compromise occurs at the carrying capacity of the environment. The **carrying capacity** is the maximum population that the environment can support for an indefinite period. The carrying capacity of the earth for humans has not been determined. Some authorities think the earth is potentially capable of supporting 50 billion to 100 billion people. Others think we already have more humans than the earth can adequately support, especially if each person is to be provided the opportunity to learn, create, and enjoy life to his or her potential.

> Populations have a biotic potential for increase. Biotic potential is normally held in check by environmental resistance so that a population's growth curve is S-shaped and levels off at the carrying capacity of the environment.

Human Population Growth

The human population has undergone 3 periods of exponential growth. *Toolmaking* may have been the first technological advance that enabled the human population to enter a period of exponential growth. Farming may have resulted in a second phase of growth; and the *Industrial Revolution,* which began about 1850, promoted the third phase.

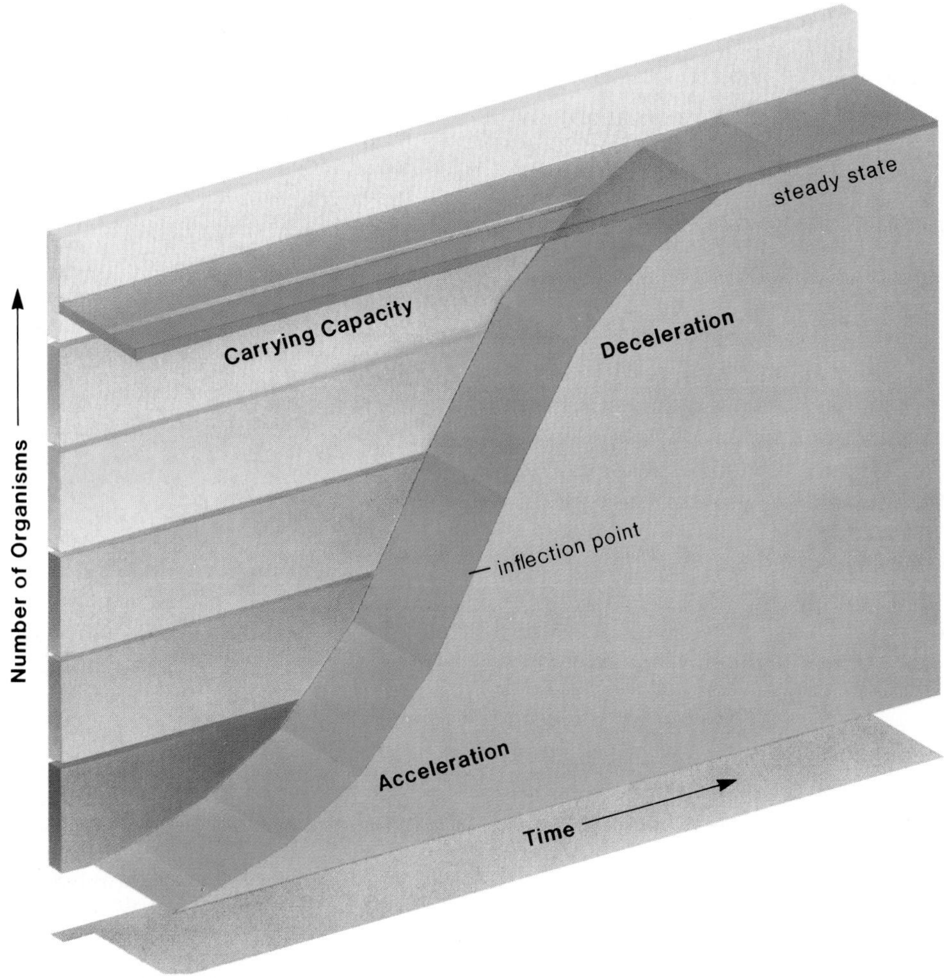

Figure 23.3

S-shaped growth curve. The population, such as fruit flies in a culture bottle, initially grows exponentially but begins to slow down as resources become limited. Population size remains stable when the carrying capacity of the environment is reached.

Countries: More-Developed versus Less-Developed

The countries of Europe and North America and also Russia and Japan were the first to become industrialized. These *more-developed countries (MDCs)* doubled their populations between 1850 and 1950. This was largely due to a decline in the death rate, the development of modern medicine and improved socioeconomic conditions. The decline in the death rate was followed shortly thereafter by a decline in the birthrate, so that populations in the MDCs experienced only modest growth between 1950 and 1975. This sequence of events (i.e., decreased death rate followed by decreased birthrate) is termed a **demographic transition.**

The growth rate for the MDCs as a whole has now stabilized at about 0.5%. The populations of a few of the MDCs—Italy, Denmark, Hungary, Sweden—are not growing or are actually decreasing in size. In contrast, there is no leveling off and no end in sight to U.S. population growth. Based on the 1990 census, it is projected that the U.S. population will reach 274.8 million in 2000, and ultimately 382.7 million by 2050. The U.S. racial/ethnic mix will grow increasingly diverse, due to a combination of factors: slow growth among non-Hispanic whites, steady growth among African Americans and native Americans, and rapid growth among Hispanics and Asian Americans. Over the next 60 years, the share of the population that is non-Hispanic white should decline steadily—from 76 percent in 1990 to 68 percent in 2010 and 60 percent in 2030. By 2050, a bare majority of Americans (53 percent) will be non-Hispanic whites.

These projections are based on these assumptions: future fertility is assumed to remain near current levels (2.12 children/female); future immigration is assumed to remain near current levels; fertility and mortality differentials by race/ethnic group are assumed to continue their current trends; life expectancy based on 1980–1990 trends is projected to improve slowly.

Most countries in Africa, Asia, and Latin America are known collectively as the *less-developed countries* (*LDCs*) because they are not fully industrialized. Although the death rate began to decline steeply in these countries following World War II with the importation of modern medicine from the MDCs, the birthrate remained high. The LDCs are unable to adequately cope with such rapid population expansion, and many people in these countries are underfed, poorly housed, and unschooled and live in abject poverty.

The growth rate of the LDCs peaked at 2.4% between 1960 and 1965. Since that time, the death rate decline has slowed and the birthrate has fallen. The growth rate is expected to be 1.8% by the end of the century. At that time, however, more than two-thirds of the human population will live in the LDCs. The reason for a decline in the growth rate is not clear, but the greatest decline is seen in countries with good family planning programs supported by community leaders.

> The history of the world population shows that the more-developed countries underwent a demographic transition between 1950 and 1975; the less-developed countries are just now undergoing demographic transition.

Comparing Age Structure

Populations have 3 age groups: dependency, reproductive, and postreproductive. One way of characterizing population is by these age groups . This is best visualized when the proportion of individuals in each group is plotted on a bar graph, thereby producing an age-structure diagram (fig. 23.4).

Laypeople are sometimes under the impression that if each couple has 2 children, zero population growth will take place immediately. However, **replacement reproduction,** as it is called, will still cause most countries today to continue growing due to the age structure of the population. If there are more young women entering the reproductive years than there are older women leaving them behind, then replacement reproduction will give a positive growth rate.

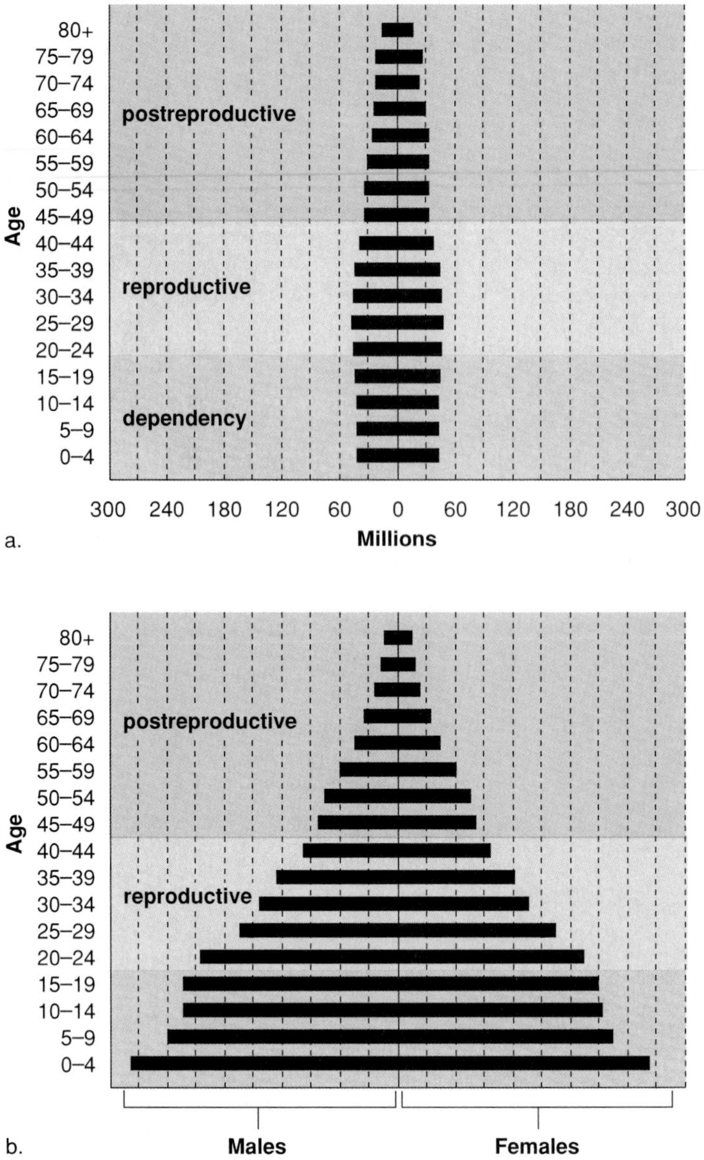

Figure 23.4

Age-structure diagram for **a.** more-developed and **b.** less-developed countries, 1989. The diagrams illustrate that the more-developed countries are approaching stabilization, whereas the less-developed countries will expand rapidly as indicated by the shape of their age-structure diagram.

Source: Data from World Population Profile: *1989, WP-89.*

Many MDCs have a stabilized age-structure diagram (fig. 23.4), but most LDCs have a youthful profile—a large proportion of the population is younger than the age of 15. Since there are so many young women entering the reproductive years, the population will still expand greatly, even after replacement reproduction is attained. The more quickly replacement reproduction is achieved, however, the sooner zero population growth will result.

Figure 23.5

Contour farming. Crops are planted according to the lay of the land to reduce soil erosion. This farmer has planted alfalfa, whose roots hold the soil, in between the strips of corn to replenish the nitrogen content of the soil. Alfalfa, a legume, also has root nodules that contain nitrogen-fixing bacteria.

Human Population and Pollution

As the human population increases in size, more energy and materials are consumed. Because the human population does not use energy efficiently and does not recycle materials (see fig. 22.12), pollutants are added to all components of the biosphere—land, water, and air.

How People Degrade Land

The land has been degraded in many ways. Here, we will discuss some of the greatest concerns.

Soil Erodes and Deserts Grow

Soil erosion causes the productivity of agricultural lands to decline. It occurs when wind and rain carry away the topsoil, leaving the land exposed and without adequate cover. The U.S. Department of Agriculture estimates that erosion is causing a steady drop in the productivity of farmland equivalent to the loss of 0.5 million hectares per year. To maintain the productivity of eroding land, more fertilizers, more pesticides, and more energy must be applied.

One answer to the problem of erosion is to adopt soil conservation measures. For example, farmers could use strip-cropping and contour farming (fig. 23.5).

Desertification is the transformation of marginal lands to desert conditions because of overgrazing and overfarming. Desertification has been particularly evident along the southern edge of the Sahara Desert in Africa, where it is estimated that 350,000 square miles of once-productive grazing land has become desert in the last 50 years. However, desertification also occurs in this country. The U.S. Bureau of Land Management, which opens up federal lands for grazing, reports that much of the rangeland it manages is in poor or bad condition, with much of its topsoil gone and with greatly reduced ability to support forage plants.

What's Causing Tropical Rain Forest Destruction

Virgin rain forests (fig. 23.6) exist in Southeast Asia and Oceania, Central and South America, and Africa. These forests are severely threatened by human exploitation (see the Ecology Focus in the Introduction, p. 4). The MDCs' demand for wood products has created a market for beautiful and costly tropical woods.

Tropical rain forests are also undergoing destruction as a result of the needs of the people who live there. For example, in Brazil, a large sector of the population has no means of support. To ease social unrest, the government allows citizens to own any land they clear in the Amazon forest (along the Amazon River). In tropical rain forests, it is customary to practice **slash-and-burn agriculture,** in which trees are cut down and burned to provide space to raise crops. Unfortunately, the fertility of the land is sufficient to sustain agriculture for only a few years. Once the cleared land is incapable of sustaining crops, the farmer moves on to another part of the rain forest to slash and burn again. In the meantime, cattle ranchers move in. Cattle ranchers are the greatest beneficiaries of deforestation, and increased ranching is therefore another reason for tropical rain forest destruction. A newly begun pig-iron industry also indirectly results in further exploitation of the rain forest. The pig iron must be processed before it is exported, and smelting the pig iron requires the use of charcoal (burnt wood).

> There are currently 3 primary reasons for tropical rain forest destruction: (1) logging to provide hardwoods for export, (2) slash-and-burn agriculture, and (3) cattle ranching. Industrialization in countries with extensive tropical rain forests no doubt will become another major reason.

Figure 23.6

Tropical rain forest sustainability. **a.** In its natural state, a tropical rain forest is immensely rich in vegetation and is breathtakingly beautiful. **b.** If the trees are not cut down, rubber tappers can earn a sustainable living by tapping the same trees year after year. **c.** Slash-and-burn agriculture is the first step toward destruction of a portion of the rain forest. With this method, agriculture cannot be sustained for more than a few years because the soil is infertile and does not hold moisture.

Losing Biological Diversity

Biological diversity is much greater in tropical rain forests than in temperate forests. Altogether, over half of the world's species are believed to live in tropical forests. A National Academy of Sciences study estimated that a million species of plants and animals are in danger of disappearing within 20 years as a result of deforestation in tropical countries. Many of these life forms have never been studied, and yet many could possibly be useful. At present, our entire domesticated crop production around the world relies on the fewer than 30 species of plants and animals domesticated during the last 10,000 years! It is quite possible that many additional species of wild plants and animals now living in tropical forests could be domesticated or at least their genes used to improve our traditional crops through genetic engineering techniques.

While it may seem like an either-or situation—either biological diversity or human survival—there is growing recognition that this is not the case. Natives who harvest rubber from rubber trees can earn a living from the same trees year after year (fig. 23.6*b*). A recent study calculated the market value of rubber and such exotic produce as the aguaje palm fruit and nuts, which can be harvested continually from the Amazon forest. It concluded that selling these products would yield more than twice the income of either lumbering or cattle ranching.

> There is much worldwide concern about the loss of biological diversity due to the destruction of tropical rain forests. The myriad plants and animals that live there could possibly benefit human beings.

Waste Disposal and Dangerous Trash

Every year, the U.S. population discards billions of metric tons of solid wastes, much of it on land (fig. 23.7). Solid wastes include not only household trash but also sewage sludge, agricultural residues, mining refuse, and industrial wastes. Some of these solid wastes contain substances that cause illness and sometimes even death; they are called **hazardous wastes.**

Hazardous wastes, such as heavy metals, chlorinated hydrocarbons (organochlorines), and nuclear wastes, enter bodies of water and are subject to biological magnification (fig. 23.8). Decomposers are unable to break down these wastes. They enter and remain in the body because they are

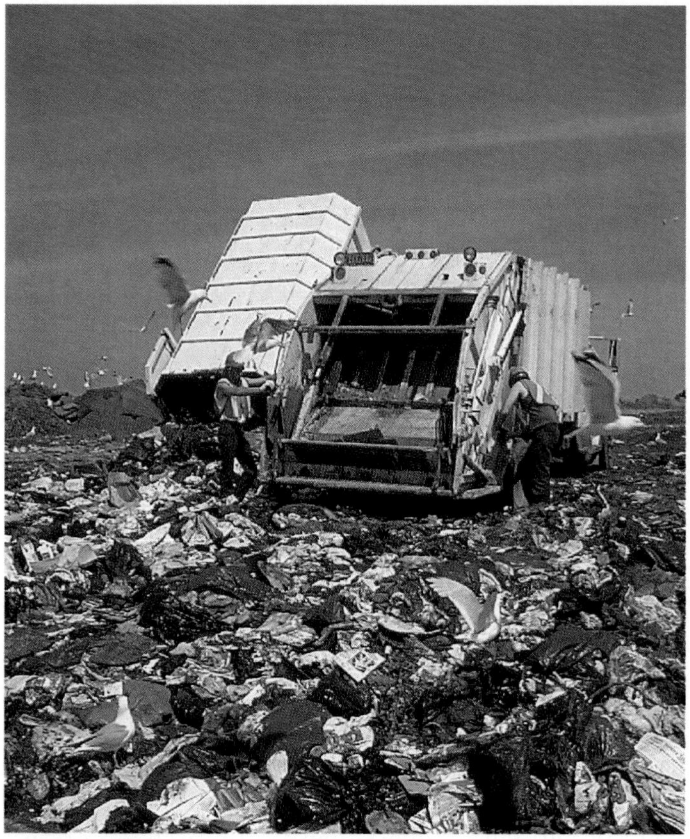

Figure 23.7

Dump sites. These not only cause land and air pollution, they also allow chemicals to enter groundwater, which may later become part of human drinking water.

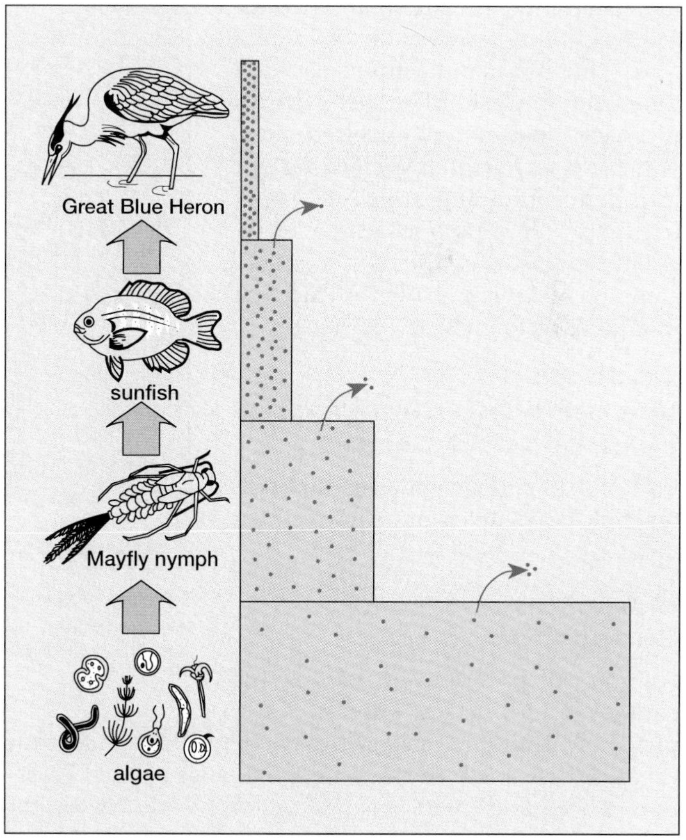

Figure 23.8

Biological magnification. A poison (*dots*), such as DDT, that is minimally excreted (*arrows*) becomes maximally concentrated as it passes along a food chain due to the reduced size of the trophic levels.

not excreted. Therefore, they become more concentrated as they pass along a food chain. Notice in figure 23.8 that the dots representing DDT become more concentrated as they pass from producer to tertiary consumer. Biological magnification is most apt to occur in aquatic food chains—there are more links in aquatic food chains than there are in terrestrial food chains. Humans are the final consumers in both types of food chains, and in some areas, human milk contains detectable amounts of DDT and PCBs, which are organochlorines.

The dumping of hazardous wastes directly endangers public health. Chemical wastes buried over a quarter of a century ago in Love Canal, near Niagara Falls, have seriously damaged the health of some residents there. Similarly, the town of Times Beach, Missouri, was abandoned because workers spread an organochlorine (dioxin)-laced oil on the city streets; this resulted in myriad illnesses among its citizens. In other places, such as in Holbrook, Massachusetts, manufacturers have left thousands of waste-filled metal drums in abandoned or uncontrolled sites, where toxic chemicals are oozing out into the ground and are contaminating the water supply. Illnesses, especially forms of cancer, are quite common not only in Holbrook but also in adjoining towns.

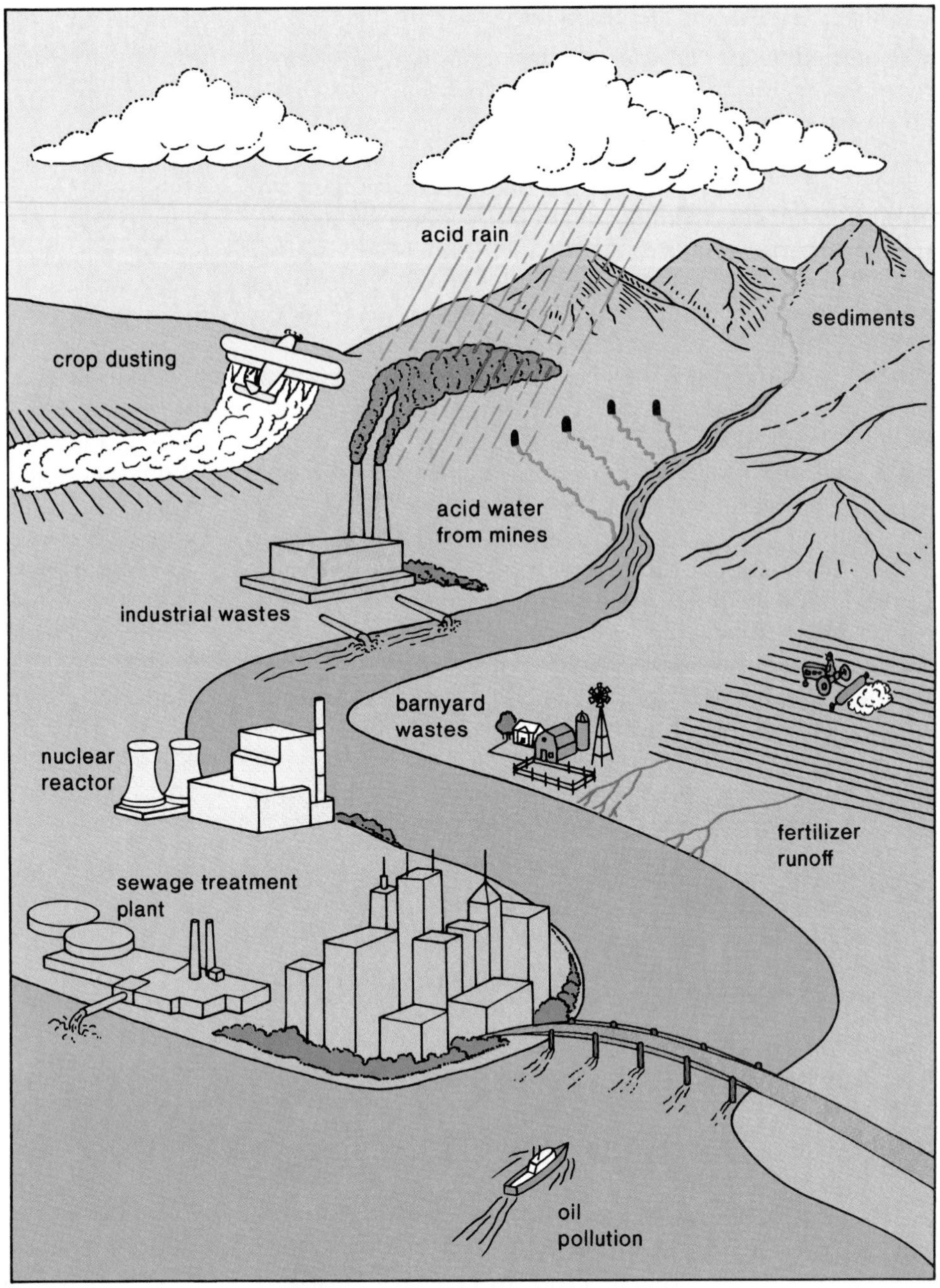

Figure 23.9

Sources of water pollution. Many bodies of water are dying due to the introduction of sediments and surplus nutrients.

Source: Adapted from U.S. Environmental Protection Agency, Office of Water Supply and Solid Waste Management Programs, "Waste Disposal Practices and Their Effects on Ground Water," Executive Summary. *Washington DC: U.S. Government Printing Office, 1977*

Water Pollution from Rivers to Oceans

Pollution of surface water, groundwater, and the oceans is of major concern today.

Many Sources Pollute Surface Water

All sorts of pollutants from various sources enter surface waters, as depicted in figure 23.9. Sewage treatment plants help degrade organic wastes, which can otherwise cause oxygen depletion in lakes and rivers. As the oxygen level decreases, the diversity of life is greatly reduced. Also, human feces may contain the microbes that cause cholera, typhoid fever, and dysentery. In LDCs, where the population is growing and where waste treatment is practically nonexistent, many children die each year from these waterborne diseases.

Typically, sewage treatment plants use bacteria to break down organic matter to inorganic nutrients, like nitrates and phosphates, which then enter surface waters. These types of nutrients, which can also enter waters by fertilizer runoff and soil erosion, lead to **cultural eutrophication,** an enrichment of bodies of water due to human activities. First, the nutrients cause overgrowth of algae. Then, when the algae die, oxygen is used up by the

Ecology Focus

Sewage Treatment

Most cities and towns take water from a nearby source and chlorinate it to kill bacteria before it is piped to offices and homes. Used water, including drainage and sewage water, is piped to a sewage treatment plant, where primary treatment removes most solids by settling, and secondary treatment, if available, removes all solids by additional bacterial digestion. Both of these sewage treatment methods release nutrients that lead to algal bloom and eutrophication. Tertiary treatment of sewage involves the removal of nutrient molecules but is very costly.

A method of sewage treatment, even if costly, that did not rely on a constant inflow of fresh water would be desirable because there are so many competing uses for fresh water. A method that uses the same water repeatedly has been developed and is illustrated in figure 23A. This system processes wastes in 4 main stages. Sewage from toilets and wastewater from other fixtures is transferred to a chamber where denitrifying bacteria consume organic compounds using nitrate as their source of oxygen. The gaseous end products, mainly nitrogen and carbon dioxide, are vented to the atmosphere. The wastewater then flows into an aerobic digestion chamber. Here aerobic bacteria continue the digestion process and convert ammonia from urine into nitrate. Thus, the second chamber replaces nitrate, which is needed as an oxygen source in the first chamber. The water is next forced through ultrafine tubular membranes that filter out bacteria and any remaining solids. Finally, the water is treated with activated carbon to improve the color and odor and with ozone to disinfect it. The water is then piped back to the toilets.

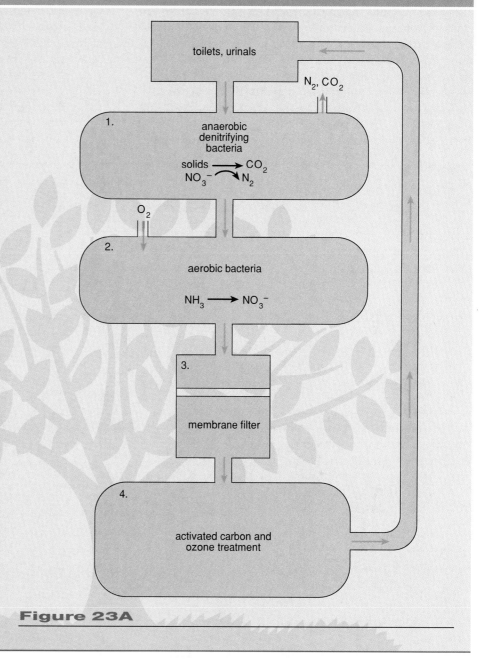

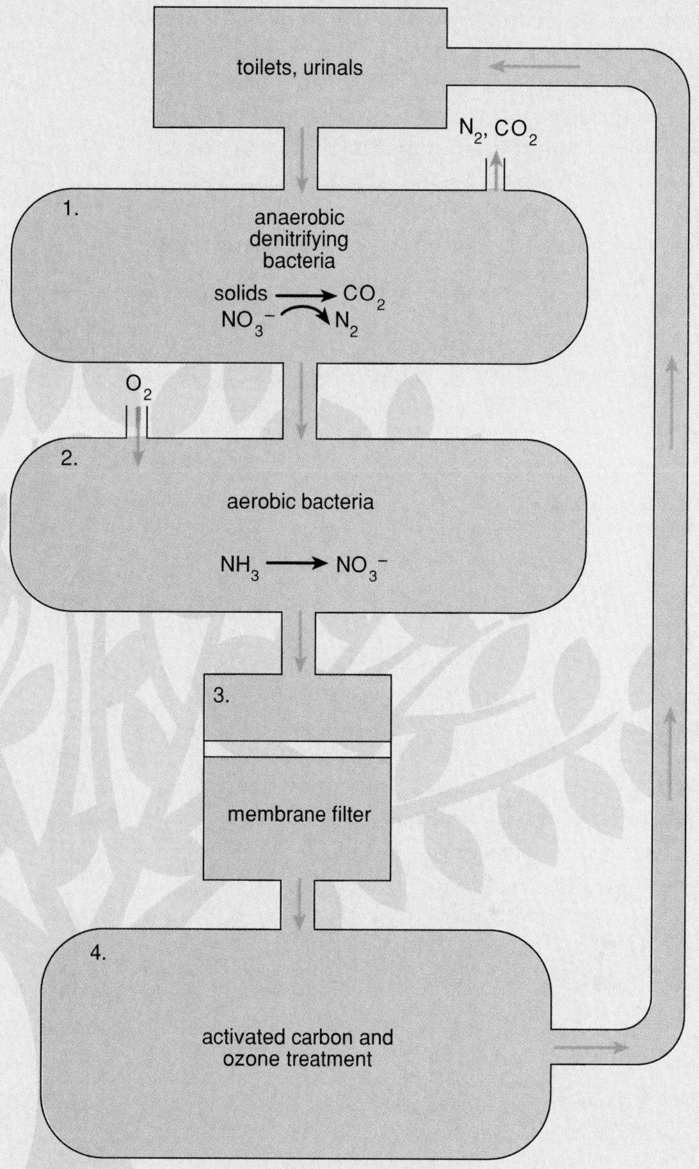

Figure 23A

decomposers, and the water's capacity to support life is reduced. Massive fish kills are sometimes the result of cultural eutrophication.

Wherever a large number of toilets are concentrated, as in shopping malls, railroad stations, or airports, a onesite method of sewage treatment can be used as described in the Ecology Focus for this chapter. Since the same water is used over and over again, this form of sewage treatment does not put a strain on natural ecological systems.

Industrial wastes include heavy metals and organochlorines, such as pesticides. These materials are not readily degraded under natural conditions or in conventional sewage treatment plants. Sometimes, they accumulate in the mud at the bottom of deltas, and environmental problems result if estuaries of very polluted rivers are disturbed. Industrial pollution is being addressed in many MDCs but usually has low priority in LDCs.

Groundwater Pollution: Farms and Industry

In areas of intensive animal farming or where there are many septic tanks, ammonium (NH_4^+) released from animal and human waste is converted by soil bacteria to soluble nitrate, which moves down through the soil (percolates) into underground water supplies. Between 5% and 10% of all wells examined in the United States have nitrate levels higher than the recommended maximum. Nitrates are converted to nitrites in the digestive tract. Nitrites are poisons because they combine with hemoglobin forming methoglobin, which has a reduced oxygen-carrying capacity.

Industries also pollute aquifers (underground rivers). Previously, industries would run wastewater into a pit from which the pollutants would seep into the ground. Wastewater and chemical wastes have also been injected into deep wells, from which the pollutants constantly discharge. Both of these customs have been or are in the process of being phased out. It is very difficult for industries to find other ways to dispose of wastes, especially since citizens do not wish to live near waste treatment plants. The emphasis today, therefore, is on prevention of wastes in the first place. Industries are trying to use processes that do not create wastes and/or to recycle the wastes they do generate.

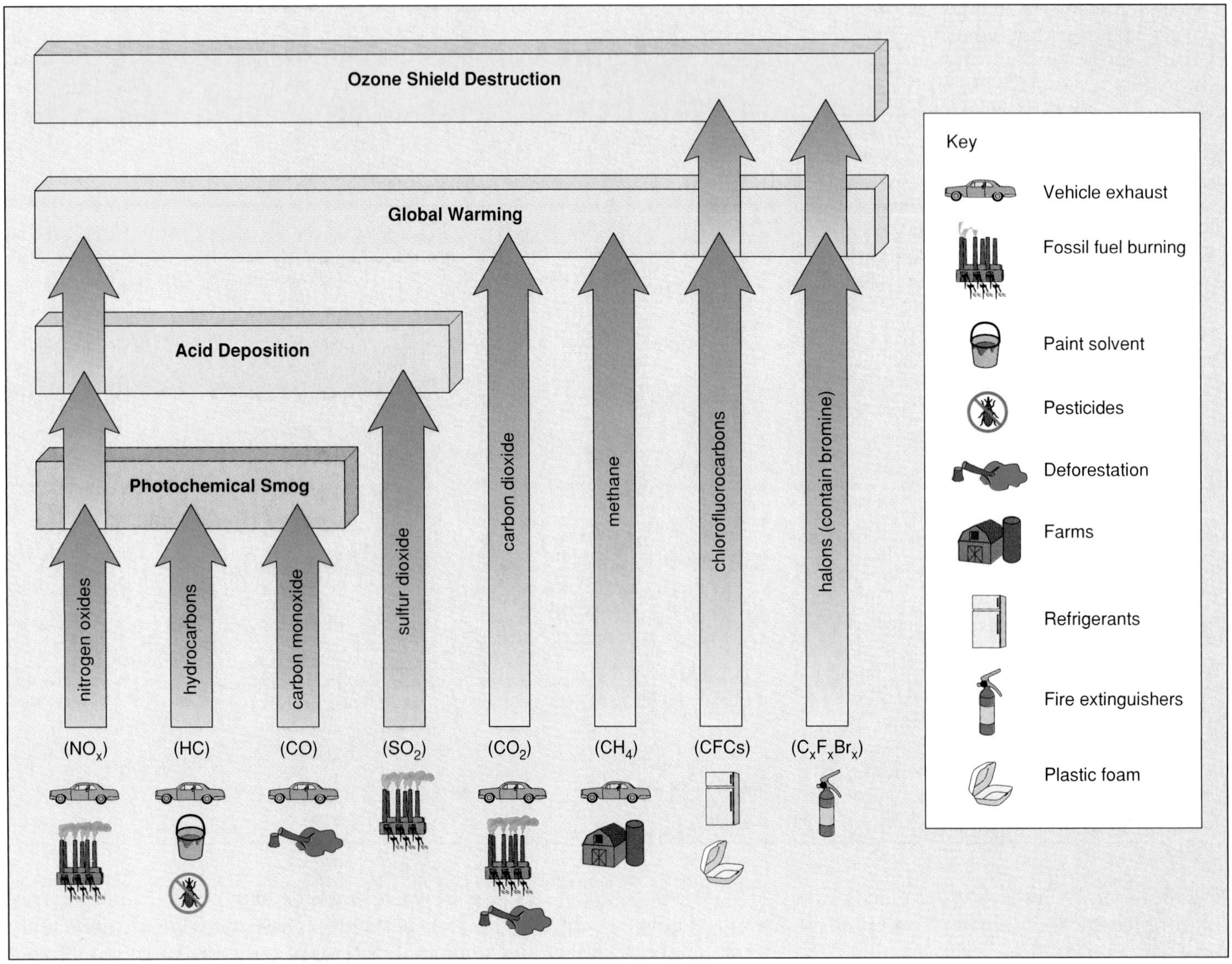

Figure 23.10

Air pollutants. These are the gases, along with their sources, that contribute to 4 environmental effects of major concern: photochemical smog, acid deposition, global warming, and ozone shield destruction. An examination of the sources of these gases shows that vehicle exhaust and fossil fuel burning are the chief contributors.

Ocean Water Pollution: The Final Dumping Ground

Coastal regions are not only the immediate receptors for local pollutants, they are also the final receptors for pollutants carried by rivers that empty at the coast. Waste dumping also occurs at sea, and ocean currents sometimes transport both trash and pollutants back to shore. Examples are nuclear wastes and the nonbiodegradable plastic bottles, pellets, and containers that now commonly litter beaches and the oceans' surfaces. Some of these, such as fishing line and the plastic that holds a 6-pack of beer, have been implicated in the death of birds, fishes, and marine mammals that mistook them for food and became entangled in them.

Offshore mining and shipping add pollutants to the oceans. Some 5 million metric tons of oil a year, or more than 1 gram per 100 m^2 of the oceans' surfaces, end up in the oceans. Large oil spills kill plankton, fish larvae, and shellfishes, as well as birds and marine mammals. One of the largest spills occurred on March 24, 1989, when the tanker *Exxon Valdez* struck a reef in Alaska's Prince William Sound and leaked 44 million liters of crude oil. During the war with Iraq, 120 million liters were released from onshore storage tanks into the Persian Gulf, an event that was called environmental terrorism. Although petroleum is biodegradable, the process takes a long time because the low nutrient content of seawater does not support a large bacterial population. Once the oil washes up onto beaches, it takes many hours of work and millions of dollars to clean it up.

> Adequate sewage treatment and waste disposal are necessary to prevent the pollution of rivers and oceans. Also, new methods are needed to prevent pollution of underground water supplies.

Air Pollution: Four Threats

The atmosphere has 2 layers, the stratosphere and the troposphere. The stratosphere is a layer that lies 15–50 kilometers above the surface of the earth. Here, the energy of the sun splits oxygen molecules (O_2). These individual oxygen atoms (O) then combine with molecular oxygen to form ozone (O_3). This ozone layer acts as a shield because it absorbs the ultraviolet (UV) rays of the sun, preventing them from striking the earth. If these rays did penetrate the atmosphere, life on earth would not be possible because living things cannot tolerate heavy doses of ultraviolet radiation. The troposphere is the atmospheric layer closest to the earth's surface; it ordinarily contains the gases nitrogen (N_2), 78%; oxygen (O_2), 21%; and carbon dioxide (CO_2), 0.03%.

Four major concerns—photochemical smog, acid deposition, global warming, and the ozone shield destruction—are associated with the air pollutants listed in figure 23.10. You can see that fossil fuel burning and vehicle exhaust are primary sources of gases associated with air pollution. These 2 sources are related because gasoline is derived from petroleum, a fossil fuel.

Photochemical Smog: Making Ozone and PAN

Photochemical smog (see the Ecology Focus in chapter 8) contains 2 air pollutants—nitrogen oxides (NO_2) and hydrocarbons (HC)—that react with one another in the presence of sunlight to produce ozone (O_3) and PAN (peroxylacetyl nitrate). Both nitrogen oxides and hydrocarbons come from fossil fuel combustion, but additional hydrocarbons come from various other sources as well, including industrial solvents.

Ozone and PAN are commonly referred to as oxidants. Breathing ozone affects the respiratory and nervous systems, resulting in respiratory distress, headache, and exhaustion. These symptoms are particularly apt to appear in young people; therefore, in Los Angeles, where ozone levels are often high, school children must remain inside the school building whenever the ozone level reaches 0.35 ppm (parts per million by weight). Ozone is especially damaging to plants, resulting in leaf mottling and reduced growth (fig. 23.11).

Figure 23.11

Effect of ozone on plants. The milkweed in **a.** was exposed to ozone and appears unhealthy; the milkweed in **b.** was grown in an enclosure with filtered air and appears healthy.

Carbon monoxide (CO) is another gas that results from the burning of fossil fuels in the industrial Northern Hemisphere. High levels of carbon monoxide increase the formation of ozone. Carbon monoxide also combines preferentially with hemoglobin and thereby prevents hemoglobin from carrying oxygen (see the Ecology Focus in chapter 6). Breathing large quantities of automobile exhaust can even result in death because of this phenomenon. Of late, it has been discovered that the amount of carbon monoxide over the Southern Hemisphere is equal to the amount over the Northern Hemisphere. The source of the carbon monoxide in the Southern Hemisphere, however, is the burning of tropical rain forests.

Normally, warm air near the ground is able to escape into the atmosphere. Sometimes, however, air pollutants, including photochemical smog and soot, are trapped near the earth due to a long-lasting thermal inversion. During a **thermal inversion,** cold air is at ground level beneath a layer of warm, stagnant air above. This often occurs at sunset, but turbulence usually mixes these layers during the day. Areas surrounded by hills are particularly susceptible to the effects of a thermal inversion because the air tends to stagnate (see the Ecology Focus in chapter 8).

Acid Deposition: From Burning Coal, Oil, Gasoline

Power plants that burn coal and oil have high sulfur dioxide (SO_2) emissions, and automobile exhaust contains nitrogen oxides (NO_2); both of these are converted to acids when they combine with water vapor in the atmosphere, a reaction that is promoted by ozone in photochemical smog. These acids return to earth as either wet deposition (acid rain or snow) or dry deposition (sulfate and nitrate salts).

Acid deposition (the return to earth of acid particles in rain or snow) is now associated with dead or dying lakes and forests, particularly in North America and Europe (see the Ecology Focus in chapter 1). Acid deposition also corrodes marble, metal, and stonework, an effect that is noticeable in cities. It can also degrade our water supply by leaching heavy metals from the soil into drinking water supplies. Similarly, acid water dissolves copper from pipes and from lead solder that is used to join pipes.

Global Warming: From Greenhouse Gases

Certain air pollutants allow the sun's rays to pass through but then absorb and reradiate the heat to the earth (fig. 23.12*a*). This is called the **greenhouse effect** because the glass of a greenhouse allows sunlight to pass through but

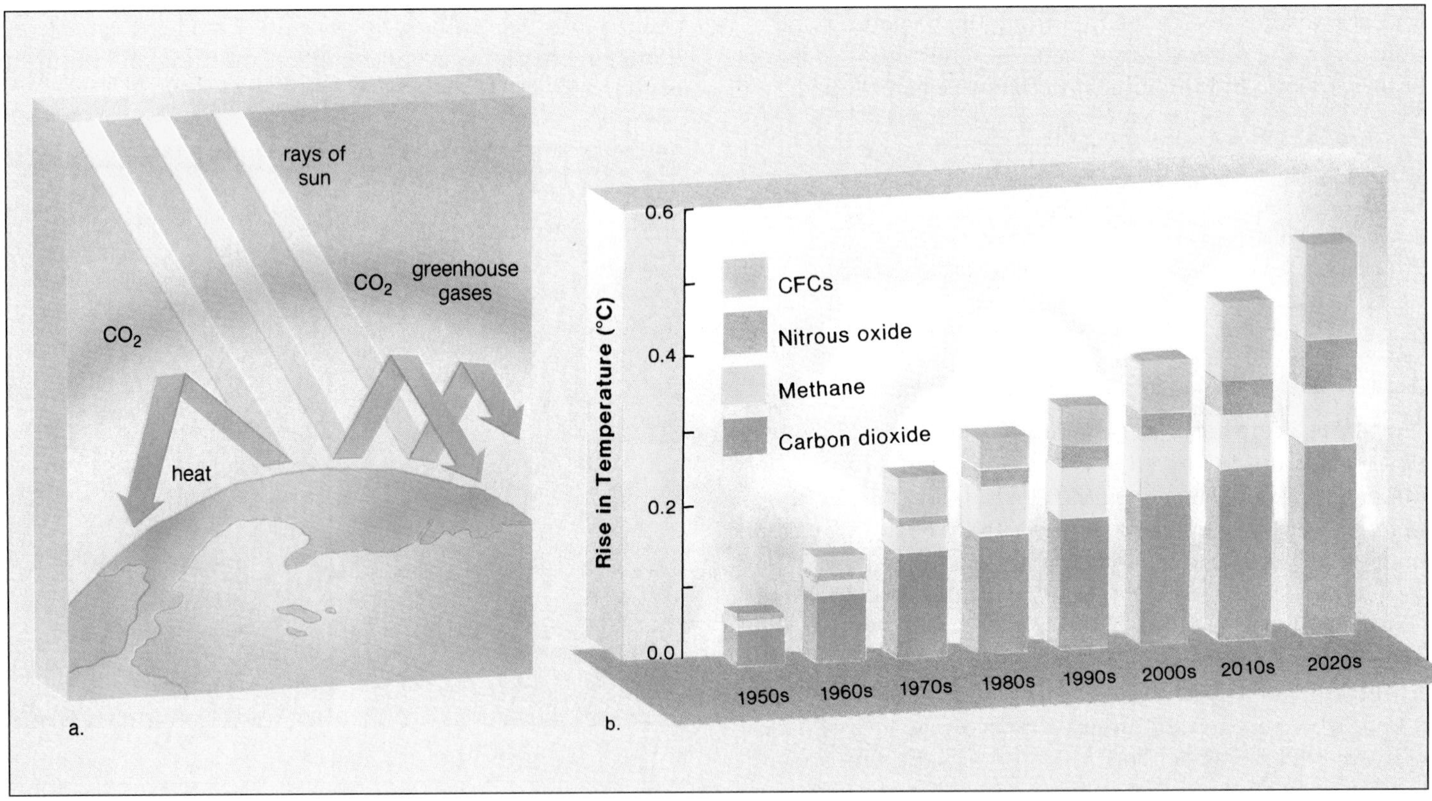

Figure 23.12

Global warming. **a.** The greenhouse effect is caused by the atmospheric accumulation of certain gases, such as carbon dioxide (CO_2), that allow the rays of the sun to pass through but absorb and reradiate heat to the earth. **b.** The greenhouse gases. This graph shows the fraction of warming caused by the gases carbon dioxide, methane, nitrous oxide, and CFCs for the decades from 1950 to 2020. There was no accumulation of CFCs in the 1950s because they were not being manufactured to any degree. By 2020, carbon dioxide and all of the other gases taken together will each contribute about 50% to the projected global warming.

then traps the resulting heat inside the structure. The air pollutants responsible for the greenhouse effect are known as the greenhouse gases. They are as follows:

Carbon dioxide (CO_2) from fossil fuel and wood burning

Nitrous oxide (NO_2), primarily from fertilizer use and animal wastes

Methane (CH_4) from biogas, bacterial decomposition (particularly in the guts of animals), sediments, and flooded rice paddies

Chlorofluorocarbons (CFCs) from Freon, a refrigerant

Halons (halocarbons, $C_xF_xBr_x$) from fire extinguishers

If nothing is done to control the level of greenhouse gases in the atmosphere, a rise in global temperature is expected. Figure 23.12 predicts a rise of over 0.5°C by the year 2020, but some authorities predict the rise in temperature could be as high as 5°C by 2050. It's possible that global warming has already begun: the 4 hottest years on record occurred during the 1980s; there has been greater warming in the winters than in the summers; there has been greater warming at high altitudes than near the equator; and the stratosphere is cooler and the lower atmosphere is warmer than before. All of these effects had been predicted by computer models of the greenhouse effect.

The ecological effects of a 5°C rise in global temperature would be severe. The sea level would rise—melting of the polar ice caps would add more water to the sea and also, water expands when it heats up. There would be coastal flooding and the possible loss of many cities, like New York, Boston, Miami, and Galveston in the United States. Coastal ecosystems, such as marshes, swamps, and bayous, would normally move inland to higher ground as the sea level rises, but many of these ecosystems are blocked by artificial structures and may be unable to move inland. If so, the loss of fertility would be immense.

There may also be food loss because of regional changes in climate. Because of greater heat and drought in the midwestern United States, the suitable climate for growing wheat and corn may shift as far north as Canada, where the soil is not as suitable.

It is clear from figure 23.12 that carbon dioxide accounts for at least 50% of the predicted rise in global temperature. Therefore, a sharp decrease in consumption of fossil fuels is recommended, and we must find more efficient ways to acquire energy from cleaner fuels, such as natural gas. We must also use alternative energy sources, such as solar, wind, and geothermal energy and perhaps even nuclear power, more aggressively. In addition to fossil fuel consumption, deforestation is a major contributor to the rise in carbon dioxide. Burning one acre of primary forest releases 200,000 kilograms of carbon dioxide into the air; moreover, the trees are no longer available to act as a sink to take up carbon dioxide during photosynthesis. Therefore, tropical rain forest deforestation should be halted and extensive reforesting all over the globe should take place.

The other greenhouse gases combined account for the other 50% predicted rise in global temperature. A complete phaseout of chlorofluorocarbon use would be most beneficial. Fortunately, in an effort to arrest ozone shield destruction, the United States and the European countries have agreed to reduce CFC production by 85% as soon as possible and to stop their production altogether by the end of the century.

Destroying Our Ozone Shield

The **ozone shield** (see the Ecology Focus in chapter 3) is a layer of ozone in the upper atmosphere that protects the earth from UV radiation. Ozone shield destruction is primarily caused by CFCs and halons (halocarbons [$C_xF_xBr_x$]). CFCs are heat-transfer agents used in refrigerators and air conditioners. They are also used as foaming agents in such products as styrofoam cups and egg cartons. In the past, they had been used as propellants in spray cans, but this application is now banned in the United States. Halons are antifire agents used in fire extinguishers.

Scientists knew that CFCs would drift up into the stratosphere, but it was believed that they would be nonreactive there during their 150-year life span. It is now apparent, however, that solar energy causes CFCs to release chlorine, which attaches to ozone molecules. Thereafter, ozone breaks down to chlorine oxide and oxygen and the ozone shield is depleted. Over the South Pole, the entire scenario is accentuated because the dry, cold air of Antarctica is filled with ice crystals on which chlorine and ozone join and react. The result is a 60% loss of ozone each spring; such a severe loss is called an *ozone hole* (fig. 23.13).

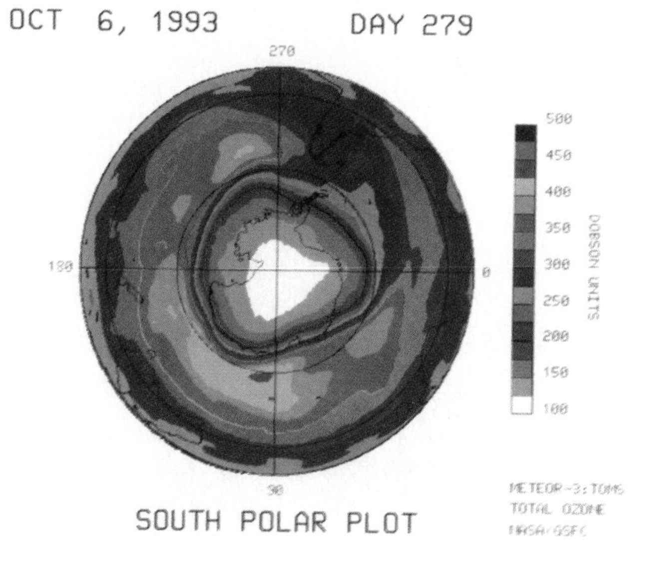

Figure 23.13

Ozone depletion. Satellites have been tracking the depletion of ozone at the South Pole of Antarctica since the late 1970s. The so-called ozone hole appears as a large white area at the center of this picture.

The Arctic and Northern Hemisphere is not as vulnerable as Antarctica. Still, scientists find that at various points during the year, the Northern Hemisphere appears to lose as much as 6% of its ozone shield.

Depletion of the ozone shield allows more ultraviolet rays to reach the earth. The incidence of human cancer, especially skin cancer, is expected to increase, and plants and animals living in the top microlayer of the oceans will begin to die. Increased UV radiation will also hasten the rate at which smog forms. Many believe that the current plans to reduce CFCs and halon emissions are insufficient and that much more stringent measures are necessary.

Outdoor air pollutants are involved in causing 4 major detrimental environmental effects: photochemical smog, acid deposition, global warming, and ozone shield destruction. Each pollutant may be involved in more than one of these effects.

A Sustainable World

Economic growth is often accompanied by environmental degradation, and there is great concern that as the LDCs become more developed, environmental degradation will increase to the point that the human population will outstrip the carrying capacity of the planet. Without economic growth, however, the demographic transition may not occur, and the sheer number of people in the less-developed countries will provoke environmental degradation to such a degree that the effects will be felt worldwide, not just in the immediate area.

The answer to this dilemma is economic growth without the side effect of environmental degradation. This is called sustainable growth. Certain MDCs, such as the Scandinavian countries and to a degree the United States, are learning to protect the environment. Energy consumption is decreasing even as economic growth continues. Industries are beginning to recycle their wastes to prevent environmental pollution. Citizens are learning to recycle their trash. More should be done, and ecologically sound practices must be exported to the LDCs very quickly. Only sustainable economic development will ensure the continuance of the world's human population.

Once zero population growth has been achieved, we can begin to consider the possibility of a steady state, in which population and resource consumption remain constant. Environmental preservation is paramount in a steady state (fig. 23.14).

What would our culture be like if we had steady-state manufacturing and a steady-state population? Perhaps it would be greatly improved. Certainly, there are no limits to growth in knowledge, education, art, music, scientific research, human rights, justice, and cooperative human interactions. In a steady-state world, the general sense of fearful competition among peoples might diminish, allowing human compassion and creativity to prosper as never before.

It is hoped that a sustainable world will result in a steady-state society, where there is no yearly increase in population or resource consumption.

Figure 23.14

In the steady state, environmental preservation is an important consideration.

Summary

Populations have a biotic potential for increase in size. Biotic potential is normally held in check by environmental resistance, thereby producing an S-shaped growth curve leveling off at the carrying capacity of the environment.

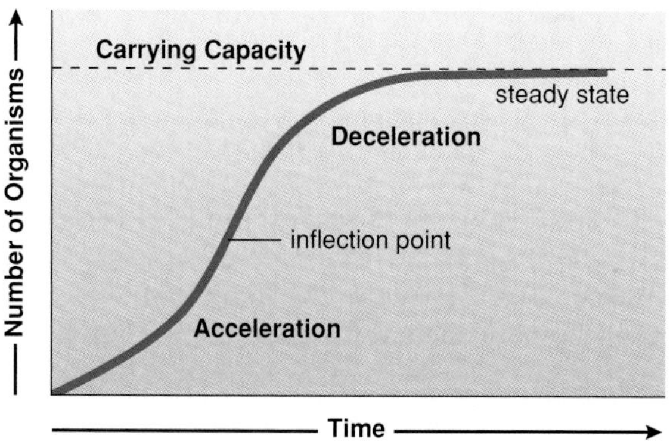

The human population is expanding exponentially; and it is unknown when growth will level off. Presently, each year exhibits a large increase, and the doubling time is now about 40 years. The MDCs underwent a demographic transition between 1950 and 1975, but the LDCs are just now undergoing demographic transition. In these countries, where the average age is less than 15, it will be many years before reproduction replacement will equal zero population growth.

On land, soil erosion sometimes reduces soil quality and leads to desertification. The tropical rain forests are being reduced in size, and the loss of biological diversity will be immense. Solid wastes, including hazardous wastes, are deposited on land and at sea. These are not biodegradable and are subject to biological magnification.

Surface waters, groundwater, and the oceans are all being polluted. Organic materials can be broken down in sewage treatment plants, but the nutrients made available to algae from this process can lead to cultural eutrophication. It is difficult to rid surface waters and particularly groundwater of hazardous wastes. The oceans are the final recipients of all the pollutants that enter water. In addition, some materials are dumped directly into the oceans, either purposefully or accidentally.

In the air, hydrocarbons and nitrogen oxides (NO_x) react to form photochemical smog, which includes ozone and PAN. Sulfur dioxide (SO_2) and nitrogen oxides react with water vapor to form acids that contribute to acid deposition. Several gases—carbon dioxide (CO_2), nitrous oxide (NO_2), methane (CH_4), and CFCs—are called the greenhouse gases because they trap heat and lead to global warming. Destruction of the ozone shield is associated particularly with CFCs, which rise into the stratosphere and react with frozen particles within clouds to release chlorine. Chlorine causes ozone breakdown.

Economic growth is required particularly in the LDCs. All countries of the world should strive for sustainable growth, which necessitates preservation of the environment. Eventually, it may be possible to have a steady state in which neither population nor resource consumption increases.

Study and Thought Questions

1. Define exponential growth. (p. 458) Draw a growth curve to represent exponential growth, and explain why a curve representing population growth usually levels off. (pp. 459–460)
2. Calculate the growth rate and the doubling time for a population in which the birthrate is 20 per 1,000 and the death rate is 2 per 1,000. (pp. 459–460)
3. Define demographic transition. When did the MDCs undergo demographic transition? When did the LDCs undergo demographic transition? (pp. 461–462)
4. Distinguish between the MDCs and the LDCs. (pp. 461–462)

   Ethical Issue Should LDCs have the same standard of living as MDCs? Why or why not?
5. Give 2 reasons the quality of the land is being degraded today. What is desertification? (p. 463)
6. Give 3 reasons tropical rain forests are being destroyed. What is another potential reason? (p. 463)
7. What is the primary ecological concern associated with the destruction of tropical rain forests? (p. 465)
8. What are the 3 types of hazardous wastes that contribute to pollution on land? (p. 465)

   Ethical Issue Do you approve of the risk/benefit (weighing the possible risk against the projected benefit of a technology) approach to pollution control?
9. What are several ways in which underground water supplies can be polluted? (p. 468)

   Critical Thinking The concept called "tragedy of the commons" says that when air, water, and land are owned in common, people tend to exploit rather than protect them. Explain by giving an example and suggest ways to overcome this tendency of people.
10. What substances contribute to air pollution? What are their sources? Which are associated with each of the following phenomena: photochemical smog, acid deposition, global warming, and destruction of the ozone shield? (pp. 469–472)

    Critical Thinking How might global warming negatively affect you personally?

Writing Across the Curriculum

1. The Mexican government at one time encouraged large families because it was believed that the greater the number of people, the greater the work force and the greater the prosperity. What's wrong with this type of thinking?
2. Sending food helps prevent present but not future famines. If we wanted to offer long-term help to the starving people of the world, what should we be doing?
3. An ever-increasing human population size puts a strain on the world's resources, yet birth control was not discussed in the Earth Summit of June 1992 in Rio de Janeiro. Speculate on the reasons why birth control was not on the agenda. Should birth control be utilized to control the size of the human population? If so, what role should the United States play in making birth control available to all?

Objective Questions

1. After a country has undergone the demographic transition, the death rate and the birthrate are both ________________ (high, low).
2. If a country has a pyramid-shaped age-structure diagram, most individuals are ________________ (prereproductive, reproductive, or postreproductive).
3. LDCs are not as ________________ as MDCs.
4. When a population is undergoing exponential growth, the increase in number of people each year is ______________ (higher, lower) than the year before.
5. The gas best associated with global warming is ________________.
6. The chemicals best associated with ozone depletion are ________________.
7. Sewage is biodegradable, but the nutrients released in the process can lead to ________________ of surface waters.
8. Pesticides and radioactive wastes are both subject to biological ___________.
9. Match the boxed terms shown in the following illustration with one of the environmental problems in the key.

Key:

sulfur dioxide	carbon dioxide
hydrocarbons	CFCs

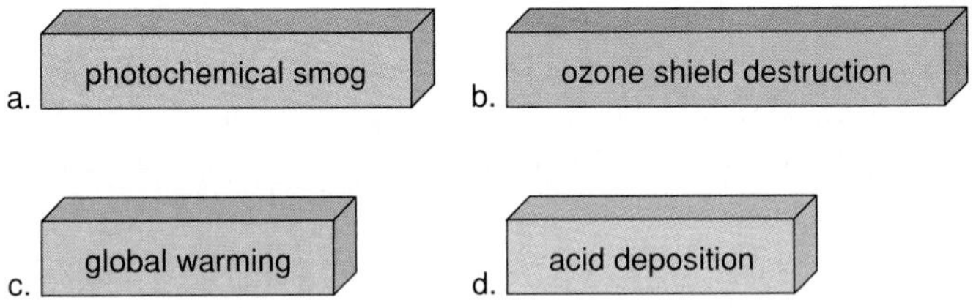

Selected Key Terms

acid deposition The return to earth as rain or snow of the sulfate salts or nitrate salts of acids produced by commercial and industrial activities on earth. 470

biotic potential The maximum population growth rate under ideal conditions. 460

carrying capacity The largest number of organisms of a particular species that can be maintained indefinitely in an ecosystem. 460

cultural eutrophication Enrichment of a body of water due to human activities, causing excessive growth of producers and then death of these and other inhabitants. 466

demographic transition A decline in the birthrate following a reduction in the death rate so that the population growth rate is lowered. 461

desertification (dez-ert″ ĭ-fi-ka-shun) Desert conditions caused by human misuse of land. 463

doubling time The number of years it takes for a population to double in size. 460

environmental resistance Sum total of factors in the environment that limit the numerical increase of a population in a particular region. 460

exponential growth Growth, particularly of a population, in which the increase occurs in the same manner as compound interest. 458

greenhouse effect Carbon dioxide (CO_2) buildup in the atmosphere as a result of fossil fuel combustion; retains and reradiates heat, creating an abnormal rise in the earth's average temperature. 470

hazardous waste Waste containing chemicals hazardous to life. 465

ozone shield A layer of ozone (O_3) present in the upper atmosphere that protects the earth from damaging ultraviolet light. Nearer the earth, ozone is a pollutant. 471

photochemical smog Air pollution that contains nitrogen oxides (NO_x) and hydrocarbons, which react to produce ozone and peroxylacetyl nitrate (PAN). 469

replacement reproduction A population in which each person is replaced by only one child. 462

Further Readings for Part Six

Bar-Yosef, O., and Vandermeersch, B. April 1993. Modern humans in the Levant. *Scientific American.*

Bazzaz, F. A., and Fajer, E. D. January 1992. Plant life in a CO_2-rich world. *Scientific American.*

Berner, R. A., and Lasaga, A. C. March 1989. Modeling the geochemical carbon cycle. *Scientific American.*

Blumenschine, R. J., and Cavallo, J. A. October 1992. Scavenging and human evolution. *Scientific American.*

Brown, B. E., and Ogden, J. C. January 1993. Coral bleaching. *Scientific American.*

Brown, L. R., et al. 1990. *State of the world: 1990.* New York: W. W. Norton and Co.

Coffin, M. F., and Eldholm, O. October 1993. Large igneous provinces. *Scientific American.*

Colinvaux, P. A. May 1989. The past and future Amazon. *Scientific American.*

Corson, W. M., ed. 1990. *The global ecology handbook: What you can do about the environmental crisis.* Boston: Beacon Press.

Cunningham, W. P., and Saigo, B. W. 1992. *Environmental science: A global concern.* 2d ed. Dubuque, Iowa: Wm. C. Brown Publishers.

Doolittle, R. F., and Bork, P. October 1993. Evolutionary mobile modules in proteins. *Scientific American.*

Enger, E. D., and Smith, B. F. 1991. *Environmental science: The study of interrelationships.* Dubuque, Iowa: Wm. C. Brown Publishers.

Francis, C. A., Flora, C. B., and King, L. D., eds. 1990. *Sustainable agriculture in temperate zones.* New York: John Wiley and Sons.

Goudie, A. 1990. *The human impact on the natural environment.* 3d ed. Cambridge, Mass.: MIT Press.

Goulding, M. March 1993. Flooded forests of the Amazon. *Scientific American.*

Greenberg, J. H., and Ruhlen, M. November 1992. Linguistic origins of native Americans. *Scientific American.*

Holloway, M. July 1993. Sustaining the Amazon. *Scientific American.*

Holloway, M., and Horgan, J. October 1991. Soiled shores. *Scientific American.*

Homer-Dixon, T. F., Boutwell, J. H., and Rathjens, G. W. February 1993. Environmental change and violent conflict. *Scientific American.*

Houghton, R. A., and Woodwell, G. M. April 1989. Global climatic change. *Scientific American.*

Jones, P. D., and Wigley, T. M. L. August 1990. Global warming trends. *Scientific American.*

Klein, J., Takahata, N., and Ayala, F. J. December 1993. MHC polymorphism and human origins. *Scientific American.*

Lents, J. M., and Kelly, W. J. October 1993. Clearing the air in Los Angeles. *Scientific American.*

Levinton, J. S. November 1992. The big bang of animal evolution. *Scientific American.*

Miller, J. T. 1992. *Living in the environment.* 7th ed. Belmont, Calif.: Wadsworth.

Milton, K. August 1993. Diet and primate evolution. *Scientific American.*

Newell, R. E., Reichle, H. G., Jr., and Seiler, W. October 1989. Carbon monoxide and the burning earth. *Scientific American.*

Nichol, J., and Newton, C. 1990. *The mighty rain forest.* London: Newton Abbott.

Odum, E. P. 1989. *Ecology and our endangered life-support systems.* Sunderland, Mass.: Sinauer Associates.

Pääbo, Svante. November 1993. Ancient DNA. *Scientific American.*

Pollack, H. N., and Chapman, D. S. June 1993. Underground records of changing climate. *Scientific American.*

Regenold, J. P., Papendik, R. I., and Parr, J. F. June 1990. Sustainable agriculture. *Scientific American.*

Rennie, J. January 1992. Living together. *Scientific American.*

Repetto, R. April 1990. Deforestation in the tropics. *Scientific American.*

Repetto, R. June 1992. Accounting for environmental assets. *Scientific American.*

Ricklefs, R. E. 1989. *Ecology.* 4th ed. New York: Chiron Press.

Robey, B., Rutstein, S., and Morris, L. December 1993. The fertility decline in developing countries. *Scientific American.*

Ross, P. E. May 1992. Eloquent remains. *Scientific American.*

Ruddiman, W. F., and Kutzbach, J. E. March 1991. Plateau uplift and climatic change. *Scientific American.*

Scientific American. September 1989. Planet earth. Special issue.

Scientific American. September 1990. Energy for planet earth. Special issue.

Seyfarth, R. M., and Cheney, D. L. December 1992. Meaning and mind in monkeys. *Scientific American.*

Smith, R. E. 1990. *Ecology and field biology.* 4th ed. New York: Harper and Row Publishers.

Storch, G. February 1992. The mammals of island Europe. *Scientific American.*

Tattersall, I. August 1992. Evolution comes to life. *Scientific American.*

Tattersall, I. January 1993. Madagascar's lemurs. *Scientific American.*

Terborgh, J. May 1992. Why American songbirds are vanishing. *Scientific American.*

Thorne, A. G., and Wolpoff, M. H. April 1992. The multiregional evolution of humans. *Scientific American.*

Toth, N., Clark, D., and Ligabue, G. July 1992. The last stone ax makers. *Scientific American.*

Vickers-Rich, P., and Rich, T. H. July 1993. Australia's polar dinosaurs. *Scientific American.*

White, R. M. July 1990. The great climate debate. *Scientific American.*

Whitmore, T. C. 1991. *An introduction to tropical rain forests.* New York: Oxford University Press.

Wilson, A. C., and Cann, R. L. April 1992. The recent African genesis of humans. *Scientific American.*

York, D. January 1993. The earliest history of the earth. *Scientific American.*

Appendix A

Table of Chemical Elements

Table of Chemical Elements

Atomic number / Atomic weight / Chemical symbol

1 1
H
hydrogen

group Ia	IIa	IIIb	IVb	Vb	VIb	VIIb	VIII			Ib	IIb	IIIa	IVa	Va	VIa	VIIa	VIIIa
1 1 H hydrogen																	2 4 He helium
3 7 Li lithium	4 9 Be beryllium											5 11 B boron	6 12 C carbon	7 14 N nitrogen	8 16 O oxygen	9 19 F fluorine	10 20 Ne neon
11 23 Na sodium	12 24 Mg magnesium											13 27 Al aluminum	14 28 Si silicon	15 31 P phosphorus	16 32 S sulfur	17 35 Cl chlorine	18 40 Ar argon
19 39 K potassium	20 40 Ca calcium	21 45 Sc scandium	22 48 Ti titanium	23 51 V vanadium	24 52 Cr titanium	25 55 Mn manganese	26 56 Fe iron	27 59 Co cobalt	28 59 Ni nickel	29 64 Cu copper	30 65 Zn zinc	31 70 Ga gallium	32 73 Ge germanium	33 75 As arsenium	34 79 Se selenium	35 80 Br bromine	36 84 Kr krypton
37 85 Rb rubidium	38 88 Sr strontium	39 89 Y yttrium	40 91 Zr zirconium	41 93 Nb niobium	42 96 Mo molybdenum	43 98 Tc technetium	44 101 Ru ruthenium	45 103 Rh rhodium	46 106 Pd palladium	47 108 Ag silver	48 112 Cd cadmium	49 115 In indium	50 119 Sn tin	51 122 Sb antimony	52 128 Te tellurium	53 127 I iodine	54 131 Xe xenon
55 133 Cs cesium	56 137 Ba barium	57 139 La lanthanum	72 178 Hf hafnium	73 181 Ta tantalum	74 184 W tungsten	75 186 Re rhenium	76 190 Os osmium	77 192 Ir iridium	78 195 Pt platinum	79 197 Au gold	80 201 Hg mercury	81 204 Tl thallium	82 207 Pb lead	83 209 Bi bismuth	84 210 Po polonium	85 210 At astatine	86 222 Rn radon
87 223 Fr francium	88 226 Ra radium	89 227 Ac lactinium	104 261 Rf rutherfordium	105 260 Ha hahnium													

Appendix B

Metric System

Standard Metric Unit	Abbreviation
Unit of length: meter	m
Unit of weight: gram	g
Unit of volume: liter	l

Length

Unit and Abbreviation (by increasing order)	Metric Equivalent	Metric-to-English Conversion Factor		English-to-Metric Conversion Factor	
Nanometer (nm)	$= 10^{-9}$m (10^{-3}µm)	—		—	
Micrometer (µm)	$= 10^{-6}$m (10^{-3}mm)	—		—	
Millimeter (mm)	= (10^{-3}) 0.001 m	mm→in:	0.039	—	
Centimeter (cm)	= 0.01 (10^{-2})m	cm→in:	0.39	in→cm: ft→cm:	2.54 30.5
Meter (m)	= 100 (10^{2})cm = 1000 mm	m→ins: m→ft:	39.37 3.28	— ft→m:	 0.305
		m→yd:	1.09	yd→m:	0.91
Kilometer (km)	= 1000(10^{3})m	km→mi:	0.62	mi→km:	1.61

How to Use This Table

To convert English to metric (mi→km):

Multiply	English unit	×	English-to-metric factor	=	metric
	5 mi	×	1.61	=	8.05 km

To convert metric to English (km→mi):

Multiply	metric unit	×	metric-to-English factor	=	English
	8.05 km	×	0.62	=	5 mi

Think Metric

Length

1. The speed of a car is 60 mph or ___________ kilometers per hour.
2. A man who is ___________ feet tall is 180 cm.
3. A six-inch ruler is ___________ cm.
4. ___________ yard is almost a meter (0.9 m).

Weight

Unit and Abbreviation (by increasing order)	Metric Equivalent	Metric-to-English Conversion Factor		English-to-Metric Conversion Factor	
Nanogram (ng)	$= 10^{-9}$ g	—		—	
Microgram (μg)	$= 10^{-6}$ g	—		—	
Milligram (mg)	$= 10^{-3}$ (0.001) g	mg→grains: approx. 0.015		—	
Gram (g)	= 1000 mg	g→grains:	15.43	—	
		g→oz:	0.035	oz→g: lb→g:	28.3 453.6
Kilogram (kg)	= 1000 (10^3)g	kg→lb:	2.2	lb→kg:	0.45
Metric ton (t)	= 1000 kg	t→tons:	1.10	ton→t:	0.90

How to Use This Table

To convert English to metric (lb→kg):

Multiply	English unit	×	English-to-metric factor	=	metric
	5 lb	×	0.45	=	2.27 kg

To convert metric to English (kg→lb):

Multiply	metric unit	×	metric-to-English factor	=	English
	2.27 kg	×	2.2	=	5 lb

Think Metric

Weight

1. One pound of hamburger is __________ grams.
2. The average human male brain weighs 1.4 kilograms. (_____ lb _____ oz).
3. A person who weighs __________ pounds weighs 70 kilograms.
4. Lucia Zarate weighed __________ kilograms (13 lb) at age 20.

Volume

Unit and Abbreviation (by increasing order)	Metric Equivalent	Metric-to-English Conversion Factor	English-to-Metric Conversion Factor
Microliter (µl)	= 10^{-6} l(10^{-3} ml)	—	—
Milliliter (ml)	= 10^{-3} l	ml→drops: approx. 15–16	—
	= 1 cm^3 (cc)	ml→tsp: approx. 1/4	tsp→ml: approx. 5
	= 1000 mm^3	ml→fl oz: 0.03	fl oz→ml: 30
		—	pt→ml: 47
		—	qt→ml: 95
Liter (l)	= 1000 ml	l→qt: 1.06	qt→l: 0.95
		l→pt: 2.1	pt→l: 0.47
		l→gal: 0.26	gal→l: 3.79
Kiloliter (kl)	= 1000 l	kl→gal: 264.17	—

How to Use This Table

To convert English to metric (fl oz→ml):

Multiply	English unit	×	English-to-metric factor	=	metric
	5 fl oz	×	30	=	150 ml

To convert metric to English (ml→fl oz):

Multiply	metric unit	×	metric-to-English factor	=	English
	150 ml	×	0.03	=	5 fl oz

Think Metric

Volume

1. One can of soda (12 fl oz) contains __________ ml.
2. The average human body contains between 10 pt and 12 pt of blood or between ________ liters and ________ liters.
3. One cubic foot of water (7.48 gal) is ________ liters.
4. If a gallon of unleaded gasoline costs $1.00, a liter costs ________.

Answers

Length: (1) 97; (2) 6; (3) 15; (4) One
Weight: (1) 453.6; (2) 3 lb, 1.3 oz; (3) 154; (4) 5.85
Volume: (1) 360; (2) 4.7, 5.6; (3) 28.4; (4) 26 cents

Units of Temperature

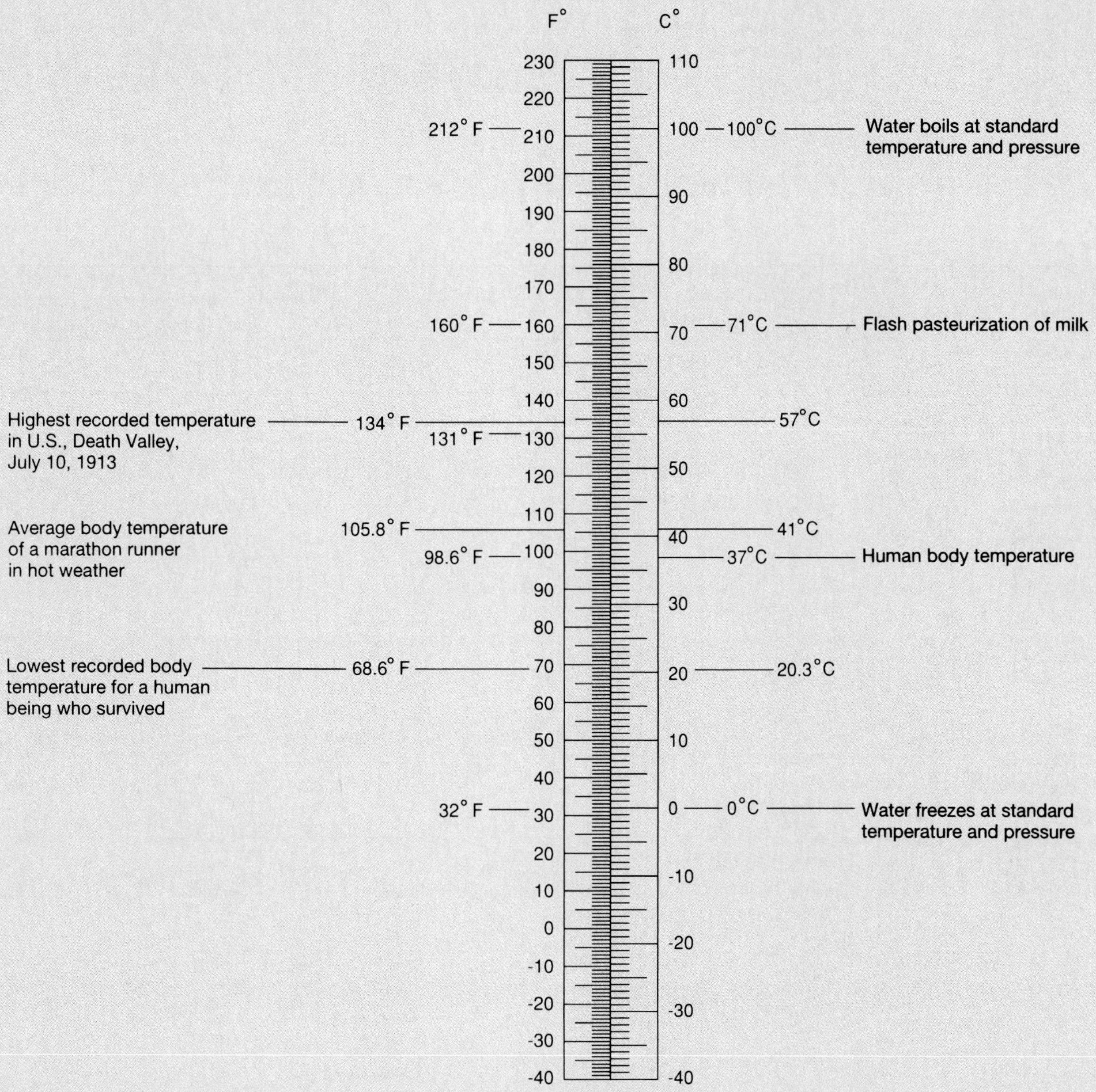

To convert temperature scales:

Fahrenheit to Centigrade $°C = \frac{5}{9}(°F - 32)$

Centigrade to Fahrenheit $°F = \frac{9}{5}\ °C + 32$

Appendix C

Drugs of Abuse

	Drugs	Often Prescribed Brand Names	Medical Uses	Potential Physical Dependence	Potential Psychological Dependence	Tolerance
Narcotics	Opium	Dover's Powder, Paregoric	Analgesic, antidiarrheal	High	High	Yes
	Morphine	Morphine	Analgesic	High	High	Yes
	Codeine	Codeine	Analgesic, antitussive	Moderate	Moderate	Yes
	Heroin	None	None	High	High	Yes
	Meperidine (Pethidine)	Demerol, Pethadol	Analgesic	High	High	Yes
	Methadone	Dolophine, Methadone, Methadose	Analgesic, heroin substitute	High	High	Yes
	Other narcotics	Dilaudid, Leritine, Numorphan, Percodan	Analgesic, antidiarrheal, antitussive	High	High	Yes
Depressants	Chloral hydrate	Noctec, Somnos	Hypnotic	Moderate	Moderate	Probable
	Barbiturates	Amytal, Butisol, Nembutal, Phenobarbital, Seconal, Tuinal	Anesthetic, anticonvulsant, sedation, sleep	High	High	Yes
	Glutethimide	Doriden	Sedation, sleep	High	High	Yes
	Methaqualone	Optimil, Parest, Quaalude, Somnafac, Sopor	Sedation, sleep	High	High	Yes
	Tranquilizers	Equanil, Librium, Miltown, Serax, Tranxene, Valium	Antianxiety, muscle relaxant, sedation	Moderate	Moderate	Yes
	Other depressants	Clonopin, Dalmane, Dormate, Noludar, Placydil, Valmid	Antianxiety, sedation, sleep	Possible	Possible	Yes
Stimulants	Cocaine*	Cocaine	Local anesthetic	Possible	High	Yes
	Amphetamines	Benzedrine, Biphetamine, Desoxyn, Dexedrine	Hyperkinesis, narcolepsy, weight control	Possible	High	Yes
	Phenmetrazine	Preludin	Weight control	Possible	High	Yes
	Methylphenidate	Ritalin	Hyperkinesis	Possible	High	Yes
	Other stimulants	Bacarate, Cylert, Didrex, Ionamin,Plegine, Pondimin, Pro-Sate, Sanorex, Voranil	Weight control	Possible	Possible	Yes
Hallucinogens	LSD	None	None	None	Degree unknown	Yes
	Mescaline	None	None	None	Degree unknown	Yes
	Psilocybin-Psilocyn	None	None	None	Degree unknown	Yes
	MDA	None	None	None	Degree unknown	Yes
	PCP†	Sernylan	Veterinary anesthetic	None	Degree unknown	Yes
	Other hallucinogens	None	None	None	Degree unknown	Yes
Cannabis	Marijuana, hashish, hashish oil	None	Glaucoma	Degree unknown	Moderate	Yes

Source: Drugs of Abuse, *produced by the Affairs in Cooperation with the Office of Public Science and Technology.*

**Designated a narcotic under the Controlled Substances Act.*

†Designated a depressant under the Controlled Substances Act.

Duration of Effects (in hours)	Usual Methods of Administration	Possible Effects	Effects of Overdose	Withdrawal Syndrome
3–6	Oral, smoked	Euphoria, drowsiness, respiratory depression, constricted pupils, nausea	Slow and shallow breathing, clammy skin, convulsions, coma, possible death	Watery eyes, runny nose, yawning, loss of appetite, irritability, tremors, panic, chills and sweating, cramps, nausea
3–6	Injected, smoked			
3–6	Oral, injected			
3–6	Injected, sniffed			
3–6	Oral, injected			
12–24	Oral, injected			
3–6	Oral, injected			
5–8	Oral			
1–16	Oral, injected	Slurred speech, disorientation, drunken behavior without odor of alcohol	Shallow respiration, cold and clammy skin, dilated pupils, weak and rapid pulse, coma, possible death	Anxiety, insomnia, tremors, delirium, convulsions, possible death
4–8	Oral			
4–8	Oral			
4–8	Oral			
4–8	Oral			
2	Injected, sniffed	Increased alertness, excitation, euphoria, dilated pupils, increased pulse rate and blood pressure, insomnia, loss of appetite	Agitation, increased body temperature, hallucinations, convulsions, possible death	Apathy, long periods of sleep, irritability, depression, disorientation
2–4	Oral, injected			
2–4	Oral			
2–4	Oral			
2–4	Oral			
Variable	Oral	Illusions and hallucinations (with exception of MDA), poor perception of time and distance	Longer, more intense "trip" episodes, psychosis, possible death	Withdrawal syndrome not reported
Variable	Oral, injected			
Variable	Oral			
Variable	Oral, injected, sniffed			
Variable	Oral, injected, smoked			
Variable	Oral, injected, sniffed			
2–4	Oral, smoked	Euphoria, relaxed inhibitions, increased appetite, disoriented behavior	Fatigue, paranoia, possible psychosis	Insomnia, hyperactivity, and decreased appetite reported in a limited number of individuals

Appendix D

Leading Types of Cancer

Cancer incidence and deaths by site are given in table 20.3. Five-year survival rates for the most common cancers are given in table D1.

Breast Cancer

Warning Signals: Breast changes that persist, such as a lump, thickening, swelling, dimpling, skin irritation, distortion, retraction, scaliness, pain, or tenderness of the nipple.

Risk Factors: Aging is a risk factor for breast cancer, as is personal or family history of breast cancer; early age at menarche;[1] late age of menopause; never had children or late age at first live birth; and higher education and socioeconomic status. International variability in cancer incidence rates correlate with variations in diet, especially fat intake, although a causal role for dietary factors has not been firmly established.

Early Detection: The American Cancer Society recommends that women have a screening mammogram by age 40; women 40–49 should have a mammogram every 1–2 years; asymptomatic women age 50 and over should have a mammogram every year. In addition, a clinical physical examination of the breast is recommended every 3 years for women 20–40, and every year for those over 40. The American Cancer Society also recommends monthly breast self-examination as a routine good health habit for women 20 years or older. Most breast lumps are not cancer, but only a physician can make a diagnosis.

Treatment: Taking into account the medical situation and the patient's preferences, treatment may require lumpectomy (local removal of the tumor), mastectomy (surgical removal of the breast), radiation therapy, chemotherapy, or hormone manipulation therapy. Often, 2 or more methods are used in combination.

New techniques in recent years have made breast reconstruction possible after mastectomy, and the cosmetic results usually are good. Reconstruction has become an important part of treatment and rehabilitation.

1. First menstrual period.

Lung Cancer

Warning Signals: Persistent cough, sputum streaked with blood, chest pain, recurring pneumonia or bronchitis.

Risk Factors: Cigarette smoking; exposure to certain industrial substances, such as arsenic, certain organic chemicals and asbestos, particularly for persons who smoke; radiation exposure from occupational, medical, and environmental sources. Residential radon exposure may increase risk, especially in cigarette smokers. Exposure to passive cigarette smoke increases the risk for nonsmokers.

Early Detection: Because symptoms often don't appear until the disease is in advanced stages, early detection is very difficult. Chest X ray, analysis of the types of cells contained in sputum, and fiberoptic examination of the bronchial passages assist diagnosis.

Treatment: Treatment options include surgery, radiation therapy, and chemotherapy. In some cancers, chemotherapy alone or combined with radiation has replaced surgery as the treatment of choice; on this regimen, a large percentage of patients experience remission, which in some cases is long-lasting.

Prostate Cancer

Warning Signals: Weak or interrupted urine flow; inability to urinate, or difficulty starting or stopping the urine flow; the need to urinate frequently, especially at night; blood in the urine; pain or burning on urination; continuing pain in the lower back, pelvis, or upper thighs. Most of these symptoms are nonspecific and may be similar to those caused by benign conditions such as infection or prostate enlargement.

Risk Factors: Aging is a risk factor for prostate cancer, with over 80% of all prostate cancers diagnosed in men over age 65. The disease is most common in northwestern Europe and North America. It is rare in the Near East, Africa, Central America, and South America. For reasons not currently known, African Americans have the highest incidence rate in the world.

Table D1

Five-Year Survival Rates by Stage at Diagnosis*

SITE	ALL STAGES %	LOCAL %	REGIONAL %	DISTANT %
Breast (female)	78	93	71	18
Lung	13	46	13	1
Prostate	76	91	80	28
Colon/rectum	57	89	58	6
Bladder	78	90	46	9
Uterus	149	183	120	40
Skin (melanoma)	83	91	54	13
Pancreas	3	8	4	1
Ovary	39	89	37	18

**Adjusted for normal life expectancy. This chart is based on cases diagnosed in 1983–87, followed through 1989.*

Source: Data from Cancer Statistics Branch, National Cancer Institute.

Early Detection: Every man 40 and over should have a digital rectal examination as part of his regular annual physical checkup. In addition, the American Cancer Society recommends that men 50 and over have annual prostate-specific antigen blood testing. If either result is suspicious, further evaluation in the form of transrectal ultrasound should be performed.

Treatment: Surgery, radiation, and/or hormones and anticancer drugs are treatment options. Hormone treatment and anticancer drugs may control prostate cancer for long periods by shrinking the size of the tumor, thus relieving pain.

Colorectal Cancer

Warning Signals: Rectal bleeding, blood in the stool, change in bowel habits.

Risk Factors: Personal or family history of cancer or polyps of the colon or rectum, and inflammatory bowel disease. High-fat and/or low-fiber diet may be associated with increased risk.

Early Detection: Digital rectal examination, stool blood test, and proctosigmoidoscopy are recommended for early detection. The American Cancer Society recommends digital rectal examination be performed annually after age 40. The stool blood test (fecal occult blood test) should be done annually after age 50. The Society recommends a proctosigmoidoscopy be done every 3–5 years after age 50. In proctosigmoidoscopy, the physician uses a long, lighted tube to inspect the rectum and lower colon. To detect cancers higher in the colon, longer, flexible instruments are being used, as well as a rigid scope. If any of these tests reveal possible problems, more extensive studies, such as colonoscopy (examination of the entire colon) and barium enema (an X-ray procedure in which the intestines are viewed), may be needed.

Treatment: Surgery, at times combined with radiation, is the most effective method of treating colorectal cancer. Combinations of chemotherapy and immunologic agents have recently been described as beneficial in postoperative patients with cancerous lymph nodes.

Bladder Cancer

Warning Signals: Blood in the urine, usually associated with increased frequency of urination.

Risk Factors: Smoking is the greatest risk factor, with smokers experiencing twice the risk of nonsmokers. Smoking is estimated to be responsible for approximately 47% of the bladder cancer deaths among men and 37% among women. People living in urban areas and workers exposed to dye, rubber, or leather also are at higher risk.

Early Detection: Bladder cancer is diagnosed by examination of the bladder wall with a cytoscope, a slender tube fitted with a lens and light that can be inserted into the tract through the urethra.

Treatment: Surgery, alone or in combination with other treatments, is used in over 90% of cases. Preoperative chemotherapy alone or with radiation before cystectomy (bladder removal) has improved some treatment results.

Lymphoma

Warning Signals: For Hodgkin disease: enlarged lymph nodes, itching, fever, night sweats, and weight loss. Fever can come and go in periods of several days or weeks. For non-Hodgkin lymphoma: enlarged lymph nodes, anemia, weight loss, and fever.

Risk Factors: Risk factors are largely unknown, but in part involve reduced immune function and exposure to certain infectious agents. Persons with organ transplants are at higher risk due to altered immune function. Human immunodeficiency virus (HIV) and human T-cell leukemia/lymphoma virus-I (HTLV-I) are associated with increased risk of non-Hodgkin lymphoma. Burkitt lymphoma in Africa is partly caused by the Epstein-Barr herpes virus. Other possible risk factors include exposures to herbicides, industrial solvents, and vinyl chloride.

Treatment: For Hodgkin disease, chemotherapy and radiotherapy are recommended for most patients. For non-Hodgkin lymphoma with localized lymph node disease, radiotherapy is recommended. Patients with later stage disease often benefit from the addition of chemotherapy.

Uterine Cancer

Warning Signals: Bleeding outside of the normal menstrual period or after menopause or unusual vaginal discharge.

Risk Factors: For cervical cancer: an early age at first intercourse, multiple sex partners, cigarette smoking, and certain sexually transmitted diseases. For endometrial cancer: early menarche, late menopause, history of infertility, failure to ovulate, tamoxifen or unopposed estrogen therapy, obesity.

Early Detection: The Pap smear, a simple procedure that can be performed at appropriate intervals by health care professionals, is done as part of a pelvic examination. The Pap smear is only partially effective in detecting endometrial cancer. Women 40 and over should have an annual pelvic exam, and those at high risk of developing endometrial cancer should have an endometrial tissue sample evaluated at menopause.

Treatment: Uterine cancers generally are treated by surgery or radiation or by a combination of the 2. In precancerous (in situ) stages, changes in the cervix may be treated by cryotherapy (the destruction of cells by extreme cold), by electrocoagulation (the destruction of tissue through intense heat by electric current), or by local surgery. Precancerous endometrial changes may be treated with the hormone progesterone.

Melanoma

Warning Signals: Any unusual skin condition, especially a change in the size or color of a mole or other darkly pigmented growth or spot; scaliness, oozing, bleeding, or change in the appearance of a bump or nodule, the spread of pigmentation beyond its border, a change in sensation, itchiness, tenderness, or pain.

Risk Factors: Excessive exposure to ultraviolet radiation; fair complexion; occupational exposure to coal tar, pitch, creosote, arsenic compounds, or radium.

The sun's ultraviolet rays are strongest between 10 A.M. and 3 P.M. Exposure at these times should be avoided, and protective clothing should be worn. Sunscreens should be used. These come in various strengths, ranging from those facilitating gradual tanning to those that allow practically no tanning. Because of the possible link between severe sunburns in childhood and greatly increased risk of melanoma in later life, children, in particular, should be protected from the sun.

Early Detection: Early detection is critical. Recognition of changes in skin growths or the appearance of new growths is the best way to find early skin cancer. Adults should practice skin self-evaluation once a month, and suspicious lesions should be evaluated promptly by a physician. A sudden or progressive change in a mole's appearance should be checked by a physician. Melanomas often start as small, molelike growths that increase in size, change color, become ulcerated, and bleed easily from a slight injury. A simple **ABCD** rule outlines the warning signals of melanoma: **A** is for asymmetry. One-half of the mole does not match the other half. **B** is for border irregularity. The edges are ragged, notched, or blurred. **C** is for color. The pigmentation is not uniform. **D** is for diameter greater than 6 mm. Any sudden or progressive increase in size should be of special concern.

Treatment: For malignant melanoma, the primary growth must be adequately excised, and it may be necessary to remove nearby lymph nodes. Removal and microscopic examination of all suspicious moles are essential. Advanced cases of melanoma are treated according to the characteristics of the case.

Oral Cancer

Warning Signals: A sore that bleeds easily and doesn't heal; a lump or thickening; a red or white patch that persists. Difficulty in chewing, swallowing, or moving tongue or jaws are often late changes.

Risk Factors: Cigarette, cigar, or pipe smoking; use of smokeless tobacco; excess use of alcohol.

Early Detection: Cancer can affect any part of the oral cavity, including the lip, tongue, mouth, and throat. Dentists and primary care physicians have the opportunity, during regular checkups, to see abnormal tissue changes and to detect cancer at an early, curable stage.

Treatment: Principal methods are radiation therapy and surgery. Chemotherapy is being studied as an adjunct to surgery in advanced disease.

Leukemia

Warning Signals: Fatigue, paleness, weight loss, repeated infections, bruising easily, and nosebleeds or other hemorrhages. In children, these symptoms can appear suddenly. Chronic leukemia can progress slowly and with few symptoms.

Risk Factors: Leukemia strikes both sexes and all ages. Causes of most cases are unknown. Persons with Down syndrome and certain other genetic abnormalities have higher than normal incidence of leukemia. It has also been linked to excessive exposure to ionizing radiation and to certain chemicals such as benzene, a commercially used toxic liquid that is also present in lead-free gasoline. Certain forms of leukemia and lymphoma are caused by a retrovirus, HTLV-I (human T-cell leukemia/lymphoma virus-I).

Early Detection: Because symptoms often resemble those of other, less serious conditions, leukemia can be difficult to diagnose early. When a physician does suspect leukemia, diagnosis can be made using blood tests and biopsy of the bone marrow.

Treatment: Chemotherapy is the most effective method of treating leukemia. Various anticancer drugs are used, either in combination or as single agents. Transfusions of blood components and antibiotics are used as supportive treatments. To prevent persistence of hidden cells, therapy of the central nervous system has become standard treatment, especially in acute lymphocytic leukemia. Under appropriate conditions, bone marrow transplantation may be useful in the treatment of certain leukemias.

Pancreatic Cancer

Warning Signals: Cancer of the pancreas is a "silent disease," one that occurs without symptoms until it is in advanced stages.

Risk Factors: Very little is known about what causes the disease or how to prevent it. Risk increases after age 50, with the most cases occurring between ages 65 and 79. Smoking is a risk factor; incidence is more than twice as high for smokers as nonsmokers. Some studies have suggested associations with chronic pancreatitis, diabetes, or cirrhosis, but these findings have not been confirmed. In countries where the diet is high in fat, pancreatic cancer rates are higher.

Early Detection: At present, only a biopsy yields a certain diagnosis, and because of the "silent" course of the disease, the need for biopsy is likely to be obvious only after the disease has advanced. Researchers are focusing on ways to diagnose pancreatic cancer before symptoms occur. Ultrasound imaging and computerized tomography (CT) scans are being tried.

Treatment: Surgery, radiation therapy, and anticancer drugs are treatment options but have had little influence on the outcome. Diagnosis is usually so late that none of these is used.

Ovarian Cancer

Warning Signals: Ovarian cancer is often "silent," showing no obvious signs or symptoms until late in its development. The most common sign is enlargement of the abdomen, which is caused by the accumulation of fluid. Rarely will there be abnormal vaginal bleeding. In women over 40, vague digestive disturbances (stomach discomfort, gas, distention) that persist and cannot be explained by any other cause may indicate the need for a thorough evaluation for ovarian cancer.

Risk Factors: Risk for ovarian cancer increases with age. The highest rates are for women over 60. Women who have never had children are twice as likely to develop ovarian cancer as those who have. Early age at first pregnancy, early menopause, and the use of oral contraceptives, which reduces the frequency of ovulation, appear to be protective against ovarian cancer. If a woman has had breast cancer, her chances of developing ovarian cancer double. Certain rare genetic disorders are associated with increased risk. With the exception of Japan, the highest incidence rates are reported from the more industrialized countries.

Early Detection: Periodic, thorough pelvic examinations are important. The Pap smear, useful in detecting cervical cancer, does not reveal ovarian cancer. Women over the age of 40 should have a cancer-related checkup every year.

Treatment: Surgery, radiation therapy, and drug therapy are treatment options. Surgery usually includes the removal of one or both ovaries (oophorectomy), the uterus (hysterectomy), and the fallopian tubes (salpingectomy). In some very early tumors, only the involved ovary will be removed, especially in young women. In advanced disease, an attempt is made to remove all intraabdominal disease to enhance the effect of chemotherapy.

Cancer in Children

Cancers in children often are difficult to recognize. Parents should see that their children have regular medical checkups and should be alert to any unusual symptoms that persist. These include an unusual mass or swelling; unexplained paleness and loss of energy; sudden tendency to bruise; a persistent, localized pain or limping; prolonged, unexplained fever or illness; frequent headaches, often with vomiting; sudden eye or vision changes; and excessive, rapid weight loss.

Some of the main childhood cancers are as follows:

1. Leukemia.
2. Osteogenic sarcoma and Ewing sarcoma are bone cancers. These may cause no pain at first, and swelling in the area of the tumor is often the first sign.
3. Neuroblastoma can appear anywhere but usually in the abdomen, where a swelling occurs.
4. Rhabdomyosarcoma, the most common soft tissue sarcoma, can occur in the head and neck area, genitourinary area, trunk, and extremities.
5. Brain cancers in early stages may cause headaches, blurred or double vision, dizziness, difficulty in walking or handling objects, and nausea.
6. Lymphomas and Hodgkin disease are cancers that involve the lymph nodes but also may invade bone marrow and other organs. They may cause swelling of lymph nodes in the neck, armpit, or groin. Other symptoms may include general weakness and fever.
7. Retinoblastoma, an eye cancer, usually occurs in children under age 4. When detected early, cure is possible with appropriate treatment.
8. Wilms' tumor, a kidney cancer, may be recognized by a swelling or lump in the abdomen.

Childhood cancers can be treated by a combination of therapies. Treatment is coordinated by a team of experts including oncologic physicians, pediatric nurses, social workers, psychologists, and others who assist children and their families.

Source: From the American Cancer Society, Inc., *Cancer Facts & Figures*, 1993. Used with permission.

Appendix E

Answers to Objective Questions

Introduction

1. vertebrates; 2. solar; 3. copy; 4. cultural; 5. multicellular; 6. control; 7. hypothesis; 8. society.

Chapter 1

1. atoms; 2. neutrons; 3. ionic, covalent; 4. hydrogen; 5. hydrogen, lower; 6. enzymes; 7. amino acids, helix (or spiral), shape; 8. glucose, energy; 9. glycerol, fatty acid; 10. DNA, nucleotides; 11. see fig. 1.7 in text.

Chapter 2

1. c; 2. e; 3. a; 4. d; 5. b; 6. cytoskeleton; 7. hypotonic; 8. active site; 9. electron transport system, cristae; 10. 2, 36; 11. a. nucleus—DNA directs; b. nucleolus—RNA helps; c. rough ER produces; d. smooth ER transports; e. Golgi apparatus packages and secretes; 12. see fig. 2.13 in text.

Chapter 3

1. epithelium; 2. simple columnar epithelium; 3. matrix; 4. loose connective, fat; 5. tendons; 6. connective, plasma; 7. striated; 8. neurons and glial cells; 9. dermis; 10. keratin; 11. constancy, tissue; 12. a. columnar epithelium, lining of intestine (digestive tract), protection and absorption; b. cardiac muscle, wall of heart, pumps blood; c. compact bone, skeleton, support and protection.

Chapter 4

1. amylase, maltose; 2. epiglottis; 3. esophagus, protein; 4. bile, emulsifies; 5. duodenum; 6. trypsin, pancreatic amylase, lipase; 7. strongly acidic, slightly basic; 8. villi; 9. glycogen; 10. essential amino acids; 11. see fig. 4.1 in text; 12. a. no digestion, enzyme missing; b. no digestion, wrong enzyme; c. digestion, both bile salts and pancreatic lipase are present.

Chapter 5

1. away; 2; aorta; 3. high; 4. lungs; 5. SA; 6. coronary; 7. blood pressure; 8. blood pressure, skeletal muscle contraction; 9. valves; 10. saturated fat, cholesterol; 11. see fig. 5.4 in text.

Chapter 6

1. plasma; 2. oxygen, fight infection; 3. oxyhemoglobin; 4. nucleus, 120; 5. fibrin; 6. neutrophil; 7. antibodies; 8. oxygen, amino acids, glucose; carbon dioxide, other wastes; 9. A, B, no; 10. Rh^-, Rh^+; 11. a. arterial end; b. venous end; c. water, oxygen, nutrients; d. water, carbon dioxide; e. plasma proteins.

Chapter 7

1. tissue fluid, subclavian; 2. purify; 3. thymus; 4. plasma, memory; 5. antibody; 6. APC; 7. lymphokines; 8. cell; 9. histamine; 10. MHC; 11. vaccines; 12. monoclonal; 13. helper (produces lymphokines), suppressor (shuts down response), memory (retains ability to kill same type of infected cell), cytotoxic (kills infected cells).

Chapter 8

1. larynx; 2. alveoli; 3. CO_2, H^+; 4. expanded; 5. bicarbonate; 6. the globin portion of hemoglobin; 7. diffusion; 8. lungs; 9. cigarette smoking; 10. bronchi; 11. see fig. 8.2 in text.

Chapter 9

1. urea; 2. bile pigments, hemoglobin; 3. urethra; 4. glomerulus; 5. water; 6. urea; 7. distal convoluted tubule; 8. ADH; 9. volume, pH; 10. dialysis; 11. see fig. 9.4 in text.

Chapter 10

1. axon; 2. sodium, inside; 3. synaptic cleft; 4. AChE; 5. muscles; 6. cranial, motor, or parasympathetic; internal organs; 7. interneuron; 8. meninges; 9. cerebrum; 10. cerebellum; 11. see fig. 10.1 in text.

Chapter 11

1. axial; 2. spinal; 3. radius, ulna; 4. synovial; 5. antagonistic; 6. tetanus; 7. myofibrils, sarcomeres; 8. creatine phosphate; 9. neuromuscular; 10. calcium; 11.a. sarcolemma; b. T tubules; c. mitrochondrion; d. calcium storage sacs; e. sarcomere; f. myofibril.

Chapter 12

From page 227:

[1]The room is a trapezoid with this shape:

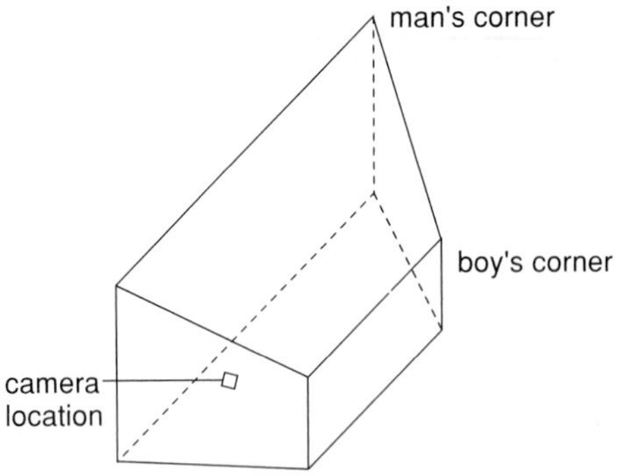

1. brain; 2. chemoreceptors; 3. rods, cones, retina; 4. color, bright (day); 5. rhodopsin; 6. rounds up (accommodation); 7. distant, concave; 8. hammer, anvil, stirrup; 9. equilibrium; 10. cochlear, cochlea; 11. see fig. 12.4 in text.

Chapter 13

1. ADH, oxytocin; 2. hormones; 3. negative feedback; 4. anterior; 5. too little, thyroxin; 6. cortex; 7. Cushing syndrome; 8. calcium; 9. pancreas, cells; 10. blood; 11. see table 13.1 in text.

Chapter 14

1. vas deferens; 2. seminal vesicles; 3. testosterone; 4. blood; 5. vagina; 6. sexual intercourse, birth canal; 7. follicle, endometrial; 8. estrogen, progesterone; 9. laboratory glassware; 10. see fig. 14.1 in text.

Chapter 15

1. nucleic acid, protein; 2. retrovirus; 3. blood; 4. cold sores, genital herpes; 5. rod, spherical, spiral; 6. PID; 7. condom; 8. gummas; 9. gonorrhea, chlamydia, syphilis; 10. yeast; 11. a. gonorrhea; b. AIDS; c. syphilis.

Chapter 16

1. sperm, egg; 2. extraembryonic, amnion; 3. cleavage; 4. implant; 5. differentiation; 6. organs; 7. second; 8. placenta; 9. head; 10. oxytocin; 11. see fig. 16.3 in text; 12. cross-linking.

Chapter 17

1. karyotype; 2. XY, XX; 3. 2N, N; 4. 24; 5. centrioles; 6. homologous chromosomes, sister chromatids; 7. spermatogenesis, oogenesis; 8. 50%; 9. c; 10. a; 11. e; 12. d; 13. b; 14. a; 15. d.

Chapter 18

1. one; 2. *w*; 3. phenotypes; 4. widow's peak; 5. 25%; 6. X; 7. mother; 8. X^BX^b; 9. recessive; 10. autosomal dominant condition; 11. a. cystic fibrosis; b. Tay-Sachs disease; c. neurofibromatosis (NF); d. Huntington disease; e. hemophilia; f. muscular dystrophy.

Chapter 19

1. C, T; 2. old, new; 3. triplet, amino acid; 4. mRNA, tRNA; rRNA; 5. tRNA; 6. translation; 7. restriction, DNA ligase; 8. bovine growth hormone; 9. retrovirus; 10. a. ACU′CCU′GAA′UGC′AAA; b. UGA′GGA′CUU′ACG′UUU; c. threonine-proline-glutamate-cysteine-lysine.

Chapter 20

1. metastasize; 2. proto-oncogene, tumor suppressor; 3. papilloma; 4. cigarette smoke; 5. tumor suppressor gene; 6. dividing; 7. chemotherapy; 8. lymphokine; 9. carcinogens; 10. A, C; 11. see fig. 20.9 in text.

Chapter 21

1. b; 2. a; 3. d; 4. c; 5. reproductive; 6. adaptation; 7. chemical; 8. erect, small; 9. Neanderthals, Cro-Magnon; 10. see table 21.1 in text.

Chapter 22

1. flow; 2. producer; 3. herbivores; 4. trophic; 5. transformed; 6. reservoir; 7. respire; 8. nitrate; 9. nitrogen gas; 10. increasing amount; 11. a. algae; b. rotifers, mosquito larva; c. water fleas, water boatmen; d. catfish, frog; e. greenbacked heron; 12. a. respiration; b. photosynthesis; c. combustion.

Chapter 23

1. low; 2. prereproductive; 3. industrialized; 4. higher; 5. carbon dioxide (CO_2); 6. CFCs; 7. cultural eutrophication; 8. magnification; 9. a. hydrocarbons; b. CFCs; c. carbon dioxide; d. sulfur dioxide.

Glossary

A

accommodation Lens adjustment to see near objects. 233

acetylcholine (ACh) (as″ ē-til-ko′ lēn) A neurotransmitter secreted at the ends of many neurons; responsible for the transmission of a nerve impulse across a synaptic cleft. 189

acetylcholinesterase (AChE) (as″ ē-til-ko″ lin-es′ ter-ās) An enzyme that breaks down acetylcholine. 189

acid A solution in which pH is less than 7; a substance that contributes or liberates hydrogen ions (protons) in a solution. 16

acid deposition The return to earth as rain or snow of the sulfate salts or nitrate salts of acids produced by commercial and industrial activities on earth. 470

acquired immunodeficiency syndrome (AIDS) A disease caused by HIV and transmitted via body fluids; characterized by failure of the immune system. 296

acromegaly (ak″ro-meg′ah-le) A condition resulting from an increase in GH production after adult height has been achieved. 253

acrosome (ak″ro-sōm) Covering on the tip of a sperm that contains enzymes necessary for fertilization. 274

ACTH *See* adrenocorticotropic hormone. 253

actin (ak′ tin) One of 2 major proteins of muscle; makes up thin filaments in myofibrils of muscle fibers. *See* myosin. 221

actin filament An extremely thin fiber found within the cytoplasm that is composed of the protein actin and is involved in the maintenance of cell shape and the movement of cell contents. 37

action potential The change in potential propagated along the membrane of a neuron; the nerve impulse. 186

active site The region on the surface of an enzyme where the substrate binds and where the reaction occurs. 39

active transport Transfer of a substance into or out of a cell from a region of lower concentration to a region of higher concentration by a process that requires a carrier and expenditure of energy. 33

adaptation A decrease in the excitability of receptors in response to continuous constant intensity stimulation; also the fitness of an organism for its environment, including the process by which it becomes fit and is able to survive and reproduce. 228, 429

Addison disease A condition resulting from a deficiency of adrenal cortex hormones; characterized by low blood glucose, weight loss, and weakness. 258

adenine (A) One of 4 organic bases in the nucleotides composing the structure of DNA and RNA. 383

adenosine diphosphate *See* ADP. 26

adenosine triphosphate *See* ATP. 26

adipose tissue A connective tissue in which fat is stored. 49

ADP (adenosine diphosphate) (ah-den′o-sēn tri-fos′fāt) A compound containing adenine, ribose, and 3 phosphates, 2 of which are high-energy phosphates. The breakdown of ATP to ADP makes energy available for energy-requiring processes in cells. 26

adrenal gland A gland that lies atop a kidney and secretes the stress hormones norepinephrine, epinephrine, and the corticoid hormones. 256

adrenocorticotropic hormone (ACTH) Hormone secreted by the anterior lobe of the pituitary gland that stimulates activity in the adrenal cortex. 253

aerobic cellular respiration The complete breakdown of glucose to carbon dioxide and water; requires glycolysis, transition reaction, Krebs cycle, and the electron transport system. 37

agglutination (ag-gloo″ tĭ-na-shun) Clumping of cells, particularly in reference to red blood cells involved in an antigen-antibody reaction. 121

aging Progressive changes over time, leading to loss of physiological function and eventual death. 332

agranular leukocyte (ah-gran′u-lar loo′-ko-sīt) White blood cell that does not contain distinctive granules. 115

aldosterone (al″do-ster′ōn) A hormone secreted by the adrenal cortex that functions in regulating blood sodium and potassium levels. 257

allantois (ah-lan-to-is) An extraembryonic membrane that serves as a source of blood vessels for the umbilical cord. 318

allele (ah-lēl) An alternative form of a gene located at a particular chromosome site (locus). 362

alveolus (pl., alveoli) Air sac of a lung. 149

amino acid Unit molecule of a protein that takes its name from the fact that it contains an amino group ($-NH_2$) and an acid group ($-COOH$). 20

amniocentesis The removal of a small amount of amniotic fluid to examine the chromosomes and the enzymatic potential of fetal cells. 324

amnion (am′ne-on) One of the extraembryonic membranes; a fluid-containing sac around the embryo. 318

ampulla (am-pŭl´lah) Base of a semicircular canal in the inner ear. 239

anabolic steroid A synthetic steroid that mimics the effect of testosterone. 261

anaphase Stage in mitosis during which chromatids separate, forming chromosomes. 352

anemia Inefficient oxygen-carrying ability by the blood due to hemoglobin shortage. 113

anterior pituitary A portion of the pituitary gland that produces 6 types of hormones and is controlled by hypothalamic-releasing and release-inhibiting hormones. 249

antibiotic A medicine that specifically interferes with bacterial metabolism and in that way cures humans of a bacterial disease. 301

antibody A protein produced in response to the presence of an antigen; each antibody combines with a specific antigen. 133

antibody-mediated immunity Body line of resistance with B cells producing antibodies. 134

anticodon A "triplet" of 3 nucleotides in transfer RNA that pairs with a complementary triplet (codon) in messenger RNA. 387

antidiuretic hormone (ADH) (an˝ tĭ-di˝ u-ret´ik hōr´mōn) Sometimes called vasopressin, a hormone produced by the hypothalamus but stored and secreted by the posterior pituitary that controls the degree to which water is reabsorbed by the kidneys. 174

antigen (ant´i-jen) A foreign substance, usually a protein, that stimulates the immune system to react, such as to produce antibodies. 133

anus Inferior outlet of the digestive tube. 73

anvil The middle bone of the 3 ossicles of the middle ear. 239

aorta (ā-or´tah) Major systemic artery that receives blood from the left ventricle. 95

appendicular skeleton Portion of the skeleton forming the upper extremities, the pectoral girdle, the lower extremities, and the pelvic girdle. 212

appendix A small, tubular appendage that extends outward from the cecum of the large intestine. 73

aqueous humor Watery fluid that fills the anterior chamber of the eye. 230

arterial duct Ductus arteriosus; fetal connection between the pulmonary artery and the aorta. 321

arteriole (ar-te´re-ōl) Vessel that takes blood from an artery to capillaries. 92

artery Vessel that takes blood away from the heart to arterioles; characteristically possessing thick elastic and muscular walls. 92

artificial insemination Introduction of semen into the vagina or uterus by artificial means. 288

aster Short microtubule that extends outward from a spindle pole in animal cells during cell division. 352

atom Smallest unit of matter that cannot be divided by chemical means. 14

ATP (adenosine triphosphate) A compound containing 3 phosphate groups, 2 of which are high-energy phosphates. It is the "common currency" of energy for most cellular processes. 26

atrioventricular A structure in the heart that pertains to both the atria and ventricles; for example, an atrioventricular valve is located between an atrium and a ventricle. *See* AV (atrioventricular) valve. 94

atrioventricular node *See* AV node. 96

atrioventricular valve A valve located between the atrium and the ventricle. 94

atrium (a´tre-um) Chamber; particularly an upper chamber of the heart lying above the ventricles; either the right atrium or the left atrium. 94

auditory canal A tube in the external ear that lies between the pinna and the tympanic membrane. 238

auditory nerve A nerve sending the signal for sound from the inner ear to the temporal lobe of the brain. 243

Australopithecine (aw˝ strah-lo-pith´ ə-sīn) The first generally recognized hominids. 436

autoimmune disease A disease that results when the immune system mistakenly attacks the body's own tissues. 143

autonomic system (aw˝to-nom´ik sis´tem) A branch of the peripheral nervous system that has control over the internal organs; consisting of the sympathetic and parasympathetic systems. 190

autosome Chromosome other than a sex chromosome. 342

AV node A small region of neuromuscular tissue that transmits impulses received from the SA node to the ventricular walls. 96

axial skeleton Portion of the skeleton that supports and protects the organs of the head, the neck, and the trunk. 209

axon Fiber of a neuron that conducts nerve impulses away from the cell body. 184

B

bacterium Unicellular organism that is prokaryotic—its single cell lacks the complexity of a eukaryotic cell. 299

basal body Short cylinder having a circular arrangement of 9 microtubule triplets (9 + 0 pattern) located within the cytoplasm at the bases of cilia and flagella. 39

basal ganglia Mass of gray matter located between the white matter of the cerebrum and the thalamus of the diencephalon; may have some control over voluntary muscle action. 196

base A solution in which pH is greater than 7; a substance that contributes or liberates hydroxide ions (OH^-) in a solution; alkaline; opposite of acidic. Also, a term commonly applied to one of the components of a nucleotide. 16

basophil A leukocyte with a granular cytoplasm, which is able to be stained with a basic dye. 115

bile A secretion of the liver that is temporarily stored in the gallbladder before being released into the small intestine, where it emulsifies fat. 75

binary fission Reproduction by division into 2 equal parts by a process that does not involve a mitotic spindle. 300

binomial system The assignment of 2 names to each organism, the first of which designates the genus and the second of which designates the species. 426

biodiversity The wide range of living things. 3

biosphere That part of the earth's surface and atmosphere where living organisms exist. 3

biotechnology Use of a natural biological system to produce a commercial product or achieve an end desired by humans. 390

biotic potential (bi-ot´ik po-ten´shal) The maximum population growth rate under ideal conditions. 460

blastocyst An early stage of embryonic development that consists of a hollow ball of cells. 322

blind spot Area of the eye containing no rods and cones where the optic nerve passes through the retina. 232

blood Connective tissue, composed of cells separated by plasma, which transports substances in the cardiovascular system. 50

blood pressure The pressure of blood against the wall of a blood vessel. 98

B lymphocyte A lymphocyte that matures in the bone marrow and, when stimulated by the presence of a specific antigen, gives rise to antibody-producing plasma cells. 133

bone Connective tissue in which the cells lie within lacunae embedded in a hard matrix of calcium salts deposited around protein fibers. 50

Bowman's capsule A double-walled cup that surrounds the glomerulus at the beginning of the nephron. 168

bradykinin (brad″e-ki′nin) A substance found in damaged tissue that initiates nerve impulses resulting in the sensation of pain. 130

breathing Entrance and exit of air into and out of the lungs. 146

bronchiole (brong-ke′ōl) The smaller air passages in the lungs that eventually terminate in alveoli. 149

bronchus (brong-kus) (pl., bronchi) One of 2 major divisions of the trachea leading to the lungs. 149

buffer A substance or compound that prevents large changes in the pH of a solution. 18

C

calcitonin Hormone secreted by the thyroid gland that helps to regulate the blood calcium level. 255

cancer A malignant tumor that metastasizes. 404

capillary (kap′ĭ-lar″e) Microscopic vessel connecting arterioles to venules through the thin walls of which molecules either exit or enter blood. 92

carbaminohemoglobin Hemoglobin carrying carbon dioxide. 156

carbohydrate One of a class of organic compounds characterized by the presence of CH_2O groups; includes monosaccharides, disaccharides, and polysaccharides. 21

carcinogen (kar-sin′o-jen) An environmental agent that contributes to the development of cancer. 406

carcinogenesis Development of cancer. 406

cardiac muscle Specialized type of muscle tissue found only in the heart. 51

carnivore An animal that feeds only on animals. 444

carrier A molecule that combines with a substance and transports it through the cell membrane; an individual that unknowingly transmits an infectious or genetic disease. 31, 365

carrying capacity The largest number of organisms of a particular species that can be maintained indefinitely in an ecosystem. 460

cartilage A connective tissue in which the cells lie within lacunae embedded in a flexible matrix. 49

cell The structural and functional unit of an organism; the smallest structure capable of performing all the functions necessary for life. 4

cell body Portion of a neuron that contains a nucleus and from which the nerve fibers extend. 184

cell cycle A repeating sequence of events in eukaryotic cells consisting of interphase, when growth and DNA synthesis occurs, and mitosis, when cell division occurs. 350

cell-mediated immunity Specific mechanism of defense in which T cells destroy antigen-bearing cells. 136

cell membrane A membrane that surrounds the cytoplasm of cells and regulates the passage of molecules into and out of the cell. 31

central canal Tube within the spinal cord that is continuous with the ventricles of the brain and contains cerebrospinal fluid. 193

central nervous system (CNS) Portion of the nervous system containing the brain and spinal cord. 193

centriole (sen′tre-ōl) A short, cylindrical organelle in animal cells that contains microtubules in a 9 + 0 pattern and that is associated with the formation of basal bodies and the formation of the spindle during cell division. 38

centromere (sen′tro-mēr) A region of attachment of a chromosome to spindle fibers that is generally seen as a constricted area. 347

cerebellum The part of the vertebrate brain that controls muscular coordination. 195

cerebral hemisphere (ser′ĕ-bral hem′ĭ sfĕr) One of the large, paired structures that together constitute the cerebrum of the brain. 196

cerebrospinal fluid A fluid found in the ventricles of the brain, in the central canal of the spinal cord, and in association with the meninges. 193

cerebrum The largest portion of the brain, consisting of the right and left cerebral hemispheres. 196

cervix Narrow end of the uterus that leads into the vagina. 278

chemotherapy The use of a drug to selectively kill off cancer cells as opposed to normal cells. 416

chlamydia (klah-mid′e ah) A tiny bacterium that causes a sexually transmitted disease, particularly characterized by urethritis. 302

chorion (ko′re-on) An extraembryonic membrane that forms an outer covering around the embryo and contributes to the formation of the placenta. 318

chorionic villi Treelike extensions of the chorion of the embryo, projecting into the maternal tissues. 318

choroid The vascular, pigmented middle layer of the eyeball. 230

chromatid (kro′ma-tid) One of the 2 identical parts of a chromosome following replication of DNA. 347

chromatin Threadlike network in the nucleus that is made up of DNA and proteins. 34

chromosome Rod-shaped body in the nucleus seen during cell division; contains the hereditary units or genes. 34

ciliary body Structure associated with the choroid layer that contains the ciliary muscle, which controls the shape of the lens of the eye. 230

cilium (pl., cilia) Hairlike projection used for locomotion by many unicellular organisms and that has various purposes in higher organisms. 39

circadian rhythm A regular physiological or behavioral event that occurs on an approximately 24-hour cycle. 263

circumcision Removal of the foreskin of the penis. 274

class In taxonomy, the category below phylum and above order. 426

cleavage Cell division of the fertilized egg that is unaccompanied by growth so that numerous small cells result. 321

cleavage furrow An indentation that deepens during cleavage, causing animal cells to undergo cytokinesis. 353

clotting Process of blood coagulation, usually when injury occurs. 118

cochlea (kok´le-ah) That portion of the inner ear that resembles a snail's shell and contains the organ of Corti, the sense organ for hearing. 239

cochlear canal Canal within the cochlea bearing small hair cells, which function as hearing receptors. 239

codon A "triplet" of 3 nucleotides in messenger RNA that directs the placement of a particular amino acid into a protein. 387

coenzyme A nonprotein molecule that aids the action of an enzyme. 40

collecting duct A tube that receives urine from the distal convoluted tubules of several nephrons within a kidney. 168

colon The large intestine. 74

columnar epithelium Pillar-shaped cells usually with the nuclei near the bottom of each cell and found lining the digestive tract, for example. 46

common ancestor An ancestor to 2 or more branches of evolution. 427

compact bone Bone that contains Haversian systems cemented together. 50

complementary base pairing Pairing of bases between nucleic acid strands; adenine (A) pairs with either thymine (T) if DNA, or uracil (U) if RNA, and cytosine (C) pairs with guanine (G). 383

complement system A series of proteins in plasma that form a nonspecific defense mechanism against a microbe invasion; complements the antigen-antibody reaction. 133

compound Two or more atoms of different elements that are chemically combined. 14

conception Fertilization and implantation of an embryo resulting in pregnancy. 318

cone Bright-light receptor in the retina of the eye that detects color and provides visual acuity. 231

connective tissue A type of tissue characterized by cells separated by a matrix that often contains fibers. 48

consumer A member of a population that feeds on members of other populations in an ecosystem. 444

control group In experimentation, a sample group that undergoes all the steps in the experiment except the one being tested. 7

convergent evolution Subjected to similar environments, genetically unrelated species evolve similar characteristics. 428

cornea The transparent membrane that forms the anterior portion of the eyeball. 230

coronary artery Artery that supplies blood to the wall of the heart. 101

corpus callosum A mass of white matter within the brain, composed of nerve fibers connecting the right and left cerebral hemispheres. 197

corpus luteum A body, yellow in color, that forms in the ovary from a follicle that has discharged its egg. 278

cortisol A glucocorticoid secreted by the adrenal cortex. 257

covalent bond A chemical bond between atoms that results from the sharing of a pair of electrons. 15

Cowper's gland Small structure located below the prostate gland in males. 275

cranial nerve Nerve that arises from the brain. 190

creatine phosphate Compound unique to muscles that contains a high-energy phosphate bond. 221

cretinism (kre´tin-izm) A condition resulting from improper development of the thyroid in an infant. 255

cri du chat syndrome A group of body malfunctions caused by a deletion of chromosome 5. 344

Cro-Magnon The common name for the first fossils to be accepted as representative of modern humans. 437

crossing-over The exchange of corresponding segments of genetic material between nonsister chromatids of homologous chromosomes during synapsis of meiosis I. 348

cuboidal epithelium Cube-shaped cells found lining the kidney tubules, for example. 46

cultural eutrophication (kul´tu-ral u″tro-fĭ-ka-shun) Enrichment of a body of water due to human activities, causing excessive growth of producers and then death of these and other inhabitants. 466

Cushing syndrome A condition resulting from hypersecretion of cortical hormones; characterized by thin arms and legs and a "moon face," and accompanied by high blood glucose and sodium levels. 258

cystic fibrosis A lethal genetic disease involving problems with the functions of the mucous membranes in the respiratory and digestive tracts. 366

cytoplasm Cellular region inside cell membrane but excluding the nucleus. 31

cytosine (C) One of 4 organic bases in the nucleotides composing the structure of DNA and RNA. 383

cytoskeleton (si´to-skel-ĕ-ton) Filamentous protein structures found throughout the cytoplasm that help maintain the shape of the cell, anchor the organelles, and allow the cell and its organelles to move. 37

cytotoxic T cell T lymphocyte that attacks and kills antigen-bearing cells; killer T cell. 136

D

data Experimentally derived facts. 7

deamination Removal of an amino group ($-NH_2$) from an amino acid or other organic compound. 76

decomposer Organism of decay (fungus and bacterium) in an ecosystem. 444

demographic transition A decline in the birthrate following a reduction in the death rate so that the population growth rate is lowered. 461

dendrite Fiber of a neuron, typically branched, that conducts nerve impulses toward the cell body. 184

deoxyribonucleic acid *See* DNA. 24

dermis (der´mis) The thick skin layer that lies beneath the epidermis. 52

desertification (dez-ert″ ĭ-fi-ka´shun) Desert conditions caused by human misuse of land. 463

detritus (di-tri´tus) Nonliving organic matter. 444

diabetes insipidus Condition characterized by an abnormally large production of urine due to a deficiency of antidiuretic hormone. 251

diabetes mellitus (di˝ ah-bē´ tēz me-li´tus) Condition characterized by a high blood glucose level and the appearance of glucose in the urine, due to a deficiency of insulin production or glucose uptake by cells. 251

diaphragm A sheet of muscle that separates the thoracic cavity from the abdominal cavity in higher animals. 150

diastole (di-as´to-le) Relaxation of a heart chamber. 96

differentiation The process and the developmental stages by which a cell becomes specialized for a particular function. 321

diffusion The movement of molecules from a region of higher concentration to a region of lower concentration. 32

digit A finger or toe. 212

diploid (2N) (dĭp´loid) The number of chromosomes in the body cells; twice the number of chromosomes found in gametes. 346

disaccharide A sugar such as maltose that contains 2 units of a monosaccharide. 22

distal convoluted tubule Highly coiled region of a nephron that is distant from Bowman's capsule, where tubular excretion takes place. 168

DNA (deoxyribonucleic acid) A nucleic acid found in the cells; the genetic material that directs protein synthesis in cells. 24

DNA fingerprinting Using fragment lengths resulting from restriction enzyme cleavage to identify particular individuals. 394

DNA ligase enzyme An enzyme that links DNA from 2 sources; used in genetic engineering to put a gene into plasmid DNA. 391

DNA probe Single strand of radioactive DNA that can be used to locate a particular stretch of DNA. 393

dominant allele The hereditary factor that expresses itself in the phenotype when the genotype is heterozygous. 362

dorsal-root ganglion A mass of sensory neuron cell bodies located in the dorsal root of a spinal nerve. 191

double helix A double spiral; describes the three-dimensional shape of DNA. 25

doubling time The number of years it takes for a population to double in size. 460

Down syndrome Human congenital disorder associated with an extra chromosome 21. 343

duodenum (du˝o-de´num) The first portion of the small intestine in vertebrates into which ducts from the gallbladder and pancreas enter. 72

dyad A chromosome with 2 chromatids held together at a centromere. 348

E

ecological pyramid Pictorial graph representing the biomass, organism number, or energy content of each trophic level in a food web, from the producer to the final consumer populations. 446

ecology (e-kol´o-je) The study of the interactions of organisms with each other and with the physical environment. 443

ecosystem (ek˝ o-sis˝-tem) A setting in which populations interact with each other and with the physical environment. 3, 443

edema Swelling due to tissue fluid accumulation in the intercellular spaces. 129

effector A structure such as a muscle or gland that allows an organism to respond to environmental stimuli. 184

elastic cartilage Cartilage composed of elastic fibers allowing greater flexibility. 50

electrocardiogram (ECG or **EKG)** A recording of the electrical activity associated with the heartbeat. 96

electroencephalogram (EEG) A graphic recording of the brain's electrical activity. 96

electron A subatomic particle that has almost no weight and carries a negative charge; orbits in a shell about the nucleus of an atom. 14

electron transport system A series of molecules within the inner mitochondrial membrane that pass electrons one to the other from a higher level to a lower energy level; the energy released is used to build ATP. 41

element The simplest of substances consisting of only one type of atom; that is, carbon, hydrogen, oxygen. 14

embryo The organism in its early stages of development; first week to 2 months. 322

embryonic development The period of development from the first week to 2 months, during which time all major organs form. 321

endocrine gland A gland that secretes hormones directly into the blood or body fluids. 46

endometrium (en˝do-me´tre-um) The lining of the uterus, which becomes thickened and vascular during the uterine cycle. 278

endoplasmic reticulum (ER) (en-do-plaz´mik rĕ-tik´u-lum) A complex system of tubules, vesicles, and sacs in cells, sometimes having attached ribosomes. Rough ER has ribosomes; smooth ER does not. 36

endospore A resistant body formed by bacteria when environmental conditions worsen. 300

environmental resistance Sum total of factors in the environment that limit the numerical increase of a population in a particular region. 460

enzyme An organic catalyst that speeds up a specific reaction or a specific type of reaction in cells. 19

eosinophil A granular leukocyte capable of being stained with the dye eosin. 115

epidermis (ep˝ ĭ-der´mis) The outer layer of skin composed of stratified squamous epithelium. 52

epididymis Coiled tubule next to the testes where sperm mature and may be stored for a short time. 274

epiglottis A structure that covers the glottis during the process of swallowing. 70

episiotomy (ĕ-piz˝e-ot´o-me) A surgical procedure performed during childbirth in which the opening of the vagina is enlarged to avoid tearing. 329

epithelial tissue (ep˝ ĭ-the´le-al tish´u) A type of tissue that lines cavities and covers the external surface of the body. 46

erection Condition of the penis when it is turgid and erect, instead of being flaccid and lacking turgidity. 274

erythrocyte (ə-rith´ro-sīt) A red blood cell that contains hemoglobin and carries oxygen from the lungs to the tissues in vertebrates. 50

esophagus A tube that transports food from the mouth to the stomach. 70

essential amino acid Nine different amino acids required in the human diet because the body cannot make them. 80

estrogen Female sex hormone, which, along with progesterone, maintains the primary sex organs and stimulates development of the female secondary sex characteristics. 279

eustachian tube Extension from the middle ear to the nasopharynx for equalization of air pressure on the eardrum. 239

excretion Removal of metabolic wastes from the body. 166

exocrine gland Secreting externally; particular glands with ducts whose secretions are deposited into cavities, such as salivary glands. 46

exophthalmic goiter (ek″sof-thal′mik goi′ter) An enlargement of the thyroid gland accompanied by an abnormal protrusion of the eyes. 255

expiration The act of expelling air from the lungs; exhalation. 147

exponential growth Growth, particularly of a population, in which the increase occurs in the same manner as compound interest. 458

external respiration Exchange between blood and alveoli of carbon dioxide and oxygen. 146

extraembryonic membrane Membranes that are not a part of the embryo but are necessary to the continued existence and health of the embryo. 318

F

facilitated transport Passive transfer of a substance into or out of a cell along a concentration gradient by a process that requires a carrier. 32

family A rank in taxonomic classification above genus and below order. 426

fatty acid An organic molecule with a long chain of carbon atoms that ends in an acidic group. 22

fermentation Anaerobic breakdown of carbohydrates that results in organic end products such as alcohol and lactic acid. 42

fertilization The union of a sperm nucleus and an egg nucleus, which creates the zygote with the diploid number of chromosomes. 317

fetal development The period of development from the ninth week through birth. 321

fibrin Insoluble, fibrous protein formed from fibrinogen during blood clotting. 119

fibrinogen Plasma protein that is converted into fibrin threads during blood clotting. 118

fibrocartilage Cartilage with a matrix of strong collagenous fibers. 50

fibrous connective tissue Tissue composed mainly of closely packed collagenous fibers and found in tendons and ligaments. 49

fimbriae (fim′bre-e) Fingerlike extensions from the oviduct near the ovary. 278

flagella (flah-jel′ah) Slender, long extensions used for locomotion by the flagellate protozoans, bacteria, and sperm. 39

focusing Manner by which light rays are bent by the cornea and lens, creating an image on the retina. 232

follicle A structure in the ovary that produces the egg and particularly the female sex hormone estrogen. 277

follicle-stimulating hormone *See* FSH. 275

food chain A succession of organisms in an ecosystem that is linked by an energy flow and the order of who eats whom. 444

food web The complete set of food links between populations in a community. 444

foramen magnum Opening in the occipital bone of the skull through which the spinal cord passes. 210

foreskin Skin covering the glans penis in uncircumcised males. 274

formed element A constituent of blood that is either cellular (red blood cells and white blood cells) or at least cellular in origin (platelets). 111

fossil Any remains of an organism that have been preserved in the earth's crust. 426

fovea centralis (fo′ve-ah sen-tral′is) Region of the retina consisting of densely packed cones that is responsible for the greatest visual acuity. 232

frontal lobe Area of the cerebrum responsible for voluntary movements and higher intellectual processes. 196

FSH (follicle-stimulating hormone) A hormone secreted by the anterior pituitary gland that stimulates the development of an ovarian follicle in a female or the production of sperm in a male. 275

fungus An eukaryote, usually composed of strands called hyphae, that is usually saprophytic; for example, mushroom and mold. 304

G

gallbladder A saclike organ associated with the liver that stores and concentrates bile. 75

gamete (gam′ēt) One of 2 types of reproductive cells that join in fertilization to form a zygote; most often an egg or a sperm. 346

ganglion (gang′gle-on) A collection of neuron cell bodies within the peripheral nervous system. 189

gastric gland Gland within the stomach wall that secretes gastric juice. 71

gene therapy The use of transplanted genes to overcome an inborn error of metabolism or to otherwise treat human ills. 396

genetically engineer To alter an organism's DNA or insert a foreign gene by a technological process; bioengineering. 390

genital herpes A sexually transmitted disease characterized by open sores on the external genitalia. 297

genital warts A sexually transmitted disease caused by papillomavirus and characterized by warts on the genitals. 297

genotype (ge′nə-tīp) The genes of any individual for (a) particular trait(s). 363

genus A rank in taxonomic classification above species and below family. 426

gerontology (jer″on-tol′o-je) The study of aging. 332

gestation period (jes-ta′shun) The period of development measured from the start of the last menstrual cycle until birth; in humans, typically 280 days. 318

gland A cell or group of epithelial cells that is specialized to secrete a substance. 46

glomerular filtrate (glo mer′u-lar fil′trāt) The filtered portion of blood contained within Bowman's capsule. 171

glomerulus (glo-mer′u-lus) A cluster; for example, the cluster of capillaries surrounded by Bowman's capsule in a nephron, where pressure filtration takes place. 169

glottis Slitlike opening to the larynx between the vocal cords. 70

glucagon Hormone secreted by the pancreatic islets (of Langerhans) that causes the release of glucose from glycogen. 259

glucose The most common 6-carbon sugar. 21

glycerol An organic compound that serves as a building block for fat molecules. 22

glycogen The storage polysaccharide found in animals that is composed of glucose molecules joined in a linear-type fashion but having numerous branches. 22

glycolysis (gli-kol´i-sis) A metabolic pathway found in the cytoplasm that participates in aerobic cellular respiration and fermentation. It converts glucose to 2 molecules of pyruvate. 40

Golgi apparatus (gol´je) An organelle that consists of concentrically folded membranes and functions in the processing, packaging, and secretion of cellular products. 36

gonadotropic hormone A type of hormone that regulates the activity of the ovaries and testes; principally FSH and LH (ICSH). 253

gonorrhea (gon″o-re´ah) Contagious sexually transmitted disease caused by bacteria and leading to inflammation of the urogenital tract. 301

Graafian follicle (graf´e-an fol´ĭ-k´l) Mature follicle within the ovaries, which contains a developing egg. 277

granular leukocyte (gran´u-lar loo´ko-sīt) White blood cell that contains distinctive granules. 115

greenhouse effect Carbon dioxide (CO_2) buildup in the atmosphere as a result of fossil fuel combustion; retains and reradiates heat, creating an abnormal rise in the earth's average temperature. 470

growth An increase in the number of cells and/or the size of these cells. 321

growth hormone (GH) Or somatotropin; hormone released by the anterior lobe of the pituitary gland that promotes the growth of the organism. 252

growth rate The yearly percentage of increase or decrease in the size of a population. 458

guanine (G) One of 4 organic bases in the nucleotides composing the structure of DNA and RNA. 383

gummas (gum´ahz) Large unpleasant sores that may occur during the tertiary stage of syphilis. 304

H

habitat The natural abode of an animal or plant species. 443

hammer The middle ear ossicle adhering to the tympanic membrane. 239

haploid (N) Half the diploid number; the number of chromosomes in the gametes. 346

hard palate Bony, anterior portion of the roof of the mouth, which contains several bones. 69

hazardous waste Waste containing chemicals hazardous to life. 465

HCG *See* Human chorionic gonadotropic hormone. 283

heart Muscular organ located in the thoracic cavity responsible for maintenance of blood circulation. 94

helper T cell T lymphocyte that releases lymphokines and stimulates certain other immune cells to perform their respective functions. 136

hemoglobin An iron-containing protein molecule in red blood cells that combines with and transports oxygen. 111

hepatic portal vein Vein leading to the liver formed by the merging of blood vessels from the villi of the small intestine. 75

herbivore An animal that feeds directly on plants. 444

herpes simplex virus A virus of which type I causes cold sores and type II causes genital herpes. 297

heterozygous Having 2 different alleles (as *Aa*) for a given trait. 363

histamine A substance produced by basophil-derived mast cells in connective tissue that causes capillaries to dilate and release immune and other substances. 130

homeostasis The maintenance of the internal environment, such as temperature, blood pressure, and other body conditions, within narrow limits. 4

hominid (hom´ĭ nid) Common term for a member of a family of upright, bipedal primates that includes australopithecines and humans. 436

Homo erectus The earliest nondisputed species of humans, named for their erect posture, which allowed them to walk as we do. 437

Homo habilis An extinct species that may include the earliest humans, having a small brain but quality tools. 436

homologous chromosome (ho-mol´o-gus kró-mo-sōm) Similarly constructed; homologous chromosomes have the same shape and contain genes for the same traits. 342

homozygous Having identical alleles for a given trait; homozygous dominant (*AA*) or homozygous recessive (*aa*). 363

hormone A chemical messenger produced in small amounts in one body region that is transported to another body region. 248

host An organism on or in which another organism lives. 294

human chorionic gonadotropic hormone (HCG) A gonadotropic hormone produced by the chorion that functions to maintain the uterine lining. 283

human immunodeficiency virus (HIV) Virus responsible for AIDS. 296

Huntington disease (HD) A fatal genetic disease marked by neurological disturbances and failure of brain regions. 368

hyaline cartilage (hi´ah-lĭn kar´tĭ-lij) Cartilage composed of very fine collagenous fibers and a matrix of a milk glass appearance. 49

hydrogen bond A weak attraction between a hydrogen atom carrying a partial positive charge and an atom of another molecule carrying a partial negative charge. 15

hydrolysis The splitting of a bond within a larger molecule by the addition of the components of water. 19

hydrolytic enzyme An enzyme that catalyzes a reaction in which the substrate is broken down with the addition of water. 77

hypertonic A solution with a higher concentration of solute and a lower concentration of water than the cell. 32

hypothalamus A region of the brain, the floor of the third ventricle, that helps maintain homeostasis. 195

hypotonic A solution having a lower concentration of solute and a higher concentration of water than the cell. 32

I

immunity The ability of the body to protect itself from foreign substances and cells, including infectious microbes. 130

immunotherapy (i-mu″no ther′ah-pe) The use of any immune system component such as antibodies, cytotoxic T cells, or lymphokines to promote the health of the body, such as curing cancer. 417

implantation The attachment and penetration of the embryo into the lining of the uterus (endometrium). 283

impotency Failure of the penis to achieve erection. 274

inflammatory reaction A tissue response to injury that is characterized by redness, swelling, pain, and heat. 130

inner ear The portion of the ear consisting of a vestibule, semicircular canals, and the cochlea where balance is maintained and sound is transmitted. 239

innervate (in′er-vāt) To activate an organ, muscle, or gland by motor neuron stimulation. 184

insertion The end of a muscle that is attached to a movable bone. 217

inspiration The act of taking air into the lungs; inhalation. 147

insulin A hormone produced by the pancreas that regulates carbohydrate storage. 259

interferon (in″ ter-fēr′on) A protein formed by a cell infected with a virus that can increase the resistance of other cells to the virus. 133

internal respiration Exchange between blood and tissue fluid of oxygen and carbon dioxide. 146

interneuron A neuron found within the central nervous system that takes nerve impulses from one portion of the system to another. 184

interstitial cell Hormone-secreting cell located between the seminiferous tubules of the testes. 274

in vitro fertilization (IVF) Union of egg and sperm in laboratory glassware. The resulting embryo may be introduced into a prepared uterus, where it develops naturally. 289

ion An atom or group of atoms carrying a positive or negative charge. 15

ionic bond A bond created by an attraction between oppositely charged ions. 15

iris A muscular ring that surrounds the pupil and regulates the passage of light through this opening. 230

isotonic A solution with the same concentration of solute and water as the cell. 32

isotope One of 2 or more atoms with the same atomic number that differs in the number of neutrons and therefore in weight. 14

K

karyotype (kar′e-o-tīp) The arrangement of all the chromosomes within a cell by pairs in a fixed order. 342

kidney An organ in the urinary system that produces and excretes urine. 167

kingdom The classification category into which organisms are placed: Monerans, Protists, Fungi, Plants, and Animals. 3

Klinefelter syndrome A condition caused by the inheritance of a chromosome abnormality in number; an XXY individual. 345

Krebs cycle A cyclical metabolic pathway found in the matrix of mitochondria that participates in aerobic cellular respiration; breaks down C_2 (2-carbon) groups to carbon dioxide and hydrogen atoms. 41

L

labia majora (la′be-ah) The outer folds of the vulva. 278

labia minora The inner folds of the vulva. 278

lacteal (lak′te-al) A lymphatic vessel in a villus of the intestinal wall of mammals. 72

lacuna (lah-ku′nah) A small pit or hollow cavity, as in bone or cartilage, where a cell or cells are located. 49

lanugo (lah-nu′go) Short, fine hair that is present during the later portion of fetal development. 327

large intestine The last major portion of the digestive tract, extending from the small intestine to the anus and consisting of the cecum, the colon, and the rectum. 73

larynx (lar′ingks) Cartilaginous organ located between pharynx and trachea, which contains the vocal cords; voice box. 148

lens A clear membranelike structure found in the eye behind the iris; brings objects into focus. 230

leukemia (loo-ke′me-ah) Cancer of the blood-forming tissues leading to the overproduction of abnormal white blood cells. 413

leukocyte White blood cell of which there are several types each having a specific function in protecting the body from invasion by foreign substances and organisms. 50

LH (luteinizing hormone) Hormone produced by the anterior pituitary gland that stimulates the development of the corpus luteum in females and the production of testosterone in males. 275

ligament Fibrous connective tissue that joins bone to bone. 49

limbic system A portion of the brain concerned with memory and emotions. 197

lipase (li′pās) A fat-digesting enzyme secreted by the pancreas. 77

lipid One of a class of organic molecules that are insoluble in water; notably fats, oils, and steroids. 22

liver A large organ in the abdominal cavity that has many functions such as production of blood proteins and detoxification of harmful substances. 75

loop of Henle U-shaped portion of the nephron tubule. 168

loose connective tissue Tissue composed mainly of fibroblasts that are widely separated by a matrix containing collagen and elastin fibers and found beneath epithelium. 49

lumen The cavity inside any tubular structure, such as the lumen of the digestive tract. 70

luteinizing hormone *See* LH. 275

lymph Fluid having the same composition as tissue fluid and carried in lymphatic vessels. 129

lymphatic system A one-way vascular system that takes up excess fluid in the tissues, filters it, and transports it to cardiovascular veins in the shoulders. 128

lymph node A mass of lymphoid tissue located along the course of a lymphatic vessel. 130

lymphocyte Specialized white blood cell, which functions in specific defense; occurs in 2 forms—T lymphocyte and B lymphocyte. 115

lymphokine (lim′fo-kīn) Molecule secreted by T lymphocytes that has the ability to affect the activity of all types of immune cells. 136

lymphoma (lim fo´mah) Cancer of the lymphoid organs such as lymph nodes, spleen, and thymus gland. 413

lysosome (li´so-sōm) An organelle in which digestion takes place due to the action of powerful hydrolytic enzymes. 36

M

macromolecule A large molecule that is formed by the joining of unit molecules; also called a polymer; for example, protein, fat, carbohydrate, and nucleic acid. 19

macrophage A large phygocytic cell derived from a monocyte that ingests microbes and debris. 131

maltase Enzyme found in the mucosa of the intestinal villi. 77

mandible The lower jaw; contains tooth sockets. 211

matrix The secreted basic material or medium of biological structures, such as the matrix of cartilage or bone. 48

maxillae Maxillary bones that contain sockets for upper teeth; the upper jaw. 211

medulla oblongata The lowest portion of the brain that is concerned with the control of internal organs. 195

meiosis A type of cell division occurring during the production of gametes in animals by means of which the 4 daughter cells have the haploid number of chromosomes. 348

meiosis I That portion of meiosis during which homologous chromosomes come together and then later separate. 348

meiosis II That portion of meiosis during which sister chromatids separate, resulting in 4 haploid daughter cells. 348

memory B cell Persistent population of B cells ready to produce antibodies specific to a particular antigen; accounts for the development of active immunity. 134

memory T cell Persistent population of T cells ready to recognize an antigen that previously invaded the body. 136

meninges mə-nin-jēz) (sing., menix) Protective membranous coverings about the central nervous system. 193

menisci (me nis´ci) (sing., meniscus) Fibrocartilage that separates the surfaces of bones in the knee. 214

menopause Termination of the ovarian and uterine cycles in older women. 283

menstruation Loss of blood and tissue from the uterus at the end of a uterine cycle. 280

messenger RNA (mRNA) Ribonucleic acid complementary to DNA; has codons that direct protein synthesis at the ribosomes. 385

metabolism All of the chemical reactions within a cell (or an organism) including break-down reactions (catabolism) and synthetic reactions (anabolism). 39

metafemale A female who has three X chromosomes. 345

metaphase Stage in mitosis during which chromosomes are at the equator of the mitotic spindle. 352

metastasis (mĕ tas´tah-sis) The spread of cancer from the place of origin throughout the body caused by the ability of cancer cells to migrate and invade tissues. 484

MHC protein A membrane protein that serves to identify the cells of a particular individual. 137

microbe Microscopic infectious agent, such as a fungus, a bacterium, or a virus. 115

microtubule An organelle composed of 13 rows of globular proteins; found in multiple units in several other organelles, such as the centriole, cilia, and flagella, as well as spindle fibers. 37

midbrain A small region of the brain stem located between the forebrain and the hindbrain. 195

middle ear A portion of the ear consisting of the tympanic membrane, the oval and round windows, and the ossicles where sound is amplified. 239

mineral An inorganic, homogeneous substance. 86

mitochondrion (mi´to-kon´dre-on) An organelle in which cellular respiration produces the energy molecule, ATP. 37

mitosis (mi-to´sis) Type of cell division in which daughter cells receive the exact chromosome and genetic makeup of the parent cell; occurs during growth and repair. 347

molecule A chemical consisting of 2 or more atoms bonded together; smallest unit of a compound that has the properties of the compound. 14

monoclonal antibody Antibody of one type produced by a hybridoma—a lymphocyte that has fused with a cancer cell. 140

monosaccharide A simple sugar; a carbohydrate that cannot be decomposed by hydrolysis. 21

morphogenesis (mor″ fo-jen´ĭ-sis) The movement of cells and tissues to establish the shape and the structure of an organism. 321

morula (mor´u-lah) An early stage in development in which the embryo consists of a mass of cells, often spherical. 322

motor neuron A neuron that takes nerve impulses from the central nervous system to the effectors. 184

multiple alleles More than 2 alleles for a particular trait. 372

muscle action potential An electrochemical change due to increased sarcolemma permeability that is propagated down the T system and results in muscle contraction. 222

muscular (contractile) tissue A type of tissue that contains cells capable of contracting; skeletal muscles are attached to the skeleton, smooth muscle is found within walls of internal organs, and cardiac muscle comprises the heart. 51

mutagen An environmental agent that induces mutations. 406

mutation A change in the genetic material. 368

myelin sheath (mi´ĕ-lin shēth) The Schwann cell membranes that cover long neuron fibers giving them a white, glistening appearance. 185

myocardium Heart muscle. 94

myofibril The contractile portions of muscle fibers. 221

myosin One of 2 major proteins of muscle; makes up thick filaments in myofibrils and is capable of breaking down ATP. *See* actin. 221

myxedema (mik″ să-de-mah) A condition resulting from a deficiency of thyroid hormone in an adult. 255

N

natural selection The process by which populations become adapted to their environment. 428

Neanderthal The common name for an extinct subspecies of humans whose remains are found in Europe and Asia. 437

negative feedback A self-regulatory mechanism that is activated by an imbalance and results in a fluctuation about a mean. 58

nephron (nef´ron) The anatomical and functional unit of the kidney; kidney tubule. 168

nerve A bundle of long nerve fibers that run to and/or from the central nervous system. 51

nerve impulse An electrochemical change due to increased membrane permeability that is propagated along a neuron from the dendrite to the axon following excitation. 185

neurilemma (nūr´ə-lem´ah) The outermost wrapping of a nerve fiber; promotes regeneration. 185

neurofibromatosis (NF) (nūr˝ o-fi-bro-mah-to´sis) A genetic disease marked by development of neurofibromas under the skin and muscles. 368

neuromuscular junction The point of contact between a nerve cell and a muscle fiber. 222

neuron (nu´ron) Nerve cell that characteristically has 3 parts: dendrite, cell body, and axon. 51

neurotransmitter A chemical made at the ends of axons that is responsible for transmission across a synapse. 188

neutron A subatomic particle that has a weight of one atomic mass unit, carries no charge, and is found in the nucleus of an atom. 14

neutrophil Granular leukocyte that is the most abundant of the white blood cells; first to respond to infection. 115

niche (nich) Total description of an organism's functional role in an ecosystem, from activities to reproduction. 443

nitrogen fixation A process whereby nitrogen is reduced and incorporated into organic compounds. 450

node of Ranvier Gap in the myelin sheath around a nerve fiber. 184

nondisjunction (non˝dis-junk´shun) The failure of homologous chromosomes or sister chromatids to separate during the formation of gametes. 356

nongonococcal urethritis (NGU) (non˝ gon-o-kok´al u˝ rē-thri´tis) An infection of the urinary tract by an organism other than *N. gonorrhoeae*. 302

norepinephrine (NE) (nor˝ ep´ĭ-nef´ren) Excitatory neurotransmitter active in the peripheral and central nervous systems. 189

nuclear envelope The double membrane that surrounds the nucleus and is continuous with the endoplasmic reticulum. 34

nucleic acid A large organic molecule made up of nucleotides joined together; for example, DNA and RNA. 24

nucleolus (nu-kle´o-lus) (pl., nucleoli) An organelle found inside the nucleus; a special region of chromatin containing DNA that produces RNA for ribosome formation. 35

nucleotide A monomer of a nucleic acid that forms when a nitrogen-containing organic base, a pentose sugar, and a phosphate join. 24

nucleus A large organelle containing the chromosomes and acting as a control center for the cell; center of an atom. 34

nutrient A portion of food usable by the body as a source of energy or of building material. 79

O

occipital lobe (ok-sip´ĭ-tal lōb) Area of the cerebrum responsible for vision, visual images, and other sensory experiences. 196

omnivore An animal that feeds on both plants and animals. 444

oncogene (ong´ko-jen) A gene that contributes to the transformation of a normal cell into a cancer cell. 409

oogenesis (o˝ o-jĕn´ ĕ-sis) Production of an egg in females by the process of meiosis and maturation. 357

optic nerve A nerve that carries nerve impulses from the retina of the eye to the brain. 231

order In taxonomy, the category below class and above family. 426

organ of Corti A portion of the inner ear that contains the receptors for hearing. 239

organelle Specialized structures within cells (e.g., nucleus, mitochondria, and endoplasmic reticulum). 33

orgasm Physical and emotional climax during sexual intercourse; results in ejaculation in the male. 275

origin End of a muscle that is attached to a relatively immovable bone. 216

osmosis (oz-mo´sis) The movement of water from an area of higher concentration of water to an area of lower concentration of water across a differentially permeable membrane. 32

ossicle One of the small bones of the middle ear—hammer, anvil, stirrup. 239

osteoblast A bone-forming cell. 209

osteoclast A cell that causes the erosion of bone. 209

osteocyte A mature bone cell. 209

otolith (o-to-lith) Calcium carbonate granule associated with ciliated cells in the utricle and saccule. 239

outer ear Portion of the ear consisting of the pinna and the auditory canal. 238

oval opening Foramen ovale; an opening between the 2 atria in the fetal heart. 321

oval window Opening between the stapes and the inner ear. 239

ovarian cycle Monthly occurring changes in the ovary that determine the level of sex hormones in blood. 279

ovary The female gonad, the organ that produces eggs, estrogen, and progesterone. 277

ovulation The discharge of a mature egg from the follicle within the ovary. 278

oxygen debt Oxygen that is needed to metabolize lactate, a compound that accumulates during vigorous exercise. 222

oxytocin A hormone released by the posterior pituitary that causes contraction of uterus and milk letdown. 251

ozone shield A layer of ozone (O_3) present in the upper atmosphere that protects the earth from damaging ultraviolet light. Nearer the earth, ozone is a pollutant. 471

P

pacemaker *See* SA (sinoatrial) node. 96

pancreas An elongate, flattened organ in the abdominal cavity that secretes enzymes into the small intestine (exocrine function) and hormones controlling blood sugar (endocrine function). 75, 259

pancreatic amylase Enzyme that digests starch to maltose. 77

pancreatic islet (of Langerhans) (lahng´ər-hanz) Distinctive group of cells within the pancreas that secretes insulin and glucagon. 259

pap smear An analysis done on cervical cells for detection of cancer. 278

parasite An organism that resides on or within another organism and does harm to this organism. 294

parasympathetic system That part of the autonomic system that usually promotes activities associated with a normal state. 193

parathyroid gland A gland embedded in the posterior surface of the thyroid gland; produces parathyroid hormone, which is involved in the regulation of calcium and phosphorus balance. 256

parathyroid hormone (PTH) A hormone secreted by the parathyroid glands that affects the blood calcium and phosphate levels. 256

parietal lobe Area of the cerebrum responsible for sensations involving temperature, touch, pressure, and pain, as well as speech. 196

parturition Birth of a human and the expulsion of the extraembryonic membranes through the terminal portion of the female reproductive tract. 328

pectoral girdle Portion of the skeleton that provides support and attachment for the arms. 212

pelvic girdle Portion of the skeleton to which the legs are attached. 213

pelvic inflammatory disease (PID) A disease state of the reproductive organs caused by a sexually-transmitted organism. 302

penis External organ in males through which the urethra passes and that serves as the organ of sexual intercourse. 274

pepsin A protein-digesting enzyme secreted by gastric glands. 77

peptidase An intestinal enzyme that breaks down short chains of amino acids to individual amino acids that are absorbed across the intestinal wall. 77

peptide bond The covalent bond that joins 2 amino acids. 21

periodontitis (per″e-o-don-ti´tis) Inflammation of the gums. 69

peripheral nervous system (PNS) Nerves and ganglia that lie outside the central nervous system. 189

peristalsis (per″i-stal´sis) A rhythmic contraction that serves to move the contents along in tubular organs, such as the digestive tract. 71

peritubular capillary Capillary that surrounds a nephron and functions in reabsorption during urine formation. 169

phagocytosis (fag″o-si-to´sis) The taking in of bacteria and/or debris by engulfing; cell eating. 115

pharynx (far´ingks) A common passageway (throat) for both food intake and air movement. 70

pH scale A measure of the hydrogen ion concentration; any pH below 7 is acidic and any pH above 7 is basic. 18

phenotype (fe´no-tīp) The outward appearance of an organism caused by the genotype and environmental influences. 363

phenylketonuria (PKU) A genetic disease stemming from the lack of an enzyme to metabolize the amino acid phenylalanine. 368

pheromone (fer´o-mōn) A chemical substance secreted by one organism that influences the behavior of another. 264

photochemical smog Air pollution that contains nitrogen oxides (NO_x) and hydrocarbons, which react to produce ozone and peroxylacetyl nitrate (PAN). 469

phylum In taxonomy, the category applied to protists and animals that follows kingdom and lies above class. 426

pineal gland A gland either at the skin surface (fish, amphibians) or in the third ventricle of the brain, producing melatonin. 263

pinna Outer, funnel-like structure of the ear that picks up sound waves. 238

pituitary gland A small gland that lies just below the hypothalamus and is important for its hormone storage and production activities. 249

placenta A structure formed from the chorion and uterine tissue through which nutrient and waste exchange occur for the embryo and later the fetus. 318

plasma The liquid portion of blood consisting of all components except the formed elements. 116

plasma cell A differentiated B lymphocyte that is specialized to mass produce antibodies. 134

plasmid (plaz´mid) A circular DNA segment that is present in bacterial cells but is not part of the bacterial chromosome. 391

platelet A cell fragment that is necessary to blood clotting; thrombocyte. 118

pleural membrane A serous membrane that encloses the lungs. 150

polar body Nonfunctioning daughter cell that has little cytoplasm and is formed during oogenesis. 357

pollution Detrimental alteration of the normal constituents of air, land, and water due to human activities. 451

polypeptide (pol″ e-pep-tīd) A macromolecule composed of many amino acids linked by peptide bonds. 21

polysaccharide A macromolecule composed of many units of sugar. 22

pons A portion of the brain stem above the medulla oblongata and below the midbrain. 195

posterior pituitary Back lobe of the pituitary gland that stores and secretes ADH and oxytocin produced by the hypothalamus. 249

postsynaptic membrane In a synapse, the membrane of the neuron opposite the presynaptic membrane. 188

pressure filtration The movement of small molecules from the glomerulus into Bowman's capsule due to the action of blood pressure. 170

presynaptic membrane In a synapse, the membrane of the neuron opposite the postsynaptic membrane. 188

primate Animal that belongs to the order Primates; the order of mammals that includes monkeys, apes, and humans. 432

producer Organism that produces food and is capable of synthesizing organic compounds from inorganic constituents of the environment; usually the green plants and algae in an ecosystem. 443

progesterone Female sex hormone secreted by the corpus luteum of the ovary and by the placenta. 279

prolactin A hormone secreted by the anterior pituitary that stimulates the production of milk from the mammary glands. 330

prophase Early stage in mitosis during which chromatin condenses so that chromosomes appear. 352

prostaglandin (PG) Hormone that has various and powerful local effects. 263

prostate gland Gland located around the male urethra below the urinary bladder; adds secretions to seminal fluid. 275

protein One of a class of organic compounds that are composed of either one or several long polypeptides. 19

prothrombin Plasma protein that is converted to thrombin during the process of blood clotting. 118

proton A subatomic particle found in the nucleus of an atom that has a weight of one atomic mass unit and carries a positive charge; a hydrogen ion. 14

proto-oncogene (pro″ to-ong′ko-jēn) A normal gene that becomes an oncogene through mutation. 409

protozoa Animal-like protists that are classified according to means of locomotion: amoeba, flagellate, ciliate. 304

proximal convoluted tubule Highly coiled region of a nephron near Bowman's capsule, where selective reabsorption takes place. 168

puberty Developmental stage after birth during which the reproductive organs become functional. 332

pulmonary artery A blood vessel that takes blood away from the heart to the lungs. 95

pulmonary circuit (pul′mo-ner″e ser-ket) That part of the circulatory system that takes deoxygenated blood to and oxygenated blood away from the gas-exchanging surfaces in the lungs. 100

pulmonary vein A blood vessel that takes blood to the heart from the lungs. 95

pulse Vibration felt in arterial walls due to expansion of the aorta following ventricle contraction. 96

Punnett square (pun′et skwar) A gridlike device used to calculate the expected results of simple genetic crosses. 363

pupil An opening in the center of the iris of the eye. 230

pus Thick, yellowish fluid composed of dead phagocytes, dead tissue, and bacteria. 131

R

receptor A structure specialized to receive information from the environment and generate nerve impulses. Also a structure found in the membrane of cells that combines with a specific chemical in a lock-and-key manner. 184

recessive allele A hereditary factor that expresses itself in the phenotype only when the genotype is homozygous. 362

recombinant DNA DNA with genes from 2 different sources, often produced in the laboratory by introducing foreign genes into a bacterial plasmid. 391

red blood cell *See* erythrocyte. 111

red bone marrow Blood cell–forming tissue located in the spaces within spongy bone. 129

reduced hemoglobin Hemoglobin that is not carrying oxygen. 157

reflex An involuntary response to the stimulation of a receptor. 191

REM sleep A stage in sleep that is characterized by eye movements and dreaming. 197

renal cortex The outer portion of the kidney. 168

renal medulla The inner portion of the kidney including the renal pyramids (loops of Henle and collecting ducts). 168

renal pelvis A hollow chamber in the kidney that lies inside the renal medulla and receives freshly prepared urine from the collecting ducts. 168

replacement reproduction A population in which each person is replaced by only one child. 462

replication The duplication of DNA; occurs during interphase of the cell cycle. 384

reproduce To make a copy similar to oneself; for example, bacteria dividing to produce more bacteria, or egg and sperm joining to produce offspring in more advanced organisms. 4

residual volume The amount of air remaining in the lungs after a forceful expiration. 152

respiratory center Group of nerve cells in the medulla oblongata that send out nerve impulses on a rhythmic basis, resulting in inspiration. 150

resting potential The voltage recorded from inside a neuron when it is not conducting nerve impulses. 185

restriction enzyme Enzyme that stops viral reproduction by cutting viral DNA; used in genetic engineering to cut DNA at specific points. 391

retina (ret′ĭ-nah) The innermost layer of the eyeball, which contains the rods and the cones. 231

retrovirus Virus that contains only RNA and carries out RNA to DNA transcription, called reverse transcription. 294

Rh factor A type of antigen on the red blood cells. 122

rhodopsin (ro-dop′sin) Visual purple, a pigment found in the rods. 234

rib Bone hinged to the vertebral column and sternum, which, with muscle, defines the top and sides of the chest cavity. 212

rib cage The top and sides of the thoracic cavity; contains ribs and intercostal muscles. 150

ribonucleic acid. *See* RNA. 24

ribosomal RNA (rRNA) RNA occurring in ribosomes, structures involved in protein synthesis. 385

ribosome (ri′bo-sōm) Minute particle found attached to the endoplasmic reticulum or loose in the cytoplasm that is the site of protein synthesis. 35

RNA (ribonucleic acid) A nucleic acid found in cells that assists DNA in controlling protein synthesis. 24

rod Dim-light receptor in the retina of the eye that detects motion but no color. 231

rough ER (endoplasmic reticulum) Endoplasmic reticulum with attached ribosomes. 36

round window A membrane-covered opening between the inner ear and the middle ear. 239

S

saccule (sak′ūl) A saclike cavity that makes up part of the membranous labyrinth of the inner ear; contains receptors for static equilibrium. 239

salivary amylase An enzyme in the saliva that initiates the digestion of starch. 77

salivary gland A gland associated with the mouth that secretes saliva. 69

SA node Small region of neuromuscular tissue in the atria that initiates the heartbeat. Also called the pacemaker. 96

saprotroph (sap′ro-trōf) A heterotroph such as a bacterium and fungus that externally digests dead organic matter before absorbing the products. 300

sarcolemma The membrane that surrounds striated muscle cells. 221

sarcoma (sar-ko′mah) A cancer that arises in connective tissue, such as muscle, bone, and fibrous connective tissue. 413

sarcomere Structural and functional unit of a myofibril; contains actin and myosin filaments. 221

sclera (skle´rah) White, fibrous, outer layer of the eyeball. 230

scrotum A pouch of skin that encloses the testes. 272

selectively permeable Indicates condition in the living cell membrane in which permeability is a regulated process. 32

selective reabsorption The movement of nutrient molecules, as opposed to waste molecules, from the contents of the nephron into the blood at the proximal convoluted tubule. 172

semicircular canal Tubular structure within the inner ear that contains the receptors responsible for the sense of dynamic equilibrium. 239

semilunar valves Valves resembling half moons located between the ventricles and their attached vessels. 94

seminal fluid The sperm-containing secretion of males; also called semen. 275

seminal vesicle A convoluted, saclike structure attached to vas deferens near the base of the urinary bladder in males. 275

seminiferous tubule (sem″ ĭ-nif´er-us tu-būl) Highly coiled duct within the male testis, which produces and transports sperm. 273

sensory neuron A neuron that takes nerve impulses to the central nervous system and typically has a long dendrite and a short axon; afferent neuron. 184

septum A partition or wall such as the septum in the heart, which divides the right half from the left half. 94

serum Light-yellow liquid left after clotting of blood; blood minus formed elements and fibrinogen. 119

sex chromosome Chromosome responsible for the development of characteristics associated with gender; an X or Y chromosome. 342

sex-linked Allele located on the sex chromosomes. 373

sickle-cell disease A genetic disorder due to the homozygous genotype of the sickle-cell gene, producing sickle-shaped cells and loss of oxygen-carrying power in the blood. 370

sickle-cell trait Characteristics due to the heterozygous genotype of the sickle-cell allele; producing sickle-shaped cells in the blood under low oxygen conditions. 370

simple goiter Condition in which an enlarged thyroid produces low levels of thyroxin. 254

sinoatrial node *See* SA node. 96

sinus A cavity, as with the sinuses in the human skull. 210

skeletal muscle The contractile tissue that comprises the muscles attached to the skeleton; also called striated muscle. 51

slash-and-burn agriculture The cutting down and burning of trees to provide space to raise crops. 463

sliding filament theory The movement of actin filaments in relation to myosin filaments which accounts for muscle contraction. 221

small intestine Long, tubelike chamber of the digestive tract between the stomach and large intestine. 72

smooth ER (endoplasmic reticulum) Endoplasmic reticulum without attached ribosomes. 36

smooth (visceral) muscle The contractile tissue that comprises the muscles found in the walls of internal organs. 51

soft palate Entirely muscular posterior portion of the roof of the mouth. 69

somatic cell In animals, any cell other than those that undergo meiosis and become a sperm or egg; body cell. 346

somatic system That portion of the peripheral nervous system containing motor neurons that control skeletal muscles. 190

species A group of similarly constructed organisms capable of interbreeding and producing fertile offspring; organisms that share a common gene pool. 426

sperm Male sex cell with 3 distinct parts at maturity: head, middle piece, and tail. 274

spermatogenesis (sper″ mah-to-jen´ ĕ-sis) Production of sperm in males by the process of meiosis and maturation. 357

sphincter A muscle that surrounds a tube and closes or opens the tube by contracting and relaxing. 71

spinal nerve A nerve that arises from the spinal cord. 190

spindle The structure that brings about the movement of chromosomes during cell division; bundles of microtubules stretch between poles that are surrounded by asters. 352

spindle fiber Microtubule bundle in eukaryotic cells that is involved in the movement of chromosomes during mitosis and meiosis. 352

spleen A large, glandular organ located in the upper left region of the abdomen that stores and purifies blood. 130

spongy bone Porous bone found at the ends of long bones. 50

squamous epithelium Flat cells found lining the lungs and blood vessels, for example. 46

stereoscopic vision The product of 2 eyes and both cerebral hemispheres functioning together, allowing depth perception. 234

sterilization The absence of living organisms due to exposure to environmental conditions that are unfavorable to sustain life. 301

sternum The breastbone to which the ribs are ventrally attached. 212

stirrup Middle ear ossicle adhering to the oval window. 239

stomach A muscular sac that mixes food with gastric juices to form chyme, which enters the small intestine. 71

stratified Layered; stratified epithelium contains several layers of cells. 46

striated Having bands; cardiac and skeletal muscle are striated with bands of light and dark. 51

subcutaneous layer (sub″ku-ta´ne-us la´er) A tissue layer found in vertebrate skin that lies just beneath the dermis and tends to contain adipose tissue. 53

succession A series of ecological stages by which the community in a particular area gradually changes until there is a climax community that can maintain itself. 443

suppressor T cell T lymphocyte that suppresses certain other T and B lymphocytes from continuing to divide and perform their respective functions. 136

sympathetic system That part of the autonomic system that usually promotes activities associated with emergency (fight or flight) situations. 193

synapse (sin´aps) The region between 2 nerve cells where the nerve impulse is transmitted from one to the other, usually from axon to dendrite. 188

synapsis (sĭ-nap´sis) The attracting and pairing of homologous chromosomes during prophase I of meiosis. 348

synaptic cleft Small gap between presynaptic and postsynaptic membranes. 188

synovial joint (si-no´ve-al joint) A freely movable joint. 213

synthesis To build up, such as the combining of 2 small molecules to form a larger molecule. 19

syphilis (sif´ ĭ-lis) Chronic, contagious sexually transmitted disease caused by a bacterium that is a spirochete. 303

systemic circuit (sis tem´ik ser´kit) That part of the circulatory system that serves body parts other than the gas-exchanging surfaces in the lungs. 100

systole (sis´to-le) Contraction of a heart chamber. 96

T

taste bud Organ containing the receptors associated with the sense of taste. 229

taxonomy The science of naming and classifying organisms. 426

Tay-Sachs disease A lysosomal storage disease that is inherited and causes neurological impairment and death. 366

tectorial membrane Membrane within the organ of Corti that transmits nerve impulses to the brain. 239

telophase Stage of mitosis during which diploid number of daughter chromosomes are located at each pole. 353

template (tem´plat) A pattern that serves as a mold for the production of an oppositely shaped structure; one strand of DNA is a template for a complementary strand. 384

temporal lobe Area of the cerebrum responsible for hearing and smelling; also the interpretation of sensory experience and memory. 196

tendon Straps of fibrous connective tissue that joins muscle to bone. 49

testis The male gonad, the organ that produces sperm and testosterone. 272

testosterone The main male sex hormone responsible for development of primary and secondary sex characteristics in males. 274

tetanus Sustained muscle contraction without relaxation. 218

tetany Severe twitching caused by involuntary contraction of the skeletal muscles due to a lack of calcium. 256

tetrad A set of 4 chromatids resulting from the pairing of homologous chromosomes during prophase I of meiosis. 348

thalamus A mass of gray matter, located at the base of the cerebrum in the wall of the third ventricle, that receives sensory input. 195

theory A concept supported by a large number of conclusions drawn by using the scientific method. 8

thermal inversion Temperature inversion such that warm air traps cold air and its pollutants near the earth. 470

thrombin An enzyme that converts fibrinogen to fibrin threads during blood clotting. 118

thymine (T) One of 4 organic bases in the nucleotides composing the structure of DNA and RNA. 383

thymus An organ that lies in the neck and chest area and is absolutely necessary to the development of immunity. 130

thyroid gland Large gland in the neck that produces several important hormones, including thyroxin, which stimulates cellular metabolism. 254

thyroid-stimulating hormone (TSH) Hormone that causes the thyroid to produce thyroxin. 253

thyroxin The hormone produced by the thyroid that speeds up the metabolic rate. 254

tidal volume Amount of air normally moved in the human body during an inspiration or expiration. 152

T lymphocyte A lymphocyte that matures in the thymus and occurs in 4 varieties, one of which kills antigen-bearing cells outright. 133

trachea (tra´ke-ah) A tube that is supported by a C-shaped cartilaginous rings; lies between the larynx and the bronchi; also called the windpipe. 149

trait Specific term for a distinguishing phenotypic feature studied in heredity. 362

transcription The process resulting in the production of a strand of mRNA that is complementary to a segment of DNA. 386

transfer RNA (tRNA) Molecule of RNA that carries an amino acid to a ribosome engaged in the process of protein synthesis. 385

transgenic organism Organism that has a foreign gene inserted into it. 394

translation The process by which the sequence of codons in mRNA dictates the sequence of amino acids in a protein. 386

triplet code Genetic code (mRNA, tRNA) in which sets of 3 bases call for specific amino acids in the formation of polypeptides. 386

trophic level (tro´fik lev´el) A categorization of species in a food web according to their feeding relationships from the first-level autotrophs through succeeding levels of herbivores and carnivores. 444

trypsin A protein-digesting enzyme secreted by the pancreas. 77

TSH *See* thyroid-stimulating hormone. 253

tubal ligation Cutting of the oviducts in females to cause sterilization. 284

tubular excretion (too´bu-lar eks-kre´shun) The movement of certain molecules from blood into the distal convoluted tubule so that they are added to urine. 173

tumor (too´mor) Growth containing cells derived from a single mutated cell that has repeatedly undergone cell division; benign tumors remain at the site of origin and malignant tumors metastasize. 404

tumor suppressor gene Gene that, when expressed, prevents abnormal cell division and cancer. 409

Turner syndrome A condition caused by the inheritance of an abnormality in chromosome number; an X chromosome lacks a homologous counterpart—XO. 345

tympanic membrane (tim-pan´ik mem´ brān) Located between the outer ear and the middle ear and receives sound waves; the eardrum. 239

U

umbilical arteries and vein Fetal blood vessels that travel to and from the placenta. 321

umbilical cord Cord through which blood vessels pass, connecting the fetus to the placenta. 319

umbilicus The scar left after the umbilical cord stump shrivels and falls off. 330

urea (u-re´ah) Primary nitrogenous waste of humans derived from amino acid breakdown. 166

ureter (u´re´ter) One of 2 tubes that takes urine from the kidneys to the urinary bladder. 167

urethra (u´re´thrah) Tube that takes urine from the bladder to outside. 167

uric acid (u´rik as´id) Waste product of nucleotide metabolism. 166

urinalysis A medical procedure in which the composition of a patient's urine is determined. 176

urinary bladder An organ where urine is stored before being discharged by way of the urethra. 167

uterine cycle Monthly occurring changes in the characteristics of the uterine lining (endometrium). 280

uterus The womb, the organ located in the female pelvis where the fetus develops. 277

utricle (u´tre-k´l) Saclike cavity that makes up part of the membranous labyrinth of the inner ear; contains receptors for static equilibrium. 239

V

vaccine Antigens prepared in such a way that they can promote active immunity without causing disease. 138

vagina Organ that leads from uterus to vestibule and serves as the birth canal and organ of sexual intercourse in females. 277

valve Membranous extension of a vessel or the heart wall that opens and closes, ensuring one-way flow; common to the systemic veins, the lymphatic veins, and the heart. 93

vas deferens Tube that leads from the epididymis to the urethra in males. 274

vasectomy Cutting of the vas deferens in males as a birth-control measure. 284

vein Vessel that takes blood to the heart from venules; characteristically having nonelastic walls. 92

vena cava (ve´nah ka´vah) A large systemic vein that returns blood to the right atrium of the heart; either the superior or inferior vena cava. 95

venous duct Ductus venosus; fetal connection between the umbilical vein and the inferior vena cava. 321

ventilation Breathing; the process of moving air into and out of the lungs. 150

ventricle Cavity in an organ, such as a lower chamber of the heart; either the right ventricle or the left ventricle; or the ventricles of the brain. 94

venule Vessel that takes blood from capillaries to a vein. 93

vernix caseosa (ver´niks ka″se-o´sah) Cheeselike substance covering the skin of the fetus. 327

vertebral column The backbone of vertebrates through which the spinal cord passes. 211

vertebrate Animal possessing a backbone composed of vertebrae. 3

vestigial structure The remains of a structure that was functional in an ancestor but is no longer functional in the organism in question. 427

villus (vil´us) (pl., villi) Fingerlike projection from the wall of the small intestine that functions in absorption. 72

virulence The ability of a microorganism to cause disease and to invade the tissues of a host. 300

vital capacity Maximum amount of air moved in or out of the human body with each breathing cycle. 152

vitamin Essential requirement in the diet, needed in small amounts. They are often part of coenzymes. 84

vitreous humor (vit´re-us hu´mor) The substance that fills the posterior chamber of the eye. 230

vocal cord Fold of tissue within the larynx; creates vocal sounds when it vibrates. 148

vulva The external genitalia of the female that surround the opening of the vagina. 278

W

white blood cell *See* leukocyte. 50

X

X-linked An allele located on the X chromosome. 373

XYY male A male that has an extra Y chromosome. 345

Y

yolk sac An extraembryonic membrane that serves as the first site of red blood cell formation. 318

Z

zygote (zi´gōt) Diploid cell formed by the union of 2 gametes; the product of fertilization. 318

Credits

Photographs

Part Openers and Table of Contents

Introduction: © Peter Arnold/Peter Arnold, Inc. **Part 1:** © Edwin Reschke; **2:** © Julie Houck/Westlight; **3:** © David Young-Wolff/Photo Edit; **4:** © Adam Smith/Westlight; **5:** © Tony Freeman/Comstock; **6:** © David Young-Wolff/Photo Edit

Introduction

Opener: © Rene Sheret/MGA/Photri; **I.3 (inset):** © MacDonald Photography/Unicorn Stock Photos; **I.3 (background):** © Thomas Kitchin/Tom Stack & Associates; **IA** (mountain gorilla): © Adrian Warren/Ardea; **IA** (deforestation): © Frans Lanting/Minden Pictures

Chapter 1

Opener: © William Martin Jr./Comstock; **1.2:** © Biomed Commun/Custom Medical Stock Photos; **1.8:** © Lawrence Migdale/Photo Researchers, Inc.; **1.12:** © Don W. Fawcett/Photo Researchers, Inc.

Chapter 2

Opener: © Bill Longcore/Photo Researchers, Inc.; **2.1a, b, c:** © David M. Phillips/Visuals Unlimited; **2.5:** © Don Fawcett/Photo Researchers, Inc.; **2.6a:** © W. Rosenberg/Iona College/BPS; **2.8a:** Courtesy of Dr. Keith Porter; **2.10b:** Courtesy of Kent McDonald, University of Colorado at Boulder; **2.11:** © John Walsh/Photo Researchers, Inc.

Chapter 3

Opener: © Prof. P. Motta/Dept. of Anatomy/University of "La Sapienza," Rome/SPL/Photo Researchers, Inc.; **3.2a–d:** © Edwin Reschke/Peter Arnold, Inc.; **3.4a–d, 3.6a–c, 3.7:** © Edwin Reschke; **3Ba:** © Steve Bourgeois/Unicorn Stock Photos

Chapter 4

Opener: © Michael Webb/Visuals Unlimited; **4.4b, 4.6c:** From R.G. Kessel and R.H. Kardon, *Tissues and Organs: A Text Atlas of Scanning Electron Microscopy,* © 1979 W.H. Freeman and Co.; **4.11:** Dr. Sheril D. Burton; **4.14:** © Robert Frerck/Odyssey Prod.; p. 81: © Kathy Hamer/Unicorn Stock Photos; **4.17a, b:** © Biophoto Associates/Science Source/Photo Researchers, Inc.; **4.17c:** Courtesy of Centers for Disease Control, Atlanta, Georgia; **4.17d:** © Ken Greer/Visuals Unlimited; **4A:** © Larry Brock/Tom Stack & Associates; **4.18:** © Michael P. Godomski/Photo Researchers, Inc.

Chapter 5

Opener: © Edwin Reschke/Peter Arnold, Inc.; **5Aa:** © Lewis Lainey

Chapter 6

Opener: © David M. Phillips/Visuals Unlimited; **6.2a:** © Lennart Nilsson: Behold Man; Little Brown and Company, Boston; **6.6:** © Boehringer Ingenheim International, GmbH, photo Courtesy of Lennart Nilsson; **6.9b:** © NIBSC/SPL/Photo Researchers, Inc.; **6.10a:** Courtesy of Stuart I. Fox

Chapter 7

Opener: © David Young-Wolff/Photo Edit; **7.1b:** © John Cunningham/Visuals Unlimited; **7.1c:** © Astrid & Hans Frieder Michler/SPL/Photo Researchers, Inc.; **7.2a–e:** © Edwin Reschke/Peter Arnold, Inc.; **7A:** © John Nuhn; **7.6c:** From J. Rini, et al., "Structural Evidence for Induced Fit as a Mechanism for Antibody Antigen Recognition," *Science,* 255: 959–965, Feb. 1992. © 1992 by the AAAS.; **7.7b:** © Boehringer Ingelheim International/photo courtesy of Lennart Nilsson; **7B:** © Custom Medical Stock Photo; **7.10b:** Courtesy of Schering-Plough. Photo by Phillip Harrington

Chapter 8

Opener: © Lynn Funkhouser/Peter Arnold, Inc.; **8.5:** © John Watney Photo Library; **8.8a:** © Dan McCoy/Rainbow; **8Ac:** © Bill Aron/Photo Edit; **8Ba,b:** © Martin Rotker/Martin Rotker Photography

Chapter 9

Opener: © Carroll Seghers/Photo Researchers, Inc.; **9.6a:** © J. Gennaro, Jr./Photo Researchers, Inc.

Chapter 10

Opener: © Bob Grant/Comstock; **10.2b:** © J.D. Robertson; **10.3:** © Linda Bartlett; **10.11:** © Tim Davis/Photo Researchers, Inc.; **10.16:** © Scott Camazine/Photo Researchers, Inc.; **10.17a:** © Frerck/Odyssey Productions; **10.17b:** © Ogden Gigli/Photo Researchers, Inc.

Chapter 11

Opener: © Prof. P. Motta/Dept. of Anatomy/University "La Sapienza," Rome/SPL/Photo Researchers, Inc.; **11.1b:** © Robert Brons/BPS; **11.11a:** International Biomedical, Inc.; **11.12:** © H.E. Huxley; **11.14b:** © Victor Eichler, Ph.D.

Chapter 12

Opener: © Arthur Sirdofsky; **12.8:** © Lennart Nilsson/The Incredible Machine; **12Bc:** Robert S. Preston/Courtesy of Professor J.E. Hawkins, Kresge; **12.14c:** © Prof. P. Motta/Dept. of Anatomy/University "La Sapienza," Rome/SPL/Photo Researchers, Inc.

Chapter 13

Opener: © Martin Dohrn/SPL/Photo Researchers, Inc.; **13.6:** © Bettina Cirone/Photo Researchers, Inc.; **13.7:** © Lester Bergman & Associates; **13.9:** © Ken Greer/Visuals Unlimited; **13.10:** © Lester Bergman & Associates; **13.11:** From Arthur Grollman, *Clinical Endocrinology*

and its Physiological Basis, © 1964, J.B. Lippincott Company; **13.12:** © Ken Greer/ Visuals Unlimited; **13.15a:** © Custom Medical Stock Photos; **13.15b:** © NMSB/ Custom Medical Stock Photos; **13.16a, b:** From B. Zitelli, *The Atlas of Pediatric Physical Diagnosis,* 2 ed. © Mosby Year Book Europe Ltd.

Chapter 14

Opener: © Prof. P. Motta/Dept. of Anatomy/University "La Sapienza," Rome/SPL/Photo Researchers, Inc.; **14.2:** © Biophoto Associates/Photo Researchers, Inc.; **14.6:** © Edwin Reschke/ Peter Arnold, Inc.; **14.11a–c:** © Ray Ellis/ Photo Researchers, Inc.; **14.11d–f:** © Bob Coyle; **14.13:** © Dan Heringa/The Image Bank

Chapter 15

Opener: © NIAID/NIH/Peter Arnold, Inc.; **15.1:** © Dr. E.R. Degginger/Color-Pic, Inc.; **15.4a:** © Charles Lightdale/ Photo Researchers, Inc.; **15.4b:** © Robert Settineri/Sierra Productions; **15A:** © F.B. Grunzweig/Photo Researchers, Inc.; **15.5a:** © Francis Leroy; **15.5b:** © David M. Phillips/Visuals Unlimited; **15.5c:** © John Cunningham/Visuals Unlimited; **15.6:** © David M. Phillips/ Visuals Unlimited; **15.8:** © Science VU/ Visuals Unlimited; **15.9:** © CNRI/SPL/ Photo Researchers, Inc.; **15.10:** Courtesy of Dr. Ira Abrahamson; **15.11:** © Janet Barber/Custom Medical Stock Photo; **15.12a,b:** © Carroll Weiss/Camera M.D.; **15.12c:** Center for Disease Control, Atlanta, GA

AIDS Supplement

Opener: © NIBSC/SPL/Photo Researchers, Inc.

Chapter 16

Opener: © Alex Bartel/SPL/Photo Researchers, Inc.; **16.1 (inset):** © David M. Phillips/Visuals Unlimited; **16.1 (background):** © David M. Phillips/ Photo Researchers, Inc.; **16.8a:** © Lennart Nilsson, A Child is Born, Dell Publishing Company; **16.9:** © Petit Format/Science Source/Photo Researchers, Inc.; **16.10:** © Claude Edelmann, Petit Format et Guigorz from *First Days of Life*/Black Star; **16.11:** © James Stevenson/SPL/ Photo Researchers, Inc.; **16.12:** © Petit Format/Photo Researchers, Inc.; **16.16:** © John Cunningham/Visuals Unlimited; **16.17:** © Richard Hutchings/Photo Edit

Chapter 17

Opener: © Tony Freeman/Photo Edit; **17.1:** © CNRI/SPL/Photo Researchers, Inc.; **17.2a:** © Jill Cannafax/EKM-Nepenthe; **17.3a,b:** Courtesy of G.H. Valentine; **17.3c:** Courtesy of Irene Uchida; **17.4:** © David M. Phillips/ Visuals Unlimited; **17.5a,b:** Photograph by Earl Plunkett. Courtesy of G.H. Valentine; **17.12a–d:** © Michael Abbey/ Photo Researchers, Inc.

Chapter 18

Opener: © Science Source/Photo Researchers, Inc.; **18.3a,b:** © Bob Coyle; **18.3c,d:** © Tom Ballard/EKM-Nepenthe; **18.3e–h:** © James Shaffer; **18.6:** Courtesy of The Cystic Fibrosis Foundation; **18.8:** © Steve Uzzell; **18.10b:** © Bill Longcore/ Photo Researchers, Inc.; **18.11:** © London Express; **18.12:** © Robert Burroughs Photography

Chapter 19

Opener: © Comstock; **19.10a:** Courtesy of Genentech, Inc.; **19.10b:** © Hank Morgan/ Photo Researchers, Inc.; **19.10c,d:** Courtesy of Genentech, Inc.; **19.12a:** © Sanofi Recherche; **19.12b:** © Applied Biosystems/Peter Arnold, Inc.; **19.12c:** © Petit Format/Nestle/Science Source/ Photo Researchers, Inc.; **19.14a:** Courtesy of LifeCodes, Inc.; **19.14b:** © Stephen Ferry/The Gamma Liaison Network; **19.15a,b:** General Electric Research and Development Center; **19.16:** Monsanto Company; **19A:** © Inga Spence/Tom Stack & Associates

Chapter 20

Opener: © Dr. Tony Brain/SPL/Photo Researchers, Inc.; **20.3a:** © Seth Joel/SPL/ Photo Researchers, Inc.; **20.3b:** © Adam Smith/Westlight; **20.8:** Howard Sochurek/Medical Images/© The Walt Disney Co. reprinted with permission of Discover magazine; **20.10:** © Wm. C. Brown Communications/Bob Coyle, photographer

Chapter 21

Opener: © Lawrence Migdale/Photo Researchers, Inc.; **21.3 (bottom left):** © Leonard Lee Rue III/Animals Animals; **21.3 (top right):** © Jeff Rotman/Peter Arnold, Inc.; **21.3 (bottom right):** © Tim Davis/Photo Researchers, Inc.; **21.3 (top left):** © Maresa Pryor/Animals Animals; **21Aa,b:** © Nadine Drabona/ Berg & Associates; **21Ac:** © Mike Douglas/The Image Works; **21Ad:** © Judy White/Picture Perfect; **21.9a:** © Linda Bartlett/Photo Researchers, Inc.; **21.9b:** © Delta Williams/Bruce Coleman, Inc.; **21.9c:** © Jocelyn Burt/Bruce Coleman, Inc.; **21.9d:** © Bachmann/Photo Researchers, Inc.; **21.9e:** © Stephen Trimble

Chapter 22

Opener: © Gregory G. Dimijian/Photo Researchers, Inc.; **22.1a:** © Peter B. Kaplan/Photo Researchers, Inc.; **22.1b:** © John Bova/Photo Researchers, Inc.; **22.1c:** © Pat O'Hara Photography; **22.3:** © Michael Gallbridge/Visuals Unlimited; **22.10:** © Jacques Jangoux/Peter Arnold, Inc.; **22B:** © Glenn Van Nimwegen; **22C:** © Joe & Carol McDonald/Tom Stack & Associates; **22D:** © Glen Van Nimwegen

Chapter 23

Opener: © Dianne Blell/Peter Arnold, Inc.; **23.1a:** © Myrleen Ferguson/Photo Edit; **23.1b:** © Wolfgang Kaehler; **23.5:** © Bob Coyle; **23.6a:** © Sydney Thomson/ Animals Animals; **23.6b:** © Michael Nichols/Magnum Photos; **23.6c:** © G. Prance/Visuals Unlimited; **23.7:** © Thomas Kitchin/Tom Stack & Associates; **23.11a,b:** Courtesy of Dr. John Skelly; **23.13:** NASA; **23.14:** © Wolfgang Kaehler

Line Art

Introduction

I.5 Copyright © Mark Lefkowitz.

Chapter 3

3.1 From John W. Hole, Jr., *Human Anatomy and Physiology,* 6th ed. Copyright © 1993 Wm. C. Brown Communications, Inc., Dubuque, Iowa. All Rights Reserved. Reprinted by permission.
3.9 From Kent M. Van De Graaff, *Human Anatomy,* 3d edition. Copyright © 1992 Wm. C. Brown Communications, Inc., Dubuque, Iowa. All Rights Reserved. Reprinted by permission.
3.11 From Kent M. Van De Graaff and Stuart Ira Fox, *Concepts of Human Anatomy and Physiology,* 3d ed. Copyright © 1992 Wm. C. Brown Communications, Inc., Dubuque, Iowa. All Rights Reserved. Reprinted by permission.

Chapter 5

5.3 Copyright © Mark Lefkowitz.
5.4 Copyright © Mark Lefkowitz.
5.9 From Stuart Ira Fox, *Human Physiology,* 4th ed. Copyright © 1993 Wm. C. Brown Communications, Inc., Dubuque, Iowa. All Rights Reserved. Reprinted by permission.
Illustration, p. 107 Copyright © Mark Lefkowitz.

Chapter 6

6.5 From John W. Hole, Jr., *Human Anatomy and Physiology,* 6th ed. Copyright © 1993 Wm. C. Brown Communications, Inc., Dubuque, Iowa. All Rights Reserved. Reprinted by permission.

Chapter 8

8.10 From John W. Hole, Jr., *Human Anatomy and Physiology,* 5th ed. Copyright © 1990 Wm. C. Brown Communications, Inc., Dubuque, Iowa. All Rights Reserved. Reprinted by permission.
8Ac From G. Tyler Miller, *Living in the Environment,* 7th ed. Copyright © 1992 Wadsworth Publishing Company, Belmont, CA. Reprinted by permission.

Chapter 9

9.2 From Kent M. Van De Graaff and Stuart Ira Fox, *Concepts of Human Anatomy and Physiology,* 3d ed. Copyright © 1992 Wm. C. Brown Communications, Inc., Dubuque, Iowa. All Rights Reserved. Reprinted by permission.
Illustration, p. 179 From Kent M. Van De Graaff and Stuart Ira Fox, *Concepts of Human Anatomy and Physiology,* 3d ed. Copyright © 1992 Wm. C. Brown Communications, Inc., Dubuque, Iowa. All Rights Reserved. Reprinted by permission.

Chapter 10

10.7 Copyright © Mark Lefkowitz.
10.10 Copyright © Mark Lefkowitz.
10.13 From John W. Hole, Jr., *Human Anatomy and Physiology,* 6th ed. Copyright © 1993 Wm. C. Brown Communications, Inc., Dubuque, Iowa. All Rights Reserved. Reprinted by permission.

Chapter 11

11.3 From Kent M. Van De Graaff, *Human Anatomy,* 3d edition. Copyright © 1992 Wm. C. Brown Communications, Inc., Dubuque, Iowa. All Rights Reserved. Reprinted by permission.
11.9 From Kent M. Van De Graaff and Stuart Ira Fox, *Concepts of Human Anatomy and Physiology,* 3d ed. Copyright © 1992 Wm. C. Brown Communications, Inc., Dubuque, Iowa. All Rights Reserved. Reprinted by permission.
11.14 From John W. Hole, Jr., *Human Anatomy and Physiology,* 6th ed. Copyright © 1993 Wm. C. Brown Communications, Inc., Dubuque, Iowa. All Rights Reserved. Reprinted by permission.

Chapter 12

12.7 From Joan G. Creager, *Human Anatomy and Physiology,* 2d ed. Copyright © 1992 Joan G. Creager. Reprinted by permission of Wm. C. Brown Communications, Inc., Dubuque, Iowa. All Rights Reserved.

Chapter 13

13.17 From Stuart Ira Fox, *Human Physiology,* 4th ed. Copyright © 1993 Wm. C. Brown Communications, Inc., Dubuque, Iowa. All Rights Reserved. Reprinted by permission.

Chapter 14

14.1a From John W. Hole, Jr., *Human Anatomy and Physiology,* 6th ed. Copyright © 1993 Wm. C. Brown Communications, Inc., Dubuque, Iowa. All Rights Reserved. Reprinted by permission.
14.3 From Kent M. Van De Graaff and Stuart Ira Fox, *Concepts of Human Anatomy and Physiology,* 3d ed. Copyright © 1992 Wm. C. Brown Communications, Inc., Dubuque, Iowa. All Rights Reserved. Reprinted by permission.
14.5 From John W. Hole, Jr., *Human Anatomy and Physiology,* 6th ed. Copyright © 1993 Wm. C. Brown Communications, Inc., Dubuque, Iowa. All Rights Reserved. Reprinted by permission.
Illustration, p. 291 From John W. Hole, Jr., *Human Anatomy and Physiology,* 6th ed. Copyright © 1993 Wm. C. Brown Communications, Inc., Dubuque, Iowa. All Rights Reserved. Reprinted by permission.

Chapter 16

16.13 From Kent M. Van De Graaff and Stuart Ira Fox, *Concepts of Human Anatomy and Physiology,* 3d ed. Copyright © 1992 Wm. C. Brown Communications, Inc., Dubuque, Iowa. All Rights Reserved. Reprinted by permission.
16.14 From Kent M. Van De Graaff and Stuart Ira Fox, *Concepts of Human Anatomy and Physiology,* 3d ed. Copyright © 1992 Wm. C. Brown Communications, Inc., Dubuque, Iowa. All Rights Reserved. Reprinted by permission.

Chapter 17

17.15 From Robert Weaver and Philip Hedrick, *Genetics,* 2d ed. Copyright © 1992 Wm. C. Brown Communications, Inc., Dubuque, Iowa. All Rights Reserved. Reprinted by permission.

Chapter 21

21.4 Copyright © Mark Lefkowitz.

Illustrators

Molly Babich: 1.12a, 2.5b, 3.2e, 3.3, 7.1a, 7.6a–b, 7.8a–b, 8.5a, 12.8b, 14.2a,b,d, 14.6a, 21.6

Todd Buck: 2.9, 16.6

Christopher Creek: 3.1, 4.1, 8.2, 9.1, 9.2, 11.3, 13.1, 18.1, and text art p. 90, p. 163, p. 179

John Frieberg/Mary Albury-Noyes: 12.7

Peg Gerrity: 21.8

Anne Greene: 12.4, 16.7, and text art p. 245

Kathleen Hagelston: 10.2a, 16.3, 16.15, 19.8, 19.9

Kathleen Hagelston/Laurie O'Keefe: 7.4

Kathleen Hagelston/Marjorie Leggit: I.1b, 9.7, 17.2b

Illustrious, Inc.: 1.13, 4.9, 4.10, 6.9a, 14.1b, 18.10a, 19.11, 23.4, 23.10, and text art p. 62, 6.63, p. 124, p. 142, p. 144, p. 150, p. 178, p. 202, p. 203, p. 221, p. 290, p. 399 (2), p. 454, p. 455 (2), p. 473, p. 474

Iverson/Hagelston: 4.3, 5.1, 11.9, 11.10, 12.1, 12.3, 17.1a

Carlyn Iverson: I.1a, 1.2b, 1.7, 1.11, 2.4, 2.13, 3.8, 3.10, 3.13, 4.4a, 5.12, 8.3, 8.4, 8.11, 8.12, 9.6b, 9.8, 9.9, 10.5, 10.15, 11.1a–b, 11.8, 12.11, 12.12, 12.13, 12.14a–b, 14A, 14.7, 14.9, 16.8b, 19.1, 19.13, 21.5, 21.7, 22.7, and text art p. 42, 44 (left), p. 174 (top)

Mark Lefkowitz/Kathleen Hagelston: 5.4, and text art p. 107

Mark Lefkowitz: I.5, 4.6a–b, 5.3, 10.7, 10.10, 21.4, 22.9, 22.11

Rictor Lew: I.2, 13.17, 16.13

Rictor Lew/Kathleen Hagelston: 3.11i, 4.2, 4.5, 9.4, 11.14, 13A

Robert Marguiles: 6.5

Steve Moon: 3.11a–h, 5.2, 16.14

Diane Nelson: 5A2, 8.10a

Laurie O'Keefe: 3.5, 6.4

Precision Graphics: 8.8b, 13.14, 18.9, 18.14

Rolin Graphics: I.4, I.6, 1.5, 1.9, 1.14. 2.2b, 2.3, 2.6b–d, 3.6a1, 3.6b2, 3.6c2, 4.12, 5.11, 6.2b, 6.8, 7.8a, 7.10a, 8.6, 8.7, 8.9, 10.3b, 10.4, 11.2a, 11.2b, 12.2, 12.5, 13.3, 13.4, 13.5, 13.8, 13.13, 13.18, 13.19, 14.4, 14.12, 15.7, 16.2, 17.8, 18.5, 18.7, 21.1, 22.2, 22.3a, 22.5, 22.6, 23.3, 23.12, and text art p. 37, p. 44 (right), p. 347, p. 154 (bottom), p. 157 (left), p. 447 (bottom)

Rolin Graphics/Kathleen Hagelston: 2.10a, 5.6, 5.10, 6.7, 6.11, 9.5, 11.4, 11.5, 11.6, 11.7, 15.6b

Shoemaker, Inc.: 1.10, 4.8, 4.13, 5.5, 6.1, 6.3, 12.6, 12.10, 17.6, 19.17, 23.9, and text art p. 26, p. 162, p. 244

Nadine Sokol: 1.3, 1.4, 1A, 1.15, 1.16, 1.18, 2.1, 2.2a, 2.7, 3A, 4.7, 4.15, 6A, 6.10b, 7.5, 7.7b, 8.1, 8A, 10.14, 13.2, 14.9, 15.2, 15.13, 15.14, A5, 17.7, 17.9, 17.10, 17.11, 17.13, 17.14, 18.16, 19.2, 19.3, 19.4, 19.5, 19.6, 19.7, 19.18, 19.19, 20A, 20B, 20C, 22.12, 22A, and text art p. 88, p. 105, p. 123, p. 190, p. 193 (2), p. 225, p. 265 (2), p. 335, p. 358, p. 401, p. 417 (2), p. 421

Nadine Sokol/Kathleen Hagelston: A5, 20.1, 20.2, 20.4, 20.5, 20.6, 20.7, 20.9

Tom Waldrop: 10.6, 10.6b, 10.9a, 14.1a, 14.3, 14.5, and text art p. 291

Yvonne Walston: 5.7, 7.3, 9.3, 10.8, 10.12

John Walters & Associates: 5.9

Index

Page numbers printed in **boldface** type refer to tables, figures, or illustrations.

A

D

E

F

G

H

I

J

K

L

M

N

O

P

T

U